S0-CDU-477

VGFT
INSTITUTE FOR INTEGRATED
ENERGY SYSTEMS (IESVic)
University of Victoria
P.O. Box 3055
Victoria, B.C. V8W 3P6 Canada

LAB

FUEL CELL HANDBOOK

FUEL CELL HANDBOOK

A.J. Appleby
Texas A&M University

F.R. Foulkes
University of Toronto

KRIEGER PUBLISHING COMPANY
MALABAR, FLORIDA
1993

Original Edition 1989
Reprint Edition 1993

Printed and Published by
**KRIEGER PUBLISHING COMPANY
KRIEGER DRIVE
MALABAR, FLORIDA 32950**

Copyright © 1989 by Van Nostrand Reinhold
Reprinted by Arrangement.

All rights reserved. No part of this book may be reproduced in any form or by any means, electronic or mechanical, including information storage and retrieval systems without permission in writing from the publisher.
No liability is assumed with respect to the use of the information contained herein.
Printed in the United States of America.

FROM A DECLARATION OF PRINCIPLES JOINTLY ADOPTED BY A COMMITTEE OF THE AMERICAN BAR ASSOCIATION AND A COMMITTEE OF PUBLISHERS:

This publication is designed to provide accurate and authoritative information in regard to the subject matter covered. It is sold with the understanding that the publisher is not engaged in rendering legal, accounting, or other professional service. If legal advice or other expert assistance is required, the services of a competent professional person should be sought.

Library of Congress Cataloging-In-Publication Data
Appleby, A. J.
　Fuel cell handbook / A. J. Appleby, F. R. Foulkes.
　　p.　　cm.
　Originally published: New York. Van Nostrand Reinhold Co., 1989.
　ISBN 0-89464-733-4
　1. Fuel cells--Handbooks, manuals, etc. I. Foulkes, F. R.
II. Title.
[TK2931.A67 1992]
621.31'2429--dc20
　　　　　　　　　　　　　　　　92-6536
　　　　　　　　　　　　　　　　CIP

10　9　8　7　6　5　4　3　2

Acknowledgments

This book represents a compilation of material over fifteen years by one of the authors (F.R.F.), who gratefully acknowledges the support of the University of Toronto Department of Chemical Engineering and Applied Chemistry, the Natural Sciences and Engineering Research Council of Canada, and the Japan Society for the Promotion of Science. Also, he must express his gratitude to his family—Irene, Bryan, and Michael—for their infinite patience and understanding during this lengthy endeavor. The final text was compiled by A.J.A., with the aid of a grant from the United States Department of Energy and from Argonne National Laboratory.

A list of some of the many people who have provided information and assistance for the preparation of this book is given below.

Dr. John Ackerman, Argonne National Laboratory
Dr. R. Anahara, Fuji Electric
Dr. A. Ascoli, CISE, Italy
James Babcock, United Technologies Corporation
Dr. Marjan Bace, Van Nostrand Reinhold
Dr. B.S. Baker, Energy Research Corporation
Dr. Heinz Behret, the Battelle Institute, Frankfurt, West Germany
E.H. Camara, Institute of Gas Technology
Dr. M.E. Charles, University of Toronto
Prof. Cha Chuan-sin, Wuhan University, Peoples Republic of China
L.J. Deggan, General Electric Company

Arnold Fickett, Electric Power Research Institute
Prof. Masamichi Fujihira, Tohoku University, Japan
Ryuzo Fukuda, Moonlight Project of the Japanese Agency of Industrial Science and Technology
Ms. Arlene Gildart, University of Toronto
Edward A. Gillis, Electric Power Research Institute
Dr. W. F. Graydon, University of Toronto
Dr. Martin Hammerli, National Research Council of Canada
James T. Hinatsu, University of Toronto
Dr. Hugh H. Horowitz, Exxon Research and Engineering Company

Diane Hooie, Gas Research Institute
Toshio Iino, Tokyo Electric Power Company
Dr. Hugh S. Isaacs, Brookhaven National Laboratory
Dr. Norburu Itoh, NEDO, Japan
Jan Jandera, Plynprojekt, Czechoslovakia
J.M. King, Jr., United Technologies Corporation
Michio Kobayashi, Tokyo Electric Power Company
Dr. Hans Kohlmuller, Siemens A.G., West Germany
Dr. Karl Kordesch, Technical University of Graz, Austria
Dr. M. Krumpelt, Argonne National Laboratory
L.J. Nuttall, General Electric Company
L. Marianowski, Institute of Gas Technology
Dr. James McBreen, Brookhaven National Laboratory
Dr. Stuart Meibuhr, General Motors Research Laboratories
Prof. Tetsuo Osa, Tohoku University, Japan
Pageworks, Ltd., Old Saybrook, CT
Dr. D. Pierce, Argonne National Laboratory
S. Pietersz, PEO, The Netherlands
Annmarie Pittman, Courtesy Associates, Washington, D.C.
J. Pottier, Gaz de France, Paris
Richard Goldstien, Electric Power Research Institute
J.W. Schmitt, United Technologies Corporation
Dr. J.W. Smith, University of Toronto
Dr. S. Srinivasan, Los Alamos National Laboratory
J.S. Stanley, U.S. Army Foreign Science and Technology Center
Dr. Dietrich Sprengel, Varta Batterie A.G., West Germany
Dr. Zenichiro Takehara, Kyoto University, Japan
Karen Trimble, Gas Research Institute
Prof. Isamu Uchide, Tohoku University, Japan
H. Van den Broeck, ELENCO n.v., Belgium
John M. Van Gelder, Monsanto Company
Dr. R. Vellone, ENEA, Italy
Prof. W. Vlelstich, University of Bonn, West Germany
Dr. Shiro Yoshizawa, Kyoto University, Japan

The authors wish to thank the following publishers for permission to reproduce, in whole or in part, the figures indicated below:

Cover illustration courtesy of the Tokyo Electric Power Company (TEPCO). It illustrates the power section arrangement of the 11 MW Goi phosphoric acid fuel cell demonstrator unit.
2–2, 2–3, 10–1, 10–2, 10–3, 12–2, 12–5, 16–4, 16–5, 16–6, 17–3, 17–7, 17–9, 18–7: Pergamon Press
2–6, 2–7, 2–8: International Society for Precious Metals
3–1, 3–2, 4–2, 6–1, 6–2: International Fuel Cells
4–1: Gas Research Institute
8–1, 8–2, 8–3, 8–4, 8–5, 17–10, 17–11, 17–12: Annual Reviews, Inc.
9–1: Gautier–Villars
12–1: The Electrochemical Society, Inc.
13–2 (D,E), 17–5: Ishikawajima-Harima Heavy Industries
15–2, 16–2, 16–3: Electric Power Research Institute
16–7, 16–10, 18–8, 18–9, 18–10: National Fuel Cell Coordinating Committee

Foreword

The principle of the fuel cell has been understood for almost one hundred and fifty years, and the first kilowatt-scale fuel cells were developed some thirty years ago. In contrast, the heat engine has been reduced to practice for more than three hundred years. There is clearly a difference in degree of maturity between the two technologies. From the thermodynamic viewpoint, the limitations of the steam engine have been understood since 1824, with the publication of Sadi Carnot's *Réflexions Sur La Puissance Motrice du Feu*. This work shows that the upper limit of efficiency of a heat engine using practical materials is on the order of 40–45%. The fuel cell directly converts the free energy of formation of a fuel into direct current electricity (i.e., work) without going through a Carnot thermodynamic cycle. It can thereby theoretically attain twice the maximum efficiency of a heat engine. For this reason, the fuel cell was considered to be the solution to the future problem of energy conversion during the 1960s and 1970s.

A natural question therefore is, "Why aren't fuel cells in widespread use?" Part of the answer is that in the atmosphere created by the success of aerospace fuel cells in the 1960s, the fuel cell concept was initially oversold. There was then no clear understanding of the nature and magnitude of the interrelated problems that had to be solved before fuel cells could be successfully applied in the commercial sector.

As we have remarked above, the fuel cell is relatively immature as a practical energy conversion device. The internal combustion engine, the electric generator, and the various types of primary and secondary batteries have all been available for much longer. Naturally, these are in a more advanced stage of development and are produced commercially in large quantities. Conversely, the fuel cell has only been in the prototype and demonstrator state until very recently, and the problem of starting up production lines is only now beginning.

Accelerated by the large amount of funding in the U.S. aerospace program, fuel cell research activity reached a high level by the mid-1960s.

By about 1970, the early problem areas had become reasonably well defined. Basically, these problem areas were fourfold:

1. Gaseous hydrogen was recognized as being the only really effective fuel for practical fuel cell use. Hydrocarbons or alcohols had insufficient reactivity even with impractically high noble-metal catalyst loadings, whereas exotic fuels such as hydrazine were hazardous and too energy-intensive from the manufacturing viewpoint. To be of commercial value within the foreseeable future, fuel cells would have to be able to operate on readily available, cheap and easy to handle fossil fuels such as natural gas and other light hydrocarbons. These would have to be converted to impure hydrogen (for example, by steam reforming) before consumption in fuel cells with a CO_2-tolerant electrolyte. Pure hydrogen would probably only become widely available as a fuel when fossil fuel reserves were depleted (i.e., sometime in the twenty-first century).
2. For fuel cells with aqueous electrolytes capable of handling reformed hydrocarbons, the only effective catalysts seemed to be expensive noble metals such as platinum, which appeared to be required in massive amounts. In the mid-1960s, these were typically in the order of 25 mg/cm^2 of electrode area, representing about 500g of Pt per kW, or, at today's prices, $10,000/kW. While this may be tolerable for aerospace and special military applications, it was completely unacceptable for general purpose commercial use. Expensive cell constructional materials and current collectors (e.g., gold-plated tantalum) also appeared to be required in the most developed type of acid electrolyte fuel cell which could operate on impure hydrogen.
3. Catalyst operating lifetime was initially quite short (hundreds, rather than thousands, of hours), due to degradation.
4. Finally, as the problem areas became more clearly understood, it became apparent that at least a ten- to twenty-year lead time would be required before practical fuel cell systems could be marketed. Many developers were unwilling to make long-term investments in a device for which the potential market was by no means guaranteed.

The combined effect of the budget reductions in the U.S. space and military programs, and the belief that no immediate commercial applications were in sight, proved to be disastrous to the continuation of private funding for fuel cell work. By 1970, with few exceptions, development funding by industry had halted, and in 1971, it was generally considered that fuel cells were another technology (like magnetohydrodynamics) that had fallen by the wayside.

In spite of these setbacks, a numerically reduced but dedicated core of true believers continued to advance the state of the art during the first half of the 1970s. Because of their efforts, during the past ten years there has been a rekindling of interest in fuel cells. The reasons for this renewed interest are as follows:

1. The U.S. Government has once again become actively interested in the development of fuel cells, largely as a result of the energy crisis, which was first brought to the public's attention in 1972 (e.g., in the *National Geographic Magazine*, November 1972; after the publication of the *Limits to Growth*, the Club of Rome Report on the future availability of natural resources, in the same year). The escalation of the price of fossil fuel in 1974, and the increasingly stringent laws on pollution and on the efficient utilization of natural gas and hydrocarbon energy resources have served to make fuel cells more attractive than was the case fifteen years ago.

2. A strong effort has been made to decrease dependency on foreign oil. To a large extent this has involved finding efficient, environmentally acceptable ways of utilizing indigenous (and available) coal reserves for large-scale power generation. Multimegawatt fuel cell power plants operating on gasified coal can help to meet this need. As a result, large fuel cell programs are now underway in Japan and Europe, as well as in the United States. The efficiency of fuel cells is almost independent of size, and their low levels of noise and of combustion pollutants should make them particularly suitable for use in small, dispersed units. Ironically, and contrary to earlier expectations, multimegawatt fuel cell units for use with hydrocarbons (reformed natural gas or other clean light fuels) are being developed first, to be followed later by multikilowatt-sized units for other applications, such as for on-site use or for propulsion. The first units will use the phosphoric acid aqueous electrolyte technology, with the fuel cell itself operating at about 200°C. Later systems will use high-temperature electrolytes (molten carbonate at 650°C, solid oxide at 1000°C), which will allow higher efficiency together with the use of non–noble metal catalysts.

3. The problem of catalyst cost for the phosphoric acid fuel cell is no longer insurmountable. A better understanding of electrocatalysis and improved techniques of catalyst preparation and electrode fabrication have made possible much lower platinum loadings without compromising performance. Noble metal catalyst loadings now have been lowered to levels of 0.5 mg/cm^2 or less. These are considered economical for commercial power plant applications, and represent costs of about \$100/kW at 1986 Pt prices. Since it is possible to recover and recycle these catalysts, their cost is no longer an overwhelming burden. It is now no longer necessary to use exotic cell construction materials: Graphite has been shown to possess the stability consistent with commercial cost-effectiveness.

4. The early problem of the short operating life of the phosphoric acid fuel cell also seems to have been largely solved. The mechanisms of fuel cell degradation processes now are well understood, and practical plant operating lifetimes in excess of 10,000 hours are being routinely demonstrated, with lifetimes of at least 40,000 hours being realistically anticipated in commercial stacks. Small laboratory cells using 1975 technology (not representative of current practice) have now operated for 100,000 hours.

The fuel cell has finally started to emerge as a new energy conversion technology. MW-sized demonstrators already have been operated successfully, and the first semi-commercial fuel cells are scheduled to begin operation in the late 1980s. The commercialization of the fuel cell should now be essentially guaranteed. The question no longer is, "Why aren't fuel cells in widespread use?", but rather, "How soon will fuel cells be in widespread use?" The artificially low price of oil since the beginning of 1986 is certainly holding up the commercialization of new, more efficient energy technologies, but it seems clear that this is just an aberration in a market where the long-term trend will be rising prices as reserves outside of the Middle East are depleted. The future of the fuel cell seems therefore to be assured for a wide variety of energy conversion applications.

In this book, we shall examine in detail the characteristics of fuel cell systems, the different fuel cell concepts and development programs, together with applications of fuel cell power plants, design considerations for practical fuel cell systems, and the current state of the art. In the very near future, *fuel cell* should be a common household expression.

NOTE

The dollar investment costs listed in this book have been extracted from a large number of sources, many published in different years. In each case we have given the costs in U.S. dollars taken directly from the source. The following tables give the relative value of the U.S. dollar for the period 1967–1988, with the value in 1982 being taken as 100. Table A shows a number of different cost deflators, taken from the most recent edition of the Electric Power Research Institute's Technical Assessment Guide[124] (TAG®, C. W. Gellings, P. Hanser and S. M. Hass, 4 volumes, 1986–1987, Electric Power Research Institute, Palo Alto, CA). The table shows that all cost deflators are not equal, and in fact a marked disparity often exists between them. For example, the TAG points out that during the period from 1980 to 1984, the GNP price index (roughly the cost of labor), changed by 24.6%, whereas the cost of structural steel, a vital part of any energy plant, changed by only 5%. This fact makes cost conversion from year to year somewhat hazardous. In individual industries, accurate conversion

TABLE A

Implicit Price Deflators for Gross National Product

Period	Gross National Product	Personal Consumption expenditures				Gross private domestic investment		Exports and Imports of goods and services			Government purchases of goods and services			
											Federal			
		Total	Durable goods	Non-durable goods	Services	Nonresidential fixed	Residential fixed	Exports	Imports	Total	National Defense	Non-Defense	State and local
1977	67.3													
1978	72.2	71.6	76.9	71.9	69.8	71.5	72.6	72.8	65.8	69.2	67.8	72.4	71.1
1979	78.6	78.2	82.1	80.0	75.6	77.8	81.4	81.6	77.1	75.4	74.2	78.0	77.7
1980	85.7	86.6	89.2	89.4	83.9	85.1	89.4	90.2	96.0	84.3	83.4	86.4	86.2
1981	94.0	94.6	95.7	96.9	92.6	93.4	96.6	97.5	101.6	93.3	92.9	94.3	93.4
1982	100.0	100.0	100.0	100.0	100.0	100.0	100.0	100.0	100.0	100.0	100.0	100.0	100.0
1983	103.8	103.9	102.1	102.0	105.7	98.9	102.2	101.4	97.5	103.4	104.0	101.3	105.0
1984	108.1	108.2	103.9	105.4	111.5	99.4	106.4	103.7	97.4	107.0	107.6	105.1	110.6
1985	111.7	111.6	104.7	107.7	116.8	100.9	108.4	102.8	95.8	110.2	111.1	107.7	116.8
1982: III	100.8	100.7	100.4	100.5	100.9	100.8	100.4	100.1	99.3	100.5	100.2	101.5	100.9
IV	101.7	101.8	100.7	101.0	102.7	100.7	99.1	100.0	99.3	101.3	102.0	99.5	102.0
1983: I	102.4	102.4	101.3	100.7	103.9	100.1	102.0	100.5	98.7	102.8	102.7	103.1	103.3
II	103.2	103.4	101.6	101.9	105.0	98.9	100.3	100.8	97.2	103.4	103.7	102.6	104.4
III	104.1	104.3	102.4	102.4	106.2	98.3	103.2	101.4	97.0	103.1	104.5	99.0	105.6
IV	105.3	105.4	103.1	103.1	107.8	98.4	103.1	102.7	97.1	104.2	105.3	100.1	106.7
1984: I	106.6	106.7	103.4	104.5	109.2	98.8	103.6	103.5	97.5	105.2	106.6	99.9	108.5
II	107.6	107.6	103.9	104.8	110.8	99.2	106.5	104.3	98.0	106.8	107.4	105.0	109.9
III	108.6	108.7	104.1	105.5	112.5	99.6	107.6	103.8	97.3	107.3	107.6	106.7	111.2
IV	109.6	109.6	104.2	106.6	113.5	100.1	107.9	103.2	96.7	108.3	108.6	107.5	112.7
1985: I	110.4	110.3	104.9	106.7	114.7	100.5	107.7	102.9	95.8	109.9	110.1	109.4	114.4
II	111.3	111.3	104.8	107.5	116.1	100.7	107.9	103.1	95.7	110.4	110.6	110.0	116.1
III	112.1	111.9	104.6	107.6	117.4	101.0	108.2	102.7	95.3	110.2	110.9	108.1	117.5
IV	113.0	113.1	104.4	109.1	118.7	101.2	109.7	102.4	96.2	110.2	112.8	104.5	119.2
1986: I	113.7	113.4	105.0	108.2	119.9	102.2	111.2	102.4	95.7	111.9	112.5	110.0	120.0

Source: Department of Commerce, Bureau of Economic Analysis. (1982 = 100; quarterly data are seasonally adjusted)

factors, made up from all typical cost components, are sometimes available. For example, the General Construction Cost (Engineering News Record), Chemical Plant Cost (Chemical Engineering), Petroleum Refinery Cost (Oil and Gas Journal), Electric Utility Plant (Handy-Whitman) and Industrial Equipment Cost (Marshall and Swift) indexes might be cited. However, even then other factors make conversion difficult, since a 1967 plant and a 1987 plant might be physically different, reflecting improved technology and often the effect of federal and local regulations, for example the addition of antipollution measures. In consequence, for an unproven technology such as a fuel cell power plant, it is customary to use the seasonally adjusted gross national product (GNP) deflator (the first column of Table A) to correct dollars of one year into those of another. The U.S. Department of Commerce, Bureau of Economic Statistics publishes quarterly raw seasonally adjusted GNP data in the "Survey of Current Business." These data are revised every July to account for various fluctuations, for the year preceding the current one by three years (i.e., 1984 data were revised in 1987). In consequence, Table A contains raw data for years after 1983. Table B shows the GNP deflator from 1967 to the second quarter of 1988, all years up to 1985 showing corrected data. The small corrections between column 1 in Table A and Table B for the years 1984 and 1985 should be noted. Table B will allow the reader to estimate in current dollars the various costs cited in the text.

TABLE B

Corrected Gross National Product Deflators, 1967–85
(1986–1987 figures not corrected)

Year (by quarters)		GNP Deflator	Year	GNP Deflator	Year	GNP Deflator
1985	I	109.7	1967	35.8	1981	94.0
	II	110.6	1968	37.7	1982	100.0
	III	111.3	1969	39.8	1983	103.9
	IV	112.2	1970	42.0	1984	107.7
1986	I	112.4	1971	44.4	1985	110.9
	II	113.4	1972	46.5	1986	114.1
	III	114.7	1973	49.5	1987	117.7
	IV	115.9	1974	54.0	1988	121.0(est)
1987	I	116.3	1975	59.3		
	II	117.3	1976	63.1		
	III	118.2	1977	67.3		
	IV	118.9	1978	72.2		
1988	I	119.4	1979	78.6		
	II	120.9	1980	85.7		

Contents

PART I	**GENERAL ASPECTS OF FUEL CELLS**	**1**
1.	**Introduction**	**3**
	A. WHAT IS A FUEL CELL?	3
	B. PRINCIPLE OF OPERATION OF A FUEL CELL	4
	C. HISTORICAL SKETCH	7
2.	**Characteristics of Fuel Cell Systems**	**15**
	A. ENERGY CONVERSION EFFICIENCY	16
	Energy Loss Mechanisms	19
	Fuel Cell Thermodynamics	26
	B. ELECTRICAL OPERATING CHARACTERISTICS OF FUEL CELLS	28
	C. AIR POLLUTION	29
	D. NOISE POLLUTION	32
	E. THERMAL POLLUTION	32
	F. VISUAL AESTHETICS	32
	G. SAFETY	33
	H. EASE OF MAINTENANCE AND LOW MANPOWER REQUIREMENTS	33
	I. RELIABILITY	34
	J. MATERIALS	34
	K. MULTI-FUEL ABILITY	34
	L. SITING FLEXIBILITY	36
	M. MODULARITY	37
	N. ECONOMIC BENEFITS	38
	O. THE POTENTIAL FUEL CELL MARKET	40

PART II FUEL CELL PROGRAMS, CONCEPTS, AND AREAS OF APPLICATION 55

3. Utility Systems 57

- A. U.S. NATIONAL FUEL CELL PROGRAM 59
- B. UTILITY PHOSPHORIC ACID FUEL CELLS 61
  - TARGET Program 61
  - United Technologies Corporation Utility Phosphoric Acid Fuel Cell Program 63
  - The 4.5 MW A.C. New York Demonstrator 65
  - TEPCO 4.5 MW A.C. Generator In Japan 69
  - UTC FCG-1 Utility Fuel Cell 73
  - Westinghouse/Energy Research Corporation Utility Phosphoric Acid Fuel Cell Program 78
- C. UTILITY MOLTEN CARBONATE FUEL CELLS 81
  - Coal-Burning Molten Carbonate Fuel Cells 81
  - Molten Carbonate Fuel Cells Vs. Phosphoric Acid Fuel Cells for Use with Coal 83
  - Utility Molten Carbonate Fuel Cells with Fuels Other Than Coal 86
  - Development Program for Utility Molten Carbonate Fuel Cells 93
- D. SOLID OXIDE ELECTROLYTE FUEL CELLS 100

4. On-Site Integrated Energy Systems and Industrial Cogeneration 105

- A. ON-SITE INTEGRATED ENERGY SYSTEMS 105
  - Gas Industry Interest in On-Site Integrated Energy Systems 109
  - The 40 kW On-Site Integrated Energy System Operational Feasibility Program 109
  - Westinghouse/ERC On-Site Fuel Cell Program 118
  - Engelhard On-Site Fuel Cell Program 119
- B. INDUSTRIAL COGENERATION 120
  - The Use of Phosphoric Acid Fuel Cells for Industrial Cogeneration 122
  - The Use of Molten Carbonate Fuel Cells for Industrial Cogeneration 127
  - Outlook For Industrial Cogeneration Fuel Cells 128

5.	**Japanese and European Fuel Cell Programs**	**131**
	A. JAPANESE FUEL CELL PROGRAMS	131
	The Moonlight Project	132
	B. EUROPEAN FUEL CELL PROGRAMS	145
	C. FUEL CELL PROGRAMS ELSEWHERE	148
6.	**Military and Aerospace Systems**	**151**
	A. U.S. ARMY PROGRAMS	153
	1.5 kW Fuel Cell Units	154
	3 kW and 5 kW Fuel Cell Units	157
	B. U.S. AIR FORCE PROGRAMS	160
	C. AEROSPACE PROGRAMS	163
7.	**Electric Vehicles**	**177**
	A. ADVANTAGES OF FUEL CELL-POWERED ELECTRIC VEHICLES	178
	Efficiency	178
	Air Quality	180
	Noise Pollution	180
	Fuel Supply	181
	Reliability and Ease of Maintenance	181
	B. REQUIREMENTS FOR ELECTRIC VEHICLES	181
	C. CHARACTERISTICS OF ELECTRIC VEHICLES (EV_s)	182
	Performance and Fuel Costs	183
	Fuel Cells Vs. Batteries	185
	D. FUEL CELL-POWERED VEHICLES	186
	Types of Fuel Cells Best Suited for Electric Vehicle Applications	186
	The General Motors Electrovan	190
	Fuel Cell-Battery Hybrids	191
	The ELENCO Alkaline Fuel Cell Program	196
	U.S. Department of Energy Program in Vehicular Fuel Cells	198
	E. RAW MATERIAL IMPLICATIONS FOR ELECTRIC VEHICLES	199
	F. OUTLOOK FOR FUEL CELL–POWERED VEHICLES	200

PART III DESIGN CONSIDERATIONS FOR PRACTICAL FUEL CELL SYSTEMS   **203**

8. Choice of Fuel and Oxidant · 205

FUELS · 205

- A. HYDROGEN · 205
- B. HYDROGEN FROM HYDROCARBONS · 208
  - External Hydrocarbon Processing · 208
  - Hydrogen from Coal · 223
  - Internal Hydrocarbon Processing · 228
  - Miscellaneous Hydrogen Generation Processes · 233
- C. HYDRAZINE · 233
- D. METHANOL · 236
- E. AMMONIA · 238
- F. FORMIC ACID AND FORMALDEHYDE · 239

OXIDANTS · 239

9. Electrodes · 241

- A. POROUS ELECTRODES · 242
  - Porous Metal Electrodes · 244
  - Porous Screen Electrodes · 246
  - Porous Carbon Electrodes · 247
  - Microporous Plastic Electrodes · 255
  - Porous Composite Electrodes · 256
- B. NONPOROUS ELECTRODES · 256
- C. ELECTRODE EVALUATION · 257
- D. ELECTRODE COST · 258

10. Electrolytes · 261

- A. AQUEOUS ALKALINE ELECTROLYTES · 261
  - Electrolyte Carbonation · 264
  - Methods of Carbon Dioxide Removal · 266
- B. AQUEOUS ACID ELECTROLYTES · 277
  - Solid Polymer Acid Electrolytes · 284
- C. OTHER AQUEOUS ELECTROLYTES · 296
- D. ORGANIC ELECTROLYTES · 297
- E. MOLTEN CARBONATE ELECTROLYTES · 297
- F. SOLID ELECTROLYTES · 303
  - Solid Oxide Electrolytes · 304
  - Beta-Alumina · 308
  - Other Solid Ionic Conductors · 310

11. Anodic Electrocatalysis — 313

- A. PRINCIPLES OF ELECTROCATALYSIS — 313
  - Function of an Electrocatalyst — 314
  - Requirements for a Fuel Cell Electrocatalyst — 315
  - Evaluation of Electrocatalysts — 318
  - Experimental Evaluation Techniques — 321
- B. ELECTROCATALYSIS OF HYDROGEN — 322
  - Noble Metal Electrocatalysts — 323
  - Poisoning of Noble Metal Electrocatalysts — 326
  - Nickel Electrocatalysts — 330
  - Nickel Boride — 332
  - Mixed Metal and Spinel Electrocatalysts — 332
  - Sodium Tungsten Bronzes — 333
  - Tungsten Trioxide — 333
  - Tungsten Carbide — 333
- C. ELECTROCATALYSIS OF HYDRAZINE — 335
  - Nickel — 336
  - Nickel Boride — 336
  - Noble Metals — 336
  - Miscellaneous Electrocatalysts for Hydrazine Oxidation — 337
- D. ELECTROCATALYSIS OF HYDROCARBONS — 337
  - Methane — 338
  - Ethane — 338
  - Propane — 339
  - Butane — 339
  - Octane — 340
  - Ethylene — 340
  - Propylene — 340
  - Methanol — 340
  - Ethanol — 350
  - Propanol — 350
  - Formaldehyde — 351
  - Acetaldehyde — 352
  - Formic Acid and Formate — 352
  - Ethylene Glycol — 354
- E. MISCELLANEOUS FUELS AND ELECTROCATALYSTS — 354

12. Cathodic Electrocatalysis — 357

- A. ELECTROCATALYSIS OF OXYGEN REDUCTION — 357
  - The Use of Platinum in the Electrocatalysis of Oxygen Reduction — 359

	Alloys of Noble Metals and Platinum	382
	Other Noble Metals	382
	Silver	384
	Nickel	387
	Carbon	387
	Perovskites	390
	Spinels and Mixed Metal Oxides	392
	Tungsten Carbide	395
	Tungsten Bronzes	395
	Metal Chelates	398
	Miscellaneous Electrocatalysts	407
B.	ELECTROCATALYSIS OF OXIDANTS OTHER THAN OXYGEN	409

13. Fuel Cell Stack Materials Selection and Design — 411

- **A. MATERIALS SELECTION** — 411
  - Aqueous Alkaline Systems — 412
  - Aqueous Acid Systems — 415
  - Use of Plastics for Aqueous Fuel Cell Components — 419
  - Molten Carbonate Systems — 426
  - Solid Oxide Systems — 433
- **B. FUEL CELL STACK DESIGN** — 440
  - Design Parameters and Correlations — 441
  - Electrical Configuration — 442
  - Manifolding Arrangements — 445
  - Pressure Seals and Pressure Control — 449
  - Stack Design Concepts — 453

14. Reactant Processing and Clean-up Systems — 455

- **A. FUEL PROCESSING** — 455
  - Hydrogen Enrichment — 455
- **B. OXIDANT PROCESSING** — 457
  - Oxygen Enrichment — 457
- **C. GAS CLEAN-UP SYSTEMS** — 458
  - Sulfur Contaminants — 458
  - High-Temperature Vs. Low-Temperature Sulfur Removal for Molten Carbonate Systems — 461
  - Hydrodesulfurization — 462
  - Other Methods of Sulfur Removal — 464
  - Other Contaminants — 464

15. Heat and Mass Transfer: System Issues   465

- A. **REACTANT SUPPLY AND DISTRIBUTION**   465
  - Feed Modes   466
  - Reactant Conditioning and Control   467
  - Examples of Feed Distribution Designs   470
- B. **ELECTROLYTE FEED AND DISTRIBUTION**   470
- C. **HEAT AND WATER REMOVAL**   471
  - Water Removal   473
  - Cooling   475
  - Passive Cooling   475
  - Active Cooling   476
  - Process Gas Cooling   476
  - DIGAS Cooling   477
  - Liquid Cooling   478
  - Comparison of Active Cooling Techniques   480
- D. **START-UP AND SHUTDOWN**   484
- E. **MECHANICAL COMPONENTS**   486
  - Blowers   486
  - Pumps   487
  - Motors   488
  - Turbocompressors   488
- F. **ELECTRICAL DESIGN AND POWER CONDITIONING**   489
  - Aspects of Electrical Design   490
  - Power Conditioning   491
- G. **BOTTOMING CYCLE SYSTEMS**   493
  - Gas Turbines   493
  - Steam Turbines   494
  - Choice of Bottoming System   495
  - Fuel Cell System Examples   496

PART IV   STATE OF THE ART   501

16. Phosphoric Acid Fuel Cells   503

- A. **UNITED TECHNOLOGIES CORPORATION FUEL CELLS**   503
  - Description of A Proposed 11 MW UTC Power Plant   510
  - Ongoing Performance and Technology Improvements   513
- B. **WESTINGHOUSE-ENERGY RESEARCH CORPORATION FUEL CELLS**   519
  - Mark II 7.5 MW Fuel Cell System   519
  - Westinghouse 7.5 MW System Description   523

C. ENGELHARD INDUSTRIES PHOSPHORIC ACID
   FUEL CELLS 528
    Fuel Cell Description 528
    Operating Conditions 528
    Electrodes and Catalysts 529
    Electrolyte Matrix 529
    Bipolar Plates 530
    Cooling Plates 532
    Fuel Processor 532
D. REQUIRED IMPROVEMENTS IN PHOSPHORIC
   ACID FUEL CELLS 533
E. COST OF PHOSPHORIC ACID FUEL CELLS 536

17. Molten Carbonate Fuel Cells 539

A. INTRODUCTION 539
B. OPERATING PRINCIPLES 539
    Internal Reforming Systems 544
C. CELL DESCRIPTION 545
D. PERFORMANCE 551
    Output 551
    Losses 553
    Performance and Lifetime Goals 555
E. PROBLEM AREAS 556
    Design and Fabrication 556
    The Electrolyte Layer 557
    Separator Plate Corrosion 559
    Seals 562
    Electrolyte Loss 562
    Contaminants 567
    Anode Creep 568
    Cathode Dissolution 570
    Other Cathode Problems 574
F. COST OF MOLTEN CARBONATE FUEL CELLS 574

18. Solid Oxide Electrolyte Fuel Cells 579

A. INTRODUCTION 579
    Principles of Operation 579
B. FEATURES OF SOLID OXIDE ELECTROLYTE FUEL CELLS 583
    Tolerance to Contaminants 584
    Typical SOEFC System Configuration 585
C. SOLID OXIDE FUEL CELL PROGRAMS 586

	D. SOLID OXIDE FUEL CELL COMPONENTS	588
	Thin Film Technology	589
	Example of the Thin Wall Concept	590
	Electrochemical Vapor Deposition	594
	E. PERFORMANCE OF SOLID OXIDE FUEL CELLS	596
	F. STATE-OF-THE-ART SOLID OXIDE FUEL CELLS	600
	Westinghouse Solid Oxide Fuel Cells	600
	Japanese Solid Oxide Fuel Cells	605
	Monolithic Solid Oxide Fuel Cells	607
	G. PROBLEM AREAS AND FUTURE RESEARCH DIRECTIONS	610

19. Conclusion   613

INTRODUCTION	613
AMERICAN ELECTRIC UTILITY POLICY	613
THE TOKYO ELECTRIC POWER COMPANY PAFC DEMONSTRATOR	616
THE ITALIAN 1 MW PAFC DEMONSTRATOR	617
THE JAPANESE 1 MW PAFC DEMONSTRATORS	617
OTHER JAPANESE PAFC UNITS	618
AMERICAN ON-SITE PAFC UNITS	620
THE PAFC ELSEWHERE	620
SPECIALIZED LOW-TEMPERATURE FUEL CELL SYSTEMS	620
FUEL CELLS FOR FUTURE TRANSPORTATION NEEDS	622
MOLTEN CARBONATE FUEL CELLS: UNITED STATES	624
MOLTEN CARBONATE FUEL CELLS: JAPAN	626
MOLTEN CARBONATE FUEL CELLS: EUROPE	629
SOLID OXIDE ELECTROLYTE FUEL CELLS	630

PART V   REFERENCES   633

Index   749

List of Tables

Table A	Implicit Price Deflators for Gross National Product	xi
Table B	Corrected Gross National Product Deflators, 1967–1985	xii
Table 2–1	Theoretical Reversible Cell Potentials ($E°_{rev}$) and Maximum % Intrinsic Efficiencies for Candidate Fuel Cell Reactions Under Standard Conditions at 25°C	18
Table 2–2	Theoretical Reversible Cell Potentials ($E°_{rev}$) and Maximum % Intrinsic Efficiencies of Selected Combustion Reactions at Different Temperatures	19
Table 2–3	Power System Exhaust Emission Comparison	31
Table 2–4	Impact of Using Different Fuels with a Phosphoric Acid Fuel Cell System Designed for Use with Naphtha or Methane	35
Table 2–5	Impact of Retrofitting a Fuel Cell Power Plant for Use With Alternative Fuels	35
Table 2–6	Value of Benefits to Be Derived from Fuel Cell Power Plants	39
Table 2–7	Projected Fuel Cell Markets	49
Table 3–1	U.S. National Fuel Cell Program Budget for FY 1983	61
Table 4–1	Tests Performed on Grid-Connected Prototype 40 kW OS/IES Fuel Cell Power Plant	115
Table 4–2	Analysis Results for Industrial Cogeneration Phosphoric Acid Fuel Cell Systems	124
Table 5–1	Fuel Cell Goals for the Japanese Moonlight Project	133
Table 6–1	Canadian Military Requirements for Portable Electrical Power Sources	152
Table 6–2	U.S. Army SLEEP Unit Goals	154

Table 6–3	Performance Qualification Testing of 1.5 kW U.S. Army Portable Fuel Cell Power Plants	155
Table 6–4	Size and Weight Data for U.S. Army 1.5 kW Fuel Cell Power Plants	157
Table 6–5	Characteristics of U.S. Army 3 kW D.C. Fuel Cell Power Plant	159
Table 6–6	Potential of Fuel Cells for U.S.A.F. Applications	162
Table 6–7	Key Parameters for U.S.A.F. Tactical Mobile Fuel Cell Power Units	163
Table 6–8	Operating Life of Aerospace Fuel Cells	164
Table 6–9	Selected Parameters of N.A.S.A. Aerospace Fuel Cells	166
Table 6–10	Estimated Requirements for Aerospace Fuel Cells	167
Table 6–11	Projected Power Demand for Aerospace Power Sources	167
Table 7–1	Weight Comparison Between Gasoline Engines and Electric Motors	182
Table 7–2	Performance Characteristics of G.M. Electrovan	191
Table 7–3	Calculated Output of Fuel Cell–Battery Hybrid Power Source	193
Table 7–4	Relative Cost and Availability of Active Elements	199
Table 7–5	Relative Cost of Fuels for Fuel Cells	200
Table 8–1	Comparative Data on Various Fuels	207
Table 8–2	Calculated Steam-Hydrocarbon Reaction Equilibria (Vol. %)	210
Table 8–3	Steam Reformation of Natural Gas	211
Table 8–4	Dry Gas Analysis for High Temperature Reformer Tests	218
Table 8–5	Average Product Compositions for Catalytic Pyrolysis	220
Table 8–6	Results of Processing No. 6 Fuel Oil Using Different Pretreatment Methods	222
Table 8–7	Percent H_2 Composition of Coal Processed by Various Methods	224
Table 9–1	Typical Physical Properties of Carbon	248
Table 10–1	Properties of Ion Exchange Membranes	289
Table 10–2	Gas Transmission Through Cationic Ion Exchange Membranes	291
Table 10–3	Resistivities of Various Doped Zirconia Electrolytes	306
Table 13–1	Physical Properties of Selected Plastics	421

List of Tables

Table 13-2	Chemical Resistance of Selected Plastics	423
Table 13-3	Permeability Ranking of Various Plastics	427
Table 13-4	Resistivity-to-Thickness Ratios of Thin Film Anodes on Doped Zirconia at 1000°C	436
Table 13-5	Selected Physical Properties of Various Interconnection Materials	438
Table 13-6	Performance of Hitachi Stack Designs for Use in Alkaline Hydrogen-Oxygen Fuel Cells	445
Table 14-1	Effect of Sulfur on Molten Carbonate Fuel Cell Performance	459
Table 15-1	Comparison of Cooling Methods for Phosphoric Acid Fuel Cells	484
Table 15-2	Comparison of Gas Turbines and Steam Turbines	496
Table 15-3	Energy Balance for Coal-Fueled 100 MW Baseload Plant	497
Table 15-4	Energy Balance for Advanced Coal-Fueled UTC Direct Molten Carbonate Fuel Cell System	498
Table 15-5	Assessment of Coal-Fueled 50 MW FCG-1 Phosphoric Acid Fuel Cell Power Plant	500
Table 16-1	Selected U.S. Patents of United Technologies Corporation That Relate to Phosphoric Acid Fuel Cells	508
Table 16-2	Comparison of UTC FCG-1 Baseline Commercial Power Plant with 4.5 MW Demonstrator Power Plant	510
Table 16-3	Design Goals of Westinghouse-ERC 7.5 MW Phosphoric Acid Fuel Cell System	527
Table 16-4	Estimated Capital Costs of Phosphoric Acid Fuel Cells	537
Table 16-5	Cost of Aerospace Fuel Cells	538
Table 17-1	Effect of Pressure and Fuel Utilization on Molten Carbonate Fuel Cell Performance	552
Table 17-2	Representative Fuel and Oxidant Mixture for Molten Carbonate Fuel Cell	553
Table 17-3	Estimated Distribution of Voltage Losses in a Molten Carbonate Fuel Cell	554
Table 17-4	Estimated Cost of ERC 50 kW Methane-Fueled Direct Molten Carbonate On-Site Fuel Cell	575
Table 17-5	Estimated Costs of ERC 2 MW Dispersed Generator Methane-Fueled Direct Molten Carbonate Fuel Cell System	576

Table 17–6	Breakdown of Cost of Electricity from General Electric 675 MW Coal-Fired Molten Carbonate Fuel Cell System	577
Table 17–7	Estimated Capital Costs of Molten Carbonate Fuel Cells	578
Table 18–1	Single Cell Voltage-Current Characteristics of Westinghouse Thin Film Solid Oxide Electrolyte Fuel Cell: Effect of Air Vs. Oxygen	597
Table 18–2	Single Cell Voltage-Current Characteristics of Westinghouse Thin Film Solid Oxide Electrolyte Fuel Cell: Effect of Fuel Composition	597
Table 18–3	Voltage Losses in 120-Cell Self-Supported Solid Oxide Electrolyte Fuel Cell Stack	599
Table 18–4	Ouput Characteristics of Japanese 132-Cell Solid Oxide Electrolyte Fuel Cell Battery	606
Table 18–5	Characteristics of Japanese Solid Electrolyte Fuel Cell Stacks used for 100-Hour Testing	606

FUEL CELL HANDBOOK

PART I

General Aspects of Fuel Cells

The Foreword briefly presented some of the inherent advantages of fuel cells that were responsible for attracting the interest of the scientific and engineering communities in the late 1950s and early 1960s. Among these advantages were high energy conversion efficiencies and low environmental pollution. Some of the problems that gradually emerged over the next twenty years as more experience was gained with the new technology were listed. These problems, coupled with the changing price of fossil fuels and the uncertainty of the marketplace, led to a decline of interest in fuel cells. Finally, over the past decade, interest in fuel cells once again has been rekindled owing to the slow, but steady, progress made by fuel cell researchers and developers and the increasing desirability of an environmentally benign, efficient energy conversion technology that can be fueled by a secure energy supply.

In Part I of this book we shall examine in more detail the nature of the fuel cell, its principles of operation, and its inherent characteristics. In Chapter 1 we explain the basic operating mechanisms of the fuel cell. We also include what we think is an interesting sketch to place the fuel cell in its historical perspective. In Chapter 2 we go on to discuss both the technical and economic characteristics of fuel cells that will ensure their eventual acceptance in our technological society. This chapter points out, however, the fact that competing technologies are also making steady progress, and that to compete in the marketplace the fuel cell power system must show continuous improvements both in efficiency and in capital cost.

CHAPTER 1

Introduction

A. WHAT IS A FUEL CELL?

A fuel cell is a device that directly converts the chemical energy of reactants (a fuel and an oxidant) into low voltage d.c. electricity. Like the familiar flashlight (zinc-manganese dioxide) primary cell and the lead-acid (lead-lead dioxide) rechargeable (secondary) battery, a fuel cell effects this conversion via electrochemical reactions. However, unlike these conventional batteries, it does not consume materials that form an integral part of, or are stored within, its structure. In the flashlight cell, the zinc negative-electrode structure is itself the fuel, and the manganese dioxide–carbon black positive-electrode structure is the oxidant. Thus, a fuel cell is a converter only, and since ideally no part of it should undergo any irreversible chemical change, it can continue to operate as long as it is fed with a suitable fuel and oxidant and the reaction products are removed.

If viewed merely as an energy conversion device, a fuel cell clearly performs the same function as a galvanic cell (primary battery) or a discharging storage battery (secondary battery). However, the design and engineering of a fuel cell are quite different from those of batteries. A primary galvanic cell, for example, must be discarded when its fuel supply (e.g., the zinc casing) is exhausted, or when the products of reaction (zinc oxychlorides or oxides) are produced in such quantity that it cannot operate further. Similarly, a secondary storage battery must be regenerated by periodic recharging by voltage reversal, normally over several hours, so that the reaction products are reconverted into the reactants.

In the case of a fuel cell, however, the reactants are stored *outside* the reaction areas (the electrodes), which ideally are invariant in composition. The reactants are fed to the electrodes only when electric power generation is required. For most terrestrial fuel cells, the oxidizing reactant is atmospheric oxygen. When its other reactant is exhausted, a fuel cell of this type is similar to an automobile with an empty gasoline tank, in that all that is required to resume operation is to refill the fuel tank. Because it is easy to make the fuel tank more or less as large as desired, a fuel cell's period of operation can be made to be much longer than that of a conventional electrochemical primary battery. For use in space, the fuel and oxidant typically will be stored as stoichiometric amounts of liquid hydrogen and liquid oxygen, as were used for the Gemini, Apollo, and Space Shuttle Orbiter missions. As indicated above for terrestrial applications, the most practical oxidant is oxygen from the air, and such a fuel cell may be fueled, for example, with natural gas from a pipeline or naphtha in a tank.

B. PRINCIPLE OF OPERATION OF A FUEL CELL

It is not within the scope of this book to review fuel cell theory in detail. For this purpose the interested reader should consult the many books and reviews that have been written on this subject* as well as the standard textbooks on electrochemistry.** Of particular value is the excellent monograph by Liebhafsky and Cairns, *Fuel Cells and Fuel Batteries*,[1613] which covers historical developments in the fuel cell field in great detail up to 1967.

The attention of the reader also is drawn to a recent comprehensive report[1989] with chapters on state-of-the-art materials problems in the five most prominent fuel cell technologies. These chapters will be referenced often throughout this text. In Part IV of this book, "State of the Art," the status of the three technologies which are the object of the most intensive R&D (research and development) effort will be reviewed. These technologies, described later, are those using phosphoric acid, molten carbonate, and solid oxide electrolytes. The remaining two technologies (cells with alkaline electrolyte and cells with solid polymer electrolyte) are for various reasons currently suitable only for special-purpose applications, particularly where pure hydrogen is available as fuel. Pure hydrogen may not become widely available until the twenty-first century, so R&D funding for the development of these latter two technologies has been less important up to the present. While their technology is adequately covered under the various topics discussed in the chapters in Parts I–III, readers are referred to Ref. 1989 for state-of-the-art development summaries for these fuel cell types, since no additional material can be added here at the time of writing.

---

* See Refs. 1, 41, 69, 70, 89, 143, 147, 196, 208, 213-215, 223, 225, 227, 305, 324, 388, 399, 451, 452, 459, 549, 560, 561, 572, 573, 575, 689, 744, 764, 770, 858, 1048, 1065, 1099, 1154, 1178, 1184, 1187, 1277, 1330, 1343, 1350, 1380, 1427, 1439, 1483, 1484, 1584, 1613, 1732, 1740, 1747, 1791, 1820, 1929, 1944, 1958, 2071, 2081, 2172, 2185, 2294, 2458, 2506, 2522, 2591, 2604, 2651, 2667, 2701, 2715, 2740, 2784.
** See Refs. 31, 250, 270, 380, 389, 390, 392, 667, 735, 736, 820, 887, 1051, 1866, 2042, 2200, 2589, 2598, 2759.

Introduction

In spite of the complexities involved in the construction and operation of a practical fuel cell system, the principle of operation of a fuel cell is readily grasped. As shown in Fig. 1–1, a fuel cell consists of a fuel electrode (anode) and an oxidant electrode (cathode), separated by an ion-conducting electrolyte. The electrodes are connected electrically through a load (such as an electric motor) by a metallic external circuit. In the metallic part of the circuit, electric current is transported by the flow of electrons, whereas in the electrolyte it is transported by the flow of ions, such as the hydrogen ion (H^+) in acid electrolytes, or the hydroxyl ion (OH^-) in alkaline electrolytes. In high-temperature fuel cells, the corresponding ionic carri-

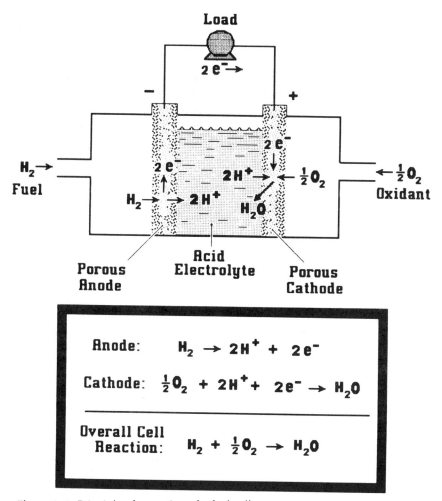

Figure 1–1: Principle of operation of a fuel cell.

ers may be the carbonate ion ($CO_3^=$) in molten carbonate electrolytes, and the oxide ion ($O^=$) in the case of the solid oxide system.

In theory, any substance capable of chemical oxidation that can be supplied continuously (as a fluid) can be *burned* galvanically as the fuel at the anode of a fuel cell. Similarly, the oxidant can be any fluid that can be reduced at a sufficient rate. For specialized systems, both reactants might be liquids, such as hydrazine as the fuel and hydrogen peroxide or nitric acid as the oxidant. Gaseous hydrogen has become the fuel of choice for most applications, because of its high reactivity when suitable catalysts are used and because of its high energy density when stored as a cryogenic liquid, such as for use in space. Similarly, at the fuel cell cathode the most common oxidant is gaseous oxygen, which is readily and economically available from air for fuel cells for terrestrial applications. When gaseous hydrogen and oxygen are used as fuel and oxidant, the electrodes are porous to permit the gas-electrolyte junction to be as great as possible. The electrodes must be electronic conductors, and possess the appropriate reactivity to give significant reaction rates.

Figure 1–1 shows a simplified schematic diagram of a hydrogen-oxygen fuel cell employing an acid electrolyte. At the anode, incoming hydrogen gas ionizes to produce hydrogen ions and electrons. Since the electrolyte is a non-electronic conductor, the electrons flow away from the anode via the metallic external circuit. At the cathode, oxygen gas reacts with migrating hydrogen ions from the electrolyte and incoming electrons from the external circuit to produce water. Depending on the operating temperature of the cell, the product water may enter the electrolyte, thereby diluting it and increasing its volume, or be lost through the cathode as vapor. Depending on the circumstances, careful water management of the electrolyte may or may not be necessary to remove product water. In all fuel cells with liquid electrolytes operating below the boiling point of water, an electrolyte circulation system incorporating an external evaporator may be necessary. For acid fuel cells employing a solid polymer electrolyte operating in this temperature range, the liquid water product may be rejected directly from the semi-solid gel electrolyte into the cathode gas chamber.

The overall reaction that takes place in the fuel cell is the sum of the anode and cathode reactions; in the present case, the combination of hydrogen with oxygen to produce water. This overall reaction may be viewed as the *cold combustion* of hydrogen with oxygen, in that it takes place at a much lower temperature than the conventional combustion of the two gases. Instead of the whole of the energy of the reaction being released as heat, as would be the case if hydrogen were burned with oxygen, part of the free energy of reaction is released directly as electrical energy. The difference between this available free energy and the heat of reaction is

Introduction

produced as heat at the temperature of the fuel cell. The theoretical standard free energy of formation of liquid water from gaseous hydrogen and oxygen at 1 atm pressure at 25°C is 1.229 V. In theory the fuel cell shown in Fig. 1-1 should therefore be capable of generating d.c. electrical energy at 1.229 V. In practice, however, on account of electrode polarization and other irreversibilities, under net flow of current the terminal voltage will be lower than this ideal value. This will be discussed in more detail in Chapter 2.

In any event, it can be seen that as long as hydrogen and oxygen are fed to the fuel cell, the flow of electric current will be sustained by electronic flow in the external circuit and ionic flow in the electrolyte. By electrically connecting a multiplicity of unit cells either in series or in parallel, it is possible to form a fuel cell battery of any desired voltage and current output. By the use of an efficient inverter it will also be possible to convert this d.c. electricity to a.c. electricity.

C. HISTORICAL SKETCH

Although the fuel cell has received much attention in recent years, the underlying concept is by no means a product of the twentieth century.[689, 1187, 1958] A more detailed review of the background of the development of the fuel cell can be found in Ref. 1613.

The forerunner of all electrochemical devices can be considered to be the various groups of electric fish that evolved millions of years ago.[2667] In these, the electric organs have evolved independently, and consist of modified muscle tissue. A Mediterranean variety, the torpedo ray, is capable of producing 300 V, and was used by Greek and Roman physicians in the first and second centuries as a type of electric shock therapy to treat ailments such as migraine headaches, epilepsy, and gout. Although electric shock therapy apparently passed out of fashion in the Middle Ages, it became popular again after the Renaissance, and was investigated by a large number of eminent biologists and physicists, including Galvani, Volta, and Faraday.

In about 1800, Volta invented the first electrochemical battery, known as the *voltaic pile*. It is interesting that he described his battery as an *artificial torpedo*, in reference to its resemblance to the column of plates in the electric organ of the torpedo ray, with which he was familiar.

The first attempt to obtain electricity directly from a fuel is usually attributed to Sir Humphrey Davy in 1802, when he reported results obtained using a cell that utilized a carbon anode and aqueous nitric acid as the cathodic reactant.[726] Twenty such cells in series were required to decompose water.

Sir William Grove, a jurist who turned his attention to electrochemical phenomena, is usually regarded as the creator of the fuel cell. Although Grove became famous in his lifetime as the inventor of a relatively conventional galvanic cell that became known as the *Grove cell*,[1034] the significance of his work on fuel cells was not recognized until much later.

As a result of insights gained through his experimentation with the electrolysis of water, Grove reasoned that it should be possible to reverse the process and generate electricity by reacting hydrogen with oxygen. In his classic experiment reported in 1839,[1033] Grove constructed what is considered to be the first known fuel cell. Although the current from this cell was rather small, Grove was encouraged by the results, and in 1842 constructed a bank of fifty such cells, which he called a *gaseous voltaic battery*, to distinguish it from his other invention, the Grove cell.[1035]

The gaseous voltaic battery consisted of a series of closed-ended glass tubes with their lower ends immersed in dilute sulfuric acid. The upper parts of the tubes alternately contained hydrogen and oxygen and had series-connected platinized platinum foil electrodes in contact with both gas and electrolyte phases. When he found that his gas battery could generate sufficient electric force to decompose water, Grove was quick to realize the beautiful symmetry inherent in the "decomposition of water by means of its composition." With reference to the dual process of first using electricity to decompose water to hydrogen and oxygen, followed by recombination of the hydrogen and oxygen to generate electricity, he went on to write, "This is, to my mind, the most interesting effect of the battery; it exhibits such a beautiful instance of the correlation of natural forces."[1035] Unfortunately Grove soon abandoned research on the gaseous voltaic battery because his device was not capable of delivering sufficent power to compete with more conventional galvanic cells. This was because of the small surface area available in the meniscus zone of each electrode in the region of *triple contact* between the gas, electrolyte, and electronically conducting phases, where all the reaction takes place. Grove however made an important contribution in realizing this, saying that what was required was "a notable surface of action."[1035]

A. C. Becquerel in 1855 was one of the first workers to attempt to develop a carbon-burning cell employing a fused salt electrolyte.[292] Becquerel's cell consisted of a carbon and a platinum electrode immersed in a fused nitrate. As pointed out by Baur and Tobler,[281] this cell actually was a nitrate-nitrate concentration cell resulting from the reduction of the nitrate ion by carbon.

No significant further attempts were made to construct a practical device until 1889, when Ludwig Mond and his associate, Charles Langer, repeated Grove's earlier work. For the first time they called the apparatus a *fuel cell*, and they attempted to make it more practical by replacing oxygen with air, and hydrogen with impure industrial gas obtained from coal by the *Mond-gas process*.[1799] Mond and Langer realized the importance of preventing the platinum black catalyst from becoming flooded with dilute sulfuric acid electrolyte, thereby losing active surface area by reducing the surface of triple contact. They avoided this problem by "using an electrolyte in a quasi-solid form, viz., soaked up by a porous non-conducting material" such as asbestos or plaster of Paris.[1799] Their device, the first to make use of a so-called matrix electrolyte, was able to develop 1.5 W at 50% efficiency. Unfortunately, they decided not to pursue the fuel cell concept further, having decided that their device was not practical because of the high cost of the platinum catalyst, which was in any case rapidly poisoned by traces of carbon monoxide in the Mond-gas, and because of the severe polarization (i.e., excessive voltage drop even at low currents) that resulted from the large concentration gradients in their porous diaphragm.

In 1896, Dr. W. W. Jacques[1302] published data on a fuel cell that he had developed for household use, based on the earlier work of Becquerel and Jablochkoff. Jacques's cell, enclosed in a massive brick structure, operated on carbon and air at a reported efficiency of 82%. Later it was revealed that the performance could be attributed to the formation of hydrogen by an electrolyte-anode reaction, and that the real efficiency was in fact much lower. Moreover, the cell required pure carbon rather than the coal originally intended. In spite of these failings, Jacques apparently was the first to envisage a fuel cell that could provide electricity for domestic use.

The development of the science of thermodynamics in the latter half of the nineteenth century led to an increase in the understanding of the relationship between electrochemical and thermal energy. This led Ostwald in 1894 to conclude that the way to solve the problem of the provision of inexpensive energy must be found in electrochemistry.[1957] He further stated that if a galvanic cell were discovered that could efficiently produce electrical energy directly from carbon and the oxygen of the air, then a technical revolution would be near.

Throughout the first half of the twentieth century several attempts were made to build fuel cells that could convert coal or carbon directly to electricity. However, because of a lack of understanding of electrode kinetics, none succeeded. One of the earliest workers to realize the importance of kinetics in the success of the fuel cell was Baur. In 1921, he built a high-temperature molten carbonate cell that operated at 1000°C using a carbon anode and an iron oxide cathode.[280] Although Baur's ideas had a solid basis, he was unable to cope with the corrosion and materials problems inherent in a high-temperature fuel cell with molten carbonate electrolyte and terminated his work.

The same period saw the introduction of the internal combustion engine, whose theoretical foundation was much better understood than that of the fuel cell. It has been stated that the lack of progress in the understanding of electrode kinetics during the first half of the twentieth century resulted from the fact that electrode discharge behavior was mistakenly interpreted in terms of equilibrium (i.e., Nernstian) thermodynamics, whereas it essentially involves irreversible, non-equilibrium processes.[389, 390, 392] J. O'M. Bockris has coined the expression "the Nernstian Hiatus" to refer to this period, and has gone so far as to blame reversible thermodynamics for the proliferation of the fuel-inefficient, polluting internal combustion engine. Regardless of the merits of this argument, it is nevertheless true that there were more electric vehicles in use in 1915 than there are at the present time.

The first major fuel cell development project that eventually led to a successful device was embarked upon in 1932 by Francis T. Bacon, an engineer associated with Cambridge University in England and a descendant of the family that produced the famous seventeenth century philosopher-scientist, Francis Bacon, Baron Verulam. Bacon reasoned that the expensive platinum catalysts used previously in acid systems by workers such as Mond and Langer would prohibit the fuel cell from becoming a commercial device. To circumvent this problem, Bacon selected the hydrogen-oxygen cell with alkaline electrolyte as a practical starting point from which to develop his fuel cell. By switching to a less corrosive alkaline electrolyte it became possible to employ relatively inexpensive nickel as the electrode material. Being an engineer, Bacon reasoned that suitably high reaction rates might be obtained with nickel electrodes by increasing the temperature and pressure. If the cell temperature was raised to about 205°C, nickel was sufficiently active catalytically to eliminate the need for any additional catalyst. At this temperature, pressurized operation was necessary to prevent the aqueous alkaline electrolyte from boiling. Bacon then discovered that further increasing the pressure increased the performance of his fuel cell, so that he finally elected to operate at about 4 MPa (~40 atm).

In order to obtain sufficiently large active electrode surface areas of three-phase contact (triple contact), Bacon had to use porous gas diffusion electrodes. Differential pressure fluctuations in the cell, however, caused difficulties in maintaining a stable gas–liquid interface inside these electrodes. Bacon solved this problem by using porous electrodes made in two layers of different porosity. By positioning the electrodes with the gas on the coarse-pore side and the electrolyte on the fine-pore side, the capillary forces in the pores made it possible to control the interface position by adjusting the pressure differential between gas and electrolyte. After more than a quarter of a century of research, Bacon was able to announce in August 1959 that he and his co-workers had developed, built, and demonstrated a 5 kW system capable of powering a welding machine, a circular

Introduction

saw, and a 2-ton capacity fork-lift truck.[28] In October of the same year, Dr. Harry Karl Ihrig of the Allis-Chalmers Manufacturing Company in the United States, demonstrated his famous 20-horsepower fuel-cell powered tractor.[689]

Developments then began to occur rapidly. Among those that received considerable publicity were several cells produced by Allis-Chalmers between 1959 and 1963. In 1964, under contract to the Electric Boat Division of General Dynamics, Allis-Chalmers produced a 750 W fuel cell system to power a one-man underwater research vessel. Running on liquid hydrazine hydrate and gaseous oxygen, this power plant was probably the first practical non-novelty application of a fuel cell as a motive power source.

It was soon discovered that the fuel cell offered a number of advantages for use as an electrical power source in aerospace flights.[689] Among these was a lack of dependence on sunlight. (In comparison, photovoltaic panels require back-up storage batteries in low-orbit applications.) In addition, fuel cells can be made to operate independently of aerodynamic forces and ambient pressure. They have high operational efficiency (low specific fuel consumption), therefore low heat rejection, and have the ability to produce potable water as a product of the cell reaction between hydrogen and oxygen stored as cryogenic liquids. The mission times of 7~14 days that were anticipated for the earlier space flights leading up to the manned lunar missions were too long for non-rechargeable primary batteries but short enough to make candidates other than solar cells attractive.

For the space application, power per unit weight (power density, kW/kg) over the given mission time (i.e., kWh/kg) is the primary consideration. The fuel cell system consists of the lightweight power-producing unit (the electrochemical cell itself), together with that of the stored fuel and oxidant. For missions over a few hours, the weight of the latter predominates over the former. For a ten-day mission, a fuel cell using cryogenic hydrogen and oxygen with 1965 technology weighed about 150 kg/kW, of which the fuel cell itself weighed 35 kg/kW, the remainder being the cryogenic fuel and oxidant with their pressurized insulated tanks and auxillary equipment. This corresponds to 1.6 kWh/kg for the 240-hour mission; in contrast, the most effective primary batteries available at that time were scarcely capable of 0.2 kWh/kg.

Accordingly, the U.S. National Aeronautics and Space Administration (NASA) decided to use fuel cells in the U.S. space program. As a result, NASA funded an extensive fuel cell research and development program during which more than 200 contracts were awarded to industry and universities for the study of the basic physics, kinetics, electrochemistry, and catalysis of fuel cell reactions; to develop methods of making electrodes, retaining electrolytes, removing by-products, and constructing workable cells; and to investigate several promising approaches to the construction of a practical power generation system.[689] The first major success of this large research effort was to be found in the Gemini series of earth-orbiting missions, in which ion-exchange membrane fuel cells developed by General Electric were used. This was followed by the fuel cell power plants for the Apollo missions, which put the first man on the moon. The Apollo fuel cells, developed by the Pratt & Whitney Aircraft Division of United Aircraft Corporation (later United Technologies Corporation [UTC]), were based on Bacon's design. New fuel cells of higher power output and lighter weight were designed and developed by General Electric and by UTC for use in the U.S. space shuttle program. United Technologies was finally awarded the Space Shuttle Orbiter contract in 1970. This massive U.S. aerospace fuel cell effort has undoubtedly provided the single most important impetus to the development of electrochemical engineering science in respect to energy conversion. Aerospace applications of fuel cells will be discussed in more detail in Chapter 6.

As a result of the success of the fuel cell as the primary electric power source for the U.S. space program, there were many predictions that it would be the solution to the world's energy problems. These hopes began to diminish somewhat as certain inherent difficulties were gradually realized. Four major problem areas almost caused the demise of the fuel cell as a potential terrestrial power source.

First, hydrogen was the only effective non-exotic fuel. Whereas it could be used with relatively inexpensive nickel catalysts in alkaline electrolyte at high temperature and pressure, the cost of the pressure vessels around the fuel cell and associated equipment would be prohibitive for ordinary applications. Second, the alkaline system required very pure hydrogen, which necessitated many production problems for use with hydrogen produced from common fuels such as natural gas or coal. Any carbon dioxide present in the hydrogen reacted with the alkaline electrolyte, early establishing that electrode kinetics in aqueous potassium carbonate were inadequate to sustain high reaction rates. This partly resulted from intrinsic kinetic and mass-transport problems, and partly from precipitation of carbonate in the electrode pores. While acid electrolytes did not suffer from these problems, for them the only effective catalyst seemed to be platinum in excessive loadings (tens of mg/cm^2 of electrode area), which were far too expensive for commercial application. Third, lifetimes were short under the conditions encountered with *dirty* commercial fuels, carbon dioxide–containing air, and with common construction materials. Finally, and in retrospect, the community of fuel cell workers themselves tended to oversell the merits of the fuel cell before really having come to terms with all the teething troubles of an immature technology. As a result of overenthusiasm, deadlines were not met, and private funding for the fuel cell greatly declined. With the major budget reductions in the aerospace program during the early 1970s, government funding for fuel cells also slowed. As a consequence, fuel cell research and development almost came to a halt because the private sector was hesitant to assume the costs previously paid by the U.S. government.

Fortunately, starting in 1967, the American Gas Association (AGA), representing the gas utility companies, decided to fund a nine-year program called TARGET (Team to Advance Research on Gas Energy Transformation), using UTC as the primary contractor and the Institute of Gas Technology (IGT), Chicago, as subcontractor. The AGA, along with a number of gas utilities, had supported IGT since about 1960 to encourage the development of the molten carbonate electrolyte fuel cell as a device which could produce electricity from natural gas without the use of noble-metal catalysts.

The objective of the TARGET program was to develop small domestic natural gas fuel cell units operating on reformed natural gas using an acid electrolyte (phosphoric acid), with the molten carbonate system as a back-up technology. The reasoning was simple: If catalyst and materials problems could be overcome, such a system could be profitable because the delivered price of gas (in terms of heating value) was about one sixth of the price of electricity for domestic use, and the fuel cell could produce both electricity and heat from gas in a ratio of 1:1 at a combined efficiency of about 80%. United Technologies considered that good engineering could make the fuel cell a success. In the end they were right, but the engineering improvements eventually took second place to a much improved understanding of materials and electrocatalytic problems.

In 1971–1972, UTC realized that economies of scale might be sufficient to reduce the per kW cost of the phosphoric acid fuel cell (PAFC) to make it economical for the production of primary electric utility power (rather than more expensive delivered electric power along with on-site heat). Accordingly, they proposed a truck-transportable 27 MW system to electric utilities. The oil embargo of 1973–1974 made many U.S. electric utilities interested in fuel cells for central power station applications. By the late 1970s U.S. federal support for large-scale fuel cell power generating systems began, ranging in current dollars from $21 million in 1977 to about $40 million in 1984.[1692] Over this same time period the total U.S. federal money for fuel cell research and development amounted to about $250 million in current dollars, or about $350 million in constant (1986) dollars. This sum excludes money spent earlier (during the 1960s) on aerospace fuel cells. The combined money spent during this same period by the utilities and manufacturers approximately matched that contributed by the government. With significant amounts of money now being spent overseas, notably in Japan and in Europe, it now appears that the fuel cell might become commercially viable by about 1990.

It is perhaps interesting to note in passing that the expected pathway for the commercialization of the fuel cell seems to be taking place in the reverse order from what was originally expected. It was considered that the first commercial fuel cells would evolve from the aerospace units, and would be in the several-kW class. These devices were expected to be used for special applications such as electric vehicles. As these smaller units proved the reliability of the technology and gained the confidence of investors and the public, the development of increasingly larger units requiring greater capital investment would then be undertaken.

The reverse in fact appears to be happening: The first commercial fuel cells will almost undoubtedly be multi-MW units that will cost hundreds of millions of dollars to develop and construct. Smaller units for special applications will probably follow. The reasons for this inverted order of emergence lie in the fact that only a large assured market can justify the major public and private expenditures necessary to ensure a proper developmental effort. Central power plant applications have not only built-in economies of scale, but are more readily seen as serving the public well-being in view of the low environmental impact and high efficiency of the fuel cell. It can also be argued that they do not come up against the resistance of strong lobby groups such as that of the automobile industry, which might prefer to use the internal combustion engine for as long as is profitable. In any event, it seems that the fuel cell finally is about to come of age.

CHAPTER 2

Characteristics of Fuel Cell Systems

Abundant energy is essential to the functioning of a modern technological society. Canadians and Americans share the dubious distinction of having the largest per capita energy consumption in the world. The Canadian Department of Energy, Mines and Resources, for example, has predicted that Canada's annual energy consumption will be about 21×10^{18} J by the year 2000, which represents a quadrupling of the value of 5.3×10^{18} J in 1970.[742] In contrast, thanks to the energy conservation efforts of the last decade, U.S. energy consumption may increase by only about 50% during the same period, going from 72×10^{18} J in 1970 to about 110×10^{18} J. In Canada, it is estimated that 33% of the energy required in the year 2000 will be in the form of electricity. Energy studies carried out by both the Canadian Federal Government and the Ontario Provincial Government[1950] confirm that electricity will become an increasingly dominant form of energy. The same is expected to be true of all countries with expanding industrial economies.

Electricity has numerous advantages at the point of use, the most important being high efficiency, ease of control, convenience, versatility, safety, and minimal public health and environmental impact. High efficiency results because electricity is energy in the form of work, the end product of energy conversion. For example, electricity can supply heat at greater than 100% efficiency by the use of the heat pump, a Carnot-cycle heat engine operating in reverse. Full electrification of the North American economy would reduce atmospheric pollution significantly, and could be coupled to the greater development of mass transit, with significant beneficial effects on cities. Unfortunately, both the generation and the transmission of electricity have significant detrimental effects that must be weighed against the obvious benefits.

How should electricity be generated? The present realistic options for large-scale use include generation from the burning of fossil fuels, from hydroelectric sources, and from nuclear energy. Fossil-fuel power plants have contributed to serious air pollution problems, including acid rain and the *greenhouse effect*, a gradual world-wide temperature increase resulting from the introduction of increasing amounts of fossil CO_2 in the atmosphere.[645, 789, 2030] The continuing demand for oil is endangering both the seacoasts and wilderness areas such as the Canadian North. Hydroelectric dams have a wide range of environmental effects, many of which are serious. More than fifty of Canada's major rivers already have been affected by large dams; and some of its finest fisheries, wildlife habitats, and agricultural lands have been lost forever. The same is true in the United States, in which no further large hydroelectric sites are available. The consequences of hydropower in other areas may have serious local effects (e.g., in the case of the Nile). In tropical rain forest areas, their effects may be global. For nuclear energy, the issues are more complex, involving many unknowns associated with a new technology requiring unprecedented caution and control.[2003]

For many applications, fuel cells promise to offer an attractive alternative to other modes of energy conversion.

A. ENERGY CONVERSION EFFICIENCY

The use of a heat engine is an essential feature of the conventional generation of electrical energy from a fuel. Heat engines operate by converting the thermal energy released by a chemical reaction (such as the burning of coal) to mechanical energy. The theoretical maximum efficiency of a heat engine is given by

$$\mathcal{E}_{max} = \frac{T_1 - T_2}{T_1} \qquad [2\text{--}1]$$

where T_1 is the absolute temperature of the hot inlet gases and T_2 that of the outlet gases. Since $T_2 < T_1$, the maximum efficiency is always less than unity. This so-called Carnot Efficiency is intrinsic to any energy converter that operates between a source and a sink temperature, such as a steam turbine, an internal combustion engine, a thermionic converter, or a magnetohydrodynamic generator. The energy conversion efficiency of a *real* system will be even lower than the Carnot Efficiency because of energy losses incurred during the various steps of the conversion process:

$$\boxed{\text{chemical energy}} \rightarrow \boxed{\text{heat}} \rightarrow \boxed{\text{mechanical energy}} \rightarrow \boxed{\text{electrical energy}} \qquad [2\text{--}2]$$

Characteristics of Fuel Cell Systems

Thus the approximate efficiencies of modern thermal power stations[742] are: nuclear 30–33%, natural gas 30–40%, coal 33–38%, and oil 34–40%.

Since the change in thermal energy between the reactants and products of a combustion reaction is given by the change in enthalpy, ΔH (i.e., the heat of combustion for the fuel), the efficiency of an energy conversion device usually is based on how much energy is produced relative to the enthalpy change for the process:

$$\mathcal{E} = \frac{\text{ENERGY OUTPUT}}{\Delta H} \qquad [2\text{--}3]$$

In an electrochemical converter, such as a fuel cell, the enthalpy change (ΔH) during combustion is not given out completely as heat. What is given out is not hot molecules in random motion, but a stream of electrons, the maximum total energy of which is given by the change in free energy (ΔG), for the combustion reaction. Random motion cannot directly perform work, since molecular motion in any one direction is exactly compensated by equal and opposite motion in another. However, motion in a unidirectional stream represents the direct production of work. In an ideal electrochemical cell, all the free energy change is available as electrical energy; thus the intrinsic maximum energy conversion efficiency of an electrochemical cell is

$$\mathcal{E}_{max} = \Delta G_T / \Delta H_o \qquad [2\text{--}4]$$

where ΔG_T is the ΔG value at T, the cell operating temperature, and where ΔH_o is the heat of combustion of the fuel under some defined standard conditions. Since there are no upper and lower temperature limits involved, the intrinsic efficiency of an electrochemical converter is not subject to the Carnot limitation imposed on heat engines.

We should however make clear that the above expression is also the upper limit for the efficiency of a Carnot heat engine operating under ideal conditions using the combustion of a fuel. This will be demonstrated in the next section. A second important point is that fuels containing hydrogen (including hydrogen itself, hydrocarbons, alcohols, and to a lesser extent coal) have two values for ΔH and for ΔG, corresponding to whether the product water is liquid or vapor. The corresponding ΔH values of the fuels are known as the *higher-heating value* (HHV) and *lower-heating value* (LHV), respectively, differing by the value of the latent heat of vaporation of water. For natural gas (methane), the LHV/HHV is 0.90; for coals of typical hydrogen and water content, it is 0.95~0.98. Unless otherwise stated, the values used in this text are higher-heating values (HHV).

As can be seen from Table 2–1, for most combustion processes, the

TABLE 2-1
Theoretical Reversible Cell Potentials ($E°_{rev}$) and Maximum % Intrinsic Efficiencies for Candidate Fuel Cell Reactions Under Standard Conditions at 25°C

Fuel	Reaction	n	$-\Delta H°$ [1]	$-\Delta G°$ [1]	$E°_{rev}$ [2]	%
Hydrogen	$H_2 + 0.5O_2 \rightarrow H_2O(l)$	2	286.0	237.3	1.229	82.97
	$H_2 + Cl_2 \rightarrow 2HCl_{(aq)}$	2	335.5	262.5	1.359	78.33
	$H_2 + Br_2 \rightarrow 2HBr_{(aq)}$	2	242.0	205.7	1.066	85.01
Methane	$CH_4 + 2O_2 \rightarrow CO_2 + 2H_2O_{(l)}$	8	890.8	818.4	1.060	91.87
Propane	$C_3H_8 + 5O_2 \rightarrow 3CO_2 + 4H_2O_{(l)}$	20	2221.1	2109.3	1.093	94.96
Decane	$C_{10}H_{22} + 15.5O_2 \rightarrow 10CO_2 + 11H_2O_{(l)}$	66	6832.9	6590.5	1.102	96.45
Carbon monoxide	$CO + 0.5O_2 \rightarrow CO_2$	2	283.1	257.2	1.066	90.86
Carbon	$C + 0.5O_2 \rightarrow CO$	2	110.6	137.3	0.712	124.18
	$C + O_2 \rightarrow CO_2$	4	393.7	394.6	1.020	100.22
Methanol	$CH_3OH + 1.5O_2 \rightarrow CO_2 + 2H_2O(l)$	6	726.6	702.5	1.214	96.68
Formaldehyde	$CH_2O_{(g)} + O_2 \rightarrow CO_2 + H_2O_{(l)}$	4	561.3	522.0	1.350	93.00
Formic acid	$HCOOH + 0.5O_2 \rightarrow CO_2 + H_2O_{(l)}$	2	270.3	285.5	1.480	105.62
Ammonia	$NH_3 + 0.75O_2 \rightarrow 1.5H_2O_{(l)} + 0.5N_2$	3	382.8	338.2	1.170	88.36
Hydrazine	$N_2H_4 + O_2 \rightarrow N_2 + 2H_2O_{(l)}$	4	622.4	602.4	1.560	96.77
Zinc	$Zn + 0.5O_2 \rightarrow ZnO$	2	348.1	318.3	1.650	91.43
Sodium	$Na + 0.5H_2O + 0.25O_2 \rightarrow NaOH_{(aq)}$	1	326.8	300.7	3.120	92.00

(1) kJ/mol (1 kJ/mol = 4.184 kcal/mol. 23.06 kcal/electron = 1 V.)
(2) Volts

Characteristics of Fuel Cell Systems

magnitude of ΔG at ordinary temperature is about 90~95% that of ΔH, so that the intrinsic maximum energy conversion efficiency of a fuel cell is very high. In theory, the combustion processes shown in Table 2-1 can be carried out in a fuel cell at room temperature, that is, as cold-combustion processes. In addition, the intrinsic efficiencies for reactions in which there is a decrease in volume in proceeding from reactants to products are greater at lower temperatures than at elevated temperatures (see Table 2-2). It is interesting to note that for combustion reactions which proceed with an increase in entropy (see Table 2-2), energy conversion efficiencies in excess of 100% are theoretically possible. In addition to converting the chemical energy contained in the reactants to free energy in the form of electricity, such reactions theoretically would absorb additional heat from the surroundings and convert this heat to additional electrical energy so long as the heat could be supplied.

Energy Loss Mechanisms

From thermodynamics, it can be shown that the heat (Q) absorbed from the surroundings by a fuel cell operating isothermally and at constant pressure is given by the expression:

$$\Delta H = Q - W_{elec} \quad [2\text{-}5]$$

where W_{elec} is the electrical work output of the cell. If the cell were operating reversibly, then it would be operating at its maximum efficiency, so that the electrical work output could be set equal to $-\Delta G$ for the reaction. Thus, from Eq. 2-5,

$$\begin{aligned} Q &= \Delta H + W_{elec} \\ &= \Delta H - \Delta G \end{aligned} \quad [2\text{-}6]$$

TABLE 2-2

Theoretical Reversible Cell Potentials ($E°_{rev}$) and Maximum % Intrinsic Efficiencies of Selected Combustion Reactions at Different Temperatures

Combustion Reaction	298K		600K		1000K	
	$E°_{rev}$ (V)	%	$E°_{rev}$ (V)	%	$E°_{rev}$ (V)	%
$C + O_2 \rightarrow CO_2$	1.03	100	1.03	100	1.03	100
$2 C + O_2 \rightarrow 2 CO$	0.74	124	0.86	148	1.02	178
$2 CO + O_2 \rightarrow 2 CO_2$	1.34	91	1.18	81	1.01	69
$CH_4 + 2 O_2 \rightarrow CO_2 + 2 H_2O_{(g)}$	1.04	100	1.04	100	1.04	100
$2 H_2 + O_2 \rightarrow 2 H_2O_{(g)}$	1.18	94	1.11	88	1.00	78

Furthermore, the heat absorbed by a reversible (ideal) cell is equal to $T\Delta S$, where ΔS is the entropy change for the reaction at temperature T. Thus for a fuel cell operating reversibly, Eq. 2–5 becomes the familiar relationship

$$T\Delta S = \Delta H - \Delta G \qquad [2\text{-}7]$$

which shows that the ideal heat absorbed by an ideal cell is just the difference between ΔH and ΔG for the cell reaction.

While the remarks at the end of the preceding section should be noted, for most fuel cell reactions ΔS is negative, signifying that heat is released to the surroundings. Since ΔH usually does not vary strongly with temperature, it follows that ΔG will become more positive as the temperature increases. At this point it is worthwhile to consider the implications of the above equation in the case of a Carnot-limited heat engine. When a fuel is burned, the reaction is spontaneous, and the free energy obtained directly as useful electrical work is zero at the temperature of combustion. Under these conditions, Eq. 2–5 shows that

$$\Delta H = Q \qquad [2\text{–}8]$$

Now, if it were possible in the limiting case to carry out the combustion process in a controlled, reversible manner, then the heat Q would be equal to the reversible heat, $T\Delta S$, and, from Eq. 2–8,

$$\Delta H = T_c \Delta S \qquad [2\text{-}9]$$

Thus to a good approximation T_c is equal to $\Delta H/\Delta S$, where T_c is the combustion temperature. T_c is therefore the maximum temperature that can be attained in a Carnot cycle involving the combustion of a fuel. Putting $T_1 = T_c = \Delta H/\Delta S$ in Eq. 2–1, assuming ΔH and ΔS do not vary significantly with temperature, and remembering that $\Delta G_{(T2)}$ is equal to $\Delta H - T_2 \Delta S$, where $\Delta G_{(T2)}$ is the value of ΔG at T_2, rearrangement shows that the maximum efficiency of a Carnot heat engine operating at a sink temperature of T_2 will be $\Delta G_{(T2)}/\Delta H$. This is exactly the equivalent of Eq. 2–4 for the maximum efficiency of a fuel cell operating isothermally at the same temperature (T_2). This point is important, since it has given rise to considerable confusion in the previous literature. A *perfect* Carnot heat engine operating under the maximum temperature limit allowed by the combustion of a fuel would have the same efficiency as that of a perfect isothermal fuel cell using the same fuel and operating at the same sink temperature.

The reason why heat engines only approach a small fraction of this limit is due to materials considerations, which require the dilution of the heat to source temperatures (T_1) within manageable limits (e.g., about 900 K for systems using a steam cycle). The efficiency of the best heat engine (e.g., a high-temperature steam cycle with a condenser) is then about 60–65% of that of the ideal Carnot cycle due to inevitable irreversible losses, giving an overall 40% for a good steam system. The important point about the fuel cell is not that it is not Carnot-cycle limited: rather it is free from the high temperature limit imposed by materials on any Carnot machine giving a similar efficiency.

Like Carnot machines, fuel cells can never approach ideal efficiencies. The heat released to the surroundings by a *real* fuel cell is always greater (i.e., is algebraically more negative) than the reversible heat, $T\Delta S$, by an amount equal to the difference between the reversible and the real electrical work output of the cell, which is given by $n \mathcal{F}(E_{rev} - E_{cell})$, where n is the number of moles of electrons transferred during one act of the overall cell reaction, $\mathcal{F}$ is the Faraday constant (96,487 coulombs/mol), E_{rev} is the reversible cell voltage, and E_{cell} is the actual cell output voltage. Thus, if a current of I amperes flows for a time of t sec, the electrical work output of a real fuel cell, given by $(E_{cell} I t)$, is less than that of an ideal cell, given by $(E_{rev} I t)$. Efficiency of fuel use in the cell is therefore less than the theoretical value. However in practice cell heat losses (i.e., waste or by-product heat) often can be utilized for other processes such as the gasification of a fossil fuel, reactant preheating, the maintenance of the cell operating temperature in the case of a high temperature fuel cell, or simply for on-site heat requirements, so that in practice most of the available energy can be put to good use.

When a net current is drawn from a practical fuel cell, the terminal voltage of the cell, E_{cell}, drops from the open circuit voltage, E_{rev}, by an amount that increases with increasing cell current output. The cell is said to exhibit *polarization*, of which there are three main types: activation polarization, concentration polarization, and ohmic polarization.

Activation polarization results when the charge transfer reaction that occurs across the electrode–electrolyte interface is slow. When this happens, a portion of the electrode potential is lost in driving the electron transfer rate to the rate required by the current demand. Activation polarization depends on factors such as the nature of the electrode material, ion-ion interactions, ion-solvent interactions, and the characteristics of the electric double layer at the electrode–electrolyte interface. Activation polarization may be reduced by increasing the operating temperature, by increasing the active surface area of the electrode (i.e., by reducing the local current density), or by increasing the activity of the electrode through the use of a suitable electrocatalyst.

The best known example of activation polarization in conventional fuel cells is that of the oxygen electrode in aqueous electrolytes, particularly in all acid-electrolyte systems and in cells using alkaline electrolytes below temperatures of at least 150°C, even when the most effective catalysts (e.g., platinum) are used. With effective catalysts, the hydrogen oxidation reaction proceeds readily so that activation polarization is negligible, that is, the hydrogen electrode operates close to thermodynamic equilibrium conditions. For example, on platinum in phosphoric acid electrolyte under practical atmospheric pressure conditions in the phosphoric acid fuel cell (PAFC), the hydrogen electrode may have an effective rate under thermodynamic equilibrium conditions of at least 10^{-2} A/cm^2. This is conventionally known as the *exchange current density* for the electrode reaction. With low-loading high-surface area platinum catalysts (0.25 mg/cm^2, 100 m^2/g), the ratio of true catalyst area to electrode geometrical area will be about 250. Hence, the electrode can sustain a current density of 2.5 A/cm^2 with hydrogen at 1 atm pressure before its potential deviates much from equilibrium. This is about ten times the typical current density used in a fuel cell due to the restrictions imposed by concentration and ohmic polarization (see following).

In contrast, the rate of oxygen reduction in aqueous media at the thermodynamic equilibrium potential is about 10^{-5} times slower than that of hydrogen, even with the best catalysts currently available. At first sight, this would appear to make the situation hopeless, since the rate close to thermodynamic equilibrium, if it could be attained, would only be about 0.25 mA/cm^2. The phenomenon which saves the situation is the fact that, in electrochemical reduction processes, the rate increases substantially as the potential is displaced in the cathodic direction (i.e., negatively, the oxygen electrode potential being considered to be positive with respect to that of hydrogen; the opposite is the case for anodic, i.e., oxidative, processes). The rate in fact increases exponentially as the potential is displaced from equilibrium, according to Tafel's law,[2397] discovered in 1905. Hence, in the case of the PAFC, an approximate potential displacement of 0.11 V (i.e., an overpotential of 0.11 V) causes the rate of the oxygen reduction reaction to increase tenfold. Thus, if the oxygen reaction is carried out at 0.25 A/cm^2 using a typical electrode, the potential is displaced from the equilibrium value by an activation polarization of about 450 mV. This is the major irreversible loss in the PAFC. The problem in other systems will be discussed later; it suffices here to say that rates for the direct oxidation of hydrocarbons or of many other carbonaceous compounds are even lower than those for the reduction of oxygen, illustrating why hydrogen is the preferred fuel.

Concentration polarization is caused by mass transport limitations on the availability of the electroactive species in the vicinity of the electrode–electrolyte interface. The electrode reactions require a constant supply of reactant species to either accept or donate electrons in order to sustain the flow of electric current. Hence, when diffusion or convection limitations reduce the availability of these species the resulting concentration gradients effectively reduce electrode activity, leading to a corresponding loss in output voltage. Similar problems can develop if the products of the electrode reaction overaccumulate near the surface of the electrode, for example by diluting the reactants. The actual physicochemical model involved depends on whether the electrode operates close to its thermodynamic equilibrium potential (as in the case of the hydrogen electrode), or under totally irreversible conditions (as for the PAFC oxygen electrode). If it is pseudoreversible, the system can be regarded as being displaced from equilibrium by the same amount as that for a reversible concentration cell. In contrast, for totally irreversible cases, the reactant concentration controls the local rate of the reaction, which is not affected by the product concentration, and the displacement in potential from equilibrium corresponds to that determined from Tafel's law. Again, a good example of the latter is concentration polarization in the case of the PAFC oxygen electrode; since the electrode reaction rate at constant potential is simply proportional to the oxygen concentration, a tenfold decrease in the latter will lead to a reduction in potential of 0.11 V at the same current density. Both these cases are limiting extremes; in practical systems, they both may be combined in a complex manner under *mixed* conditions.

For the generalized *reversible* cathodic overall process,

$$sR + ne^- = qP \qquad [2\text{--}10]$$

where s molecules of reactant R and n electrons (e⁻) yield q molecules of product P, the change in cathode (i.e., fuel cell positive) potential is given by the Nernst equation:

$$E = -\Delta G/n\,\mathcal{F} = E° + \{2.303RT/n\,\mathcal{F}\}\log_{10}\{[R]^s/[P]^q\} \qquad [2\text{--}11]$$

where R and P are the concentrations (activities) or pressures of the reactant and product, and $E°$ is the standard reduction potential at the particular temperature T in Kelvins (i.e., $E°$ is the value of E when R and P are unity). Hence, increasing reactant concentration increases potential, and increasing product concentration decreases it. For an anodic process, the signs are reversed, so that $E°$ becomes more negative as reactant concentration increases. However, since the output voltage of a fuel cell is the *absolute* difference between the cathode and the anode potentials, increasing pressures or concentrations of reversible reactants at both anode and cathode would increase overall cell potential and vice versa; for products the effect would be opposite. As remarked above, the Nernst equation is not obeyed by the irreversible oxygen reduction process in aqueous media, however the hydrogen process (for which $E°$ is nominally put equal to zero to provide a measuring scale) does obey it on effective catalysts. In high-temperature fuel cells, both anodes and cathodes are considered reversible.

As an illustration, Fig. 2–1 shows typical overpotential/log current density relationships in a diagrammatic form for the PAFC. Due to its low exchange current, the Tafel polarization region for the oxygen electrode is several decades long before the rate is sufficient to deplete the oxygen reactant concentration. When this occurs, a rapid deviation from the linear potential (V)/ log current density (i) relationship occurs, leading to the limiting current region where the reaction is entirely controlled by the maximum surface concentration allowed by the rate of oxygen diffusion to the active part of the electrode. With state-of-the-art high-surface area electrodes this occurs at current densities over 1 A/cm^2 at 1 atm oxygen pressure. In contrast, the hydrogen electrode, with its much larger exchange current, shows no Tafel region at all, and its limiting current is slightly higher than that for oxygen due to its improved diffusion characteristics. At a current density of 0.25 A/cm^2, the cell potential corrected for internal resistance losses is about 0.7 V, all the loss being on the oxygen side. This corresponds to a theoretical reversible open circuit value of about 1.2 V.

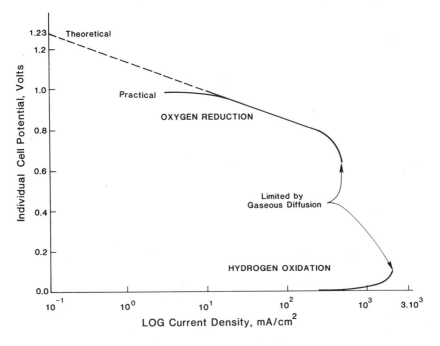

Figure 2–1: Typical polarization characteristics of an aqueous electrolyte (phosphoric acid) fuel cell.

In order to avoid or minimize both activation and concentration polarization, cells are constructed so that the reactants can be supplied to as large an area of electrode (in terms of real cm^2 per geometric cm^2) as possible. This is one of the major reasons why fuel cell electrodes are normally made highly porous. Forced convection of reactants also may be helpful in avoiding concentration gradients.

In this connection, a very important point is the difference in the characteristics of cells using pure reactants and those that must use impure reactants. Cells of aerospace type operating on pure hydrogen and oxygen essentially can be operated *dead-ended*, that is, the reactants are at constant pressure everywhere in the cell. Product water then can be removed from the electrolyte via a separate evaporator. Such cells therefore have a theoretical open-circuit potential close to the standard value for the particular temperature and pressure conditions of operation; and apart from a small continuous or intermittent bleed to remove accumulated trace impurities, they can consume the entire amount of reactants supplied. In contrast cells operating on impure fuels and oxidants, for example hydrogen-CO_2 mixtures derived from reformed hydrocarbons and ambient air, clearly cannot be operated dead-ended, since the unreacted impurities must be removed continuously from the cell. Nor can they consume more than a certain percentage of the hydrogen fuel (generally 80–90%; the percentage of the oxygen consumed in the air is much less important), otherwise the resulting polarization would become too great close to the fuel cell exit. In consequence, the theoretical open-circuit voltage of such cells when operating under normal conditions of high fuel conversion is considerably less than the value under standard conditions, because of the effect of pressure on ΔG or alternatively on the open-circuit potential. Such practical cells are therefore penalized from the efficiency viewpoint not only because they do not operate under standard ΔG conditions at high fuel utilization, but also because they cannot consume all the fuel supplied. Efficiency will be a descending function of utilization from the viewpoint of the voltage developed by the cell, and an ascending function of utilization from the viewpoint of percentage of fuel that can be consumed. The efficiency will therefore pass through a maximum optimum value as a function of utilization. The hydrogen apparently wasted at the anode exit can however be used in the fuel processing system, for example, to supply the heat of reforming, so that it does contribute to the efficiency of the overall system (hydrocarbon fuel to electricity).

Concentration polarization due to concentration gradients also can occur in the electrolyte itself. This can be reduced by using highly concentrated solutions to ensure an adequate supply of ions to the electrodes, and in some cases by using forced convection of the electrolyte.

Ohmic polarization, also known as resistance polarization, results from electrical resistance losses in the cell. These resistances can be found in the electrolyte (ionic), in the electrodes (electronic and ionic), and in the terminal connections in the cell (electronic). In order to minimize these losses, electrolytes of the highest conductivity possible are chosen, and fuel cell electrodes are designed to have high electronic conductivity and are closely spaced to minimize electrolyte resistance.

Fuel Cell Thermodynamics

At constant pressure, the heat absorbed per mole of reaction for a real fuel cell can be expressed as

$$Q = \Delta H + W_{elec} \qquad [2\text{--}12]$$

where Q is the heat in J/mol and W_{elec} is the electrical work delivered by the cell in J/mol. Since, for essentially all fuel cell reactions, heat is evolved, then Q will have a negative value (i.e., will be a heat loss). This expression also can be expressed as

$$Q = T\Delta S - n\mathcal{F}\{\Sigma|\eta| + IR_{int}\} \qquad [2\text{--}13]$$

where $\Sigma|\eta|$ is the sum of the activation and concentration polarizations, I is the total cell output current, and R_{int} is the sum of the cell internal and other ohmic resistances. The rate of heat loss per mole in the cell also can be expressed as an equivalent power loss (in watts), given by

$$P_{heat\ loss} = I\left\{\frac{T\Delta S}{n\mathcal{F}} - \Sigma|\eta|\right\} - I^2 R_{int} \qquad [2\text{--}14]$$

The expressions just given are important for calculating the cooling requirements of the fuel cell stack.

The efficiency of a fuel cell system normally is expressed in terms of its *heat rate*. This is defined as the ratio of the HHV of the fuel used in the system to the electrical energy output. The lower the heat rate, the more efficient the system. Following normal power plant usage, in the United States it is expressed in BTU/kWh (3413 BTU = 1.00 kWh$_{th}$). It is expected that the initial design heat rate for the first generation of PAFC central power station fuel cell systems operating on reformed hydrocarbon fuels will be about 8300 BTU/kWh (8755 kJ/kWh, 2.43 kWh$_{th}$/kWh),[1087] although it could be as low as 7900 BTU/kWh (8330 kJ/kWh, 2.31 kWh$_{th}$/kWh),[692] the latter corresponding to a heat efficiency (HHV) of 43.2%. The best estimate in 1983 was about 8300 BTU/kWh (41.1%) by the end of life of the fuel cell.[177,1087] Not only are these heat rates better than those for any

Characteristics of Fuel Cell Systems

fossil steam units in the United States; but, unlike fossil steam or gas turbine units, they do not increase substantially under part-load operating conditions.

Since $E_{cell} < E_{rev}$, the fuel cell *voltage efficiency* is given as

$$\mathcal{E}_v = \frac{E_{cell}}{E_{rev}} \qquad [2\text{-}15]$$

where E_{rev} can be defined in various ways, depending on the circumstances. For example, since E_{rev} is numerically equal to $-\Delta G$ for the overall process expressed in eV, E_{rev} (cf. Eq. 2–7) falls with the operating temperature of the cell. Similarly, it varies according to whether the reaction product (typically water) is in the form of vapor or liquid water. Since it also depends on the concentrations of the reactants and products, it depends on the definition used for the partial pressure of reactants and product water, that is, whether these are defined under fuel cell entry or exit conditions. All practical fuel cells liberate water from the cell as vapor, and therefore strictly speaking, they operate on the lower free energy value of the fuel (i.e., they do no work by condensing the liberated water vapor). In consequence, the theoretical open-circuit potential under exit reactant conditions in the PAFC operating at 1 atm pressure with normal reformate fuel gases and utilizations is about 1.10 V at 200°C (473 K), and varies only by a negligible amount (about 20 mV) at an operating pressure of 8 atm. In the atmospheric-pressure molten carbonate fuel cell (MCFC) operating at 650° C (923 K), the corresponding value has fallen to close to 0.9 V, to reach only about 0.8 V in the solid oxide fuel cell (SOFC) operating at 1000°C. These changes result from the variation of ΔG with temperature, and are one of the unfortunate consequences of reversible thermodynamics (Eq. 2–7). In consequence, it has become habitual to use the maximum attainable value of E_{rev} under ordinary conditions (1.229 V with hydrogen and oxygen in their standard states, with liquid water as product at 25°C, the value being corrected to the initial partial pressures of the reactants) even though in practice this is somewhat meaningless.

There is also the possibility that in some cases part of the reactants may participate in a nonproductive side reaction. While this does not occur with hydrogen and oxygen, unless gas cross-over occurs, it is seen, for example, in the autodecomposition of hydrazine to nitrogen and hydrogen in hydrazine fuel cells, where it results in incomplete conversion of reactants to desired products. This is accounted for by the *Faradaic efficiency*, $\mathcal{E}_F$. The Faradaic efficiency is also known as the *current efficiency*. Fortunately, for most practical fuel cell systems, the Faradaic efficiency is very close to unity.

The overall free energy conversion efficiency in the cell is given by

$$\mathcal{E}_{overall} = \mathcal{E}_{max} \cdot \mathcal{E}_v \cdot \mathcal{E}_F \qquad [2\text{--}16]$$

This expression must be divided by $\Delta G/\Delta H$ for the cell reaction to obtain the efficiency in terms of the heating value of the fuel consumed in the cell. This in turn must be corrected for the utilization of fuel in the cell, as explained earlier, and finally it must be multiplied by the efficiency of any transformations required in the fuel pretreatment system, including parasitic power losses (e.g., in steam-reforming natural gas to produce the required hydrogen), and that for any electric transformations required, such as d.c.–a.c. conversion.

In general, the 35~40% upper performance for the HHV efficiency of most systems limited by Carnot cycles[1169] is the minimum efficiency for a practical fuel cell system. While this is lower than the usual part-load efficiency of the cell, it results from the power/voltage characteristics of a fuel cell, since the maximum power is obtained at close to 50% of the open circuit voltage when the voltage/current curve is linear. Free energy efficiencies as high as 60~80% have been obtained for fuel cells,[2122] and, in theory at least, there are no reasons why the maximum intrinsic energy conversion efficiencies indicated in Table 2–2 cannot be closely approached under optimum operating conditions and with steady improvements in electrocatalysis.

B. ELECTRICAL OPERATING CHARACTERISTICS OF FUEL CELLS

Many of the operational characteristics of fuel cell systems are superior to those of conventional power generation systems. Conventional heat engines operate most efficiently at full load, and show a rapid fall-off in efficiency at part load. Like that of a battery, the voltage (i.e., the cell efficiency) of a fuel cell is higher at part load than at full load. However, in most cases the characteristics of the fuel cell itself reflect only part of those of the overall system, particularly when the system includes a fuel processor (e.g., a steam reformer and its associated subsystems). The latter tends to have a lower efficiency at part load, due to heat transfer losses and parasitic power requirements. Consequently, the overall efficiency of a fuel cell unit incorporating a fuel processor tends to be approximately independent of load, but may start to fall off at loads of less than 25% of full load.

Such a fuel cell system is therefore able to follow changes in the power load accurately and without economic penalty on account of its relatively flat heat rate curve. For example, the 8300 BTU/kWh heat rate of United

Technologies Corporation's 11-MW phosphoric acid fuel cell essentially will be constant between 50 and 100% of rated power; furthermore, the plant can operate at as low as 30% of rated power.[1616,2397] Similarly, the 40% overall electrical efficiency of the on-site integrated energy fuel cell system being developed by the U.S. gas and electric utilities is constant over a range of from 50 to 100% of its rated power output.[2728] This unit, rated at 40 kW, can sustain transient outputs of up to 56 kW for five seconds.

In addition to its relatively constant efficiency over a wide range of operating conditions, a fuel cell system also has a very fast reactive power response.[692,1616] For example, the 4.5 MW (a.c.) demonstrator units in New York and Japan were designed to respond to load transients in only 0.3 seconds. In order to simplify the control system and lower expenses, the 11 MW units now under development are being designed to respond to each 1 MW load change in one second, which is still a very fast response time. This fast power response should permit a utility to lower its need for on-line spinning reserve, since, when required, such reserve can be provided almost instantaneously by the rapid reaction time of the fuel cell.

On account of their fast reactive power response capabilities, fuel cells can provide alternatives to shunt reactors and capacitors, since they are able to react to and dampen system voltage surges and swings such as may occur during switching operations or lightning storms. Similarly, if a short circuit occurs in part of a series-connected fuel cell stack array, the short circuit current generated will be limited, thus reducing the necessity for high-current circuit breakers. This feature of this type of generator may enable a utility system containing a significant number of fuel cells to better survive transmission cascading, voltage disturbance and associated frequency regulation that often have resulted in widespread blackouts in the past.

C. AIR POLLUTION

The major source of air pollutants is the consumption of fuel for heat and energy. Coal and its by-products, natural gas, fuel oil, and gasoline are the chief contributors to community air pollution, which has been demonstrated to have adverse biological effects.[2638] These effects are produced not only by the gases and particulates that issue directly from smokestacks and automobile exhausts, but also by secondary pollutants derived from complex chemical interactions between emitted gases and atmospheric constituents over a relatively long time-scale.

Vegetation also is extremely sensitive to toxic air pollutants.[2445] Sensitivity to specific pollutants varies widely among species and sometimes among strains within a single variety. The syndrome produced by a given pollutant may be influenced by the concentration of the pollutant, the length of exposure, climatic conditions, as well as by the characteristics of the exposed vegetation. Chlorotic and/or necrotic lesions characteristic of injury from a specific type of toxicant may be produced by short-term, high-concentration exposure of sensitive crops. Less specific symptom expressions, such as growth suppression, abscission of plant parts, pigment alterations, and distortion of growth tend to be produced by long-term, low-concentration exposure to phytotoxic pollutants.

Sulfur dioxide, fluorides, nitrogen dioxide, particulates, and acid aerosols threaten certain types of vegetation where coal is burned in large quantities. Combustion of fuel oils may produce some of the same types of toxic pollutants, and natural gas under certain combustion conditions may contribute significantly to the atmospheric concentration of nitrogen oxides. Nitrogen dioxide is well known to be toxic, and has been shown to be a key factor in the reaction of certain hydrocarbons in the presence of sunlight to produce ozone and peroxyacyl nitrates to yield highly toxic oxidative air pollutants.

In addition to the more obvious hazards outlined above, there also is the possibility that air pollution may bring about more subtle, far-reaching changes on the earth's climate, including the so-called greenhouse effect.[645, 789, 2030]

Most of these air pollutants are emitted from fossil fuel–burning power plants and from automobile internal combustion engines. For nuclear power plants, air pollution aspects appear minimal, as long as no major malfunctions occur. The greatest problem with nuclear power in the future will involve the disposal of highly radioactive wastes.[1226, 2003]

In contrast to the above, the impact of fuel cells on the environment should be minimal. Fuel cells generate clean power. Table 2–3 shows the measured exhaust emissions from the 1 MW prototype natural gas–burning fuel cell developed and tested by United Technologies Corporation.[692, 1400] It is immediately evident that its emission level was much lower than that permitted by the U.S. Environmental Policy Act. Furthermore, a study funded by the U.S. Environmental Protection Agency (EPA) showed that not only are fuel cells themselves more efficient and environmentally benign than conventional power generators, but also that the same advantages are apparent from the resource extraction stage to the final end use of the electricity they generate.[1324] The latest available data were obtained with the UTC 4.5 MW (a.c.) natural-gas PAFC prototype unit tested at the Tokyo Electric Power Company (TEPCO) during 1983–1985. Emissions design values were NO_x 20 ppm, SO_x 0.013 g/GJ. Measured values were < 10 ppm, and zero (i.e., unmeasureable) respectively.[1088] We should note that SO_x emissions standards for new fossil fuel plants in developed countries are about 200 g/GJ, more than four orders

TABLE 2-3

Power System Exhaust Emission Comparison[692, 1400]
(kg/1000 kWh)*

	TYPE OF POWER STATION				
Contaminant	Gas-fired	Oil-fired	Coal-fired	FCG-1 Fuel Cell	EPA Limits
SO_x	—	3.35	4.95	0.000046	1.24
NO_x	0.89	1.25	2.89	0.031	0.464
Particulates	0.45	0.42	0.41	0.0000046	0.155

*1 kg/1000 kWh = 0.646 lb/million BTU

of magnitude higher than that from the fuel cell. Finally, the fuel cell NO_x emissions, while very low, result from the burner in the fuel processing system. In second-generation fuel cell systems with improved cathode chemistry (cf. Chapter 17), NO_x emissions should be negligible.

In the shorter term, the low-emission characteristics of fuel cell power plants will enable some utilities to defer or offset the costs of installing additional scrubbers or of emission control equipment. Indeed, in many jurisdictions, re-equipment with fossil fuel plants (such as fuel cells) that are below the federal or state emission limits will enable emissions from existing facilities to be increased proportionately. In the longer term, the environmentally benign characteristics of fuel cells should help to reduce the social and material costs of pollution that normally are associated with economic development, particularly in frontier regions such as the Canadian North.

A decision to substantially reduce pollution involves a substantial commitment of resources. Cost-benefit analyses have been made that provide a systematic framework to estimate the cost to society of committing these resources to pollution abatement and of its benefits to society. For example, in an early study made in 1971, it was estimated that the benefits of lowering air pollution by 50% in the largest North American cities would result in benefits in excess of $8 billion per year at a cost of only about $1.3 billion per year,[1578] both expressed in 1971 dollars. In addition to the savings in the economic cost of illness and mortality, there also would be savings in cleaning, maintenance, and life (e.g., of structures and buildings); savings corresponding to animal and plant life; and beneficial psychological effects including improvements in visibility, in the number of sunny days, and in the general quality of life.

D. NOISE POLLUTION

Noise interferes with sleep, tranquility, communication, and even with the ability to think rationally. Since the processes that take place in a fuel cell are electrochemical reactions, ideally there would be no moving parts to generate noise and vibration. In practice however, pumps, blowers, transformers, and other auxiliary devices are generally required. Despite this, fuel cells operate much more quietly and smoothly than most other power-generating devices. Any potential noise problems (e.g., those from turbocompressors) can be readily solved by the use of soundproofing or baffles. The noise level at 30 meters from a fuel cell power plant has been measured to only about 55 decibels.[692] In this regard, fuel cell generators are ideal for dispersed urban or suburban power generation.

E. THERMAL POLLUTION

Because fuel cells are more efficient than Carnot-limited power generators, especially in small units, they produce less waste heat. Small amounts of heat would be produced even if the fuel cell system does not have a separate fuel processing subsystem to generate hydrogen from a fossil fuel. This will depend on the future availability of hydrogen fuel, or alternatively on the development of integrated fuel processing (e.g., internal reforming of methanol or natural gas in high-temperature fuel cells (see Chapters 3, 14, 17). When the waste heat that is generated is not utilized for auxiliary processes or on-site heating, it normally can be rejected directly into the air rather than into a body of water.

F. VISUAL AESTHETICS

Although the fuel processing subsystem of a fuel cell generating station may tend to resemble a miniature refinery, it does not have a striking visual impact. In addition to there being no smokestacks or obvious cooling towers, it is possible to enclose the low profile subsystems within buildings, as is being proposed for urban use in Japan. With the possible exception of photovoltaic generators, a fuel cell generating system is probably the only type of power plant that can be located without intrusion in downtown, residential, or scenic areas. This results from its low chemical, thermal and acoustic emissions as well as from its small visual impact. In this connection, fuel cell generators will be the only power systems that will receive a license to operate in Manhattan in the future.[113]

G. SAFETY

Fuel cells certainly are as safe as coal–, oil–, and natural gas–burning central power stations, and they should in the long run be safer than nuclear power stations on account of the possiblility of disaster in the latter—resulting from systems or materials failure, neglect, and more particularly by sabotage or acts of war.[1226, 2003] In this connection, distribution systems using dispersed power generation will always be safer and inherently more reliable than those using centralized generation.

H. EASE OF MAINTENANCE AND LOW MANPOWER REQUIREMENTS

Since the power generating processes that occur in a fuel cell are essentially mechanically static with auxiliary circulation devices operating under relatively benign conditions, they can be designed so that maintenance should be required only at infrequent intervals. Above all, they can be used for unmanned operation with automatic or remote dispatch of electricity generated, which is vital for the economics of dispersed generators, for which manpower costs would otherwise predominate.

Depending on system requirements, the overall hardware can be as complex or as simple as required. For example, pressurized units with computerized feedback or feedforward subsystems between the heat- and energy-integrated fuel cell stack and fuel processor will be quite complex. Such systems are necessary to obtain the highest possible efficiency and rise-time for electric utility use. On the other hand, atmospheric pressure units for continuous use in, for example, developing countries can have more modest characteristics. These can indeed be designed so that only the simplest tools and skills will be needed for maintenance activities, provided that plug-in replacement modules (e.g., for the fuel cell stack) are available at short notice. The latter will be an important feature because the limited availability of technically trained manpower for the installation, operation, and maintenance of power supply equipment is a characteristic common to both frontier and rural areas, where the use of transported skilled manpower will add significantly to operating cost. Indeed, the lack of skilled personnel is often cited as the reason for the failure of rural electrification programs involving the use of small, conventional on-site generators. Reference 2538 mentions a case in which more than one third of the installed capacity in one country was out of operation for a considerable period of time because of the lack of repair capability. Thus, fuel cell power systems which require few technically trained personnel, or which permit less expensive, more rapid training of indigenous manpower, should gain wide acceptance in developing and rural areas.

I. RELIABILITY

The U.S. aerospace program has proved the reliability of fuel cells for steady, uninterrupted power generation. Design reliabilities of greater than 95% have been achieved consistently under the stringent conditions of space flight. No fuel cell battery in NASA's program has yet failed. (The often-cited failure on board Apollo 13 in 1970 was that of a cryogenic oxygen tank, not of a fuel cell battery.) The design reliabilities required for terrestrial use are much lower than those needed for man-qualified aerospace applications. It is expected that the first generation of commercial fuel cell power generating systems will have an on-stream availability of about 90%.[692]

The high reliability of fuel cell systems is largely attributable to their lack of moving parts operating under extreme conditions, their ease of maintenance, and their modularity. The latter is discussed in more detail below.

J. MATERIALS

Like other emerging technologies, fuel cells do tend to be materials-limited. However, their problems in this respect are certainly much less critical than those associated with magnetohydrodynamics, nuclear fission, and nuclear fusion. This topic is considered in detail in Chapter 13, Section A.

K. MULTI-FUEL ABILITY

Although the first generation of electric utility fuel cells can consume only hydrogen as fuel, the hydrogen does not need to be pure and may be from a variety of potential sources, including steam-reformed natural gas or light petroleum distillates such as JP-4 or naphtha, or from steam-reformed alcohols such as methanol. These fuel cells also can use the hydrogen contained in synthetic fuels such as remotely-manufactured high, medium, or low heating–value coal gas. This versatility of the fuel cell will enable it to be readily adapted to future changes in energy sources, and should ensure that it will not become redundant or obsolete upon the unavailability of certain fuels. In addition, large additional capital investments are not required to switch from one steam-reformable fuel to another.

Characteristics of Fuel Cell Systems

This aspect of fuel cell operation has been studied by UTC under contract to the Tennessee Valley Authority (TVA). Tables 2–4 and 2–5 show some of the results.[166] Table 2–4 shows the effects on plant cost, and efficiency of using alternate fuels in a phosphoric acid fuel cell power plant that was designed for use with natural gas or naphtha. The baseline configuration includes fuel cleanup, a steam-reformer, a shift reactor to reduce CO to less than 1.5% to prevent anode catalyst poisoning, and the heat exchanger equipment required to integrate the waste heat produced from the fuel cell power sections to the fuel treatment subsystem.

It is noteworthy that the electrical energy conversion efficiency of the plant (input fuel to a.c. power) is almost independent of the fuel used, since the heat requirements for the fuel processor derive from the waste heat from the power section (steam and anode tail-gas) and not from the burning of extra fuel. In this connection, we should remember from Table 2–1 that upgrading the HHV of propane or methane (about 111 kJ/equiva-

TABLE 2–4

Impact of Using Different Fuels with a Phosphoric Acid Fuel Cell System Designed for Use with Naphtha or Methane(166)

	Naphtha or Methane	Medium BTU Coal Gas*	Methanol	Hydrogen
Cost Impact, % above or (below) baseline	0	(1) ~ 5	1	(11)
Efficiency, %	41	41	41.5	42

*depends on specific gas composition

TABLE 2–5

Impact of Retrofitting a Fuel Cell Power Plant for Use With Alternative Fuels[166]

Fuel	Heat Rate (kJ/kWh)	(BTU/kWh)	Capital Cost (% change)
Methane (baseline)	8808	8350	0
Medium BTU Gas			
• with methane	8818	8360	+0.4
• without methane	8808	8350	+4.3
Methanol	8660	8210	+1.0
Hydrogen	8544	8100	+1.1

lent) or of methanol (about 121 kJ/equivalent) to that of hydrogen (143 kJ/equivalent) requires a net energy input. If hydrogen were available on site as a fuel (e.g., as a byproduct from the chlor-alkali industry), a total capital savings of about 11% could be realized.

There are special cases where a more efficient and cheaper plant would result if it were designed for one fuel. Two of these are a dedicated methanol-fueled system (which has the advantage of simplified, low-temperature reforming), and a plant designed only for use on pure hydrogen, should this be available. For the methanol plant, an overall HHV efficiency of about 49% should be possible, with a capital savings per kW of about 15%, resulting from the higher efficiency. If a complete plant were designed for use with hydrogen only, then much more significant savings could be realized, because the entire fuel processing sections and water recovery subsystems could be eliminated. In this latter case, a plant cost reduction of at least 25% and an absolute HHV efficiency of almost 50% may be expected. Another case is a dedicated coal-gas plant. In the latter, the integrated steam reforming unit is unnecessary and the fuel cell stack can be directly supplied with remotely-manufactured low-CO gas, or can eventually be integrated on site with a coal gasifier.

Table 2-5 shows the effects of retrofitting a plant that has been designed for use with natural gas to enable it to run on alternative fuels. As indicated in this table, and concluded by the TVA/UTC study, the equipment modifications required to retrofit a phosphoric acid fuel cell power plant to an alternate fuel are not extensive. Furthermore, the use of coal-derived fuels will have no impact on power rating, and only a very slight impact on heat rate or capital cost.

L. SITING FLEXIBILITY

Owing to the modular construction of a fuel cell power plant, the amount of land required for the installation of a single unit will be small. For example, the 11 MW utility design being developed by United Technologies Corp. will occupy only about 0.5 hectares (1.2 acres).[1616] In addition, the unit can be made to be independent of a water supply. These desirable characteristics permit a fuel cell power plant to be located in remote, relatively inaccessible sites that might be unsuitable for conventional generation units. In addition, its environmental acceptability and lack of concerns for safety will allow it to be sited close to the point of use; for example, in urban areas, where its waste heat may be used in cogeneration applications.

This dispersed siting cabability of the fuel cell not only permits a reduction in transmission and distribution losses, but also provides for active and reactive power control at the actual load center and enables a utility to defer transmission and distribution investments. The lack of siting restrictions coupled with the fact that on-site generation will reduce the need for transmission corridors together with their right-of-way, results in highly acceptable land-use characteristics for fuel cells.

M. MODULARITY

Both of the 4.5 MW (a.c.) PAFC systems that have been tested in the United States[113] and in Japan,[1088] as well as the future commercial systems now in the planning and development stage,[1087] have been designed so that many major components of the various subsystems can be pre-assembled in modular form at the factory. Each UTC fuel cell power module is expected to be capable of generating about 11 MW (a.c.) based on the 1983 design characteristics.[1087] Many of the advantages of fuel cell power systems are attributable to their modularity, which derives from the inherent scalability of individual fuel cells and fuel cell stacks. Unlike most technologies, the results obtained from small bench-scale fuel cells having electrode areas of 10~100 cm^2 directly can be applied to estimate the performance of prototype-size units having electrode areas of 1 m^2. Since a large fuel cell power plant uses more, not bigger, modules, it is found that small cycling plants of only a few MW will be as efficient as large baseload plants in the several hundred MW class. This means that a fuel cell power plant can achieve high efficiency, independent of plant scale. This characteristic lowers technical risk and capital cost burdens.

Because large parts of the system, including the cell stacks (the only item involving appreciable technological risk) will arrive at the site pre-assembled and pre-tested, construction times can be very short. Installations are of a repetitive nature, so that site preparation can be standardized, allowing low costs per MW compared with those for conventional power plants. In essence all that will be required is the laying down of electrical and other services and concrete pads for the system units.

The high reliability of a fuel cell system also stems largely from its modularity. Because the modules are repetitive, any problems that arise will tend to be similar instead of one-of-a-kind, permitting the rapid build-up of highly competent well-trained personnel. Similarly, testing can be done on a repetitive basis, which enables rapid feedback of any potential problems to the manufacturer or designer in time to be useful.

The use of modular units permits a site layout that can be designed to permit the replacement of complete modules, which not only allows for a more economical use of spare parts, but also minimizes lost output. Also, by replacing spare modules, a plant could be operated at full power during periods of routine maintenance, should this be necessary. Even without spare parts, plants could be designed so that only partial shutdown will be necessary in the event of failure.

N. ECONOMIC BENEFITS

It has become axiomatic that both power generation systems and transmission networks should rely on large-scale systems to achieve economies of scale. Because of this, build-up of capacity tends to occur in quantum leaps in anticipation of future demand. Hence, large amounts of system capacity must remain idle while the demand matures. This problem is exacerbated by the fact that in developing rural and frontier areas demand growth is unpredictable, especially in the early stages of power availability. Because of the overriding importance of the efficient use of capital, there is a need for an approach permitting capacity addition that is in phase with demand growth, thereby allowing the capital input and power capacity in a system to be utilized at all stages of its development.

The use of fuel cells can provide such an improved use of construction capital because system capacity can be added in small increments that are consistent with current demand growth. On account of their modular nature, the lead times for fuel cell system construction will be short, typically less than three years from the initial planning stages to final installation. This is in sharp contrast with the 10- to 14-year lead times required for conventional power plants.* Their short lead times, combined with their high efficiency even at multikilowatt-class ratings, will permit fuel cells to be added annually to the generating mix, whereas conventional large central station systems would not be fully utilized for five to ten years after building commences. This may result in capital charges that are 30 to 50% lower for fuel cells[1400] than for conventional central station plants.

The ability to add capacity in small increments, as required by increased demand, minimizes the penalties and business risks that result from inaccurate load forecasting and provides a significant degree of financial flexibility for planners. For example, by gradually adding a few smaller units per year, a relatively constant annual cash flow can be achieved. During short periods of relatively high interest rates it is possible for a utility to limit its short-term debt by delaying the installation of one or two

---

*Simple-cycle combustion turbines also can be constructed within 2~3 years, however they are very inefficient (typically 25–28% HHV) when used with hydrocarbon fuels in units in the 1–10 MW class. Since they are combustion machines producing relatively high nitrogen oxide levels, their environmental acceptability is also much less than that of fuel cells.

small incremental fuel cell units. This type of planning flexibility is possible with fuel cell power plants, whereas making even modest changes in the installation schedule of a large baseload conventional power plant is very capital-intensive.

The Fuel Cell Users Group (FCUG), which had members representing about fifty utility organizations, has worked closely with the Electric Power Research Institute (EPRI) to determine in more detail the benefits to be obtained from the use of fuel cells.[166] The results of these efforts, obtained after close consultation with 25 U.S. utilities, representing about 15% of total U.S. electrical generating capacity, are summarized in Table 2-6. This table indicates in a quantitative manner the savings in 1982 dollars that may be made by installing fuel cells instead of conventional power plants. Another report has estimated that the use of fuel cell power plants in dispersed siting applications could result in savings of up to $154/kW in deferred transmission and distribution costs, up to $26/kW in reduced line losses, $10~20/kW in reactive power control, and up to $200/kW in environmental credits.[2118] The most complete and up-to-date assesssment of fuel cell benefits to utilities was published in April 1986.[122] This report was prepared under EPRI sponsorship for FCUG by Temple, Barker and Sloane.

Because of high construction costs, budget constraints, and uncertainties in load growth, few U.S. utilities are currently building new power plants. In spite of this, many projections indicate that utilities will require additional generation capacity, especially in urban and suburban areas, in the late 1980s and early 1990s.[1616] If these predictions materialize, utilities

TABLE 2-6

Value of Benefits to Be Derived from Fuel Cell Power Plants[166]*

Benefit	Value ($/kW capacity)†		
	Low	High	Average
Air Emission Offset	0	217	11
Spinning Reserve	0	33	8
Load Following	0	52	7
VAR Control	11	22	12
Transmission and Distribution Capacity	0	247	59
Transmission and Distribution Energy Losses	8	35	15
Cogeneration	0	442	56

*Average of values selected by 25 utility planners
† 1982 dollars

will have no time to install conventional large-scale power plants. Fortunately, it appears likely that the first commercial fuel cells will become available at about this time. Fuel cells may therefore be the solution for future utility load growth.

O. THE POTENTIAL FUEL CELL MARKET

Due to the high cost of the fuel, fuel cell systems using premium fuels (natural gas or light distillates) will be suitable only for the intermediate duty and peaking electric utility markets. The electric utility baseload will continue to be provided in the forseeable future by plants using the cheapest sources of energy, namely coal, nuclear, and hydropower. While baseload plants eventually may use fuel cells operating on coal gas, the first units to be introduced will be intermediate duty electric utility units, or on-site natural gas cogeneration units. Market surveys have therefore concentrated on these two classes of fuel cell systems up to the present time.

For electric utilities, fuel cells must compete with other established technologies using the same premium fuels. These technologies are the simple-cycle gas turbine, which is characterized by low initial cost, but low overall thermal efficiency; and the higher cost, higher efficiency combined-cycle gas turbine. The latest issue of the EPRI Technical Assessment Guide (EPRI-TAG),[124] which lists the characteristics and costs of different generating technologies, uses as a general guide $310 (January 1986)/kW as the capital cost of a simple-cycle gas turbine, which in large units (>100 MW) may have an HHV efficiency approaching 30%, or 25% in small units (5 MW). For interested readers, a current listing of heat rates of commercial simple- and combined-cycle gas turbines is given annually in Ref. 123. The combined cycle unit has a capital cost quoted in the EPRI-TAG[124] of $650/kW, again in January 1986 dollars. Such machines incorporate an additional steam cycle to recover waste heat from the gas turbine exhaust, and are typically available in sizes above 100 MW, though some smaller systems exist.[123]

The efficiency characteristics of fuel cells compared with other generators, as a function of load and unit size, are shown in Fig. 2–2.[177] For the thermal machines, the characteristics are for units available today, whereas for the PAFC, the efficiencies as a function of load are those expected for UTC's 1983 design for an 11 MW unit,[1087] three of which are combined to give the 33 MW unit quoted. The characteristic of the molten-carbonate fuel cell units as a function of load are those projected for internal-reforming (IRMCFC) natural-gas systems (see Chapters 3, 14, 17). It is clear that the fuel cell systems possess the great advantage of higher efficiencies, particularly in small sizes and at intermediate loads, which will result in the operating credits discussed earlier in this chapter. These, combined

Characteristics of Fuel Cell Systems

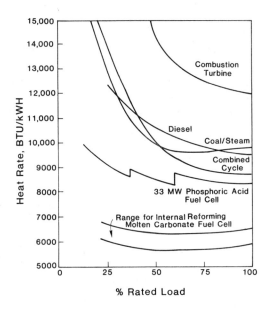

Figure 2–2a: Heat-rate (higher heating value) as a percentage of rated load for different generation technologies.

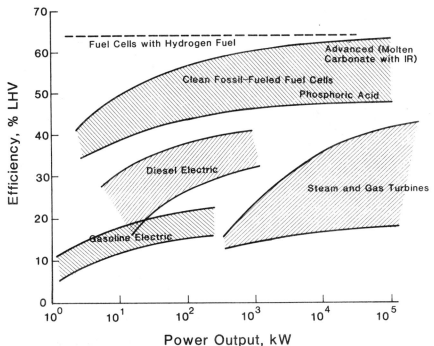

Figure 2–2b: Lower-heating-value efficiency of different technologies as a function of scale.

with the dispersibility of the PAFC, should make it highly competitive with the combined cycle unit, which is less efficient even at full load. The future IRMCFC should be much better.

Unfortunately, considerations of fuel efficiency alone only tell part of the story, since the cost of the fuel is only part of the cost of operating a generator. The fixed costs (interest on capital and depreciation, together with other costs such as liability insurance), as well as the operating and maintenance (O&M) costs (man-hours of work and cost of replacement parts) must also be taken into account. For generating technologies available today, the corresponding fixed costs, including O&M costs, are shown as the intercept on the y-axis of Fig. 2-3. They are taken to be 20% of the per

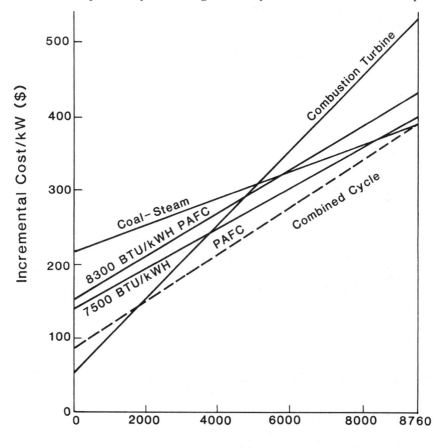

Figure 2–3: Illustrative screening curves for different power plants. assumptions: Capital costs as in Ref. 183 (20% annual capital charges): Coal cost, $2/MMBTU; clean hydrocarbon cost, $4/MMBTU.

Characteristics of Fuel Cell Systems

kW capital cost of each technology on an annual basis, which has been an electric utility industry rule of thumb in recent years. Capital costs are taken from the 1982 edition of the EPRI-TAG.[183] The x-axis shows the number of hours over the year, and the straight lines therefore show the cost of generating electricity for each installed kW, taking into account the fuel used (natural gas and light distillate at $4/MMBTU and coal at $2/MMBTU, corresponding to $25/bbl for distillate and $40/ton for delivered coal). As can be seen from the plot, the capital cost of the 8300 BTU/kWh heat rate PAFC has been taken to be $770 (1981)/kW, whereas that of a hypothetical 7500 BTU/kWh heat rate PAFC is reduced to 90% of this figure, because of the improved heat rate using the same hardware. The 1981 combustion turbine (CT) and combined cycle (CTCC) capital costs were taken to be $300/kW and $450/kW respectively, whereas the large coal plant is assumed to cost $1100/kW. The combined cycle plant was assumed to be 40% efficient and the coal plant 35%.

In this simplistic example, it can be seen that the costs of generating electricity over the whole 8760 h year will be the same for the CTCC and coal plants, since both screening curves intersect at $390/kW for the complete year, yielding an electricity cost of 4.45¢/kWh. Operating the coal plant for less than the whole year will result in higher costs than those for a CTCC plant operating under the same conditions, because of the difference in fixed charges between the two technologies. For example, a coal plant operating only over six months (4380 h) per year would cost $305/kW, whereas a combined cycle plant operating over the same number of hours would cost only $240/kW. Due to its low efficiency (and therefore high fuel cost) the simple CT would cost about $500/kWy when operated over the whole 8760 h, therefore the utility would only use it on a permanent basis (i.e., as baseload) if an emergency such as a sudden surge in capacity arose because of a major breakdown in a baseload plant using cheaper fuel at higher efficiency. However, the simple combustion turbine has very low fixed costs, and it is capable of producing lower-cost electricity than the CTCC out to about 2000 h/y, where its screening curve intersects that for the latter.

An older coal or hydropower plant will have lower fixed costs than those for the coal plant shown in Fig. 2–3, and a nuclear plant will have lower fuel costs. All the above will therefore produce the cheapest electricity on a 8760 h utilization basis, and will provide the baseload power. This will provide the production level set by the minimum demand of a typical utility, as shown in Fig. 2–4. The intermediate duty, in this scenario, can be met by the CTCC units, up to perhaps 5000 h/y, and the peak demands, over a total period of 2000 h/y, can be met by simple CT units. The latter, because of their low cost and short installation time, can be used for emergencies and in cases where demand temporarily exceeds supply.

Of course, not all utilities are as described in this simple scenario, since

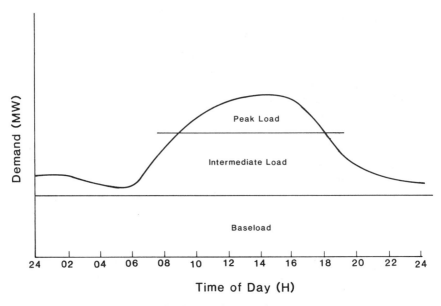

Figure 2–4: Typical utility daily electrical demand.

different local conditions of supply, demand and load-sharing prevail: for example, some utilities are predominently hydropowered, such as in Canada and in the Pacific; some have peak demand in winter; whereas for others the peak load occurs with summer air conditioning. However, this mix of generating capacity is common enough to show the outstanding attraction of the CTCC system for many utilities.

The CTCC, while not dispersible, is available in much more manageable units (100⁺ MW) than a central coal steam plant, and its emissions are much lower than those of a traditional coal plant when it uses natural gas or oil-based fuels. Its outstanding advantage, however, is that it can be combined with a coal gasifier that is scaled to its size. The coal gasifier (see Chapter 14) partially oxidizes coal with air (or oxygen) and steam to give a mixture of H_2, CO, and CO_2 containing sulfur compounds derived from the coal. The latter are relatively easy to remove, so the overall product is a clean gas that can be burned at high efficiency in the CTCC. The gas is produced at about 60% thermal efficiency from the coal, but the gasifier produces waste heat that can be used to raise more steam. When perfected, the overall system may be able to produce electricity with low emissions (e.g., 25 ppm NO_x with very low sulfur) from coal at 45% efficiency, although the Cool Water prototype unit in Southern California[631, 1155] currently operates at lower efficiency values than this. The system can be run as a baseload unit, with extra CTs (and CTCCs) installed on site as required to provide

Characteristics of Fuel Cell Systems

peaking and intermediate duty application using natural gas fuel. When system demand increases, these can be upgraded to baseload duty by the addition of extra coal gasification facilities in which waste heat from the gasifiers is used in a further steam cycle (integrated gasifier combined cycle, IGCC). Altogether, the concept offers flexibility (modularity in units of 100+ MW), relatively low potential cost to the utility, and good efficiency.

When we compare the PAFC with the natural gas–fueled CTCC in Fig. 2–3, we immediately see that it is at a disadvantage. It has a higher capital cost, and its improved efficiency (even at 7500 BTU/kWh in a hypothetical improved version, compared with 8530 BTU/kWh, or 40%, for the CTCC) is not sufficient to make it more economical than the CTCC at present fuel prices. Indeed, compared with the CTCC, it must be sold on the strength of credits against its capital cost. These credits include those in Table 2–6, and these will in many cases swing the balance by lowering the intercept for the PAFC in Fig. 2–3.

One of the best-known studies on utility fuel cell markets, the Energy Management Associates (EMA) report prepared for EPRI in 1983,[(1972)] has often been quoted (see for example, Ref. 177). This study assumed that the baseline cost of the 8300 BTU/kWh heat rate PAFC would be $770 (1981)/kW, and it looked at the requirements of typical utilities throughout the United States in the time period 1985–2005, taking into account the credits that could be attributed to the fuel cell for various applications, types of utility, and geographic and local conditions. It considered a range of learned-out fuel cell capital costs below $770/kW, together with a range of credits from $0/kW to $300/kW, the average being $166/kW. Finally, it assumed a natural gas cost of $4 (1981)/MMBTU. It used three different scenarios—optimistic, expected, and pessimistic, depending on business climate—estimated the total added and replacement generating capacity market for each. It concluded that according to the *expected* scenario (final fuel cell cost $600/kW, credits $166/kW), 44,500 MW of 8300 BTU/kWh PAFC capacity might be installed by the year 2005, most of the capacity being put into service during the last few years of the twenty-year period. This capacity would represent about 6.6% of total new capacity (including baseload), or about 33% of the total intermediate and peaking load of about 130,000 MW in the year 2005, the remainder being simple-cycle CTs. The *optimistic* scenario, based on a different installed capacity estimate, projected 104,000 MW of PAFC capacity by the year 2005, whereas the *pessimistic* scenario showed only 5700 MW. This wide dispersion showed that unknowns made the market very unpredictable, although a sensitivity study appeared to indicate that capital cost, rather than fuel cost, would be the key factor determining the market. For example, a $150/kW cost reduction, or increase in credits, could double the market in the expected scenario.

The EMA study had two vital flaws: First, it did not specifically consider the combined cycle as a competitor in the intermediate-duty market; and second, it did not consider how the fuel cell cost would be driven down to the \$770/kW or \$600/kW level where it could compete in the market. Any new product will have a high cost when it is first introduced into the marketplace and is in limited production; cost goes down as the volume of production rises and as both production and product improvements are made. The cost–cumulative production relationship for a product is usually called the *learning curve*, or *experience curve*. This curve is a logarithmic expression, and it is conventionally referred to as the percentage of the previous value to which the cost will fall if total production doubles. While the various fuel cell developers have determined their own learning curves for fuel cell systems, these are of course proprietary. However, a study conducted for EPRI (which has remained unpublished in its entirety) has shown that, by comparison with similar technologies in the chemical, electrochemical, and chemical engineering industries, the global learning curve for the PAFC can be considered to have a factor of about 0.87; that is, every doubling in production will result in a cost equal to 0.87 of the previous value. This study, which has a number of rather sensitive implications with regard to the potential marketing strategy to be used by a developer, was first summarized briefly in Ref. 959, and then given in more detail for the first time in Ref. 177.

The 0.87 learning curve can be closely approximated by the expression:

$$C_p/C_i = (P_c/P_i)^{-0.2} \qquad [2-17]$$

where C_p is the unit cost (cost/kW) at cumulative production equal to P_c, and C_i is the unit cost at some initial production point, P_i. The Ref. 959 study took the P_i value as 11 MW (i.e., the cost of the first 11 MW unit of UTC FCG-1 type, Ref. 1087), and gave it an estimated cost of \$3800/kW (1982 dollars). In contrast, the cost of the 4.5 MW New York PAFC demonstrator was \$7800/kW expressed approximately in 1978 dollars (\$35 million total construction cost, about half of the total project cost), of which the stack alone cost \$2000/kW. This cost is accurately known, since the project was largely publically supported (DOE 48%, EPRI 25%, UTC 20%, Consolidated Edison 7%). Construction of the next 4.5 MW PAFC unit for TEPCO was a private effort, wholly funded by the utility and UTC, and its exact cost is not publicly available. However, it is known that the total cost in 1980 dollars was about \$25 million, or \$5500/kW, which is much less than that of the New York demonstrator. Hence, \$3800 (1982)/kW for the first unit of FCG-1 type is not unreasonable; indeed, in the learning process that has gone on since 1983, many potential cost reductions already have been identified, and this figure is now a high estimate.

The learning curve for this technology is shown in log-log form in Fig. 2-5. After a total production run of about 3850 MW (350 11-MW units), the

Characteristics of Fuel Cell Systems

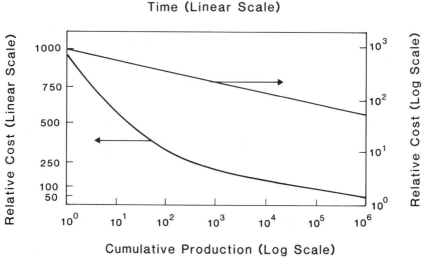

Figure 2–5: Learning curve ($\log_{10}$ cost as a function of $\log_{10}$ cumulative production). In many cases, the slope is close to -0.2, indicating a fall in cost of 13% for each doubling of production (a so-called "0.87" learning curve). Time is linear since growth of production is generally exponential.

price should be below $1000/kW, and general market penetration without special incentives can then take place. However, in the initial period, where the price is being driven down from $3800/kW to $1000/kW by the rate of production, incentives such as special tax credits, cogeneration credits and subsidies will be needed to enable utilities to purchase it. Reference 959 suggests that in the most probable scenario penetration to 1600 MW, total capacity will take about 12 years starting at the beginning of 1987, finishing at the end of 1998. In the first and second years, one 11-MW will be sold in each year; in the third, 2, followed by 3, 4, 6, 8, 11, 16, 22, 29, and 39 units in later years, to give a total of 1562 MW of total penetration over this 12-year period. The assumptions of this penetration model are given in Ref. 177: It suffices to state here that natural gas prices were again taken to be $4 (1982)/MMBTU, as in Fig. 2–3 and in Ref. 1972. However, this time a 7%/y fuel cost escalation rate was used.

This study showed that in many markets (though of low total volume), break-even might occur at a capital cost as low as $1500/kW–$1800/kW (depending on fuel cost, hence year, in the time-frame 1993–1998) in electric utility applications with cogeneration (without tax credit), and at about $1000/kW in some stand-alone generation markets about the year 2003. This study therefore shows what might be the driving force to achieve the lower cost for final breakthrough into the wider markets predicted by the EMA study. The penetration curve in Ref. 959 (log annual production as a function of time) is shown in Fig. 2–6. The points refer to production

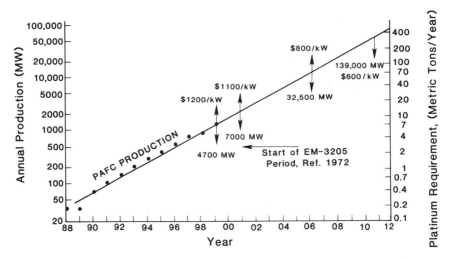

Figure 2–6: Model for 8300 Btu/kWh (HHV) efficiency PAFC penetration into the world market, based on mature production costs and marketing assumptions in Ref. 1972 for U.S. electric utilities (44,000 MW over approximately a 10-year period, assumed to be between the years 2001 and 2011). Total world market assumed to be three times U.S. electric utility market (177,182). Points are initial annual rates for early worldwide production, assumed equal to three times those for the electric utility requirement given in Ref. 182. Annual market penetration rate is 37%. Arrows indicate cumulative production and unit costs as a function of year. Costs assume a 0.87 learning curve (See Fig. 2–5). Right axis shows annual platinum requirement, assuming present loading of 5 g Pt/kW (present world platinum production is about 100 metric tons/yr).

in the time-frame to 1998 referred to above. In this period, the annual growth rate is about 38%. This was not considered to be sustainable in the long term due to the effect of the normal S-shaped curve on market growth as the market approaches saturation (i.e., PAFC production approaches the growth of appropriate generating capacity). Hence, nominal maximum growth was assumed in about the years 2003 and 2004, each with about 2300 MW of electric utility production per annum, which is then followed by a period of decline (cf. Marchetti's functions for market penetration, Ref. 1681). The production rate is considered to decline at 38%/y for four years until 2008, when the total U.S. electric utility market for this expected case is 14,000 MW, at a cost of $775/kW. It should be emphasized that confirmatory costing studies given in Ref. 177 support the learned-out costs given in Refs. 959 and 1972. For optimistic and pessimistic scenarios (not shown on Fig. 2–6), the total U.S. electric utility markets in this study were proposed as 30,000 MW and 4000 MW respectively in the year 2008.

The results of the studies in Refs. 959 and 1972 are shown in Table 2–7. The contrast between the expected scenarios for the two models for the 2005–2008 time-frame is striking: By this time, the learned-out cost of the Ref. 959 study is about the same as the baseline cost in Ref. 1972, in which not only has no learning-curve restraint been placed on market penetration since an unrealistic constant cost was assumed, but no upper limit of S-curve type has been placed on the market. If, as an exercise, we remove the latter constraint from Ref. 959, growth will continue as shown by the dashed line in Fig. 2–6, the total market reaching the expected 44,000 MW–Ref. 1972 scenario between the years 2008 and 2009, when the PAFC cost will be \$600/kW, and the production rate about 14,000 MW/y. This scenario shows that 70% of the capacity would be installed in the final three years, 50% being in the last two years. This model is unrealistic for two reasons: First, platinum catalyst availability (based on present cell technology) during the years of massive PAFC production from 2005 (see following), and the simple lack of a market to fill. The latter will occur since competing technologies that would have been available more rapidly at more competitive costs would already have been used to fill the available niches, and the growth of new capacity in this time-frame would be insufficient to provide such a large market for the fuel cell.

The latest study for EPRI, by Decision Focus Incorporated (DFI),[432] attempts to resolve the above conflicts by introducing more realistic economic and market considerations, taking into account both the learning curve and the extrapolated characteristics of competitive technologies, particularly the improved CTCC. For the latter, the example used was the state-of-the-art future General Electric (GE) Frame 2000 Model F (135 MW), with the high turbine inlet temperature of 1260°C (2300°F), at a cost of \$650 (January 1986) /kW, with a further cost where necessary of \$130/kW for

TABLE 2–7

Projected Fuel Cell Markets[432, 959, 1972]

	EPRI[959]					EMA[1972]	DFI[432]	
Scenario	1988	1993	1998	2003	2008	2005	2010	
							7800 Heat Rate	7200 Heat Rate
Optimistic	60	500	3,400	14,000	30,000	104,000	22,300	
Expected	50	300	1,600	6,000	14,000	44,500	14,800	68,000
Pessimistic	40	175	650	1,850	4,000	5,700	7,100	39,000

catalytic NO_x reduction. The HHV heat rate of this generator under standard conditions at full load is calculated to be 7500 BTU/kWh (45.5% efficiency, equivalent to 50.5% LHV). The possibility of further future improvement to 7250 BTU (HHV)/kWh, at some increase in capital cost, was also considered. Again, a fuel (natural gas) price of $4 (January 1986)/MMBTU was used (present depressed end-1986 prices lie between $0.61 for old contracts and $4.00, with a mean about $2.50–2.70).

The study showed that it was immediately apparent that any 8300 BTU/kWh PAFC, in spite of all the possible dispersion or other credits, would hardly compete in the stand-alone electric utility context against such a machine. However, a reduction in the heat rate of the fuel cell system to 7800 BTU/kWh substantially improved its competiveness provided that the capital cost was reasonable. The baseline figures taken for the latter (again in January 1986 dollars) were $1300/kW when out of the prototype stage, at the beginning of mass production, $920/kW for 1000 MW, and $690/kW for mature production (several thousand MW/y). Taking into account inflation, these are about 60% of the costs assumed in Ref. 959, but they do reflect many cost-cutting approaches and improvements since 1982, in particular in the cell stack. These include improved structural configurations allowing lower IR-drop and higher current density at the same or higher cell voltage, as discussed in Chapter 16. The 7800 BTU/kW heat-rate value, obtained by fuel processing system improvements, turbocompressor efficiency improvements, as well as improvements to the cell stack, also helps to reduce residual cost by 6% compared with that of the 8300 BTU/kWh baseline system, allowing a unit giving 11.7 MW rather than 11.0 MW from the same hardware.

On this basis, the U.S. electric utility market to the year 2010 is predicted to be 22,300 MW (optimistic high-growth), 14,800 MW (expected), or 7,100 MW (low growth) for the 7800 BTU/kWh, $690/kW fuel cell unit, the major competitor being a CTCC of GE Frame 2000 Model F type. Again, most of this market opens up after the year 2000. On this basis, the fuel cell would take 8% of a total market of about 185,000 MW of intermediate load and peaking capacity; the CTCC taking 63%; the simple CT, 26%, and the simple-cycle CT with storage of compressed air produced under off-peak conditions, 3%. Other conventional storage technologies (e.g., pumped hydro) showed no growth beyond present plans, and new technologies (e.g., storage batteries for peaking) showed no penetration at all. However, premium fuel prices in 2010 may be much higher than supposed in the study, which would make storage technologies more attractive.

The most striking aspect of the study was the effect of system heat rate and capital cost on market penetration. Improvement of fuel cell heat rate from 7800 BTU/kWh to 7200 BTU/kWh (increase in unit power from 11 MW to 12.7 MW), assuming the baseline (7500 BTU/kWh) CTCC, would

Characteristics of Fuel Cell Systems

increase the fuel cell market share to 68,000 MW (37%) to the year 2010 at $690/kW. This would fall to 39,000 MW (17%) if the 7250 BTU/kWh CTCC becomes available. However, the fact that cost, rather than efficiency, is a major driving force is shown by the effect of reduction of the cost of the baseline 7800 BTU/kWh unit from $690 (1986)/kW to $590 (1986)/kW, equivalent to about $700 (1981)/kW, close to the Ref. 1972 value. The market then increases from 14,800 MW to 43,000 MW (23%). Correspondingly, an increase in cost by $100/kW will reduce the total market to only 3,000 MW, which is too small to be economically realistic. In all cases, the fuel cell's intrinsic credit offsets are the most important factor in swinging the balance against the CTCC. The baseline DFI conclusions are included in Table 2–7.

The latest analysis, which is not really inconsistent with Refs. 959 and 1972, shows that mature fuel cell system cost must cost less than about $700/kW, and preferably under $600/kW, to achieve an effective stand-alone electric utility market penetration. For this application, heat rate is however important: It must be comparable to, or lower than, that of the advanced CTCC, otherwise the fuel cell system will play only a minor part in electric utility generating technology. It is certain that system improvements will make a 7800 BTU/kWh PAFC possible with cell performance being currently demonstrated (Chapter 16); indeed, in 1981 it was suggested that 7900 BTU/kWh would be possible by simple system improvements alone.[692] Further improvements, such as burning the cathode exhaust in the reformer, could result in 7700 BTU/kWh at the conventional 0.73 V UTC cell potential.[177] A system heat rate of 7200 BTU/kWh is not unreasonable at a nominal stack lifetime (e.g., 10,000 h) with some stack performance improvements, for example in IR-drop, and by the use of a low nominal current density (216 mA/cm^2) to raise cell voltage somewhat, provided that this is not inconsistent with materials problems due to corrosion (Chapter 16). However, when the fuel cell is used for cogeneration applications with a balanced heat-to-power ratio, an electrical output of 40% efficiency (8500 BTU/kWh) will be quite satisfactory and economical. Hence, the PAFC should make its first (and largest) impact in the cogeneration, rather than the stand-alone utility, market.

Instead of the highly-improved PAFC, a stronger contender for a simple, less costly generator with a heat rate of 7000 BTU/kWh or less may be the IRMCFC (Chapters 3, 14, 17), which when commercialized after the year 2000 should take a large share of the market, provided that its development problems can be overcome. Some of the reasons for this will be discussed later, after the market for the fuel cell for other than U.S. electric utility applications is considered.

The on-site market must compete with the cost of electricity as delivered to the customer, rather than with the bus bar–cost to the electric utility. In addition, a heat credit is taken for the cost of the fuel that otherwise would be burned in the place of the fuel cell waste heat. A higher capital cost is therefore possible for the fuel cell, which is a necessary condition since small (200 kW) dispersed units will inevitably be more expensive than large substation units in the multi-MW class. It has been suggested that a capital cost of $2500 (1985)/kW may be economical in this context.[696] A current projection for the U.S. on-site market towards the year 2000 is about 1000 MW/y,[1157] ultimately representing about 25–30% of the U.S. electric utility market.[177] In a recent report, the On-Site Industrial Users Group estimates the commercial on-site U. S. market to be ultimately 18,000 MW (at $1000/kW in 1985 dollars).[115] In Japan, the electric utility market over the period up to 2005 has been estimated at 25,000 MW,[1854] and the on-site market may add a further 40%.[177] First estimates for the potential market in Europe put it on the same order as that in Japan.[2159] The on-site market everywhere will clearly be about a third of the total market, and that for industrial cogeneration in the United States alone may be potentially as much as 50,000 MW.[177]

Since all the above assessments were generated on the basis of EMA report assumptions for the *most probable* case,[1972] it is not unreasonable to suppose that a minimum estimate of the total world market for stationary fuel cells up to the year 2010 will be about three times the yearly production on Fig. 2–6, with a number of parallel production lines in different countries. Figure 2–7 is a suggested scenario. PAFC production is shown to follow a penetration plot similar to the EPRI Ref. 959 scenario, with the IRMCFC following some five years later in development.

The PAFC paves the way for the IRMCFC (or other advanced high-efficiency fuel cell) until about 1997, at which time total world PAFC production is about 2500 cumulative MW, representing only a small fraction of the potential world market. By this time cumulative MCFC production will have reached 200–300 MW, and a cost in the $1200/kW–1300/kW range at this production level.[1981] The detailed justification of this cost is given in Chapter 17.

At this point, assuming the same 0.87 learning curve for both the IRMCFC and for the PAFC, rapid market penetration of the efficient IRMCFC takes place (heat rate nominally 6500 BTU/kWh, 7000 BTU/kWH at a lower capital cost). IRMCFC production crosses the PAFC plot in 2001, after 3400 MW have been produced. Its expected cost will then be $680/kW, well within the DFI model assumptions,[432] and it then goes on to capture the market identified by DFI. After about 2004, its growth becomes limited by that of the electricity market as a whole, and by 2010 it has filled the market niches identified by DFI. This scenario of course excludes any baseload (coal-gas) applications, which would represent a further future market.

Characteristics of Fuel Cell Systems

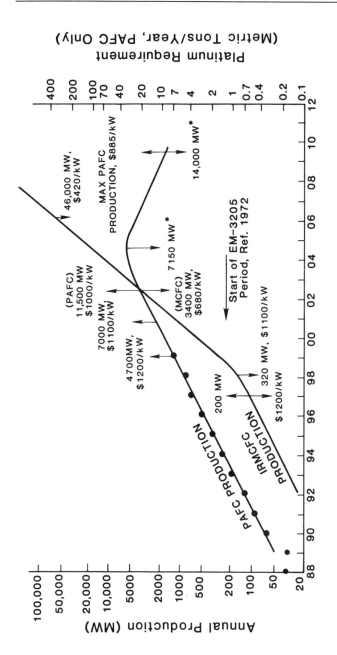

Figure 2-7: As Fig. 2-6, showing effect of the IRMCFC on PAFC market penetration. The total IRMCFC market is assumed to be about three times the PAFC market (corresponding to the effect of system efficiency improvement given in Ref. 432). IRMCFC commercialization is assumed to be five years after that of the PAFC, with the same learning curve. IRMCFC cost is $1200 per kW after 200 MW cumulative production (1982). Asterisks show total installed PAFCs in U.S.-Electric Utilities using "expected" assumptions and 7800 BTU/kWh heat rate in Ref. 432.

Finally, the turn-around point for PAFC production on Fig. 2–7 is not arbitrary, as the right-hand axis, showing platinum requirements based on present technology, indicates. The maximum PAFC production level will require a new supply of almost 40 metric tons of platinum per year. This represents 50% of current (1986) production, and any quantity much in excess of that figure will be very difficult to obtain at economic prices. Thus, unless there is a breakthrough in electrocatalysis within the next decade, the rate of production of the PAFC will be limited by platinum supply considerations.[178]

Finally, the most effective strategy for market penetration must be considered. Penetration of a new market at the high-cost end of the learning curve, trying to find special niches where high-value technology can compete, is not necessarily the best economic strategy, though this choice must be left to the individual developer. In most cases, however, early units are best subsidized and sold at a loss, later units in much larger numbers being sold at a profit. This type of strategy is shown in the curve in Fig. 2–8.[178]

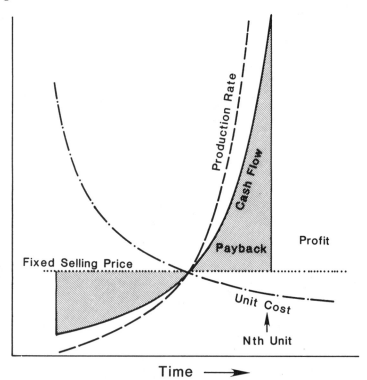

Figure 2–8: Unit cost, production rate, and cash flow as a function of time (schematic), showing a possible selling strategy yielding a profit starting with the nth unit.

PART II

Fuel Cell Programs, Concepts, and Areas of Application

In the previous chapter, fuel cells were discussed mainly from the viewpoint of central utility power generation, for which the power ranges vary from hundreds of kilowatts to hundreds of megawatts. However, fuel cells also appear attractive for many applications that require much lower power levels. As discussed earlier, the initial impetus for fuel cell research came from the fact that fuel cell systems, characterized by high energy densities (kWh/kg), but relatively low power densities (kW/kg), appeared to be ideal for applications that required longer duty cycles than those obtainable from storage batteries. As will be discussed in the following chapters, fuel cells promise to meet the requirements of a sizable number of both military and nonmilitary applications, in many cases displacing units such as engine-generator sets.

A brief look at the U.S. National Fuel Cell Program budget for fiscal year 1983 (Table 3-1) provides some idea of the scope of the ongoing interest in fuel cells in the United States. The fiscal year 1984 DOE fuel cell budget was increased to about $40 million.[1692] During the period from 1976 to 1984 the U.S. Federal Government spent over $250 million (about $350 million total in 1986 dollars) on fuel cell research and development, in addition to the funding allocated in the 1960s for aerospace fuel cells. It has been estimated that the utilities and manufacturers approximately have matched federal expenditures.[1692] Furthermore, there is now considerable government support for fuel cell development elsewhere, particularly in Japan and Europe, which indicates the promise of this emerging energy conversion technology.

CHAPTER 3

Utility Systems

Methane is the fuel that most electric utility companies would prefer to use in fossil plants because of its ease of handling, low environmental impact, and combustion characteristics that minimize capital equipment requirements. It is particularly attractive for use in combined-cycle (combustion turbine with steam bottoming cycle, CTCC) plants, which will be discussed in further detail later. Its cost compared with other fuels (particularly coal) is of course important, but by 1986 long-term contracts for natural gas were quite competitive in the United States ($0.61 to $4.00, averaging about $2.50 per MMBTU, see Chapter 2).

It has been estimated that methane will be available continuously, either as natural gas beyond the end of this century, or as synthetic natural gas (SNG), until well into the twenty-first century.[2118] This conclusion, combined with legal and regulatory incentives*, has provided a major driving force for the development of utility fuel cell power plants. For example, over 300 electric utilities that appear to be near-term potential users of fuel cells have been identified by the Fuel Cell Users Group (FCUG). These factors, combined with the realization that high-efficiency fuel cell technology can reduce the economic impact of high priced synthetic, conventional, and nonconventional fuels, have resulted in the undertaking of a comprehensive mission-oriented program in the United States to develop and commercialize fuel cells for utility use.[1678]

---

*For example, utility fuel cells are exempt from the Power Plant and Industrial Fuel Use Act of 1978 (PIFUA), which restricts the use of natural gas and petroleum fuels in conventional power plants. Also, they comply admirably with the Clean Air Act, which regulates the maximum amounts of nitrogen oxides, sulfur dioxide, and total suspended particulates (TSP) permissible, especially for urban use.[262]

A major application for utility fuel cells will be as dispersed power generators that provide an alternative method for meeting increased loads in congested urban areas with restrictive pollution standards and limited potential for new transmission rights-of-way. At present, utilities are being forced to build large central station plants that are more remotely located, along with the corresponding transmission facilities, in some cases underground, to supply the needs of local areas. By using fuel cells as dispersed generators at selected distribution substation locations, power in required increments can be provided to areas of growing demand, at the same time meeting local municipal requirements and codes.[1524]

In remote or developing areas not connected to a utility grid, fuel cells are able to supply electricity at a lower cost than diesel generating sets. For example, for sparsely populated areas located more than about three kilometers from a transformer substation it should be cheaper to install a dispersed fuel cell generating plant than to invest in a long distribution line.[322]

It is equally probable that utility fuel cell plants also can be used for urban plant replacement. In cities, many utilities have old plants that must be upgraded because of age, pollution restrictions, or uneconomic operation. The upgrading of an old plant now usually must include the addition of an extensive stack gas cleanup system for pollution control. The difficulties of retrofitting stack gas cleanup equipment on an existing site and problems of removing byproducts from the cleanup process have been shown to be significant.[1379] This retrofitting cost, combined with increased operating costs, including a reduction in plant efficiency, can add substantially to the per kWh cost of service. Replacement of an old plant at an existing urban site by currently available conventional plants will with good reason frequently encounter opposition on the grounds of atmospheric emissions, noise, vibrations, or thermal discharge into waterways. For these reasons, it is generally mandated that the replacement of old plants will require the construction of a new plant at a site outside the urban area where the power is needed. This can be an expensive option, particularly if transmission lines must be run underground or if the utility has a limited service area with few available sites.

Since we have shown that fuel cell power plants are highly compatible with urban area siting requirements, they can be installed at existing old plant locations. The short lead times associated with fuel cell planning and construction can permit the retirement of obsolete units and their replacement with fuel cells to be accomplished very rapidly, thereby reducing the time interval in which capital is tied up in nonproductive facilities, and minimizing the need for power purchases from other utilities during the construction phase.

A. U.S. NATIONAL FUEL CELL PROGRAM

Electric utility interest in fuel cells dates back to the late 1960s and early 1970s. In 1971 the Edison Electric Institute (EEI), United Technologies Corporation (UTC), and a group of electric utilities began to study the benefits of fuel cells to the electric utility industry.[2118] Sufficient interest was generated to initiate the Fuel Cell Generator-1 (FCG-1) program in 1972 between UTC and nine electric utilities. This ultimately led to the 1 MW demonstrator power plant in the UTC Power Systems Division facility in South Windsor, Connecticut in 1977 and the two 4.5MW (a.c.) demonstrators in New York and in Goi, Japan (Chiba Prefecture), to be discussed later in this chapter.

The recognition that commercialization of utility fuel cell power plants would need assistance from potential users early in the development of the technology led to the formation of the FCUG in April 1980. The FCUG started with 37 charter members, with both large and small utilities including investor-owned, municipal, cooperative, federal and foreign utility organizations; including the three major utility trade associations (the American Public Power Association, the EEI, and the National Rural Electric Cooperative Association). By 1983 the membership had grown to more than 65 members.[1971] The purpose of the Fuel Cell Users Group was to expedite the near term commercialization of fuel cell power plant technology by sharing information on fuel cell developments within the utility industry, by coordinating and helping members identify, sponsor, support, and participate in fuel cell research and development, by educating potential users to assist fuel cell developers in defining the needs of the utilities, and by determining the potential market for the technology. The FCUG was divided into three technical and three supporting subcommittees.

In 1982 the System Planning Subcommittee of the Fuel Cell Users Group and the Electric Power Research Institute (EPRI) sponsored the *Fuel Cell Power Plant Applications Study*, which was the first coherent attempt to determine the potential applications for 10 MW PAFC power plants over the time period 1986~2005.[1971] The 25 of the 65 members of the FCUG who took part in this study represented about 15% of the total U.S. electricity generation capacity. The purpose of the study was to provide tools that would enable quick, efficient development of generation expansion plans, to use real utility data, and to provide a better economic determination of the benefits and applications of fuel cell power plants. While it cannot be pretended that the economic assumptions used were entirely realistic (e.g., the study assumed that the per kW cost of fuel cells would be the same on introduction and in mature production), the study allowed a first cut at the market potential after the year 2000.

The major conclusion was that most of the 25 participating utilities would expect to start adding fuel cells to their systems after about 1995, when many of their present plants should be retired. By 2006 the same 25 utilities estimated that they may require an additional 9000 MW of power. The general conclusions of of this study,[1972] extrapolated to include all the U.S. Electric Utilities, have been given in Chapter 2, Section O.

In response to the interest in utility fuel cells, the National Fuel Cell Coordinating Group (NFCCG), in cooperation with the U.S. Government, fuel cell developers, and users groups, formulated a U.S. National Fuel Cell Program. In addition to extensive private funding, the U.S. Government is coordinating the overall plan and is providing funding for the program via the Department of Energy (DOE). During the early years of fuel cell research, much of the U.S. Government fuel cell research effort was administered by the Environmental Protection Agency (EPA) under its mandate to promote the development of environmentally acceptable energy systems. With the creation of the Energy Research and Development Administration (ERDA), later the DOE, the EPA's only present connection with fuel cells is through its work in coal gasification systems.[1321, 1323]

The DOE supports research and development on the phosphoric acid fuel cell (PAFC, *first generation technology*), the molten carbonate fuel cell (MCFC, *second generation technology*), and on the solid oxide fuel cell (SOFC, *third generation technology*), as well as in various areas of more fundamental fuel cell–related research.[769, 2652] The DOE coordinates and apportions contracts to major developers such as UTC, Westinghouse Electric Corporation, and Energy Research Corporation (ERC; Danbury, Connecticut). In addition to the DOE, EPRI (the research and development organization representing the majority of U.S. electric utilities since 1972) is supporting the development of utility fuel cell power plants.[859, 860] The objective of EPRI's fuel cell program is:

> To develop and assist the commercial introduction of a fuel cell power plant technology that will provide the electric utility industry with a power generation option that is highly efficient, environmentally acceptable, and compatible with available utility fuels while fully realizing the economic and operating advantages of an essentially modular power generator.

EPRI has been allocating 60~70% of its fuel cell budget towards the development of the first generation PAFC system.

The level and distribution of effort of the National Fuel Cell Program can be learned from its budget, as given in Table 3–1 for fiscal year 1983.[1061, 1576] It can be seen that well in excess of $60 million per year of public funding

Utility Systems

TABLE 3-1

U.S. National Fuel Cell Program Budget for FY 1983 [1061,1576]

Agency	$ Millions	Main Areas of Interest
Dept. of Energy	26.8	Mostly PAFC with some MCFC support
Dept. of Defense	4.6	40 kW PAFC unit field tests (4 units); 1.5~5 kW portable units
Gas Research Institute	19.5	40 kW unit field tests with DOE (49 units)
NASA	1.6*	Advanced space units (alkaline, SPE**, some PAFC)
Electric Power Research Institute	8.7	4.5 MW demonstrator with ConEd.; PAFC and MCFC for utility applications

* R&D budget only; applications budget extra
** solid polymer electrolyte

alone was expended at that time on the U.S. National Fuel Cell Program. This gives some idea of the importance attached to the development of this technology.

B. UTILITY PHOSPHORIC ACID FUEL CELLS

TARGET Program

TARGET (Team to Advance Research for Gas Energy Transformation), was a cooperative undertaking initiated in 1967 between 32 U.S. gas companies and Pratt & Whitney Aircraft Division of United Aircraft Corporation (now UTC).[518] The goal of TARGET was to develop an economical natural gas fuel cell electric power supply for homes and small commercial and industrial users. A major incentive for the program was the gas industry's realization that the trend towards increased use of electricity was resulting in a gradual shift from gas-heated homes to all-electric homes,[950] in spite of the large disparity between delivered gas and electricity prices (about a factor of six in terms of heating value in 1967). In the hope of realizing larger profits, the gas industry decided instead to try to capture a share of this expanding electricity market and compete with electric utilities in selling on site power for commercial, industrial, and residential use.[1208] The TARGET fuel cell was to be scaled for typical maximum domestic demand (12.5 kW$_e$), and would use reformed natural gas with a CO_2-rejecting fuel cell (the PAFC or the MCFC). The ultimate objective was to develop a "total energy and environmental conditioning" package, in which the system would provide not only electrical energy but also heat, would humidify and purify the air, and possibly would be used even as a waste processor. Pratt & Whitney was chosen as the prime contractor because of its longstanding corporate interest in fuel cells, dating from 1959.[105]

Starting in January 1967, the sponsoring gas companies provided up to $7 million annually in current dollars in a joint venture with Pratt & Whitney for the development of the PAFC-based system. Pratt & Whitney in turn, used the Institute of Gas Technology (IGT), the gas industry's research facility in Chicago, as a subcontractor, mostly for work on the MCFC as a backup system. The TARGET program was the largest single research effort ever undertaken by the U.S. gas industry up to that time. One Canadian gas company, Canadian Western Natural Gas Co., Ltd. of Alberta, also participated.[1401]

The program consisted of a nine-year plan comprising three three-year phases (1967-1975), broadly consisting of research and evaluation, development and testing. During the last phase, 60 experimental 12.5 kW phosphoric acid PC-11 (Power Cell-11) fuel cell power plants were extensively field tested in the United States, Canada, and Japan. These power plants were installed and maintained by utility personnel and operated unattended for about three months at each site in a wide range of environments under actual service conditions. The power plants, which were operated singly or as multiple units in parallel, provided on-site power to a variety of residential, commercial, and industrial buildings and also were operated in parallel with the utility network at two electric utility substations.[1400] As a result of these trials, essential data were obtained that defined the technical, economic, and business factors affecting the operation of a fuel cell energy service. In addition, potential problems were identified in relation to building and utility codes, insurance, reliability and maintenance, response to peak load demands, and other factors.

The PC-11, consisted of a steam reformer, a stack (battery) of phosphoric acid matrix-type fuel cells, and an inverter to convert d.c. power to line-quality a.c. The fuel cell and reformer were housed in a single unit, the inverter being a separate, slightly smaller unit. The combined units were initially approximately the size of a household furnace and weighed about 725 kg (1600 lb). However, a later model was reduced to about one quarter the original volume and weighed 500 kg (1100 lb). It generated utility-quality a.c. electrical power from natural gas and air with a rated output of 12.5 kW at 120/240 V single phase or 120/208 V three phase. The heat required to drive the reforming reaction was supplied by burning unconsumed fuel exhaust from the fuel cell anode, and it could be operated on

Utility Systems

natural gas, propane, and light distillate liquid fuels. Tests showed that its exhaust pollutant level was at least one to two orders of magnitude lower than that of a conventional heat engine–driven electrical power plant, and that its noise level was less than that of a home air conditioner.

The PC-11 made its debut in May 1971, when the world's first natural gas fuel cell–operated home was unveiled in a condominium residence in Wethersfield, Connecticut. In August of the following year Canada's first PC-11 fuel cell power plant was installed in a single family dwelling in Calgary. A testing program involving six of the Pratt & Whitney 12.5 kW units also was carried out by Hydro-Quebec in Quebec City during 1972~1973. The aims of these tests were to permit electrical engineers to acquaint themselves with the new type of electrical generator, to study the maintenance problems associated with it, to test the compatibility of fuel cells with an electrical network, and to determine the actual performance characteristics of the device.[304]

Through a licensing agreement, a program complementary to TARGET also was started in Great Britain by Energy Conversion, Ltd. (ECL), a venture derived from a consortium of three British companies along with the National Research and Development Corporation (NRDC), for the purpose of developing commercial natural gas fuel cells.[96] Under the agreement, there was to be a free interchange of research and development results between ECL and TARGET.

The goal of the TARGET program was to ultimately produce a fuel cell at a cost of about $150 (1967)/kW (about $500/kW in 1986 dollars) that could be leased by the gas utilities to homeowners. The fuel cells were intended to be stand-alone (i.e., not grid-connected) so that the homeowner would rely exclusively on the gas utility for all energy needs. This required the unit to possess very high reliability. Electrical efficiency was to approach 40% and the required economic life of the fuel cell stack was estimated to be about 40,000 hours. The PC-11 was only a prototype unit that could not have achieved these goals even in large-scale mass production, due to its high platinum content and short stack lifetime. Technical improvements and economies of scale were needed to bring its per kW cost down to acceptable levels. It did however prove the dispersed fuel cell concept.

United Technologies Corporation Utility Phosphoric Acid Fuel Cell Program

United Aircraft and its successor, UTC, have been involved in the development of fuel cell technology for almost 30 years. In 1962 they supplied the first commercial fuel cell, a 500 W gasoline-consuming device, to Columbia

Gas Systems for use in their Stanton, Kentucky pumping station, and in 1966 they supplied the fuel cells to NASA for the Apollo spacecraft. They also have delivered hydrocarbon-air systems to the U.S. Army Electronics Command and a 3.75 kW natural gas fuel cell to Columbia Gas. Since the early 1970s, UTC have been involved in a number of private and government-funded fuel cell programs having the goal of developing a commercially acceptable phosphoric acid fuel cell for electric utility use.[1086, 2600]

The experience obtained in the TARGET program showed that the subsystems of the PAFC could profit from economies of scale. Accordingly, in 1971 UTC collaborated with the EEI and ten independent utilities to study the viability of fuel cell power plants in larger units for electric utility generation.[976] More specifically, this study was undertaken to evaluate the potential benefits of fuel cells for dispersed power generation, a concept in which a fuel cell is installed either in place of a normal power substation, on the site of the power demand itself, or at some point between the two. This preliminary study defined the research, technology, and engineering development required to bring an initial electric utility-sized fuel cell power plant into service. The outcome of the effort was a decision to proceed with a program for the development of a modular 27 MW PAFC power plant, to be designated the FCG-1 (Fuel Cell Generator-1, a code name which has since been applied to other later units, in particular the 11 MW design of the early 1980s). The 27 MW unit was intended to be large enough to generate electric power for a community of 20,000 people. The FCG-1 program commenced in January 1973, but was not announced publicly until December of the same year.[106]

It was foreseen that utilities would use these units for baseload, peaking, and stand-by power generation, depending on their individual requirements. Electric utility participation and funding were based on initial provisional orders for 56 power plants. The participating companies and the number of units tentatively ordered by each were: Boston Edison Co. of New York (1 unit), Consolidated Edison Co. of New York (10 units), Consumers Power Co. of Michigan (1 unit), New England Electric System (1 unit), Niagara Mohawk Power Corp. of New York (4 units), Northeast Utilities of Connecticut (10 units), Philadelphia Electric Co. (2 units), Southern California Edison Co. (15 units), and Public Service Electric and Gas Co. of New Jersey (12 units). Public Service Electric already previously had spent $900,000 in fuel cell research and had tested the first fuel cells operated in parallel with a utility system. Each utility was to contribute $500,000 per unit, with UTC putting up $14 million of its own money.[1208]

In addition, the utilities were to pay the total cost for fabricating and testing (and ownership) of a demonstrator fuel cell estimated at that time to cost an additional $7 million (1974). This unit would become, as time elapsed, the 4.5 MW (a.c.) New York demonstrator. To avoid conflicts of interest with the TARGET program, only technical developments were to be shared between the two programs; all economic and applications-oriented data were to be kept separate.

As a part of this program, the 1 MW PC-19, the world's first MW-class fuel cell, was built at South Windsor, Connecticut, and was tested between January and June of 1977. Costs for this unit were shared between DOE (or its predecessor, ERDA) and EPRI (which took over joint electric utility research funding from the EEI in 1972–1973). They were shared in such a way that cell-stack technology was DOE supported, whereas utility input (systems and the fuel cell–grid interface) was provided by EPRI.

During its testing period, the PC-19 generated 698,000 kWh during 1069 hours of on-line operation connected in parallel to the Connecticut Light and Power system. The unit fulfilled all its test objectives, and showed the feasibility of technology scale-up to utility size.[1086] Compared with the PC-18, the technology had been advanced to allow the use of a low platinum loading in the fuel cell stack, but the PC-19 was still not cost-effective from the viewpoint of materials lifetime. However, the PC-19 became the forerunner of the 4.5 MW (a.c.) demonstrators in New York and Goi, Japan.

The 4.5 MW A.C. New York Demonstrator

In February 1977, DOE and EPRI issued a request for proposals to the electric utility industry to select a host to install and operate a 4.5 MW (a.c.) demonstrator fuel cell power plant as a prototype for the proposed FCG-1. With funding from UTC, DOE, EPRI, and the selected host utility, this initial demonstrator was to operate on reformed naphtha.[976, 1580, 2609] This plant has often been referred to as the "4.8 MW" unit, from the expected d.c. power output of its fuel cell stacks. However, the parasitic losses in the system (about 4% from pumps and 2% from d.c.–a.c. conversion) would be expected to reduce net output to 4.5 MW a.c., so that we will henceforth refer to the unit as the 4.5 MW system.

In July 1977, Consolidated Edison was selected as the host utility. The purpose of the plant was to demonstrate the installation and operating characteristics of a utility fuel cell power plant manned by utility personnel in an urban utility environment. The construction site chosen was a three-quarters acre (0.30 hectare) site located at East 15th Street and the Franklin D. Roosevelt Drive in lower Manhattan, opposite an older ConEd thermal power plant. From the viewpoint of obtaining an operating license, the site had to be considered as one of the most difficult in the United States. The original plan was that this unit would be constructed rapidly, and would operate in late 1978, the test being completed in 1979. Unfortunately, such optimism proved to be unfounded, as the following describes.

The actual construction cost of the system was about $35 million (1978), or $7800/kW, the stack alone costing $2000/kW. While these figure may seem high, it is gratifying to know that the next unit (the Tokyo Electric Power Company 4.5 MW demonstrator) did cost considerably less. By the time the Manhattan project was completed, its total cost (including construction) had become about $70 million, of which DOE provided 48%, EPRI 25%, UTC 20%, and Consolidated Edison 7%.

Field construction of the facility was begun in November 1978.[977] By mid-1979 the design of the demonstrator was completed and much of the equipment fabricated.[1083] By mid-1980 all the major items of equipment had been installed at the site,[978] and in January 1981 Consolidated Edison's 19-member operating team began their training.[979]

From the beginning, the New York demonstrator plant was plagued with recurring problems. It is instructive to review the history of this plant, not only to illustrate the activities involved in setting up a large new fuel cell system, but also to illustrate the need for patience associated with the development of an emerging technology under real-world conditions.

In addition to the usual difficulties that might be expected with the installation of any new-technology concept, such as contractual disputes with subcontractors and late deliveries of items, difficulties arose with the local city authorities. Since they were hesitant about the construction of a new kind of power plant in downtown Manhattan, the responsibility of public safety was placed on the shoulders of the New York City Fire Department. The Fire Department's first reaction was that the plant was essentially a refinery, which therefore would not be permitted in New York City. After arbitration, it was willing to accept the arguments put forward to let it go ahead as an experimental power plant, provided that it could pass a number of stringent and unorthodox inspection and safety tests, including a demanding water pressure test.[1692] These not only resulted in long delays because they required that plumbing be modified and parts of the plant redesigned, but also resulted in freeze damage to some of the one-of-a-kind brazed flat-plate heat-exchangers. The latter, which were much more compact than conventional tube-and-shell exchangers, had been made in 1978 after many difficulties by the Harrison Radiator Division of General Motors Corp. They were damaged, in some cases irreparably, because the period of testing coincided with the coldest December night in New York in a decade. Insofar as was possible, the heat exchangers were repaired, replaced by spares or substituted by less effective tube-and-shell heat exchangers during the winter of 1980-1981.

In November 1981 the PAC (Process and Control) test was begun in order to check out the chemical engineering system before installation of the fuel cell stacks. The PAC validation test and other test sequences laid down in the contract were considered to be very important because their successful completion was a key step in UTC's contractual obligations with the nine utility companies before they would commit themselves to purchasing commercial FCG-1 power plants. Supplemental testing of components was also necessary to provide experience to project the operational, maintenance, and economic characteristics of a fuel cell operating on the grid. During this period, a number of components (particularly valves, sensors, etc.) required replacement and in some cases relocation. In addition, the inverter required repair.

Testing of the Thermal Management System (TMS) in December 1981 was delayed because of contamination of the piping by contaminants, which finally were cleared by early March of 1982.[925] Hold 2, involving the operation of the High Flow Turbocompressor using the auxiliary burner with naphtha fuel, finally was achieved by mid-January, 1982. Hold 4, heat up of the fuel processing system, was achieved by February 6, 1982.

Problems then were encountered with the reformer. Five of the 37 tubes burned out from poor flame patterns resulting from clogged (carbonized) torch nozzles. These were a result of using high–flash point naphtha instead of the design low–flash point material. In addition, a retrospective modification of the reformer burner-head design was not immediately apparent. High–flash point naphtha had been mandated by the NYC Fire Department for safety reasons, which partly involved the question of the volume of the on-site naphtha tank, which was larger than the gasoline tank allowed in the average city service station.

The second problem was more serious: In order to make the reformer truck-transportable, it had been shortened by about 6 inches (15 cm) from the original design specification, and the air-supply tube to the showerhead burner had been relocated at an offset angle to further reduce height. In consequence, not only was the air supply to the burner maldistributed, but the reformer tubes were too close to the hot zone. This required the addition of a ring to raise the reformer head to the appropriate height, together with that of a modified (vertical) air-supply tube.

On November 14, 1982, the reformer had been repaired and the PAC test re-initiated.[926] Then on November 30, the Fuel Processing System and the Hydrodesulfurizer System were started up, only to be shut down because of problems with the High Flow Turbocompressor. Finally, testing started to proceed relatively smoothly, and on March 30, 1983, the plant produced its first hydrogen, and completed the testing of the Power Conditioning System. On April 13, the reformer was transitioned to process gas, and on April 25 the auxiliary burner was checked out to process gas and was placed on standby. On May 13, the fuel processing plant was run on a simulated 25% load for 3 minutes, and on June 8 for 52 minutes. On June 13, standby-to-load transition was verified, and on June 14 the Thermal Management System burner was cleared to process gas. On June 16, the plant reached standby with the transitioned TMS burner, and the plant then was shut down manually, the PAC test being thereby completed on June 17, 1983.

By the spring of 1984 the fuel cell stacks finally were in place, only to reveal an unexpected source of difficulty which had been considered, but rejected as improbable, during the period during which they had been in storage in South Windsor. The stacks destined for the 4.5 MW (a.c.) New York demonstrator used technology dating to 1975 (the so-called *conventional* technology). Their construction was such that the amount of electrolyte per active cm^2 was limited. The difference between this form of construction and that of the later *ribbed substrate stack* will be discussed in Chapter 16.

It was discovered that these early stacks, which had been in storage for almost seven years, had a limited shelf life resulting from the slow migration of the small inventory of phosphoric acid electrolyte from the electrolyte matrix to the somewhat porous graphite bipolar plates between each cell. The design changes that have been made in the ribbed substrate stack have and will continue to prevent this from happening in future; these changes were incorporated into both the gas industry 40 kW units and in the TEPCO 4.5 MW (a.c.) demonstrator.

The major syndrome noted in the New York stacks was that many of them showed unacceptable cross-leaks between anode and cathode gas due to lack of intervening electrolyte. As expected, the stacks that had been manufactured first were the worst affected. Attempts were made to add more acid to the stacks; however this proved not to be an easy operation, since they were stored hot containing highly concentrated phosphoric acid (about 100% H_3PO_4), which cannot be cooled to less than about 35°C to avoid the risk of freezing, with resulting damage to the electrode microstructure. Removal of the stacks from the heated pressure vessels was therefore considered to be hazardous. An attempt was made to dilute the acid in the stacks (i.e., increase its volume and avoid freezing problems) *in situ* by operation at low current density on pure hydrogen at their normal storage temperature, with success in many cases. Some stacks, however, had such heavy cross-leaks that this operation was unsuccessful.

At first, it was hoped that a full complement of twenty stacks might eventually be made available, using two spare stacks that were in storage along with a TEPCO-technology stack that had been purchased separately by the Tennessee Valley Authority (TVA). Another, less preferred option was to rearrange the whole system, making one useful pallet of the ten best stacks, to yield 2.25 MW (a.c.). In order to do this, it was hoped that the stacks in which the acid had been successfully diluted might be saved by acid addition to their matrices via a Teflon-tube acid sprinkler device located at the top of each stack. Due to problems resulting from the

Utility Systems 69

blockage of holes in the acid-addition tube, suitable flow rates could however only be obtained by using unacceptable pressures, even with addition of relatively dilute acid.

At this point, the possibility of operating the original stacks had to be definitively abandoned, since the old-technology stacks could then only be saved by disassembly and rebuilding. This had not been done before, and the operation was therefore considered to be not cost-effective.

The only possible alternative was to replace the stacks by a new series incorporating the (then) latest technology, at an estimated cost of $11 million (1984). This would have caused further delay (12–18 months). However the project was already well over budget, the New York fuel processing system had been so modified as to be less than reliable, and the TEPCO demonstrator was then operating. Consequently, the only logical step was to terminate the project and clear the site, which was done by December 1985. A report is available on the Consolidated Edison 4.5 MW unit, including the testing of its hydrogen-generation system.[113, 1088]

In spite of not having produced electric power, the Manhattan plant passed all the other tests required to confirm the system viability. Most important, and in spite of the responsibilities and complexities involved in the codes at a major urban site, the plant showed that it was possible for a fuel cell power generating station to comply with the demands of the local authorities. After extensive extra testing under nondesign conditions, which resulted in considerable damage and long delays, it received its license to operate, and as a result fuel cells will be the only new type of fuel-burning generator that will be allowed in Manhattan in the future.

In retrospect, it may have been better to install the demonstrator elsewhere than in Manhattan, which would have allowed the possibility of start-up 3~4 years earlier, however a license to operate in the most demanding site in the United States would not then have been obtained.

It is certainly fortunate that the next 4.5 MW unit was ordered and was successfully operated close to schedule by the Tokyo Electric Power Company, otherwise the U. S. electric utility fuel cell might well have been abandoned because of the discouraging experience of New York.

TEPCO 4.5 MW A.C. Generator In Japan

The Tokyo Electric Power Co., Inc. (TEPCO), with a generating capacity (July 1980) of 31,835 MW, is the largest of the nine Japanese electric utility companies,[1435] and the largest privately owned electric utility company in the world. About 40% of its capacity uses imported natural gas (as LNG) from Indonesia.

Persuaded by the advantages of fuel savings and low emission characteristics of the PAFC, TEPCO executed a contract with UTC in February

1980 to demonstrate UTC's Mark II–4.5 MW a.c. PAFC power plant. TEPCO joined the National Fuel Cell Coordinating Group on September 16, 1980. By then they had eight full-time engineers and about fifty part-time engineers and technicians on the project. By 1981, TEPCO had invested over five billion yen (~$25 million) in this project. In August 1980, construction of the demonstrator plant was started at TEPCO's 1760-megawatt LNG-fired thermal power station at Goi, in Chiba Prefecture, near Tokyo.[1436] While the construction and licensing procedure was very simple compared with the New York plant, special precautions were needed for possible seismic activity, high tides and oil spills. As in the New York unit, the power section consisted of 20 cell-stack assemblies (CSAs), each rated at 280 V at 850 A d.c., connected in series in two parallel strings to supply the power conditioner with 2800 V, 1700 A d.c. at full load (4.8 MW d.c.). The power conditioner (after parasitic losses) was designed to deliver 4.5 MW three-phase 50 Hz a.c. power into the TEPCO 66 kV network. A report on the operation of the TEPCO unit is available.[1088]

As indicated in the previous section, in comparison with that of the New York demonstrator, the TEPCO demonstrator had an improved power section that used the newly-developed ribbed substrate structure,[277,447,1084] giving sufficient reservoir capacity to prevent the acid inventory problems encountered in New York. Parts of the other subsystems also were redesigned by UTC for the Goi plant, based on the learning experience in Manhattan. The Process and Control test of the subsystems, excluding the fuel cell stacks, was commenced in October 1981 and all tests were completed by October 1982, when the cell-stack assemblies were installed.

During the period October 1982 ~ March 1983, the system was leak-tested and adjustments and modifications were made as necessary. In particular, burst disks on the CSAs were replaced by relief valves, a stack coolant drain and a nitrogen supply line to aid purging were installed, the auxiliary burner liner was replaced, and controls and the power conditioner were adjusted and given necessary maintenance. Finally, changes were required in the water-treatment plant supplying purified recovered water to the cell stacks for cooling purposes, which is then used as steam in the reformer.

This problem had already been uncovered in New York, and it is instructive to discuss it in some detail. This type of modification philosophy shows the unforeseen difficulties that may occur if any part of a new technology is designed or modified to give an apparent improvement that extends beyond more than one engineering discipline.

Because of its high thermal conductivity, copper piping sheathed with Teflon to protect it from acid was the material of choice in the cooling plates between every five cells in the fuel cell stacks (four cells per cooler had been used in the New York stacks). Other supply and manifolding parts of the piping system were stainless steel and mild steel. Even using water of the highest purity with ppb levels of dissolved oxygen, enough corrosion was observed to result in clogging of parts of the cooling system under certain conditions. This resulted from the use of dissimilar metals in the system.

The provisional cure used in Japan consisted of replacing as much of the external piping as possible with stainless steel, adding extra ultrafine filters to remove particulates, and improving the deaeration system. The oxygen specifications were tightened to <10 ppb, with water turbidity at <10 ppb, pH 6.7, and conductivity at $< 4 \times 10^{-7} \Omega^{-1} cm^{-1}$. Future designs will require the use of an entirely stainless steel cooling system, including the piping in the intercell cooling plates.

Overall operational testing of the plant at about half nominal electrical load (2.0 MW) started on April 7, 1983. During this phase, half-load operation was necessary because of certain system problems. These included the fact that one of the specified Harrison Radiator brazed flat-plate heat exchangers was not available and had to be replaced by a conventional tube-and-shell exchanger and because the reformer had not yet been modified to the original specifications (i.e., the addition of the 15 cm–high ring to raise the position of the burner away from the reformer tubes and the relocation of the air supply tube in New York). Modifications to the reformer were found necessary and were made in May~July, 1983. They included conversion of the start-up burners from naphtha to natural gas, spacer ring installation, flow distribution optimization, and reduction in the maximum operating temperature. At the same time some damaged insulation in the reformer dome was replaced.

In August 1983, Phase-I operation of the plant was resumed and continued in order to verify plant characteristics. A total of 16 generating test runs, totaling 196 h, with 89 h on hot standby, and yielding 371,000 kWh, were made between August 1983 and February 1984. During this period, the voltage of a few CSAs fell from 280 V to 200 V; the problem was traced to peeling of the Teflon coating inside the air manifold, impeding air flow. This was simply corrected.

In February 1984, the plant operated for the first time at 4.5 MW on natural gas containing 88.4% methane, 6.1% ethane, 3.8% propane and 1.7% butane. Gross a.c. power was 4519 kW, of which 277 kW was used for power station requirements. Heat rate based on gross power and 38.59 MMBTU/h HHV fuel input was 8540 BTU/kWh, or 9098 BTU/kWh based on net power (40% and 37.5% efficiency respectively). These exceeded design values. Net efficiencies at part load were slightly lower (34.8% at 2.0 MW, 35.2% at 3.0 MW, and 35.8% at 4.15 MW), reflecting the rather high accounting of parasitic power, which included the experimental control-room facility which would not be required in a commercial plant.

A number of problems were observed during this period. Start-up time was 6~8 h, rather than 4 h, due to the use of an incorrectly dimensioned nitrogen blower. Plant maintenance was difficult, because its compact design gave poor accessibility. Reliability of valves, instruments, and of the water treatment plant was less than adequate. Accordingly, the plant was shut down for servicing from March to May, 1984.

During this period, the reformer received new ceramic tube caps and some new insulation, the CSA manifolds were inspected, and fuel manifold gasketing was checked for leaks and repaired where necessary. A gas sensor on-line calibration system and a boost pump to allow better flow at high loads were installed. The power conditioner, computer, water treatment system, recycle blowers, and feedwater pumps were refurbished; and parts replaced where necessary. Finally, the turbocompressor was equipped with silencer baffles.

Phase-II generating tests took place from June 1984 to December 1985 to assess reliability, transient response, emissions, and general characteristics. Reliability improved dramatically: The longest run was then 500 h, and manpower went down significantly. No performance deterioration was seen in 2800 h of operation and hot standby, though a planned outage to repair several reformer tubes and replace reformer catalyst took place in the spring of 1985. By December 1985, the plant had accumulated 2400 load hours and had produced 5,428,000 kWh.

This plant therefore operated successfully on reformed methane for two years, proving the technology and that the system was able to achieve the promised reliability and efficiency, and that maintenance by ordinary utility personnel was possible.[1088] In comparison with the ConEd unit, the TEPCO demonstrator was built within budget and operated (with the equipment to the standard then available) only 38 months after the order was signed. Total building cost was about $25 million (1980), or about 60% of that of the New York unit, allowing for inflation.

Considering the problems that had occurred with similar subsystems in New York, completion of the experiment in Goi must be considered as a remarkable achievement, which leaves in no doubt the possibilities of a properly designed fuel cell operating in a utility context. However, future plants would require a more open structure for improved maintenance, simple tube-and-shell instead of compact formed-plate heat exchangers, an improved reformer burner and general reformer design, improved-quality inverter parts, including gate-turn-off (GTO) thyristors to increase efficiency from about 94% to a future 98.5%, and a simplified microprocessor-control system. The preceding were the lessons learned from the TEPCO Goi experiment.[1088] A photograph of the Goi plant is shown in Fig. 3–1, and a simplified flow-sheet is shown in Fig. 3–2.

TEPCO expects fuel cells to supply 14~20% of the increase in their future marginal demands, mainly as individual units of 30~250 MW size.

Utility Systems

Figure 3–1: Goi (Japan) 4.5 MW ac UTC unit (1983).

They foresee the necessity that fuel cells in Japan should be able to operate on coal-derived fuels and also that such power plants may have to be installed within multistory buildings in order to reduce siting costs.

UTC FCG-1 Utility Fuel Cell

As mentioned earlier, the full scale UTC FCG-1 power plant was originally planned as a 26-megawatt system in 1972. However, based on the background and experience gained with the 1-MW and 4.5-MW demonstrators, the present program has been changed to develop an 11-megawatt power plant as the final commercial version.[1086] The preliminary version of this power plant (called by UTC the *generic* 11 MW unit) was published in 1983 in a study funded by EPRI.[1087]

This unit has often been loosely called the *PC-23*, but the latter was UTC's internal code name for the final utility version, which differs in many details and was first announced in a UTC technical brochure in

74   Chapter 3

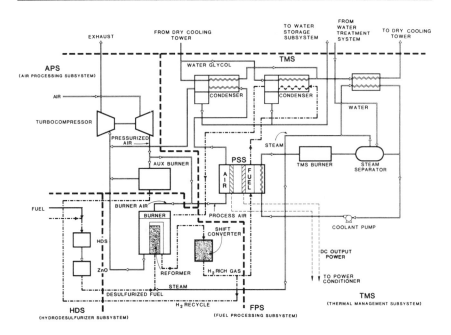

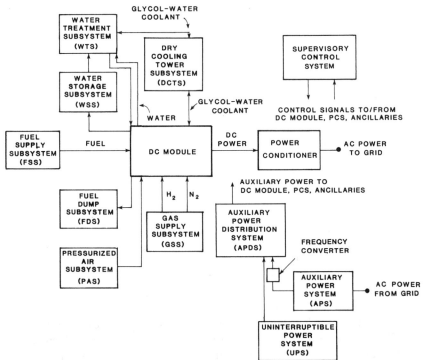

Figure 3–2: Flowsheet of Goi plant.

1985,[116] after UTC had announced a joint venture to develop, manufacture, and market this system with Toshiba in January 1985. This joint venture (International Fuel Cells, IFC) is essentially aimed only at a product for the electric utility PAFC market, although further options are possible in the future. The consortium will use proprietary UTC-developed cell component and stack technology, which remains the property of UTC. The partners will collaborate in developing the rest of the system, with assistance from Bechtel, Inc.

The 11 MW system as described in Ref. 1088 is an upgraded version of the 4.5 MW demonstrators. The higher output is achieved by operating the fuel processing system at higher pressure, so that the cell stacks operate at 8.2 atm (120 psia), rather than at 3.4 atm (50 psia). This allows the reformer to process more than twice as much gas, making the system more cost-effective by more than doubling output. The use of a higher pressure goes hand-in-hand with a higher cell operating temperature, since the fuel cell itself produces the steam output required in the higher pressure reformer. Both of these in turn increase the cell voltage from an average value of 0.65 V to 0.73 V, allowing an efficiency increase of 0.73/0.65, or 12.3%. The HHV heat rate is thus reduced from the value of 9300 BTU/kWh in the 4.5 MW demonstrators to 8300 BTU/kWh, and the total output increases by approximately the product of the efficiency ratio and of the pressure ratio (ignoring changes in parasitic losses) from 4.5 MW a.c. to about 11 MW a.c.

Operation at increased temperature and pressure has required the use of improved materials technology, particularly in the power section and in the reformer. The sensor system and the power conditioner also have been upgraded. Most important, the concept of a system consisting of plug-in truck-transportable modules constructed with an aerospace compactness has been modified to allow for ease of maintenance after the experience with the 4.5 MW demonstrators. The cell stacks in their pressure vessels are now separate units, rather than packed in a trailer-sized container, and the cells have been increased in area from 0.34 m^2 (3.7 ft^2) in the 4.5 MW units to 0.93 m^2 (10 ft^2). The various parts of the system are spread out in such a way as to allow easy access. An important aspect of the design that has been retained is that many of the components, such as the modular d.c. fuel cell stacks, the different parts of the fuel processing system, and the inverters, will still be factory-assembled and then transported to the site by common carrier, so that on-site assembly will be rapid.

The engineering design of the FCG-1 was aided by the use of a computer model that has been developed under contract to DOE by Mueller Associates, Inc., of Baltimore, Maryland.[2106] The desired operating conditions of the fuel cell stack are fed into the model, which then carries out material and energy balances throughout the system, permitting an evaluation of the effects of variations in pressure and temperature, concentrations of hydrogen, oxygen, carbon dioxide, carbon monoxide, and contaminants; hydrogen and oxygen utilizations; and the effect of stack ageing. In this way, optimum system performance and economics can be determined.

Two major applications foreseen for the commercial 11 MW power plant are dispersed power generation and urban plant replacement, both of which have already been discussed in some detail. A survey of 26 older urban plants in the U.S. indicated an average heat rate of 15,600 BTU/kWh (16.45 MJ/kWh). Replacing such old units with 11 MW fuel cells would result in fuel savings of approximately 40% in BTU terms. Thus the use of FCG-1 power plants will allow a utility to better utilize existing sites, reduce fuel consumption, and provide power in an environmentally acceptable manner. Manhattan is again an excellent example, where meeting increased demand using new plants outside the island would require installing new transmission lines under the Hudson River at an unacceptably high cost.

The 11 MW power plant also can be fitted into a system operating on centrally processed fuels, which will usually involve a plant supplied by either coal or heavy liquid hydrocarbons to provide a range of synthetic gases and/or clean liquids of varying quality. The synthetic gases range from low BTU gas (heating value of 120~150 BTU/scf), to an intermediate gas (300~500 BTU/scf; in contrast, natural gas is 1000 BTU/scf). The potential liquid products might range from desulfurized fuel oil to clean light naphtha or methanol. In its present configuration, the 11 MW plant can operate on synthetic natural gas, light naphtha, and methanol. Intermediate BTU gas containing carbon monoxide, carbon dioxide, and hydrogen with little or no methane would be even easier to handle because steam reforming can be eliminated, and the only fuel processing requirement would be catalytic conversion of carbon monoxide by the water–gas shift reaction to produce additional hydrogen. This configuration will result in a simpler dedicated power plant with reduced capital cost. Low BTU gas containing significant quantities of nitrogen can be handled in a similar fashion to that of the intermediate gas except that the lower partial pressure of hydrogen in the shift gas will slightly reduce fuel cell efficiency and power output. These losses will be much less in the fuel cell than if the gases had been used in a combustion machine, such as a gas turbine. The fuel cell therefore has particular advantages if only low BTU gases are available.

As has been outlined above, UTC's phosphoric acid utility fuel cell program has consisted of a series of sequential efforts to demonstrate technical, operational, and commercial feasibility of the PAFC power plants.[646,1086] After demonstrating operational feasibility with the Goi 4.5 MW demonstrator, the next step has been the attempt to establish commercial feasibility, including proof of component cost-effectiveness and lifetime. In this connection, UTC has been conducting a number of related technology component programs for the PC-23 under the sponsorship of DOE, the Tennessee Valley Authority (TVA), the Niagara Mohawk Power Corp., and the Northeast Utilities Service Co.[1334] The system design work has been divided into four areas: (a) system simplification studies to improve reliability and manufacturing costs; (b) preparation of a model specification describing the physical and operational characteristics of the power plant; (c) an expansion and confirmation of this model specification with system schematics, component descriptions, and realistic performance descriptions to provide the basis for mechanical design; and (d) a preparation of preliminary cost estimates. As a result of these studies, three main goals were identified that must be met by the final form of the 11 MW unit in order to achieve commerical viability:[1086]

1. The ultimate installed cost in volume production should be less than $1000/kW (1982 dollars).
2. Volume production units should have maximum heat rates of about 8500 BTU/kWh, based on the higher heating value for natural gas, with preferred heat rates of about 8000 BTU/kWh or less.
3. The availability characteristics of the power plant must be as good as or better than those of competitive systems.

However, these 1982 goals have turned out to be moving targets, and, as indicated in Chapter 2, Section O, future heat-rate and cost improvements will be needed to reach the largest possible market. The definitive UTC (now IFC) PC-23 first commercial PAFC power plant design is now established,[116] but the hope that units would be commercially available in 1986[1971] did not materialize. Part of the problem is dependent on the effective collapse of OPEC and of the relatively low cost of oil (and of natural gas in the United States) in 1986. Utility economists naturally feel that investment in an energy efficient, though comparatively high capital–cost technology that is as yet incompletely proven, despite its promise of ubiquitous siteability and other advantages, is not yet justified.

There is every incentive only to look at present fuel costs, and forget the *aberration* of 1974. The feeling is that fossil fuel prices will continue to rise in the long term, since readily available reserves will eventually be exhausted. However, the general attitude is a lack of interest in these arguments, since they are outside of the usual time-scale of predictions that the market economy is used to considering; for example, commodity futures markets only consider trends over about 18 months.

The final argument, which does contain a great deal of truth, is that the most extensive reservoir of world oil happens to be the geographic area with the lowest production costs: namely Saudi Arabia and the Gulf States. This of course ignores the strategic implications that are important for a country such as the United States, which currently obtains its nondomestic oil requirements from strategically-reliable producers with high production costs, such as Canada, Mexico, Britain, and Norway. The future cost of hydrocarbon energy is, to say the least, unpredictable, and a degree of political contingency would be prudent. Similarly, the price of coal cannot be regarded as permanent: Minehead coal prices increased by a factor of three in current dollars from $0.33/MMBTU to $1.00/MMBTU between 1970 and 1974 as a result of supply and demand. It is not inconceivable that this will occur again in the future.

Marketing the fuel cell to electric utilities and to other users will require a strategy on the part of the developer. Naturally, the production cost of early units off the line will be high, and their overall reliability may be doubtful. Users will prefer to wait for the operational experience of others to accumulate, and then buy later units at a lower price. The strategy must take into account this psychology; early units must have a guarantee of support, and must be available at realistic capital costs for the application. Such an approach may involve a consortium or condominium of utilities, as was envisaged with the proposed 27 MW unit in 1972. Developers, particularly the pioneer of the electric utility fuel cell (IFC), need to develop such a strategy to give an initial impetus to marketing, some of the problems of which have been discussed in Chapter 2, Section O.

Westinghouse/Energy Research Corporation Utility Phosphoric Acid Fuel Cell Program

Westinghouse Electric Corporation's Advanced Energy Systems Division is undertaking a program to design two 7.5-megawatt air-cooled phosphoric acid utility fuel cell power plants with funding from DOE/NASA, EPRI, and the Empire State Electric Energy Research Co. (ESEERCO).[508, 1864] Westinghouse is working in this program with Energy Research Corporation (ERC) to develop utility fuel cells utilizing licensed ERC gas-cooled (DIGAS®) phosphoric acid fuel cell technology.

The approach being taken by Westinghouse and ERC is to use a single baseline fuel cell stack consisting of pressurized 12" x 17" fuel cells that can be used for multiple applications, including utility power plants for package sites, for substations, and for urban use (for such applications, 5~15 MW pressurized modules that can be combined to give plants up to the 100 megawatt range). Alternative service-sector uses might be in unpressurized or pressurized (depending on output) on-site–integrated energy systems for hospitals, shopping centers, and apartment buildings in the 100 kW to 10 MW range. Cogeneration plants for industrial and commercial applications (municipal heating, pulp and paper, food processing, waste treatment, metals recovery, etc.) in the 1 MW class or greater, together with specialty chemical processing applications (1 MW to 100 MW) constitute other potential markets.

By focusing development and test efforts on a single design, the result should be a unit of maximum reliability and cost effectiveness due to the economies in large-scale manufacturing. The system designs into which this baseline fuel cell will be integrated vary only in the total power level, reject heat utilization, and possibly in power conditioning for specialized applications. Until the fuel cell market becomes established, the d.c. power modules will be kept distinct from the fuel processing systems. Early commercial plants will be designed more for reliability and maintainability than for optimum thermal integration with fuel processors to allow the best possible system heat rates. In other words, the approach will be maximum cost-effectiveness rather than maximum efficiency.

The Westinghouse/ERC philosophy is to develop prototypes that are as close as possible to the final commercial product, with flexible features that permit easy adaptation to future customer needs. Westinghouse believe that fuel cells can achieve their potential only if they are pre-tested to the maximum extent possible before leaving the central factory. Their market studies indicate that, provided heat rates can be lowered to less than 8000 BTU/kWh and costs kept below about $1000/kW (1983 dollars), by the year 2000 there should be a U.S. market for the PAFC lying between 700 and 1200 MW/year, mostly for plants sized between 5 and 50 MW, of which cogeneration will be an important segment.[1864] A 7.5 MW plant was chosen in the development program as a basis for modularity, because above this size, any gains in heat rates are relatively small and it becomes difficult to keep the hardware small enough to be shippable by truck.

The Westinghouse phosphoric acid fuel cell program consists of two complementary programs: a Technology Development Program, and a Utility Prototype Power Plant Program.[851, 2732] The Technology Development Program started in May 1982, and is based upon ERC technology and on advances made from the NASA-funded on-site–integrated energy system program (to be discussed in the next chapter). This has resulted in a baseline fuel cell stack referred to as the Mark II Design, which will be developed further for high pressure operation.

The aim of this program is to provide the basic technology needed to demonstrate the performance, endurance, and economical manufacturability of the gas-cooled PAFC. The program has been planned in stages: first, the demonstration of subscale cells; followed by that of nine-cell stacks of 10 kW, 25 kW, and finally a 375 kW plant module. In Phase I, emphasis has been on the development of the Mark II pressurized stack design having a heat rate of < 9000 BTU/kWh. Testing of 2 kW and 10 kW stacks as well as development of the 2.5 kW and 30 kW pressurized facilities were scheduled for 1983, with 2kW, 25kW, and 100 kW stack testing in 1984. In 1985, 100 kW stacks and a 375 kW module were originally scheduled for testing, to be followed in 1986 by further construction and testing, and of 375 kW modules and plant manufacture. The objective was to design and operate two 7.5 MW prototype power plants on two host utility sites to demonstrate commercial feasibility by late 1986 to mid-1987. The first proposed plant (1986) was to be for electric generation only, whereas the second plant (1987), would allow cogeneration. The first should allow the same heat rate as UTC's proposed 11 MW unit (8300 BTU/kWh), in spite of having a lower unit cell potential (0.68V vs 0.73 V). This would be achievable by the use of a better integrated use of waste heat in the system, including obtaining extra work by condensing steam.[177,1865] However, this plant, unlike UTC's, would have no waste heat suitable for cogeneration. The cogeneration plant, without the steam bottoming cycle, was proposed to have a heat rate of 8900 BTU/kWh.[177,1865]

The status of the Westinghouse program in mid-1986 may be summarized by the fact that great progress has been made in establishing the material requirements for a pressurized air-cooled stack based on the original licensed ERC technology. Due to a lack of hands-on knowledge, the difference between operation under atmospheric and pressurized conditions was not originally appreciated. Pressurization implies operation not only under higher reactant (hydrogen and oxygen) pressures, but also under higher product (water vapor) pressure. As we will discuss later in Chapter 16, the required oxygen for the corrosion of carbon structural components at the fuel cell cathode (which is at the highest electrochemical potential, therefore under the most oxidative conditions) must be supplied by water vapor.

Hence, corrosion is strongly pressure-dependent. In addition, pressurized cells appear to develop anomalous corrosive degradation modes under certain operational conditions. Many of these factors, which have been discussed at some length in Ref. 177, were either unknown or were discounted at the start of the Westinghouse program, which has had to develop the know-how that UTC had acquired some years previously by the same method, that is, by experience. The latest Westinghouse 10 kW pressurized short stacks, using their latest materials, now appear to be achieving satisfactory lifetimes. These units will be used as building blocks for the 375 kW pressurized stack.

Due to a diminution in utility interest that has resulted from present cash-flow and fuel price conditions, Westinghouse was unable to secure the announced intention of Southern California Edison for the installation of a 7.5 MW prototype unit in Long Beach. This decision was announced in May, 1985. Since that time, Westinghouse has been concentrating on the development of a smaller 1.5 MW prototype.

C. UTILITY MOLTEN CARBONATE FUEL CELLS

A large-scale program also has been in progress in the United States since 1976 to develop large-scale utility MCFC power plants. This program has involved close cooperation between many public and private institutions. Like the phosphoric acid fuel cell program, it also is managed by the U.S. DOE and EPRI. The original objective of the program was to allow the earliest possible utilization (perhaps by 1990) of cost-effective, clean, coal-burning MCFCs to reduce dependence on foreign oil, as a result of the events of 1973–74.

Coal-Burning Molten Carbonate Fuel Cells

The original incentives for undertaking the development of MCFCs derive from their ability to operate (at least indirectly) on coal. First-generation PAFCs are usually considered to be more appropriate for higher quality fuels such as reformed natural gas, naphtha, or methanol, since it was known that PAFCs could only consume hydrogen. Carbon monoxide is a poison to the platinum anode catalysts in the PAFC at concentrations that depend on the operating temperature. Under normal utility fuel cell conditions, CO must be removed to a level of less than about 1–1.5% using a water–gas shift converter. This extra fuel processing step results in a net cost and efficiency penalty.

At the high operating temperatures (~650°C) of molten carbonate fuel cells, anode kinetics are rapid and do not require noble metal catalysts that are easily poisoned at lower temperatures. Furthermore, carbon monoxide in coal-derived fuel gas is readily shifted to hydrogen and carbon dioxide under the *forced* conditions of the MCFC, where one of the reaction products (hydrogen) is electrochemically consumed *in situ*, and one of its reaction products (water vapor) serves to drive the shift process. In principle, this permits the use of a wide range of fuels such as coal, distillates, and residuals in a suitable fuel processor, and in particular it allows the integration of the fuel cell with a coal gasifier.

In 1978, under contract to NASA, UTC prepared a report that was to allow a selection of the technology that would best utilize the vast reserves of U.S. coal for the generation of electricity. Various competing advanced technologies were considered. The resulting study, known as ECAS (Energy Conversion Alternatives Study), was the most thorough of its kind to date, and used as a basis, a 650 MW baseload electric generating plant supplied by low BTU coal gas.[295, 490]

Plant efficiency studies for various advanced technology coal utility generators were examined, including MHD and combined cycle gas turbine–low BTU gasification systems. These indicated that for a competitive cost of electricity, a system incorporating an MCFC might have the highest overall efficiency, on the order of 47~50% expressed as input coal to a.c. output. This result of course, depends on certain extrapolated efficiency, cost, and lifetime assumptions for the MCFC.

In contrast, efficiencies were only 32~33% for conventional coal-fired plants with flue-gas cleanup systems.[733] In addition, the emissions from MCFC plants would be intrinsically extremely low, so that such units would make it possible to easily utilize coal without the usual gaseous and particulate emissions. The only particulates would be those that result from coal handling and ash disposal, the remainder being removed during gasification. Typically cited emissions for a molten carbonate system were SO_2 = 0.12 lb/million BTU; NO_x = ~0; TSP = 0.01 lb/million BTU.[733] These emission considerations would be especially important in environmentally constrained locations. The NO_x concentration will be particularly low because of the electrochemical scrubbing characteristics of the MCFC cathode, through which all (or almost all) of the stack gas would exit.

The study selected by EPRI and General Electric (GE) as a result of the ECAS conclusions was for a 650 MW system burning Illinois No. 6 coal with an overall coal–a.c. efficiency of 50%. The ECAS design assumed the use of a fluidized bed gasifier and a fuel cell stack tolerating anode gases containing sulfur concentrations given by the EPA limit.[1112] Subsequent studies showed that the MCFC sulfur tolerance was much lower,[2104, 2105, 2179, 2293, 2617] necessitating more complete sulfur removal, which fortunately had been shown to be feasible. These subsequent studies, sponsored by EPRI, were made for a plant consuming 10,000 tons of coal per day utilizing three state-of-the-art gasifiers (Texaco entrained-bed oxygen-blown, Texaco air-blown, and British Gas Council oxygen-blown slagging gasifiers). They appeared to confirm that molten carbonate fuel cell power plants will be more efficient and will have a lower environmental impact than that of gas turbine power plants. At the same time, they should be able to generate electricity from coal at competitive prices. The latter point is important, since the annual fixed cost of a coal-fired central station, which depends on capital cost, is in practice generally more important than the annual fuel cost. The GE/EPRI study[421] concluded that a coal-to-a.c. efficiency of 50% was indeed possible, at a capital cost of $1000/kW (1980). Efficiency could rise to as high as 55% if the somewhat modest performance that GE assumed for the pressurized MCFC is corrected to the latest performance data. General Electric continued to perform further similar studies for the DOE, with the conclusion that the modularity of the MCFC would allow coal-gasifier plants to be built in units of as small as 150 MW.[1977] They would thus be very attractive to utilities since they would permit increase of baseload capacity in incremental quantities. An excellent summary of

Utility Systems

MCFC coal-gasifier plants, based on all available information, has been published by Argonne National Laboratory (ANL).[1521]

Four major areas of application have been considered for molten carbonate fuel cells as follows:[378]

1. 150~650 MW coal-fired baseload central power plants. This application is for purely baseload electric power generation and is one of the main thrusts of the DOE program. As coal gradually becomes the primary fossil fuel, the use of plants based on this principle should increase. The low pollution and high efficiency of the molten carbonate fuel cell are the main incentives in this application.
2. 10~25 MW plants using petroleum or coal-derived liquids for dispersed baseload application, principally for load leveling use. EPRI studies have indicated that availability of such fuels for load leveling should occur well into the 1990s and beyond. Since the molten carbonate fuel cell should have an efficiency of about 45% with these fuels, it should allow energy savings compared with conventional oil-fired plants.
3. 1~50 MW units fired by natural gas, SNG, petroleum, coal, and coal-derived liquids for industrial cogeneration. In this application, which is discussed in greater detail in Chapter 4, the cogeneration of electricity and heat using a molten carbonate fuel cell may enable greater than 90% utilization of the heating value of the fuel. The high grade heat produced can be used for district heating or industrial process heat, a large energy-demand sector.
4. 10~100 MW coal-fired plants using the economy of scale of a central gasifier, associated with dispersed fuel cell units.

Molten Carbonate Fuel Cells Vs. Phosphoric Acid Fuel Cells for Use With Coal

It is generally conceded that for operation on coal-derived fuels the MCFC will be preferable to the PAFC.[733] The MCFC operating at about 650°C, has a higher cell efficiency than the PAFC in its present form, which operates at a maximum temperature of about 210°C. This results from the lower polarization of the oxygen cathode, which more than offsets the 0.2 V reversible thermodynamic loss (see Chapter 2) at its high operating temperature. A well-designed MCFC can be expected to operate at about 0.1 V higher than current-technology PAFCs under design operating conditions (about 200 mA/cm^2 for the PAFC, 160 mA/cm^2 for the MCFC). This should improve d.c. system efficiency by a factor of approximately 1.14.

The higher temperature of the MCFC permits the use of bottoming cycles to produce more electricity from its high-grade waste heat. MCFCs can use both gas and steam turbines in the bottoming cycle to recover process energy and improve the overall electrical efficiency. For example, GE's calculations referred to in the previous section[421] indicate that for a coal-fired MCFC system with an integrated gasifier operating on Illinois basin coal with a higher heating value of 12,235 BTU/lb (28.4 MJ/kg) the output from the fuel cell itself is about 40%, with an additional 4% gain achieved using a gas expander-generator, and a further 12% gain from a steam turbine generator. Deducting 7% for parasitic losses yields an overall electrical efficiency of 49%.[2005]

It should not be inferred from the previous data that PAFCs cannot be used with coal-derived fuels; just that they will probably be less efficient than MCFCs for such applications, when the latter are available for commercial application. Studies made to evaluate the use of state-of-the-art PAFCs operating on gasified coal indeed show that their heat rates are not as low as those achievable with MCFCs. For example, a study by Kinetics Technology International Corporation (KTI) of Pasadena, California, evaluated the use of U.S. coals or lignites with a commercially available fixed-bed air-blown Wellman-Galusha coal gasifier to be used in a 44~50 MW plant using four of the 11 MW modular FCG-1 PAFCs being developed by UTC.[493,1836] A heat rate of 11,100 BTU/kWh was calculated for the baseline plant, or 9600 BTU/kWh if credit was taken for the by-product tars, oils, and coal fines; if they could be sold *over the fence*, such as for boiler heat.[493] Where permitted, the plant could burn the by-product tar and oil, and use the reject thermal energy to generate auxiliary steam for additional electricity generation in an expansion steam generator. The earlier study[1836] used a heat rate of 10,880 BTU/kWh, compared with a baseline value of about 12,100 BTU/kWh. The later report[493] did not specifically consider this option because of the environmental problems it would incur, but an optimized plant operating in this mode should have a heat rate of about 9800 BTU/kWh. However, a more interesting result was obtained if a KGN gasifier was used, which can recycle the oils and tars, allowing higher overall coal conversion and a system HHV heat rate of 8239 BTU/kWh, or 41.4% efficiency.[493] While this heat rate is still significantly higher than that for an integrated coal-fired MCFC system, it is a considerable improvement over a low-emissions conventional coal plant.

If the Westinghouse PAFC were used in an identical system to that in Ref. 3034, the heat rate would be 11,500 BTU/kWh with by-product credits, and 12,800 BTU/kWH without the credit.[492] These heat rates are higher than those for a UTC-based system, reflecting the lower cell voltage and smaller by-product steam output for the coal gasifier when Westinghouse cells are used. Because the Westinghouse cells would require a larger balance-of-plant per kW, its total capital cost would be higher.

Another study made by Westinghouse for its planned 7.5 MW PAFC integrated with the Westinghouse oxygen-blown fluidized bed coal gasifier indicated a calculated heat rate of 9110 BTU/kWh.[2707] In this innovative system the anode exhaust, which contains 16 volume percent hydrogen, would be burned with air and the resulting hot gases expanded through a turbine driving an a.c. generator producing additional electricity. The system also would use both high and low temperature shift reactors. By incorporating these bottoming cycle steps a significantly greater net electrical output can be achieved. Integrating the system by recycling the anode exhaust gas instead of burning it to drive a turbine might achieve an even lower heat rate of only 8520 BTU/kWh, but then such a system would be complex and difficult to operate. The calculated HHV efficiencies of the non-integrated and integrated systems correspond to 37.5% and 40.0%, respectively.

Another study was made by the Tennessee Valley Authority (TVA), in which the modifications necessary to allow the use of the UTC 11-MW FCG-1 PAFC with coal-derived fuels were examined.[1266] In this case, the gasifier is assumed to be remote, that is, not integrated with the fuel cell power plant; and the assumed heat rates are for gas-to-a.c. power, not coal-to-a.c. (the BTU conversion efficiency for coal to gas might typically be 60–65%). The most extensive modifications to the FCG-1 would be required when medium-BTU gas containing no methane is the coal-derived fuel. Although the modified plant still has the same power capability as the original, and only a slightly different heat rate at 100% rated power (8350 BTU/kWh vs. 8300 BTU/kWh for the baseline plant), to adapt the plant to run on both baseline fuel (naphtha) and coal gas results in an increased plant cost of 4~5% above that of the baseline plant. This is mainly as a result of necessary additions such as shift converter beds, heat exchangers, start-up electric motors, associated controls, line blinds or isolation valves, along with modifications to the auxiliary burner, anode recycle blower, and controls schedules. However, if the plant were designed to run exclusively on the coal-derived fuel, then the cost would be 3.3% *lower* than the baseline design, because several equipment items could be deleted, such as the reformer, hydrodesulfurizer, heat exchangers, a fuel pump, and various associated controls.

In spite of the possibilities of PAFC-based units, the higher intrinsic cell efficiency of the MCFC still makes it look more attractive for use with either remote or on-site fuels produced by coal gasification. A cost of electricity analysis made by GE for a 675 MW integrated coal-gasifier MCFC power plant showed that the cost of electricity will not be strongly dependent on the cost of coal.[2097] Instead, the reliability (on-stream capacity factor), endurance (the effect of cell life on long-term cost), and performance (polarization) of the fuel cells are more important. The capacity factor was found to have the greatest influence on the cost of electricity, and improvement to give greater than 70% on-stream time could lead to significant per kWh cost reductions.

Increasing the cell life to beyond about 40,000 hours would not have much effect on the cost of electricity, whereas failure to achieve 40,000 hours may significantly increase operating costs. Cell performance (polarization) also is important, but no more so than is cell lifetime. The actual cost of the cells and their replacement have been estimated to represent a considerable percentage (30%) of the fixed costs. Hence, it was concluded that it would be more important to improve cell cost and endurance than to improve cell operating characteristics. Thus, emphasis on a reduction of fuel cell cost cannot be overemphasized, and probably will be the single most important factor determining the commercial success or failure of the technology. A recent study by ANL appears to strongly underline the above conclusions, indicating that minimum electricity cost might be achieved at some expense of efficiency by minimizing fuel cell capital and replacement cost. For example, using a smaller percentage of gasifier output in fuel cell modules operating at high current density might be advantageous from the cost viewpoint.[1787] In such a plant, the fuel cell would act as a supplemental cycle in an integrated-gasifier combined cycle (IGCC) plant.

A study made by the Gas Research Institute (GRI) on fuel cell capital costs required to penetrate the U.S. utility market indicated that a 50% efficient coal-fired fuel cell system would be viable at a capital cost of about $800/kW (1980), whereas a 40% oil-fired fuel cell system could not be allowed to exceed a capital cost of about $300/kW (1980),[378] based on relative coal and oil costs at that time. These figures require updating to correspond to present economic conditions.

In spite of the above considerations, the technology of the PAFC can be said to be essentially proven, which is not yet true for the MCFC, whose commercial development is about five years behind that of the PAFC.

Utility Molten Carbonate Fuel Cells with Fuels Other Than Coal

Although the main impetus for the development of MCFCs has been their ability to more effectively use the coal-derived fuels which might be readily available in the future, conventional fuels such as reformed natural gas, light hydrocarbons, or alcohols could be used advantageously at efficiencies as high as 60% in this high-temperature cell. Such high efficiencies will permit rejection of the relative small amount of high temperature waste heat to the air, with an attendant increase in system simplicity, for plants not making use of bottoming cycles. Alternatively, their high-grade waste heat can be used in cogeneration schemes.

There is a strong possibility that the MCFC may have the highest efficiency of any fuel cell, unless further technological breakthroughs occur in the future. The improvement in cell potential, hence efficiency, in the MCFC compared with that of the PAFC has already been discussed. For a solid oxide fuel cell (SOFC) operating at 1000°C, the reaction kinetics are even faster than in the case of the molten carbonate fuel cell. However, on account of the lower theoretical open circuit potential and of the relatively high ohmic resistances encountered in the present-generation SOFC (Chapters 2 and 18), these fuel cells operate at a unit cell voltage under atmospheric pressure conditions similar to that of the PAFC under the same conditions. Due to their construction, there is as yet no evidence that they can take advantage of the higher cell potential that would result if they were pressurized. Although the waste heat available from an SOFC is very high grade and can be used very effectively in a bottoming cycle, it is generally maintained that a SOFC-based system will not be able to achieve overall efficiencies that approach those obtainable from a suitable MCFC.[177]

Recently there has been much interest in developing dispersed molten carbonate fuel cell power plants for operation on natural gas. In addition to the recognition of the cell high efficiency of the molten carbonate system using hydrogen fuel, this interest has arisen largely from the fact that the price of natural gas has not escalated as much as was predicted some years ago. It now seems probable that natural gas will be available well beyond the end of the century; indeed, there may be an almost limitless supply (though at high production cost) if the Gold theory of the non-biogenic origin of natural gas[984] is shown to be correct.

Because of the realization that natural gas should be relatively inexpensive for some time to come, utilities have been reluctant to invest in high capital–cost coal gasifier plants, using either combined cycles as at Cool Water in Southern California,[631, 1155] or with fuel cells and bottoming cycles. As a consequence, the developers of the MCFC are now considering the use of these power plants with natural gas. In addition to its inherently higher cell efficiency, an added advantage of the MCFC over the PAFC is that its operating temperature is high enough to allow *internal reforming* of the natural gas. In this concept, a suitable reforming catalyst is located in the fuel flow section of the anode compartment of the fuel cell, and the waste heat from the cell reaction is directly used to reform a desulfurized methane–steam fuel gas mixture. This arrangement does away with the complex and costly separate steam reforming unit. Also, since carbon monoxide can be consumed in the MCFC via the water–gas shift reaction, the catalytic shift convertor required for the PAFC system can be eliminated.

These practical aspects of internal reforming are secondary to great kinetic and thermodynamic advantages. First, the temperature of the MCFC is somewhat low compared with that in a normal commercial methane reforming unit. In the latter, however, the heat of reforming is supplied by a flame, and the catalytic reaction takes place in tubes within the reformer vessel, the reaction rate being limited by heat transfer from the flame zone through the tube walls. The system is limited by heat transfer, the local rates of the reforming reaction are high, and the effective reforming temperature is high (at least 750°C, and more typically over 800°C). In the internal reforming MCFC, the ratio of the surface area of the heat-transfer plate (the bipolar plate between the cathode of one cell and the anode of the next) to that of the reforming catalyst volume is high, so that the system is no longer heat-transfer limited. In spite of the lower temperature (under 650°C), the chemical reaction rates for reforming are adequate since the reaction is driven (as is that for the internal water–gas shift, as was described earlier). That is, the hydrogen produced in the process is consumed in the fuel cell as it is produced, the product being water, the reactant for the reforming process. Consequently, the reaction goes beyond the equilibrium value of about 85% conversion at 650°C, essentially to completion. In fact, the cell behaves to within a few millivolts on the methane–steam input mixture as it would on the equivalent fuel reformed externally. In this connection, while the steam required as a reactant is produced within the fuel cell, some steam (typically at least corresponding to a steam/carbon ratio of 0.5) must be added to the methane input gas, otherwise cracking with irreversible carbon deposition will take place within the anode manifold and in the flow channels.

Second, in internal reforming the waste heat in the fuel cell is used to upgrade the HHV of methane fuel (the equivalent of 1.153 V [see Table 2–1]) to that of hydrogen (1.48 V). The heat output of the cell falls accordingly, by about 0.33 V. However, performance of the cell on methane–steam and on the equivalent hydrogen-containing reformate is identical. Therefore, from the thermodynamic viewpoint, the cell acts as a black box, consuming methane, but producing the same cell potential as that for hydrogen. If the cell potential is 0.73 V, the efficiency for methane used in the cell is therefore 0.73/1.153, or 63.3%. If 90% of the methane can be consumed in the cell, and if parasitic power and d.c.–a.c. conversion losses amount to 6% (as in the PAFC), then the overall system efficiency will be 55.6%, equivalent to a heat rate of 6140 BTU/kWh. In contrast, the UTC 11 MW PAFC operating at the same cell potential, but suffering from the thermodynamic problems of a low cell temperature and the difficulties resulting from heat and mass transfer in its external-reforming system, can only manage 8300 BTU/kWh, even with the best-optimized thermally-integrated subsystems. In addition, it should be remembered that this MCFC, unlike the PAFC, would be an atmospheric pressure unit. It could therefore be produced in relatively small sizes for on-site use. Finally, if it could incorporate a bottoming cycle, or be pressurized using a turbocompressor driven by waste heat and/or waste anode exhaust, its heat rate would be even lower: For example, pressurization to 5 atm could add 0.1 V to cell performance, yielding a theoretical heat rate of 5400 BTU/kWh, provided that the cells could then supply enough heat for internal reforming. The potential interest of the internal-reforming MCFC over that of the PAFC is therefore clear: Not only will it be much more efficient, but its system simplicity may also make it cheaper.

System analyses carried out at ANL indicate that a conventional natural gas-fired MCFC with external reforming and a system operating pressure of 6.8 atm can achieve an overall electrical efficiency of about 50% at 58% fuel utilization using off-the-shelf components for everything but the fuel cell stack.[1516-1518] At greater than 58% fuel utilization, the overall efficiency declines, because there is then not enough energy left in the anode exhaust stream to reform all the methane, and unreacted fuel must be burned to provide heat for the reforming reaction. About 75% of the net electrical output from this baseline system comes from the fuel cells, and 25% from a turbocompressor bottoming system. Its flow-sheet is shown in Fig. 3–3.[1517] Incoming natural gas is mixed with a small stream of reformed gas (needed to provide the hydrogen necessary for the hydrodesulfurization process) and directed to the desulfurizer (hydrodesulfurizer–zinc oxide absorber). The clean gas then is reformed at 800~950°C and fed to the anode compartment of the fuel cell stack. After exiting the anode compartment, the stream is regeneratively cooled to eliminate most of the water and is fed to the reformer burner, where the unreacted fuel provides heat for the reforming reaction. The combusted gas, containing carbon dioxide, then leaves the reformer, is mixed with incoming fresh, preheated air, and is fed to the cathode compartment of the fuel cell. The hot cathode exhaust gas is then passed through a turbocompressor to generate extra electricity before exhausting from the stack.

The net efficiency for the system is calculated first by multiplying the fuel cell d.c. output by 0.96 to take account of the d.c.–a.c. conversion efficiency of the power conditioning subsystem, then adding the difference between the gas turbine output and the air compressor input, and finally subtracting the power consumption of the two blowers. Increasing the pressure of the system from 6.8 to 10 atm has been shown to decrease the overall system efficiency by about 2% as a net result of the harmful effect of increased pressure on the reforming process (whose equilibrium com-

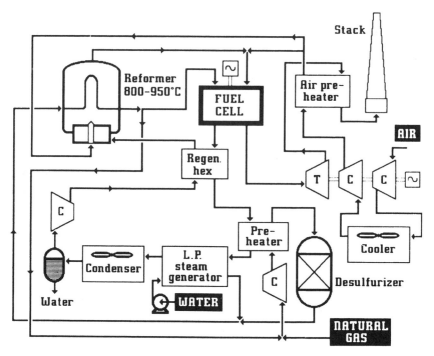

Figure 3-3: Simplified flow-sheet for natural gas-fired molten carbonate fuel cell system with external reforming.[1517]

position decreases with pressure in non-driven reforming), which overrides the beneficial effect of pressure on the fuel cell performance. This baseline molten carbonate fuel cell system is of a conventional design, a bottoming turbine providing some of the power. Such a pressurized system appears to be most feasible for power plants in the 10~100 MW range.

A second system, termed a *sensible heat reforming system*, is shown in Fig. 3-4.[1517] In this configuration, only recycled fuel cell waste heat is used for reforming. The reformer reactor has been replaced with a packed bed catalytic reformer. A recirculation blower recycles about 90% of the hot anode exit gases through the catalytic reformer. As well as heat, the anode recycle stream also supplies the water vapor required for the reforming reaction. In the configuration shown, a water electrolyzer is used to generate the hydrogen needed for the cleanup (hydrodesulfurization) of any organic sulfur compounds in the fuel gas.

As in the previous case, the overall electrical efficiency of the system is determined by taking the sum of the fuel cell and turbocompressor outputs

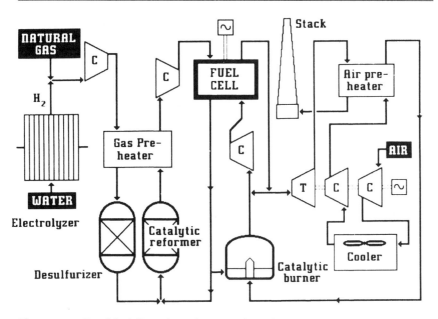

Figure 3-4: Simplified flow-sheet for natural gas-fired molten carbonate fuel cell system with sensible heat reforming.[1517]

minus the various parasitic losses, including that of the electrolyzer. The system analysis indicates a net system electrical conversion efficiency of as high as nearly 58% at 92% fuel utilization. There is, however, a risk of damaging the fuel cell at fuel utilizations greater than about 90% because minor flow field variations within the cell can lead to local depletions of fuel (i.e., hydrogen), which can initiate anode corrosion reactions. At very high fuel utilizations the system efficiency starts to level off due to the extra power requirements of the air blower, which starts to override the d.c. efficiency improvement resulting from greater fuel use in the fuel cell. In practice, it may therefore be preferable to operate a system of this type at about 85% fuel utilization, yielding a very high 56% efficiency value. Because the system requires a special catalytic reformer and high temperature blower, neither of which is an off-the-shelf item, it would be more expensive than the previous baseline case. This pressurized system would again appear to be most feasible for use with power plants in the 10~100 MW range. If the turbocompressor is eliminated it also might be feasible for use in smaller power plants of somewhat lower efficiency.

The third and most advanced system considered uses direct internal reforming and would operate at atmospheric pressure. In this configuration, the reforming catalyst is placed directly into the fuel cell, thereby eliminating the need for gas recirculation required for sensible heat reforming. This results intrinsically in a more efficient system, since simple driven reforming can take place as just indicated.

The system flow-sheet is shown in Fig. 3-5.[1517] Incoming natural gas is mixed with hydrogen coming from a small catalytic reformer and is desulfurized. Some steam also is injected into the clean fuel to preheat the gas and to stabilize it from carbon formation. Next the fuel gas goes directly to the anode of the fuel cell. After leaving the anode compartment, the spent anode exhaust gas is cooled to remove most of the water vapor that is formed in the fuel cell reaction. The stream then is burned in a catalytic combustor with excess air to generate the carbon dioxide–oxygen mixture required for the cathodic process, and is then fed to the cathode compartment of the fuel cell. The cathode exhaust then passes through two heat exchangers and is exhausted from the system. Even without a bottoming gas turbine, a net system electrical efficiency of about 52% appears to be attainable for this system with 80% fuel utilization. The system has been designed without a bottoming turbine because an analysis indicates that its addition would provide little more than sufficient power to run the compressors and auxiliaries. Although the efficiency continues to increase

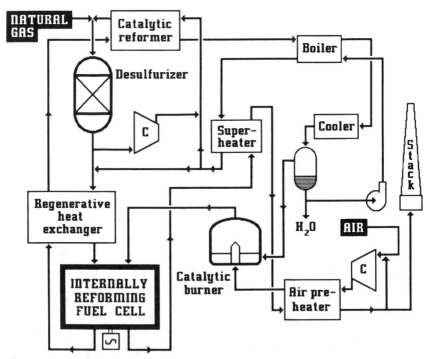

Figure 3–5: Simplified flow-sheet for natural gas-fired molten carbonate fuel cell system with internal reforming.[1517]

Utility Systems

with increasing fuel utilization (reaching ~54% at 92.5% fuel utilization), there again may be a risk of damaging the fuel cell anode at fuel utilizations greater than about 90%. The 5~10% efficiency gain that may be attainable by operating the system at 10 atm is probably not sufficient to justify the capital expense of a turbocompressor. Because such an internal reforming system is simple, and does not use a bottoming system, it would seem to be best suited for small on-site generators of less than 1 MW power output.

In view of the preceding, it would appear likely that a 2~10 MW molten carbonate fuel cell operating on natural gas, using either an internal or external reformer, will be able to achieve an overall system efficiency of about 50% without using bottoming cycles. With a plant of 50 MW or greater, it should be cost-effective to add a bottoming cycle consisting of either a steam or gas turbine, which would raise the overall system efficiency to close to 60%.

With a 600 MW molten carbonate fuel cell it would be worthwhile to employ both steam turbines and gas turbines, with the result that operation on coal gas with an integrated gasifier should yield an overall coal-to-a.c. efficiency of close to 50%, as described earlier for the General Electric design.[421, 2005] With optimum integration of the coal gasifier system, efficiencies approaching 60% may even be possible.[1519] These values are considerably higher than the values of 45~48% that will be obtainable from an advanced methane-fueled combustion turbine combined cycle (CTCC) plant.

Development Program for Utility Molten Carbonate Fuel Cells

In 1975 DOE and EPRI started to fund the development of the MCFC.[2023] The original program, administered by the DOE (then ERDA), in conjunction with EPRI, private companies, and a number of electric utilities, had the objective of developing and building a multimegawatt utility MCFC plant by about 1990.[17] In early 1980, contracts were awarded by DOE to GE and UTC for the development of a coal-fueled baseload MCFC system. The purpose of these contracts was to obtain enough cell and stack materials and manufacturing development data to provide a basis for the construction and testing of prototype stacks, as well as to develop detailed reference designs and establish cleanup requirements for the cell fuel and oxidant feedstock streams. In the same time-frame, EPRI supported work at three developers: GE, to define utility plant parameters;[421] UTC, to establish the viability of the MCFC system compared with other fuel cell concepts,[1404-1406] then to examine pressurized laboratory stack development;[2101] and ERC, to investigate an alternative cell design concept.[125]

In addition, the Gas Research Institute (GRI) supported work on the MCFC systems for cogeneration. In the original program, contracts were awarded to IGT to study contaminant effects; to ERC, for alternative cell designs; to EIC Corp. and Physical Sciences Inc., to study electrode kinetics; to the ANL, to develop rigid sintered electrolyte structures; to Oak Ridge National Laboratory, to study ion transport mechanisms; to Montana Energy and MHD Research and Development Institute (MERDI), to develop fibrous ceramic materials; and to various utilities such as Pacific Gas and Electric (PG&E), Southern California Edison (SCE), and others. The emphasis of all the funding organizations was stack development and the scale-up of component manufacturing processes.

The 1980 DOE program consisted of four phases: (a) to develop a reference power plant design, (b) to develop component and stack designs, (c) to develop plans and fabrication facilities needed to progress to prototype stack testing, and (d) to verify cell operation on coal-derived gas. The first phase consisted of an *Exploratory Development* phase, in which a reference design was to be developed in sufficient detail to establish a set of requirements for the fuel cell stack, for example, determination of the user requirements such as size, materials used, reliability, duty cycle, site requirements, cost of electricity, construction lead time, and other parameters. This design would then define a complete set of fuel cell requirements for inlet and outlet temperature, pressure, flow rates, gas stream compositions, frequency of start-up, cool-down, load change, and transients. Also included in the initial phase was the development and testing of an initial stack design meeting these requirements. Cell development activities included those problems that had been identified up to that time, such as electrolyte management, anode sintering, cathode performance, thermal cycling, together with cell and stack scale-up.

The second part of the program was to be a *Technology Development* phase, involving optimization of the reference design and demonstration of the compatibility of the coal gasifier, fuel purification, and fuel cell subsystems. Other tasks included determining the effects of off-design conditions on the system, defining interface operating conditions and requirements for each subsystem, and constructing a gasifier–stack test facility to test individual components. The results were then to be used to refine stack and reference power plant design, to establish fuel cell performance and reliability, and to support multimegawatt level testing.

The third phase of the program would be *Engineering Development*, involving verification of subsystem compatibility, system operation, and control. This phase would provide the database to verify fuel cell performance and reliability at the high confidence level required to proceed to the demonstration phase. To this end a multimegawatt test facility was to be built to permit large-scale, integrated testing of all the major demonstration power plant subsystems, to aid in resolving any power plant scale-up difficulties in advance of actual demonstration plant construction.

Utility Systems

The final part of the program was to be a *Demonstration* phase, to demonstrate the technical feasibility of two full-scale commercial size, cost-effective power plants: a cogeneration system of 5~10 MW, depending on the industry selected; and a baseload demonstration power plant, initially to be the smallest independent power train for a commercial plant, probably of the order of about 200 MW.

Most of the cell and stack development was to be conducted early in the first phase of the program, with continuation at a reduced level well into the final phase. This would mostly involve life testing, and the development of materials and manufacturing methods. The time-frame designed for the program envisaged the testing of a 500- to 700-cell prototype stack in 1983, the building and testing of a several-megawatt pilot plant in 1985, the demonstration of a small cogeneration plant or electric utility plant using volatile fuels in 1987~1989, and the demonstration of a coal-fueled baseload plant in about 1990.[17] Funding constraints have of course prevented this planned progress.

The ultimate goal of this ambitious MCFC program was to build a 650 MW coal-fired utility power plant incorporating a coal gasifier by about the year 1990. Although such a plant is still desired by DOE, recent budget cutbacks have made it impossible to adhere to the above schedule for the development of the MCFC system. The program budget rose from about $1.5 million in 1975, to $4 million in 1978, $9 million in 1979, and about $11 million in 1980, and has been holding relatively steady over the past five years at $9~10 million per year in current-year dollars. In comparison, the total DOE annual fuel cell budget during the same period was $30~40 million in current-year dollars, much of which went to the New York PAFC demonstrator and to the 40 kW PAFC on-site systems.

Budget cutbacks, combined with the desire to also develop the natural gas MCFC, have resulted in a branching of the program into two parts. One will be for the development of a coal-fueled system, the other for that of a natural gas system.[19,1519] The first would involve the development of a full-size, roughly 500 kW stack operating on coal gas; the second, the development of a 10~50 MW dispersed natural gas power plant. The major developers were originally UTC and GE for the pressurized coal-gas system, with ERC receiving support for materials development parallel to its EPRI contract to develop an alternative stack.

After the DOE budget reduction, combined with interest in the utilization of natural gas, and with GE's corporate decision to leave the program (as explained later), the largest fraction of the budget has now gone to the technology leader, UTC. United Technologies and ERC are therefore now the main DOE contractors for MCFC development, UTC receiving about twice as much support as ERC for its stack development effort. The organization monitoring DOE work (funded from the Morgantown Energy Research Center, METC) is ANL, which also conducts some in-house laboratory work.

Department of Energy philosophy has been to carry out systems designs in parallel with technology development to ensure relevant technology goals and a commercially attractive end product. The philosophy has been to use or adapt commercially available, off-the-shelf technology as much as possible, and to carry out no extensive development for components other than those for the fuel cells and stacks. It was anticipated that full-scale stack development could begin in 1986; and the first 0.74 m^2 (8 ft^2) 20-cell short stack was started up in September, 1986. Progress in MCFC development by UTC throughout the period of the DOE program will be summarized in Chapter 17. To aid in predicting cell performance and requirements, DOE has made use of sophisticated computer models that will simulate the behavior of MCFC electrochemistry, heat- and mass-transfer, and materials decay over a wide range of operating conditions.[2693] Analogous computer programs also are used by DOE for PAFC and SOFC systems.[2794]

Energy Research Corporation has been working on MCFC development in-house since 1969, and with external support since 1976. In their original DOE work (1976–1980) the major focus was to provide cost effective and functionally sound technology. Emphasis was therefore on the development of basic component materials, fabrication techniques, and cell designs, with some work on performance modeling and system design.[1703] EPRI awarded a contract to ERC in 1979 to develop a new thin thermally-cyclable cell concept, and to examine the possibility of using new techniques for fabricating thin layers of cell components.[125] Although not chosen as a major contractor for the DOE stack development program in 1980, DOE has continued to fund ERC in parallel with EPRI's project for cell component development up to the time of writing.

Since 1982–1983, ERC has concentrated on developing and demonstrating the advantages of the internal reforming MCFC (IRMCFC; or in ERC's nomenclature, the *direct fuel cell* [DFC]) for use with natural gas,[1980, 1981] under EPRI contracts. The latter include work on the design of a cost-effective stack, again in parallel with DOE component development. In addition to the use of such a fuel cell for on-site applications (~50 kW), ERC, in collaboration with Fluor, Inc. (its parent company) also has examined its application to simplified atmospheric pressure–dispersed power plants in the 2 MW class.

The simplest configuration studied was for a molten carbonate fuel cell power plant operating at atmospheric pressure on odorant-free natural gas.[711] At an operating current density of 160 mA/cm^2, and an estimated unit cell voltage of 0.73 V, with 90% overall methane utilization (83% per pass), and 45% fuel recycle, the heat rate for such a plant was calculated to be 6450 BTU (HHV)/kWh, corresponding to an overall efficiency of about 53%. This plant was designed by Fluor and ERC with emphasis on utilizing commercially available components as far as possible (all were available from manufacturers' catalogs except the high-temperature recycle blower and of course the fuel cell stacks). In the original concept, the stacks were to be the same as those for the ERC-designed 50 kW on-site unit, so that only one production line would be needed for both applications. The unit was designed on four shop-assembled and truck-transportable 12 ft x 40 ft skids: two for the fuel cells, one for auxiliary equipment, and one for the power conditioner. The overall footprint for the power plant would only be about 3400 square feet. The objective of the design was to determine the probable per kW cost of a limited production run of 100 2-MW units for this simple non-optimized configuration based on the use of multiple 50 kW stacks with their large overall piping requirement. This was calculated to be about $1200/kW (1982),[711] which would be competitive for many applications. Even limited production of such a system could therefore be profitable to a developer, and could create sufficient cash-flow to allow further optimization (e.g., the use of a much larger cell size to minimize the number of stacks and peripherals). Optimized units of this type should eventually be produced at a much lower cost in mature series-production, which would permit reaching a much larger future market segment. This has already been discussed in more detail in Chapter 2, Section O.

Although ERC has run a small internal reforming single cell for over 7000 hours, internal reforming must still be considered to be in the laboratory stage, since the present life goal for IRMCFCs is 25,000-40,000 hours. These advanced versions of the MCFC will probably follow the development of nonreforming stacks by about two to three years.

It should finally be pointed out that the MCFC also can be operated on methanol-water fuel, which internally-reforms so readily on the metal anode parts that no special reforming catalyst is necessary. If methanol fuel becomes widely available in the future, the simple MCFC with no fuel processor will allow its conversion at over 50% efficiency. This alternative might be very attractive for many utility and on-site applications.

General Electric developed utility MCFCs under contract to DOE, starting in 1976. They were intended for a 675 MW coal-fired central power station, using an oxygen-blown Texaco coal gasifier and the Selexol gas purification process,[733, 1095] in an extension of the work discussed earlier.[2005] A simplified flow-sheet for this plant is illustrated in Fig. 3–6. In this system, electric power is obtained both from molten carbonate fuel cells and from steam turbine and gas turbine expander bottoming cycles. Steam is injected into the fuel gas upstream of the fuel cell anodes to suppress carbon formation. General Electric developed parametric optimum design curves correlating fuel cell current density, polarization, life,

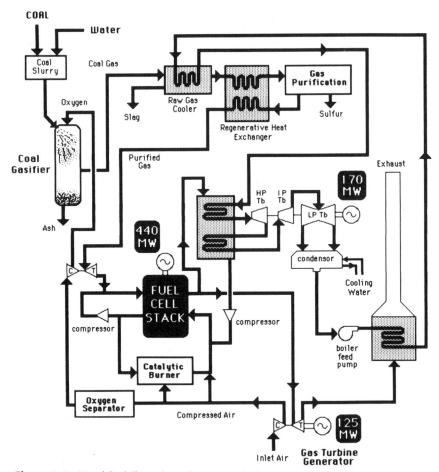

Figure 3–6: Simplified flow-sheet for General Electric coal-fired 675 MW molten carbonate fuel cell central power station.[733]

and cost of electricity for this system. Further studies, carried out under EPRI contract,[421] on the integrated MCFC coal gasifier designs have been referred to earlier in this chapter. This work was followed by an EPRI-sponsored investigation of new methods for the fabrication of thin electrolyte layers.[486] The final GE work under the 1980 DOE stack development program (for which UTC was the other contractor) consisted of the development of internally-manifolded 0.93 m² (1 ft²) active area stack components and some new materials concepts.[117] In parallel, new materials developments and possible IRMCFC systems were funded under a PG&E program.[1207] Some aspects of GE's work are discussed in more detail in Chapter 17.

Utility Systems

In the spring of 1984, after many years of conducting fuel cell and other work in the energy field, GE made the corporate decision that there was not sufficient profit either at the present time or in the near future to stay active in the new-technology energy conversion market. They therefore completed their contractual obligations and, after many years of pioneering research and development, they closed down their Advanced Energy Systems Division, abandoning their work in areas such as sodium-sulfur batteries, fuel cells, solar energy, and wind turbines. They retained only their business in conventional steam and gas turbines, which is profitable and which requires little development effort. In addition to abandoning MCFC technology, they also sold off their interests in Solid Polymer Electrolyte technology for both fuel cells and advanced electrolyzers to the Hamilton Standard Division of UTC.

The strong point of GE MCFC work was the development of many innovative methods and advanced processes for the manufacture and production of materials (Chapter 17), which deserve consideration by other developers.

With the withdrawal of GE, United Technologies Corporation now is the prime DOE contractor for the MCFC stack development program. UTC has continued to follow DOE/METC's emphasis on the development of utility MCFC systems intended to operate on gasified coal. Their stack design is relatively conventional (i.e., it follows the general lines of their cross-flow PAFC stack, and incorporates external manifolding. More details on this design will be given in Chapter 17.

The development of the molten carbonate fuel cell now is in a pilot scale phase. Up to mid-1985, five subscale 20-cell pressurized stacks of conventional design (0.94 dm^2, 1 ft^2 cells) were operated under DOE funding. The last two of these operated for almost 4000 and 5000 hours respectively. EPRI (along with DOE) and Niagara Mohawk had each tested earlier versions of these short stacks incorporating manufacturable electrolyte components, in mid-1981 and mid-1980 respectively. EPRI intended to scale this system up vertically to 100 cells in 1982; but the project was abandoned when it became clear that this was premature, and that further short-stack testing (funded by DOE) would be required to resolve a number of difficulties, particularly that of electrolyte management on vertical scale-up (Chapter 17).

In contrast, no major problems were expected with horizontal scale-up,[1519] and this has been confirmed in the latest work at UTC. The next step (starting September 1986) was the test of a pressurized 20-cell stack using cells with 0.74 m^2 (8 ft^2) active area (i.e., approaching the proposed size of commercial prototype cells). This stack may be capable of a lifetime of perhaps 15,000 hours. A $2.2 million (1985) DOE program was started at the end of 1985 to manufacture the components for this stack, which ran for about 1000 hours at the end of 1986. Closedown was voluntary (i.e., it resulted from interruption in funding), and a later restart was anticipated (December 1986).

While it is a nonprofit research institute, and not a potential fuel cell developer, the Institute of Gas Technology (IGT, Chicago) must be mentioned because it is the pioneer of MCFC cell and component development in the United States, having continuously worked on the technology since 1959. Many of the basic concepts of the MCFC have resulted from IGT studies, including the basic patents on the composition of the electrolyte,[236, 1709] on stable anode structures,[1686] on the principle of internal reforming,[1618] and on corrosion-resistant wet seals.[555] The Institute of Gas Technology worked on the MCFC as a subcontractor to UTC under the TARGET program from 1967–1975, and later as a subcontractor to GE's DOE-sponsored studies. Since 1977, IGT has been supported by EPRI to perform research on the basic electrochemistry of the molten carbonate fuel cell;[120] on the effect of contaminants, particularly sulfur, on MCFC reactions;[2425] and on new materials, components and internal reforming catalysts. These will be considered in more detail in Chapter 17. In view of its strong MCFC background, IGT has had much influence on new programs on fuel cell technology, particularly outside of the United States. Recently, IGT has announced its intention to form a joint venture to manufacture the MCFC with a group of industrial partners, including a major Japanese corporation. The results of this development are awaited with interest.

Japanese work on the MCFC has recently become important. Japan's program now aims to have two 1-MW demonstration stacks in operation by 1994–1995. Similarly, the Netherlands aims to have a 200–300 kW demonstrator by 1992. In late 1986, DOE issued a request for proposals for an approximately $35 million–development and testing program for a protoype pressurized utility stack (500 kW~1 MW), to be developed over a three-year period followed by a further 18 months of testing. This stack development, which may use as many as 500 cells with areas as large as 0.93~1.4 m^2 (10~15 ft^2), follows the proposed plan for the mid- to late-1980s.[1517] The preceding are considered in more detail in Chapters 5 and 17.

D. SOLID OXIDE ELECTROLYTE FUEL CELLS

As their name implies, solid oxide electrolyte fuel cells (SOFCs) utilize a solid oxide, usually doped zirconia, as the electrolyte. When this material is heated to about 1000°C, it becomes sufficiently conductive to oxide ions (while remaining nonconductive to electrons) to serve as a solid-state electrolyte.

There are several features of SOFCs that may make them very attractive for utility and industrial applications.[843, 844, 1160, 1252] In the first place, their high operating temperature ensures that all fuel compositions, provided they are protected from cracking in the manifolding by the use of excess steam, will spontaneously internally reform and then oxidize rapidly to chemical completion. This eliminates the need for specialized (and possibly expensive) catalysts of the IRMCFC and for a fuel processing system. A further advantage of the SOFC over the MCFC is that carbon dioxide is not a reactant at the cathode and a product at the anode, since ionic transport occurs as $O^=$ ions, rather than as $CO_3^=$ ions. Hence, no CO_2 recycling system is necessary.

Experiments have shown that solid oxide electrolyte fuel cells operate equally well on dry or humidified hydrogen or on carbon monoxide fuel, as well as on mixtures of the two. In addition, its high operating temperature and its lack of reliance on high-activity catalysts that might be easily poisoned are said to give the SOFC a high tolerance to impurities such as sulfur. It has been shown, for example, that the presence of 50 ppm of hydrogen sulfide in the fuel lowers the cell operating voltage by only about 5%, apparently without causing any permanent damage to the cell, since the original performance can be restored when the impurity is removed from the fuel.[494] This tolerance to impurity-containing fuels may make it desirable for operation on heavy fuels that have received little upgrading, such as fuel oil. More importantly, it might also be able to use untreated coal gas. However, according to Chapter 6[495] of Ref. 1989, the extent of sulfur tolerance is far from understood.

Recent system analyses by the ANL have indicated that small, direct reforming solid oxide electrolyte fuel cells operating on natural gas may be able to achieve overall system efficiencies of about 44~45% (HHV) without bottoming cycles or a desulfurization system.[1518] For the thermodynamic reasons discussed earlier, this is somewhat lower than the achievable values for MCFC systems (~50%). However, it can be argued that SOFCs may have lower capital costs than MCFC systems since fewer components will be required in the power plant. It will be appreciated, however, that this will be highly dependent on the costs of the fuel cells themselves, and at present it is by no means evident that a ceramic SOFC technology can compete with a sheet-metal MCFC from the viewpoint of per kW cell costs. This aspect of the SOFC remains to be demonstrated.

The SOFC stack will produce very high grade waste heat, making the system attractive for small utility and industrial cogeneration applications where absolute electrical efficiency may be less important than heat/power ratio. Heat of this quality is advantageous even for applications that do not require very high temperature, since it permits a reduction in the size and cost of heat recovery equipment. Because of the quality of its reject heat, for large utility applications it would certainly be economical to obtain higher overall efficiencies by incorporating steam turbine or gas turbine bottoming cycle systems.

The conventional design for a SOFC has been one in which a large number of very small cylindrical cells are joined end-to-end, electrically connected in series to build up a stack of cells of a desired power output. In such systems, the fuel electrodes were on the interior of the tubular cell arrays, with the oxygen cathodes on the outside. While ceramic technology can make such systems feasible, at least on a laboratory scale, a major problem has always been the susceptibility to thermal stresses of the large number of ceramic-to-ceramic contacts. This question will be reviewed in more detail in Chapter 18.

Starting in 1980, Westinghouse Electric Corporation developed a new cell design, inverted from the viewpoint of the electrodes and reactants, in which all electrical connections are on the outside of each tubular cell in the reducing fuel atmosphere. Cell-to-cell interconnection can thereby be accomplished with low cost metallic (nickel sponge) components.[1252, 2171] Densely packed bundles of long single cells (21 cm or more) are connected in series or parallel using the nickel connectors, which are ductile at the elevated operating temperatures of the cell and eliminate many ceramic-to-ceramic contact points.

Argonne National Laboratory also has been working on a SOFC of a new design.[844] The ANL cell, called by them the *monolithic fuel cell*, is intended to use the thin ceramic layer components of existing SOFCs in a strong, lightweight honeycomb structure that should allow much higher energy and power densities than conventional configurations of Westinghouse type.

The technology of material preparation of the thin film deposition techniques involved in the fabrication of SOFC components is now in a reasonably advanced stage, largely through research efforts undertaken by Westinghouse Electric Corporation. However, analyses of the performance of SOFC systems still contain a large element of uncertainty, because long-term thermal and electrical performance of full-size stacks have yet to be demonstrated. For this reason, the level of funding and the state of the art for the SOFC third-generation technology are not at the level of the second-generation technology MCFC. However, the initial performance of small-scale cell groups now looks sufficiently promising that the detailed design, engineering, and construction of generator-sized hardware will be undertaken.

The majority of research on the SOFC has been conducted by Westinghouse in the United States and by Brown, Boveri, and Cie in Germany; with more recent contributions from the Ministry of International Trade and Industry (MITI) in Japan, and Dornier Systems in Germany. In the United States, the SOFC program has been centering around Westinghouse work in a DOE-funded program to demonstrate a self-supporting SOFC module of about 10 kW size.[19] Currently, modules of 24 tubes producing about 300 W are being made available. Over the past several years the level of funding for this program has been relatively constant at about $2 million per year, with a lower level of supporting research directed at an understanding of fundamental cell phenomena and the development of alternative designs. Further work on the SOFC will be discussed in more detail in Chapter 18.

CHAPTER 4

On-Site Integrated Energy Systems and Industrial Cogeneration

A. ON-SITE INTEGRATED ENERGY SYSTEMS

Buildings require a combination of electrical and thermal energy.[1662] Electricity operates lighting, motors, appliances, and air-conditioning equipment, whereas heat is needed for space heating, cooking, domestic hot water, process needs, and absorption air conditioning. Natural gas-fueled on-site phosphoric acid fuel cell (PAFC) power plants, known as *on-site integrated energy systems* (OS/IES), may be the best near-term option for the on-site supply of the electrical and thermal energy requirements of buildings. With such systems, electricity generated by the fuel cells, which must compete in cost with *delivered* electric utility power, not bus-bar power, provides the electrical needs of the building after d.c.–a.c. conversion, while by-product heat from the fuel cells provides a significant portion of the building's thermal needs, the remainder being supplied directly from the combustion of natural gas or via electrically-powered heat pumps. The combined output of electricity and thermal energy from such a fuel cell power plant results in the utilization of approximately 80% of the total energy content of the natural gas fuel, compared with conventional electrical generators which typically deliver only about 35% of the fuel energy (in the form of electricity) to the user.[2121]

Because an on-site fuel cell power plant does not require an external water supply for cooling, its siting is not limited by water availability. In fact, such power plants can be sited within existing buildings or even as rooftop installations.[863] Indeed, the low emissions and minimal noise from on-site fuel cells permit siting at any location; and their rapid load response, good quality electrical power, and high efficiency at part load enable them to satisfy a wide range of building needs. Furthermore, the fuel cell systems do not have to be autonomous, since they can be grid-connected, operating part of the time in an electric utility cogeneration mode, transferring surplus electricity to local electric power companies. In this way, both the reliability and use of the fuel cell can be maximized, leading to the lowest possible operating cost.

NASA has studied the economic competitiveness of fuel cell on-site integrated energy systems with heat recovery against conventional, non-cogenerating energy systems using life-cycle cost and simple payback as criteria for comparison.[418] Between 50 and 70 different systems were simulated and analyzed for use with buildings such as apartment houses. The results of the study showed that fuel cell on-site integrated energy systems will be competitive over a wide range of delivered fuel costs, including considerably higher costs than those currently prevalent. The results also indicated that grid-connected systems, which permit the fuel cell power plant to be arbitrarily sized, were more competitive than autonomous stand-alone systems, which must be sized for building peak load. However, even stand-alone systems were shown to be able to compete against conventional systems. For grid-connected systems, it was shown that the optimum output capacity of the fuel cell power plant is usually considerably less than the peak electric load. Climatic variations had little impact on the results, though variations in building energy demands did. It was shown that the cost of the fuel required to produce the electricity sold back to the grid was not, within wide limits, a strong function of the buy-back price of the electricity. The reason for this was that if buy-back electricity prices were low compared with fuel costs, the fuel cell operator then had the option of not selling electricity at all; the unit as a whole would still be competitive without this feature. In contrast, if electricity buy-back costs were high compared with fuel costs, selling to the utility would constitute an excellent technique for overall savings to further offset fuel cell fixed costs. All of the above was of course conditional on the fuel cell system having appropriate capital costs, lifetime and reliability.

An electrical load profile analysis made by Westinghouse Electric Corporation and Energy Research Corporation (ERC) for an Albany, New York 48-unit apartment complex with a peak electric load of 220 kW showed that the electricity consumption is less than 50% of the peak value for more than 80% of the time.[1159] The high part-load efficiency of fuel cells makes them admirably suited to the steep energy profiles associated with this type of application, since the fuel cell units can be designed to operate at high power densities (low efficiency) during the relatively short times at or near peak load, thereby minimizing the capital cost of the hardware, and at lower power densities (higher efficiency) the remainder of the time, thereby minimizing fuel consumption.

This type of operation is sometimes preferable to that in which the fuel cell is designed to operate at a relatively constant output to satisfy the baseload demand of the building, with a tie line to the local utility grid to provide the peak requirements, since the demand or stand-by charges for such a system often are very high, resulting in high per kWh utility electricty costs. This results from the fact that the OS/IES peaking requirement will probably occur when the utility load is also at or near its peak. In addition, the load factor on the utility distribution equipment that must be installed to supplement the OS/IES will be low, giving high effective per kWh costs.

In another DOE-supported study made by Westinghouse [1419] for the same 48-unit apartment building, the performance and cost of several 100 kW natural gas OS/IES PAFC systems were compared with those for conventional natural gas appliances for heating, cooking, and domestic hot water, with utility electricity for air conditioning and other electrical requirements. The fuel cell systems required from 37 to 44% less primary energy than did the conventional system to satisfy the energy requirements of the building. The capital costs of the proposed OS/IES's would, however, be 57 to 70% higher than the cost of the conventional system, which was estimated to be $98,700 (1979). These additional costs would require payback periods of from 4.7 to 5.6 years to cover the additional investment.

A similar study made for the DOE by Mathtech, Inc. compared the use of on-site fuel cells with that of conventional all-electric and gas-electric systems for a multifamily low-rise apartment building, a retail store, and a hospital, all located in Washington, D.C.[2637] An important factor imposed in this study was that the on-site fuel cell systems had to provide electric service at a reliability equivalent to that of an electric utility. The results of the study indicated that in all cases, the fuel cell systems could reduce the building energy consumption by 10 to 50%. The system life-cycle costs for the on-site PAFC systems would be 0 to 30% higher than those for conventional apartment building energy systems, 13% lower to 26% higher than those for conventional energy systems for retail stores, and 5 to 40% lower than those for conventional hospital energy systems. The economics for the apartment building and retail store were marginal because of the relatively high fixed and operating-and-maintenance costs of the fuel cell systems, which were not completely offset by the savings in energy costs. In the case of the hospital, however, the life-cycle costs using the fuel cell systems were attractive because of the high relative importance of energy costs as a component of overall energy system life-cycle costs.

Although the results of the aforementioned study indicated that the economics for the use of on-site fuel cells were not attractive for the specific apartment building considered under the given criteria, it should not be inferred that this is a general conclusion with regard to apartment buildings. Each case must be studied in detail on account of the wide range of differing factors involved. To aid in such studies, a computer model has been developed by Arthur D. Little, Inc. to evaluate the optimum integration of an on-site fuel cell system with the heating, cooling, and ventilation subsystems of a residential apartment and retail building.[1589] This model allows the user to group components and subsystems in any practical combination with fuel cells of any type and to analyze the technical and economic performance of the resulting system. For applications in buildings where the ratio of thermal energy demand to electrical energy demand is greater than about four, the coupling of an on-site fuel cell to a heat pump looks especially attractive from the standpoints both of cost and fuel savings.[415, 824]

In addition to residential buildings, retail stores, and hospitals, on-site fuel cells may also be attractive for other less frequent building applications. A study made by Westinghouse, for example, indicated that the use of a fuel cell OS/IES operating on arctic grade diesel fuel would be ideally suited as the electrical and thermal energy source for remote DEW Line radar stations. The use of such a unit with an overall (electrical plus thermal) efficiency of about 75%, could result in a 25% improvement in fuel consumption compared with that for diesel-electric generators.[2376] On-site fuel cell systems also appear to be well suited for operating large computer installations, because the quality of the power is not affected by problems such as blackouts, thunderstorms, and line surges.[683]

A major market for on-site fuel cells will probably be in Europe, especially West Germany.[987, 2159] Almost half the primary energy consumption in West Germany comes from oil, of which 96% is imported. Because more than half of this oil is used for space heating, fuel cells should prove highly attractive to regional energy planners who seek to reduce their dependence on imported oil for this application. It has already been stressed that fuel cells offer the considerable advantages of efficiency in small units, fuel supply flexibility, and favorable environmental characteristics (chemical and acoustic emissions) compared with any competitive dispersible systems, such as diesel units. Fuel cells might be especially attractive for use in the district heating mode, which has been much further developed in Europe than in North America. Already, about 7% of West Germany's residential heat comes from district heating schemes, and this percentage is expected to increase to 15 to 20% in the future. In Denmark, all new thermal power plant capacity must be linked to district heating. Thus the European situation may offer a unique opportunity to the emerging U.S. fuel cell industry by means of near-term demonstration projects. On-site fuel cells also might find applications in small, less developed countries, especially those consisting of many islands where a large national electric grid is not possible.[1860] Small units could serve in rural electrification schemes in countries such as India,[1970] using biogas fuels to provide power in currently non-electrified villages for irrigation pumps and lighting. Waste heat can be used to improve biogas generator output. There are currently some 300,000 such villages in India.

Gas Industry Interest in On-Site Integrated Energy Systems

Because the U.S. supply of natural gas presently somewhat exceeds demand, the U.S. gas industry is again searching aggressively for new markets and therefore is supporting new fuel cell development,[1576] as in the TARGET program of 1967–1975. New fuel cell–related markets that are predicted for natural gas include fleet and private vehicles, industrial cogeneration, and the use of natural gas along with coal for electrical generation. The commercialization of gas-fired on-site fuel cell systems is seen by the gas industry as being advantageous not only to its own members, but also to customers and to the nation as a whole. Advantages to the gas industry include increased sales, diversification into a new energy service business, and the increased value added to the gas sold. To the customer, major factors will be the convenience of having all energy needs supplied by one contractor, lower total energy consumption and energy cost, and a means of overcoming the investment and ownership barriers inherent in advanced-technology energy conversion equipment.[2083] Overall, there will be a net energy saving, less reliance on imported fuel and lower environmental emissions.

A study sponsored by the U.S. Department of Energy concluded that the on-site integrated energy systems that may be installed over the period 1985–1999 could result in a net U.S. national energy savings of about 0.4 quadrillion BTU (4.2×10^{17} J), or an average energy savings equivalent to about five million barrels of oil per year.[1063]

The 40 kW On-Site Integrated Energy System Operational Feasibility Program

As discussed in Chapter 3, the U.S. gas industry initially became interested in the concept of on-site fuel cell systems in the 1960s.[1575] This resulted in the TARGET program over the period 1967–1976, during which 60 experimental 12.5 kW *total energy* PAFC fuel cell systems were operated throughout the United States, Canada, and Japan. At that time, limitations occurred in the form of far from cost-and-lifetime effective equipment. Perhaps more important, institutional barriers existed, including very negative electric utility attitudes that severely hindered the introduction of natural gas-fired on-site fuel cell power plants. These factors, coupled with the energy crises of the mid-1970s and predictions of the early unavailability of natural gas as a fuel, made it difficult to justify the on-site fuel cell at that time.

Outlook for the supply of natural gas is now much better, largely because of the economic incentives for exploration due to the deregulation of gas prices and the reduction of BTU input per constant dollar of GNP by almost one third since 1970. The latter has occurred because of market forces resulting from the trend from energy-intensive industry to other sectors, coupled with the rising real cost of energy. As an example, wholesale long-term contract prices were $0.20/MMBTU in 1967 (about $0.65/MMBTU in 1986 dollars), compared with $6/MMBTU at the height of OPEC's influence, and about $2.50/MMBTU in 1986. The latter cost, not much different from that for delivered coal, must be considered to be below the long-term trend. Gas exploration activities are therefore once again at a high level and there is optimism that the development of processes to recover gas from eastern Devonian shales and western tight sands will ensure a firm supply of methane in the United States for many years to come. The Gold theory of its non-biogenic origin[984] has already been alluded to.

In addition, there now is a greater emphasis than before on higher energy end-use efficiency and lower environmental emissions, which has resulted in a shift towards dispersed power production and an increasing acceptance of cogeneration by electric utilities and federal and state regulatory bodies. This shift has resulted in legislation, such as PURPA, the Public Utility Regulatory Policies Act of 1978,[1887] that encouraged cogeneration using natural gas. This act also forced electric utilities to purchase power from independent producers, making the grid-connected fuel cell selling back surplus power attractive.

For these reasons, about 30 major gas utilities again have become determined to commercialize the on-site natural gas fuel cell, and have embarked on a program with the GRI and DOE to develop and commercialize natural gas-fueled PAFCs for use as on-site integrated energy systems. Forty kilowatts was initially selected as an optimum size,[1063] to achieve further economies of scale compared with the 12.5 kW TARGET cell. This choice means that the device will be aimed for the commercial and residential apartment market, rather than that for single-family dwellings. Unlike the original TARGET cell concept, units were intended to be grid-connected if the application required it. The units were specified for totally automatic hands-off operation, with high reliability and a minimum of required maintenance.

This project was known as the On-Site Integrated Energy System Operational Feasibility Program and was an outgrowth of the UTC/TARGET private venture, the role of the TARGET gas utility team being taken over by the Gas Research Institute,[1082] founded as the gas utility research funding agency in the mid-1970s. To help implement the plan, the gas industry formed an On-Site Fuel Cell Users' Group, made up of gas and combination gas and electric utilities; and GRI, which was to examine the policy issues related to the commercialization of natural gas on-site fuel cells.[2083] An important assignment of group members was to conduct

On-Site Integrated Energy Systems and Industrial Cogeneration 111

market and business assessment analyses of the role of fuel cells in their respective territories. These included the definition of market and business opportunities, identifying barriers and problems to be overcome, and proposing actions and solutions required to bring about successful commercialization of on-site natural gas fuel cell systems.

The initial objective of the program was to investigate the practical potentialities for use of on-site natural gas fuel cell power plants in multifamily residential, commercial, and light industrial buildings by installing and testing about fifty 40 kW units, designated PC-18 by UTC, at 20~30 selected commercial and light industrial locations in the United States and Japan.[862,863,1407,1421,1662,1857,2019,2121,2774] United Technologies, as the prime contractor and manufacturer of the units, provided technical assistance for their installation and operation. The program included technology development, fabrication of prototype power plants, and extensive field testing under many different conditions. The initial cost of the field test project, which was coordinated by the Gas Research Institute, was announced in 1981 as approximately $40 million (1981 dollars), with GRI and DOE funding $16.1 million and $9.2 million, respectively. In addition, each participating utility was expected to supply about $500,000 towards the total cost of the project.

The early field test operating goal was a minimum of 8000 hours for each power plant to obtain detailed operational data for both the equipment and the entire system. In addition to the numerous units evaluated at sites in the United States, two Japanese utilities (Osaka Gas and Tokyo Gas) have tested three units. The National Research Council of Canada also purchased one unit for testing, and UTC has tested several units at its Connecticut plant.

The overall goal of the Operational Feasibility Program was to establish the national acceptance of on-site fuel cell energy systems as a viable new business option for the utilities, manufacturers, and sponsoring agencies. To achieve this goal, the program has had two major thrusts: a field test program to verify operational feasibility by showing that the 40 kW units can meet utility performance standards under actual field operating conditions in a variety of attractive early-entry markets; and a business assessment by the potential gas utility users showing that on-site natural gas fuel cells can represent a significant and viable private commercial market for the fuel cell supplier, thereby leading to future commitments from the fuel cell manufacturers.

The 40 kW OS/IES program began with an Engineering and Development Program, which was completed in 1982.[1027] During this phase of the program, 7300 hours of power plant operation confirmed that the specifications set by the gas industry could be met. All major components performed at or above specifications, and any deficiencies in ancillary components were identified and corrections made. After this part of the program was completed, six pioneering units were constructed and placed in service: two in Japan, one in Mexico, and three in the United States.

These six original prototype units were each housed in a cabinet approximately nine feet long, five feet wide, and six feet high, weighing approximately 8000 pounds.[2121] Grid-connect (power conditioning) units were in separate housings, approximately four feet wide, two feet deep, and four feet high. Each of the power plants produced 220 V a.c. grade electricity and recoverable thermal energy. The units were fully automated and could respond to a fixed or variable electrical load demand. A detailed description of the system is given in Ref. 1733, and technical specifications will be discussed in detail in Chapter 16. Figure 4-1 shows the arrangement of the components within the unit,[1083] and Fig. 4-2 is a photograph of a typical 40 kW system.

The first prototype unit to be tested in the United States was sponsored by Northwest Natural Gas Company in cooperation with its neighboring gas utilities. It was installed in *Rawlinson's New System Laundry*, in Portland, Oregon,[2121] and started operation on April 1, 1982. The cost of the unit's installation alone, with the necessary instrumentation for testing, was about $140,000. The laundry, which services hotels, restaurants, and commercial establishments, operates nine hours per day, six days per week, using approximately 30~70 gallons of hot water per minute for washing. During operating hours the entire electrical output of the fuel cell (about 32 kW) was used in the plant. During off-peak hours the electricity generated by the system was diverted to the local utility grid after the requirements for night lighting and other miscellaneous uses were satisfied. The power plant supplied high grade process heat to the laundry at a temperature of 132°C and lower grade heat at a supply temperature of 71°C, the return water being at 27°C. Prior to the installation of the fuel cell system, the laundry used a 200 hp gas-fired boiler to generate steam for drying, pressing, and heating of wash water. Electric power came from the local electric company's grid system. The use of the on-site fuel cell system resulted in a net energy savings of over $300 per month.

Another of the early 40 kW units (again with a grid-connect unit) was installed in late 1982 by Northeast Utilities, an original member of the TARGET program, and a long-term active fuel cell supporter. The unit was installed at the Southern New England Telephone's Rockville Central Office building in Vernon, Connecticut,[897] which contains computer controlled telephone switching equipment for about 20,000 telephone customers. It is representative of a large number of the Bell Telephone System's 26,000 buildings throughout the United States. The fuel cell was connected to operate in parallel with the electric utility grid, while the

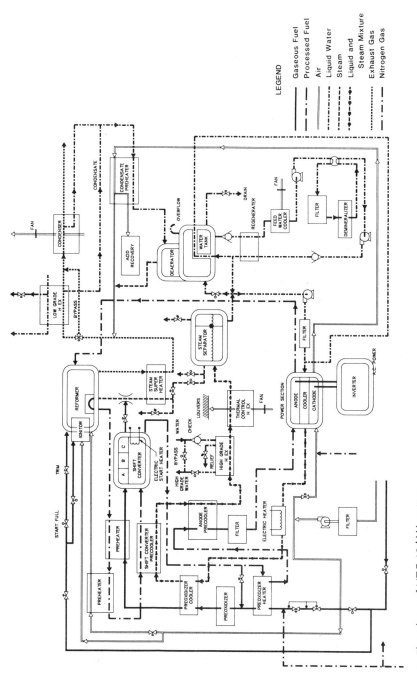

Figure 4-1: Flowsheet of UTC 40 kW system.

Figure 4-2: Photograph of 40 kW system.

thermal energy output of the system was used to supply most of the space heating needs of the building.

Economic studies have shown that a fuel cell power plant that is able to operate in parallel with the electrical grid network can be economical at twice the installed cost per kilowatt of a conventional power plant that serves only a grid-isolated load (see Ref. 696). The unit was therefore used to acquire technical information and operating experience relative to microprocessor-based control systems operating in parallel with the electric grid. In addition, it confirmed the design specifications of the power plant and developed operational and maintenance cost data for a one-year test period. Table 4-1 shows some of the tests that were performed with this unit.

The results of these early field tests showed that on-site fuel cell technology had reached the stage at which total energy supply systems are able to meet the commercial requirements for automatic operation and economic performance. However, the tests also showed that the earlier units did not meet the commercial requirements for reliability and cost. To remedy these deficiencies, in 1981 the Gas Research Institute began an On-

TABLE 4-1
Tests Performed on Grid-Connected Prototype 40 kW OS/IES Fuel Cell Power Plant[2516]

Audio Noise Test

Electrical Testing: Isolated Operating Mode
1. No-load electrical characteristics
2. Zero to full-load electrical characteristics
3. Greater than full-load protection characteristics
    a. Three-phase over-load test
    b. Unbalance load test

Electrical Testing: Grid Connect Operating Mode
1. Minimum to full-load electrical characteristics
2. Protection characteristics
    a. Single-, two-, and three-phase open line to grid
    b. Grid over-voltage trip condition
    c. Grid under-voltage trip condition
    d. Automatic return to grid-connect mode from trip

Thermal Energy Recovery Test

Site Technology Development Program, which represented a major expansion of the gas industry's efforts to commercialize on-site natural gas fuel cells.[1407]

In January, 1981, UTC made a proposal to NASA to take part in the design, manufacture, fabrication, and support of 45 field test power plants.[2019] The contract was awarded in the summer of 1981 and the first unit was delivered for field testing in November, 1981. The Department of Energy also joined this program, since it conforms to the government's long-term policy of encouraging the efficient and clean utilization of natural resources. The DOE program is managed by NASA's Lewis Research Center in Cleveland, Ohio, which lends support primarily for technology development, as opposed to commercialization.

The overall goal of the Technology Development Program is to reduce costs to threshold levels at which private sector energy service can begin.[1027] The program involves coordination of projects involving component technology, systems technology, and fabrication processes, the studies being aimed at selecting the optimum power plant specifications and size to enable viable commercial business to begin. These aims would be achieved by: (a) improving the reliability and capability of the basic system components (i.e., the cell stack, with a goal of doubling power density), the fuel processor, and the inverter; (b) further lowering the system cost by examining the use or adaptation of off-the-shelf industrial components for ancillaries and controls with a high level of reliability; and (c) matching power plant design concepts to the needs identified by the gas utilities during the Field Test Program.[1407] This provided for the delivery of 45 plants in the United States and Japan for one year of field testing, and four for installation and testing at U.S. Department of Defense facilities.[1027] This action was initiated early in 1982, the first units being delivered in 1983, with completion of planned testing by the end of 1985.[1857]

To select the candidate buildings in which to install the fuel cell power plants, the gas utilities had to determine the energy needs of buildings of various types. Hence, a survey was made to determine the annual thermal and electrical energy use and equipment characteristics of nearly 500 buildings.[1662] This was followed by detailed monitoring and hourly analysis of the thermal and electrical loads in approximately 100 buildings across the United States for a one-year period. The results of these analyses have been used both to select the most effective sites for the field test power plants, and also to develop models and correlations that can be used to assess alternative on-site fuel cell systems. In buildings selected for field testing, monitoring was continued after power plant installation to verify the analytical predictions.

The methodology for site selection and for the field test portion of the 40 kW Fuel Cell Operational Feasibility Program was coordinated by Science Applications Inc. (SAI) of La Jolla, California.[690] The test units were to be installed in a variety of multifamily residential, commercial, and industrial installations over a wide geographic range to provide data on the variations in energy load demand, use patterns, weather, transportation, installation, maintenance, operation, product support, and other factors. The final selection of the test sites was made jointly by the GRI and DOE.

The candidate sites selected for the installation of the 40 kW power systems included: low-rise apartment buildings of 25 to 50 units; office buildings and banks of less than 16,000 square feet; hospitals with fewer than 100 beds; nursing homes with 50 to 80 beds; laundries; schools of less than 50,000 square feet; light industries; groupings of single family homes; refrigerated warehouses of 50,000 to 80,000 square feet; and health spas of less than 25,000 square feet.[690] The best sites for an 80 kW system (two 40 kW units) include: 24-hour restaurants of less than 7000 square feet; motels under 80 units; ornamental nurseries; 24-hour grocery stores of less than 8000 square feet; meat packing plants of less than 12,000 square feet; apartment buildings with swimming pools and air conditioning (40 to 50 residential units); motor vehicle parts manufacturing plants of less than

4000 square feet; refrigerated warehouses of from 50,000 to 80,000 square feet; health and sports clubs of less than 12,000 square feet; and auto and tire supply stores of less than 2500 square feet.[203]

At about the midpoint of the program (end of February 1985), 41 power plants had been delivered to 36 sites, one was operating in-house at UTC, and a total of 97,000 hours of operation had been reached, of which 50,000 hours were accumulated between October 1984 and February 1985. The longest total period of operation for a single unit was then 7000 hours at UTC, followed by 6470 hours at Southern California Gas. More than 1000 hours of operation before stoppage for maintenance or repair had been recorded over 30 times on different units, and continuous runs of over 2000 hours had been recorded twice. A 24-cell short-stack with on-site technology had exceeded 25,000 hours of continuous operation.[112,177] A complete report on the program, which successfully terminated in 1986, is now available.[114].

The results of a study made by TRW Energy Systems Group of Redondo Beach, California, indicated that the combination of five 40 kW systems in parallel would be economically feasible to service the electrical and thermal demands of a large office building at the McClellan Air Force Base.[924] The installation of such a system could be expected to save 1600 MWh of electricity per year and offset 10^4 therms of natural gas currently consumed in the building heating, ventilating, and air-conditioning (HVAC) system.

The On-Site Fuel Cell Users Group feels that from the marketing standpoint, the product is now ready for commercial field testing, and that the market potential for the small on-site integrated fuel cell system now depends more on marketing research and planning than it does on technical feasibility.[1857] Thus, the ambitious goal of the Technology Development Program and the Field Test Program was to reach the threshold of private sector commercial service by about 1986–1987.[1027] A report on the potential commercial use is now available.[115] It shows that the feasible U. S. market is 18,000 MW, or 10% of the total commercial market for electricity, provided the cost is $1000/kW (1985).

However, some lessons are already apparent. The average cost of manufacture of the later 40 kW on-site units was $12,500/kW (1985),[177] whereas it is estimated that a minimum price for commercial application will be $2500/kW (1985),[696] with the hope that this will eventually fall to $1500/kW, then to $1000/kW, to make the system widely attractive in the marketplace. In this respect, the tendency is again to aim for an economy of scale in the systems supplying fuel to the fuel cell stack and converting the d.c. stack output to a.c. In other words, this means increasing the total unit size. At the same time, it is hoped that the power density of the fuel cell stack will be increased with no change in cell potential or efficiency, thereby reducing the cost of this key component. The present philosophy is to orient the program towards a 200 kW stack that initially will use a new-technology version of the TEPCO 0.34 m^2 (3.7 ft^2) cells derated to operate at atmospheric pressure, so that the power of the complete stack falls from 275 kW at 3.4 atm to 200 kW at 1 atm. Later systems presumably will use cells of the same area, but each will have a higher power output, giving fewer cells per stack. Some of these options are reviewed in Chapter 16. From the learning curve philosophy outlined in Chapter 2, Section O, cost should descend to the expected levels as production rises.

As would be expected with any new technology, reliability and maintenance problems have occurred. These have however been progressively reduced or eliminated as later units have been produced. As has been already described in Chapter 3 for the TEPCO 4.5 MW demonstrator, the major problem is in the design and materials in the pressurized piping and manifolding in the fuel cell cooling plates. For these, a major design change is intended, with replacement of the serpentine Teflon-coated copper cooler tubes by a parallel-flow stainless steel system that not only will result in the reduction of the possibility of bimetallic corrosion, but also will be less susceptible to blockage by corrosion products. This redesign will decrease purity requirements in the treatment system for recovered water for cooling and for use in the steam reformer, and will improve reliability. In particular, it will greatly reduce the need to have a frequent chemical cleaning schedule requiring a down-time of one to two days to remove accumulated metal hydroxide from the cooling system using complexing agents such as EDTA. Such cleaning has been required at 1500 to 2000 hour intervals, the longest continuous run without cleaning (up to the end of October 1984) being 2165 hours at Consumers' Power in Jackson, Mississippi. It is hoped that with the anticipated modifications cleaning intervals will be increased to at least 8000 hour (i.e., yearly intervals). Most recently, cooling system manifolding changes have increased this maintenance interval substantially.

Westinghouse/ERC On-Site Fuel Cell Program

In parallel to their joint program to develop PAFC systems for electric utility use, Westinghouse Electric Corporation and ERC also have shown interest in the development of an OS/IES incorporating a pressurized module of four gas-cooled PAFC fuel cell stacks with a nominal output capacity of 400 kilowatts.[507] Energy Research Corporation has indicated that its air-cooled phosphoric acid fuel cell systems might be especially attractive for use with on-site integrated energy applications because air-cooled systems inherently should be simple, trouble-free, and inexpensive. Their coolant circuit does not introduce any materials compatibility, materials joining, or electrical isolation problems into the system, which can result in the corrosion problems discussed earlier; and also does not require high pressurization, which can cause catastrophic failure if the cell stack is flooded with water.[1159]

Such attributes are very desirable for OS/IES systems since these should be capable of unattended operation with minimal maintenance and risk, at the same time being economically competitive with conventional generators. Furthermore, if the stack is cooled by cathode exhaust recirculation rather than by a once-through air flow, then waste heat from the fuel cells can be delivered to the thermal load at a temperature approaching that of the cell, allowing greater system efficiency. This cooling system also will result in more uniform cell temperatures, will reduce acid loss by evaporation from the cells, and will facilitate water recovery. The Westinghouse/ERC on-site fuel cell program is receiving support from both DOE-sponsored contracts and from the ERC in-house technology development program.

Engelhard On-Site Fuel Cell Program

Engelhard Industries Division of Engelhard Minerals and Chemicals Corporation are engaged in a five-segment DOE-funded program under contract to NASA Lewis Research Center to develop on-site integrated energy systems based on Engelhard's dielectric liquid-cooled PAFCs.[505, 1325, 1364-1366] In October 1978, Engelhard began a 20-month program under joint DOE–NASA sponsorship to improve the performance and cost of their PAFC technology, including electrodes, catalysts, matrixes, bipolar plates, and cooling hardware. The goal of this program was to examine the parameters that might lead to the development of a cost-effective on-site integrated energy system for residential and commercial buildings. Unlike the UTC and ERC on-site units, which are intended to operate on reformed natural gas, Engelhard has focused its efforts on the use of reformed methanol as an alternative fuel, as a result of its accumulated experience with methanol in its military fuel cell development program. One of Engelhard's goals for the on-site building application has been the energy-efficient integration of the fuel cell power system with building HVAC systems. They have developed guidelines to facilitate this integration.[914]

In January 1981, their program entered another phase, which was to construct a breadboard 50 kW system module for OS/IES applications. A series of multicell stacks as well as two advanced 5 kW stacks were built and tested, and development work on the isothermal steam reforming of methanol was conducted.

The Engelhard cooling approach employs forced circulation of a dielectric liquid (an oil such as Therminol 44) through the metal piping of the cooling plates in the stack, which are similar in principle to those in the UTC OS/IES PAFC stacks. Liquid cooling is particularly attractive for OS/IES applications because of the high quality of the heat that the cooling fluid can transfer from the stack to the HVAC system of the building. In addition to providing relatively high heat transfer temperatures, the use of a dielectric liquid coolant also permits the use of small heat exchangers, requires only minimal pumping power, and offers good compatibility with efficient absorption-type air conditioners. Finally, in the event of a total cooling system failure, it may not lead to the total loss of a stack and its peripherals by acid leaching and its attendant corrosion implications.

In addition to its commitment to using dielectric liquid cooling, Engelhard is emphasizing maximum heat utilization, high system and component reliability, and detailed economic analyses early in the program to ensure its system's commercial viability.[2757] Economic assessments have indicated that connection of the system to the electric grid is generally favorable, and providing that the fuel cell system cost goals are met, on-site building applications may represent a sizable commercial market for fuel cells.

B. INDUSTRIAL COGENERATION

The first part of this chapter has dealt with cogenerative on-site integrated energy systems that supply the electrical and thermal requirements primarily of residential, commercial, and light industrial buildings. In this section, fuel cell systems using cogeneration in heavier industrial applications, such as in the chemical processing and metallurgical industries, will be discussed.

For these applications: The thermal output of the fuel cells would normally be used to supply industrial process heat (most often in the form of steam); and the fuel (pure or impure hydrogen) generally would be supplied from an internal by-product stream from some chemical process in the plant. Such industrial systems would possess all the benefits usually associated with fuel cells, such as resource savings, low emissions, modularity, and regulatory acceptability (e.g., the U.S. National Energy Act and the Energy Tax Act of 1979, which gave investment tax credits and other economic advantages to cogenerators). In addition, the use of fuel cells using process fuel to supply electricity and process heat are seen as bringing important benefits to industrial cogeneration systems.[2308, 2635,]

The economic viability of fuel cells used for this application is quite sensitive to the value of the thermal energy produced by the system, since revenue derived from this heat would offset the cost of both the fuel cells and of the heat recovery equipment required by the system.[1952] Similarly, the cost of the electricity produced by the fuel cells will depend on the value of the cogenerated steam. For example, in 1976, Westinghouse Electric

Corporation made a study of the concept of using a central coal gasifier to generate fuel gas, which then would be piped to a number of industrial cogeneration PAFCs located up to several hundred miles away.[2305] The fuel cell systems were envisaged as producing high pressure process steam (at 3 MPa or greater, about 30 atm, 450 psia) by burning the anode vent gas in a boiler, and low-pressure process steam (at about 0.5 MPa, 5 atm, 75 psia) from the fuel cell waste heat. Baseline selling prices of $8.80 per thousand kg of steam and 3.02¢ per kWh for the electricity produced were assumed, with all costs in 1976 dollars. Economic analysis indicated that if no revenue from the sale of steam could be produced, production cost of electricity would then be 4.16¢ per kWh, but that with steam revenue at $5.69 per thousand MJ ($6/MMBTU), the electricity would then cost only 2.44¢ per kWh to produce. The study concluded that sale of reject heat from fuel cells would therefore seem to be worth from $100 to $400 per kW in 1976 dollars,[2118] in other words, it would largely offset the capital and O & M cost of the fuel cell for this application..

There is evidence that fuel cells may indeed be even better suited to the requirements of the chemical industry than to those of electric utilities.[1630] Electric utility companies needing high quality, high voltage a.c. power, must be concerned about siting regulations and system compactness, as well as about finding local uses for the waste heat produced by the fuel cell. In contrast, the requirements of the chemical industry are almost the opposite. It would prefer high current density, low voltage d.c. power that can be easily coupled with electrolytic cells, does not have siting problems, and has use for waste heat. The ideal modular fuel cell size for such applications is probably about 250 kilowatts. In 1979, the U.S. inorganic chemicals and metals processing industries alone consumed about 16,500 MW.[1366] For these industries, d.c. power costs can represent as much as 75~80% of product cost. It has been estimated that the hydrogen generated by the U.S. chlor-alkali industry, which consumes about 3200 MW of d.c. power, could produce from 15 to 20% of the total d.c. power needs of the industry.[75]

On-site fuel cells also appear to be attractive for numerous applications in the metallurgical and metal working industries, largely because these industries require low-voltage, high-amperage direct current power.[1337] Examples are the electrolytic production of aluminum and magnesium, electroplating, galvanizing, and plasma spraying. The use of fuel cells for these applications would eliminate the need for power rectification. Other industrial processes such as the electrolytic production of chlorate or the reduction of iron ore using blast furnaces produce large quantities of hydrogen or carbon monoxide (which can be shifted to hydrogen). After suitable cleanup, these gases could be used as fuels for fuel cells which can produce electrical energy for other uses in the plant. If cogeneration is incorporated into these fuel cell systems, high rates of return on investment may be projected.[1337]

Although it may generally be desirable to have an on-site chemical process that produces a stream of by-product pure or impure hydrogen that can be directly used in the fuel cell, it should not be inferred that only those industries that produce hydrogen or some other fuel can benefit from the use of cogenerative fuel cells. In terms of tonnage, hydrogen is the largest chemical commodity produced in U.S. industry, after iron (steel) and chlorine, which it exceeds on a molar basis. Approximately 3×10^{12} cu ft (7.2 million metric tons) of hydrogen are produced on an annual basis in the United States. Almost all this hydrogen is produced and consumed within the plants themselves, for use in chemical processing where appropriate. Many of the chemical industries have an excess of hydrogen, however (e.g., in the dehydrogenation of hydrocarbons to alkenes, and in the caustic-chlorine industry). In such cases, they can afford to use the excess hydrogen as fuel, or simply flare it. In this connection, a study was made by Engelhard Industries to examine the economics of retrofitting fuel cell cogeneration systems in chemical industries that generate by-product hydrogen.[1998]

Results indicated that retrofitted fuel cell cogeneration systems are economically viable when the hydrogen purge streams are either flared or vented, but are less attractive for cases where hydrogen is used as a higher-value chemical feedstock. Generalizations could not be made for plants where the hydrogen purge streams are burned for their fuel value. The incentive for fuel cell retrofit was definitely lowered for industries requiring conversion of direct current to alternating current. A sensitivity analysis showed that the value-in-use of the hydrogen as currently consumed in the plant was the major determinant in the economic viability of the system, rather than the installed capital cost of the fuel cell. Another important factor was the relative price of electricity to that of heat: With everything else constant, a 50% increase in the price of electricity could change an uneconomical retrofit to an economic one.

The Use of Phosphoric Acid Fuel Cells for Industrial Cogeneration

Several studies have dealt with the use of PAFCs for industrial cogeneration. Most have been based on the use of a fuel cell system with a similar specification to that of UTC's 4.5 MW demonstrator as the basic modular unit, since the characteristics of this system are well known. In one EPRI-supported study by Mathtech, Inc. in Washington, D.C., 4.5 MW PAFC modules were considered for industrial cogeneration in 11 industries, based on 4-digit Standard Industrial Classification (SIC) groups.[2635] The economic performance and energy consumption pattern of each fuel cell cogeneration system was evaluated under two basic operating strategies: (a) *Economic Dispatch*, in which the plant electrical output is designed to minimize utility system production costs, and thermal energy is produced as a by-product; and (b) *Thermal Dispatch*, in which the system operating level is determined by the on-site demand for thermal energy, with electrical energy being produced as a by-product. In each case the fuel cell cogeneration system (which might include a supplemental boiler) was compared with so-called conventional systems, which included a fuel cell with identical electrical generating capacity but without thermal recovery, as well as one or more fossil-fired conventional boilers with an assumed thermal efficiency of 83%.

The energy saving implications of each system were evaluated in terms of the overall utilization of total (electrical + thermal) energy. The economic evaluation of each system was reported in terms of an *incremental break-even capital cost* (ΔBECC). This represented the amount by which the capital cost of the cogenerating fuel cell could be permitted to increase above that of a similar fuel cell producing only electric power before the life-cycle cost of the cogeneration system would exceed that of the conventional fuel cell system.

The results, shown in Table 4–2, indicated that the energy utilization efficiencies of the cogeneration fuel cell systems averaged about 77%, excluding the chlor-alkali, cyclic crudes, and paving mixture cases, which showed rather low values. This compared with an average value of 63% for conventional fuel cell systems. For the economic dispatch configurations, the ΔBECC values ranged from about \$70/kW to \$400/kW. The cogeneration configuration used in Table 4-2 was based on taking 20% of the heat as steam at 177°C and the rest as hot water at 71°C. Even better results were obtained when some of the fuel was burned to obtain additional thermal energy. For these particular industrial applications, it was found that MCFCs were not significantly more attractive economically than PAFC systems because the amount of higher temperature heat provided by the MCFC was relatively small on account of its higher electrical efficiency. Consequently, its total available heat was also less than that from a PAFC. The overall results of the study showed that industrial cogeneration was technically and economically feasible for the applications considered, that the break-even capital costs for such fuel cells were about \$100 to \$400 per kilowatt higher than those for electric-only fuel cells, and that cogeneration resulted in significant energy savings over conventional fuel cells with separate heat production systems for all applications.

Several other applications studies for PAFC cogeneration systems will now be briefly discussed to indicate their potential range of applicability.

TRW Energy Systems of Redondo Beach, California, under contract to

TABLE 4-2
Analysis Results for Industrial Cogeneration Phosphoric Acid Fuel Cell Systems[(2635)]

SIC	Industry	Fuel Cell Size* (MW)	Economic Dispatch			Thermal Dispatch		
			ΔBECC ($/kW)	% Average Efficiency		ΔBECC ($/kW)	% Average Efficiency	
				Cogen. F.C.	Conv. F.C.		Cogen F.C.	Conv. F.C.
2062	Cane Sugar Refining	18.0	302	77	68	271	81	68
2063	Beet Sugar	13.5	344	78	65	380	81	65
2085	Distillery	9.0	232	71	65	145	79	65
2261	Finishing Plant, Cotton	18.0	340	74	57	398	79	57
2262	Finishing Plant, Synthetics	27.0	301	69	52	317	76	52
2600	Pulp and Paper Mill	18.0	367	80	69	403	81	69
2812	Chlor-Alkali	612†	71	45	40	23	45	40
2824	Non-Cellulosic Fibers	90.0	244	63	48	267	69	48
2865	Cyclic Crudes & Intermediates	104†	75	45	40	30	45	40
2911	Petroleum Refining	13.5	407	83	80	487	83	80
2951	Paving Mixtures	698†	72	43	40	41	45	40

* Except as noted (by †), fuel cells were sized to meet the peak requirements for thermal energy at temperatures below 100°C.
† Fuel cell sized to meet peak requirement for thermal energy at temperatures between 100° and 177°C.

the U.S. Department of Energy, investigated the use of cogeneration fuel cells (UTC 4.8 MW units) in an aluminum plant.[922, 924] The fuel cell system could be used to replace one of the two aluminum lines in a Reynolds Metals Company electrolytic aluminum plant in Alabama. The line operates at about 50,000 amperes, requiring a voltage of 610~640 V d.c. The conceptual fuel cell system was configured into two parallel strings of ten series-connected cell stack assemblies. Since the output voltage from the fuel cell considered was only 560~600 V, a series-type boost regulator using plant a.c. power was incorporated into the design to achieve the desired voltage level. In practice, there would be no difficulty in adapting a fuel cell system to give the voltage required.

The waste heat from the fuel cells would be utilized in the cryolite recovery process, the boiler steam currently used being replaced with 152°C steam generated by the fuel cells. The fuel cells would provide about 80% of the steam requirements directly, utilizing about 90% of the fuel cell waste heat. In addition, the fuel cell anode exhaust gas could be used for carbon dioxide injection into the cryolite precipitation step. This proposed installation would displace 3.8×10^4 MWh of electricity per year, resulting in a net yearly fuel saving of almost \$300,000 (1982 dollars).

Westinghouse Electric Corporation also has studied the use of fuel cell cogeneration in an aluminum production plant, but using a high temperature solid oxide electrolyte fuel cell system planned to become commercial in about 1990.[839] Like TRW, Westinghouse concluded that substantial savings could be realized by the introduction of a fuel cell cogenerator. The benefit analysis for the proposed system was most sensitive to the cost of conventionally-supplied electricity, and least sensitive to the cost of the actual fuel cell; the sensitivity to the cost of the natural gas used in the fuel cell being intermediate.

In 1980, Holmes and Narver, Inc. of Orange, California, investigated the use of a UTC 4.5 MW unit adapted for cogeneration, in a brewery in Los Angeles.[244, 2308] The results of the study indicated that the use of the cogeneration fuel cell system would yield an energy savings equivalent to 23,000 barrels per year of No. 6 fuel oil, and an additional savings equivalent to \$244,000 (1980) per year in investment for pollution control of NO_x and SO_x, as required by the special regulations for the Los Angeles basin area.

An EPRI-supported study conducted by R.M. Parsons Company of Pasadena, California, evaluated the use of a 50 MW integrated coal gasifier/PAFC system to provide electricity, heat, and process steam for an industrial park outside Willmar, Minnesota.[1030] The system was to operate on 740 tons of North Dakota lignite per day. The fuel cell cogeneration system was compared with other alternate small-scale plants for power generation; namely, coal gasification–combustion turbine plants, atmospheric fluidized bed boilers, and conventional direct-fired boilers. The much higher overall efficiency of the fuel cell plant (56.1%) compared with those of the other plants (~43.6%), would result in similar bus-bar power costs, even though the fuel cell plant had a much higher capital cost. This was considered to be a favorable indication for the applicability of cogenerative fuel cell systems, since future improvements could be expected for the fuel cell, but not for the mature conventional technology systems considered.

A DOE contract to Burns and McDonnell Engineering Company of Kansas City, Missouri, identified the pulp and paper industry as a promising application for fuel cell cogeneration.[684, 685] This study evaluated the feasibility of using waste heat from a natural gas–fired 4.5 MW UTC PAFC cell to replace heat currently generated by natural gas–fired boilers in a paperboard mill in California. In the proposed installation, the fuel cell would produce 8.8 MBTU/h of 35 psia steam plus 11.2 MBTU/h of 93°C hot water. The electricity from the fuel cell would be sold to the local utility. This use of fuel cell waste heat could allow a 22% reduction in the fuel consumed by the mill, with considerable savings for NO_x, SO_x, and particulate emissions control equipment. The high efficiency of the fuel cell would allow a net annual energy savings equivalent to 37,000 barrels of crude oil.

A study jointly sponsored by DOE, UTC, Gilbert/Commonwealth, and the Public Service Electric and Gas Company of Newark, New Jersey, investigated the integration of a UTC 4.5 MW PAFC with a large water pollution control facility in New Jersey.[1131, 1132] Methane-rich gas from the anaerobic digestion of raw sewage would be used as feedstock for the fuel cell, and heat rejected from the fuel cell would be used to sustain high levels of bacterial action in the digesters, which have an optimum operating temperature of 35°C. The electrical output of the fuel cell would be sold to the electric utility. The results of the study indicated an increase of 21~27% in the net fuel utilization efficiency of using cogeneration compared with the use of a fuel cell producing electric power only. The fuel cell cogeneration system would be able to deliver 30 million kWh of electrical energy annually (about half the needs of the pollution control facility) while providing over 90% (~74 × 10^{12} J) of the annual plant thermal energy requirements. It was estimated that the cost savings resulting from the use of this system might be of the order of $589,000 per year (1982 dollars). This use of a cogeneration fuel cell would save about 26,000 barrels of No. 6 fuel oil per year for on-peak electric utility power, together with 5900 tons of coal per year for off-peak power.

A recent DOE study performed by Los Alamos National Laboratory examined the use of fuel cells to consume excess hydrogen produced in the chlor-alkali industry, the d.c. electricity produced providing some of the needs of the plant.[68] The report concluded that this approach was feasible, provided that the fuel cells produced enough excess process steam. The PAFC was therefore to be preferred over hydrogen alkaline fuel cells such as the Alsthom-Occidental system (see Chapter 5), which would be satisfactory but for its low operating temperature (65°C).

The Use of Molten Carbonate Fuel Cells for Industrial Cogeneration

In contrast to the results of the Mathtech study reported above,[2635] there is evidence that MCFCs may be more effective than PAFCs for industrial cogeneration.[431] The advantages of molten carbonate systems derive from their high efficiency, which should be relatively independent of power plant size and load, and their high operating temperature (~650°C), which will allow the production of high grade by-product thermal energy that is suitable either as high temperature process heat or for use with bottoming cycles for additional electric power generation. Advanced MCFC systems can convert 70~80% of fuel input energy to electricity plus high temperature (400°C) steam.[1112] For industrial processes that require power-to-heat ratios as high as 2.5, molten carbonate systems may be capable of fuel utilization levels as high as 80~90%.[2005]

It should also not be forgotten that while MCFC systems are usually accepted as being more efficient than PAFCs, and will therefore have less available waste heat, even though it is of higher quality, it may be relatively easy to design a less efficient MCFC producing more waste heat if industrial conditions warrant such an approach. By operating the fuel cell at higher power density, its capital cost could be considerably reduced, especially in systems with simplified chemical engineering (e.g., the IRMFC) in which stack cost promises to be a substantial proportion of total capital cost. Such a system could have about the same efficiency as that of the PAFC, a lower capital cost, and about the same amount of waste heat (corresponding to a power/heat ratio of about unity).[181] This should make it extremely attractive for many cogeneration applications, including those considered rather negatively in the Mathtech study.[2635]

Some years ago, NASA undertook a broad DOE-funded screening study, known as the Cogeneration Technology Alternatives Study (CTAS), to advance the use of coal and coal-derived fuels, and to a lesser degree distillates, in industrial cogeneration applications.[519] Emphasis was on the use of high-sulfur coal, minimally processed coal-derived liquids, and low- or intermediate-BTU coal gas as fuels for industrial cogeneration. Contracts were awarded to UTC and to General Electric, and to the NASA-Lewis Research Center to manage the study and carry out independent analyses.

CTAS evaluated a number of energy conversion systems, including steam turbines, diesels, open cycle gas turbines, combined gas turbine –steam turbine cycles, closed cycle gas turbines, Stirling engines, PAFCs, MCFCs, and thermionic converters. The various systems were examined for potential application to about 85 key industrial processes. The results of the study indicated that PAFC-based systems had the potential for cogeneration fuel savings and yielded very high values of emissions savings; but, because of the relatively expensive fuels they required, the cost savings and return on investment were not attractive for many of the industries considered.

Conversely, MCFCs not only yielded fuel energy savings and emissions credits, but also looked attractive from the standpoint of cost savings through the use of coal-derived fuels and had returns of investment that appeared promising. The MCFC cogeneration fuel cell attained a sufficiently high score for each parameter studied in a sufficient number of the industries considered to be more attractive than any of the other systems.

The Institute of Gas Technology also has studied the use of molten carbonate fuel cells for industrial cogeneration, with the objective of saving natural gas, which accounts for approximately 40% of industrial primary energy use.[431] The IGT study concluded that cogenerative natural gas–fueled MCFCs would yield 30~50% primary energy savings over a central non-cogeneration power plant with an on-site boiler. In addition, replacement of central-station baseload electricity by on-site natural gas MCFCs would be achieved with only a small increase in total natural gas consumption in most parts of the United States. As high as 80% overall efficiency might be achieved in several of the instances considered.

In most cases, about one third of the energy savings was derived from the higher electrical conversion efficiency of the fuel cell vs. the baseline conventional power plant, and about two thirds from the utilization of the by-product thermal energy from the fuel cell. Both the energy savings and fuel utilization were found to be significantly higher than those achievable with the other gas-fueled cogeneration systems, including advanced gas turbines and PAFCs. In addition, the average emissions were less than 1% of those from conventional power plants associated with boilers.

Outlook For Industrial Cogeneration Fuel Cells

At present it is difficult to accurately evaluate the savings that might be realized in industrial cogeneration. This is because (a) economic viability often strongly depends on the relative projected escalation rates of electric power costs compared with those for natural gas, (b) the capital costs of commercial fuel cells are not yet certain, and (c) dependable statistics are not yet available for maintenance costs and reliability of commercial industrial cogenerative fuel cells.[685] However, a number of points are

On-Site Integrated Energy Systems and Industrial Cogeneration 129

already certain. Cost sensitivity studies made by Arthur D. Little, Inc., for a number of naphtha-fueled PAFC industrial cogeneration fuel cell systems indicated that:

a. Fuel cell costs sometimes may dominate the economics of fuel cell cogeneration.
b. Fuel cell cogeneration systems are cost-competitive with conventional non-cogeneration systems when the cost of purchased electricity and the electric load factor are high.
c. The economics of industrial cogeneration are very sensitive to the relative values of fuel and electricity.
d. The capital cost of the fuel cell system can be reduced without increasing the total annual costs by relying on a utility connection for unexpected outage requirements.
e. Fuel cell system peripheral equipment can be a major cost item in small scale industrial cogeneration systems (less than 500 kW) due to lack of economy of scale.

Because the rate of payback on a cogeneration investment tends to be longer than that usually desired by many industrial users, ownership by utilities should tend to be encouraged. As for other fuel cell systems, the ultimate economic viability of their use in industrial cogeneration may need to be settled by the initial subsidized installation of a number of power plants. In particular, the economic viability for a system involving fuel cell retrofit will require a site-specific case-by-case approach.[1998]

In spite of the above uncertainties, there is optimism for the successful application of fuel cell systems for industrial cogeneration. This was recognized as far back as 1978 when EPRI started to develop a methodology to assess cogeneration options, to identify promising candidates for industrial cogeneration, and to identify research and development needs and priorities for cogeneration.[1617, 2635] At that time, environmentalists were attacking the U.S. electric utilities for not encouraging cogeneration systems, so that they were then known by the euphemism *DEUS* (Dual Energy Use Systems, Ref. 2636). In November of 1982, a Potential Industrial Fuel Cell Users Group (PIFCUG) was formed to stimulate efforts for developing fuel cells for industrial cogeneration and other applications.[1631] By 1983, this group had grown to 22 members, and in the same year a chlor-alkali section of the PIFCUG was formed to promote and facilitate the development of cogeneration fuel cell systems for specific use in the chlorine industry.[75]

CHAPTER 5

Japanese and European Fuel Cell Programs

A. JAPANESE FUEL CELL PROGRAMS

In addition to the fuel cell activities being carried out in the United States, an important parallel effort is now being made in Japan by both government agencies and by the private sector.[904, 2333] The major incentive for Japanese fuel cell development is that country's lack of self-sufficiency in energy supply. In spite of economic recession and energy conservation, it has been estimated that by the year 2000, Japan's peak electricity demand (GW) and total electricity generation (GWh) will be respectively 2.1 and 2.4 times greater than the 1982 values.[1165] To respond adequately it will be necessary for the Japanese to develop new electric generating capacity that is cost effective, efficient, reliable, environmentally acceptable, and fuel flexible.

It is expected that fuel cells will be introduced commercially into the Japanese utility market by the late 1980s, natural gas being the most attractive fuel for dispersed use because of the country's fuel distribution system. About 40% of the total generating capacity of the Tokyo Electric Power Company (TEPCO), the world's largest private utility, is natural gas fueled. As an approximate breakdown, in 1983, 65% of TEPCO's 23.7 GW of thermal plants were fueled by natural gas, and the remaining 5.8 GW were nuclear or hydroelectric plants. Similar percentages hold for other Japanese electric utilities. Oil-fired power plants in Japan that were commissioned before 1959 are scheduled to be retired by the year 2000. After 1995 it is considered probable that these plants will be replaced with natural gas fuel cell power plants at the rate of about 500 MW/year. It is also expected that dispersed fuel cell power plants will be able to replace as much as 26% of the total electric power demand, or 25,000 MW to the year 2005.[1854] The potential demand for industrial on-site cogeneration fuel cells in Japan is believed to be about 2600 MW. These plants are attractive for use in the steel and nonferrous metals industries; in ceramics and stone quarrying operations (e.g., cement, glass, and lime); and in the rubber, fibers, machines, food, and pulp and paper industries.[1165] The potential market for fuel cell cogeneration plants in the commercial and residential sectors amounts to about 4000 MW. Potential users in this area include hotels; heated swimming pools; hospitals; restaurants; newly developed towns; and, in particular, office buildings. The total cogeneration demand may reach 40% of the electric utility demand, or 10,000 MW, by the year 2005.[177]

By 1990~1995, it is anticipated that PAFCs will comprise a few percent of the electric utility market. By the year 2000, it is expected that MCFCs will be commercially on-stream. At this point, the fuel cell share of the Japanese added utility capacity (i.e., the market for new and replacement electric utility generation units) should increase to about 10%. By the twenty-first century it is expected that as much as 7~8% of the total Japanese electric power supply may be generated by fuel cells (see Ref. 1854).

The Moonlight Project

The Japanese government's contribution to fuel cell research and development is included in the national *Moonlight Project*, directed by the Agency of Industrial Science and Technology of the Ministry of International Trade and Industry (MITI).[2551] The objective of the Moonlight Project is to develop higher efficiency energy conversion systems than those that were available in 1974 in order to reduce dependence on fuel imports.

Originally, the fuel cell projects of the Japanese Government were started in 1974 as part of the *Sunshine Project*, to address the problems caused by future energy shortages or embargoes. This project was sponsored by the Agency of Industrial Science and Technology. However, the entire fuel cell program was transferred to the Moonlight Project in 1981. Over the period 1981–1986, the Moonlight Project was budgeted to spend $44 million (based on the 1985 value of 250 Yen = $1.00) to seed a national program to develop energy-efficient fuel cell systems operating on a variety of fuels such as natural gas, methanol, and coal-derived gas. Of the $44 million, $30 million was intended for the development of PAFC systems to be used for both small scale dispersed generation as well as for large scale central generation. The fuel cell objectives of the Moonlight Project up to 1986 were:

a. To design and construct in 1984~1985 and demonstrate in 1986 two 1-MW PAFC systems operating on reformed fuels.
b. To develop by 1986, a 5~10 kW molten carbonate fuel cell stack that can operate on coal-derived gas with an overall efficiency of over 45%.

Japanese and European Fuel Cell Programs

c. To develop and demonstrate in 1986, a 1 kW solid oxide electrolyte fuel cell stack operating on coal gas and having an overall system efficiency of 50%
d. To develop in 1984, a 1 kW hydrogen-air alkaline electrolyte fuel cell stack having an overall system efficiency of 45%.

Table 5–1 summarizes these Moonlight Project fuel cell system objectives.[904, 1921] In contrast to the slow-down and stretching out of objectives that has occurred in the United States since the late 1970s, all the goals of the Japanese program are on schedule at the time of this writing (mid-1986).

TABLE 5-1

Fuel Cell Goals for the Japanese Moonlight Project[904, 1921]

	1 MW Phosphoric Acid Fuel Cell System	
	Dispersed Generation (low-pressure type)	**Central Generation** (high-pressure type)
Main contractors	Mitsubishi, Fuji	Hitachi, Toshiba
Fuel	natural gas	natural gas
A.C. electrical output	1000 kW	1000 kW
Overall efficiency (HHV) (with cogeneration)	40% (60–65%)	42% —
Cooling method	water-cooled	water-cooled
Operational mode	unmanned, automatic	one-man control
A.C.–D.C. converter	self-commutated	self-commutated
Fuel processor:		
Type	catalytic combustion	pressurized combustion
Efficiency	85–90%	85–90%
Fuel cell stack:		
Operating pressure	4.87 atm	6.8–7.7 atm
Operating temperature	190°C	205°C
V-I characteristics	200 mA/cm² @ 0.70 V	220 mA/cm² @ 0.72 V
Platinum loading	4 g/kW	4 g/kW
Contaminant levels	NO_x < 20 ppm < 0.12 g/kWh SO_x < 0.1 ppm	NO_x < 20 ppm < 0.12 g/kWh SO_x < 0.1 ppm
Noise level at plant boundary	45 db at plant boundary	55 db

(continued)

Table 5-1 (continued)

	Dispersed Generation (low-pressure type)	Central Generation (high-pressure type)
Main contractors	Mitsubishi, Fuji	Hitachi, Toshiba
Power response:		
Cold start to standby	< 4 h	< 4 h
Normal operating range	25–100% rated power	25–100% rated power
Min. to max. power	< 1 min	< 1 min
Normal plant shutdown	1 h	1 h
Emergency stop	< 1 min	< 1 min
Predicted module life	40,000 h	40,000 h

Molten Carbonate Fuel Cell

Power output	5–10 kW
V-I characteristics	150 mA/cm^2 @ 0.80 V
Electrode area	~ 2000 cm^2
No. of cells/stack	several tens
Overall efficiency	> 45%

Solid Oxide Electrolyte Fuel Cell

Power output	0.5–1.0 kW
V-I characteristics	200 mA/cm^2 @ 0.7 V
Life	1000 h

Alkaline Fuel Cell

Power output	1 kW
V-I characteristics	100 mA/cm^2 @ 0.75 V
Electrode area	1000 cm^2
Life	40,000 h

As the preceding indicates, the major emphasis is on phosphoric acid fuel cell (PAFC) development. This is contracted through the New Energy Development Organization (NEDO), to four major electric manufacturing companies, namely Toshiba Corporation, Hitachi Ltd., Fuji Electric, and Mitsubishi Electric Corporation (MELCO). In addition to work under the Moonlight Project, a considerable amount of PAFC field testing and development work is being undertaken by the Japanese private sector, in particular by gas companies and electric utilities. The joint project between the Tokyo Electric Power Co. and UTC involving testing of the Goi 4.5 MW (a.c.) demonstrator plant has already been described in detail. A number of Japanese-developed dispersed PAFC systems in the 4~200 kW range are about to be field-tested at the time of writing. These will be described later in this chapter.

The Japanese gas utilities are also supporting the development of on-site natural gas fuel cell systems to promote the use of energy technologies that will substitute for oil.[2189] Since 1969, they have been importing LNG as a major primary energy feedstock. For example, in 1980 they imported 3.42 million tons of LNG, and it is estimated that by 1990 the figure will reach about 11.2 million tons. A major effort is being made by these utilities to develop appliances and equipment that will contribute to the expansion of the natural gas market. Japan has a total of 250 gas utilities, but more than 60% of the nation's total gas supply is provided by the Tokyo Gas Co. and the Osaka Gas Co.

In 1972, these two utilities joined the TARGET program; and in 1973, each tested two of United Technologies' 12.5 kW PC-11 TARGET fuel cell units in Japan. This was the first Japanese fuel cell test program that was widely publicized, and it provided a great stimulus to the development of indigenous fuel cell technology.

As members of the On-Site Fuel Cell Users Group, the Tokyo Gas Co. and Osaka Gas Co. have collaborated with the U.S. Gas Research Institute (GRI) in its program to field-test UTC's 40 kW PC-18 on-site fuel cell system. The two gas utility companies have invested over $7 million (1982 U.S. dollars) in this project. In March 1982, each purchased one 40-kW unit from UTC and began field tests. The unit purchased by Tokyo Gas was installed at a swimming club in Yokohama, whereas that purchased by Osaka Gas was used in the *Royal Host* restaurant at Sakai, on the outskirts of Osaka. Both were arranged to welcome visitors as part of a promotion program.

The two units were operated in the grid-isolated mode close to their maximum capacity of 40 kW. Operating experience with these units has generally confirmed the specified plant efficiencies and other operating characteristics. The NO_x content in the exhaust was determined to be only about 9 ppm, more than an order of magnitude below that of turbines or any other possible competing power sources.

The development of the molten carbonate fuel cell under the Moonlight Project up to 1985 has been conducted by the Government Industrial Research Institute, Osaka (GIRIO) for the evelution and survey of new materials; by the Central Research Institute of the Electric Power Industry (CRIEPI) for system studies; and by the Toshiba Research and Development Center, the Toshiba Corporation, the Hitachi Research Laboratory, and Hitachi Ltd., who are responsible for cell and stack development.[1449] In addition, various other companies also have been involved in cell and stack development. These include Fuji Electric Co., Ltd.; Mitsubishi Electric Corporation (MELCO); Ishikawajima-Harima Heavy Industries Co. Ltd., (IHI); and Matsushita Electric Industrial Co., Ltd. Also reported interested in this development, are Sanyo, Kawasaki Heavy Industries, and NNK.

In Phase I of the molten carbonate fuel cell project, extending from 1981~1986, molten carbonate fuel cell stacks of several kW output were built and tested. The goal was to demonstrate 1 kW stacks by 1985, and 10 kW stacks by late 1986. The target output of these stacks was to be 150 mA/cm^2 at about 0.8 V, using electrodes with areas around 2000 cm^2. If the evaluation of these 10 kW stacks by CRIEPI (see following) was favorable, then Phase II would be undertaken, namely, research and development aimed at a 1000 kW–scale demonstration plant.

Operation of the 10 kW (nominal) stacks from the different developers took place over 500 hours in late 1986–early 1987. While operation took place at a fuel utilization of only 40% (which would yield impractically low efficiencies in a commercial stack), the results were successful. The reason for the choice of this low utilization was the use of machined stainless steel anode bipolar plate components, which would corrode if higher utilizations were used, and which must be nickel-clad in practical stacks (see Chapter 13). In contrast, oxidant utilization was maintained at a practical value of 40% in each case, and the imposed mean current density was 150 mA/cm^2. Hitachi's stack used 30 cells, arranged in three blocks, with an active area of 3600 cm^2, yielding 12.4 kW at an average cell potential of 0.76 V. The stack developed by IHI had 29 cells of the same active area in a single block (12.5 kW at 0.8 V per cell). The Fuji Electric system used 24 2500-cm^2 cells in three blocks (7.1 kW at 0.79 V per cell). Mitsubishi Electric's had 40 cells of area 2000 cm^2 in two blocks (9.1 kW at 0.76 V per cell). Finally, Toshiba's used the smallest cells (57 1600-cm^2 cells in one block, giving 10.4 kW at 0.76 V per cell). Details of the cell technology approaches used will be given in Chapter 17.

Funding of Phase II, an approximately $100 million, eight-year program involving five developers (Toshiba, Hitachi, Fuji Electric, MELCO and IHI), started in 1987.

The development of the high temperature solid oxide electrolyte fuel cell for the Moonlight Project has been primarily undertaken by the Energy System Division of the Electrotechnical Laboratory in Ibaraki Prefecture. The objective is to investigate the basic characteristics of an yttria-stabilized zirconia electrolyte–based system via fabrication and power generation tests.[1846] The first stage of the program was to develop fuel cell stacks consisting of small series-connected cells leading up to a 500~1000 W demonstration battery by the end of 1986.

The following paragraphs briefly outline current Japanese fuel cell development activities.

The MITI's Agency of Industrial Science and Technology (AIST) is conducting in-house work on both alkaline fuel cells and on SOFCs as part of the Moonlight Project. Again within the Moonlight Project, the Government Industrial Research Institute of Osaka (GIRIO) is collaborating with Toshiba Corp. and Hitachi, Ltd. to study and further develop the basic electrochemical and materials technologies for MCFC systems. The GIRIO is conducting similar work on SOFCs with the Electrotechnical Laboratory.

The electric power companies are working with MITI, through the Central Research Institute of the Electric Power Industry (CRIEPI), the Japanese equivalent of EPRI in the United States, in a program for the construction, operation, system interconnection, and evaluation of the 1 MW PAFC power plants being developed by the electric machinery manufacturers for the Moonlight Project. Since 1978 CRIEPI and a utility, the Chubu Electric Power Co. of Nagoya (CHEPCO), together with the Fuji Electric Co., Ltd., and Hitachi, Ltd., have studied options for connecting large-scale fuel cells to the power distribution networks. The main role of CRIEPI in the Moonlight Project has been to promote the development of large-scale power system technology. The organization also entered into agreements with Hitachi Ltd. and Fuji Electric Co. to carry out joint research and evaluation of molten carbonate fuel cell technology. After CRIEPI constructed the first large Japanese test facility for MCFCs, experimental testing started in July, 1983. This facility was used for the interim evaluation of Moonlight Project 1 kW molten carbonate fuel cells during 1985–1986.

Fuji Electric Co. and Fuji Electric Corporate Research and Development Ltd. have been involved in work on fuel cells since 1961, which saw the start of research and development on alkaline hydrogen-oxygen fuel cells and on MCFCs.[2774] In 1972 they developed a 10 kW alkaline hydrogen-oxygen shipboard fuel cell system for the Japanese Defense Agency. Work on PAFC technology started in 1973. In association with CRIEPI, they developed a 2 kW hydrogen-air alkaline fuel cell system and a 3 kW inverter system in 1977. Since 1978 they have further developed their alkaline hydrogen system under the Moonlight Project. In parallel to this work, they also developed a 1 kW alkaline fuel cell system operating on by-product hydrogen in 1983–84.[2402] This system employed an oxygen electrode utilizing a silver catalyst supported on a furnace black substrate and a zirconium-containing Raney nickel anode. From 1979–1981 Fuji undertook studies for CRIEPI on the use of fuel cell systems as substitutes for thermal power plants and as power sources for buildings. They also developed a 15 kW alkaline fuel cell stack for the Central Electric Power

Council. In early 1983, 10,000 h of endurance were recorded at 70°C on alkaline cells, using 5 mg/cm² cathode catalyst. One kilowatt stack goals (1000 cm², 75 mW/cm², 2000 h operation at 50% efficiency) were completed in 1984. Fuji's AFC now uses low-loading platinum on carbon cathode catalyst with edge-collected electrodes, and in its latest version it is a 94 V, 80 A (7.5 kW) unit with 162 cells in three blocks. The size of the unit is 60 x 70 x 150 cm, with a weight of 380 kg. It is intended as an uninterruptible or emergency power source, either alone or in a battery hybrid configuration.

In June of 1980, Fuji launched a joint project with the Kansai Electric Power Co., Inc. (KEPCO, the electric utility for the Osaka area) to verify the feasibility of a fuel cell power plant in the electric power system.[2774, 2775] During October 1981, Fuji designed and built a complete 30 kW air-cooled natural gas–fired PAFC system. After undergoing three months of testing at Fuji's Kawasaki factory, it was installed in April 1982 at the Kansai Electric Power Company's Sakai-ko Thermal Electric Power Station near Osaka for on-site testing. From December 1982 until December 1983 the plant was grid-connected, and it performed very satisfactorily under continuous long term operation at various load levels. Although the initial voltage-current characteristics showed some decline with time, replenishing the phosphoric acid in the stacks after 1000 hours of operation stabilized their performance. Throughout the test period the performance of the reformer also was stable. The NO_x content in the exhaust gas from the reformer was found to be only 0.06 g/kWh under full load conditions, which was significantly better than even the design value of 0.24 g/kWh. The experimental value corresponds to approximately 50 g/MJ (close to 0.1 lb/MMBTU), or 50 ppmv in the exhaust gas. These are well within any presently mandated norms.

The 30 kW Kansai system consisted of a fuel processor built by the Japan Gasoline Co., the d.c. power section, a power conditioner, and a control and protection system. The air-cooled power section was composed of four 7.5-kW stacks and operated at 190°C slightly above atmospheric pressure to ensure good air flow. The fuel cells used state-of-the-art components, which will be discussed in detail in Chapter 16. The experimental demonstrator system required four hours for start-up from ambient and thirty minutes for shutdown. Its overall HHV efficiency at the a.c. bus bar was 30%, which is considerably less than if the total system had been fully thermally integrated.

At present, Fuji is participating in the Moonlight Project to develop, construct and install a 1 MW dispersed PAFC generating plant along with the Mitsubishi Electric Co. (MELCO). This plant is designated by NEDO as being of *low pressure–low temperature* (LPLT) type, in which the fuel cell stacks are designed to operate at 190°C and at 3.4 atm. These conditions were chosen by NEDO at the start of the development program as a system benchmark reflecting the operating conditions in the UTC 4.5 MW demonstrators. The perception was that such operating conditions might be most appropriate for units in the 1~10 MW class, however this view may change as a result of practical operating experience. The other Japanese 1 MW demonstrator, being developed by Toshiba Corporation and Hitachi, Ltd., is in contrast of *high pressure–high temperature* type (HTHP), and is planned to operate at about 205°C, and 8.2 atm (i.e., UTC 11 MW conditions) at a slightly different cell power density. Its operating conditions are considered to better match uses in the 10~100 MW class. Both systems use cells with areas similar to those of the 4.5 MW UTC demonstrators, namely around 3600 cm^2. Their technology will be discussed in more detail following, and in Chapter 16.

For both systems, the contract with NEDO involves an interesting work-sharing arrangement, in which one company develops fuel cell stacks, and the other partner a fuel processing system and peripherals, both under NEDO funding. However, both partners must carry out an in-house program to develop its own complementary part of the total system under in-house funding, thus allowing a wide choice of combinations to construct the most effective final system. For example, Fuji Electric Co. has been working on the reformer and system design for the 1 MW LPLT unit, including ancillary equipment, under NEDO funding; whereas Fuji Electric Corporate R&D, Ltd. has developed an alternative stack technology with in-house support. In contrast, MELCO is responsible for NEDO-supported stack technology for the 1 MW LPLT unit, and is developing its own system and peripherals.

As indicated in Chapter 4, the chemical industry produces large quantities of by-product fuel in the form of pure or impure hydrogen. Since it is experienced with the techniques required to operate fuel cells, Fuji believes that this sector will comprise one of the biggest potential users of fuel cells outside of the electric utilities. In order to penetrate this market, Fuji believes that it will be necessary to bring down the cost of fuel cell stacks to reasonable levels.[2402] For low-cost alkaline fuel cell stacks, the specific goals are for the development of air-cooled cells with active electrode areas of about 1000 cm^2, operating on carbon dioxide–free hydrogen and carbon dioxide–free air at 0.7 V, 150 mA/cm^2, 65°C, and 1 atm pressure. For air-cooled PAFCs operating on hydrogen-rich gas and air at 0.65 V/cell, 150 mA/cm^2, 180°C, and 1 atm, the active electrode area should be about 1500 cm^2/cell. These electrode areas were chosen because they are the optimum size for mass production utilizing the manufacturing techniques Fuji had available in 1983, and because small sized cell stacks allow easy replaceability, hence favorable economics. Low pressure and temperature operation result in a performance penalty, but should also lead to increased reliability and simplicity, together with lower cost. Fuji is hoping for costs of $200~$300/kW in 1983 dollars.

New developments at Fuji include a 50 kW system using natural gas or LPG fuel supported by the Tohoku Electric Company (northern part of the island of Honshu), to be delivered in mid-1987. This will operate at 2 atm, using a very small 60% efficient turbocompressor with a flow of 157 m^3/h, which is being developed by Ishikawajima-Harima Heavy Industries (IHI). This contrasts with the 70% efficient off-the-shelf Shimazu turbocompressor used in the Fuji-Mitsubishi LPLT 1 MW unit. In addition, Fuji is developing a 200 kW methanol-fueled PAFC under a NEDO contract awarded in May 1986 ($10 million over five years, with a large cost-share by the contractor). This will be an atmospheric pressure unit, and is intended as a primary generator for use on remote islands. Unlike the other Fuji PAFC cells developed since the Kansai unit, which are all of UTC ribbed-substrate type, this will use the unique Kureha KES-1 integrated substrate (Ref. 177: also see Chapter 16).

Fuji is also being supported by NEDO to develop an atmospheric-pressure air-cooled methanol 4 kW PAFC unit for use in conjunction with photovoltaic panels in remote applications such as ski resorts. The methanol unit is intended as a photovoltaic backup under poor weather conditions, and its waste heat will be used to keep the solar panels snow-free. Fuji is also involved in a 6 kW in-house PAFC project to produce a compact methanol-fueled unit, and has performed studies with Takenaka Engineering on the feasibility of placing 1 MW PAFC units in large buildings. This study concluded that the PAFC was advantageous compared with its competitors if its cost was under ¥250,000/kW (about $1400/kW; see Chapter 2, Section O) for a fuel cost of $3.40/MMBTU. The low noise of the PAFC would also make it desirable over diesel generators or gas turbines in other applications, such as for offshore drilling rigs where the crews work and sleep in shifts. Under MITI contract, Fuji and TEPCO have examined large (300 MW) methanol-powered PAFC projects: As an example, a flow diagram of a 49% efficient LPLT 300 MW methanol unit is shown in Fig. 5–1. The HPHT consortium has received similar contracts to study large natural gas powered PAFC units.

Fuji's in-house stack design for the 1 MW LPLT unit will be described in Chapter 16, and its MCFC work in Chapter 17.

In 1979, Mitsubishi Electric Corporation (MELCO) started a research and development program on PAFC systems, and they then built a complete 500 W system, including a fuel processor and inverter. In late 1981 they constructed a water-cooled 50 kW prototype (Mark I) scaled-up 112-cell pressurized PAFC system operating on reformed natural gas with cells of approximately the same size as those of the UTC 4.5 MW demonstrators (3600 cm^2). Like Fuji's air-cooled Kansai unit, this was really only

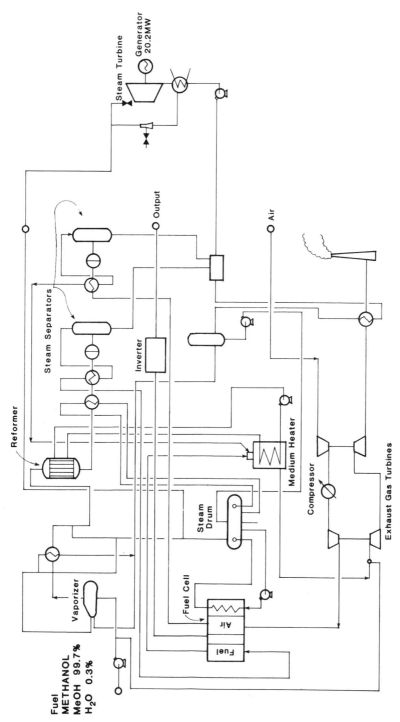

Figure 5-1: Schematic flow-diagram of proposed 300 MW Fuji-TEPCO methanol unit (49% HHV efficiency).

a breadboard system, no attempt having been given to optimum system integration to allow highest efficiency. The overall efficiency was 31% (similar to that of Fuji's 30 kW unit), however the fuel cell module efficiency of 47% was state-of-the-art.[1421] This system operated during 1982.

On the basis of their operating experience with the Mark I cell, a more advanced Mark II model with higher performance and a more elaborate system design was built, with operation beginning in 1983.[1422] Mitsubishi's Mark II 50-kW PAFC system used a stack of 121 water-cooled 3520 cm^2 cells operating on steam-reformed and CO-shifted natural gas at 190°C and 4.4 atm. The fuel cell module efficiency was 51%, with an overall system efficiency of 35% at fuel and air utilizations of 75% and 50%, respectively. The total catalyst loading was 1.05 mg of catalyst/cm^2.

Having shown their ability to design and operate successful PAFC systems, in January 1982 they began contract research with NEDO on the Moonlight Project, using the results of their experience to design and develop the cell technology for the 1 MW LPLT-type PAFC dispersed generating plant. As indicated above, a parallel stack technology for this plant was developed by Fuji Electric Corporate R&D, Ltd.; Fuji Electric Co. being responsible for the development of the fuel processing system.[1423] Detailed system design of the 1 MW plant was carried out on an individual basis by both Mitsubishi and Fuji. Fuji Electric already had a 1/10-scale fuel reformer on test in 1983, scale-up to the full size being completed in 1985. In the meantime, both organizations were verifying and testing cell and other components for the unit.

The operating point for each unit cell is targeted for 200 mA/cm^2 at 0.75 V. By 1983, an 11-cell 3240 cm^2 stack had achieved 200 mA/cm^2 at 0.66 V, with no performance decay after 6000 hours of testing.[1423] The system will consist of 25 kW stacks, each containing 56 cells of area 3540 cm^2. Electrospray coating techniques are used to lay down the uniform matrix layers and catalyst loadings. Dispersed platinum catalyst is used for the cathodes and 10% ruthenium–90% platinum for the anodes, the ruthenium helping to increase the resistance of the anode to carbon monoxide poisoning. The output of the reformer (NiO catalyst) will contain 76% hydrogen, which will be increased to 78~79% hydrogen after passing through a shift converter (CuO catalyst). The chemical engineering and electrical systems part of the plant at KEPCO had been completed in the summer of 1986, and process and control (PAC) testing had been started later that year. Operation of the whole unit started in 1987, with completion of testing expected by the end of 1988. In a change from the original plan, the stacks will consist of two 250-kW Fuji units and two similar 250-kW MELCO units. Due to the lack of exact matching between the two sets of stacks, the cooling and steam systems are separate, as are the parallel d.c. systems. The final goal is a

20 MW power plant operating at a hydrogen utilization rate of 80%, an air utilization rate of 60%, and an overall system efficiency of 42.3% (HHV).

Like Fuji Electric, Mitsubishi also believes that industrial clients as well as utilities will be significant users of fuel cells in the future, and it has conducted a study to integrate a fuel cell system into a petrochemical operation. Recent MCFC developments at Mitsubishi are the construction and testing of a successful internal-reforming natural gas–fueled stack (see Chapters 3 and 8), partly based on licensed Energy Research Corporation technology, which successfully operated over 4000 hours.

In 1981 Toshiba Corporation built and tested 20 kW and 35 kW pressurized PAFC stacks. Along with Hitachi, Ltd., they were selected in 1982 by NEDO to develop the 1 MW HPHT PAFC system under the Moonlight Project proposals. The situation is similar to that between Fuji and Mitsubishi: Toshiba has the NEDO contract for HPHT system development, and Hitachi that for the cell stacks. However, Toshiba is at the same time developing its own stack under in-house funding (Chapter 16) and Hitachi is similarly developing its own systems technology. For this project, Toshiba developed an innovative reforming system that uses exhausted air (of low oxygen concentration) from the fuel cell cathode that is burned in the (catalytic) reformer combustor, which reduces the air required for compression.[56] Simulation studies have suggested that this technique may increase overall plant efficiency by more than 1%. In 1982, Toshiba fabricated a 1/4 scale–50 kW burner to establish the best catalysts and catalyst packing methods. Based on the results of these experiments, a full-scale 50 kW exhausted-air combustion reformer was built and tested in 1983 in cooperation with Mitsui-Toatsu Chemical, Inc. The burner catalyst bed was packed in layers, those at the top containing a platinum catalyst, the lower layers having one based on cerium. Hydrogen was burned in the upper layer to produce the high temperature required to completely burn methane in the lower layer.

At the end of 1986, the status of the 1 MW HPHT plant at the Chubu Electric Power Company (CHEPCO) near Nagoya was similar to that of the KEPCO LPLT plant: PAC testing started in the summer of 1983, and the stacks were expected to be installed later in the summer. As in the case of the KEPCO unit, stack technology would be shared, so that the unit would contain two 250-kW Toshiba stacks and two 250-kW Hitachi stacks. While production of electricity from both units was planned for September 1986, with operation through September 1987 and teardown and site clearance by March 1988, at the time of writing (December 1986), teething troubles similar to those in the 4.5 MW UTC demonstrators had not allowed this ambitious schedule to be kept.

As part of the Moonlight Project, Toshiba, like the other Japanese corporations listed, had developed MCFC stacks in the 5~10 kW class by 1986. As part of this effort they developed a nondestructive testing method using an ultrasonic pulse echo technique to evaluate the mechanical strength of hot-pressed electrolyte tiles.[1449] However, more recent work described in Chapter 17 makes the hot-pressing method obsolete.

Hitachi, Ltd. commenced basic fuel cell research in 1960. In the period 1960~1977 they carried out work on alkaline hydrogen-oxygen fuel cell systems, culminating in 1977 with the construction of a 1 kW system that was tested in collaboration with the Chubu Electric Power Co., Inc. In 1970 they began work on alkaline hydrazine fuel cells. This work led to the development of a 3 kW hydrazine-air unit for the Japanese Defense Agency.[2333, 2373] This power plant, which is used to power field communications and signal systems, has been manufactured by the Shin-Kobe Electric Machinery Co., Ltd., a member of the Hitachi group. Starting in 1978 they developed direct methanol fuel cells, first using sulfuric acid as electrolyte. From this experience, they developed a 48 W direct methanol-air fuel cell with a proprietary platinum-ruthenium anode catalyst applied to an ion exchange membrane, which gives an output of 60–80 mA/cm^2 at 0.4 V/cell.[111] Commercialization of this system was planned for 1985~1986, for example, in conjunction with a lead acid battery as a golf cart power source or as a small portable generator for other intermittent uses.

Hitachi's work on PAFCs started in 1975. In 1978~1979 they built a 100 W reformed natural gas PAFC system, which was tested in a program with the Chubu Electric Power Co., Inc., following Hitachi's work on the 1 kW AFC referred to above. In 1980 they developed and demonstrated a 1 kW PAFC stack employing electrodes of practical size. In 1981 they constructed and demonstrated an experimental 20 kW pressurized system.

As already indicated, Hitachi is working with Toshiba as part of a team to develop the HPHT type 1-MW PAFC system for the Moonlight Project. NEDO is supporting Hitachi's cell technology, which uses high surface area dispersed colloidal platinum-containing binary catalysts, ribbed carbon substrates containing chopped carbon fibers for improved tensile strength, glassy carbon separators, and matrixes made from a combination of silicon carbide and metallic phosphates with good affinity for phosphoric acid and excellent thermal stability.[56] Hitachi's technology is reviewed in more detail in Chapter 16. A high pressure stack with more than 15 cells (3600 cm^2 area) was first tested during 1983.

Since 1978, Hitachi has been working with Toshiba and GIRIO on MCFC systems research for the Moonlight Project.[1449, 2475] More recently, this work has been supported by NEDO.

Sanyo Electric Co., Ltd. started fuel cell research in 1959, and they have developed hydrogen-air, methanol-air, and hydrazine-air fuel cells.[2166]

They have been working on PAFCs for on-site integrated energy systems since 1972, and, in conjunction with Mitsubishi Electric Corporation, they developed 2 kW and 10 kW stacks consisting of 36 and 180 cells, respectively, operating on hydrogen (either pure or from reformed gas) and air. At 180°C, these generated 60 mA/cm^2 at 0.63 V with cathode and anode platinum loadings of about 1.0 and 0.6 mg/cm^2 respectively.

In 1980, Hitachi entered into a cross-licensing agreement with Energy Research Corporation (Danbury, Connecticut) to accelerate the development of air-cooled phosphoric acid power plants. They have built and tested a four-stack 50 kW air-cooled PAFC system that operates on reformed natural gas using a steam reformer developed jointly with the Toyo Engineering Corporation (TEC). The fuel cell modules, of dimensions 30 x 42.5 cm (ERC technology) operate at 180~190°C and at atmospheric pressure. At a total platinum loading of 0.7 g Pt/cm^2 the cells can generate 200 mA/cm^2 at about 0.65 V. The performance characteristics of this 50 kW unit are being used to design a larger scale power plant having optimal overall efficiency and economy. The present plan is to have a 200 kW air-cooled unit of a design similar to the Westinghouse version of the ERC technology[177] operating at a TEPCO site by June 1987. This will be followed by a larger system (600 kW, incorporating three 200-kW units). As the characteristically well-coordinated research and development effort of the Japanese starts to pick up momentum and develop focus, it is becoming increasingly clear that in the near future Japan will become a world leader in fuel cell technology.

B. EUROPEAN FUEL CELL PROGRAMS[2333]

The extensive European developments in the various fuel cell technologies up to 1968 are reviewed in Ref. 1613. Since that time, however, European activity has not been very extensive until the period 1985–1986, when fuel cells were seen as an alternative to nuclear power in some countries. This has become particularly true since the Chernobyl incident in April 1986, which accelerated new programs in both the Netherlands and Italy, and caused some rethinking of fuel cell work in Sweden. France, which has been committed to the rapid growth of nuclear power since 1975, has very little fuel cell work at present, except some exploratory work for aerospace applications. In the United Kingdom, British Gas was considering some advanced fuel cell demonstration work in 1986.

In West Germany Siemans AG has developed both a 225 W ethylene glycol-air alkaline (6 M KOH) fuel cell employing a Pt-Pd-Bi catalyst and a 7 kW alkaline fuel cell. The latter is described in detail in Ref. 394. Siemans is also developing the solid polymer electrolyte (SPE) cell[176] for military applications under license from General Electric (the technology is now owned by the Hamilton Standard division of UTC). The SPE acid electrolyte will be described in more detail in Chapter 10. In the early 1970s, AEG Telefunken developed a 700 W 40-cell phosphoric acid fuel cell stack using tungsten carbide (40 mg/cm^2) as anode catalyst and platinum (1 mg/cm^2) as cathode catalyst. A 0.5 mm–thick, 400 cm^2 resin-bonded fiber sheet was used as the electrolyte matrix. This sheet weighed only 16 g but absorbed 45 g of acid. At about the same time, Brown, Boveri, & Cie AG worked on high-temperature doped zirconia solid oxide electrolyte fuel cells (Chapter 18).

In the Netherlands and Belgium, earlier research from the 1960s and 1970s culminated in the formation of ELENCO, a joint venture between SCK/CEN (the Belgian Atomic Energy Commission), Dutch State Mines, and the Brussels corporation Bekaert to apply alkaline hydrogen-oxygen fuel cells to vehicular transportation. This and further details of cell design are discussed in Chapter 7. It is ELENCO's belief that its circulating electrolyte alkaline fuel cell (which requires periodically changing a carbon dioxide filter and the alkaline electrolyte) is superior to the acid fuel cell for transportation applications on account of its higher efficiency, higher output, and fewer corrosion problems. This system,[394, 2557, 2562, 2563] which has received about $12 million in current dollars from the Belgian and Dutch governments since the early 1970s, is expected to become commercially viable by about 1990 for hydrogen-fueled vehicles. The monopolar edge-collected circulating electrolyte alkaline cells operate at 65°C and use inexpensive components; for example, injection-molded plastic cell frames and manifolding. Details of modules, business plans, and production capability are given in Chapter 7. ELENCO proposes that the hydrogen for vehicle operation will be supplied by chlorine-caustic plants.

The Royal Institute of Technology in Stockholm, Sweden has developed a 50 W methanol-air fuel cell, and its technology was instrumental in developing a circulating-electrolyte alkaline fuel cell system in the form of 10 kW modules during the late 1960s and early 1970s by ASEA in Sweden.[1627] The objective was a 1 MW alkaline hydrogen-oxygen fuel cell unit as a power plant for a small submarine. The Institute is presently interested in developing large alkaline fuel cell power plants that operate on hydrogen produced by the conversion of biomass and peat, using a modified steam-iron process known as the *Ferrugas Process*. These developments are reviewed in Refs. 394 and 1627. Earlier estimates of the cost-effectiveness of large (600 MW) coal plants with alkaline fuel cells carried out by the Royal Institute of Technology showed that capital costs would be little more than $1000/kW (1978) for plants with coal-to-a.c. efficiencies of about 45% when all excess steam was recovered and used in bottoming cycles.[1625] In this system, CO_2 removal would be by scrubbing with NaOH,[1622] which would be regenerated by electrodialysis.[1624] Such plants (based on the use of coal or biomass) are considered to be a solution to electricity generation problems in Sweden in view of that country's adoption of a measure to eliminate nuclear power by the year 2000. However, the estimated efficiency of these systems is surprisingly high in view of the calculated heat rates of coal-PAFC plants quoted in Chapter 3.[492, 493, 2707] This is because the AFC operates at temperatures that are too low to generate steam from its waste heat, whereas the PAFC, operating at 190°–205°C, produces useful steam which can itself be used in the plant, thereby increasing overall efficiency. The philosophy behind the European preference for alkaline fuel cell systems for terrestrial power applications is discussed in Chapter 10.

In the United Kingdom, Johnson Matthey is developing 500 kW fuel cells for military applications. Small portable 250 W units operating on reformed methanol also are being designed. Little information is available about their characteristics, though the units include a palladium hydrogen diffusion membrane in conjunction with an acid (SPE?) fuel cell. Some of these systems are perhaps intended for submarine applications.

In France, the French Petroleum Institute (IFP) and the Compagnie Generale d'Electricité (CGE) developed various AFCs for military and transportation applications in the late 1960s and early 1970s, but are no longer active in this area. Their work is reviewed in Ref. 394.

An inexpensive AFC technology was developed by Alsthom starting in about 1967. An agreement was signed with Peugeot to develop the technology in 1968, and when this lapsed in 1970 a joint program was undertaken with Exxon until the mid-1970s. Alsthom's program, under the direction of B. Warshawski, was oriented around the use of thin, lightweight cells made from cheap plastic components. Unlike ELENCO's cells, Alsthom's cells have always been of bipolar construction, with ultra-thin structures to give minimum IR-drop, light weight, and the lowest possible cost. The first cell packages constructed in 1968 used 50 micron stainless steel bipolar plates with flow channels pressed in a finely-ribbed herringbone pattern and a central diaphragm. The fuel and oxidant were hydrazine and hydrogen peroxide carried in the flow-by mode in KOH, and the catalysts were silver and cobalt-tungsten oxides attached to the bipolar plates with a conducting epoxy material.[689] These devices were capable of 1 kW/kg, 1 kW/L, which were then unspectacular power densities. Later, the system switched to methanol–air to allow more general use, and then to hydrogen–scrubbed air using circulating KOH and conventional gas-diffusion electrodes when the reactivity of methanol in a buffered CO_2-rejecting electrolyte was shown to be insufficient.

The bipolar plate now consists of 30% conductive carbon black and 70%

polypropylene, and still has the pressed herringbone pattern for gas distribution. The pattern is 0.4 mm wide and 0.5 mm deep, with a web thickness of 0.35 mm. The active area is 400 cm^2. Each bipolar plate is incorporated in a 40% talc-filled polypropylene frame incorporating internal gas and electrolyte manifolding. The whole 900 cm^2 unit can be produced in one single injection molding operation using 12 injection points to the mold. Both hydrogen and air electrodes are calendered low-loading platinized carbon-Teflon, and are glued by a thin layer of epoxy to the bipolar plate. A spacer frame is inserted in each 3.3 mm thick cell. The process–air cooled system can operate using 2M sulfuric acid at 65°–80°C instead of KOH, but performance then falls by 100 mV at the same current density.

Stacks are assembled in subgroups of 100 cells (3.1 kW, 33 cm long) to keep electrolyte circulating currents (shunt currents) to a minimum. All components are epoxy-cemented. The cost is estimated to be \$30/m^2 or less than \$30/kW without catalyst. The platinum loading is 0.3 mg/cm^2 at both electrodes, the goal being to reduce this by a factor of two. Performance with 4M KOH at 65°C is 0.72 V at 150 mA/cm^2, with a goal of 0.8 V. Component lifetimes of 8000–10,000 h have been demonstrated. Further details are given in Ref. 394.

Present catalyst cost (at \$500/troy ounce) is therefore \$90/kW. Cost could be decreased, and performance substantially increased, if the newly-developed pyrolyzed macrocyclic cathodes (Chapter 12) are used. Production capacity was 400 electrodes/day in 1984. During the early 1980s, Alsthom and Occidental Petroleum were hoping to use this technology in caustic-chlorine plants.[118,790] However, despite their low cost, studies have shown that they would be less attractive for this application than the PAFC system, since their low operating temperature (65°C) cannot supply process steam for the plant.[68] For the same reason, their use was discounted for electric utility fuel cell generation from reformed natural gas or light distillates in an earlier study for EPRI, since they could not supply excess steam for reforming, which would therefore require burning more fuel, resulting in a higher (and uneconomic) system heat rate.[786]

Studies at Sorapec in France are directed towards the development of AFCs for terrestrial and space applications. The former would use low-loading platinum electrodes, the latter electrodes with pure noble metal catalysts in high loadings.[2151]

C. FUEL CELL PROGRAMS ELSEWHERE

Very little is known about contemporary fuel cell work in the Soviet Union and other Warsaw Pact countries. In the Peoples' Republic of China, the most interesting development has been the design and construction of a man-rated fuel cell similar to that used in the Apollo program, though of smaller size (500 W, rather than 1.5 kW), which has been tested in a simulated space environment but which has not yet been flown in a manned capsule[583]. The use of small dispersed fuel cells has received a good deal of attention in solving the power problems in developing countries, such as in non–grid-connected villages,[1970] as a recent UNESCO meeting testifies. The proceedings of this meeting provide a good review of this interest.[173, 180]

CHAPTER 6

Military and Aerospace Systems

Since 1965, the United States Department of Defense (DOD) has been involved in the development of military fuel cells for various applications. These have included mobile power generators, transportation vehicles, airfield lighting, communication and transmitter devices, satellite power systems, prime mover remote site generators, submerged vehicles, and buoy transmitters.[262] Among the main reasons for military interest in fuel cells are their low detectability (low noise and infra-red signature), increased force readiness, greater reliability, lower maintenance requirements, cogeneration potential, decreased reliance on petroleum fuels, reduced vulnerability to energy supply disruptions, and cost savings on fuel through more efficient power generation.

With regard to the last advantage, it should be noted that the U.S. Department of Defense presently consumes about 2% of all U.S. national energy, and that in 1979 it paid $2.0 billion for its energy requirements. Studies by the DOD have indicated that major fuel savings are possible through the use of fuel cells. For example, it has been estimated that their use at remote sites could reduce energy consumption by as much as 50% in many cases.[262]

The fuel cell falls into the military category of a *silent power source*. Although the silent nature of the fuel cell is only one of its many advantages, it is very significant for fuel cells in military applications. Its low infra-red signature is equally important for battlefield applications. Because of these qualities, together with its high efficiency, high energy density, long life, simplicity, and reliability, the possibility that various fuel cell systems may eventually replace the entire U.S. military standard line of gasoline engine-driven generator sets has been considered.[1185]

The same features also encourage consideration of the fuel cell for vehicular auxiliary power units, either alone or in conjunction with batteries or other energy storage devices. The U.S. Department of Defense estimated in 1970 that during the period to 1990 there should be a decrease in the percentage of military gasoline-driven auxiliary power sources from

65% of the total to 12%; of diesel-driven power sources, from 34% of the total to 28%, an increase in the number of turbine power sources from 1% of the total to 30%; and an increase in other power sources, which would be primarily electrochemical, from almost 0% of the total to 30%.[1028] The U.S. Navy also predicted a requirement for advanced electrochemical technology devices to keep up with the fast moving pace in the electronics field, the need for nonpolluting underwater propulsion sources, for new weapons requirements, and for electrically powered vehicles of various types.[785] Thus electrochemically generated power has been recognized as a vital component of military systems.

A great many military applications call for portable power equipment, which is classified into two types: *tactical power sources* and *general purpose power sources*. Tactical power sources provide power for the advance combat units operating in contact with the enemy. Unlike the stationary civilian power plants so far considered, tactical military fuel cell power units would be small (typically varying from 30 W to 10 kW), would be expected to endure numerous start–stop operations, and would be specified to allow storage and use under severe climatic conditions. For these combat applications, they must be very rugged and possess very low detectability characteristics. General purpose power sources provide power for the supporting units in the forward area where security is not so vitally dependent on non-detection. Typically, for fuel cells they would vary in size from 1.5 to 30 kW. Most proposed military fuel cell applications to date have been in the tactical power sources category, such as manpack units for field radios. Table 6–1 lists typical military requirements for such portable electrical power sources.[1644] As fuel cell technology matures, the

TABLE 6–1

Canadian Military Requirements for Portable Electrical Power Sources[1644]

	Tactical	General Purpose
Power Range, kW	0.03–10	1.5–30
Noise Level: inaudible distance, m	0–4.5	9–30
Desired Power Density, W/kg	40–110	110
Desired Fuel/Oxidant	Common/Air	Common/Air
Desired Fuel Consumption, kg/kWh	0.2–0.3	0.2
Reliability, %	95–98	95
Desired Life, h	8,000	8,000
Operating Environment	–40°C to +52°C; all humidities; up to 2400 m altitude	

unique characteristics of the fuel cell systems may also qualify them for use in the higher power rating ranges for general purpose power sources.

A. U.S. ARMY PROGRAMS

The United States Army has stressed the need for silent power sources in tactical operations.[262] Past practice usually required running long cables to remotely-located generator sets to isolate the noise. In contrast, silent power sources can be located close to the equipment that they service. A further advantage they possess is that their reject heat can be used to heat shelters and trailers.[54] Tactical systems of up to 3 kW are required for backpacks; and for operating portable power tools, portable searchlights, electrically heated and cooled clothing, and weapons that operate electrically.

Another requirement is for continuous, reliable (no-break) power. No-break power is especially important for front-line medical requirements and other critical operations in fire control, data processing, and communications. The U.S. Army has foreseen batteries and fuel cells as providing the power sources for many of these applications.[54]

Another important Army criterion for fuel cell power systems is that they must be capable of operating on a wide range of both current logistic fuels and future synthetics. This will allow increased potential for standardization and inter-operability, and ultimately will decrease dependence on petroleum-based fuels. In conjunction with U.S. Army Regulation 70-64 ("Design to Cost"), which took effect on February 1, 1980, the U.S. Army now is obliged to consider *cost* ("life cycle management") as a factor of equal importance with technical viability for all military systems.[310] Thus, the long-term cost-effectiveness of the fuel cell, deriving largely from its high efficiency and low maintenance requirements, may be estabished as one of the most important factors for its widespread introduction into military service. It should also be remembered that effective fuel costs for many military applications can be very high, for example when fuel has to be supplied to the field in small quantities by helicopter.

For example, a conceptual design and cost estimate made for the U.S. Army by Westinghouse Electric Corp. indicated that switching to air-cooled PAFCs operating on arctic grade diesel oil in order to power and provide heat for remote radar stations on the Distant Early Warning (DEW) line for ballistic missile detection, would achieve a 25% improvement in fuel consumption over the diesel-electric generators currently employed for this electricity-intensive application. The fuel cell units would operate at 75% overall electrical, plus heat efficiency,[2376] but would be able to supply a larger percentage of the heat content of the fuel as electricity than the diesel system. The consequences would be large life-cycle savings, because of the logistic cost of supplying fuel to remote regions.

The U.S. Army Mobility Equipment Research and Development Command (USAMERADCOM) at Fort Belvoir, Virginia, has been pursuing a program to obtain a family of general purpose silent, lightweight electrical energy plants (SLEEP) in the 1~10 kW range, having characteristics similar to those shown in Table 6–2.[1185] These SLEEP units were to have aural non-detectability at 100 m and minimum infra-red detectability.

The 1.5, 3.0, and 5.0 kW fuel cell power plants were recently at the most advanced stage of engineering development. These systems were all PAFC units which would operate on steam-reformed, 58%-by-weight methanol-in-water–fuel and air. Methanol was chosen as being the best approach to silent power plants since it is a liquid at all ambient temperatures, is easily stored and transported, can be readily steam reformed over commercial catalysts at low temperatures (230~315°C), and can be used in PAFCs in commercially available form without any complex impurity removal.[6] The development of the 1.5 kW unit was contracted to United Technologies Corp. (UTC), whereas the 3.0 and 5.0 kW systems were developed under contract to Energy Research Corporation (ERC).

1.5 kW Fuel Cell Units

In 1980, UTC was awarded a contract to design and fabricate 17 1.5-kW PAFC units, which were to be ready for the field by 1988.[1551] Of the 17 contracted units, nine were to provide a.c. output and eight d.c. output; for example, for charging Nickel-Cadmium battery packs. In addition to the fuel cell units themselves, UTC was to supply the necessary spare parts

TABLE 6-2

U.S. Army SLEEP Unit Goals[1185]

Power Ratings, kW	0.5, 1.5, 3.0, 5.0, 10
Electrical Output	All standard a.c./d.c.; 50, 60, & 400 Hz
Power Density: W/kg	44–73
kW/m^3	8.8–17.7
Fuel	Multifuel*
Fuel Consumption, kg/kWh	0.21–0.28
Life, h	≥ 5000
Noise Level	Inaudible at 100 m

*It is desirable that these units operate on logistically available military fuels; e.g., JP-4, compression ignition turbine engine (CITE) fuel, and combat gas.

and software, such as training and operating manuals. The original budget was $6 million (1980 dollars), but this had exceeded $9.5 million by 1982 in that year's dollars. After that time, a further overrun occurred. Table 6-3 contains a list of the rigorous tests that these units were to pass, allowing

TABLE 6-3

Performance Qualification Testing of 1.5 kW U.S. Army Portable Fuel Cell Power Plants[1551]

Test	Objective of Test
1. Start and stop.	Determine any abnormal start and stop conditions.
2. Battery charging system.	Determine adequacy of battery charging at −65°F and +125°F as well as at normal ambient temperature.
3. D.C. control test.	Determine ability of set to operate with other than nominal rated control voltage.
4. Fuel consumption.	Provide a measure of full load fuel consumption.
5. High temperature.	Capability of unit to perform at +125°F.
6. Transient response.	Capability of a.c. power units to maintain frequency during load changes.
7. Voltage and frequency drift.	Measure of output voltage and/or frequency drift when ambient temperature changes from −25°F to +35°F over eight hours.
8. Long term frequency and voltage stability.	Measure output for four hours at constant ambient temperature.
9. Voltage waveform-harmonic analysis.	Document voltage waveform quality of a.c. power units.
10. Voltage modulation.	Determine magnitude of voltage envelope.
11. Voltage dip and rise.	Determine variations in output voltage during transient loads.
12. Short circuit test.	Determine ability of unit to withstand a direct short circuit across output terminals.
13. Circuit interruptor.	Determine operation and time constant for output circuit breaker.

(continued)

Table 6-3 (continued)

Test	Objective of Test
14. Panel instrumentation.	Determine accuracy of unit output meters.
15. Controls.	Determine consistency of control knob rotation.
16. Voltage adjust range.	Ability of output voltage to be varied without excessive voltage drop.
17. Altitude operation.	Operational capability at 8000 ft above sea level and 95°F ambient; and at 5000 ft and 107°F.
18. Inclined operation.	Ability of set to operate normally when inclined in any direction up to 15° from horizontal.
19. Extreme cold start and stop.	Determine ability of unit to endure five consecutive starts at −65°F using external batteries for start power.
20. Moderate cold start and stop.	Same as (19) but at −25°F using internal batteries.
21. Audio noise.	Noise level in all directions.
22. Electromagnetic capability.	Effectiveness of EMI shielding.
23. Rain.	Capability of unit to operate when exposed to intense rain.
24. Humidity.	Check for deterioration when exposed to high humidity ambient.
25. Lifting and towing.	Ability of unit to be lifted and towed.
26. Endurance.	Reliability of unit when operated continuously for 100 H (2 units) and for 6000 H (2 units).
27. Ripple voltage.	Measurement of a.c. component of d.c. output.
28. Railroad humping.	Typical RR hump test: imparts a 20 kg shock load in three directions at the same time.
29. Drop.	Simulate a drop of a packaged unit from an aircraft.
30. Vibration.	Determine if power unit can withstand dynamic vibrational stresses.
31. Reliability.	Point estimate of mean time between failure derived from endurance testing.

Military and Aerospace Systems

Table 6-3 (continued)

Test	Objective of Test
32. Maintainability.	Point estimate of maintenance ratio, derived from endurance testing.
33. Storage at 155°F.	Conducted concurrently with humidity test: checks for unit deterioration incurred due to storage at high temperature.
34. Salt fog.	Checks for unit deterioration caused by salt fog conditions similar to exposure to ocean spray.
35. Human factors.	Formal examination of control placement, operation, and other elements relating to machine–human interface.
36. Reverse polarity.	Check to see if unit is damaged by reversed polarity hookup of start batteries.

an idea of the reliability that can be expected from military fuel cells.[1551] United Technologies was able to meet all specifications except those of size and weight, for which the original goals were found to be unrealistic. This is illustrated in Table 6–4. Thus, these units, which originally were specified to be two-man lifts (i.e, two-man portable units) are now considered to be four-man lifts. The power plants were to be self-contained, completely foolproof packages with various warning lights to indicate any malfunction. Due to the cost overrun and Army policy changes (see following), only one unit was eventually delivered.

3 kW and 5 kW Fuel Cell Units

In 1978, Energy Research Corp. (ERC) built a 1.5 kW indirect methanol demonstrator fuel cell system for the U.S. Army.[3,1700] This unit was able to

TABLE 6–4

Size and Weight Data for U.S. Army 1.5 kW Fuel Cell Power Plants[1551]

	Original Specifications	Achieved Values	
		A.C. Units	D.C. Units
Weight, lb	150	315	300
Size, ft^3	6	11.5	11.5

meet all military non-detectability requirements, at the same time exhibiting an LHV thermal efficiency of almost three times that of comparably-sized gasoline-engine generator sets (about 22% vs. 7%). Consequently, in 1979 USAMERADCOM contracted ERC to develop corresponding 3 and 5 kW units. Two of the 3 kW prototypes were to be scheduled for delivery to the Army by the end of 1983. ERC delivered one a.c. and one d.c. unit, plus one breadboard unit. All these ran on methanol-water mixture. In 1985, the army decided that pure methanol was a more desirable battleground fuel, so a further breadboard unit was delivered which incorporated a condenser for water recovery. Finally, the whole program was terminated in late 1985, after the Army had decided that methanol (and gasoline) were undesirable fuels for the battlefield, and that only diesel would be used in the future. Since it was impossible to run small PAFC systems on reformed diesel fuel, the program terminated at that point. The Army's present (late 1986) interest in fuel cells is therefore zero.

The reformed-methanol units were based on ERC's air-cooled (DIGAS®) PAFC technology, and they operated on the standard pre-mixed 58%-by-weight methanol in water.[4] The DIGAS® technology[177] is described in Chapter 16. Reformate supply to the fuel cell anodes was about 1.5 times stoichiometric, the excess stack exhaust being burned in the reformers to provide the heat of fuel evaporation and of reaction. While this was more than twice the theoretical heat requirement, it reflects the problem of heat loss in a small, thermally simplified reformer unit. The latter were of tube-and-shell type operating in the 200~345°C temperature range, with annular catalyst beds to minimize infra-red signature and give the maximum efficiency consistent with design. The fuel cells operated at 185°C at about 0.60 V per cell and 160 mA/cm^2. Internal electrical heaters were used to maintain the operating temperature of the fuel cell stacks at low loads.[5] During the start-up procedure, the fuel cell stacks required about 20 minutes heat to reach operating temperature.[4]

The 3 and 5 kW units were designed for maximum interchangeability of parts to allow for best cost and logistical optimization. For example, both used stacks with 557 cm^2 (0.6 ft^2) cells, the 3 kW unit (capable of generating 4.5 kW d.c.) having 80 cells, whereas the 5 kW unit (maximum power, 7.5 kW d.c.) contained two 70-cell stacks in parallel. Output was regulated either to constant d.c. voltage at standard military output (28 V) or was inverted to a.c. by the use of solid state electronic devices. Overall power plant operation was controlled by microprocessors and was fully automatic. As mentioned above, the thermal efficiency of the devices (based on the LHV of methanol) was about 22%, which is equivalent to 19.3% HHV.

Prototypes of both units were operated well in excess of 6000 hours with no drop in performance, and delivered units were operated for 400 hours. Table 6–5 indicates some of the characteristics of the 3 kW unit.[5] The stack

Table 6-5

Characteristics of U.S. Army 3 kW D.C. Fuel Cell Power Plant[5]

	Percent of Rated Power (3 kW)				
	Hot Idle	25	50	75	100
Load:					
Current, A d.c.	—	27.0	54.5	80.2	107.7
Voltage, V d.c.	—	28	28	28	28
Stack:					
Current, A	29.2	35.6	44.2	64.3	92.4
Voltage, V	52.8	52.4	50.7	47.5	42.9
Power, kW	1.54	1.86	2.24	3.05	3.96
Temp., °C	185	185	188	191	196
Internal Power, W	595	580	545	532	527
Reformer:					
Temp., °C	313	310	304	285	263
Conversion, %	100	100	100	100	99.8
CO in reformer gas, %	2.5	2.4	2.3	2.0	1.2
Fuel Consumption, kg methanol/kWh	—	1.63	1.00	0.73	0.77

design originally employed the cathode air stream for stack cooling, corresponding to the original DIGAS® concept.[177] However, the design was later modified to incorporate a separate manifold for the cooling air loop, thereby permitting lower stoichiometric air flow through the cathode, which is important both for water recovery from this stream as well as to ensure multiple restart capability for the unit when wet fuel combustion products are used to heat the stack.[6] This design allowed the use of pure methanol, rather than methanol-water mixture, as fuel.

More recently, ERC built a 20 kW indirect methanol fuel cell power plant for the Los Alamos National Laboratories,[6] which was delivered in 1984. This self-contained unit has two 100-cell stacks and will be used to obtain baseline performance data for simulated vehicle loads.

In addition to the development of the SLEEP fuel cell power plants, the U.S. Army also examined the possibilities for low power, long duration units for applications such as tactical surveillance or communications equipment, which would require high energy density systems of up to several hundred–watts output.[191] For example, the Army designed and built in-house, a small prototype 30 W hydrogen-air solid polymer electrolyte (SPE) fuel cell that incorporated a small hydrogen generator operating on calcium hydride and water.[1189] There also has been some interest in using battery-fuel cell hybrids to power materials handling equipment, such as fork lift trucks, that can meet noise and pollution standards and can be used in difficult locations such as ammunition magazines, outdoor areas, loading zones, and warehouses,[435] where diesel systems might be inappropriate.

In addition to the above activities, the U.S. Army supported fundamental and applied research programs aimed towards a better understanding of catalyst–electrolyte interactions, as well as of the mechanisms and kinetics of fuel cell processes.[23,27,1191] For example, work was contracted to ERC in 1981 to explore the use of aqueous trifluoromethylsulfonic acid (TFMSA) using small atmospheric pressure fuel cell stacks operating at 65°C, 0.65 V/cell, and 200 mA/cm^2 using hydrogen. Other work has been contracted to UTC and to Giner, Inc. (see Ref. 177).

The longer-term objective of this more fundamental research effort was to develop thermally integrated indirect methanol–air and direct alcohol–air fuel cell systems.[1318] Most recently, the Army's electrode development program focused on carbon monoxide-tolerant anode electrocatalysts such as platinum-ruthenium alloys and platinum-doped tungsten carbide, which facilitate low temperature start-up on reformer gas; and on the effects on electrode degradation of on-off and/or freeze-thaw cycling to very low temperatures (+180°C to –55°C).[23]

B. U.S. AIR FORCE PROGRAMS

The U.S. Air Force has a wide variety of electric power requirements ranging from milliwatts to hundreds of kilowatts. Air Force fuel cell applications range from ground utilizations in which storage batteries will probably be unacceptable (such as supplying the lighting for runways and taxi strips varying in length from 1200 m to 3700 m), to ground support, munitions, and life support devices, together with uses in spacecraft, aircraft, and missile systems. Fuel cells appear attractive for many of these applications, particularly those related to aircraft systems where reliability, weight, and volume are the most important factors.[360]

Although the involvement of the U.S. Air Force with fuel cells has been largely related to its previous aerospace experience, at the present time it is mainly interested in their ability to lower logistical and fuel costs at its numerous remote sites and mobile facilities, for which it sees their high reliability and low maintenance costs as additional advantages.[262] The Air Force has made a comparative study of power generating systems for its terrestrial power requirements, in which nineteen energy conversion

Military and Aerospace Systems 161

systems and two energy storage systems were evaluated. The results of this study indicated that fuel cells can satisfy many of the Air Force's needs, as shown in Table 6–6.[1070]

The U.S. Air Force, through the U.S.A.F. Aero Propulsion Laboratory, is currently pursuing the development of large PAFC units for use at remote sites and in mobile systems, as well as of small SPE fuel cells for mobile communications systems.[262] In addition to this involvement, it coordinated the participation of the Department of Defense in the DOE's National Fuel Cell Program to demonstrate the operational feasibility of UTC 40 kW On-Site/Integrated Energy System (OS/IES) fuel cells. Several of these 40 kW units were tested at five military installations.[114]

The Air Force also supported the Advanced Energy Systems Division of Westinghouse Electric Corporation to investigate the feasibility of applying PAFC technology to several ground electric power and heating requirements.[.2043] Two tactical mobile applications where fuel cells were recently considered as replacements for inefficient small gas turbines having practical thermal-to-electrical efficiencies of only 4 to 8% are (a) for Forward Air Controller Radar Power, and (b) for Tactical Aircraft Ground Support Power. Tactical Aircraft Ground Support Power units are used to provide cooling-air during aircraft maintenance (7~8 kW) and to provide electrical power during aircraft maintenance (50~60 kW). Compared with the gas turbines currently employed, it is expected that each fuel cell–powered Radar Power Unit will save 65% on fuel costs and result in a total 20-year life cycle cost saving of $640,000. Similarly, each Tactical Aircraft Ground Support Power unit should save 87% on fuel costs and yield a net 20-year total life cycle cost saving of $630,000.[2043] The only foreseeable disadvantages of the fuel cell units are that their weights and volumes will be substantially larger than those of the turbine sets they will replace. Table 6–7 gives details of some of the key parameters for these units.

The Air Force also had a program with Energy Research Corporation to develop a 5 kW stationary fuel cell power plant for use at remote locations,[6] prototypes of which were delivered in 1984. This power plant operated on steam-reformed methanol, and was essentially a modification of ERC's 3 and 5 kW fuel cells originally developed for the U.S. Army. Because it was intended that this unit would be installed at remote sites, minimum maintenance and low fuel consumption were key requirements in its design. Unlike the first portable units built for the Army, the 5 kW Air Force version incorporated internal water recovery to supply the water required for reforming. This decision was made because of the high logistical cost of supplying fuel (in the form of 58%-weight methanol-water mixture) to a remote site, alluded to earlier. A heat pipe condenser was used to recover the water into a holding tank, from which the water then was metered along with pure methanol into a mixing tank. However, the

TABLE 6-6

Potential of Fuel Cells for U.S.A.F. Applications[1070]*

System	Continuous						8 Hours			1 Hour			
	10 kW	50 kW	250 kW	750 kW	10 MW	50 MW	10 kW	50 kW	10 MW	10 kW	50 kW	10 MW	50 MW
Fuel Cells	•	•	•	•	•		•	•	•	•	•	•	•
Diesels	x	x								x	x		
Stirling engines	•	•	•	x			•	•		•	•		
Solar turbine			x	x					x				
Gas turbine			x	•	•					•		x	x
Wind turbine				x	•				•		•		
Nuclear steam turbine					•								
Photovoltaics	x	x	x										
MHD					•								•
Coal-fired steam turbine				•	•								•

*Key: • = Most promising
x = Promising

TABLE 6-7

Key Parameters for U.S.A.F. Tactical Mobile Fuel Cell Power Units[2043]

	Radar Power	Maintenance Power
Operational Mode:	Intermittent	Intermittent
Physical Parameters:		
Primary/Secondary fuel	JP-4/DF-2	JP-4/DF-2
Fuel consumption at rated power, L/h	22.3	25.4
Volume, m^3	6.4	7.7
Footprint, m^2	3.5	4.2
Weight, kg	1840	2280
Performance Parameters:		
Mean time between failures, h	2620	2000
Availability, %	99.3	98.6
Rated electrical output, kW	60 @ 400 Hz	60 @ 400 Hz
Heat rating, kW	none	42.5
Electrical generation efficiency, %*	28.9	25.5

*Based on JP-4 heating value of 120,000 BTU/gal (33,440 kJ/L)

Air Force has recently taken the same attitude as the Army, and has decided that methanol is not an appropriate fuel. As a result, the emphasis is now on large units (up to several hundred kilowatts) using reformed diesel fuel for remote site use. These are currently (1986) being developed by ERC.

C. AEROSPACE PROGRAMS

Fuel cells have played a major role in the programs of the U.S. National Aeronautics and Space Administration (NASA) for spacecraft power generation since the start of manned spaceflight.[2256, 2257] They originally were chosen as primary electrical power sources for these applications because conventional batteries could not meet the energy density requirements of the Gemini and Apollo space missions. Accordingly, the decision was made to develop fuel cells for manned orbital applications that were to be powered by cryogenic hydrogen and oxygen stored in pressurized insulated tanks.

As was indicated in Chapter 1, NASA's mission to land a man on the moon was the key factor for the development of the fuel cell from a laboratory concept to a reliable multi-kW power source, at the same time enormously advancing electrochemical and electrochemical engineering science. The massive infusion of fuel cell development funding for the Gemini and Apollo missions, which reached a maximum of more than $20 million in current dollars during the mid-1960s, resulted in a fuel cell practical life increase from about 100 hours to several thousand hours in a period of less than ten years. This is illustrated in Table 6–8.[297]

The Gemini missions employed three General Electric 32-cell acid fuel cell stacks with low-operating temperature SPE electrolytes that generated a total output of about 1 kW, whereas the Apollo program utilized three 1.42 kW high temperature, high pressure alkaline (Bacon) fuel cells developed by UTC, designated PC-3A-2. No further units of this type were manufactured after the end of the Apollo program. The Apollo fuel cell is described in detail in Refs. 394 and 689.

The major NASA use of fuel cells at the present time is to provide in-flight and in-orbit electrical requirements for the Orbiter vehicle in the space shuttle program.[2257] Parallel contracts for the Orbiter hydrogen-oxygen fuel cells were awarded to General Electric for SPE[176] membrane fuel cells (see Chapter 10) and to United Technologies Corporation for alkaline capillary matrix fuel cells. The UTC system was selected by NASA in 1970, the GE system being retained as a backup in case technical deficiencies were revealed during development.

The UTC power plant used on the shuttle flights, designated PC-17C, consists of three low temperature (about 82°C gas inlet, 89°C gas outlet) alkaline fuel cell units operating at 4.0–4.4 atm pressure, each rated at about 12 kW in their original form (18 kW maximum power) with two parallel-connected stacks.[394] Each stack weighed about 23 kg, with a total original system weight of 90 kg. In contrast, the weight of each of the three 1.42-kW Apollo modules was 110 kg. More recently, the PC-17C has been upgraded to three-stack 18 kW units as Orbiter mission power requirements have increased, which allow a more constant voltage to be maintained under high-load conditions. Unlike the fuel cells used in the Gemini and Apollo programs, which had back-up Nickel-Cadmium battery systems, all the power for the Space Shuttle Orbiter comes from its fuel cell stacks.

TABLE 6–8

Operating Life of Aerospace Fuel Cells[297]

Year	Hours
1962	500
1969	1,000
1973	3,000
1980	10,000

Each PC-17C stack consists of 32 bipolar cells with an active area of 465 cm^2 giving nominal power at 27.5 V (0.86 V/cell at 470 mA/cm^2). Unlike the Apollo units, the electrolyte is totally immobilized in a thin reconstituted asbestos matrix, and the cell must contain a sintered nickel electrolyte reservoir plate (ERP) to compensate for volume changes. This is in contact with the anode, and is pierced to allow hydrogen to pass through. This 1970 technology is the precursor of the PAFC ribbed substrate technology (see Chapter 16). The ERP accounts for 50% of the total stack weight, and lighter components are used in UTC's later developments of the PC-17C system for NASA. In contrast, the PC-17C uses lightweight gold-plated magnesium bipolar plates. Active water rejection is via the hydrogen feedback loop, and dielectric liquid cooling within the cell stack is used. These subsystems and controls account for almost half the total system weight. The electrodes represent 1960s technology, consisting of Teflon-bonded pure noble metal blacks (anode 10 mg/cm^2 80% Pt, 20% Pd on an Ag-plated Ni screen; cathode 20 mg/cm^2 90% Au, 10% Pt on an Au-plated Ni screen).[394]

During the first four shuttle test flights, which lasted from 2~8 days, the average vehicle power requirement was 15.5 kW, which was about 2 kW above the design prediction.[1835] The highest power level required (29 kW) was generated during lift-off, while the lowest levels (about 10 kW) occurred during crew sleep periods. On Missions 5, 6, and 7 the alkaline fuel cell units again performed flawlessly, delivering a total of 1.89, 1.57, and 1.90 MWh respectively during each flight, the average power requirement being about 14 kW. The performance of the Orbiter fuel cells has equalled or exceeded that predicted, and none of the fuel cells has so far exhibited any degradation. One of the fuel cells, Power Plant 102, had accumulated 757 hours of operating time by the end of Mission 7. The design life of the PC-17C is 2000 hours, with refurbishment between missions, although three stacks have been bench-tested to 10,000 hours of continuous operation.[394, 2256] A photograph of the PC-17C, which is 35 cm x 38 cm x 101 cm, is shown in Fig. 6–1.

The ongoing work in the alkaline fuel cell technology program is focused on increasing the life and power capabilities for future space shuttle operations. Table 6–9 gives a comparison of some of the design parameters of the fuel cell systems used in the Gemini, Apollo, and Shuttle programs.[1737, 1835] A review of fuel cells and batteries for aerospace applications has been published.[1154]

Future aerospace applications will require fuel cell systems capable of much greater power output than has been the case in the past. These new applications should include those for unmanned platforms in low earth orbit (LEO) and in geosynchronous orbit (GEO), permanently-manned space stations in LEO, vehicles for orbital transportation from LEO to GEO,

Figure 6–1: Space shuttle fuel cell unit (UTC PC-17C).

TABLE 6–9

Selected Parameters of N.A.S.A. Aerospace Fuel Cells [1737]

Program	Current Density mA/cm²	Mission Energy kWh (per module)	Mission Duration, h	Life, h
Gemini	36	65	360	1000
Apollo	68	115	192	400
Shuttle	172	2600	168*	2000

*nominal

and complementary military applications in the areas of surveillance, command, and space weapons. Primary power will be provided either by deployable photovoltaic systems, solar-thermal systems, or from a small nuclear reactor such as GE's SP100 unit (originally rated at 100 kW, with the potential to reach 300 kW), or a combination of all three. In 1986, the safety problems involved in launching nuclear reactors into orbit were not entirely clear. All the above systems would require on-board storage for peaking duty and (for the solar systems) during eclipse periods. Tables

Military and Aerospace Systems

6–10 [1737, 2229] and 6–11 [297] list the fuel cell requirements for some of these applications.

NASA's fuel cell development effort over the next several years originally was intended to be directed towards the development of a regenerative energy storage (RES) concept for use with large orbital power platforms needed for orbital construction, manufacturing, and scientific activities. This would consist of a system that is capable of being recharged,

TABLE 6–10

Estimated Requirements for Aerospace Fuel Cells[1737, 2229]

Mission	Crew	Operation Life, y	Electrical Power, kW
Space station	12	≥ 10	25
Exploration modules	?	5	0.1–5
Space base	50–100	≥ 10	100–200
Mass module	6	≥ 2	12–20
Shuttle	12	7 days	5–20
Lunar rover	1	≤ 1	0.3–0.6
Lunar base	?	> 5	≥ 15
Skylab	3	1	7–8
Orbital power platform	?		100–500

TABLE 6–11

Projected Power Demand for Aerospace Power Sources[297]

Use	Year	Power
Orbiter, Space Lab	1980	30 kW
Materials Processing	1982–83	40 kW
Science & Technology Payloads	1982–83	60 kW
Life Sciences	1984–85	80 kW
SEPS	1987	115 kW
Construction	1987	350 kW
Space Manufacturing & Processing	1988–89	240 kW
Pharmaceuticals	1989	115 kW
Space Power Technology Demonstration	1989	1.5 MW
Public Services	1990	160 kW
Space Station	1992–93	1.0 MW
Advanced Space Propulsion	1994–95	11.5 MW
Radars	1995	5.0 MW
Materials Processing	1995–96	18.0 MW
SPS Geo Demonstration	1996–97	45 MW

like a battery, using photovoltaic panels delivering enough energy to satisfy mean power requirements after conversion losses, allowing for the time when the orbiting platform will be wholly or partly eclipsed. The product of recharging would generally be stored hydrogen and oxygen produced by a separate charging unit (electrolyzer), which will be consumed in the primary fuel cell units. Other concepts have however been proposed (see following). The main technical objectives for the fuel cell power plants would be long cyclic life with on-orbit maintainability. Initial units were expected to be in the 50–100 kW peak load class, but a modular approach should eventually enable power plants of almost any output to be built.

NASA worked on LEO energy storage applications from 1979 under a combined program involving the Johnson Space Center in Houston (a mission center) and the Lewis Research Center in Cleveland (a technology center). The aim of this program was the development of a regenerative fuel cell having a 40,000-hour life for a technology readiness demonstration by the end of 1986.[2256]

As indicated above, this would have used hydrogen-oxygen fuel cells for darkside power and solar cells for sunlight power. During sunlight periods the solar cells also were to be used to drive specialized electrolyzer cells to decompose fuel cell product water. This type of fuel cell–electrolysis cell combination is known as a *regenerative fuel cell* (RFC), although that term should be restricted to a fuel cell unit that is truly regenerative; that is, one which itself can operate in both discharge and charge modes. However, development of such a system will lead to problems of optimized electrode area and number of active cells, which ideally would be different in the charge and discharge modes.

In addition, if alkaline electrolyte systems are used, materials problems exist which render the development of a truly regenerative cell with a long lifetime difficult on account of the different materials requirements for the charge and discharge electrode environments. In contrast, the technology for the long-lifetime alkaline-electrolyte RFC system using separate charge and discharge cells already exists. The same problem may not occur in acidic (fluorinated solid polymer) electrolyte cells, which can operate in both charge and discharge modes.[176] However, SPE cells incur a voltage penalty on discharge compared with alkaline fuel cells, and the different electrical requirements on charge and on discharge have, in any case, generally led to the conclusion that separate charge and discharge units will be necessary in practice. The choice of a separate alkaline fuel cell–electrolyzer system with optimized units, or an SPE charge–discharge unit with lower overall efficiency but perhaps lighter weight, will depend on mission requirements.

Although the overall efficiency of the hydrogen-oxygen fuel cell–

Military and Aerospace Systems 169

electrolyzer combination is only 50~60%, if the device is thermally integrated with the spacecraft it could also provide a useful source of heat at 80~95°C, which should make it superior to proposed competing battery systems as a total energy source. The combined long life and unrestricted charge–discharge characteristics of a typical RFC system will result in a lighter overall system weight than all but the most optimistic advanced secondary battery systems. Thus, the calculated system weights for a 100 kW power plant suitable for a ten-year orbit life have been estimated as: RFC = 6000 kg; nickel-cadmium batteries (50% depth of discharge) = 7000 kg; nickel-cadmium batteries (25% depth of discharge) =16,000 kg,[1737] where the ten-year orbit lifetime of the 50% depth of discharge nickel-cadmium battery system cannot be guaranteed. It has been predicted that the energy densities for some RFC configurations may be as high as 200 Wh/kg, compared with less than 30 Wh/kg for nickel-cadmium batteries. In addition, it now seems likely that the useful operating life of the RFC may be limited more by auxiliary equipment, such as pumps, than by the electrochemical cells themselves.[1824] Although the majority of the effort has been spent on the hydrogen-oxygen system, with which there is the most experience, and EPRI has also supported work on hydrogen-chlorine and hydrogen-bromine regenerative systems,[309, 1596, 2392] which may also find use in future space applications.

Studies have indicated that for longer missions (> one year) the acid solid polymer electrolyte regenerative fuel cell may be most appropriate, whereas for shorter missions (< 5000 hours) the alkaline capillary matrix fuel cell system with a separate electrolyzer unit would appear to be best. In order that the most effective technology will be available when required, NASA has supported the development of both types of RFC system. The Lewis Research Center coordinated basic cell technology and alkaline cell development; while the Johnson Space Center was responsible for the SPE technology, interim breadboard evaluations, and the engineering *proof of technology readiness.*

The acid SPE regenerative fuel cell technology being pursued is based on small G.E. fuel cell–electrolyzer components, now being developed by UTC's Hamilton Standard Division. Earlier programs supported by the U.S. Navy to produce oxygen for submarine life-support systems using SPE electrolyzer cells with Nafion membranes[176] demonstrated 35,000 hours life at 82°C and indicated a potential electrolyte membrane life in excess of 100,000 hours. This Navy experience has provided a strong technology base for the acidic RFC system, which has demonstrated > 40,000 hours life.

The acid RFC *feasibility breadboard demonstrator* chosen for endurance testing consisted of an 8-cell stack of 1000 cm^2 SPE fuel cells in combination with an electrolysis section comprising a 22-cell stack with 210 cm^2 cells.

The fuel cell unit operated at 112 amperes, 6.5 V, and 71°C; while the electrolysis unit operated at 24 amperes, 36 V, and 23°C; although the cell voltages required to operate the electrolyzer at current densities of 200 and 800 mA/cm^2, presumably at a higher temperature, have been reported as being 1.59 and 1.78 V, respectively.[1824] The hydrogen was stored at 8.8 atm and the oxygen at 7.8 atm.

This breadboard unit was delivered to the Johnson Space Center for testing in February, 1983. By July, the unit had accumulated 473 LEO cycles, each cycle consisting of operating the electrolysis subsystem for 54 minutes followed by the fuel cell subsystem for 36 minutes. No permanent degradation in cell performance has been observed with either the fuel cells or the electrolysis cells. Although no measurable water loss from the system was determined, apparently 4% more gas was produced than consumed. This was presumably due to greater coulombic losses from hydrogen diffusion across the SPE membranes in the fuel cells. The next proposed step was to replace the fuel cell stack with an advanced design consisting of 10~15 cells of 800 cm^2 area. This was to test open-ended endurance on the LEO operational duty cycle to establish a better lifetime data base. The breadboard was also to be used as a test bed to verify advanced cells and components.

The alkaline RFC technology was based on an improved version of UTC's earlier alkaline fuel cells for the Space Shuttle Orbiter.[2456a] As described earlier, their stack lifetimes have already been established as over 10,000 hours.[394, 2256] It was determined that stacks containing cells of 500 cm^2 area should be suitable for use up to approximately 100 kW. Desirable performance goals would be 40,000-hour life at 110 mA/cm^2 and, for high current density applications, 3000-hour life at 0.9 V and 1100 mA/cm^2.[2456a] Orbiter hardware has demonstrated in excess of 3500 hours under a simulated LEO test profile, and it is likely that the minimum goal of 20,000 hours is achievable.

The electrolysis technology for the alkaline units was being developed by Life Systems, Inc. of Beachwood, Ohio, who manufactures submarine life-support electrolysis systems for the U.S. Navy. They were evaluating the endurance capability of state-of-the-art electrolysis cells of about 90 cm^2 area. This hardware had achieved over 13,000 hours of successful testing, and was finally scaled up to the 0.10 m^2 area needed for a 100 kW system.

The alkaline breadboard, delivered to the Johnson Space Center in January, 1984, contained a 30-cell alkaline electrolysis unit nominally rated at 1.5 kW integrated with an Orbiter power plant. Applied voltages of 1.52 and 1.82 V were required to drive the electrolyzers at current densities of 200 and 800 mA/cm^2, respectively.[1824] In April 1984, the 1.5 kW electrolyzer was replaced by a 3 kW unit consisting of 6 cells, each of 930 cm^2 area. This provides a better power match for the Orbiter power plant, and is considered to be full-size space station hardware.

Military and Aerospace Systems

As of July 1983, endurance testing of the electrolysis unit alone at 82°C, ambient pressure, and 160 mA/cm^2 using 90 cm^2 single cells, had surpassed 22,000 hours in the LEO regime with no voltage degradation. Similarly, a 6-cell fuel cell stack of Orbiter-sized hardware had accumulated 6800 hours of LEO cycle endurance testing at 60°C, 4.1 atm, and 215 mA/cm^2 with a voltage degradation of less than 1 µV/hour. A 20,000 hour endurance test of a complete electrolysis subsystem was begun in April 1984. All tests of RFC equipment were terminated when funding ran out. No degradation beyond that just indicated was observed.

The four main technology requirements for these RFC energy storage systems were (a) a 40,000 h life, (b) a two-year minimum life for individual components, (c) greater than 50% overall electrical efficiency, and (d) an optimum voltage range between 100 and 240 V. The program concentrated on improving both component life and electrical efficiency.

However, in May 1986, the program was changed drastically, when NASA decided not to consider RFCs for the space station as an economy measure. The main reason for this was that the projected power requirements of the future manned station were starting to change as its size shrunk along with NASA budgets. At that time, it was proposed to have 25 kW of installed deployable photovoltaic panels, together with 50 kW of dynamic solar power (i.e., a solar thermal system incorporating heat storage for eclipse periods, peaking, and heating). The proposed polar orbiting platform was also to have 25 kW of photovoltaic power, and for it nickel-hydrogen batteries had been identified as the most effective electrical energy storage system.

The choice of the nickel-hydrogen system over nickel-cadmium for aerospace use is an advantageous one, since Ni-H$_2$, first described in 1972[2351] offers both an improved and highly reliable cycle life and better energy density than Ni-Cd. As its name implies, the battery has a standard aerospace Ni positive and a rechargeable hydrogen negative, the latter being a standard Teflon-bonded fuel cell electrode. The battery resulted from the discovery that hydrogen reduced charged nickel positives only slowly, so that self-discharge is acceptable.

Generally, early hydrogen electrodes were 3 mg/cm^2 pure platinum black on a nickel screen, but they now tend to use low-loading platinum on carbon catalyst. Unlike AFC oxygen electrodes, the hydrogen negative cycles indefinitely at negligible overpotential. For aerospace use (e.g., in communications satellites since the late 1970s), the battery has been assembled as a monopolar array of parallel-connected bicells; each consisting of two central back-to-back Ni positives contacting on each side a nylon or reconstituted asbestos matrix containing immobilized 6M KOH electrolyte, a hydrogen electrode, and finally a porous plastic gas diffusion screen. The whole array is enclosed in a lightweight Inconel 718 cylindrical pressure vessel with hemispherical end caps containing the appropriate insulated current feed-throughs. The system cycles between approximately 7 atm and 41 atm pressure.[773] The arrangement is highly reliable, since it lacks the possibility of corrosion due to shunt currents in a bipolar array, although these are now proposed, including arrangements with centralized pressurized hydrogen storage.[774] The system is capable of very high charge-discharge rates to more than 4000 cycles, has overcharge protection, and is capable of 45 Wh/kg on the single-cell level at a practical C/2 (2-hour) discharge rate. Mounted in a satellite, it and the structural support system necessitated by its rather large volume allow a practical 32 Wh/kg,[773] which is about 50% better than that of an aerospace Ni-Cd battery pack under real cycling conditions. Consequently, for LEO applications involving short discharge times (i.e., storage of few kWh), it and the RFC would have comparable weights.

In fact, the $Ni-H_2$ aerospace battery grew out of a proposed true RFC concept for geosyncronous orbital use in the late 1960s.[1428, 1429] This consisted of an AFC with high-loading noble metal electrodes in a lightweight pressure vessel divided internally in such a way that both hydrogen and oxygen could be stored separately. This required a 50% larger (i.e., 50% heavier) pressure vessel than a $Ni-H_2$ cell of equivalent Ah capacity basis, after correction for any volume differences for the components. In addition, it required provision for water storage, which is not necessary in the $Ni-H_2$ system in which the cell reaction is:

$$NiOOH + \tfrac{1}{2} H_2 \rightarrow Ni(OH)_2$$

In other words, no free water is involved in the reaction and the electrolyte is invariant in volume. Finally, the pressurized fuel cell operated at 0.9 V, whereas $Ni-H_2$ is a 1.2 V system. Overall, the RFC Ah/kg value was at best comparable to that of $Ni-H_2$, and its poorer charge-discharge voltage efficiency required a heavier solar panel array for charging.

The RFC only becomes of real interest for applications where a large number of kWh must be stored for each kW of power output. For example, pressurized storage of hydrogen and oxygen in optimized lightweight containers corresponds to about 450 Wh/kg for the reactants and containers. Hence, the RFC concept is viable only if the power system weight is small compared with the stored reactant weight (i.e., if storage is for hours, rather than for minutes) with the present state-of-the art RFC equipment. This can however be expected to change in the future as lighter weight components are developed and as requirements change. For example, the space station is already growing in power requirement, the proposed PV array having been upgraded first to 37.5 kW, and most recently to 50 kW, which may make the RFC more attractive than batteries for eclipse storage. Finally, electrolyzers are already required for life-support systems, and may in future be used to manufacture cryogenic oxygen-hydrogen fuels in space (see following). In this case, the fuel cell will be the logical choice for eclipse and peaking power (e.g., for Air Force SDI applications).

Another area of interest to NASA is on-board power for future orbital transfer vehicles (OTV), which will be used to ferry equipment between space stations in low earth orbit and satellites in geosynchronous orbit. Fuel cells are the prime candidates for OTV power because they are already flight-qualified and can operate using the boil-off from the cryogenic hydrogen and oxygen propulsion reactants.[2257] The latter will be preferred to, for example, unsymmetrical dimethyl hydrazine–N_2O_5, because of their higher specific impulse. In addition, since it is easier to ship water from earth to LEO than it is to ship cryogenic propellents, these will be manufactured in orbit using electrolyzers, to which FCs will make a complementary package.

Preliminary power requirements for an OTV have been identified as a nominal range of 0.5~2.0 kW at about 28 V, with 3.5 kW peak power, and a required operating life of about 2500 hours. These requirements probably will be met with Orbiter alkaline fuel cells scaled down to 232 cm² (0.25 ft²) hardware. These requirements have already been addressed in NASA's advanced lightweight fuel cell program at UTC, which developed and lightened the Space Shuttle Orbiter cell over the period 1975–1979.[1706] The main changes were the elimination of the sintered nickel electrolyte reservoir plates (ERPs), and their replacement by Ni-plated polysulfone plates, replacement of the 10 mg Pt/Pd anode by an anode with 0.5–0.75 mg/cm² Pt supported on carbon, and improvements in water rejection to reduce subsystem weight. The 0.75 mm thick ERP contacts the cell anode and the electrolyte is 8M KOH immobilized in asbestos. On the opposite side of each anode gas chamber (which contains a porous Teflon screen) is a passive water-removal assembly consisting of first an asbestos matrix containing KOH of a higher concentration than that in the actual cell matrix, a further ERP, a porous Teflon membrane maintained by a Ag-plated Ni screen, and finally a further Teflon screen occupying the space for the passive evaporation of water from the electrolyte. This contacts a nickel-foil separator, which then contacts a cooling plate containing dielectric liquid. The d.c. cell and cooling plate frames are resin-impregnated zirconia mats, and the water-rejection cell frame is resin-impregnated fiberglass. A cross-section of this complex, though lightweight, assembly has been given in Ref. 176, and a photograph of the components is shown in Fig. 6–2.

Because of the use of the asbestos component in the water-rejection cell, the system must be monopolar, the electrodes being 30 cm × 7.5 cm with edge-collection of current along their long sides. Performance of the system is the same as that of the Space Shuttle Orbiter cell after 1000 h at

Figure 6–2: UTC advanced lightweight alkaline fuel cell components (1979).

82°C, 4 atm (0.86 V, 460 mA/cm^2; 1.01 V, 70 mA/cm^2). The target performance of 0.9 V at 1.1 A/cm^2 was exceeded at 127°C and 10 atm, and reached 0.98 V at 149°C, 17 atm and 1.1 A/cm^2 (1.10 V, 100 mA/cm^2). Polarization curves are given in Ref. 394. Performance degradation at 149°C was however rapid, and was 60–70 mV (7%) over 1000 h at 120°C. This was attributed to dissolution and degradation of the asbestos matrix, which must be replaced with, for example, potassium titanate (see Chapter 13) if higher performance is to be obtained. The cell was also examined with low-loading Pt-carbon cathodes, but these suffered from catastrophic corrosion at temperatures much above 80°C.[394]

The 30-cell 2 kW stack reached its target power density of 550 W/kg at a nominal 2 kW rating for the 8 kg unit, compared with 275 W/kg for that of the PC-17C stack, overall system nominal rating being 250W/kg, again twice that of the Orbiter design at an efficiency about 7% higher, which is important in reducing cryogenic fuel weight. At full target power (6.75 kW) at a current density of 1.1 A/cm^2 at 0.9 V, the system gives the remarkable figures of 1.8 kW/kg for the cell stack or 840 W/kg for the complete system. The corresponding PC-17C figures at beginning of life are 400 W/kg at stack level, 200 W/kg for the complete system, at 0.75 A/cm², 0.83 V. Since 1979, even thinner, lighter cells, (e.g., using porous carbon ERPs at the anode side) have been designed for Air Force applications. A recent request for proposal[126] cites 6.3 kW/kg of short-term power as the ultimate goal for the cell stack of an auxiliary power unit (APU) for the future National Space Transport System (NSTC). The monolithic solid oxide fuel cell may also be able to achieve or exceed these goals (see Chapter 18), and it has been mentioned as the potential APU for the National Aerospace Plane.[2585]*

In Chapter 5, some military and aerospace applications of fuel cells in countries outside of the United States have already been mentioned. These include the Chinese Apollo-type system,[583] as well as potential uses of the AFC as a submarine power plant in Sweden.[1627] SPE development at Siemens in West Germany is believed to be directed towards similar goals. Finally, Breelle in France has alluded to the use of alkaline fuel cells in the European Space Agency's Hermes refurbishable mini-orbiter,[454] which apparently will require a 2~3 kW cryogenic H_2–O_2 AFC. Several developers are reportedly undertaking work on high-loading, Teflon-bonded noble-metal electrodes for this application (cf. Ref. 2151).

---

*In 1987, UTC announced that they had developed an advanced lightweight alkaline system with a gold-plated magnesium co- or counterflow bipolar plate, a graphite ERP, and a 50 micron matrix, with a total active component weight of 1.66 kg/m² (0.34 lb/ft²). At 150° C and H_2/O_2 at 13.6 atm, it is capable of 1A/cm² at 1.0 V, 5A/cm² at 0.8 V, or 9A/cm² at 0.72 V. Thus, at 0.8 V, the stack components are capable of 24 kW/kg. The system uses active cooling (water evaporation in the reactant flow field) and external water removal. A total system, weighing 90 kg should produce 300 kW continuous power. Performance is illustrated in Figure 10–3, page 295.

CHAPTER 7

Electric Vehicles

The transportation sector accounts for 17~20% of the total energy consumed in North America. In the United States, the transportation industry alone accounts for about 20% of the GNP, split roughly equally between passenger (mostly private automobile) and freight.[2371] In addition to the importance of the whole of the transportation sector in terms of the energy it consumes and the total percentage of the GNP that it represents, it has been estimated that vehicular power plants account for approximately 60% of atmospheric pollution.[2334] In view of these facts, there has been widespread interest in the replacement of internal combustion (IC) engines in transportation by electric traction, which should be more fuel efficient and have less environmental impact.

The interest in electrical propulsion for vehicles has not grown out of any sudden realization that an advantage might be gained in performance. Rather, political, economic, and social pressures are forcing an evaluation of alternatives to the internal combustion engine.[1770] There is also nothing new about electric automobiles. In the fifth edition of *Self-Propelled Vehicles*, 1907, the author, James E. Homans, stated:

> Vehicles propelled by electric motors, whose energy is derived from secondary batteries, are much preferred by many authorities on account of the combined advantages in point of cleanliness, safety, and ease of manipulation. When well constructed and well cared for, they are also less liable to get out of order from ordinary causes. Among their disadvantages, however, may be mentioned the fact that the storage batteries must be periodically recharged from some primary electrical source, which fact greatly reduces the sphere of efficient operation.[2680]

As might be inferred from the preceding passage, the battery-powered car was at one time an equal entity in the competition with the gasoline buggy. In the United States, in 1900, there were about 2000 of each, together with

2000 steam cars. By 1915, there were almost 40,000 battery driven vehicles, but by then a much larger number were fueled by gasoline.[2334] By 1915, most battery-powered vehicles were light trucks and delivery vans. In fact, in the latter category, battery power enjoyed a 4-to-1 preference over piston engines. The batteries used were almost entirely of the lead-acid type, which then, as now, were relatively inexpensive.

It is therefore obvious that electrochemical power was at one time a significant and competitive technology in the propulsion field. Electric vehicles succumbed to competition from the internal combustion engine because of the disproportionate gains in performance, economy, and flexibility (i.e., range) that came to be associated with the latter. One of the more important technological innovations that favored the internal combustion engine was the development of the self-starter, which greatly increased its ease of use and consequent popularity. On a day-to-day basis, the range and performance of the electric automobile was competitive with that of the horse-drawn buggy or delivery van: Neither could finally compete with the flexibilty of the IC engine, which allowed the individual driver more freedom of travel than did the railroads. In contrast, the fuel cell may, in future, offer the automobile the range flexibility of the IC engine, along with the advantages of silence and lack of pollution of traditional electric propulsion. In addition, and more important, it could offer a much more advantageous utilization of primary energy than has been the case for traditional IC power sources. A summary of its potential advantages, both to an individual and to society as a whole, is given in the following section.

A. ADVANTAGES OF FUEL CELL–POWERED ELECTRIC VEHICLES

Efficiency

The efficiency of a fuel cell differs in two important respects from that of a heat engine: (a) It does not decrease sharply as the size of the unit is decreased, and (b) fuel cell efficiency does not appreciably change if the fuel cell operates at part load. The latter characteristic is especially important in applications with a variable duty cycle, such as in transportation, particularly under urban conditions. In the average power ranges of most interest for vehicle operation (10–100 kW), fuel cells have at least twice the efficiency of internal combustion engines and at least one and a half times the efficiency of diesel engines.[1735]

Care must be exercised when an attempt is made to evaluate fuel cell efficiency compared with that of secondary batteries. The major problem

Electric Vehicles 179

is to determine how far back to go in the energy transformation chain starting with the source of primary energy. For example, if the electricity used to charge the battery is of nonfossil (e.g., hydro or nuclear) origin, it is difficult to put it in the same frame of reference as fuel of fossil origin consumed in the fuel cell. In an economy where the majority of electricity is of nonfossil origin, electricity might be considered as primary energy, and would be directly compared with primary fuel from coal, oil, or natural gas.

Thus electricity may be available from the terminals of a storage battery at 60% efficiency (including line losses of 10%, charger efficiency, and charge-discharge efficiency of the battery). Compared with this, electricity may be available at the terminals of a fuel cell using methanol at 45% efficiency. However, the methanol must be made from coal at 65% efficiency, or from natural gas at 80% efficiency. Hence, if coal is regarded as the primary energy source, the overall coal-motive power efficiency using the fuel cell converter is about 30%, which appears to be half that attainable when nonfossil electricity is regarded as primary energy. However, in a coal-based primary energy economy, electricity is produced from coal using conventional power plants at only 35% efficiency. Hence, the coal-motive power efficiency using rechargeable batteries will then only be 21%, much less than that using coal-based methanol and fuel cells. Finally, if the methanol (a typical coal-based synthetic fuel) had been burned in an IC engine with an average urban-cycle conversion efficiency to shaft power of 12.5%, then the coal-motive power efficiency will be only 8%. In this instance, the advantage of the fuel cell is clear.

In a coal-based primary energy scenario, synthetic hydrogen could be produced from coal at the same efficiency as methanol, and used in the fuel cell at 50% efficiency, giving almost 33% overall efficiency. However, if the hydrogen had to be made from primary hydroelectric power by conventional electrolysis at 75% HHV efficiency, the overall primary electric-motive power efficiency will be about 37.5%, appreciably less than that of the battery storage system calculated on the same basis. Overall, therefore, whether fuel cells or batteries (or a hybrid of both) will be the best choice depends on the energy economy and the economics of the energy supply chain. However, for practical purposes it will also depend on the mission requirement: The battery is a short-range device, whereas the fuel cell allows flexible autonomy.

In any vehicle powered by an electrochemical energy source (including both fuel cell- and battery-powered vehicles), the efficiency with which the energy is converted to effective power in the drive train is dependent on the power requirement (i.e., the rate at which the energy is used). However, this is true to a much less extent than is the case for IC engines. Batteries tend to be worse than fuel cells in this regard. In addition, for the electrochemical generators, energy available goes down as load *increases*; whereas for the IC engine it goes down as load *decreases*. The electrochemical converters are therefore particularly appropriate to part-load, stop-start, urban use, where the IC engine is least desirable and least clean-burning.

To determine the thermal efficiency of a storage battery, detailed knowledge must be obtained for the primary energy to charger efficiency, for the charger efficiency, for the battery round-trip energy storage efficiency, and for the voltage capability of the battery in providing current at given load levels (i.e., in the driving cycle). In addition, if a significant amount of time elapses between the end of the charge period and the beginning of the discharge period, another factor, self-discharge due to chemical recombination of reactants, may occur with some storage battery systems, reducing the amount of energy remaining. This also must be taken into account in the calculation of the complete energy storage efficiency of batteries.[892] The overall result is that in practical driving cycles under real-world conditions, battery efficiencies tend to be less than expected. In the example just quoted for the nonfossil electric economy comparing batteries with 60% overall efficiency with hydrogen-based fuel cells at 37.5%, the battery efficiency for urban driving may in practice be much closer to that of the fuel cell, because of the better electrical characteristics of the latter under urban load conditions.

Air Quality

The levels of emissions from fuel cell–and battery-powered vehicles will be orders of magnitude lower than those from conventionally powered vehicles, even those equipped with catalytic converters.[110] The only potentially noxious emissions from battery electric vehicles consist of small quantities of ozone from the electric motors. However, it should not be forgotten that emissions may be involved with battery-powered vehicles at the site of the power plant, rather than in the urban environment. In this respect, fuel cell vehicles will have an advantage in a coal-based economy, since clean gas from coal is required for the production of synthetic fuel.

Noise Pollution

Electrically-driven vehicles powered by batteries or fuel cells are much quieter in operation than other types of vehicles, although some noise can be expected from the electrical components.[2119] This is caused by current surges when the silicon-controlled rectifiers used to invert d.c. power to three-phase a.c. power of controlled frequency and voltage are cycled on and off by control systems. This noise seems to be as fundamental to a control system using pulsed power as exhaust noise is to an internal combustion engine, but it is at a very much lower level than the latter.

Fuel Supply

The use of electricity to power vehicles will lessen dependence on imported or indigenous petroleum-based fuels, since both fuel and electricity can be generated from a variety of fossil fuels and also from nuclear, hydroelectric, and eventually from solar sources in suitable locations. Fuel cells can operate effectively on a variety of specific fuels and will present new options in regard to considerations of fuel supply.

Reliability and Ease of Maintenance

Electrochemical power sources, particularly batteries, are intrinsically simpler and may potentially be more reliable than IC engines. Similarly, electric motors require virtually no maintenance over lifetimes of 50,000 hours or more. This is equivalent to 1.5–2 million kilometers in typical urban use. An excellent recent book reviewing power sources, including fuel cells, for electric vehicles is now available.[1760]

B. REQUIREMENTS FOR ELECTRIC VEHICLES

Analyses of the energy and power requirements of electric highway vehicles indicate that a full-sized electric family automobile of average performance and with a range of about 350 km would require a power source with an energy density of about 220 Wh/kg and a power density of 130~220 W/kg.[547, 2388] Other types of vehicle with lower performance specifications will have correspondingly lower specific energy and power density requirements for the power source. Small urban vehicles, for example, will be competitive with energy and power densities as low as 45~90 Wh/kg and 65~90 W/kg, respectively.[2217] For a battery-powered family car, the battery life should be equivalent to about 150,000 km of operation.[1385] Any routine servicing that involves cell-by-cell activity is undesirable for an electrochemical power system on account of the large number of cells involved. For this reason maintenance activities should be limited to only once every 1~2 years. In addition, it should be noted that this argument also holds against refuelable metal-air batteries (fuel cells using replacable solid metal anodes) which must be replaced in each cell at regular intervals, for example every 1000 km. Such systems include the aluminum-air fuel cell system that has been suggested for use in vehicles.

When specifying the output of an electrochemical power source, the power requirements of the accessories should also be taken into consideration.[1124] Lights, horns, radios, windshield wipers, defrosters, and the like, consume about 0.15 kW. Power steering requires about 0.18 kW. Typical air-conditioning systems require about 4.5 kW, and are therefore probably impractical in the average electric vehicle for urban use. Similarly, automobile heaters produce 6~8 kW$_{th}$, which is a large amount if it is to be supplied by stored energy in a low-temperature battery, whose waste heat output is relatively small and is at too low a temperature to be effective. However, it will be possible to use the waste heat of certain high-temperature batteries and of fuel cells operating at 65°C or more for vehicle heating.

C. CHARACTERISTICS OF ELECTRIC VEHICLES (EVs)

As indicated in Table 7-1,[1124] the present generation of electric motors tends to be rather heavier than internal combustion engines for a given power rating. This does not mean that high-performance, lightweight a.c. induction motors have not been built. The 75 kW (100 hp) electric motor used by General Motors in their Electrovair II weighed only 0.8 kg/kW and occupied only 0.0235 m^3.[95] Motors for special applications such as electric torpedoes are even lighter. There seems therefore to be no reason why electric motor design should be a limiting factor for EV applications. The control systems for electric vehicles tend to increase in weight if instant acceleration response is required, though this is expected to be reduced in the future by the use of new power electronics, which will lead to an increase in control system efficiency. On the other hand, advantageous

TABLE 7-1

Weight Comparison Between Gasoline Engines and Electric Motors[1124]

Horsepower	(kW)	Gasoline Engine, lb	Electric Motor, lb
25	(19)	100	150
50	(37)	200	500
75	(56)	300	750
100	(75)	350	850
150	(112)	450	1100
200	(149)	500	—
300	(224)	650	—

characteristics of electric motors and their associated systems are that they are very rapidly and easily started and controlled compared with IC engines, particularly under cold weather conditions.

Performance and Fuel Costs

The performance of most electric vehicles, in particular those powered by heavy lead-acid batteries, has been generally less than that obtainable from internal combustion engines. The trend towards smaller cars with smaller engines has, however, shown that the public may be willing to accept vehicles with lower performance than manufacturers previously considered desirable. In addition, the fact that fuel cell and battery performance has constantly (if unevenly) shown improvement indicates that electric vehicles will eventually become more acceptable. Tapley drag data show that 11 kW (15 hp) is sufficient to power an urban vehicle,[2117] and that only 21 kW (28 hp) is required to maintain a constant road speed of 100 km/h under normal atmospheric conditions for the majority of present-day North American automobiles.[1124] For subcompacts, the figure will be in the 15 kW range. In general, the propulsive energy requirements for vehicles range from 0.15 to 0.35 Wh/kg^{-1} km^{-1}.[545, 1124, 2096]

The operating cost of a gasoline-powered vehicle is very sensitive to the cost of fuel. A study of the cost trends for both electric power and gasoline has clearly shown that the cost of electricity has been decreasing throughout this century, while the cost of gasoline has been increasing for more than three decades.[1124] Because of the limitation on readily available petroleum reserves with low costs of exploitation, it is expected that the cost of gasoline will continue to rise in the long term. The fall in oil price that occurred in 1986 can be regarded as being within the overall noise in historical prices (Fig. 7–1). However, the trend in the cost of electricity, while still rising, should be somewhat more favorable because of the shift to new, improved, and larger power generating plants, particularly nuclear plants. Recent experience in the United States has shown that the latter can be built for $1000/kW (1984) if the design and construction teams are not constrained by unnecessary demands, and have previously built power plants. The data given below were obtained from EPRI. The best example is the McGuire-2 plant at Duke Power in North Carolina, which was ordered in 1969 and completed in 1984 at a cost of $932/kW (1984), interest on construction capital being 33% of the total cost. The cost of this plant is less than half of the U.S. norm of $2779/kW (1984), including interest charges during construction of about 32.5%, for plants ordered since 1966. Plants such as Shoreham in Long Island, which cost $5192/kW (1984) (of which 35% was interest) show up as being beyond any reasonable engineering estimate for a properly-designed plant on an acceptable

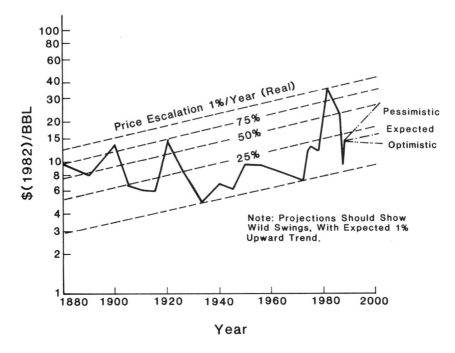

Figure 7-1: Historic escalation of U.S. crude oil prices (log cost in $ (1982)/BBL as a function of time). *Source*: Polydyne, Inc., San Mateo, California: American Petroleum Institute; Energy Information Administration Annual Reports to Congress; Pacific Gas & Electric Co.; *Wall Street Journal*; U.S. Bureau of Labor Statistics.

site. If future plants can be standardized so that engineering and licensing can be made routine, as in the French nuclear program, safe nuclear power using PWR or CANDU reactors should be able to stabilize electricity costs in future. However, the long-term physical, physiological and psychological effects of the Chernobyl accident may change this perception in the future.

Taking this into account, the future operating cost of a battery-powered vehicle should be much more sensitive to the expense and life of the batteries than to the cost of electricity. For fuel cell vehicles, the situation is different. While they of course require a fuel, it need not be petroleum-based. Although the nonconventional (i.e., synthetic) fuels such as methanol or hydrogen that can most readily be used by fuel cells still are relatively expensive, they can use these fuels much more efficiently than IC engines use gasoline. Nevertheless, it may be expected that the operating cost of fuel cell–powered vehicles will be more dependent on fuel cost than that of battery-powered vehicles on electricity cost. Like battery-powered vehicles, however, the operating cost of fuel cell–powered vehicles will certainly be sensitive to the cost and life of the installed fuel cell units.

Fuel Cells Vs. Batteries

The advantages for using fuel cells instead of storage batteries for powering electric vehicles are as follows:

1. The energy density of a fuel cell is greater than that of a battery. On an overall system basis, an increase in fuel cell energy density can be achieved merely by increasing the size of the fuel tank. Taking into account efficiency of use, methanol in a tank for use in a fuel cell represents 1900 Wh/kg; whereas gasoline for an IC engine corresponds to 900 Wh/kg. In contrast, lead-acid batteries are less than 40 Wh/kg, advanced batteries with the most optimistic specifications only being 200 Wh/kg.
2. A fuel cell can be recharged much faster than a battery since it is only necessary to refill the fuel tank, as for an IC engine. Since classical secondary batteries require much longer times to recharge, a battery leasing and exchange system would be required to obtain the same mission flexibility for highway vehicle operations.
3. A liquid fuel distribution network already is present in the form of the extensive network of gasoline stations. These stations could be converted to handle fuels such as methanol, hydrazine, or possibly even liquid hydrogen without too much difficulty. The comparable so-called recharging network that would be required for the large-scale use of battery-powered vehicles with interchangeable battery packs, available like 19th century stage horses, is not currently available.

The disadvantages of fuel cells include problems of the complexity of the currently available systems, together with their volume, weight, cost, and lifetime. The resolution of these problems is anticipated as time and understanding progresses. Already, it is possible to design an alkaline fuel cell (AFC) system using hydrogen fuel that would be competitive in weight and volume with an IC engine, except for the problem of the storage of hydrogen [393] in the vehicle. This has been reviewed in Ref. 172. In the case of methanol, a suitable fuel processing system to give hydrogen—not methanol—storage, is the problem. Both difficulties lie outside the development of the fuel cell itself, and indications are that rapid progress will be made to allow the early development of both system concepts.

Regardless of whether fuel cells or batteries are used, most of the energy that will eventually be required for electric vehicles probably will be supplied either directly or indirectly by electric utilities. For a rechargeable battery system, this is obvious. For fuel cells the most desirable fuel (hydrogen) will probably be economically produced by electrolysis, using relatively low cost energy from nuclear, hydroelectric, and ultimately solar power plants. In fact, if future electric vehicles are powered by fuel cells or by batteries, the overall effect on the power generation industry may well be about the same.[1708]

D. FUEL CELL-POWERED VEHICLES

The feasibility of operating vehicles with fuel cells has been demonstrated during the last twenty years. The first practical vehicle was the Allis-Chalmers fuel cell–powered tractor in 1959, which inspired the well-known Soviet electrochemist A. N. Frumkin to start a fuel cell program in the U.S.S.R..[394] Since then, experimental electric vehicles have been built and powered with hydrogen-oxygen,[90, 91, 92, 1213, 1620, 1705] hydrazine-oxygen,[2474] and hydrazine-air [93, 94, 955, 1420] fuel cells. Various companies have been involved in these efforts, including ASEA of Sweden; Shell Thornton, Ltd. (Shell Thornton Research Center) of the United Kingdom; and ELENCO in Belgium and the Netherlands.[2557-2560, 2562, 2563] In the United States, General Motors Corporation, Union Carbide, and several other companies have been involved.[525] In addition to use in automotive applications, fuel cells also have been considered for use in powering railway trains,[1349] submarine tankers,[1537] and submarines.[97, 107, 632, 779, 953, 1627]

Types of Fuel Cells Best Suited for Electric Vehicle Applications

For applications in the small vehicle sector of the transportation market, where the impact on total energy use will be greatest, the two most attractive fuel cell options are the hydrogen-air alkaline fuel cell (AFC) and the methanol-air acid electrolyte fuel cell.[161, 163, 172, 1486, 1735, 2559]

The advantages of using hydrogen as a vehicular fuel are its high energy density, no need for on-board fuel treatment, and the absence of carbon dioxide emissions at the site of use. The disadvantages include the problems involved in its distribution and storage and the uncertainty of its economics. The advantages of AFCs are that they require little or no platinum electrocatalyst (see Chapter 11), that they have very fast start-up times of approximately five minutes (largely because they are capable of yielding up to 50% of their rated power at ambient temperature), that their structural corrosion problems are relatively few, and that they have high efficiencies. The advantage of not requiring platinum catalyst will be very important if fuel cell vehicles are to have a major market impact: Present platinum production is 80 metric tons per year, equivalent to only 800,000 20-kW automobiles with the present state-of-technology loading of 5 g/kW in acid cells. This represents less than 3% of world annual automobile production.[178]

AFCs have been developed to a very high stage of reliability and sophistication as a result of the U.S. aerospace program. For example, recent space shuttle cells have demonstrated performances as high as 750 mA/cm^2 at 0.83 V, and energy densities of 400 W/kg at the stack level, or 200 W/kg for the complete system,[2142] whereas the more advanced small (2 kW nominal, 8 kg) NASA lightweight cell stack will produce 1.1 A/cm^2 at 0.9 V, or 1.8 kW/kg for the stack and 840 W/kg for the complete system.[394,1706] While these systems, described in Chapter 6, are pressurized H_2-pure O_2 units using noble metal catalysts, derated nonpressurized versions with carbon-transition metal catalysts, offering 300 mA/cm^2 with air are now becoming available. These are discussed in Chapter 12. We should point out that aerospace cells still use platinum catalysts only because these were qualified for manned flight in the late 1970s, based on technology that was available in the late 1960s. (See footnote, p. 175.)

Some disadvantages of AFC systems are that certain engineering aspects such as water management can be complex, that their lifetimes with non-noble cathode catalysts are still somewhat uncertain, and, perhaps most important for practical purposes, that they are intolerant to carbon dioxide, which results in the formation of carbonate in the electrolyte, with loss of performance. AFCs therefore cannot be directly used with reformed carbonaceous fuels on account of the large amounts of residual carbon dioxide in the hydrogen produced. Pure hydrogen must therefore be produced from these gases for AFC use. It is also necessary to remove essentially all the carbon dioxide from the air, which contains 350–400 ppm CO_2. As strange as it may seem, to date, no carbon dioxide removal systems have been demonstrated that are able to meet the criteria of cost, efficiency, weight, and volume required for vehicular applications.[622] In spite of these disadvantages, if pure hydrogen were readily available and easily could be stored on board the vehicle, the alkaline fuel cell would be an obvious choice for vehicular applications. State-of-the-art AFC technology has been reviewed in Ref. 394.

The primary advantages of methanol as a fuel are that it can be synthesized from coal and that its storage and distribution are similar to that of gasoline. Also, synthesis methanol is a clean fuel that requires no desulfurization, and which can be readily steam-reformed at only about 200–250°C. At these temperatures, the CO content of the reformate is low enough to be tolerated by a PAFC, so that a separate water–gas shift reactor is unnecessary. The disadvantages of methanol include the uncertainty of the economics and the generation of carbon dioxide, which precludes the use of the AFC. Reformed methanol yields a mixture containing about 25% carbon dioxide and 75% hydrogen, so that methanol fuel requires the use of acid electrolyte fuel cells with the current state-of-technology.[622] PAFCs have the advantages of being tolerant to carbon dioxide, of having long demonstrated life, of having simple water management, and of being technically in the most advanced state. Their disadvantages include the necessity of using platinum electrocatalysts (common to all acid electrolytes, including the SPE; slow start-up (~15 minutes, on account of the relatively high operating temperature of ~200°C), possible difficulties in building shock- and vibration-resistant stack and component structures, and the rather large volumes and weights of present systems designed for stationary applications.

A third possibility for use in vehicular applications is the solid polymer electrolyte (SPE) fuel cell, which utilizes an acid electrolyte based at present on perfluorosulfonic acid ion exchange membranes (see Chapter 10), typically Nafion® (Du Pont de Nemours). The state of the art of these systems is reviewed in Ref. 176. As has been discussed in Chapter 6, SPE fuel cells were originally developed by General Electric for use in the Gemini space program. The apparent advantages of SPE fuel cells for vehicular applications should be their fast start-up, since they can produce appreciable power at temperatures close to 0°C, their high power density, their reasonable compatibility with reformed hydrocarbon fuels, and their ruggedness.

Solid polymer electrolyte fuel cells do have a major disadvantage. This has usually been expressed as a tendency of the solid electrolyte membrane to *dry out* if care is not taken with water management. However, it goes much beyond this simple statement, since drying out of the membrane, which occurs in certain water vapor pressure ranges, leads to an almost complete loss of ionic conductivity, which will be discussed in more detail in Chapter 10. Because of the chemistry of phosphoric acid, it is possible to operate the PAFC at very low water contents (e.g., that corresponding to a water vapor pressure of 0.2 atm at 200°C), and still obtain good conductivity. Under the same conditions, the SPE membrane would be completely dry, nonconducting, and probably cracked. As a result, the maximum operating temperature of current-technology SPE cells is about 85–90°C under atmospheric pressure conditions, and perhaps 105°C at 4 atm.[176] Under these conditions, their anode carbon monoxide tolerance is low (ppm levels, compared with 1.5% at 190–200°C with the PAFC). Changing the ionic conductivity parameter will require a major change in their chemistry, which will require further research.[176] In the meantime, they need high platinum loadings, and will require a methanator or other chemical reactor to reduce CO levels from a methanol reformer to acceptable levels. Finally, their operating temperature (less than the boiling point of water at operating pressure) means that they cannot use waste heat to produce steam for the reforming reaction. This will inevitably involve a system efficiency disadvantage, since extra fuel will then have to be burned to raise the steam for reforming.

In 1981, the Los Alamos National Laboratory sponsored a study by General Electric to show the feasibility of adapting the SPE fuel cell for use in an indirect methanol-fueled power plant for vehicular applications.[1900] This system used cracking, rather than steam reforming, for methanol processing, and required a methanation pretreatment system to reduce CO content in the fuel to acceptable levels. To obtain the highest possible power density, the system was designed to take advantage of the SPE membrane's ability to operate under differential pressure conditions. The best performance was estimated to be with the anode stream at 10 atm, and the excess air supply at 3 atm to avoid excessive compression work. In the relatively small vehicle unit considered, the air supply system was to use a reciprocating compressor. While they stretched the state of the art, the general conclusions appeared favorable, particularly with regard to overall volume energy density. Even so, it appears unlikely that the SPE fuel cell will be viable for use in vehicular transportation until Nafion®, which currently costs \$400/m^2, with little hope of cost reduction, can be replaced with a considerably less expensive SPE membrane. In addition, the noble metal catalyst loadings of state-of-the-art cells must be lowered significantly.[622] These loadings currently are about ten times greater than those in the PAFC because of the difficulty of obtaining an optimized gas–catalyst interface with a solid polymer membrane.[176]

Indirect ammonia–air fuel cells also have been considered for use as power sources for electric vehicles.[375, 1620, 2139, 2142, 2210] In the proposed indirect ammonia–air AFC system studied at the Lawrence Berkeley Laboratory anhydrous liquid ammonia would be catalytically cracked, the heat for the cracking process being derived from the combustion of anode vent gas. The use of liquid ammonia as a method of on-board storage of hydrogen and the subsequent recovery of the hydrogen content by catalytic cracking results in a hydrogen storage system that contains about 5% by weight hydrogen, which is 2.5 to 6 times higher than possible with metal hydride storage systems.[393] Using currently available materials and technology, calculations indicated that the proposed fuel cell system, operating at 80°C, should be able to generate 20 kW continuous power output, or 50 kW peak power. A peak power density of 141 W/kg and a continuous fuel efficiency of 3.03 kWh/kg of ammonia was envisaged. The latter corresponded to a conversion efficiency of 62.8%, based on the lower heating value of ammonia. Because of carbonation of the electrolyte, the system proposed will require a complete electrolyte replacement (i.e., 2.5 L of 35 weight percent KOH for a 20 kW fuel cell) every 500 kWh, or about every 5000 km for a four-passenger vehicle. In addition, means would be required to remove product water vapor from the anode exit gas before it is burned to provide the energy for cracking. This may need a technological advance; for example, the use of a membrane separation system to remove excess water vapor from the anode exit stream so that this tail gas can be combusted. In addition, since this system has never been built, there are no established manufacturing costs for the proposed hardware. An economic analysis is therefore difficult. The system is described in detail in Refs. 375 and 2139.

As a result of the above discussion, in general, it can be concluded that the alkaline hydrogen–air fuel cell is the most promising system for use as a power source for electric vehicles, especially if the most important criterion is the highest possible efficiency.[1733, 1735, 2559] A discussion of some of the alkaline-hydrogen systems that have been proposed is given below. A discussion of recent suggestions for fuel cell vehicle systems is given in Ref. 172.

The General Motors Electrovan

To demonstrate the technical feasibility of a fuel cell–powered automotive vehicle, in 1967 General Motors converted a 1966 GMC Handivan to operate on hydrogen–oxygen fuel cells.[1705, 2733] The fuel cells used in the Electrovan, which were manufactured by Union Carbide, could operate on both cryogenic hydrogen and compressed gaseous hydrogen. However, they used liquid, not atmospheric, oxygen. They powered a 60 kg, 27 cm diameter × 36 cm long, 93 kW (125 hp) 4-pole, 3-phase induction motor designed to operate up to 13,700 rpm with oil cooling.

The fuel cell system consisted of 32 modules, each having a dry weight of 19 kg, connected in series to provide an open circuit voltage of 520 V, a continuous power output of 32 kW, and a peak power output of 160 kW. The system used recirculating potassium hydroxide electrolyte at 65°C. The final weight of the Electrovan was 3230 kg, of which 1790 kg could be directly attributed to the fuel cell and electric drive systems. In comparison, the standard van weighed only 1480 kg, including a 395 kg power train. The Electrovan was therefore very heavy, although 450~680 kg was attributable to overdesign and to the requirement for test instrumentation. It also must be remembered that the Electrovan utilized 1967 technology.

In spite of the weight, the Electrovan was able to perform comparably with the standard van. Table 7–2 shows some of the performance characteristics of the vehicle.

The Electrovan was certainly characterized by a number of important

TABLE 7-2

Performance Characteristics of G.M. Electrovan[1705, 2733]

Acceleration from 0 to 100 km/h, s	30
Top speed, km/h	112
Fuel consumption, g hydrogen/kWh:	
at 32 kW	45
at 160 kW	86
Range, km:	
cryogenic fuel	160–240
compressed gases	70–100
Energy consumption, kWh/km	~0.6
Overall thermal efficiency, %:	
at rated output	> 50
at peak power	~30
Auxiliary power consumption, kW	3.0
Parasitic power consumption due to shunt currents in electrolyte flow path, kW	2.4

problems. These included excess weight and volume, short lifetime, costly components and materials, a lengthy and complicated start-up procedure, electrolyte water balance and carbonation, system complexity, temperature control, critical gas–electrolyte pressure balance during transient conditions, and finally, safety. The latter included electrolyte and hydrogen leaks, the high system voltage, and the hazards in event of collision. Most of the above can however be attributed to the relatively primitive technology used, and to the circulating alkaline electrolyte.

Despite the problems, the investigators concluded that "... the rate of progress in this field and the strong advantages of fuel cells are sufficient incentives to maintain this effort."[1705] Unfortunately, the automotive industry then undertook little further work to devise an improved fuel cell–powered automobile.

Fuel Cell–Battery Hybrids

Fuel cell–battery hybrid power sources appear to be well suited to overcome both the so-called battery problem (low energy density) and the fuel cell problem (low power density) without having to resort to the addition of an internal combustion engine. During periods of high power demand, such as acceleration, high-power-density batteries will perform most of the

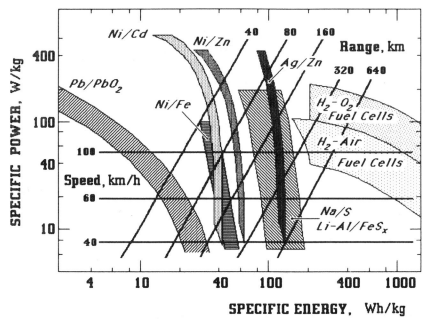

Figure 7-2: Ragone plot for various battery and fuel cell systems.[1486]

work. During periods of low power demand, such as cruising, it will be provided by the fuel cells. The batteries will then be recharged by the fuel cells during periods of low power operation.

Hybrid operation will therefore permit the most efficient use of the inherently high energy density of the fuel cell and the high power density and energy-interchange characteristics of the battery. Figure 7-2 illustrates this by showing a Ragone plot; that is, a plot of power density vs. energy density, for a number of battery and fuel cell systems.[1486]

Other benefits may also be associated with the use of a fuel cell–battery hybrid,[1733] an electric-electric system that is inherently much simpler than the corresponding mechanical-electric hybrid. For example, the battery allows rapid fuel cell start-up and protects the fuel cells against cell reversal during this operation. Batteries also have much longer life under the shallow cycling conditions characteristic of hybrids than they do under deep discharge, as occurs in a conventional traction-battery electric vehicle. A fuel cell–battery hybrid power plant offers the best of all worlds: good performance, long range, fast refueling and long life.

One hybrid design study was carried out for a system employing an indirect ammonia-air MCFC in conjunction with a molten lithium-lithium salt secondary battery system.[892] This study concluded that the use of such a hybrid power source would provide (a) full range capability, (b) excellent

Electric Vehicles

acceleration characteristics, and (c) operation on conventional hydrocarbon fuels. However, whether the use of a thermal-cycling MCFC would be technically suitable for use in anything but the largest vehicles is open to question. Similar calculations have been made for an urban car powered by a hydrazine-air fuel cell and a lithium-tellurium battery.[545] Table 7-3 indicates the expected output characteristics for such a system.

More recently, Engelhard Industries developed a fuel cell–battery hybrid fork lift truck for the United States Army,[435] whose studies have shown that fork-lift trucks are ideally suited for fuel cell–battery hybrid power sources.[1190] The power plant for this 4000-pound-rated lift truck consisted of a 4.3 kW indirect methanol-fueled PAFC unit in tandem with four lead-acid starter (SLI) batteries for peak power demand and start-up duty. Engelhard compared the economics for a 5000-pound-rated fork-lift hybrid system with those of a conventional battery-powered fork-lift truck. Their analysis directly compared the capital costs, operating costs, and cumulative costs in constant dollars for the two systems over a ten-year life cycle. The results showed that, depending on the local cost of electricity, operating cost savings of 30~60% could be realized with the fuel cell system, and that the capital costs of the fuel cell–battery hybrid system would be about 15% lower than for a conventional traction battery–charger- system.

Union Carbide Hybrid City Car

Almost twenty years ago a fuel cell–battery hybrid city car was built and tested as a personal venture by Karl Kordesch of Union Carbide.[689,1479,1480] This car was a 900 kg vehicle based on a 1961 Austin A-40 two-door, four-passenger sedan. The six-kilowatt AFC used Union Carbide components

TABLE 7-3

Calculated Output of Fuel Cell–Battery Hybrid Power Source[545]

Weight of 4.56 kW hydrazine–air fuel cell, kg	70
Weight of 18.55 kW Li-Te battery, kg	28
Weight of fuel and tanks, kg	83
Total weight of hybrid power source, kg	**181**
Power density of power source, W/kg	128
Energy density of power source, Wh/kg	708
Range of vehicle at 6.0 km/kWh, km	770

similar to those used in the GMC Electrovan, with an air-breathing cathode, thereby eliminating the need for liquid oxygen tanks, piping, and the associated controls. On CO_2-scrubbed ambient-pressure air, the polarization of this cathode under heavy load was quite high, but this could be tolerated because peak power demands were satisfied by lead-acid batteries. There is no question that this low-cost project was a milestone in the development of fuel cell–powered vehicles because (a) the vehicle was operated by the designer on public roads as a functional means of transport over a four-year period (1971~1975), during which it was driven about 21,000 km; and (b) the control of the power plant was reduced to an easily-operated fail-safe system.

The electric motor and lead-acid battery were installed under the hood of the vehicle. The motor was a series-wound Baker d.c. unit with two field windings, similar to that used in fork-lift trucks. It was rated at 7.5 kW continuous, 20~25 kW peak power, 4000 rpm maximum speed, and weighed 82 kg. The power train was the standard four-speed manual transmission fitted to the automobile.

The seven 12-V, 84 ampere-hour lead-acid batteries were installed above and beside the motor and connected in series. The total battery weight was 150 kg. A charging unit installed in the car could recharge the batteries from a regular household 117 V a.c. supply, and was automatically disconnected at 10~15% overcharge. An ampere-hour counter installed under the instrument panel registered the instantaneous state of charge of the batteries at all times.

The AFC system was installed in the trunk and consisted of 120 cells in 15 modules arranged in three blocks operating at a nominal 90 volts and producing about 6 kW. The fuel cells were supplied with pure hydrogen from a group of high pressure tanks on the roof, and with scrubbed air by means of a blower. Air was passed through a Plexiglas trough filled with soda lime to remove carbon dioxide. Recirculating aqueous potassium hydroxide electrolyte was the fuel cell electrolyte: It was pumped from a reservoir under the cells and remained there when the system was not operating. This feature allowed the fuel cells to be shut down completely without parasitic currents, with no reactants required to maintain the electrode potentials, and with no hydrostatic pressure on the electrodes. The fuel cells weighed 59 kg, and the accessories (pumps, tanks, blowers, etc.), which all were commercially available items, weighed 23 kg.

The hydrogen supply was stored in six steel cylinders, at 11 MPa (1900 psi), on the roof. The designer noted that the day-to-day transportation of gasoline in normal automobile tanks appeared to be at least as dangerous as the transport of hydrogen gas in steel high-pressure cylinders.

To start the vehicle, all the operator had to do was to turn the key switch. The automobile could be driven immediately on the batteries while the fuel cell was being brought on line by means of a sequence switch device. Conversion from battery to fuel cell power required 1~2 minutes. Emergency shutdown was automatically achieved by throwing a single switch.

Speed control was achieved by means of a gas pedal-type foot lever that closed a series of relays in sequence, which was combined with a normal clutch and gearshift. In operation the car behaved very much like its gasoline-powered predecessor: Inside, it even sounded the same, since most of its internal noise was from the gearbox, which was retained.

The panel instrumentation consisted of meters that displayed the motor voltage and total current, together with the current being drawn only from the fuel cells. The panel also carried the sequence switch and indicator lights, below which was the ampere-hour counter showing the battery condition. A pressure gauge overhead monitored the hydrogen supply, while gauges mounted on each side of the fuel cell, which were visible in the rear-view mirror, indicated air differential pressure and hydrogen supply pressure.

The vehicle could accelerate from zero to 20, 30, 50, 60, and 80 km/h in 2.5, 7, 12, 20, and 35 seconds, respectively. The driving range was about 320 km, which is a significant improvement over even the latest generation of vehicles powered by traction batteries. Kordesch then estimated operating costs at 0.3¢/km, based on a bulk purchase price of 65~90¢/kg for hydrogen ($5.55–$6.55/MMBTU in 1970 dollars). Based on the delivered price of individual cylinders, the cost would be as much as 3¢/km in 1970, equivalent to about 10¢/km in 1986. This corresponds to present delivered hydrogen costs in individual cylinders of about $100/MMBTU (1986), whereas gasoline at 90¢/gal is equivalent to $6.70/MMBTU.

Routine maintenance procedures for the car included battery water replenishment and periodic replacement of the soda lime carbon dioxide scrubber approximately once every 1600 km. Indicator particles in the soda lime changed color when it was exhausted, giving a clear visual indication of the need for replacement. Periodic flushing of the fuel cells with water was found to be desirable to remove the build-up of dried electrolyte at the gas outlets. The manifolds could be flushed with tapwater.

Few problems were encountered with the system. The motor and its control circuitry operated without attention since their original installation. The batteries had a life expectancy of at least two years, and the fuel cells demonstrated lifetimes of thousands of hours of continuous operation. Experience indicated that even longer lifetimes could be expected under conditions of intermittent use.

The only serious problem reported was the development of leaks in the fuel cells, believed to be caused by the use of a mixture of materials with different thermal expansion rates (nylon, Plexiglas, and epoxy). Kordesch suggested that this problem could be solved by using one material throughout for construction, such as polysulfones. However, the leaks did not prove catastrophic either to the fuel cells or to the environment. For example, potassium hydroxide did not attack the vehicle body and could be washed off readily with water.

The significant feature of this vehicle was its practicality. It was essentially the project of an individual (though very knowledgeable) enthusiast, with access to the right materials. Kordesch produced a low-pollution vehicle with a useful performance and range, with an easy-to-operate failsafe system. The technology employed was quite old even in 1970, and present-day electrode construction would significantly improve vehicle performance and range. The use of ammonia as fuel with a thermal cracker to convert it to hydrogen can be considered as a possible means of reducing the fuel tank volume.

The ELENCO Alkaline Fuel Cell Program

A considerable amount of fuel cell research was conducted in both the Netherlands and Belgium during the early 1960s. In the Netherlands this was mostly connected with the pioneering MCFC studies carried out by Broers and his co-workers at TNO, referred to in Chapter 17. In contrast, the Belgian work concentrated on the AFC, and was carried out by the ABEPAC group at SERAI.[2558] In the first half of the 1970s the latter work was continued on a limited scale at SCK/CEN (the Belgian Nuclear Energy Research Center). Following this, a collaborative industrial program was initiated by SCK/CEN, the Belgian company Bekaert, and the Dutch company Dutch State Mines (DSM), who created a new company known as ELENCO (Electrochemical Energy Conversion). The three founding partners and the Belgian and Dutch Governments have made available total annual funding of about $2.5 million to develop hydrogen-air AFC systems for vehicular applications at ELENCO.

ELENCO made the initial decision to develop the alkaline system over acid-electrolytes cells for several reasons: First, they felt that the future energy situation in Europe, unlike that in the United States, would be more conducive to the use of high grade hydrogen as fuel. This naturally gives an advantage to the higher-performance AFC system, which also has the advantage of cold-start capability. ELENCO also felt that acid systems (i.e., the PAFC) might soon be replaced by the MCFC and SOFC systems in utility applications, whereas this was unlikely to happen to the AFC because of its unique suitability for transportation. Finally, they felt that the material and temperature constraints for AFCs were much less severe than those for their competitors. After screening the potential areas of use for the AFC, ELENCO decided to concentrate its initial attention on vehicular applications.

Electric Vehicles

The ELENCO fuel cells employ thin flat gas diffusion electrodes using one or two layers of Teflon-bonded carbon-supported platinum catalyst on a nickel screen for both anodes and cathodes. A combined loading of 0.6 g Pt/cm^2, with the majority of the catalyst on the cathode, was reported in 1981 and 1983,[2560, 2561] More recently, the loading has been stated to be 0.7 mg/cm^2,[2562] with an introductory commercial value of 0.15 mg/cm^2, which it is hoped can be divided by a further factor of three in the future.[394] As indicated in Chapter 5, which briefly discusses the inexpensive cell construction materials, the system uses a monopolar configuration with edge collection and is built into fuel cell stacks with 24-cell modules, with an active unit electrode area of about 300 cm^2 (17 cm x 17 cm). The fuel cells operate at about 65°C using circulating 6 M potassium hydroxide electrolyte. Each module has a nominal power output of about 400–500 W and a weight of about 4 kg, lighter modules being under continuous development. Present power density is 100 W/kg, the ultimate goal being 150 W/kg.

The majority of performance and life testing has been carried out on stationary 0.5 and 5 kW units. In 1981 the specific power density of the cells was about 70 mW/cm^2, and the performance degradation (caused by interaction between the platinum catalyst and the carbon substrate) was considered to be reasonable during 5000 hours of continuous operation. A degradation of 4% per 1000 h has been reported.[394, 2562] It was expected that by 1985 power densities of 125 mW/cm^2 would be achievable with lifetimes in the order of 10,000 hours. The fuel cell modules have been designed to deliver tens of kilowatts and are connected electrically in series and parallel strings yielding a total power plant voltage compatible with electric vehicle traction system requirements; for example, 96 V, 144 V, 216 V, or more, depending on the type of vehicle. Minimum weight and volume and fast maintenance have been important design criteria. The latest 20-module system, which has been tested in a vehicle, is 10 kW, 336 V (0.7 V/cell).

The company has built an automated pilot plant for electrode manufacture, based on an original manufacturing technique that was developed and patented by SCK/CEN in the early 1970s. This plant can manufacture about 250,000 pairs of electrodes per annum, or more than 5000 kW of fuel cells.[394, 2562] ELENCO believe that their electrode manufacturing concept has solved the problem of electrode cost reduction through mass production. They project that by the late eighties they will reach fuel cell unit costs that are acceptable for fleet vehicles, such as city buses and fork-lift trucks, as well as for certain light commercial vehicles. Their aim is a production cost of $200–300/kW for the total alkaline unit. While this cost currently is too high for passenger cars, fuel cells for this application should be possible in a further stage of development of their technology, possibly after the year 2000, depending on the overall fuel situation. However, their present costs may be acceptable for many stationary fuel cell applications (i.e., in the 100 kW to 10 MW range) provided lifetime can be extended to 25,000~50,000 hours. Although this may be a problem in regard to catalyst development, it is considered to be achievable if enough money and time can be devoted to it. Finally, ELENCO's philosophy is that the AFC using hydrogen fuel is the type best suited for vehicles, since it will allow the use of catalysts other than platinum. Since vehicle applications call for lifetimes of about 5000 hours, rather than 50,000 hours, component lifetimes are less important than they are for developers of the PAFC or other fuel cells for stationary applications.

ELENCO is currently working on the development of a hybrid fuel cell–lead-acid battery power plant for use with fleet vehicles such as buses and vans. Their cost estimates are about $6 per unit cell, corresponding to about $200 per kilowatt output.[1933] For a city bus they estimate a total operating cost of $1.50~$1.80 per km (1990 dollars), compared with a figure of $1.50/km for a diesel bus (1979 dollars). A calculation of favorable life-cycle costs for various city vehicles, including garbage trucks, has recently been published.[2563]

U.S. Department of Energy Program in Vehicular Fuel Cells

In 1978, the U.S. Department of Energy initiated a program entitled *Development of Fuel Cell Technology for Vehicular Applications*, involving the collaboration of Brookhaven National Laboratory, Los Alamos Scientific Laboratories, the U.S. Army Mobility Research and Development Command, and Energy Systems Research Group Inc.[1733] Various fuel cells and battery–fuel cell hybrid combinations were examined for this application, and it was concluded that if hydrogen fuel were readily available, the alkaline hydrogen–air fuel cell would constitute a compact power plant with fast start-up capabilities making it suitable for use as a power source for transportation.

As part of this effort, workers at Brookhaven National Laboratory have worked on developing improved alkaline hydrogen–air fuel cells for vehicular applications.[1735] Using Union Carbide Corporation dual porosity nickel electrodes developed for air-depolarized chlor-alkali cells and metal-air batteries, a number of single cell alkaline fuel cells and a 0.5 kW 16-cell hydrogen-air AFC battery module were built and tested. The units were monopolar bicells, and are shown in Ref. 394. The cathodes were catalyzed with platinized carbon (0.2 mg Pt/cm^2) and the anodes with palladium; the active electrode area was 289 cm^2. Using 50% potassium hydroxide electrolyte, the stack was operated at 65°C at a current density of about 100 mA/cm^2 for 11 weeks (40 hours of operation per week).

Process air was scrubbed of carbon dioxide in a scrubber-demister unit containing 50% potassium hydroxide before being fed to the cells.

Power densities of 100 mW/cm^2 were obtained at a unit cell voltage of 0.65 V. At 70°C values as high as 126 mW/cm^2 could be obtained at the same voltage. Operation on oxygen increased the unit cell voltage by about 50 mV. These values should be compared with the value of about 27 mW/cm^2 obtained in 1969 by Kordesch with his fuel cell–battery hybrid car. Power densities in the order of 180 mW/cm^2 were projected for 1985.

A degradation in anode performance was observed after about 8 weeks of operation. This was caused by sloughing of the carbon layer on the hydrogen electrodes, and may have been due to dilution of the potassium hydroxide electrolyte in the pores of the electrode as a result of operation at current densities in the range of 100 to 250 mA/cm^2, since the electrodes were only designed for operation at 50 mA/cm^2. Plans were to improve the performance of these electrodes and to further lower the noble metal content. The use of a flexible epoxy seal (3M Corporation Scotch Weld Structural Adhesive 2216A and 2216B) to join the cell components to the polysulfone frames permitted trouble-free thermal cycling of the 0.5 kW stack.

E. RAW MATERIAL IMPLICATIONS FOR ELECTRIC VEHICLES

The active materials required for the large scale production of rechargeable batteries include elements such as lithium, sodium, sulfur, chlorine, lead, graphitic carbon, nickel, cadmium, and zinc, most of which are relatively abundant, as indicated in Table 7-4.[524] Although there is some question as to whether the ultimate reserves and production of materials such as lead,

TABLE 7-4

Relative Cost and Availability of Active Elements[524]

Element	Pb	Zn	Ni	Cd	Li	Na	Ag	S
Electrochemical equivalent, g/Ah	3.8	1.2	1.1	2.1	0.25	0.9	1.0	0.6
Amount in earth's crust, g/tonne	18	145	88	0.17	72	31,100	0.10	572
Numerical order of amount in earth's crust	36	23	24	66	27	6	67	15
Cost, relative to Pb = 1	1	26	17	25	167	27	264	0.17

nickel, and zinc will allow their use for batteries powering the entire vehicle population of North America, and ultimately that of the world, they are certainly sufficiently available for use in fuel cell–battery hybrid electric vehicles. In any case, range limitations do not make a conversion of the whole of the vehicle fleet to rechargeable battery drive desirable, even with the most ambitious performance projections. The materials that have been used for the construction of fuel cell batteries may however present some difficulties. Elements such as tantalum, niobium, and tungsten that are used in experimental cells cannot be used in large scale production on account of availability and cost. As we have already stressed, the possible fuel cell requirements for noble metal catalysts such as platinum will make a serious impact upon the rate of production of these materials. The development of cheap non-noble electrocatalysts with wide availablity will have an important effect upon the use of fuel cells for electric vehicles.

Availablity of catalysts also concerns availablity of fuels. For example, hydrazine is the best liquid fuel from the viewpoint of storage and reactivity (though not in regard to toxicity). However, the production cost of hydrazine, as well as the energy required for its manufacture, as is indicated in Table 7–5.[2604] With hydrogen, the problem is in distribution and storage, rather than fuel cost as such. Breakthroughs in the synthesis, distribution, and availability of these fuels must be made before the fuel cell can compete on a large scale with the gasoline engine.

F. OUTLOOK FOR FUEL CELL–POWERED VEHICLES

If low cost hydrogen becomes readily available, the AFC will certainly be an attractive candidate for use as a power source for vehicular transporta-

TABLE 7–5

Relative Cost of Fuels for Fuel Cells[2604]

Fuel	Usable Energy, MJ/kg	Relative Cost/MJ
Hydrogen:		
95% pure at plant	118.3	1.0
99% pure in cylinders:	120.0	7.4
Propane (LPG)	47.4	0.5
Gasoline	45.1	0.8
Methanol	21.8	3.3
Ammonia	20.9	3.6
Hydrazine hydrate	12.0	16.8

tion, because of its easy start-up, high performance, and lack of need for materials of low availablity. While there has been little recent development activity in alkaline fuel cells as such in North America, the significant improvements made in recent years in air-depolarized alkaline electrodes for use in the chlor-alkali industry and in certain metal-air systems (e.g., aluminum-air [520,675,676,2174,2175]) can be directly utilized in alkaline fuel cells. Long-lived alkaline fuel cells having specific power densities of 50~60 W/kg are well within the range of near-term technology.[1735] Although this power is more than adequate for cruising, an auxiliary power source such as a battery will be necessary for acceleration. Hence, if this power density cannot be improved, fuel cell–battery hybrid systems will probably be the most effective power sources for electric vehicle applications. Such hybrid power plants offer good performance, long range, and fast refueling, and have the advantage of requiring no particular breakthrough in battery development. However, if the alkaline fuel cell can be made to operate at 200 mA/cm^2 at 0.7 V, which should be readily attainable using recent technology, much higher power densities might be made available using lightweight hardware weighing an average of 2 kg/m^2, which will allow 700 W/kg for the fuel cell stack. Such power densities may allow the use of the fuel cell without a hybrid traction battery.

Although many of the technical obstacles that previously impeded fuel cell development have now been overcome, the economic viability of this technology is still uncertain, so that it appears likely that the first fuel cell–powered vehicles will have to be subsidized fleet vehicles such as city buses. Although the initial costs of electric vehicles using batteries or fuel cells will be higher than those using internal combustion engines (ICEs), for fleet vehicles at least, operating costs may be significantly lower and the service life much longer than for ICEs.[1487] Further work is required to assess initial costs, fuel costs, and maintenance costs.

Finally, the so-called monolithic SOFC (Chapter 18), which may be eventually capable of very high power densities (optimistically 8 kW/kg and 4 kW/L at maximum power) is a possible future contender for vehicular use. Even if these high values are not attained, it is not impossible that the system might reach 1 kW/L under nominal power conditions, so that a 1 cu ft (27 L) cell stack weighing perhaps 20 kg could yield 30 kW (40 hp), from either hydrogen fuel or internally reformed methanol-water mixture.[182] In other words, the motive power of an automobile may in future be obtained from a ceramic multilayer fuel cell that differs little in total volume from that of a present-day ceramic catalytic muffler. The main problem with this exciting prospect (as well of the obvious ones of uncertain manufacturing feasibility and cost; see Chapter 18) is the high operating temperature of the cell, which would produce no power below 800°C and full power when heated to 1000°C. However, rapid crack-free thermal cycling of a very thin ceramic structure with components of matched thermal expansion coefficient is not inconceivable, even though a battery hybrid of small capacity (2–3 kWh) may be needed to give instant drive-away capability and to improve fuel economy on very short trips where the use of fuel to start the cell from cold might be wasteful. Alternatively, the fuel cell might be maintained hot overnight (in a garage) by resistance heating from the utility grid. Otherwise, heat-up from cold might be achieved by burning some fuel in excess air, and passing the hot gases through the cathode gas channels. This exciting development, if eventually successful, could have many broad applications where small size and weight is more important than the highest efficiency.

PART III

Design Considerations for Practical Fuel Cell Systems

In Chapter 2 we discussed the basic operational principles and mechanisms that limit fuel cell performance. Although the schematic unit shown in Figure 1-1 embodies the essential elements of a single cell, it may perform inadequately for a number of reasons. For example, if the overpressures of the reactant gases are too high (i.e., they exceed that of the electrolyte by too large an amount), the gases may bubble through the porous electrodes into the electrolyte. This will result in the loss (and possible recombination) of reactants in the cell, leading to low coulombic efficiency and possible structural damage. It would also reduce the effective area of the three-phase gas-solid-liquid interface, thereby resulting in a much lower cell power density. Conversely, if the gas pressures are too low compared with that of the electrolyte, electrode flooding may occur. This will result in lost electrolyte, reduction in the superficially active area of the electrode, and again, destruction of the three-phase interface. If the cell employs alkaline electrolyte and if either fuel or oxidant gas contains carbon dioxide, insoluble carbonates may form in the electrode, again leading to loss of effective active area. By locally changing wetting characteristics, the presence of precipitated carbonate will again change the nature of the three-phase boundary.

Taking the example of an acid electrolyte, in the configuration shown, hydrogen ions flowing from anode to cathode must cross a wide interelectrolyte gap, so the cell would suffer ohmic and possibly ionic concentration polarization losses. The latter is especially true if the electrolytes are dilute or otherwise have limited buffer capacity.[314,1649] In addition, product water formed as a consequence of the overall cell reaction may further dilute the electrolyte, affecting its conductivity, viscosity, wetting characteristics, and transport properties, further affecting performance. In addition, forming a stack or battery of cells to attain an acceptable system voltage may lead to a far from compact system.

As is customary in engineering, in the design of a fuel cell and of its associated systems, the solution of one problem often results in the emergence of another; thus the procedure becomes one of optimization (i.e., of compromise) so that a solution results that approaches the design goal as closely as possible without creating more problems than it solves.

Because each type of fuel cell has its own particular set of engineering problems with its own set of solutions, it is difficult to formulate a general description of the design process. The following chapters in this section deal in a general way with some of the major issues that must be faced in designing and optimizing fuel cells.

CHAPTER 8

Choice of Fuel and Oxidant

FUELS

A. HYDROGEN

In Chapter 2 it has been pointed out that hydrogen is currently the only practical fuel for use in the present generation of fuel cells. The main reason for this is its high electrochemical reactivity compared with that of the more common fuels from which it is derived, such as hydrocarbons, alcohols, or coal. Also, its reaction mechanisms are now rather well understood,[168,169] and are characterized by the relative simplicity of its reaction steps, which lead to no side products. In both respects, hydrogen is more desirable than hydrocarbons or alcohols (e.g., methanol), whose catalytic reaction rates are not only very slow at temperatures under 300°C, but also lead to the formation of poisoning side-products. At higher temperature, particularly at 1000°C in the SOFC, the oxidation rates of these compounds may become sufficiently rapid, but they tend to irreversibly decompose, yielding carbon. This reaction can be prevented by the addition of steam to the fuel gas, when the carbonaceous material is decomposed by the reforming reaction to H_2, CO, and CO_2. Hence, hydrogen is still the effective fuel even under these conditions.

When ordinary fuels are transformed into hydrogen, the energy density of the product is always less than that of the original fuel, because of the necessity to provide water for the reforming process, whose oxygen appears in the reaction products as CO and CO_2. However, in some cases, this water can be provided by condensation of the reaction product in the fuel cell itself. If the reformate mixture can be converted to pure hydrogen, or if the latter is available as fuel from some other source, then the energy density of the fuel used in the cell is much greater than that of hydrocarbons on a weight basis. Under practical fuel cell conditions, hydrogen has an energy density of about 20 kWh/kg, whereas methane has a value of 10 kWh/kg if water for steam-reforming is recovered from the fuel cell. In contrast, the energy density of methanol-water mixture for use in simple fuel cell systems is only about 2 kWh/kg. However, on a volume basis, gaseous or liquid hydrogen has only about one third of the energy density of gaseous or liquid methane. Similarly, liquid hydrogen has about two thirds the energy density of liquid methanol. Hydrogen is therefore not an easy fuel to handle and store, especially in view of its very low boiling point of 20.4 K, which means that the irreversible work of liquefaction of hydrogen represents about one third of the heat content of the gas using standard methods of liquefaction.[1018]

Pure hydrogen is attractive as a fuel, because of its high theoretical energy density, its innocuous combustion product (water), and its unlimited availability so long as a suitable source of energy is available to decompose water. One of the disadvantages of pure hydrogen is that it is a low density gas under normal conditions, so that storage is difficult and requires considerable excess weight compared with liquid fuels. Reviews of various methods of gaseous hydrogen storage are quoted in Refs. 172 and 393. It is also highly flammable, and as supplied, it is relatively expensive (cf. the costs quoted in Chapter 7 for Kordesch's hydrogen-fueled vehicle, Refs. 689, 1479, 1480).

Its cost and the problems of storage have been major drawbacks to the use of hydrogen in the past. However, as hydrocarbon fuels become relatively more expensive as low-cost reserves diminish, and as the atmospheric CO_2 level rises, increasing the mean temperature as predicted by the greenhouse effect,[645, 789, 2030] hydrogen will become increasingly more attractive as a universal energy carrier and fuel. During the 1970s, a great deal of attention was paid to the concept of a *hydrogen economy* in which hydrogen, generated from water using nuclear, fusion, or solar energy, will become the all-purpose fuel of the future.[99, 104, 393, 762, 1018, 1604, 2197] If such a concept materializes, delivered or *merchant* hydrogen will undoubtedly become relatively cheaper than at present, on account of the economies associated with large-scale production. New or improved methods of producing hydrogen are currently being actively investigated.[100, 102, 103] Instead of the potentially dangerous production of hydrogen from nuclear energy, and the somewhat uncertain problems of solar hydrogen production (including that from hydropower), it is still conceivable that hydrogen produced from coal could solve the greenhouse effect atmospheric CO_2 problem, at least as an interim measure. The key would be to dispose of CO_2 at a central point of hydrogen production, such as in deep strata (e.g., depleted hydrocarbon deposits), or in the ocean depths. This concept of course requires solution of the problem of hydrogen transport from coal-producing areas to those consuming energy.

Table 8-1 presents some of the properties of hydrogen. Methane,

TABLE 8-1

Comparative Data on Various Fuels [1018]

	H_2	CH_4	CH_3OH	NH_3
Molecular weight	2.016	16.04	32.04	17.03
Freezing point, °C	−259.2	−182.5	−97.8	−77.7
Heat of fusion, J/g	58.2	58.6	3.10	332.1
Boiling point, °C	−252.77	−161.5	64.7	−33.4
Heat of vaporization, J/g	445.6	510.0	1,100	1,371
Critical temperature, °C	−239.9	82.1	240	132.3
Critical pressure, atm	12.8	45.8	78.5	111.3
Critical density, g/L	31	162.5	272	235
Liquid density, g/L at temp. of (°C)	77 (−252.77)	425	792 (20)	674 (−33.4)
Vapor density, g/l at temp. of (°C)	1.3 (−252.77)	1.8	—	0.89 (−33.4)
Gas density, g/L at temp. of (°C)	0.082 (25)	0.7174 (25)	—	0.71 (21)
Molar volume @ STP, L/mol	22.420	22.360	0.0396	22.094
Liquid–gas expansion ratio	865	650	—	—
Net heat of combustion at 25°C				
BTU/lb	51,756	21,522	8,572	7,990
kJ/mol	241.8	802.5	638.5	316.3[1]
Specific heat at STP, (J mol^{-1} K^{-1})	28.8	34.1	76.6	36.4
Flammability limits in air, %	4~75	5~15	6~50	15~28
Flammability limits in oxygen, %	4~95	5~61	—	15~79
Stoichiometric mixture, %	29.50	9.47	12.24	21.81
Maximum flame temp., °C[2]	2,200	1,980	—	—
Autoignition temp. in air, °C	571	632	470	651
Thermal conductivity, (0°C) (J cm^{-1} s^{-1} K^{-1}) × 10^4	15.9	2.94	20.9[3]	2.00
Viscosity, μP (0°C)	84.11	102.4	80.80	92.6
Diffusivity in air, (0°C), cm²/s	0.611	—	0.162[4]	0.198
Solubility in water, (0°C), cm³/100 g solvent	2.1	3.3[3]	—	117 × 10³

[1] 15°C
[2] air at 18°C
[3] 20°C
[4] 25°C

methanol, and ammonia fuels also are included for comparison and for later reference.[1018]

B. HYDROGEN FROM HYDROCARBONS

External Hydrocarbon Processing

As already indicated, the difficulties of direct anodic oxidation of hydrocarbons are normally circumvented by externally processing the hydrocarbon fuel to give a hydrogen-rich mixture. Fuel processing of any type is intrinsically undesirable, since it increases overall system cost, complexity, and weight; at the same time decreasing efficiency in most cases. However, unless suitable electrochemical catalysts can eventually be developed to allow direct conversion of hydrocarbon fuels at acceptable rates and reasonable cost, fuel processing of some type must remain an integral part of hydrocarbon fuel cell systems.

The trend will be towards processing heavier distillate fuels such as No. 2 and No. 4 fuel oils and coal-derived liquids. Although the technologies for processing gaseous fuels (such as natural gas) and light liquids (such as naphtha, methanol and ethanol) are relatively mature, much work remains to be done for heavier fuels.[2783] The latter are difficult to desulfurize, and then to reform, since they tend to crack in the vapor state. There are three basic modes by which hydrocarbon fuels can be processed to produce hydrogen, namely steam reforming, partial oxidation, and pyrolysis. These are discussed in the sections that follow.

Steam Reforming

In the steam reforming process the hydrocarbon is reacted with either steam, or in some cases a mixture of steam and air, to produce a hydrogen-rich product gas. Strictly speaking, *steam reforming* refers to the endothermic reaction of the carbonaceous fuel with steam only, whereas reaction with a mixture of steam and air is more properly called *autothermal reforming* In contrast, in the process termed *partial oxidation*, the hydrocarbon is reacted with air. This is discussed later.

Taking the case of methane, steam reforming produces a maximum of four moles of hydrogen gas per mole of carbon after water-gas shift to yield largely CO_2 in the product gas, whereas partial oxidation produces a maximum of only three moles, and autothermal reforming between three and four moles. Steam reforming is therefore inherently the most desirable of the three processes, since it is the most efficient. Although at present, steam reforming can only be easily used to process light hydrocarbons up

Choice of Fuel and Oxidant

to about the molecular weight corresponding to naphtha (C_6~C_8), efforts are being continuously made to extend it to higher fractions. Autothermal reforming can accommodate somewhat heavier hydrocarbons and has, for example, been demonstrated to operate satisfactorily with No. 2 fuel oil.

Catalyzed Steam Reforming

Light Fuels

Excess steam and hydrocarbons in a ratio of 2~3 moles of water per mole of carbon are reacted over a catalyst in a temperature range that is normally between 450°C (inlet) and 800°C (hot zone) to obtain optimum rates. In practice, the temperature in a typical catalyst-filled reformer tube supplied with external heat by burning waste fuel rises from about 430°C at the inlet to about 760°C at the outlet.

A mixture containing hydrogen, carbon monoxide, carbon dioxide, methane, and steam is produced. Assuming for the moment that all the carbon in the product gas is in the form of CO (i.e., at the high temperature limit, where the water-gas shift process is negligible), the ideal reaction for a saturated hydrocarbon can be written:

$$C_nH_{2n+2} + nH_2O \rightarrow (2n+1)H_2 + nCO.$$

For an unsaturated hydrocarbon, it will be:

$$C_nH_{2n} + nH_2O \rightarrow 2nH_2 + nCO$$

Taking the common example of methane, the reaction is

$$CH_4 + H_2O \rightarrow 3H_2 + CO.$$

By way of example, a typical product mixture obtained from reacting vaporized JP-4 fuel at 800°C and 1 atm pressure would be approximately 48% hydrogen, 32% water, 12% carbon monoxide, 7% carbon dioxide, and a trace of methane.[793] Table 8–2 shows the effects of temperature and pressure on the calculated overall equilibrium concentrations.[2702] A relatively low operating temperature might at first sight seem attractive in that it lowers the possibility of carbon deposition on the catalyst via the reaction

$$CH_4 \rightarrow C + 2H_2,$$

which may limit catalyst life. However, this reaction only tends to become important in the case of heavy hydrocarbons. Conversely, carbon formation via the Boudouard reaction,

$$2CO \rightarrow CO_2 + C,$$

TABLE 8–2

Calculated Steam-Hydrocarbon Reaction Equilibria (Vol. %) [2702]
Feed: JP-4* ($C_nH_{2n} + 3nH_2O$)

	Before Water Removal				After Water Removal			
	1 atm		20 atm		1 atm		20 atm	
	527°C	827°C	527°C	827°C	527°C	827°C	527°C	827°C
H_2	34.7	47.9	12.1	42.4	58.8	70.5	31.1	66.8
CO	2.5	12.0	0.4	9.9	4.3	17.7	1.0	15.6
CO_2	12.6	8.0	9.4	8.4	21.7	11.8	24.2	13.2
CH_4	8.2	0.01	17.0	2.8	14.2	0.015	43.7	4.4
H_2O	42.0	32.1	61.1	36.5	0	0	0	0

*Typical composition: $C_{8.53}H_{17.59}$

is much less at higher temperatures. Of more practical importance, the yield of hydrogen is greater at higher temperatures. Although the product gas has a relatively high carbon monoxide content as reformer temperatures increase, this can easily be reduced to very low levels with the formation of more hydrogen by the use of a low temperature water-gas shift reaction (see following), so that the reaction essentially goes to completion in two steps.

Operation at high pressure somewhat lowers the concentration of carbon monoxide but also increases the formation of methane and lowers the yield of hydrogen, since the volume of gaseous products is greater than that of the reactants, forcing the process to the left according to le Chatelier's principle. In any event, operation at high pressure is usually undesirable on account of increased system complexity, although there will often be some fall in cost by reduction in reactor and piping volume. In practice, this will result in a compromise for maximum cost-effectiveness.

For highest efficiency, it is necessary to increase the yield of hydrogen to the maximum value, at the same time decreasing that of carbon monoxide, which has a poisoning effect on noble metal fuel cell anode catalysts. The carbon monoxide is therefore reacted further with additional steam via the water-gas shift reaction,

$$CO + H_2O \rightarrow H_2 + CO_2.$$

At this point, the composition of processed JP-4 would typically be 59% hydrogen, 21% water, 1% carbon monoxide, 18% carbon dioxide, with a

Choice of Fuel and Oxidant

trace of methane.[2049] The water-gas shift reaction is a reversible, exothermic reaction for which practical conversion is limited by thermodynamic equilibrium. This reaction sometimes is carried out in two stages. For example, the dry product gas from the steam reforming of naphtha contains about 70% hydrogen, 11% carbon monoxide, 16% carbon dioxide, and 3% methane.[1081] When this gas is passed over a high temperature iron-chromium shift catalyst at 400°C the carbon monoxide content is lowered from 11% to 4.5~5.0%. Passing it next over a second-stage low temperature copper-based shift catalyst at 200°C further reduces the carbon monoxide content to 0.25~0.35%.[1081]

Even further reductions in carbon monoxide content are possible with the use of a methanation catalyst, which converts any remaining carbon monoxide to methane:

$$CO + 3H_2 \rightarrow CH_4 + H_2O.$$

Table 8-3 shows data obtained by the Institute of Gas Technology for the processing of natural gas.[1762] After methanation, the overall product gas had a composition of 78% hydrogen, 19.7% carbon dioxide, 0.3% methane, 2% water, and 8 ppm carbon monoxide, corresponding to only about 3×10^{-4} volume percent CO. Another possible method of lowering the carbon monoxide in the hydrogen-rich stream is by selective oxidation of carbon monoxide to carbon dioxide without attacking the hydrogen in the fuel gas mixture.

If the reforming and shift reactions are combined, the overall steam reforming process for a paraffin can be expressed as

$$C_nH_{2n+2} + 2nH_2O \rightarrow (3n+1)H_2 + nCO_2,$$

TABLE 8-3

Steam Reformation of Natural Gas[1762]
Feed: 95% CH_4 + 5% higher hydrocarbons
$H_2O/CH_4 = 7.3/1$

	Space Velocity (scm/m³ catalyst/h)	Temperature (°C)	Catalyst	ppm CO in Product
Reformer	250	800	Girdler G-56B	150,000
Shift Reactor	500	500	Girdler G-66	2,000
Methanator	1,000	1,000	Ru on alumina	8

Total system ΔP = 10 cm water

and for an olefin as

$$C_nH_{2n} + 2nH_2O \rightarrow 3nH_2 + nCO_2$$

Although the processes involved in catalytic steam reforming are well understood, they are in fact quite complex and present interesting engineering problems, particularly when they are to be used with small lightweight power plants for mobile applications. In general, the fuel cell anode exhaust gas is burned to supply the heat of reforming, with a burner being needed to bring the reformer rapidly to its operating temperature. This means that a series of pumps, vaporizers, valves, and temperature sensors are required for start-up alone. The burner should be silent, efficient, and able to withstand abuse. Pot burners are silent but cannot be oriented, whereas pressure jet burners are noisy and have small orifices which may become blocked. Aspirator burners have easily varied outputs but require a small flow of relatively high pressure air. In addition, they tend to be noisy.

Once at operating temperature, the problem of uniform temperature distribution in the reactor must be considered. This depends on the design of the burner supplying the heat of reforming and on the heat transfer surfaces between it and the catalyst bed. The use of small fluidized beds may be helpful, but these require gas for fluidization. In addition to carbon formation and sulfur poisoning, other problems commonly encountered in steam reforming include premature reaction before the fuel-steam mixture reaches the catalyst bed, as well as incomplete conversion.[1176]

Heavier Fuels

The steam reforming of simple alkane premium fuels such as natural gas is technically highly advanced. Understandably, there is a great desire to use such fuels as feedstocks for processes where they have greater value, such as for chemical applications. Instead of methane and propane, it would be advantageous to use heavier and cheaper fuels in fuel cell systems. Steam reforming of these alternate fuels is one of the major research challenges in the fuel processing area.

A process that has been thoroughly investigated by UTC is the steam reforming of naphtha.[1081] UTC ran tests in which naphtha (Shell Tolusol-5) was steam reformed for almost 4200 hours with steady fuel conversion rates of 84~86% expressed as carbon converted to carbon monoxide and carbon dioxide, the remainder being converted to methane. No original naphtha or hydrocarbon intermediates were found in the product gas, which had a typical dry gas composition of 70% hydrogen, 11% carbon monoxide, 16% carbon dioxide, and 3% methane. The CO content was further lowered to 0.25% using a two-stage shift reactor. During the test the

Choice of Fuel and Oxidant

system was given 34 start-up and shutdown cycles with no loss in performance or change in pressure drop.

More recently, UTC has examined three types of steam reformer for processing No. 2 fuel oil containing 3200 ppm sulfur.[333] The reactor studied most completely was an autothermal reforming system known as an *adiabatic reactor*. Adiabatic reforming achieves high temperature by the addition of air to the steam–fuel reactor feed,[332] where the enthalpy of reforming is supplied by partial combustion. When coupled with reactant preheating, the heat of combustion is sufficient to raise the catalyst bed into the temperature range for significant activity and to supply the endothermic heat for steam reforming the remaining fuel to hydrogen. Because the reactor runs adiabatically, that is, at a uniform temperature with no hot-spots, the ceramic-lined walls can be expected to have long life. The elimination of pyrolytic carbon deposition, which reduces reactor life, is possible in this case because of the short residence time of the unreacted feed prior to steam reforming in the catalyst bed. Using advanced nickel and metal oxide catalysts, greater than 98% conversion has been obtained at exit temperatures of 925~950°C with O_2/C molar ratios of about 0.4. It should be noted that the above reactor cannot make use of the heat content of the fuel cell anode exhaust stream except for preheating, so that the system as a whole will have a slightly lower efficiency than that of a conventional steam reformer. Burning the anode exhaust steam in the adiabatic reactor is self-defeating, because it results in the recycling of a large amount of CO_2, which drastically reduces H_2 partial pressure.

UTC also studied a *hybrid reformer* using a two-step process involving a primary and secondary reformer; and a *cyclic reformer*, in which heat is generated in the first cycle by combusting anode exhaust gas and is stored in the heat capacity of the bed. In the second cycle, this heat is used to supply the endothermic heat of reforming. Based on cost alone, the cyclic reformer and the hybrid reformer appeared to be about 25% and 15% cheaper, respectively, than the adiabatic reformer. Tests also were conducted with the adiabatic reactor operating on coal-derived liquid fuels.[334] Carbon-free operation comparable to that on No. 2 fuel oil was obtained with hydro-treated distillates from the H-Coal Process and light organic liquids from the SRC-1 Process. However, the use of heavier coal-derived fuels resulted in carbon deposition in the reactor.

In a program sponsored by the U.S. Department of Energy, the Institute of Gas Technology studied the use of *hydrogasification* as a means of steam reforming No. 2 and No. 4 fuel oils.[554, 1740] Hydrogasification refers to the noncatalytic addition of hydrogen to the fuel oil at high temperature and pressure. Using Ni-MoO_3 catalysts, methane was found to be the major gaseous product, along with the deposition of some by-product coke. IGT also evaluated the use of proprietary sulfur-resistant catalysts for reforming unleaded gasoline and diesel oil.[1303] At 860°C and 8 atm pressure, the gasoline underwent 72% conversion to a product consisting of 62.9% hydrogen, 9.2% methane, 13.2% carbon dioxide, and 8.8% carbon monoxide. At 965°C hydrocarbon conversion increased to 98%.

The Jet Propulsion Laboratory (JPL) at Pasadena, California, under contract to EPRI, conducted work on the autothermal reforming of No. 2 fuel oil using small reactors employing commercial alumina-supported nickel catalysts.[1174, 1175, 1177] Typical No. 2 fuel oil compositions were 86.2% carbon, 12.9% hydrogen, and 0.28% sulfur; corresponding to 82% paraffins, 10% olefins, and 8% aromatics. Temperatures as high as 1100°C were developed in the reactor via the fuel-rich combustion process. Hydrogen was produced both by fuel decomposition and by reaction of carbon and carbon monoxide with steam. At the high reactor temperatures used, the catalyst was more tolerant to sulfur and there was less tendency for soot formation than at lower temperatures. A minimum steam-to-carbon ratio of about 0.5 appeared to be necessary to avoid soot formation. Using a catalyst bed consisting of three different catalysts in series, 2.1 moles of hydrogen gas per mole of carbon were obtained. About 80% of the hydrogen was formed by partial oxidation and 20% by steam reformation. Some typical experiments at an air-to-carbon molar ratio of 2.5 and a steam-to-carbon ratio of 0.6 yielded a product gas containing 24% hydrogen, 21% carbon monoxide, 4.8% carbon dioxide, and 0.2% hydrocarbons.

The Jet Propulsion Laboratory also has investigated the steam reforming of No. 2 diesel oil, under support of the U.S. Army, who for logistic reasons preferred to replace methanol with diesel oil for use in their portable fuel cell before the mid-1980s (see discussion in Chapter 6).[1174] A commercial diesel oil containing >98% paraffins, <1% each of olefins and aromatics, and 0.3% sulfur was used. Using alumina-supported nickel catalysts, a steam-to-carbon molar ratio of 5, and a catalyst bed temperature of 700~1040°C, a 113-hour run produced gas of composition 69% hydrogen, 21% carbon monoxide, 10% carbon dioxide, and 0.1% methane with no soot formation.

Under contract to EPRI, Kinetics Technology International Corporation (KTI) of Pasadena, California, evaluated two high temperature steam reforming processes proposed for use with a 5 MW utility PAFC system designed to operate on No. 2 fuel oil.[554, 2462, 1675] This work was an extension of earlier EPRI-sponsored work by Catalytica Associates.[691] The system heat rate target was 9.5 MJ/kWh (9000 BTU/kWh). The first fuel treatment system evaluated was the French Grande Paroisse Fluidized Bed Process, which is able to handle feedstocks as heavy as No. 2 fuel oil containing up to 4.0% sulfur. Using a nickel-based catalyst with 60~630 μm particles, this process can operate at steam-to-carbon ratios as low as one, compared with

Choice of Fuel and Oxidant

values of 3~4 for conventional processes, thereby giving a higher efficiency, hence a lower system heat rate. Another major advantage of a fluidized bed process is that carbon deposition is not as serious a problem as it can be with a fixed bed reactor. In the fluidized bed, carbon-precursors are rapidly re-gasified as they circulate through different parts of the bed where the partial pressures of the steam and hydrocarbons are not favorable for carbon deposition.

Another process evaluated was Toyo Engineering's THR (Total Hydrocarbon Reforming) Process.[1172, 1676] This is known as a high temperature steam reforming (HTSR) process because it operates at higher than conventional temperatures to allow conversion of heavier fuels. The Toyo THR reactor is similar to a conventional tubular steam reformer, but is characterized by the use of two separate catalyst beds. The catalyst used in the first part of the reactor is calcium (or beryllium) oxide–aluminum oxide that is totally resistant to sulfur and highly resistant to carbon deposition, but which has a relatively low reforming activity. The catalyst used in the second part of the reactor uses the same material as a support for nickel to increase the reforming activity to a level that is still consistent with high resistance to sulfur poisoning. This system appears promising for processing fuels of No. 2 fuel oil type. At reactor temperatures of 820~980°C feedstock sulfur levels of 1500~2300 ppm can be tolerated, although twice as much catalyst is required as for processing naphtha.[1172, 1675, 1676, 1785, 2462]

Unfortunately, analyses of the above processes indicated heat rates of about 10.6 MJ/kWh (10,000 BTU/kWh) and 12.7 MJ/kWh (12,000 BTU/kWh) for the Toyo and Grande Paroisse processes, respectively, using the PAFC technology available at the time of the study (roughly equal to the 4.5 MW demonstrators). Neither process could therefore meet the desired value of 9.5 MJ/kWh, the reason being the excessive partial oxidation of the fuel. However, it was proposed that a heat rate of 9.4 MJ/kWh (8930 BTU/kWh) might be achievable without a bottoming cycle through the use of a hybrid system combining a Toyo HTSR with an autothermal reformer (ATR).[2462, 2623] The use of such a combined system transfers duty from the HTSR system, which has a rather low thermal efficiency of 60~71%, to the ATR system, which has a thermal efficiency close to 100% on account of its much higher operating temperature and its use of an adiabatic reactor. In this hybrid system, the heat content of the fuel cell anode exhaust stream is not wasted, since it can be used in the HTSR, thereby improving overall efficiency. It was estimated that the potential cost for such a fuel processor might be of the order of $136/kW for the hundredth unit.[1785]

Engelhard Industries Division of Engelhard Minerals and Chemicals Corporation of Edison, New Jersey, have investigated the autothermal steam reforming of No. 2 fuel oil under contract to the U.S. Department of

Energy.[2756] They found that rhodium-platinum catalysts appear to be able to steam reform light olefins with little coke formation. How cost-effective they will be, remains to be seen however.

Methanol

Engelhard has also studied low temperature (205°C) steam reforming of both pure and technical grade methanol.[2755] Using a Cu–ZnO reforming catalyst, 100% methanol conversion could be maintained for several thousand hours. Using their Pd–Al_2O_3 catalyst, initial conversions of 100% were obtained with pure methanol, but at faster catalyst ageing rates than those of Cu–ZnO. Fuel purity appeared to be critical for high initial catalyst activity, higher alcohols being temporary (i.e., reversible) poisons for Cu–ZnO catalysts.

Since late 1979 until 1986, ERC, under contract to the U.S. Army MERADCOM, has developed portable methanol reformers to provide fuel for the Army's 3 and 5 kW portable air-cooled PAFC power plants.[4] They redesigned their earlier tube and shell reformer intended for a 1.5 kW fuel cell, to give a reactor with an dual catalyst bed around an annulus, the burner being inside the annulus. This design gives better thermal efficiency and a smaller infra-red signature than earlier units. The reformer operates at 232°C, and is fed with a premixed fuel consisting of the standard 58 weight % methanol in water. A commercial shift catalyst is used in the reactor bed. After a rather large initial fall in effectiveness, the activity of this catalyst stabilized for more than 1600 on-off cycles during more than 2400 operating hours.

Also under contract to the U.S. Army, JPL investigated steam reforming of methanol contaminated with various higher impurities such as gasoline, ethanol, 1-propanol, and 1-butanol.[2607] The results showed that the conversion efficiency of methanol was somewhat lower in the presence of the contaminants, but no soot appeared to be formed in the catalyst bed. In addition, increasing the steam-to-carbon ratio or the temperature improved methanol conversion efficiency.

Under an EPRI contract, Exxon Research and Engineering Co. made a detailed study of steam reforming of methanol, as well as that of naphtha.[788] Hydrogen production from methanol also has been studied by German,[400] Japanese,[2423] Russian,[2039] and other [1119, 1746, 2055, 2627] workers. Steam reformer designs have been patented by UTC [2377, 2491, 2640] and by ERC.[238] The U.S. Army has reported details of a hydrocarbon fuel processor running on common hydrocarbon fuels.[553] Exxon Research and Engineering Co. have patented cobalt and molybdenum oxide catalysts for the water-gas shift process.[2238] More recently, Los Alamos examined a compact methanol reformer design intended to supply an ERC 20 kW PAFC

Choice of Fuel and Oxidant

as a vehicle power source.[1194] It was clear that a considerable reduction in overall volume could be achieved if internal heat-transfer could be improved compared with that in a reformer of annular type. The latter had such a poor catalyst utilization that over 220 kg of commercial Cu–ZnO was required in a unit rated at 20 kW. The improved Los Alamos reformer design has reduced this by a factor of ten by the use of a design which has some of the features of an adiabatic reformer, but also contains an internal fan to improve heat transfer. This design also may be advantageous in other applications. Other work recently conducted at Los Alamos has thrown light on the operation of the Cu–ZnO reforming catalyst, which apparently functions via a H_2–H_2O redox mechanism, the rate of which depends on the relative gas partial pressures. Only part of the bed will therefore operate at maximum activity at any given time.[1194] This analysis suggests that reforming rates would be improved by the use of multilayer catalysts, each with redox properties optimized for the various stages of reforming. The mechanism should also apply to hydrocarbons and higher alcohols, and may suggest methods of improving the high-temperature reforming of these compounds.

Since, steam reforming, followed where necessary by water-gas shifting, offers the highest yield of hydrogen per mole of hydrocarbon reactant, it is the method of choice in applications where size and weight are not rigid constraints and where there is a supply of pure water available. The latter can be added to the fuel stream via metering from a fuel cell product water recovery system, or in the case of ethanol or methanol it can be premixed with the fuel.

Uncatalyzed Steam Reforming

In the 1970s, a short experimental program was conducted by ERC to demonstrate the feasibility of operating a reformer at 1100°C without using a catalyst.[793] Such a system might have a distinct advantage over previous reactor designs because its small thermal mass would allow faster system start-up. In addition, the absence of a catalyst would simplify the problem of sulfur removal from the fuel. In this connection it should be remembered that the PAFC anode can tolerate about 50 ppm of total sulfur (as H_2S and COS) under normal operating conditions,[2146] whereas a typical steam-reforming catalyst can only tolerate a fraction of a ppm.

Two test runs were made using a reactor in the form of a silicon carbide tube, one at approximately 25% of rated flow and the other at 50% of rated flow. The results, shown in Table 8-4, indicated that JP-4 fuel was completely steam reformed to lighter components in the high temperature reactor without the use of a catalyst, although the composition predicted

TABLE 8–4

Dry Gas Analysis for High Temperature Reformer Tests [793]

Run	Mole Fractions			
	CO_2	CO	CH_4	H_2
1. 25% Flow	0.024	0.268	0.052	0.656
2. 50% Flow	0.024	0.282	0.132	0.560
Equilibrium values at 1100°C, 1 atm.	0.062	0.252	0.000	0.686

by calculations of chemical equilibrium was not attained. This probably indicated that the fuel mixture had too short a residence time in the hot zone of the reactor. However, the test demonstrated the feasibility of the concept. It was observed that in a practical operating device, care must be taken not to expose the reacting gas to any cold surfaces before conversion was complete, to avoid carbon by-product formation. In addition, a longer residence time would be required.

The chemical vapor–deposited silicon carbide reactor did not show any deterioration during the tests, and it was concluded that it would be an excellent construction material. The results were used for the preliminary design of two portable hydrogen generators. The first had a production capacity of 0.2 kg of hydrogen per hour, sufficient for a 500 W fuel cell, whereas the second was designed for a 5 kW unit. Both systems were designed to steam reform JP-4 with subsequent separation of pure hydrogen for the fuel cell by the use of a pressurized silver-palladium hydrogen diffusion diaphragm separator.

Partial Oxidation

Hydrocarbons may easily be broken down using a partial oxidation process in which the fuel is incompletely burned with oxygen or air. The reaction products are ideally hydrogen and carbon monoxide:

$$2CH_4 + O_2 \rightarrow 2CO + 4H_2$$

The use of pure oxygen for this purpose is undesirable because of its cost and the weight and volume required for its production and/or storage. The partial oxidation process produces a fairly low quality product in terms of heating value per unit volume when the reaction is carried out

Choice of Fuel and Oxidant

with air at a temperature of 875~975°C, because of the high nitrogen content. Overall, it only contains about one third as much hydrogen by volume as that produced by steam reforming.[1988] A typical product mix might contain 14% hydrogen, 19% carbon monoxide, 3% carbon dioxide, 60% nitrogen, and 4% methane.[2450]

Advantages of partial oxidation over steam reforming, at least for certain applications, are that water is not required for the process, the equipment is compact, allowing rapid start-up, and the high-temperature reactor can tolerate the presence of many impurities, including sulfur. In fact, partial oxidation may be the only reasonable method of treating very heavy fuels with high sulfur contents.[1599] One disadvantage is its thermal inefficiency, since it is entirely exothermic and cannot make use of any waste heat, for example from fuel cell anode exhaust. Consequently, after shifting, it will produce less hydrogen from the fuel than steam reforming followed by shift, in the ratio $(2n+1)/(3n+1)$ for the general saturated hydrocarbon C_nH_{2n+2}. In addition, care must be taken to select and maintain the appropriate conditions within the reactor (especially the oxygen-to-carbon ratio and the temperature) to prevent carbon deposition. Since the process is exothermic, high temperatures will be encountered in parts of this system, giving possible materials problems, although Inconel 600 and Incoloy 800 have performed satisfactorily, at least for short periods of time. Provided that its sulfur content is low, the product gas is suitable for direct use in the MCFC, which has the capability of internal shifting. However, it requires upgrading, such as by two-stage shifting, for satisfactory use in the PAFC. For low-temperature acid fuel cells (e.g., SPE and fluorinated sulfonic acid cells operating at a maximum of about 90°C), a methanation step will be necessary to reduce the CO content to acceptable levels.

Partial oxidation, while less complex than steam reforming from the system viewpoint, still requires several subsystems. Heat exchangers are required to bring air from ambient temperature to 785°C, at which point fuel is added. The mixture passes into the reactor, which is maintained at about 970°C. The products of the reaction are sent through a cleanup unit to remove sulfur and lead compounds which can affect Ag-Pd diffusers (for pure hydrogen production) or fuel cell anodes. Partial oxidizers have found application in supplying hydrogen to some systems until the primary steam reformer can be started and brought up to operating temperature.[1232]

Pyrolysis (Thermal and Thermal-Catalytic Cracking)

Hydrocarbons can be broken down to hydrogen and carbon by heating at elevated temperatures in the absence of air:

$$C_xH_{2y} \xrightarrow[\text{catalyst}]{\text{heat}} xC + yH_2$$

This process may be conducted with or without a catalyst, although in the latter case higher temperatures are required to obtain the same yields.[793] Furthermore, uncatalyzed thermal cracking tends to result in the production of methane, which is difficult to crack further.[550] For these reasons, the majority of work has involved catalyzed pyrolysis.

The U.S. Army MERADCOM has studied the catalytic pyrolysis of a range of fuels.[550,551] The hydrogen-producing reaction was carried out in a reactor packed with catalyst (see following), together with inert packing material (alumina) at 850°C to 1200°C. During the reaction the heat supplied resulted in the conversion of the fuel to carbon and hydrogen, the carbon being deposited inside the reactor. The latter had no moving parts and was capable of operating over a fairly wide temperature range on a variety of liquid or gaseous hydrocarbons, proving to be both simple and versatile. The process was quite efficient from the conversion viewpoint: At fuel space velocities on the order of 230 cm^3 of fuel per 1000 cm^3 of reactor volume per hour, it was capable of conversions in excess of 95%, with a product stream purity of over 90% hydrogen.[551] The thermal efficiency was however rather low, even compared with partial oxidation, since the heat of combustion of the carbon to CO is lost. However, in contrast to partial oxidation, anode exhaust gas from the fuel cell can be used to provide the enthalpy for the endothermic cracking process. Once cracking range had been achieved in the range 850~1200°C, temperature had much less influence on rate than did fuel space velocity. Table 8–5 gives product compositions averaged over a range of fuels.

TABLE 8–5

Average Product Compositions for Catalytic Pyrolysis[551]*

Fuel Flow Rate (cm^3/min)	% H$_2$	% CO	% CH$_4$	% H$_2$O
2.6	95.3	2.8	1.3	0.7
4.8	92.6	3.6	3.7	0.1
7.7	89.3	2.1	8.5	0.2
12.0	87.4	1.6	10.8	0.1

* Approximate compositions of fuels used:

Combat gasoline	$C_{7.36}H_{13.5}$
Sunoco 190 gasoline	$C_{7.32}H_{15.15}$
JP-4 Jet fuel	$C_{8.53}H_{17.59}$
Diesel fuel	$C_{15.0}H_{24.76}$

Choice of Fuel and Oxidant

The best catalysts proved to be nickel, cobalt, chromium, and iron; nickel being finally chosen because of its stability and availability. Although sulfur compounds attack nickel under reducing conditions, this was shown not to be a problem because the reactor design was such that most of the sulfur appeared to be deposited with the carbon in a noncatalyzed alumina-packed initial zone of the graded bed. Thus the fuel entered a zone where pure thermal cracking could first occur, giving carbon, some hydrogen, and lighter hydrocarbons; after which the gas moved to a nickel-catalyzed zone in which catalytic cracking took place. This bed design resulted in negligible plugging of the reactor, the elimination of severe longitudinal temperature gradients, and a better quality product gas.

The endurance of these graded beds was only several hundred hours. Experiments confirmed that sulfur would not be a major problem for a 1000~1500 hour design life, a possible exception being for continuous operation on 0.6% sulfur diesel fuel. Results indicated that the nickel catalysts were not only resistant to catastrophic activity loss by poisoning, but also had some self-cleaning capability and could withstand considerable abuse without degradation in performance. Like sulfur, heavy metal contaminants such as lead were deposited in the noncatalyzed upper zone of the reactor, and posed no problems with regard to catalyst poisoning.

For continuous use with a fuel cell, some provision must be made for removing carbon particles from the bed. This may be effected by alternately exposing it to air, then to fuel. The air cycle oxidizes the carbon to carbon monoxide and carbon dioxide, providing heat for subsequent cracking in the fuel cycle. Practical systems have been built, in which two parallel catalyst beds are arranged so that one bed is being cleared when the other is operating.[796]

The advantages of catalytic pyrolysis are that it can tolerate fuel additives or impurities such as sulfur and lead, water is not required, it is simple, and it can yield a hydrogen stream of purity greater than 95%. However, it has several major problems. These involve the optimum method of engineering an integrated cracker unit as part of the overall fuel cell system, as well as materials degradation in the extreme environment in the reactor. Perhaps the most important drawback of pyrolytic reactors is the limited amount of operational experience that has been obtained with them as practical devices.

Thermocatalytic cracking was the fuel conditioning process chosen for use with a family of general purpose, tactical PAFCs,[956, 2050] which could operate directly on impure hydrogen fuel.[35] They could tolerate large amounts of impurities such as nitrogen, carbon dioxide and methane, the only adverse effect on performance resulting from lowered hydrogen partial pressure. In addition, these PAFCs could tolerate up to about two volume percent carbon monoxide before significant (and reversible) anode performance loss due to poisoning was observed. United Technolo-

TABLE 8-6
Results of Processing No. 6 Fuel Oil[1] Using Different Pretreatment Methods[(554)]

Composition of Product Gases, mol%	Partial Oxidation (O_2-blown)	Catalytic Cracking	Pyrolytic Cracking	HDS–HC Reforming[2]	HG–Steam Reforming[3]	Autothermal Reforming (Air-blown)
H_2	46.5	46.5	43.4	48.8	54.5	28.4
H_2O	—	35.0	40.7	30.8	29.0	19.8
CO	47.3	13.5	13.8	13.7	12.1	13.1
CO_2	4.3	3.0	1.4	6.7	4.4	7.2
CH_4	0.4	0.8	0.3	—	—	0.1
N_2	1.4	1.2	0.4	—	—	31.4
Thermal Efficiency	52~65	52~61	67 (max)	61~79	65~80	80

[1]84.6% C, 11.3% H, 3.5% S
[2]Hydrodesulfurizing–Hydrocracking Reforming
[3]Hydrogasification–Steam Reforming

Choice of Fuel and Oxidant

gies developed a hydrocarbon catalytic cracker employing nickel catalyst deposited on a porous ceramic substrate,[2250] and the U.S. Army has reported details of a thermocatalytic cracker that converts JP-4 fuel to hydrogen at temperatures lower than 900°C.[552]

Table 8–6 presents a comparison of the product compositions to be expected when No. 6 fuel oil is processed according to the various techniques discussed above. The process selected for a given application will depend largely on the relative importance of factors such as desired weight and volume, cost, reliability, efficiency, ease of maintenance, and pure water availability.

Hydrogen from Coal

Coal Gasification

As discussed in Chapter 3, a large-scale program to develop MCFCs and SOFCs has been under way in the United States since about 1980. These fuel cells are intended to operate on hydrogen produced by the gasification of typical coals, such as Illinois No. 6. In coal gasification, slurried coal is reacted with oxygen and steam to produce a mixture consisting primarily of carbon monoxide and hydrogen. The main types of coal gasifiers under study include those using high-temperature fluidized beds, low-temperature fixed or fluidized beds, together with high-temperature entrained flow systems and molten salt reactors.[427]

Entrained flow gasifiers (e.g., Texaco) are now commercially available, and are used in the Cool Water prototype integrated coal-gasifier combined cycle (IGCC) generating plant in Southern California.[631, 1155] Both high-temperature fluidized bed and entrained flow reactors produce a high-quality gas, together with a low output of hydrocarbons and tars. However, they may not result in the highest system efficiencies when used with fuel cells, and these reactors have yet to be fully integrated into fuel cell system design concepts. The Rockwell molten salt gasifier, using molten sodium carbonate, has been shown to produce clean gas on a pilot scale, but its development has yet to go beyond this stage. Low-temperature fixed or fluidized bed reactors are economically attractive but they produce relatively high levels of light hydrocarbons and tars which must be upgraded for fuel cell use.

High- and low-temperature reactors differ in their oxygen requirements and in their efficiency in converting coal to BTUs in the form of gas. Coal undergoes less partial oxidation in the low-temperature type, giving a higher energy conversion efficiency, but the output gas may have a methane content of 10~15%. Such reactors may be very effective when used with the MCFC incorporating internal reforming capability, or with the SOFC. The molten salt gasifier, which produces little or no methane or tars, can be regarded as a catalytic low-temperature system.

Oxygen can be supplied pure, as oxygen-enriched air, or as ambient air. The use of pure oxygen or oxygen enrichment requires a dedicated energy-intensive liquid oxygen plant and is clearly more costly than gasification with air. However, it yields higher BTU product gas with little or no nitrogen content (see Table 8–7), which can increase fuel cell efficiency sufficiently more than to compensate for the energy requirements for oxygen separation from air. As an example, a coal-water slurry at 60°C, together with oxygen at 150°C and 50 atm, are fed into a typical entrained bed oxygen-blown gasifier (Texaco) operating at 40~45 atm and 1350~1400°C.[378] Typical oxygen/water/coal feed ratios are 0.84/0.44/1.00. As in all coal gasification systems, gas cleanup subsystems to lower hydrogen sulfide and carbonyl sulfide (COS) contents to acceptable levels are required. For use in the PAFC,[2146] and possibly the SOFC, cleanup to 50~80 ppm (or to the highest level that emissions regulations permit, if below this range) will suffice. However, in the MCFC, cleanup to less than 1 ppm is required.[2104, 2105, 2243, 2293]

The major problem with coal gasifiers is one of system efficiency. If a greater yield of hydrogen could be achieved, the overall fuel cell system efficiency would correspondingly improve. Preheating of the gasifier feed using fuel cell waste heat helps in this respect, but a decrease in the carbon-to-oxygen feed ratio may result, leading to some soot and methane formation, but a higher coal-product BTU conversion. The presence of soot in the exit gas requires the use of special (and unproven) heat exchangers while the presence of methane lowers the efficiency of conventional fuel cell subsystems unless further processing is used. As indicated in Chapter 3, internal reforming MCFCs and SOFCs can consume the methane, giving the highest overall system efficiency due to the lower degree of coal oxidation.

Since 1972 Westinghouse Synthetic Fuels Division (Madison, Pennsylvania) has developed a technically and economically viable pressurized fluidized bed coal gasifier.[2707] Under contract to the U.S. DOE and to EPRI,

TABLE 8–7

Percent H_2 Composition of Coal Processed by Various Methods[378]

Air-Blown Coal Gasification	16~30
Oxygen-Blown Coal Gasification	36~50
Partial Oxidation	45
Reformed Naphtha	53
Reformed Methane	60

Westinghouse evaluated this system for use with a 7.5 MW PAFC system using the DIGAS-cooled fuel cell concept in conjunction with a technology transfer agreement with ERC. Their gasifier is an efficient, single-stage, air- or oxygen-blown reactor capable of converting coal to low- or medium-BTU gas. In the process, coal is ground, dried, sized, and fed to the gasifier along with recycled char fines from the particulate recovery system. Steam produced by heat recovery is fed with oxygen to the combustion zone of the reactor, which combusts and gasifies the coal at 1010~1038°C. The process takes place with a high carbon conversion efficiency of 97~98%, no tars or oils being present in the product gas. The latter contains from about 5 mole percent hydrogen when the process is air-blown, to about 30 mole percent hydrogen when oxygen-blown. Following a shift reaction, the hydrogen content of the air-blown and oxygen-blown product gases is upgraded to 30 and 50 mole percent, respectively. Assuming a hydrogen utilization at the PAFC anode of 0.8~0.9, these hydrogen concentrations lower the overall fuel cell voltage from that obtained with pure hydrogen gas by only 38~45 mV and 25~35 mV, respectively.

The Westinghouse gasifier is claimed to be simple, controllable and safe to operate on all types of coal, including reactive western lignites, sub-bituminous coals, coking eastern coals, high-ash coals, and run-of-mine coals. The production cost (1980 dollars) of medium-BTU gas using this process was estimated to be about \$3.70/MJ (\$3.90/10^6 BTU). When used with the 7.5 MW PAFC system with an additional steam reforming step, the expected heat rate based on the gas produced has been shown to be 8.99 MJ/kWh (8520 BTU/kWh). Without steam reforming a value of 9.61 MJ/kWh (9110 BTU/kWh) was obtained.

Detailed evaluations of coal gasification technology have been made by TRW Inc.,[2500] Physical Sciences Inc.,[1603] and the Indian Institute of Technology.[240] A recent comprehensive review of gasification technologies, including research needs for future developments, is now available.[1990, 2276]

Coal-Assisted Electrolysis

Couglin and Farooque of the University of Connecticut have reported a novel electrolytic method of gasifying coal.[679, 680, 829] In this process, finely ground coal is slurried in an electrolyte solution of 3.7~5.6 M sulfuric acid and electrolyzed using platinum or graphite electrodes. The primary reactions that take place are:

Anode: $C_{(s)} + 2H_2O_{(l)} \rightarrow CO_{2(g)} + 4H^+ + 4e^-$

Cathode: $4H^+ + 4e^- \rightarrow 2H_{2(g)}$

Cell: $C_{(s)} + 2H_2O_{(l)} \rightarrow 2H_{2(g)} + CO_{2(g)}$

The thermodynamic standard reversible potential for the overall cell reaction at 25°C is 0.21 V (assuming coal to be chemically equivalent to graphite, as a first approximation), compared with a value of 1.23 V for conventional water electrolysis. Correspondingly, the practical cell voltage for the process, including overpotentials on non-optimized electrodes, was found to be 0.8~1.0 V at 25°C at the current densities used. This may be compared with a value of about 2 V for conventional water electrolysis under the same non-optimized conditions. In effect, part of the energy required to electrolyze water to produce hydrogen is supplied by oxidation of the coal. This reaction depolarizes the anode, thereby lowering the electrical energy requirement.

This process was examined using three coals, one lignite, and one char, ranging in sulfur contents from 0.3~3.6% and in heating values from 19.1~31.9 MJ/kg (8240~13,740 BTU/lb). Using coal ground to a particle size of <250 μm as a slurry in 3.7 M sulfuric acid electrolyte, a cell voltage of 0.83 V at 25°C was obtained using platinum mesh electrodes at a current density of 0.23~1.38 mA/cm^2 geometrical area. As expected, the reaction rate increased both with increasing cell voltage and with increasing temperature. For example, at a cell voltage of 1 V, the apparent activation energy over the temperature range 25~114°C was 38~50 kJ mol^{-1} K^{-1}. At 1.2 V and 114°C, 60 mA/cm^2 was obtained. Lignite was found to be 2~3 times as reactive as bituminous coal, whereas solid graphite and graphite felt electrodes, which are known to be chemically inert, performed similarly to platinum electrodes. Reducing the particle size of the coal to 60~70 μm significantly increased the reaction rate by increasing particle surface area. Sulfuric acid appeared to be the most effective electrolyte, while phosphoric acid, although more stable at higher temperatures, gave much lower performance. At low temperatures trifluoromethanesulfonic acid yielded reaction rates comparable to those using sulfuric acid, but it could not be used at temperatures greater than about 80°C. A further advantage of sulfuric acid was its higher conductivity than that of the other acids: At optimum concentrations at 40°C, the conductivity of sulfuric acid is about 1.8 and 3.7 times that of TFMSA and phosphoric acid, respectively. The same process was found to apply in alkaline solution, but carbonation rules out serious consideration of alkaline media.

The potential advantages of coal-assisted electrolysis are that it can be made to proceed at relatively low temperatures, and that essentially pure hydrogen is automatically separated from carbon dioxide and carbon monoxide. At the cathode, hydrogen was produced at 100% current efficiency, and 85~90% of the anode gas consisted of carbon dioxide with 10~15% carbon monoxide. Mass spectrometric analyses of both anode and cathode gases showed no traces of hydrogen sulfide or sulfur dioxide, even though the coal contained sulfur. This was presumably oxidized to sulfuric acid. Other gaseous impurities such as hydrocarbons, tars, and particulates were also absent, so that gas cleanup subsystems required with other methods of coal gasification should not be required.

However, the experimental technique is certainly not free from problems, and is a long way from being a practical process. A preliminary estimate of its energy efficiency at 1 V and 23°C showed it to be only about half that for conventional coal gasification when the energy necessary to produce the electricity is taken into account. However, operation at higher temperature, yielding a lower cell voltage will significantly improve efficiency. In addition, the energy losses involved in gas separation and gas cleanup are avoided.

A further disadvantage is the fact that not all the coal will react, so a use must be found for the by-product material. The coal has been demonstrated to gradually lose reactivity as it is consumed. Thus at 114°C the cell voltage had to be gradually increased from an initial value of 0.8V to about 1.1V after 300 hours of cell operation to maintain current density. During this period only 20% of the coal was consumed. Loss of reactivity may be attributed to the formation of oxygen-containing functional groups on the surface of the coal particles or to the build-up of a tar-like coating on the coal. If the former is the explanation, it should be possible to maintain reactivity by operation at much higher temperatures (200~600°C), where such oxygen-containing compounds are known to decompose. This will however result in great challenges in respect to materials and the problem of the choice of electrolyte and its change with time. At low temperature, sulfuric acid will remain essentially invariant, since tars, oils, and ash can probably be easily separated, while sulfur in the coal is converted to sulfuric acid. At high temperatures, however, ash products are likely to dissolve in the molten salt electrolytes that would be necessary. The design of reactors for these conditions must not be underestimated.

The process is pointless if the electricity consumed has to be derived from coal (or other fossil fuel). It therefore should be compared with conventional electrolysis if electricity of nonfossil origin (nuclear or solar, including hydro) is available. However, when coal-assisted electrolysis was compared with advanced SPE conventional water electrolysis it was found to have almost the same total energy consumption per unit of hydrogen produced. The relative energy usage (REU), giving the ratio of the free energy consumed using ordinary water electrolysis to that consumed using coal-assisted electrolysis to produce the same quantity of hydrogen, was shown to be given by

$$REU = E^*/(E + 1.02),$$

where E^* is the cell voltage for conventional electrolysis and E is that for coal-assisted electrolysis, the effective free energy of the coal being given by 1.02 eV. Using $E^* = 2$ V and $E = 0.9$ V gives REU = 1. However, carrying out coal-depolarized electrolysis at 100°C should make it possible to lower the voltage using properly optimized electrodes to about $E = 0.4$ V, giving REU = 1.4. It should therefore be possible to produce hydrogen by coal-depolarized electrolysis using about 30% less free energy than is required for conventional water electrolysis.

However, we should remember that the free energies of coal and electricity are not equivalent, since electricity can be used at 100% efficiency to give work, whereas coal can only be converted into electricity at a maximum efficiency of about 50% (state of the art being closer to 35–40%). In consequence, 1.02 in the above equation should be divided by at least 2 to give a meaningful comparison of energy input, giving a much more favorable REU. In consequence, the viability of coal-depolarized electrolysis will depend much more on economics than on efficiency calculations. These will include the capital cost and lifetime of the reactor, and the value of by-products. If the process can be made practical, the economics of electrolytic hydrogen production will favor coal-assisted electrolysis when electricity costs are high and coal costs are still comparatively low.

Other work on coal-assisted electrolysis has been conducted more recently by Bockris and coworkers.[492, 1837] Under the experimental conditions used in sulfuric acid electrolyte, the Fe^{2+}/Fe^{3+} redox couple appears to be the catalyst involved in coal oxidation,[750] and much of the coal is converted to oils at the anode, presumably via the free-radical Kolbe process. In principle, this should provide an attractive way to upgrade coal to a clean liquid product, at the same time producing hydrogen as a by-product.

Internal Hydrocarbon Processing

Studies supported by EPRI have indicated that metallic components in the MCFC anode chamber at 650°C catalyze the shift reaction, so that the cell consumes carbon monoxide in the presence of water vapor under the same kinetic conditions as hydrogen.[12, 1112] In the conceptual, conventional MCFC system, hydrocarbon fuel would be reformed with steam in an external fuel processor to produce the hydrogen and carbon monoxide anode feedstock. Because of thermodynamic limitations, complete conversion of a hydrocarbon fuel such as methane is not possible with an external reformer: Typically, conversion is 85–95%, depending on temperature. The concept of internal reforming has already been introduced in Chapter 3. If direct internal reforming (DIR) can take place in the fuel cell, then the heat from the exothermic fuel cell reaction can supply the reaction heat without the need for a supply of external heat:

Choice of Fuel and Oxidant

$$CH_4 + H_2O \rightarrow 3H_2 + CO$$
$$3H_2 + 3CO_3^= \rightarrow 3H_2O + 3CO_2 + 6e^-$$
$$CO + CO_3^= \rightarrow 2CO_2 + 2e^-$$
Net: $$CH_4 + 4CO_3^= \rightarrow 2H_2O + 5CO_2 + 8e^-$$

Electrochemical hydrogen consumption accompanied by water vapor production, shifts the reforming reaction to the right, so that it is *driven*, giving a higher conversion and an improved system efficiency. Besides higher efficiency and reduced capital and (perhaps) operating costs, a further advantage of DIR systems is facilitated heat removal, especially if operating current density be increased from the present value of 160 mA/cm^2 to future values of 200~300 mA/cm^2.

Since carbonate in the vapor phase or in the form of films of molten salt may affect the activity of the reforming catalyst, it may in practice be necessary to use indirect internal reforming (IIR), whereby the gases are recycled to a separate catalyst-containing chamber in thermal contact with the fuel cell. The actual reforming would then take place in this chamber, which would resemble the cooling plate (one for every five or six cells) in the liquid-cooled PAFC. However, in this concept, reforming is not driven *in situ*, but via an anode recycle loop.

Energy Research Corporation has evaluated the potential benefits of internal reforming on the performance of a 10 MW natural gas–fueled MCFC under contract to GRI.[1979] Analysis showed that direct internal reforming appeared more attractive than indirect internal reforming because of the parasitic power consumption and capital cost of the high temperature recycle blower required for the latter. In addition, compared with external reforming, DIR might increase overall system efficiency by as much as 15% and save up to 80% in heat exchanger surface area.

To verify these conclusions ERC operated a small laboratory, direct internal reforming molten carbonate fuel cell (IRMCFC) on natural gas at 650°C and a steam-to-carbon ratio of 3:1.[1979] At 160 mA/cm^2 and a cell voltage of 0.71 V (30 mV lower than anticipated) this cell operated for 700 hours with no drop in performance, although there was some evidence of a partial loss of both activity and structural integrity of the reforming catalyst, due to some electrolyte attack on its alumina support. When the process was scaled up to a 300 cm^2 bench scale cell under EPRI support, methane conversion at open circuit was 82%, compared with the theoretical value of 85%. On load however, complete conversion was obtained at hydrogen utilization rates greater than about 50%, when the driven reaction could take place. After 7000 hours of operation the reformer and anode still gave reasonably stable performance, with a slight decay that was identified as being due to the presence of traces of sulfur.[1965] Results are shown in Fig. 8–1; the latest data show excellent cell performance (Fig. 8–2).

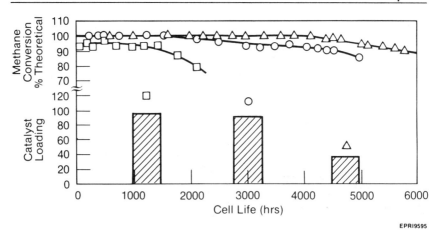

Figure 8–1: Efficiency of Energy Research Corporation internal reforming molten carbonate fuel cells as a function of life, showing results of progressive technology improvements allowing reduction in catalyst loading. Falloff in upper plot after 4200 h due to slight sulfur poisoning of reforming catalyst.

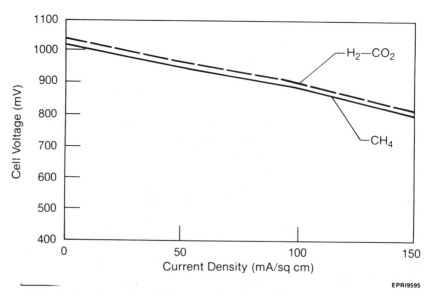

Figure 8–2: Performance of ERC internal reforming stack 10–1. V-i plot under constant flow conditions on methane and reformate, corresponding to 60% fuel utilization at 150 mA/cm².

Choice of Fuel and Oxidant

In older internal reforming systems, serious corrosion problems often occurred, since high operating temperatures were required to obtain good results,[1988] and performance with hydrocarbon fuels was never good. It now appears that many of these problems have been overcome by the use of improved reforming catalysts on supports that cannot be attacked by carbonate (e.g., lithium aluminate or magnesia rather than alumina) and by the use of a fine-pore nickel gas-diffusion sheet between the reforming catalyst in the gas channels and the anode catalyst containing electrolyte.[182] The nickel sheet is not wetted by carbonate and hence acts as an effective electrolyte barrier.

Another early form of internal reforming fuel cell relied on the use of a solid anode as a hydrogen-diffusion membrane. The fuel was reformed with steam in a catalyst chamber in contact with the palladium or silver-palladium foil acting on the other side as an anode. The foil allowed only hydrogen to pass through to the electrolyte side. Fused alkali could therefore be used as electrolyte if carbon dioxide was removed from the air supply. Pratt and Whitney Aircraft used this concept in 1964 [890] to react liquid hydrocarbons at a temperature of 260°C. Subsequently, this type of internal reforming fuel cell was operated at temperatures in excess of 400°C [1340] using a palladium foil anode and a nickel-based reforming catalyst. While cells of this type can produce very high current densities due to the very effective catalysis of the oxygen electrode in molten KOH, they have not found any practical application because of the high cost of the hydrogen diffusion membrane, which also requires a differential pressure gradient to drive hydrogen to the anode at appreciable rates. However, the development of new non-noble alloys that can dissolve hydrogen may result in further future interest in cells of this type.

However, the major advantage of the IRMCFC lies in its system simplicity, therefore potentially lower cost than that of the PAFC. In addition, since it can be regarded as a black box, consuming methane directly without any external heat requirements or losses, it has a higher system efficiency for a given cell voltage than that of the PAFC. For example, based on state of the art performance, the simple system shown in Fig. 8-3[1982] should offer a cell potential of about 0.68 V (atmospheric pressure operation) at 90% utilization and 150 mA/cM^2. This yields an overall HHV efficiency of 53% before auxiliary power requirements are taken into account, or about 49% net. Only one small heat-exchanger (fuel-to-air) is used. This should be compared with the complex 40 KW atmospheric pressure PAFC (Fig. 4–1), which had nine heat exchangers associated with the electrical part of the unit and gave an efficiency of only 37% at about 0.625 V unit cell potential. The more complex IRMCFC system with three heat-changers (none of which are fuel-to-air), shown schematically

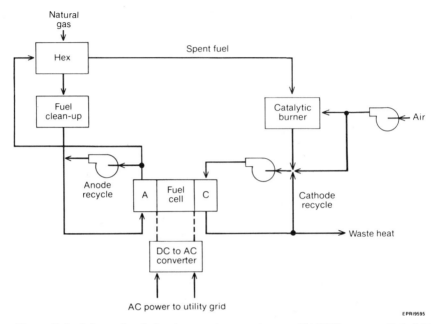

Figure 8–3: Schematic of simple, one heat-exchanger IRMCFC system, Ref. 192 (0.68 V at 90% fuel utilization, 160 mA/cm². Net efficiency would be 49% HHV (7000 BTU/kWh).

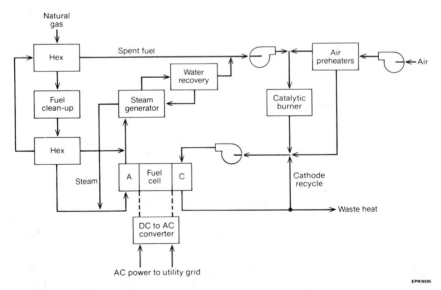

Figure 8–4: As Fig. 8–3, three heat-exchanger unit (0.73 V under same conditions as Fig. 8–3 due to addition of anode exit gas condensor). Net efficiency: 52.5% HHV (6500 BTU/kWh, Ref. 1983).

Choice of Fuel and Oxidant

in Fig. 8-4[1983], is capable of 0.73 V under the same conditions, since it incorporates a water-recovery system to yield a richer cathode gas mixture. Its efficiency would be about 53% net. Finally, the more complex scheme with an air turbine bottoming cycle (Fig. 8–5)[1207] would have an HHV efficiency of 56% assuming only 0.7 V at the same current density and utilization.[182]

Miscellaneous Hydrogen Generation Processes

A number of miscellaneous fuels and processes have been used to generate hydrogen for use in fuel cells. The Osaka Gas Co. of Japan has developed a process whereby aqueous solutions of ammonium ion are electrolyzed to produce fuel cell hydrogen.[1811] Ammonia also has been cracked to generate hydrogen.[1363, 2250] As part of the utility fuel cell program in Sweden, the Swedish Royal Institute of Technology has developed a modified steam-Fe processor for the conversion of biomass and peat to hydrogen.[1626] Licentia Patent-Verwaltungs-G.m.b.H. reported equipment allowing decomposition of organic liquids to hydrogen.[1640] A patent has been issued for a method of producing hydrogen from scrap aluminum by reacting it with strong alkali solutions.[2793] A Kipp-type generator has been used to generate hydrogen from lithium hydride for a low-power (30 watt) fuel cell [933] and from sodium aluminum hydride for use in a 60 W manpack fuel cell.[2171] Patents also have been issued for hydrogen generators that make use of the reaction between metal hydrides and steam or water vapor.[184, 2441] Seawater can also be reacted with magnesium foil to produce hydrogen for low-power fuel cells used for navigational purposes such as buoys.

C. HYDRAZINE

The catalytic difficulties encountered in the anodic oxidation of hydrocarbons have prompted the investigation of *composite fuels* as hydrogen carriers—in particular, hydrazine, ammonia, and methanol—in descending order of reactivity between hydrogen and the hydrocarbons. Hydrazine in particular exhibits some very desirable characteristics, and has been found to be a promising fuel in fuel cells.[2584] It possesses high solubility in aqueous electrolytes at normal temperatures and is very readily oxidized in both acidic and alkaline media at low negative potentials:[1072]

acid: $N_2H_5^+ \rightarrow N_2 + 5H^+ + 4e^-$ $\Delta E° = -0.23$ V

alkaline: $N_2H_4 + 4OH^- \rightarrow N_2 + 4H_2O + 4e^-$ $\Delta E° = -1.16$ V

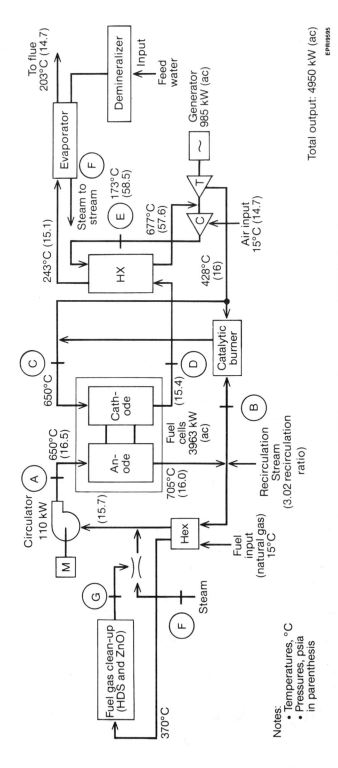

Figure 8–5: IRMCFC system with air-turbine bottoming cycle.[1207] Net efficiency: 56% HHV (Ref. 182).

Choice of Fuel and Oxidant

The theoretical energy density for liquid hydrazine monohydrate is 3560 Wh/L, compared with 590 Wh/L for gaseous hydrogen at 200 atm pressure. Since the practical operating potential for cells using both fuels is the same (about 0.7~0.8 V), hydrazine is therefore about six times more compact than compressed hydrogen. The practical specific energy densities on a weight basis (including storage tanks) might be 700 Wh/kg for hydrazine monohydrate, but only 375 Wh/kg for hydrogen stored at 2% loading in a modern pressure container. In terms of energy density, hydrogen only becomes attractive compared with hydrazine when it is in liquid form.

Hydrazine can be used as a fuel in various ways. The simplest is to dissolve it in aqueous electrolyte as hydrazine hydrate ($N_2H_4 \cdot H_2O$), thereby somewhat simplifying the construction of the fuel cell anode and permitting the use of nonporous flow-by electrodes with suitable catalysts (Chapter 9). Cells of this design may offer savings in cost and volume over cells with gas diffusion electrodes. This advantage is at least to some extent offset by substitution of electrolyte circulation for that of gas, which increases system complexity. Hydrazine is interesting in being the only liquid fuel whose reaction products do not react with alkaline electrolyte.[1453] Because any advantages of CO_2 tolerance are offset by the greater polarization at both the hydrazine anode and oxygen cathode in acidic media, alkaline hydrazine fuel cells have usually been preferred.

As a fuel, hydrazine can be used either directly or after decomposition. Since it is a liquid and has a very high reaction rate in alkaline electrolytes, most of the large corporations that conducted fuel cell research in the 1960s and 1970s attempted hydrazine fuel cell development.[689] However, the high cost of hydrazine (about 15~20 times the cost of hydrogen for the same energy output) prohibits its use in any systems except those used for military applications. Another disadvantage is that it is highly toxic by ingestion, inhalation, and skin absorption, as well as being a strong eye and skin irritant. Human tolerance is about 1 ppm in air.[1108] It also poses a severe explosion hazard when exposed to heat or by reaction with oxidizing materials.

Hydrazine usually is synthesized from ammonia via an energy intensive process, which makes it a costly chemical. However, since its constituents are both extremely abundant in nature, attempts have been made to develop cheaper production methods. The Institute of Chemical Technology in Darmstadt, Germany, worked on a direct electrolytic synthesis in which the anodic oxidation of amide anions is followed by the desired N–N coupling.[65,271] A photochemical process has also been described.[217] Even so, energy input for hydrazine production greatly exceeds energy output in the fuel cell, so that with the present state of technology hydrazine makes no sense for use as a general energy vector. In fact, its theoretical reversible potential is considerably more negative than that of hydrogen, but in practice the reversible hydrogen potential is the best that can be achieved in aqueous solution, representing an irreversible energetic loss. For these energetic reasons, and because of its high toxicity, hydrazine-fueled cells are unlikely to be widely used.

D. METHANOL

Methanol is an attractive liquid fuel because it is relatively cheap when manufactured from natural gas or from coal (in both cases, it should cost about the same as hydrogen on a BTU basis). Its biomedical and environmental hazards are well known, and have been discussed in the literature.[1810] It is readily available today, is soluble in aqueous electrolytes, and is easy to handle. Unfortunately, it is not particularly reactive on any known catalyst under aqueous conditions, and its direct conversion, even at high overpotentials, usually requires large amounts of catalysts containing alloyed noble metals. The electrochemical oxidation of methanol produces carbon dioxide:

$$CH_3OH + H_2O \rightarrow 6H^+ + CO_2 + 6e^-$$

Even though oxidation rates are greater in alkaline electrolytes (e.g., KOH) than in acids, the former are of limited usefulness because of carbonate and bicarbonate formation, necessitating periodic electrolyte renewal. Neutral electrolytes have rather low ionic conductivities, but more important, they show much greater concentration polarization [314,1649] for the conducting ions than strong acids or alkalis. In consequence, they can produce an accumulation of OH^- ions at the cathode and H^+ ions at the anode, reducing cell voltage by as much as 800 mV at practical current densities. This is shown by substituting the possible hydrogen ion concentration change in the Nernst Equation for initially neutral aqueous solutions, with H^+ concentration equal to about 0.1~1 mol/L at the anode, and OH^- at the same concentration at the cathode, equivalent to H^+ values of 10^{-14}~10^{-13} mol/L. Such electrolytes are clearly not suitable for use with methanol. However, CO_2-rejecting buffer solutions containing a very soluble salt, (in particular concentrated cesium carbonate–bicarbonate [534, 535, 542, 1017]) can somewhat reduce this problem, although the high cost and low availability of cesium does not render their use very practical. Fuel cell performance using buffers of this type is discussed in more detail in Chapter 10.

For these reasons, acid electrolytes, notably sulfuric acid in earlier work, have been most commonly used for methanol oxidation. Direct methanol

Choice of Fuel and Oxidant

fuel cells have therefore required acid-resistant materials. Since the alcohol dissolved in the electrolyte is much less reactive than hydrazine, it can pass unreacted over to the air electrode, even though the latter is protected by a membrane. This results in both coulombic and voltage losses.

Platinum anode catalysts also become poisoned by methanol partial oxidation products, although less susceptible catalysts, such as Pt-Sn, have been developed.[83,1753,1755,1758,1759] The search for a more effective methanol oxidation catalyst was actively pursued by several investigators during the early 1970s, notably at Shell Research, Ltd. in England,[83,1753,1755,1758,1759] and at Hitachi in Japan.[2423] While Shell's research was abandoned in about 1978, Hitachi in 1985 was on the point of commercializing a small direct methanol fuel cell unit utilizing a sulfonic acid electrolyte and an Asahi Glass Flemion or Du Pont Nafion SPE membrane to reduce the problem of methanol crossover.[2423] This will be used in conjunction with a lead-acid battery in a golf cart, and will operate at 0.4V, 80 mA/cm^2.

As discussed earlier, methanol is very readily steam-reformed at low temperature, and is therefore very suitable for indirect use in acid fuel cells. External steam-reforming allows much better use of the fuel cell hardware and catalysts than does the direct methanol system, since power density and cell efficiency are both much greater. To avoid carrying water for steam reforming, there has been some interest in methanol cracking for mobile applications. A supported nickel-copper catalyst at 200°C seems to suffice, the reaction apparently going to completion with no by-product formation.[217,2627] However, the cracking reaction leads to the formation of CO and hydrogen, so a low-temperature shift converter will be necessary before the gas mixture is used in a fuel cell.

It may be possible to incorporate such a reactor into a PAFC or other relatively high-temperature acid electrolyte fuel cell so that the cracking reaction can directly use waste heat. Since the Ni-Cu catalyst is acid-sensitive, the reaction zone would have to be incorporated within cooling plates distributed throughout the cell stack, similar to those for the IIR methane MCFC described above. Such a system would have to incorporate a subsidiary shift process, for which water vapor and a suitable catalyst would be necessary. It may be possible to consider direct cracking combined with shift in the anode chamber of the PAFC, analogous to direct methane reforming in the MCFC, if acid-resistant (e.g., noble metal or mixed oxide) catalysts could be developed. Internal steam reforming of methanol in the PAFC has been examined.[2086] In contrast to the MCFC, water is produced at the cathode of acid fuel cells by reaction between the carrier H$^+$ ion and molecular oxygen. However, the methanol reforming reaction can still be driven by the local anodic oxidation of the hydrogen produced catalytically. Ideally, the PAFC temperature should be around 225°C for effective steam reforming of methanol. However, new catalysts may be developed that allow this process (or a combination cracking and shift process) to proceed effectively at 200°C or less. This could result in a greatly simplified, low-cost methanol-PAFC system. The development of new proton-conducting inorganic electrolytes that operate at higher temperatures than the PAFC will make such systems even more flexible.[176]

Finally, methanol-steam mixtures are immediately internally reformed in the high-temperature MCFC and SOFC systems, without the necessity of catalysts. Again, such systems will be simple and efficient, and could be used in many terrestrial and mobile applications when and if methanol is available. We will return to this in Chapters 17 and 18.

E. AMMONIA

Ammonia is readily available worldwide, and costs only slightly more than methanol on a BTU basis. Some properties of anhydrous liquid ammonia are as follows:

Heating value	6.23 kWh/kg
Specific gravity (21°C)	0.60
Vapor pressure (-30°C)	126 kPa (1.3 atm)
(+52°C)	2100 kPa (21.4 atm)

Anhydrous ammonia is therefore easy to handle under low pressure. The ammonia manufacturing process is such that the purity of bulk commercial ammonia is 99.5%, the main contaminant being water, which of course would not be detrimental to fuel cell operation. Ammonia is only flammable with difficulty and has a relatively low toxicity, though at high concentrations it strongly irritates mucous membranes. However, it can be detected readily by smell at very low concentrations. It is therefore probably at least as safe, if not safer, than gasoline, its lack of combustion being offset by its irritant action in case of spills. As a fuel, ammonia would be stored as the pure liquid in a lightweight vessel, and therefore is much easier to handle and transport than hydrogen.

Ammonia has vapor pressure-vs.-temperature characteristics (see preceding) almost identical with propane, thus it can be fed to a fuel cell in the vapor phase except at ambient temperatures below -30°C. Like hydrazine, ammonia may be readily cracked into hydrogen and nitrogen (although, at 700°C), without the disadvantage of producing components that react with alkaline electrolyte, as is the case with processed hydrocarbons. On the other hand, cracked ammonia is difficult to use with acid electrolytes, since the small amount of ammonia in the feed stream reacts with acid electrolytes in an analagous manner to CO_2 with alkaline electrolytes. Because of its water solubility, it can be circulated in an alkaline electrolyte, though since its reactivity is very low, its direct oxidation is not normally very effective.

In general, ammonia fuel cells have received little serious attention and usually are not considered suitable for economic power production except as specialized remote, low-power systems in which the availability of ammonia makes it unusually attractive. However, vehicle applications may be of future interest.[375, 2139]

F. FORMIC ACID AND FORMALDEHYDE

Formic acid–air fuel cells with acid electrolyte have not been proposed commercially but apparently could be developed without major difficulty.[2186] Formate-air cells are available.[2591] The latter are attractive for remote use as refuelable air batteries in which a mixture of solid potassium hydroxide and potassium formate is dissolved in ordinary water to form the stationary electrolyte. The carbonate mixture is thrown away at the end of use. After rinsing the air electrode, the system is recharged with fresh fuel-electrolyte mixture. Formaldehyde electrodes and cells with non-platinum metal catalysts are still in the development stage.[2187]

OXIDANTS

Although pure oxygen is desirable for best cathode performance, air is usually employed as the favored oxidant in terrestrial fuel cells because of its availability. Performance on air is obviously less than on pure oxygen, owing to the usual effect of partial pressure both on cell potential at a given current density, and on the currents that can be supported as the diffusion limit is approached. If ambient air is used in an alkaline electrolyte system, a reduction in performance is eventually observed due to carbonation of the electrolyte by the 380 ppm of atmospheric CO_2.

While a small amount of carbonate in the electrolyte (up to perhaps 2N concentration) does not significantly reduce fuel cell performance, cathode performance deteriorates as the carbonate content increases beyond this level. The effect is particularly marked at high current densities. Physical examination of such cathodes always reveals some mechanical blockage with carbonate; thin, highly-hydrophobic electrodes being superior to thick electrodes in this respect. Spinel catalysts (mixed transition–metal oxides; see Chapter 12) have been found to be superior to platinum-based catalysts in respect to carbonate blockage, and it now is possible to make alkaline air cathodes that are more resistant to loss of performance than earlier designs. However, if operation over many thousands of hours is required without changing the electrolyte, and particularly if immobilized or quasi-immobilized electrolyte is used, some form of carbon dioxide or carbonate removal is required. In the past, this was considered to be one of the greatest disadvantages of alkaline fuel cell systems, but the newer methods of CO_2 removal discussed later can greatly reduce the magnitude of the problem.

In contrast, the great practical advantage of acid electrolytes is that they reject CO_2, allowing the use of unscrubbed air and of fuel gases containing significant amounts of CO_2. In acid cells, the use of air instead of oxygen merely corresponds to a dilution effect, and performance usually suffers by about 80 mV with well-designed electrodes with little gas-phase concentration polarization. This millivolt value is called the *oxygen gain*, and its measurement is a valuable diagnostic test for a properly made electrode. The small loss due to the use of air instead of oxygen is usually acceptable because an air-breathing fuel cell is not burdened with the cost of oxygen, and the extra weight and complexity of oxygen storage or enrichment systems. Although there have been no reports of difficulties encountered with impurities which may be present when filtered city air is used in acid fuel cells, there is some indication that inert impurities found in cylinder-grade oxygen may gradually lower fuel cell performance.[1336]

Hydrogen peroxide has sometimes been used as an oxidant for certain applications. It is highly reactive under normal conditions and can be added in the liquid phase to a circulating electrolyte fuel cell with simple, suitably catalyzed flow-by electrodes. For example, very high performance hydrazine–hydrogen peroxide systems have been developed for lightweight cells using circulating KOH electrolyte with a porous plastic membrane separator between the anode and cathode compartments, notably by Alsthom in France.[689] Hydrogen peroxide has also been used (or suggested) for certain submarine applications where pressure and buoyancy problems are factors, such as for torpedoes.

CHAPTER 9

Electrodes

The main task of electrode design is to ensure that all reactants have ready access to a sufficiently large catalytically-active interface between the electrolyte and the current-carrying electrode. In the normal case of a gaseous reactant, three physical steps must be completed before electron transfer (oxidation or reduction) can occur. These are:

1. Dissolution of gas molecules at the interface between the gas and the electrolyte, which will be facilitated by the presence of the largest possible interfacial area.
2. Diffusion of dissolved gas molecules from the gas–electrolyte interface to the electrolyte–electrode interface, which will be most rapid if the thickness of the interfacial electrolyte film is minimal.
3. Physical contact with the electrode with (where appropriate) adsorption onto the electrode surface followed by charge transfer, which will be promoted by the presence of a large area of material of the highest electrocatalytic activity on the surface of the electrode.

Charge transfer is followed by the flow of electrons to or from the reaction site through the electronically-conducting electrode material and current collector, together with a corresponding flow of ions in the electrolyte. Finally, reaction products must be eliminated. If this occurs via the gas phase, they must travel in the opposite direction to that of the gaseous reactants, following the reverse sequence of events.

In the electrolyte, highly effective migration of ions (e.g., hydrogen or hydroxyl ions), together with the transport of appropriate molecules such as water, must occur to and from the bulk of the electrolyte and the electrode–electrolyte interfaces. This requires that electrodes should be thin, to reduce the length of the reaction site to bulk solution pathways. Several different electrode configurations, depending on the fuel cell type, have been developed to deal most effectively with the above problems.

A. POROUS ELECTRODES

To obtain high current densities, in general the gas-diffusion electrode must be a thin, porous, and high-surface area body activated by the presence of a suitable catalyst, which must make electronic contact with the remainder of the electronically-conducting structure. The electrode must provide a large number of suitably active reaction sites where both the reactant gas and the electrolyte can come into contact. It must also maintain a stable interface between the electrolyte and the gas, so that the number of reaction sites remains as high as possible as a function of time. In addition, the electrolyte must not leak through into the gas supply chamber, and gas must be prevented from being forced through into the electrolyte. Finally, the electrode structure must provide a conductive path for the flow of electrons to and from the reaction sites.

In early fuel cell work porous electrodes were essentially judged on an empirical basis in terms of performance, but there is now a substantial theoretical foundation for the relationship between performance and electrode structure. Many aspects of this relationship have been studied for electrodes intended for use in aqueous electrolytes, including the general theory of porous gas diffusion electrodes;[607, 725, 1356, 1689, 2324] the theory of current distribution in porous diffusion electrodes;[813, 1612, 1664] the theory of mass transport in porous diffusion electrodes;[8, 695, 945, 946, 1295, 2028, 2107] the role of the electrical double layer in porous electrodes;[2300] theoretical evaluation of the performance of hydrophobic porous electrodes[605] and of anodically dissolving solid organic reactant porous electrodes;[1523] the modeling of flooded hydrophobic porous electrode performance;[2611] the mechanisms of current generation in hydrophobic porous gas diffusion electrodes;[763] studies of the structure of hydrophobic gas diffusion electrodes;[972] a review of skeleton structure electrodes;[997] the use of electrical analogs to interpret diffusion and discharge processes in electrodes;[2162] and the theory of partially immersed electrodes[1348] and flow-through porous electrodes.[207] A number of investigations has been made concerning the maintenance of the three-phase interface in gas diffusion electrodes, such as studies of gas bubbling and electrolyte weeping in porous semi-hydrophobic electrodes;[245, 1780] gas blow-through in porous gas diffusion electrodes;[221, 1505] the wetting of diffusion electrodes;[1047] and even the effect of the gravitational field strength on the wetting angle at the three-phase interface.[1943]

Kinetic parameter studies with electrodes of this type have included the determination of the double layer capacity in porous gas diffusion electrodes;[1004, 2460] the theory of transient behavior in porous electrodes;[737, 993, 2038] a study of polarization vs. time for hydrophobic porous gas diffusion electrodes;[2394] methods for determining the processes occurring inside electrode pores;[1522, 2395] and a study of polarization in flooded porous electrodes.[1779] The theory of electrode contact resistance;[313,359,752,894] methods of measuring IR-drop [52, 1110, 2704] and cell impedance;[469] and special alloys for reducing contact resistance[592] have been considered for electrodes in general. Methods of determining the IR-drop[989] and the effective electrolyte resistivity[1888, 2797] inside porous gas diffusion electrodes also have been reported. In addition, methods of determining the depth of gas penetration in the pores of porous gas diffusion electrodes;[1763] the penetration of gas and electrolyte into hydrophobic gas diffusion electrodes;[2612] the structural characteristics of porous electrodes (using pulsed potentiostatic methods);[1893] and the pore structure of Teflon membranes [2296] have been studied. Studies dealing with the effects of various factors on the performance of porous gas diffusion electrodes include: the effect of electrode porosity on performance,[831] the effect of current density on the moisture content,[2621] and structural changes [1489] in PTFE-bonded porous electrodes.

With regard to the fabrication of porous gas diffusion electrodes for use with aqueous electrolytes, a number of papers have been published that provide general guidelines. In the majority of cases, they refer to electrodes that have a hydrophilic porous structure that is made sufficiently wetproof by the use of colloidal Teflon (polytetrafluorethylene, PTFE), the latter being the most stable and effective hydrophobic agent known. Descriptions have included the optimization of structural parameters,[1357, 1654] how to determine the optimum PTFE content for hydrophobic electrodes,[716] improving performance by the use of small outer pores and large inner pores,[2711] designing for minimum IR-drop,[2052] the use of honeycomb-type structures,[1539, 1540] factors involved in the deposition of rhenium and rhenium oxide catalysts,[1130] and the use of plate-type porous gas diffusion electrodes.[1098] Preparation methods have been disclosed for microporous PTFE sheet,[1583] hydrophilic porous PTFE sheet,[1927] catalyzed porous plastic electrodes,[1805] and hydrophobic porous gas diffusion electrodes consisting of silicon-based substrates.[1511, 1855, 2346] Operational guidelines for hydrophilic porous hydrogen electrodes also are available.[2064]

One of the most important results of the study of porous gas diffusion electrodes has been the discovery that over 90% of the reaction in porous electrodes occurs over a very narrow band of electrolyte located at the meniscus with the electrolyte.[387,531,972,2064] The thickness of this electrolyte band is in the order of only 10^{-7} to 10^{-6} cm.[1621] For the sake of comparison it is interesting to note that for a smooth, partially immersed electrode, the thickness of the active meniscus zone is much greater, being about 3×10^{-4} cm.[1665] To be effective even at high current densities, this implies that the catalyst need only be present near a meniscus edge in a porous electrode. This has been a most important discovery because, coupled with improved quality control of electrode porosity, it has enabled great reductions in required catalyst loadings, so that the catalyst is only present where it is needed.

The theory of porous gas-diffusion electrodes, which refers to electrodes under mixed activation and diffusion control, will not be given here. This may seem anomalous, but the whole art and science of making successful fuel cell electrodes is to make them free of all diffusion control under normal current density conditions, so that they can be regarded as a high area convoluted surface under activation control. Porous gas-diffusion electrode theory is only of academic interest, and interested readers are directed to the references given above. Some practical aspects for making successful electrodes are given in the following sections.

Porous Metal Electrodes

Porous metal electrodes represent the earliest concept used within the modern era of fuel cells. Basically, they are simply metal powder sinters in which the powder itself serves as both catalyst and support. They may be constructed on a metal screen to give mechanical strength during the fabrication process, although its presence is not really necessary after sintering. Such electrodes contain no wet-proofing agent and rely on surface tension, that is, pore-size variation between components, with or without differential gas–electrolyte pressure gradients, to maintain the three-phase boundary. Sintered nickel electrodes of this type were first used in the Bacon high-temperature (200~280°C) alkaline system,[28, 394] however their major use is now for the sintered Ni or Cu anode of the MCFC. Both the Bacon and certain MCFC cathodes have typically been a modification of this type of electrode, since both consisted of lithium-doped NiO, which gives greatly improved electronic conductivity over that of the parent oxide. In the case of the Bacon electrode, it was sintered, lithiated, and then oxidized. In the case of the MCFC, which uses a lithium-containing carbonate electrolyte, in-$situ$ oxidation of the sintered structure in the fuel cell itself is all that is necessary.

The classical porous metal electrode was the porous nickel structure employed by Bacon in his pioneering work on alkaline fuel cells.[28, 689] The same electrodes were subsequently used by UTC for the Apollo fuel cell system for the moon landing missions. The anode, which contained relatively low surface area sintered carbonyl nickel powder (about 0.5 m^2/g), relied on the high activity of nickel for hydrogen oxidation in molten KOH. The cathode had a similar initial structure, and gave an excellent performance owing to the ability of KOH to catalyze the oxygen electrode reaction at high temperatures. This will be discussed in more detail in Chapter 10.

Both electrodes consisted of two layers, the layer on the electrolyte side having a smaller pore size than the layer on the gas side. When a small pressure differential exceeding the capillary pressure in the large pores was applied between the gas and the electrolyte, the liquid was expelled from the larger pores on the gas side. However, the gas could not pass through into the electrolyte because of the higher capillary pressure in the small pore layer, which is inversely proportional to the capillary pore radius.

Bacon electrodes were prepared from a mixture of coarse nickel powder with ammonium bicarbonate as a pore-forming agent. The mixture was compacted into a flat plaque, which was fired for 30 minutes at 850°C in a reducing atmosphere. During this process, the ammonium bicarbonate evaporated and the nickel particles sintered, leaving a highly porous rigid structure.[28,212] Favored carbonyl nickel powers for these applications were Inco 287 and 255. A layer of finer nickel powder was then sintered on one surface to make the fine pore layer of the dual porosity electrode.

Dual porosity electrodes of this type were also occasionally made from materials other than nickel. For example, the fine pore electrocatalyst might be deposited onto larger pore substrates such as carbon, boron carbide, metallized Teflon, and porous metals. Sometimes three or four graded pore layers of this type have been applied.

The effects of sintering [1410, 2011] and of sintering temperature [840] on the performance of porous metal gas diffusion electrodes have been described, as have the operational mechanisms of graded porosity electrodes using high-surface area Raney nickel catalysts.[521, 1344, 2063] Various disclosures have been made for the preparation of porous metal gas diffusion electrodes, including those using Raney nickel;[317, 318, 878, 1501, 1503, 1635, 2583] Raney silver;[1926] hydrophobic sintered silver;[1905] and sintered porous iron.[78] Porous metal gas diffusion electrodes also have been made by incorporating nickel or silver fibers into a screen.[62]

An improvement in porous metal oxygen electrodes was pioneered by Esso Research and Engineering Company,[265, 266, 808] based on studies conducted by Cahan at the University of Pennsylvania.[387, 531] The electrodes were made by the sputtering and vapor deposition of very light loadings of platinum onto porous nickel. Typical performance for oxygen reduction at 0.60 V in 6 M potassium hydroxide at 60°C for sputtered electrodes with loadings of only 0.14 mg Pt/cm^2 was about 30 mA/cm^2. For vapor deposited electrodes at the same cell potential, only 0.012 mg Pt/cm^2 was required at 50 mA/cm^2, and only 0.104 mg Pt/cm^2 at 100 mA/cm^2. Furthermore, these electrodes were apparently free from diffusion limitations even at high current densities. Structures of this type pointed the way to making the platinum electrocatalyst totally accessible to the reactants in electrodes of other types, particularly those with low-loading carbon-supported platinum. These would particularly include the state-of-the-art structures developed since 1970–1975, in which platinum catalyst loadings vary from 0.5 mg/cm² to 0.1 mg/cm².

Except for the MCFC, for which no lyophobic agents exist, porous metal gas diffusion electrodes are now outmoded; two reasons being the high cost of producing electrode structures of the required rigid specifications of pore size and layer thickness, and the difficulty of maintaining the correct gas–electrolyte pressure differential in the fuel cell. Since these electrodes are not wet-proofed, they tend to become too wetted for optimum performance when used with caustic electrolytes. Even for the MCFC, where the three-phase boundary is maintained by the difference in pore characteristics between the electrodes and matrix, it is well recognized that the performance of both single- and dual-layer structures is still not optimized.[2243] A model for porous metal gas diffusion electrodes has been proposed by Burshtein.[521]

Porous Screen Electrodes

This family of porous electrodes may be regarded as the direct ancestor of PAFC and many AFC electrodes with carbon supported catalysts. It is also the type of electrode used in the 1970-technology Space-Shuttle Orbiter AFC, which has high loadings of noble-metal blacks. The electrodes employed a stable conductive screen as a support, to which was applied a mixture of hydrophilic catalytic material and hydrophobic material. The hydrophilic catalytic material was pure platinum black, or noble metal alloy, which, in addition to being the primary electronic conductor, also ensured that there was a continuous film of electrolyte across the electrode. The hydrophobic material was colloidal polytetrafluoroethylene (PTFE, Teflon®). The principles involved in the three-phase structure of these electrodes will be described in the following section on carbon electrodes.

Porous screen electrodes were first developed in the mid-1960s by the American Cyanamid Corporation.[657] The conductive support for use in alkaline solution was woven nickel wire gauze or screen, which led to the term *screen electrode*. For use in acid solution, the screen was platinum. The electrodes were normally supplied with a platinum black loading of 40 mg/cm². These electrodes were designated AB40 (alkaline electrolyte) and AA40 (acid electrolyte); later, lower loadings became available, such as 20 mg/cm² in the AB20 electrodes.

Porous screen electrodes may be prepared by first mixing the catalyst or catalyst and a substrate with PTFE emulsion,[1880] generally Du Pont's Teflon 30 (see next section). Using the technique developed by American Cyanamid, the mixture was cold pressed onto both sides of the support–current collector screen using a pressure of 70~210 kg/cm². Next the electrode was hot pressed at 300~700 kg/cm² at 315°C for 7.5 minutes to partially sinter the Teflon. Lastly, if desired, Teflon gaskets to give edge seals and a Teflon backing layer to give additional hydrophobicity on the gas side could then be applied. The pressing steps probably caused the porosity of the electrode to vary with thickness because of the resistance of the screen to material flow: The resulting catalyst compaction could adversely affect the electrode performance. To offset this, a more open final structure after pressing could be achieved by incorporating ingredients such as nickel or Baymal into the electrode, which could be leached out after pressing.[932] In electrodes of this type the catalyst forms porous clumps that appear to be wetted by the electrolyte, associated with unwetted areas which form the gas diffusion pores.

Derivatives of the original screen electrodes have been successfully used in many types of fuel cells with alkaline or acid electrolytes. They can be designed for use with a freely circulating electrolyte, but they have been much more frequently employed with the electrolyte immobilized in a thin matrix of stable porous or fibrous electronically-insulating material. Asbestos in mat form is suitable for this application, and has usually been employed for alkaline electrolyte, in which it can be used up to about 90°C as in the Space-Shuttle Orbiter fuel cell discussed in Chapter 6. The matrix should ideally have a mean pore size less than that of the wetted pores in the electrodes, so that the electrolyte will always wet the matrix, preventing gas-crossover and giving minimum ionic resistance. The matrix should be in contact with an electrolyte reservoir to compensate for volume changes due to dilution, concentration, or evaporation. If the matrix-electrode system is properly designed from the viewpoint of differential porosity, electrolyte management can be maintained at all times and there will be no tendency for the electrodes to flood.

Porous Carbon Electrodes

Carbon has long been a favored material for fuel cell electrodes on account of its low cost and unique properties. Although carbon is a nonmetallic substance with a wide range of crystalline and amorphous structures, structures other than diamond show overlap of the valence and conduction bands, giving many of the properties of a metallic conductor. Table 9-1 lists some typical physical properties of carbon.[1485]

When assessing the chemical stability of carbon it is important to characterize the material clearly: Not only do different forms of carbon exhibit large variations in chemical stability, but pretreatment history also has a strong effect. Reversible thermodynamics tells us that carbon should be oxidized to CO_2 or its derivatives (HCO_3^-, $CO_3^=$, depending on pH) at only about 50 mV above the hydrogen potential in aqueous media. This is illustrated by the potential-stability diagram for carbon as a function of pH

TABLE 9-1

Typical Physical Properties of Carbon [1485]

Triple point	~4020K and ~110 bar
Normal melting point	~3820K
Normal boiling point	4470~5070K
Specific gravity:	
diamond	3.515
graphite	2.266
amorphous carbon	1.8~1.9
C-C bond length	0.154 nm
Electrical resistivity, $\mu\Omega$ m	
carbon, petroleum coke base	35~46
carbon, anthracite base	33~66
carbon, lampblack base	58~81
carbon, lampblack graphitized	46~66
carbon fiber	> 600
glassy carbon	30~50
graphite, petroleum coke base	8~13
graphite fiber	~42

(Pourbaix diagram, Ref. 2044), shown in Fig. 9-1. In practice, however, the oxidation of carbon is kinetically so slow that carbon is considered to be very resistant to chemical attack except in oxidizing environments at elevated temperatures. The following chemical characteristics of carbon are important indications in regard to its use in fuel cells: Hot sulfuric acid (an oxidizing agent) can react with carbon to produce carbon dioxide and sulfur; alkali hydroxides generally do not react with carbon except in the presence of oxidizing agents at elevated temperatures; graphite can react with concentrated phosphoric acid in the presence of oxidizing agents such as chromium trioxide at 80~100°C; at temperatures above about 500°C (in the presence of nickel) or 1100°C (presence of platinum) hydrogen reacts with carbon to form methane; sulfur dioxide can react with carbon to produce COS, CO, CO_2, and S at 700°C; oxygen can react with carbon at temperatures of 500°C to produce CO and CO_2; and, at elevated temperatures both CO_2 and water vapor can react with carbon, the latter giving CO and H_2. Hence, the fact that carbon can react in these powerful oxidizing environments under extreme conditions means that it will react, though slowly, at the oxygen electrode in aqueous fuel cells.

Kordesch has pointed out [1485] that much confusion exists in the literature on carbon electrodes because workers have failed to be specific on the

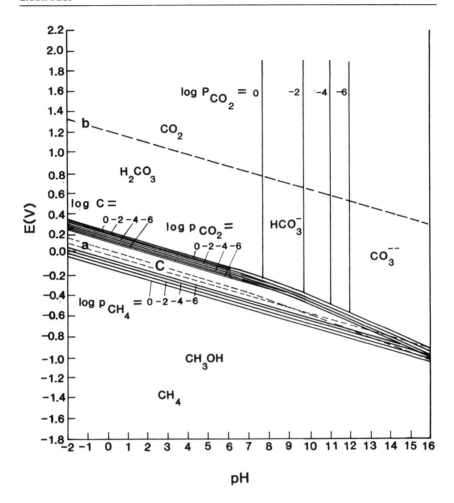

Figure 9–1: Pourbaix diagram for carbon in aqueous media at 25°C, showing zones of stability of elemental carbon and oxidized and reduced species at different concentrations as a function of pH. Lines (a) and (b) are the lines separating zones of stability for hydrogen and water and water and oxygen, respectively.

materials they employed. As might be expected, there are many forms of carbon, including natural graphite (macrocrystalline, mesocrystalline, microcrystalline); artificial graphite; pyrographite; flexible graphite, sometimes used for gaskets, seals, and manifold separators; carbon fibers, cloths, and felts; graphite whiskers; carbon blacks, such as lampblack, industrial blacks (channel blacks, oil furnace blacks, thermal blacks), and acetylene blacks (used as catalyst carriers); and activated or *adsorbing* carbons or charcoals. Activated carbons, and furnace and acetylene blacks have traditionally been the most important for use in fuel cell electrodes. Active carbons have been used in AFC electrodes, whereas furnace blacks and acetylene blacks have been found to be good catalysts supports for PAFCs and more recently, AFCs. These applications are reviewed in Refs. 177 and 394.

All these materials have different particle size and porosity distributions as well as a wide variation in other properties. Factors such as as electrode life, performance, catalyst adherence, surface area, chemical stability, and electrical and thermal properties depend strongly on the carbon material used and how it has been previously processed.

As amorphous (microcrystalline) carbon is heated, its disordered lattice approaches the more structured hexagonal planar structure of graphite in the following sequence:

carbon black → graphitized carbon black → graphite .... → diamond

Diamond (tetragonal) is placed apart: Graphite is only converted to diamond under extreme conditions of temperature (giving atomic mobility) and pressure (to achieve the volume reduction in the graphite–diamond transformation). At atmospheric pressure, diamond is converted into graphite if heated in the absence of oxygen to over 2000°C, since it has a stability of about +12.6 kJ/mole compared with graphite. For many carbons, care must be taken not to equate *heat treatment* with *graphitization*, since heat-treatment may imply nothing more than a partial structural ordering. *Carbonization* refers to the conversion of an organic material to a coke, which is about 95% carbon. It is usually carried out by heat treatment at temperatures below 1000°C. The conversion of the coke to the more stable and structured graphite is a solid-state process that gradually takes place between 1000~3000°C, with the appearance of progressively more order, and a lattice spacing more closely approaching that of pure graphite.

Soft carbons refer to those materials that can be graphitized by heat treatment below the melting point of carbon and at ambient pressure; whereas *hard* carbons such as cellulosic carbon, vitreous carbon, acetylene black, and some carbon fibers are nongraphitizing, retaining an imperfect structure even after prolonged heat treatment at high temperature. Among the graphitizable carbons, *pyrolytic* carbons, formed by the pyrolysis of methane or heated graphite, are notable for their high purity, regular structure, and near theoretical density (2.20 g/cm^3). The residence time at a given temperature largely determines the extent of graphitization. The strength of the carbon increases with the degree of graphitization, finally passing through a maximum. Increased graphitization also increases chemical stability and lowers electrical resistivity. Other properties that undergo changes with graphitization include the lattice parameters, diamagnetism, Hall coefficient, thermoelectric power, and electron paramagnetic resonance.

In order to obtain stable carbon structures it is necessary to remove oxidizable components (mostly hydrogen) from the carbon. In order to optimize stability and surface area, carbon processing for use in electrodes may involve treatment with steam and/or carbon dioxide as well as heat treatment. This removes the least stable areas of the carbon particles as CO_2–CO or CO respectively, improves corrosion resistance, and increases surface area. Different end results are obtained, depending on whether these *activation* processes are fast or slow. When no further weight losses occur during treatment the carbon has attained a stable structure under the temperature and atmospheric conditions used. For fuel cell electrode applications, active carbons of surface areas of 300~500 m^2/g to more than 1000 m^2/g have been employed for AFC electrodes. For PAFC systems, certain furnace blacks in the 70~300 m^2/g range or acetylene black (70 m^2/g), with or without a heat or chemical treatment, have been favored. While uncatalyzed high-surface carbons can be used as moderately good cathodes (but not as hydrogen anodes) in the AFC,[132,149,155,394] carbons used in the PAFC require catalysis, normally with platinum.[2048]

Porous carbon electrodes usually contain a highly conductive supporting material to provide effective current collection and to give strength to the structure. Up to about 1975, nickel screens (gold-plated in the case of the Space-Shuttle Orbiter cathodes) were used for this purpose in AFC electrodes, just as platinum or gold-plated tantalum screens had been used in the earliest pure noble metal black electrodes intended for acid electrolytes. However, for acid electrolytes, either carbon or graphite felt from a resin precursor, or carbon or graphite cloth that has been made rigid by a resin treatment followed by pyrolysis, are now favored by all developers. This carbon-based material is conventionally termed the *backing*, the *backing layer*, or the *substrate*. It may in some cases contain ribbed gas channels, and serve as an electrolyte reservoir, or it may be flat and partially wet-proofed, so that it remains dry. In the latter case, the gas-distribution ribs are in the solid carbon–graphite bipolar plate beween each cell. These options are discussed in further detail in Chapter 16.

The active zone of the electrode, containing the gas diffusion pores and the active catalyst, is applied to the substrate. For the PAFC, the catalyst zone typically contains a loading of 0.2~0.5 mg of high surface-area Pt/cm^2 in a usual loading of 10% by weight on the carbon black support. Generally, the phase boundary in the catalyst layer is maintained by the use of a mixture of the catalyst (e.g., Pt on C, in the preceding) with Teflon. The latter is typically added to the catalyst in the form of a colloidal aqueous suspension (see following) to give a mixture containing 35~40 wt % Teflon. This mixture is normally produced in the form of an aqueous paste, latex-like dough or ink; and is rolled, spread, printed, or sprayed onto the substrate; in some cases with application of pressure. Some developers reportedly prefer the use of a dry colloidal mix, each having his own proprietary recipe and preparation technique. The green electrode is heated to about 350°C for a few minutes in the absence of air, sometimes under pressure. This operation removes the wetting and stabilizing agents present in the original Teflon suspension, but more important to electrode operation is the fact that it sinters the Teflon. The net result is a structure in which the Teflon has been partially depolymerized, melted, and selectively wets parts of the catalyst structure on a rather random basis.[177, 1433, 2613] The areas where the carbon carrying the catalyst has become wetted by Teflon will be hydrophobic, producing the gas pores, whereas those that are not covered by a Teflon film will become the hydrophilic electrolyte pores.

Sometimes, the gas side of the electrode is made more hydrophobic than the rest of the active layer, or is even a porous layer of pure Teflon. Since a poorly wetted interface will result in an unacceptably high electrical resistivity and will have low catalyst utilization due to lack of electrolyte, whereas a more hydrophilic interface may flood, great care is required in regard to Teflon loading and distribution and electrode pretreatment. The previous use of the word *recipe* is not unreasonable, since the whole operation is on the skilled artisan level, with a definite analogy to fine cuisine. However, the long-term stability of the three-phase interface in hydrophobic porous Teflon-bonded carbon electrodes is more difficult to achieve than in porous metal electrodes, where the interface can be maintained by the use of graded porosities and pressure differentials. Because of this, some consideration is now being given to the use of graded porosity structures in carbon electrodes.

Both the carbon-based support and the high surface-area catalyst, (e.g., platinum black) must be stable for the duration of the mission under the electrode operating conditions, in particular at the air cathode. Since the rates of corrosion of both carbon materials and platinum are strongly potential-dependent, anodes generally show no problems, whereas cathodes tend to operate under the corrosion limits of the materials used. Temperature and water vapor pressure at a given temperature (i.e., total system pressure) also have a strong effect on support corrosion and catalyst corrosion and recrystallization. For example, the PAFC air cathode in an electric utility cell typically operating at 4.8~8.2 atm pressure at 200~210°C, and at 0.78 V vs. SHE in concentrated phosphoric acid, is definitely pushing available carbon support materials, as well as platinum catalysts, close to their operating limits.[175, 177, 357, 2144] The situation in the

AFC cathode is even worse, since carbon-based cathode catalysts cannot be operated for extended periods at temperatures higher than about 80°C, although pure noble metals can be used at much higher temperatures.[394] Losses in active surface area as a result of platinum recrystallization and graphite substrate corrosion are a major problem in the life of porous carbon gas diffusion electrodes in acid electrolytes. Such area loss is a major obstacle to further increasing performance, and will be discussed in more detail in later chapters.

For use in the PAFC or in other acid fuel cells, (e.g., those employing trifluoromethanesulfonic acid or its higher, less volatile homologs; Chapter 10), only all-carbon electrodes using low-loading noble metal catalysts on carbon–graphite supports or backing are practical from the standpoints of cost and durability. The only (impractical) alternatives from the corrosion viewpoint are the electrodes used in early work (see preceding), which had pure noble-metal catalysts on noble-metal screens.

The earliest successful porous carbon electrodes were those developed by Union Carbide Corporation,[1477] who pioneered some of the principles of the catalyzed electrodes in use today. These electrodes were typically made by mixing a finely divided carbon, such as lampblack, with a binder and pressing the mix into the required shape, followed by baking, which removed some of the binder and resulted in a porous structure. Since the carbons used were not very active electrocatalytically, noble metal catalysts were introduced (e.g., by impregnation) during the manufacturing process. Wet-proofing for electrolyte retention was achieved in the earliest electrodes by dipping them in wax, but later, hydrophobic materials (first polyethylene, finally PTFE) were employed.

The most popular of the PTFE materials used was in the form of a colloidal suspension, produced by E. I. duPont de Nemours and Co., Inc. (Wilmington, Delaware) under their Teflon® trademark. According to this company's technical literature,[1485]

> Teflon 30 TFE resin aqueous dispersion is a negatively charged hydrophobic colloid containing TFE resin particles 0.05 to 0.5 microns in size, suspended in water. The dispersion contains 59 to 61% solids (by weight) and is stabilized with 5.5 to 6.5% (based on weight of resin) non-ionic wetting agent. The pH as supplied is normally about 10. The viscosity at room temperature is approximately 15 centipoises. The melting point is $327 \pm 10°C$. Teflon 30 TFE resin dispersions are applied as coatings, cast as films, or impregnated into a variety of porous structures by conventional dipping or flowing procedures. After application, the dispersion is heated to about 120°C to remove the water. Water-repellent structures with useful properties can be obtained by simply applying pressure to the coated

materials, such as by calendering or molding. If a strong, continuous film is to be obtained, however, the deposited polymer must be sintered after drying. Sintering is usually done at 360~370°C.

In the Union Carbide work, careful grading of the amount of wet-proofing agent could be used to control the localization of the reaction zone. One important advantage of this technique is that it allowed the catalyst to be restricted to the reaction site instead of being spread throughout the complete structure, thereby allowing potential cost savings. However, electrodes made in this fashion could be complex, therefore expensive, and could consist of as many as seven distinct layers which required many fabrication steps.

Several theoretical studies were reported that specifically concern porous carbon electrodes. These include a general discussion of the fabrication process for porous carbon electrode;[2249] a study of gas transport in porous carbon electrodes;[2731] and the determination of the wetting characteristics of porous hydrophobic carbon electrodes.[2011] Papers dealing with the preparation of porous carbon substrate materials include: the preparation of porous carbon sheets [615,1582] and vitreous carbon foams [1688] for PAFC electrode substrates, a method of making carbon-coated oxide powders for use as electrode substrates,[905,906] preparation of nickel-coated porous carbon electrode substrates,[316] preparation of graphite molding particles,[1205] and a method for coating porous carbon with pyrolytic graphite for PAFC substrates.[889,2705,2706] A method also has been developed for coating graphite electrodes with thin layers of silicon carbide.[2347] Coating with Pt_3Si has been claimed to lower the overvoltage of carbon electrodes.[1326] Many methods have been developed for the preparation of catalyzed [30, 235, 243, 364, 445, 898, 899, 903, 1263, 1440, 1571, 1586, 1636, 1745, 1778, 1847, 2165, 2262, 2280, 2597] and uncatalyzed [64,1821,2181,2625,2742] hydrophobic porous carbon gas diffusion electrodes. Methods to deposit platinum [2018] and tungsten carbide [576] on hydrophobic porous carbon electrodes are also available. An electrostatic method of binding metal complexes to graphite electrodes has been published.[1959,1960] Some newer developments in the preparation of hydrophobic porous carbon electrodes include: preparation from carbon fibers,[746, 1570, 2231] preparation from graphite paper substrates,[256, 443, 996, 1481] and the use of screen printing techniques.[995, 2493, 2494] Reports dealing with the preparation of catalyzed [2477] and uncatalyzed [1368] hydrophilic porous carbon gas diffusion electrodes also are available. Recently there has been considerable interest in producing composite porous electrode structures in which porous carbon is bonded to some other material for the purposes of cost-saving or improved electrode performance, particularly with the idea of incorporating an electrolyte reservoir in the PAFC. In particular, UTC has developed this type of structure in their PAFCs as described in detail in Chapter 16. Some of the work published to date includes: the bonding of hydrophobic porous carbon layers to hydrophilic electrolyte reservoir matrixes for PAFCs,[446, 523, 616, 745, 1025, 1561, 1562] laminar layers of different carbons,[579, 581, 1508] laminar layers of catalyzed porous graphite–resin composites having different porosities,[1212] preparation of hydrophobic porous carbon electrode composites bonded to plastics [1217] and metals,[983, 1830] the bonding of highly porous carbon gas distribution layers to porous carbon electrode layers for use in phosphoric acid fuel cells,[702] and preparation of porous gas diffusion electrodes of mixed carbon–plastic compositions.[1448] A special adhesive paste for joining carbon-graphite materials also has been formulated.[1507] A detailed description and discussion of many of the above methods can be found in Kordesch's review of carbon materials,[1485] which also provides a listing of the names and addresses of worldwide suppliers of the products required to make carbon electrodes (we should note that some of these are no longer available; see Chapter 16). Teflon-bonded catalyzed carbon electrodes are commercially available from Prototech, Inc. in Newton Highlands, Massachusetts.

Microporous Plastic Electrodes

Another design that found some favor up to the 1970s was developed in England by Shell Research, Ltd. It was based on the use of a thin microporous plastic (polypropylene) sheet that was readily available commercially for use as a separator in lead-acid batteries. It was generally suitable for use below about 80°C in both acid and alkaline media. The separator had a uniform pore size of about 5 microns and a thickness of about 0.8 mm. The electrode was prepared by first coating one side of the plastic with a thin sputtered metal layer (normally silver for alkaline solution and gold for acid) to render it electrically conductive. The film thus produced was then thickened by electroplating to give a porous, highly conductive layer. The noble metal catalyst layer, usually platinum or palladium black or a noble-metal alloy, was then applied on top of this either by electrodeposition or in a Teflon binder, as appropriate. The total thickness of the conductive and catalyst layers, making up the gas side of the electrode, was only a few microns. A small pressure differential (about 0.07–0.2 atm) was applied to prevent the gas side from being flooded by electrolyte.

The concept was basically close to that of Cahan's electrodes [387, 531] as developed by Exxon,[265, 266, 808] discussed earlier in this chapter. Typically, precious metal loadings on the order of 1 mg/cm^2 were found to be adequate, giving a low-cost structure. These electrodes indeed confirmed that the electrode reaction takes place within a zone of a few microns thickness, and many variations of this design were examined. For example, instead of metallized microporous plastic, porous nickel sheet was also used as a conductive electrode support for use in alkaline electrolyte, a thin layer of platinum black and colloidal Teflon being applied to the gas side. However, it was generally found to be necessary to apply a third, pure Teflon layer to prevent slow leakage of electrolyte into the cathode gas channels. This phenomenon was known as electrode *sweating* or *weeping*, and was particularly associated with the surface tension changes that take place on KOH concentration at the AFC cathode at high current densities (i.e., concentration polarization associated with water transport from anode to cathode).

Porous Composite Electrodes

A recent trend has been to bond electrode structures directly onto the electrolyte matrix, particularly in the case of the PAFC. Going beyond this, attempts have been made to develop composite electrode structures using bonded layers of materials that improve their cost, performance, manufacturability, or structural integrity. Typical examples include porous metal sheet bonded to plastic laminates;[1848] electrodes using porous plastic substrates;[1858] and the bonding of plastic [398] and glass fiber, or asbestos [397] fuel matrixes to porous electrodes. However, most of these are only appropriate for use below 80°C, often only for alkaline systems.

Some examples of composite structures are given in the section on porous carbon electrodes, although recent work has tended towards simplification in the direction of a modernized American-Cyanamid electrode structure, since many of the earlier concepts were found to be unnecessary. Recent advances in materials science indicate that composite electrode-matrix combinations, involving cathode-matrix–anode-electrolyte reservoir sandwiches, perhaps using successive layers of ionically and electronically conducting polymers, will become increasingly important in the future. Some aspects of this work are discussed in Refs. 177, 394, and 1989.

B. NONPOROUS ELECTRODES

In fuel cells other than those using hydrogen and oxygen, the fuel may in some cases be supplied as a gas or liquid dispersed or dissolved in the electrolyte. The electrode does not then have to be of the porous three-phase boundary type, since both fuel and electrolyte will be present at the same face of the electrode. However, the maximum effective catalyst surface, implying a large roughness factor, will still be required for optimum performance. Even so, this concept may greatly simplify electrode construction, and reduce both the cost and the volume of the flow-by fuel cell system compared with those of a structure using gas circulation. For example, a very thin, ribbed metal sheet may be used on which the catalyst is simply applied to one surface.

Nonporous electrodes have generally been metallic. Examples include titanium electrodes coated with cobalt-nickel spinel catalysts,[2372] spinel electrodes,[1450] nickel-catalyzed electrodes on mild steel substrates,[192] titanium-based ruthenium oxide electrodes,[2470] and pressed electrodes using mixed catalyst powders.[1987] The use of stainless steel cathode current collectors has also been described.[2747]

C. ELECTRODE EVALUATION

Electrodes are evaluated by their performance during short-term polarization measurements, as well as by conducting life tests. The latter may be carried out under constant electrical resistance, constant voltage, constant current, or constant power conditions. However, constant current conditions are experimentally simplest and are therefore most commonly used. Reference electrodes are used where necessary for cell endurance testing to separate anode and cathode polarization and performance loss. Voltage free of IR-drop is measured to estimate any electrolyte conductivity changes (e.g, due to drying out of electrodes or electrolyte loss) during the test. Ohmic drop measurements may also serve to detect relative changes in contact resistance as well as changes in electrode wetting properties or reactant distributions that may result in non-uniformity of cell current flow. If performed correctly, surface area measurements and crystallite size determinations on electrocatalysts before and after life tests can give valuable information on structural changes as a function of time. The relationship between performance and reactant utilization is usually obtained during the course of life tests.

In general it has been found that under moderate current load, the voltage decline with hydrogen–air alkaline fuel cells and molten carbonate fuel cells is very gradual and involves a very slow, practically linear, decay rate over thousands of hours. With state-of-the-art electrodes, PAFCs at 200°C often show a voltage reduction of as much as 3~5% from the initial value during the first 100~1000 hours, after which the decline is much slower. In general, PAFC voltage decay is entirely at the cathode and is logarithmic with time.[177] The decay mechanism has been identified as recrystallization of the platinum-based catalyst.[177] It should be remarked that if this slow logarithmic rate dependence is masked in the PAFC, particularly in stacks, by a rather rapid linear loss in potential (e.g., 5 mV/

1000 h), then other problems can be diagnosed, for example loss of electrolyte or severe corrosion of carbon–graphite structural parts (bipolar plates), usually associated with gas leakage between cells. Oxygen cathode life in the PAFC has now reached 100,000 hours using 1975 technology,[177] and current technology shows very low decay. Well-constructed stacks now show a very small loss in performance (about 20 mV between 1000 and 10,000 hours, with the same loss projected between 4000 and 40,000 hours). System heat rates are quoted for end-of-life performance, based on these projections.

Hydrogen electrodes based on catalyzed carbon anodes show no performance degradation either in the PAFC or (generally) in the AFC, even up to 150°C.[394, 1706] However, when easily oxidized metals such as Raney nickel with a sintering inhibitor (e.g., titanium) are used as a substitute for carbon-supported platinum or palladium in the AFC, care must be taken not to overpolarize by running the cell at a very high current density, especially with an inadequate hydrogen supply. If the anode reaches the oxidation potential of nickel (about 110 mV vs. HE in alkaline solution, [2044]) its structure will be irreversibly damaged.

In contrast to hydrogen, direct organic oxidation usually results in performance decay of the most effective electrocatalysts. When attempts were made to directly electro-oxidize hydrocarbons, periodic voltage cycling was sometimes observed,[932] along with rapid decay. This has been attributed to poisoning of anode reaction sites by hydrocarbon oxidation intermediates, leading to increased overpotentials. The more positive potential then allows intermediate oxidation and partial reaction site cleanup, so that a cyclic process is superimposed on performance decay. Due to the lack of interest in the direct hydrocarbon anode since 1970, this work has not been followed up in detail. It may however suggest ways of maintaining the activity of anodes for direct organic oxidation, such as by the use of pulsing at periodic intervals.[384]

Recent advances in electrode performance and progress in the development of low-cost catalysts still in the laboratory stage are discussed in Chapters 11 and 12.

D. ELECTRODE COST

The cost of fuel cell electrodes is dependent both on materials cost and on cost of manufacture. In recent years manufacturers have become hesitant to quote the learned-out production costs of electrodes under high-volume production conditions, since they bear no relation to those in the prototype or limited production stage. In any case, mass-production costs are carefully-guarded trade secrets. In limited production, PAFC cathodes containing 5 g of platinum; 40 g of Teflon; 50 g of carbon support; and 200~1000 g of carbon-graphite substrate, depending on design, may cost $2000/m^2. The platinum materials cost is about $80, that of the Teflon about $1, and that of the support and the substrate (based on graphite cost), less than $1 and about $3~$15, respectively. However, the experimental substrate itself may cost $200/m^2, as purchased from an outside vendor. The reason for the large markup is that labor input is very high in limited production; and that in mass production the aim will be to reduce manufacturing costs to no more than 2~3 times the cost of the primary materials, apart from the special case of the platinum catalyst (cf. the discussion in Chapter 2, Section O). Thus, it is reasonable to consider a final electrode cost of $95~130/m^2, including $80 in platinum, as a reasonable learned-out mass-production cost for the electrode and substrate, with all components manufactured in-house. It should be noted that exclusive of the noble metal, the most expensive electrode component will be the substrate. The corresponding cost of anodes will be about $40/m^2 less because of their lower platinum content. To estimate the total cost of the cell, we must add the bipolar plate separating individual cells and that of the matrix. Materials cost for the latter will be about $2/m^2; and that of a flat 1.5 mm thick graphite bipolar plate (to be used with the thicker, more expensive ribbed substrates) about $50/m^2; that of a ribbed plate for use with the flat thin substrates being about $75/m^2. Multiplying these by a factor of two to give estimated cost of manufacture, and adding the cost of the electrodes with appropriate substrates, gives a mean cost for the total cell of $180~$200/m^2 without the catalyst. This corresponds to $115~$125/kW, again without the catalyst, at typical utility PAFC performance levels (about 150 mW/cm^2). As has been already remarked, these electrodes will have lifetimes of at least 40,000 hours.

These are definitely cost effective, and it is easy to foresee new manufacturing methods or shortcuts that will further reduce costs; for example, the fabrication of the two substrates and the bipolar plate in one piece.[177] These will be discussed in Chapter 16.

Present alkaline fuel cells operating at 75 mW/cm^2 (half the power density of the PAFC) over 5000 hours [2558] are predicted to cost $200~300/kW.[2560] These figures are not out of line with those just estimated for the PAFC. Although the material costs of present AFC electrodes would not appear to be cheaper than some of those developed earlier (e.g., the pure carbon cathodes developed in the mid-1970s; Refs. 132, 149, 155), it should be remembered that their lifetime can be as much as an order of magnitude longer, which is very significant from the viewpoint of overall life-cycle cost.

The cost of hydrazine or other specialized electrodes is not readily available, but their catalyst costs will certainly be higher than those of

PAFC or AFC hydrogen-air electrodes.[788] For acid direct methanol systems, catalyst cost is likely to be rather high on a per-kW basis even if low-loading electrodes are used, since typical power densities at present are low. For example, the Hitachi system [2423] operates at about 0.4 V and at only 80 mA/cm^2. With a 1 mg/cm^2 platinum loading on a thin porous carbon sheet, this corresponds to almost \$700/kW for the anode catalyst alone. If direct hydrocarbon oxidation is considered, high loading platinum anodes (9 mg/cm^2 and higher) were always necessary, even at 200°C, to obtain power densities similar to those quoted previously for methanol (about 30 mW/cm^2 maximum). Even at this elevated temperature, performance often could not be effectively sustained for more than a few hundred hours and electrocatalyst costs were in the thousands of dollars per kilowatt. Unless an electrocatalysis breakthorough occurs, direct hydrocarbon oxidation electrodes will be impractical for anything other than specialized low-power applications.

Before leaving the subject of electrode costs in low-temperature fuel cells, the special case of the SPE system should be mentioned. At the present time, the membrane alone costs about \$300/kW calculated on the basis of an output of 1.5 kW/m^2, and the fabrication techniques are such that a total platinum loading of about 8 mg/cm^2 is currently necessary.[176] Some of the reasons for this will be discussed in Chapter 10. The platinum loading alone corresponds to about \$900/kW, and both platinum loading and membrane cost must be brought down before the system can find broader use. While platinum can be recovered and reprocessed at the end of the useful operating life of the electrode, its use can only be economically justified if the initial platinum investment cost is a relatively small fraction of the total system cost, since interest on investment is part of the annual running cost.

Low electrode costs are certainly possible for the sintered metal structures used in MCFCs. In 1966 the Institute of Gas Technology quoted probable electrode costs of \$20~40 per kilowatt for molten carbonate fuel cells then operating at 55 mW/cm^2 for "thousands of hours".[1231] This is equivalent to \$40~80/m^2 per square meter in 1986 dollars. Present MCFCs operate at about 135 mW/cm^2 and have probable electrode costs of no more than \$20~30 per kilowatt, or \$30~40 per square meter. These cells should be capable of economic lifetimes of 25,000 to 40,000 hours. Noble metals are neither necessary nor suitable for use in the MCFC, although thin platinum films have been examined in the SOFC in early work performed by Westinghouse Electric Corporation.[2673] They have now been replaced by cheaper materials for which the cost of fabrication will exceed materials costs. This work will be reviewed in Chapter 18.

CHAPTER 10

Electrolytes

The optimum electrolyte for a fuel cell depends on a number of factors. As well as cost of materials, these include: the nature of the fuel and oxidant, the most effective cell operating temperature, the cell materials and their corrosion properties, the electrode performance as a function of reactants and electrolyte, and electrolyte stability and conductivity as a function of operating conditions. The latter will include temperature and the conditions for engineering requirements such as heat and reaction product removal. All the above are of course related in the final analysis to system and operating costs. Methods for evaluating electrolytes [2072] and reviews of fuel cell electrolytes[209, 873, 2781, 2782] have been published.

Fuel cell electrolytes tend to fall into groups in which the current is primarily carried by one of the following four ions: hydroxyl (OH^-), protons (H^+), carbonate ($CO_3^=$), or oxide ($O^=$). The major reason for selecting electrolytes on this basis is that the transport ion must be a product of one of the half-cell reactions, and must be present in the highest possible concentration to avoid polarization due to mass-transport.[314, 1649] For almost all practical fuels and oxidants, the reactions involve -H oxidation at the anode and =O reduction at the cathode, typically as H_2 (molecular hydrogen or dihydrogen) and as O_2 (molecular oxygen or dioxygen), but occasionally as other -H and =O carriers discussed in Chapter 8 (e.g., N_2H_4, HCOOH, H_2O_2, or even HNO_3, etc.). Favored media are therefore concentrated aqueous or molten hydroxides, strong acids (including solid polymer electrolytes), molten (and occasionally concentrated aqueous) carbonates, and conducting solid-phase oxides.

A. AQUEOUS ALKALINE ELECTROLYTES

Alkaline electrolytes have excellent electrochemical properties, but they react with CO_2 derived from carbonaceous fuels; therefore the most developed alkaline electrolyte fuel cells employ hydrogen as the fuel. They also can be used in hydrazine fuel cells and in certain systems employing methanol or formate. In the latter cases, the electrolyte does not remain invariant, but is more or less rapidly converted to carbonate, and requires periodic or continuous recycling or replacement.

Aqueous alkaline electrolyte systems have a low activation energy for the cell reactions, that is, they show a relatively small temperature coefficient of performance. They therefore have high power output even at below-ambient temperatures. They possess the further advantage of good compatibility at cathode potentials with many conventional materials (such as carbon, silver, and plastics to 80°C, and with nickel to 200–250°C). A wide range of highly active electrocatalysts (noble metals, catalyzed carbons, Raney alloys) is available for use in alkaline media, since they allow the formation of stable passive films on many materials at oxygen cathode potentials. They can also operate effectively on non-noble metal catalysts at the oxygen cathode below 80°C, such as carbon-supported transition-metal macrocyclics and their derivatives (Chapter 12); and can easily use oxidizable fuels such as hydrogen and hydrazine. The main, indeed only, disadvantage of alkaline electrolytes is that carbonaceous compounds, particularly hydrogen that contains CO_2 derived from conventional fuel processing operations, cannot be used directly as fuels on a continuous basis. Similarly, ambient air supplied to the cathode requires a carbon dioxide removal system.

The development of electrodes requiring low or zero loadings of noble metal catalysts has been a promising feature in the development of alkaline fuel cells. For example, air electrodes using uncatalyzed high surface-area carbon can operate satisfactorily at high current densities in alkaline electrolytes,[132, 149, 155, 1476] although at polarizations that are about 100 mV greater than those with noble metals. Carbon-based hydrogen electrodes perform satisfactorily with noble metal (platinum or palladium) catalyst loadings as low as 0.1 mg/cm^2. These are stable indefinitely, even at 120°C, since carbon corrosion is negligible at anode potentials, although it becomes severe at the cathode at over 80°C.[394] There are indications that different types of thin porous metal anodes using Raney alloys (i.e., very high surface-area nickel stabilized against recrystallization by titanium) can match the output-per-unit-weight performance of low noble-metal loading carbon anodes.[638, 2710] The life expectancy of low-cost cells employing carbon cathodes with non-noble catalysts together with high surface-area transition metal anodes is promising.

Long-term corrosion tests in cells with alkaline electrolyte using sintered metal (Bacon-type electrodes; cf. Chapter 9) showed that it is wise not to exceed a temperature of about 200°C if long life for the oxygen electrodes is to be achieved. To obtain the same performance as the Bacon cell at a lower pressure, the fuel cells used in the Apollo service module were operated at 260°C, and were suitable only for the duration of the typically 11-day mission. Under these conditions, the oxygen electrode in KOH becomes reversible, since molecular oxygen can dissolve chemically in the melt to produce peroxide and superoxide ions:

$$O_2 + 2OH^- \rightarrow O_2^= + H_2O$$
$$O_2 + O_2^= \rightarrow O_2^-$$

This mechanism was first shown in Ref. 130, 1653. These ions are readily reduced.

Work in the mid-1970s was directed towards the use of temperatures below 90°C for aerospace use, to obtain lifetimes in excess of 10,000 h using pure noble metal (Au-Pt) cathodes, as in the Space Shuttle Orbiter fuel cell (see Chapter 6 and Ref. 394). These cells use electrolyte immobilized in an asbestos matrix, the corrosion of which becomes serious at temperatures exceeding 90~110°C, depending on the grade of asbestos used.[1706] However, other materials, such as Teflon-bonded potassium hexatitanate,[1706] allow operation at higher temperatures. For aerospace cell operation, such matrixes may be as thin as 0.05 mm.

In all alkaline systems, the favored electrolyte is usually aqueous potassium hydroxide, which has a higher conductivity than the cheaper alternative, sodium hydroxide. A further advantage is the fact that the conductivity of concentrated aqueous KOH rises rapidly with temperature. Unlike acid cells, which are constructed using a thin (0.25 mm) electrolyte matrix, for which the resistance is very small, electrolyte resistance has often been a controlling factor in AFCs intended for use on scrubbed air, since these have used circulating electrolytes with a wide (2~3 mm) interelectrode gap. A further important advantage of KOH is that carbonate ion is more soluble than in NaOH, so there is less risk of precipitation of carbonate in electrode pores if the fuel and oxidant are not entirely CO_2-free. However, the latest cells have a gap of only 0.05 mm (p. 175).

The usual electrolyte concentration for use close to ambient temperature is about 6 M (27 weight percent), corresponding to that of maximum conductivity. However, concentrations as high as 12 M (46%) or even higher will be present in cells operating at higher temperature, the value depending on the vapor pressure of product water. The higher concentrations do provide somewhat improved oxygen cathode performance, particularly if pure carbon cathodes are used, since their oxygen reduction overpotential is pH-dependent, becoming much less as pH increases.[132,149,155] Even with carbon-supported platinum cathode catalysts, for which the overpotential is almost pH-independent, a small increase in performance is achieved in very concentrated solution at constant temperature,[131] in spite of the fact that oxygen solubility falls rapidly with increasing KOH concentration.

In the fuel cells used in the Apollo missions mentioned earlier, KOH concentrations of 75~85% were used as the electrolyte at 260°C.[394] The reason for this choice, which corresponds to commercial reagent-grade melted KOH pellets, was to give the lowest product water pressure (i.e., overall system pressure) consistent with high activity, in order to reduce pressure vessel weight. This made it possible to operate the cell at only four atm using a noncirculating electrolyte. Under Bacon cell conditions at 200°C using a lower KOH concentration, it was necessary to operate at pressures as high as 45 atm to prevent the circulating electrolyte from boiling.[28] However, under all these high-temperature conditions, the water vapor pressure of the electrolyte varies rapidly with water content, so if cell temperature and pressure are controlled, the concentration of the electrolyte automatically remains approximately constant. However, these comments only apply to cells developed in the past for specialized applications; when air is used in more conventional alkaline cells, it is not practical to operate at pressures much above atmospheric because, in contrast to the 200°C PAFC, the required compression work cannot readily be provided by the waste heat from cells operating at about 80°C.

The electrolyte may freely circulate in parallel through all the cells in the battery; this was the arrangement in Bacon's 200°C system [28] and it has been followed in many low temperature hydrogen-oxygen and hydrogen-air alkaline fuel cells. This arrangement has the advantage of easy removal of waste heat from the electrolyte in an external heat exchanger, as well as that of product water using an evaporator, as in the Siemens 80°C system, which operates on pure hydrogen and oxygen at 2 atm.[394] On the other hand, with parallel electrolyte circulation, electrical losses resulting from shunt currents between cells are inevitable, and can result in serious corrosion problems. However, careful design can reduce losses from this cause to an acceptably low level. Cells for aerospace use (UTC Apollo and space shuttle systems, Allis-Chalmers system; Ref. 176) have always used stationary electrolyte to give maximum system simplicity, reliability, and weight reduction. This lead is likely to be repeated in new designs for terrestrial applications.[394, 2335]

Electrolyte Carbonation

As we have repeatedly stressed, for economical operation in terrestrial applications fuel cells must operate with air as the oxidant. The greatest problem with alkaline electrolytes is that they react with traces of carbon dioxide that are present even in so called pure hydrogen fuel and scrubbed air to precipitate insoluble carbonates. If the hydrogen fuel for the cell is not of electrolytic origin (e.g., if it is technical grade manufactured from natural gas, hydrocarbons, or coal), it will contain ppm traces of CO_2, as will air scrubbed with normal absorbants such as soda lime.

In alkaline fuel cells the water produced during cell operation is formed at the anode by reaction between H_2 and the OH^- transport ion. It is removed either by controlling the vapor pressure in the subsystem for water removal or by controlling the dew point of hydrogen in the anode gas recirculation loop in systems such as the space shuttle cell, which uses stationary electrolyte. In the former approach, the loss of water can be arranged to be from special cells in the electrolyte circulation system, as in the Siemens unit,[394] or from the built-in cell electrolyte reservoir if circulation is not used.[176] The control devices used are designed to prevent the cells from operating wet, (i.e., with excess water), or dry, with a deficiency of water. Water management should normally be controlled via the physical properties of the electrolyte, such as by water vapor pressure vs. temperature as a function of concentration.

As potassium carbonate builds up in the electrolyte, the water vapor pressure above it will change. If the control device cannot differentiate between, and compensate for, an increase in vapor pressure resulting from carbonate accumulation from one due to other causes, the net result will be a change in electrolyte volume as potassium hydroxide is slowly converted into potassium carbonate. This may result in the drying out or wetting of the electrodes, giving loss of active electrode area, that is, increased polarization. Loss of electrolyte also can lead to gas crossover, burn-through of cells due to catalytic recombination, and irreversible losses in catalytic activity.

In addition to changes in electrolyte volume, carbonation also can produce other undesirable effects. It could certainly lead to a decrease in the conductivity of the electrolyte, as well as increase the possibility of the precipitation of solids.[2477] For example, the conductivity of 27% KOH solution at 97°C is about $1.69 \, \Omega^{-1} cm^{-1}$, but with the conversion of 30% of the potassium ion to carbonate it falls to about $1.06 \, \Omega^{-1} \, cm^{-1}$,[2455] giving increased ohmic losses in the cell. Similarly, certain electrode regions become inactive as a result of precipitation of solids from the electrolyte, which may occupy electrode pores, deactivating part of the electrode area. In particular, under certain operating conditions solid $(KOH)_2(K_2CO_3)_3$ may be formed in addition to K_2CO_3.

Other effects also can result from the conversion of KOH to K_2CO_3 in the fuel cell. For example, for many catalysts in alkaline solution the open circuit voltage can be approximated [168] by the reversible potential for the process

$$O_2 + H_2O + 2e^- \rightarrow HO_2^- + OH^-,$$

at least in the absence of side-reactions. This reaction often may be combined with an anodic process to give a mixed potential (e.g., anodic oxidation of a catalyst support combined with cathodic oxygen reduction on the catalyst). However, the presence of carbonate ion may change the overall process, particularly from the viewpoint of electrode lifetime. This is demonstrated by the fact that in one case, cell degradation rates increased ten- to twentyfold in the presence of 0.1% carbon dioxide or carbon monoxide in oxygen used as a reactant.[2455] Carbon monoxide present in the oxygen stream, in effect is equivalent to carbon dioxide since it can be readily oxidized at the fuel cell cathode to carbon dioxide. Adverse polarization effects caused by the presence of carbonates have been found to be worse when cells are run at high current densities (600 mA/cm^2). In such cases, estimates of percentage voltage loss resulting from the long-term formation of carbonate ranged from 20 to 80%.[944] Even so, these results cannot be taken to be definitive, since improved electrode formulations and structures give very different results. Since alkaline fuel cell electrolytes can undergo carbonation in as little as 500 h,[788] efficient scrubbing is essential. Since the effects resulting from carbonate formation must be avoided in all but short duration applications, Alsthom et Cie. in France (see Chapter 5), patented an alkaline fuel cell electrolyte containing lithium hydroxide, lithium borate, lithium citrate, and lithium acetate that they claimed had improved properties in this respect.[530] Ion-selective electrodes that rapidly measure the partial pressure of oxygen and carbon dioxide in gases have been developed in Hungary.[759] These could allow control of the CO_2 scrubbing process.

Methods of Carbon Dioxide Removal

As long as carbon dioxide is present in the oxidant of an AFC, carbonation of the electrolyte will eventually occur. Electrode structures now exist which allow the carbonate formed to migrate from the electrode surface into the bulk of the electrolyte, where it does relatively little harm.[1476] As the concentration of carbonate builds up in such a system, routine servicing to replace the electrolyte becomes necessary. Alternatively, an electrochemical device to maintain carbonate at a maximum level of about 2 N can be built into the electrolyte circulation system, for example by operating a separate smaller fuel cell under very high current density conditions, so that this subsystem fuel cell anode operates at low pH, liberating CO_2 from the electrolyte (see following; Ref. 1078), or by electrodialysis.[1301] For this reason (and despite their other disadvantages), circulating electrolyte systems have been favored for both AFCs and other similar electrochemical energy converters that use air electrodes, such as metal-air batteries.[144, 520, 675, 676, 2174, 2175] While it may be necessary to eliminate the carbon dioxide scrubber in smaller units where weight and volume are important, it is generally agreed that removal of carbon dioxide from air is necessary for larger units to give maximum life and reliability.

There are various options for removing carbon dioxide, including expendable absorbers, regenerable absorbers, electrochemical removal, membrane methods, and molecular sieves. Regenerable systems are preferred, provided that they do not involve excessive penalties in cost, weight, volume, or parasitic power consumption. A further consideration is the level to which carbonate can be tolerated in the electrolyte, since most regenerable systems will not achieve terminal concentrations as low as those for expendable systems. For reformer gas, complete removal of carbon dioxide after the gas has entered the fuel cell stack is impractical owing to the high (20%) concentrations of carbon dioxide present in the fuel; and preremoval of at least 90% is required before a regenerable system can be considered. Since there is not a great difference in system cost between 90 and 100% CO_2 separation, complete preremoval from reformate is preferable.[788] Methods of CO_2 removal from reformate have been reviewed, and over 22 processes identified.[394, 974] Conversely, in the case of removal from air, continuous elimination of carbon dioxide from unscrubbed air is possible by the use of a regenerative system in the fuel cell.

Expendable Absorbers

The three most common expendable absorbers are soda asbestos, lithium hydroxide, and soda lime (a mixture of sodium hydroxide and calcium oxide). Soda asbestos has a tendency to become blocked if the reactant air has a high humidity, and is thus unsuitable for many applications. Lithium hydroxide is somewhat more active than soda lime and, volume for volume, has 20% more capacity for absorbing carbon dioxide. Efficient (82%) absorbents containing mixtures of calcium oxide and lithium hydroxide also have been developed.[1536] However, lithium hydroxide is not only more expensive than soda lime, but is generally available only from the United States. For these reasons soda lime containing an exhaustion indicator has been the most common external CO_2 scrubbing agent.

One kilogram of soda lime can clean 1000 m^3 of air, taking the carbon dioxide concentration from 0.03% to 0.001%.[1054] This corresponds to 135 to 250 kWh under normal fuel cell operating conditions (oxygen utilization between 20 and 30%). To obtain the maximum capacity from soda lime prior to carbon dioxide breakthrough, a relative humidity of greater than 50% must be maintained. However, providing water for the dry air humidification can sometimes present problems. In a well-designed absorber, carbon dioxide breakthrough occurs abruptly, so that the exact level of electrode tolerance to concentration is relatively unimportant. The minimum tolerance concentration usually is taken to be 10 ppm.

Instead of soda lime, aqueous caustic alkali (NaOH, KOH, or LiOH) can also be used to remove carbon dioxide, using a wet scrubber consisting of a small packed column. Adsorption approximates to a steady state process, the important variables being the available wetted surface area and the required gas flow rate. The liquid flow rate is not critical provided that it is sufficient to continuously wet the the packing without flooding. The concentration of carbonate in the solution affects the level to which carbon dioxide can be removed, but tests have indicated that, with a well-designed tower, fairly high carbonate concentrations can be tolerated before the equilibrium value of carbon dioxide in the gas phase rises above 10 ppm. If electrodes are used which will tolerate a higher level of carbon dioxide in the reactant air without detrimentally affecting performance or life, then the volume of the packed column at a given air flow rate can be reduced. Siemens A.-G. have developed a wet air-scrubber using porous sintered PVC plates containing KOH solution.[1531]

Another wet scrubbing approach for AFCs is to use the electrolyte itself as the scrubbing medium. This dispenses with the separate scrubbing circuit and since it does not allow carbonate to build up inside the cathodes, it has an advantage over the direct use of ambient air. However, the electrolyte must be considered as an extra expenditure, since it will adsorb virtually all the carbon dioxide in the air and therefore has to be replaced relatively frequently. For this reason it is important to try to reduce the quantity of air required above the stoichiometric amount. It has been shown, for example, that for a 750 W fuel cell, 90 L/min of air (about five times the stoichiometric amount) can be scrubbed to about 10 ppm in a fairly small volume using a flow of about 2 L/min of 30% aqueous KOH at 70°C.[2134] Solutions containing up to one molar carbonate have been used, but they do lower scrubber efficiency. Thus the absorption tower must be dimensioned to allow for the maximum expected carbonate concentration in the electrolyte.

In some systems it is sufficient to let the intake air pass through a labyrinth of deflecting plates wetted with caustic to remove most of the carbon dioxide. For example, in one version of a hydrazine fuel cell battery, the air electrodes had a small portion of carbonate-free, KOH-wetted, porous nickel exposed to the incoming air. A 2 cm-wide strip cleaned carbon dioxide from the air sufficiently to prevent damage to other portions of the electrodes.[845]

A major disadvantage associated with wet scrubbing with the electrolyte when using a dissolved fuel such as hydrazine, is fuel carry-over via the scrubber. Although tests showed that normal carry-over from anodes

Electrolytes

to cathodes was not sufficient to affect air electrode performance, a fuel loss of some 20% was observed when a hydrazine concentration of 1.5 M was used.[2134] This could probably be reduced by using a less concentrated fuel supply, provided that the fuel electrode performance were not affected. The use of separate fuel and electrolyte circuits would eliminate this loss, but fuel savings would have to be balanced against the extra weight, volume, and complexity involved.

United Technologies has used packed beds of dry sodium hydroxide particles of ≤10 μm diameter, uniformly mixed with 10~60% by volume of Teflon particles of the same size, to remove carbon dioxide from air to a level below 0.25 ppm.[1359] Consideration has been given to the use of partly hydrophobic carbon particles as a more efficient packing material than that conventionally used in carbon dioxide absorption towers.[974]

Regenerable Absorbers

Reduction of atmospheric carbon dioxide concentrations to desired levels can easily be achieved by the use of molecular sieves, but these have the drawbacks of requiring dry air (since water is preferentially absorbed) and they also require considerable energy for regeneration. One such system is Union Carbide's *Pressure Swing Absorption* (PSA) process,[974] which can be used to treat reformer product gas to separate almost all impurities from the hydrogen, giving a resulting product containing 99.99+% hydrogen.

Monoethanolamine also is commonly used to absorb carbon dioxide, but such systems also require considerable energy (in the form of steam) for regeneration. Two other disadvantages of systems of this type are the constant vaporization loss of monoethanolamine reactant, which may become a fuel cell catalyst poison, and its inability under normal conditions to reduce carbon dioxide content below about 50 ppm. Siemens A.-G. has developed an improved system that employs a mixture of monoethanolamine and butanol.[1125] The use of monoethanolamine scrubbers was examined by the CGE (France) for the removal of carbon dioxide from methanol reformate for a 500W AFC system. Regeneration was via waste heat from the reformer.[1301]

The Fluor Corporation's *Fluor Solvent Process* is based on the absorption of carbon dioxide in propylene carbonate.[788] This process operates at very high pressures of 4600 kPa (45 atm) and requires methanation to remove final traces of carbon dioxide. While not appropriate for small fuel cell applications, it can be used to supply pure hydrogen from a large reformer unit for use in AFC systems.

Exxon has developed the *Catacarb* and *Axorb* additive processes for the absorption of carbon dioxide.[788] These processes use the additives to increase the rate of the reaction of carbon dioxide with alkali to form carbonate. The systems employ steam stripping and regeneration towers, and are able to remove 97% of the carbon dioxide from reformer output gas initially containing 20 mol % carbon dioxide.

Finally, the *Rectosol* and *Selexol* pressure-swing and temperature swing processes for removal of acid gases from reformate or coal gas should be mentioned,[974,1794,1868] although they are more commonly used for H_2S and COS rather than CO_2 removal.[394] The first uses methanol as the solvent, the second dimethylether polyethylene glycol. All the above processes are reviewed in Refs. 974, 1794, 1868.

Electrochemical Removal

The sulfate cycle is a partially electrochemical process that removes carbon dioxide and generates oxygen. The carbon dioxide is removed by absorption in caustic soda solution which is then treated with sulfuric acid. The resulting sulfate solution is electrolyzed in a three-compartment cell to regenerate sulfuric acid at the anode, and caustic soda at the cathode by membrane electrodialysis. The catholyte is recycled for further absorption of carbon dioxide. Cell lifetimes of 3000 h or better are obtainable with a suitable selection of membranes, using liquid differential pressures on the order of only 20~30 kPa (3~4 psi).[2785] The principal disadvantages of this system appear to be the high overvoltage required to electrolyze the solution, the reduced reactivity of the hydroxide in the presence of the sulfate, and the inefficiency caused by the neutralization of unreacted hydroxide by acid every time it is returned to the electrolyzer. Typically, cell voltages of 4~7 volts are required. If five times the stoichiometric amount of air were processed with a 50% utilization of sodium hydroxide, some 10% of the gross output of a typical fuel cell would be consumed by the regeneration system. This cannot be claimed as an energy-efficient process for electrolyte regeneration.

An electrochemical regeneration method which can be used directly in an AFC system has been devised by Pratt and Whitney Aircraft.[1078] In this system, carbonate ions that have accumulated in the electrolyte are converted to carbon dioxide at a fuel cell anode by cell operation at a high current density. The carbon dioxide is then swept from the anode gas cavity by the hydrogen, and can be rejected by bleeding off part of the fuel stream.

The basic principle of operation is as follows: Both hydroxyl ions and carbonate ions formed at the cathode migrate to the anode where the hydroxyl ions are consumed in the power production process which occurs within the anode pores:

$$2OH^- + H_2 \rightarrow 2H_2O + 2e^-$$

If the anode is run at a sufficiently high current density, the OH⁻ ions will be consumed in the anode pores so fast that the steady state hydroxyl ion concentration in the pores, $[OH^-]_{pores}$, will be lower than that in the bulk electrolyte solution. Consequently, because of the water dissociation equilibrium—

$$H_2O \rightarrow H^+ + OH^-; \quad K_w = [H^+][OH^-] \approx 10^{-14}$$

—the concentration of H⁺ ions within the anode pores will be much greater than that in the bulk electrolyte solution:

$$[H^+]_{pores} = K_w / [OH^-]_{pores} \gg [H^+]_{bulk} = K_w/[OH^-]_{bulk}$$

The primary and secondary dissociation of carbonic acid can be represented as follows:

$$H_2CO_3 \rightarrow H^+ + HCO_3^-; \quad K_1 = [H^+][HCO_3^-]/[H_2CO_3] \approx 4.3 \times 10^{-7}$$

$$HCO_3^- \rightarrow H^+ + CO_3^=; \quad K_2 = [H^+][CO_3^=]/[HCO_3^-] \approx 4.7 \times 10^{-11}$$

Accumulated carbonate in the electrolyte sets up the following hydrolysis equilibrium:

$$CO_3^= + H_2O = HCO_3^- + OH^-$$

$$K_{h1} = [HCO_3^-][OH^-]/[CO_3^=] = K_w/K_2 = 2.1 \times 10^{-4}$$

If the concentration of OH⁻ ions inside the anode pores is low enough, this reaction proceeds to the right and forms bicarbonate ions. The carbonate and bicarbonate will coexist in equal proportions when the ratio $[HCO_3^-]/[CO_3^=]$ equals unity. This occurs when $[OH^-] = K_{h1} = 2.1 \times 10^{-4}$; that is, when the OH⁻ concentration in the anode pores is about four orders of magnitude lower than in the bulk solution. Since pH $\approx -\log[H^+]$, this corresponds to a pH of $-\log(K_w/[OH^-]) = 10.3$.

Similarly, as bicarbonate in turn accumulates, a further hydrolysis reaction can occur, generating free carbonic acid:

$$HCO_3^- + H_2O \rightarrow H_2CO_3 + OH^-$$

$$K_{h2} = [H_2CO_3][OH^-]/[HCO_3^-] = K_w/K_1 = 2.3 \times 10^{-8}$$

The bicarbonate and free carbonic acid will coexist in equal proportions when the ratio [H_2CO_3] / [HCO_3^-] equals unity. This occurs when [OH^-] = K_{h2} = 2.3 × 10^{-8}, that is, when the OH^- concentration in the anode pores is about eight orders of magnitude lower than in the bulk solution. This is followed by decomposition of the carbonic acid, liberating carbon dioxide at the anode:

$$H_2CO_3 \rightarrow H_2O + CO_2\uparrow$$

The key to this method of electrolyte regeneration is clearly the establishment of a very low hydroxyl ion concentration inside the fuel cell anode under regeneration conditions; meaning, creating a sharp hydroxyl ion concentration gradient between cathode and anode working surfaces. As mentioned in the preceding, this condition can be brought about by operating the fuel cell at a high current density. Thus, a cell operating at a constant current density of 100 mA/cm^2 began to show performance degradation from carbonate build-up after about 7000 h of operation. After 10,000 h, the current density was increased to 400 mA/cm^2 for 100 h and then returned to 100 mA/cm^2. The cell voltage increased by 190 mV, returning almost to the performance level at the beginning of the test. At 15,000 h the regeneration procedure was repeated, and the performance again showed improvement. This simple regeneration procedure doubled the useful life of the cell from 10,000 h to more than 20,000 h.

The preferred Pratt & Whitney system was to use a circulating electrolyte system having special cells incorporated into the regular fuel cell stack designed to operate on hydrogen and unscrubbed air at high current density. These special cells would contain membrane diffusion barriers to accentuate the development of a sharp OH^- or pH gradient, and would be capable of continuous electrolyte regeneration. The overall efficiency penalty resulting from their use was reported to be less than 1% for a power plant operating on normal ambient air and clean hydrogen. Furthermore, this regenerative system was stated to require only a few percent of the total cell area in the fuel cell stack. For a power plant rated at several kW or less, a single cell was stated to be sufficient for the regeneration. Experiments showed that the regenerator cells were of reasonable size and did not seriously penalize overall efficiency, even at input rates equivalent to using air containing 3000 to 4000 ppm of carbon dioxide.

The above concentration is comparable to 1 volume % carbon dioxide in the hydrogen entering the fuel cell stack. The hydrogen effluent from a simple reformer contains at least 20% carbon dioxide and will therefore still require purification before entering the fuel cell, even if an electrochemical regenerator is used. However, it may be possible to use such a system to separate hydrogen of the purity needed from reformer effluent using inexpensive selective organic membranes rather than a costly and energy-intensive palladium-silver separator. It may be feasible also to use the effluent from a thermal cracking hydrogen generator directly, since the cracker exhaust contains only small concentrations of carbon monoxide and carbon dioxide.

We should finally point out that a high current density electrolyzer anode could also be used to reject CO_2 instead of the high-current density fuel cell proposed. Such a cell might be even more effective on a volume basis, since electrolyzers can run at much higher current densities than fuel cells, so many fewer cells per stack would be required. The disadvantage would be the fact that electrolyzer cells (as in the sulfate process discussed earlier) are parasitic users of power, whereas the proposed Pratt & Whitney decarbonation cells still produce power, though under very inefficient conditions because of their built-in concentration polarization requirements. The choice in any given system will depend on overall efficiency, and ultimately on cost-effectiveness. However, it should be borne in mind that in all cases the mechanical, efficiency (IR-drop due to wide inter-electrode gap), and cost penalties resulting from the circulating electrolyte must be considered.

Exxon Research and Engineering Co. has investigated the use of an electrochemical decarbonatation scheme for use with alkaline methanol-air fuel cells.[788] In this process the anolyte and catholyte were separated by a membrane and circulated along the length of several cells. The lower pH (acidity) generated at the anode by OH^- consumption eventually results in the evolution of the carbon dioxide formed from the oxidation of methanol. The alkalinity is then restored by mixing the anolyte and catholyte. Potassium fluoride should be added to avoid nonconductive solutions at the anode, and to raise the pH for carbon dioxide evolution.

A large-scale absorption tower approach using NaOH as the scrubbing agent, with regeneration of the reagent by electrodialysis, has been patented by the Royal Institute of Technology in Stockholm, Sweden.[1622, 1624]

Life Systems, Inc. of Cleveland, Ohio has examined electrochemically regenerable carbon dioxide absorber concepts utilizing regenerable absorption beds.[2225, 2727] The cells they propose incorporate advanced lightweight plated anode current collectors, internal liquid cooling, and lightweight cell frames, and are designed to meet the carbon dioxide removal requirements of 1.0 kg/day; that is, the requirements for one person in a submarine life-support system.

Soviet workers have examined the removal of carbon dioxide from air using a proposed hydrogen-air carbon dioxide concentrator cell employing aqueous cesium carbonate electrolyte and a carbonyl nickel anode.[1974] Concentrated cesium carbonate also has been studied as a potential CO_2-tolerant highly-buffered alkaline electrolyte.[534, 535, 542, 1017]

Finally, electrochemical reduction mechanisms of carbon dioxide have been studied using metal porphyrin catalysts [2406] and on rhodium electrodes.[2299]

Membrane Removal

Alsthom/Exxon (France-USA) researched and developed a novel carbon dioxide removal technique in the 1970s, in which carbon dioxide is transferred from the fuel (from a reformer) through a selective membrane into the fuel cell exhaust air.[788] The transport rates through the membrane that were demonstrated were several times higher than the best literature data available. The only additional energy required by the system was required to pump the fuel and air along the opposite sides of the membrane. In a related development, Exxon Research and Engineering Co. patented a membrane for carbon dioxide removal.[308] It consisted of carefully graded layers of hydrophobic and hydrophilic materials, the latter being treated with pH buffering catalysts such as sodium arsenate or boric acid to ensure that carbon dioxide was converted to bicarbonate, which could then decompose to carbon dioxide in the exhaust air stream. Exxon claimed that the carbon dioxide content of reformer gases containing 23.6 and 10.1 mole percent carbon dioxide could be lowered to 5~12 and 0.4~3.5 mole percent, respectively, by using such membranes.

General Electric Co. has developed extremely thin (< 50 nm) polymer film membranes that are chemically reversible to carrier gas species.[1399] The company has reported that these membranes display nearly perfect selectivity for carbon dioxide as well as for other reactive gases. In a similar development, the Monsanto Co. (St. Louis, Missouri) has commercialized a tubular pressurized membrane separator system capable of separating gases such as carbon monoxide from hydrogen, or carbon dioxide from oxygen.[1801] It is claimed that the hollow fiber selectively-permeable polymer tubes can operate efficiently in the presence of various contaminants such as hydrogen sulfide, ammonia, and hydrocarbons for periods of up to five years.

As summarized, the major disadvantage of alkaline electrolyte fuel cells over those employing acid electrolytes is their inability to operate on gas containing carbon oxides or oxidizable carbon compounds, at least over long periods. However, the competition beween alkaline and acidic fuel cells will in future be closely related to cell cost per unit of output. If an alkaline cell can produce twice the power density of an acid cell at the same cost per square meter of cell stack, then the more expensive reformer and cleanup system required by the alkaline cell can be justifed. Electrode kinetics are definitely improved in alkaline systems, particularly at the cathode. A great future advantage of the alkaline cell will be its lack of reliance on platinum as a cathode catalyst, which will open up the world's transportation market, in which acid cells can never seriously compete using present catalysts because of the constraint of availablity, hence of market-driven cost.[178] Strongholds for alkaline cells will be found in future liquid fueled systems, and above all in electric vehicles. In the latter case, they may be used in parallel with secondary batteries to combine large capacity (Wh/kg) with high current density (W/kg). Similarly, promising experience has been obtained with the parallel operation of simple hydrazine-air fuel cells with nickel-cadmium batteries in powering communications equipment.

While most of the work of the U.S. Army was on acid cells for use with reformed methanol, it maintained some interest in alkaline systems. The U.S. Navy has sponsored work on fuel cells only for the Deep Submergence Search Vehicle System. The U.S. Air Force continues to support alkaline hydrogen-oxygen fuel cells (regenerative and high-current output studies), while NASA has been concerned with alkaline systems for the space shuttle and manned orbiting-laboratory projects (see Chapter 6).

Thus systems destined for commercial use have tended to be based on the PAFC, while alkaline work has been funded mainly by government. On the other hand, renewed interest in commercializing alkaline fuel cells for use in electric vehicles has been shown by Brookhaven National Laboratory [394] and by ELENCO in Belgium (Chapter 7), while Occidental Petroleum has continued to support the French Alsthom AFC system, which would use surplus hydrogen and produce d.c. electricity in caustic-chlorine plants. In addition, the alkaline cell is of great interest in Sweden. These developments are discussed in Chapter 5.

It may be interesting at this point to briefly examine why the Europeans have held onto the philosophy of the AFC for general use, whereas in the United States more attention has been focused on the PAFC since 1970. Looked at from the aspect of the fuel cell itself, materials problems are fewer in alkaline solution, and satisfactory alkaline fuel cells can be made from simple, inexpensive materials (for example, plastic cell frames) together with non-noble catalysts (for example, carbon at the cathode, Raney nickel at the anode). In contrast, acid fuel cells required exotic materials in the late 1960s, including noble metal catalysts. Finally, performance was much better in AFCs. This is illustrated by the oxygen electrode data shown in Fig. 10–1 [177] for 96% phosphoric acid at 165°C and for 30% KOH at 70°C, both using the same low-loading (0.25 mg/cm^2) carbon-supported platinum electrodes under pure oxygen at atmospheric pressure. To show the effect of changing the nature of the acid, results are also shown in a dibasic fluorinated sulfonic acid (see following) at 110°C. The latter is about 40 mV better than phosphoric acid, despite being at a lower temperature. However, the spectacular improvement due to the lower Tafel slope (Chapter 2) in KOH, even at 70°C, is apparent: A difference of about 120 mV is observed at 200 mA/cm^2.

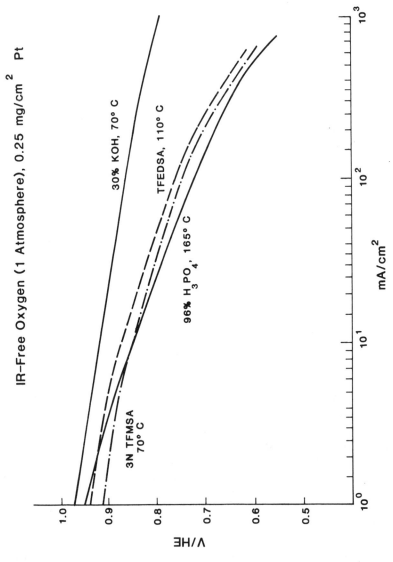

Figure 10-1: Oxygen reduction (1 atmosphere O_2) on 0.25 mg/cm² supported pt electrodes (10 wt% on carbon) in aqueous acid and alkaline electrolytes.

Electrolytes

These characteristics stimulated European interest in the AFC, but the assumption had to be made that hydrogen fuel would be available for the system. At 0.77 V overall cell potential, hydrogen can be consumed in the AFC at an HHV efficiency of 0.77/1.48, or 52%, whereas the PAFC operating at 0.65 V under the same current density conditions will allow only 44% efficiency. If hydrogen is available, the use of the AFC is clearly indicated. However, if the hydrogen must be manufactured at 65% efficiency from another fuel (natural gas or coal), which requires a heat input and steam, then the effective efficiency of fuel use in the AFC is only $0.65 \times 52 = 33.8\%$, since the AFC operating at 70°C can provide no waste heat for use in the on-site fuel treatment process. In contrast, a PAFC operating at about 200°C produces waste heat in the form of steam. This not only allows it to be pressurized, thereby increasing cell performance from 0.65 to 0.73 V, but also provides steam for fuel processing. The use of this steam effectively increases fuel processing efficiency to 85%, so the efficiency of fuel use becomes $(0.73/1.48) \times 0.85$, or about 43%. Overall, the PAFC is therefore a much more efficient generator of electricity from fossil fuel than the AFC, in spite of its lower cell potential. These systems aspects of the fuel cell–integrated fuel processor were largely ignored in Europe, but were strongly stressed in the United States, particularly at UTC. However, the inexpensive, non-noble metal alkaline system is certain to become more important as fossil fuels become more scarce, and as hydrogen becomes more widely available as a fuel.[393, 394]

B. AQUEOUS ACID ELECTROLYTES

The principal advantage of acid electrolytes over alkaline electrolytes is that they are carbon dioxide–rejecting, thus acid fuel cells can use unscrubbed air or fuel gases containing 30% or more carbon dioxide from a hydrocarbon fuel processor. Purification is only required to keep anode poisons such as sulfur compounds and CO at acceptable levels. When reformate is used, the percentage of carbon dioxide increases as hydrogen is consumed during passage through the fuel cell, making it necessary to vent gas containing some hydrogen at low partial pressure from the anode exhaust (Chapter 2). In general, only 80~90% of the hydrogen is used in the cell, whereas alkaline cells can use 99% of pure hydrogen feedstock. (We should note that the same is true of acid cells if they also use pure hydrogen.) However, the lower fuel utilization in acid cells using reformate is not serious from the viewpoint of overall system efficiency, since anode tail-gas can be burned to provide useful heat for the reforming reaction. If reformers are designed to produce pure hydrogen (e.g., by pressure-swing adsorption), the tail-gas from the absorption process is burned to provide reformer heat, so that the percentage of hydrogen available for use in the fuel cell is the same in both cases. As we have just pointed out, acid cells appear to have a major disadvantage over alkaline cells: that of a higher Tafel slope for the oxygen reduction reaction (Fig. 10–1). This is a result of the difference in reaction mechanism, which is at present not entirely understood.[168] In reactions requiring the transfer of several electrons for completion, of which the four-electron oxygen process is typical, theory [168] shows that the Tafel slope should decrease from a value of about $2.303(2RT/F)$ in increments or fractions of $2.303RT/F$, according to where the rate-determining (i.e., slowest) step is situated in the overall step-wise reaction sequence. In the preceding expressions, R is the Gas Constant, F is the Faraday, and T is the absolute temperature.

Neglecting any effects resulting from complex adsorption processes, the simple theory[389] gives the slope as $2.303RT/\alpha_c F$, where α_c, the *transfer coefficient* for the reduction reaction, is given by $[(\gamma/\nu) + r\beta]$, where γ is the number of electrons transferred before the rate-determining step(s), ν is the number of times the rate-determining step occurs for one act of the overall reaction, β is the *symmetry factor* for the rate-determining step (usually assumed to have a value of about 0.5), and r is the number of electrons transferred in the rate-determining step. The only permitted values of r are 1 (if the rate-determining step is an electron transfer step) and 0 (if the rate-determining step is a purely *chemical* step). If the first electron transfer step is rate-determining, $\gamma = 0$ and $r = 1$; whereas if the second is rate-determining, $\gamma = r = 1$. If a chemical step following a primary electron transfer is rate-determining, $\gamma = 1$, and $r = 0$, and so on. Consequently, at 25°C (298 K), the Tafel slope should vary from about 120 mV/decade for reactions in which a first electron transfer is rate-determining, to 60 mV/decade for a rate-determining chemical step following an electron transfer; then to 40 mV/decade for a rate-determining second electron transfer, to 30 mV/decade for a rate-determining chemical step following two electron transfers. The latter two very low slopes imply a very efficient process in which earlier steps have been catalytically strongly accelerated: even lower slopes might occur if catalytic means could be found to push the rate-determining step farther back towards the end of the process, but this is not within the means of our present knowlege of electrocatalysis.

The theory as currently understood [168,389] is defective, since Tafel slopes for oxygen reduction often do not show the precise values expected or the correct temperature coefficients;[673, 2764] although this may be an artifact resulting from the complications of adsorption. This observation is particularly true for the reduction in phosphoric acid on platinum, for which a first electron transfer appears to be rate-determining.[168] The Tafel slope is indeed about 120 mV/decade at 298 K, but it is virtually independent of temperature, so that at 200°C (473 K) it deviates considerably from the expected value of 190 mV/decade. On carbon-supported high surface-area platinum in phosphoric acid, the slope is even less, typically about 110 mV/decade, and is again independent of temperature.

In aqueous KOH at 80°C the corresponding slope is about 50 mV/decade, corresponding to a second rate-determining electron transfer. The practical implication of this is that the oxygen electrode is about 80~100 mV less polarized in KOH at 70~80°C than it is in phosphoric acid at 200°C, and the change in voltage with current density is correspondingly much lower in KOH. Performance of the PAFC can be improved to the same levels as in KOH by pressurizing the cell's fuel and air supply, which is practical since waste heat from the 200°C PAFC is of high quality. In contrast, we have already noted that a KOH cell operating at 70~80°C (the maximum temperature allowed for the use of carbon and other low-cost materials in this medium), cannot profitably be pressured. Another major point is that a tenfold pressurization would increase PAFC oxygen cathode performance by the same amount as the Tafel slope, viz., 110 mV. For KOH electrolyte, the corresponding increase would be only 50 mV.

In summary, at normal levels of pressurization (currently a maximum of 8.2 atm at International Fuel Cells, IFC) the higher temperature PAFC recovers the performance that it has lost, compared with the low-temperature AFC, which cannot be pressurized. Even if the latter could be pressurized, the increase in voltage (as distinct from possible improvements in current density) would hardly be worthwhile, since the improvement in overall system efficiency would be minor compared with that available on pressurizing the PAFC. Where pressurization can be used (i.e., in large stationary units), the acid-electrolyte PAFC has an advantage in being able to operate on impure hydrogen. For unpressurized use (e.g., small units, mobile applications), the AFC is advantageous provided that pure hydrogen is available.

The two acid electrolytes which initially received most attention were sulfuric acid and phosphoric acid. Sulfuric acid is limited to temperatures below 100°C, since it is catalytically reduced at the hydrogen anode at higher temperatures, but in other respects it is an excellent high conductivity electrolyte. However, when fuel gases containing carbon monoxide are used with platinum catalysts, it is necessary to raise the cell temperature to 130~165°C (depending on the CO concentration) to prevent poisoning of the anode catalyst. This is one the main reasons for the original choice of phosphoric acid by UTC for the small, nonpressurized TARGET-type units (12.5 kW) in 1967 (Chapter 3).

Among acids that are reasonably strong, phosphoric acid has a high temperature stability that is unique for an inorganic aqueous system. Orthophosphoric acid systems are not invariant, but lose water at elevated temperatures to form relatively stable condensed compounds.[875] These condensed acids (metaphosphoric, pyrophosphoric) not only are stronger acids than orthophosphoric acid, but also have a higher conductivity. Indeed, this conductivity rises with temperature as further water loss and condensation takes place, and at 200°C the conductivity of 100% H_3PO_4 is close to that of 6M KOH at room temperature. Similarly, changes in relative humidity at a given temperature do not give rise to large conductivity changes. The acid is now known to be unique in its property of being able to operate at very low relative humidities at high temperatures (200°C) without loss of ionic conducting properties. A further advantage is its very low (although not negligible) volatility at normal operating temperatures. The properties of phosphoric acid as an electrolyte have been reviewed.[2192]

Phosphoric acid electrolyte fuel cell technology has progressed more rapidly over the past decade than that of any other type of fuel cell. It is particularly attractive for air-breathing fuel cells, and since it can use dilute and impure hydrogen streams, it can be matched to many different types of fuel source. As indicated in the preceding, its prime advantage in air-breathing cells lies in reactant-product-thermal control. The condensed acid always exhibits adequate ionic conductivity under practical conditions so that there is no effective upper air flow limit to cause excessive water removal and cell failure due to dehydration, loss of conductivity or gas crossover, as in other aqueous or solid polymer electrolyte (SPE) systems (see following).

Similarly, the entire gas flow rate–humidity–stack temperature interrelationship for moisture control is open-ended, so that stack temperature control need not be precise. Stacks have been designed to float as a function of load, and some have been designed to let the cells within a stack operate at different temperatures. Large phosphoric acid fuel cell stacks have been built which are cooled by process air; others have been liquid or evaporatively cooled. In this last method the cell waste heat vaporizes water to supply the steam reformer, thereby improving overall system efficiency by waste heat recovery. The concentration of the phosphoric acid initially used to fill cells is usually between 85 and 98% by weight, but the effective concentration under operating conditions will depend on the temperature and humidity. The condensed acid may have a concentration (expressed as % H_3PO_4) of up to 103~105% under unpressurized operating conditions, and of about 95~98% under pressurized conditions. Cells have also been constructed that were initially filled with polyphosphoric acid.[1611]

The main disadvantage of the acid cells employed in early work was the corrosion of most of the common cell construction materials and catalysts, especially under cathode conditions. For this reason, early workers employed costly noble metal catalysts. Rapid progress in reducing the cost of

Electrolytes

acid cells has been made, and the unexpected stability of carbons, particularly graphite, under cathode conditions even at 200°C in phosphoric acid, was the major breakthrough (dating from about 1970) that made the PAFC feasible.[170, 177, 618, 1414, 2360] Machined—followed by molded—graphite parts were developed, and the latter now are used routinely for stack separator plates and electrode substrates. Similarly, carbon blacks that have undergone heat- or chemical treatment have proven to be excellent catalyst supports. The minimal thermal control requirements of the PAFC also have helped to reduce costs.

Noble metal catalyst loadings have been lowered from several mg/cm^2 in 1970 to $0.25\ mg/cm^2$ or less for the hydrogen anodes used in liquid acid electrolytes. Similarly, cathode loadings have been reduced to $0.3\sim0.5\ mg/cm^2$. Catalysts such as those based on tungsten carbide have even shown some promise in eliminating the need for platinum metals at the anode,[352, 406] but for the oxygen cathode in acidic electrolytes, an effective substitute for platinum has yet to be found. Organometallic electrocatalysts such as derivatives of the transition metal macrocyclics [177, 335, 2261] may point the way towards a future platinum substitute for the oxygen cathode, but these catalysts have yet to achieve anything approaching the long-term stability (40,000 h) required for practical use in acid fuel cells.

A recent tendency in PAFC development has been to increase the cell operating temperature to 200°C or even to 210°C, in order to obtain better cell performance. However, as discussed in the next chapter, concentrated phosphoric acid is measurably volatile and may even become unstable at temperatures above 200°C. This has necessitated the use of more dilute acid (i.e., higher humidity conditions) which accelerates carbon corrosion (corresponding to the attack of carbon by water vapor), at the same time making water management somewhat more difficult. It has long been known that phosphoric acid is not the most effective acid electrolyte for oxygen reduction on platinum. Indeed, it is not as effective as sulfuric acid, and is much inferior to HF and even $HClO_4$;[1613] however, all these common acids are unstable or volatile at high temperature. From 1967 onwards, there has been a search for new acid electrolytes that might offer better performance than phosphoric acid at the high temperatures required for utility fuel cell operation.

Much of the recent work has centered around organic sulfonic acids such as trifluoromethansulfonic acid (TFMSA) and its monohydrate (TFMSA-MH). A 60% aqueous solution of this acid has been shown to support current densities greater than $400\ mA/cm^2$ at 25°C and 0.6 V in a hydrogen-air fuel cell system using phosphoric acid hardware.[938] Such performance is analogous to that obtainable with HF.[1613] Aqueous TFMSA-MH solutions have conductivities that are lower than similar solutions of sulfuric acid but higher than solutions of phosphoric acid.

Considerable work was conducted at the American University in Washington, D.C., on evaluating the fuel cell–related properties of TFMSA-MH and TFMSA.[874, 875, 2190, 2191] The corrosiveness of TFMSA towards most fuel cell materials appeared to be 1~2 orders of magnitude lower than that of phosphoric acid.[2191] However, the volatility of TFMSA-MH appears to be acceptably low for fuel cell operation only up to about 150°C. The material wets Teflon at acid concentrations greater than about 60%, so that it cannot be used with conventional electrodes under these conditions. On the other hand, it has been found that for the direct oxidation of hydrocarbons on platinum catalysts, TFMSA-MH electrolyte at 95~135°C supports higher oxidation rates than hydrofluoric acid, sulfuric acid, and phosphoric acid, the order of increasing reactivity being propane > ethane > n-butane > methane.[24, 26] Workers at the American University found that hydrogen and propane oxidation rates were an order of magnitude higher in TFMSA-MH than in phosphoric acid.[25] Similarly, workers at New York University found steady state propane oxidation currents to be twice as high in TFMSA-MH as in 85% phosphoric acid; and the U.S. Army found that propane oxidation rates in TFMSA-MH at 115°C and 135°C were 200 and 1000% higher, respectively, than in 85% phosphoric acid at 135°C.[255]

The durability and performance of fuel cell materials and catalysts in TFMSA-MH and TFMSA at 25°C and 70°C were evaluated by ERC.[938] Workers at the University of Ottawa have measured the density, electrical conductivity, freezing point, and viscosity of TFMSA-water solutions over the complete composition range.[677] Work carried out at the Argonne National Laboratories showed that at 104°C and 140°C, direct hydrocarbon oxidation proceeds much faster in TFMSA than in 95~97% phosphoric acid at the same temperature.[11] ERC also tested hydrogen-air fuel cell stacks using 6 M TFMSA at 60°C and found better performance than similar cells using phosphoric acid.[940] In fact, fuel cell performance using TFMSA at 60°C appeared to be better than that obtained using phosphoric acid at 180°C.[939] ERC also compared the performance of fuel cells using phosphoric acid, sulfuric acid, TFMSA, and tetrafluorobenzenesulfonic acid.[231, 232] The dissociation constant of TFMSA in sulfuric acid [2156] and in various non-aqueous solvents [896] has been determined, and the electrochemical properties of the acid and its salts in non-aqueous systems have been reviewed.[1257] Methods for the preparation of these and of other partially fluorinated alkane sulfonic acids have been discussed.[503, 511, 512]

Other synthetic organic acids being studied as possible improvements over TFMSA-MH have included methanedisulfonic acid;[502] ethanesulfonic acid; ethane-1,2-disulfonic acid; propane-1,2,3-trisulfonic acid; chlorotrifluoroethanesulfonic acid, and what was originally reported to be 1,2,2-trifluoroethane-1,1,2-trisulfonic acid.[875] Methanesulfonic acid also appeared at one time to be an attractive substitute for phosphoric acid. Workers at Pennsylvania State University found that this acid was superior to 85% phosphoric acid at 135°C for both direct and indirect hydrocarbon oxidation.[209] Since it contains no fluorine, its monohydrate does not wet Teflon. However, the acid is volatile and the long-term stability of its CH_3- group in regard to oxidation is doubtful.

Emphasis subsequently was placed on dibasic or higher synthetic acids whose volatility would be low. The aim was to synthesize fully fluorinated materials for maximum acidity and stability, but to avoid the presence of CF_3- groups that were suspected of wetting Teflon. Two acids that appear attractive as fuel cell electrolytes are difluoromethanediphosphonic acid and 1,2-tetrafluoroethanedisulfonic acid.[161, 2644] They were synthesized with EPRI support and were investigated by ECO, Inc.; Lawrence Berkeley Laboratory; and the United States Army. These acids appeared to be completely nonvolatile and not to wet Teflon. They were tested at temperatures up to 236°C. Eco, Inc. considered that oxygen reduction in these two acids occurs at more than 100 mV higher than in phosphoric acid under the same conditions, corresponding to a tenfold increase in the catalytic activity of the oxygen cathode. This increased activity does not seem to be a function of the hydrogen ion activity, and may be related to specific adsorption characteristics in fluorinated acids. Work has shown that tetrafluoroethane disulfonic acid is indeed stable, but that the stability of difluoromethanediphosphonic acid (which was prepared for the first time in 1980; Ref. 688) at temperatures over 150°C in the presence of water vapor is doubtful.[2146, 2322] In this connection, dibasic organic acids, particularly carboxylic acids, with both acid groups attached to the same carbon are generally unstable. In contrast to the preceding, the feasibility of employing substituted carboxylic acids as fuel cell electrolytes does not appear promising, on account of their high volatility, even at room temperature.[875]

The initial promise of the higher fluorinated sulfonic acids, the simplest stable member of which is 1,2-tetrafluoroethanedisulfonic acid, which is chemically closely related to Nafion (see following), has been frustrated by the realization that its monohydrate becomes a very poor ionic conductor as the water content becomes low (i.e., at temperatures over about 110°C[177, 2143]). Even so, it is more active than phosphoric acid for oxygen reduction by about 20 mV, but in a practical fuel cell this advantage is lost because of higher IR-drop. This results from the fact that the H_3O^+ ion is the only conducting species. Since this species is relatively immobile, the proton can move only if water molecules are present, enabling a Grotthus chain conduction mechanism to be set up. As is now apparent, Nafion behaves in the same way. It is also clear that the high temperature conductivity properties of phosphoric acid, which involve the creation of conduction chains containing $H_4PO_4^-$ ions and neutral molecules (or their polymers) are altogether exceptional. The relative conductivities at high temperature are shown in Fig. 2.15–2 in Ref. 177.

The development of acid electrolyte fuel cells has progressed rapidly in the past few years: major commercial fuel cell development in the United States mostly involves 200°C PAFC systems. Even though platinum supply ultimately may limit the market, a broad range of future applications may be expected to open up now that noble metal loadings have been lowered to less than a total of 1 mg/cm^2 on a routine basis (corresponding to 5g/kW, or \$80/kW), and lifetimes in the tens of thousands of hours are being demonstrated. In view of the high cost of research and development and the low levels of production, the involvement of industry depends to a great extent upon government procurement. Outside of DOE's interest, the U.S. Army had up to late 1985 leaned largely to acid cells using reformed methanol and hydrocarbons (Chapter 6).

Solid Polymer Acid Electrolytes

Solid polymer electrolytes (SPEs) are usually understood to consist of a polymer network to which are attached functional groups capable of exchanging cations or anions. The polymer matrix generally consists of a cross-linked organic polymer such as polystyrene, polyethylene, partially fluorinated polyethylene, fluorinated polyethylene, or other PTFE analogs. In general the material is acid, with a sulfonic acid group incorporated into the matrix. All SPE membranes of practical importance are therefore acidic and conduct hydronium ions. An alkaline system containing suitable groups that would act as an OH$^-$ ion conductor would be highly desirable, but no such stable system has been synthesized. All acidic SPEs are gels requiring the presence of water molecules as well as protons for conductivity. The ratio of H_2O to H^+ for effective conductivity is typically about 3:1. For an OH$^-$ conducting polymer (if it existed), the ratio would probably be about the same. The conductivity of such polymeric materials therefore depends on the local water vapor pressure, and therefore on temperature. Overall, the majority of such SPEs can be regarded simply as acid electrolytes in which the anion is immobilized by the polymer structure. Hence, for the most part their properties are simply those of the involatile sulfonic acid electrolytes described in the preceding section. For this reason, we have included them here in the category of acid electrolytes.

The apparent advantages of using acid SPEs are numerous. The solid electrolyte membrane is simpler and more compact than other types of electrolyte, with the possible exception of the matrix PAFC structure, which in any case requires an electrolyte reservoir to allow for electrolyte volume expansion changes and to give long life. We should remember that when the SPE was developed in the late 1950s, other systems often required electrolyte recirculation and containment, together with relatively complex instrumentation. The SPE membrane itself can act as an electrode support and a major cell structural component. Furthermore, in early approaches there were no severe requirements for a well-defined electrode structure from the viewpoint of porosity, pore volume, and pore-size distribution, as was the case for a cell using liquid electrolyte and porous gas diffusion electrodes. However, this has been shown to be an oversimplification, since the electrode-electrolyte interfaces in liquid-electrolyte cells have allowed great reductions in catalyst loadings as the nature of the porous interface became better understood; whereas the two-dimensional solid polymer-catalyst interface, which wastes catalyst surface area, has so far proved to be intractable to much reduction in catalyst loading without drastic loss of performance. Even so, it is certain that bonded SPE membrane–electrode assemblies are amenable to mass production and can be fabricated into very thin (0.025 cm) sheets, giving low IR-drop under normal conditions.

The solid matrix of the membrane is invariant (although membrane volume varies somewhat with water-vapor pressure), since product water at the cathode side can be rejected either by static capillary wick systems [176] at low temperature, or in the vapor phase at higher temperature. This feature is important for zero-gravity aerospace applications in which cavitation and handling problems can be encountered with circulating liquid electrolytes, although the characteristics of the matrix PAFC are similar. However, and in contrast to the PAFC, the ability of the SPE to function at ambient (or less than ambient) temperature and pressure may be a factor that increases reliability and decreases complexity and therefore cost for certain applications. In addition, creep of the gelled electrolyte is usually negligible in SPE systems, and corrosion is minimal since pure water (rather than an acid leach) is the only reaction product.

For an acid ion exchange membrane used as an electrolyte, a number of properties are desirable.[1666, 1791] These include: high ionic conductivity with zero electronic conductivity; low gas permeability; dimensional stability (resistance to swelling); high mechanical strength; minimal water transport; high resistance to dehydration; oxidation, reduction, and hydrolysis; high cation transport number; surface properties allowing easy catalyst bonding; and homogeneity. Since no single membrane will possess all these properties, a compromise structure will result.

The progress in SPE fuel cells can be seen by considering the work carried out by the General Electric Company during its program to develop membrane cells and electrolyzers.[1094, 1557, 1741] The earliest membranes developed by General Electric for fuel cell applications were made by condensation of phenolsulfonic acid and formaldehyde. These membranes were brittle, cracked when dried, and rapidly hydrolyzed to sulfuric acid. The next series of SPE membranes was based on partially sulfonated polystyrene. The properties of these membranes were still unsatisfactory, and lifetimes of only about 200 h at 60°C were obtained. These were followed by the D series of membranes, made by cross-linking styrene-divinylbenzene into an inert fluorocarbon matrix, followed by sulfonation. The D membranes were the first to have acceptable strength in both the wet state and the dry state, and were used in seven Gemini space missions, starting in 1964.[176] However, lifetimes were still only about 500 h at 60°C, because of degradation resulting from attack on the weak alpha C-H bond in the polymer structure. Improved cross-linking increased life to about 1000 h.

The next series of membranes, designated the S series, was made from homopolymers of alpha-β,β trifluorostyrene sulfonic acid.[586] These showed excellent chemical and thermal stability but still left much to be desired from the viewpoint of physical properties. They were improved by blending fluorocarbon sulfonic acid polymer with polyvinylidene fluoride using triethyl phosphate plasticizer, giving membranes capable of a 2000 h life at 80°C. Grafting trifluorostyrene into an inert fluorocarbon matrix further improved the lifetime to about 3000 h.

At this point (about 1968) DuPont developed their Nafion series of SPE membranes, which consists of polymers of perfluorosulfonic acid, sometimes reinforced with Teflon.[2465, 2767] The polymers are made by reacting tetrafluoroethylene with SO_3 to form a cyclic sultone (see Fig. 10-2, Ref. 2770), which forms a rearranged product that is then reacted with $(m+1)$ molecules of hexafluoropropylene epoxide, where $m > 1$. When the product of this reaction is heated with sodium carbonate, a sulfonyl fluoride vinyl ether is formed. This is copolymerized with tetrafluoroethylene to form XR resin, which can be melted and made into sheets or tubes. Base hydrolysis of these sheets or tubes yields Nafion®. The Nafion series of SPE membranes may be regarded as sulfonated analogs of Teflon, even though they contain ether linkages (see Fig. 10-2). On the molecular level, the polymer has a tubular structure, in which the $-SO_3^=$ groups are on the inner surface of the tubes, which provide the hydrophylic conduits for conduction, the outer parts of the tubes being hydrophobic fluorinated material. The tubular structures shrink and rearrange as the water content decreases (see Fig. 4.6-2 in Ref. 176).

These polymers are generally unaffected by both strong acids and oxidants and are highly conductive to hydronium ions when hydrated to about 20% by weight. They can be fabricated in 120 cm-wide sheets of indefinite length with typically 7 mils (175 microns) thickness. Commercial products are denoted by numbers (e.g., Nafion 1170), where the first two

Tetrafluoroethylene reacts with SO_3 to form a cyclic sultone. After rearrangement, the sultone can then be reacted with hexafluoropropylene expoxide to produce sulfonyl fluoride adducts, where

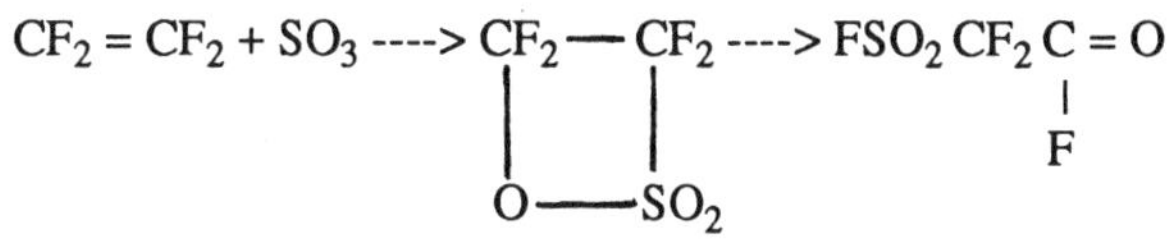

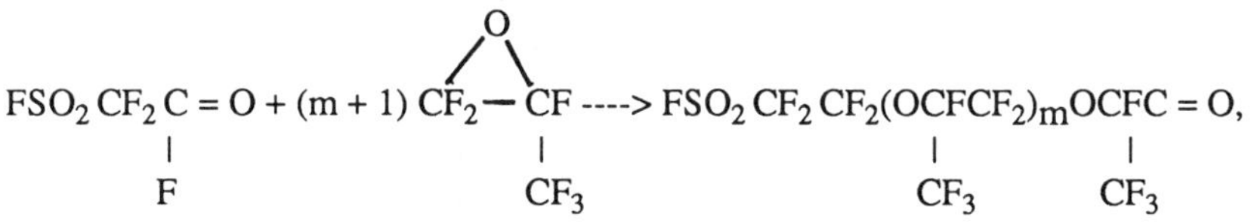

$$CF_2 = CF_2 + SO_3 \longrightarrow \underset{\underset{O \longrightarrow SO_2}{|\qquad\quad|}}{CF_2 - CF_2} \longrightarrow FSO_2 CF_2 \underset{F}{\overset{|}{C}} = O$$

$$FSO_2 CF_2 \underset{F}{\overset{|}{C}} = O + (m+1) \underset{CF_3}{\overset{|}{CF_2}} \overset{\overset{O}{\diagup\,\diagdown}}{-} \underset{}{CF} \longrightarrow FSO_2 CF_2 CF_2 (O \underset{CF_3}{\overset{|}{C}} FCF_2)_m O \underset{CF_3}{\overset{|}{C}} FC = O,$$

$m \geq 1$. When these adducts are heated with sodium carbonate, a sufonyl fluoride vinyl ether is formed. This vinyl ether is then copolymerized with tetrafluoroethylene to form XR resin:

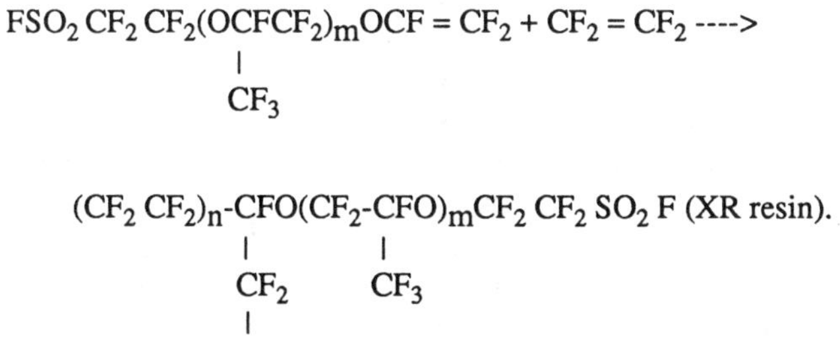

$$FSO_2 CF_2 CF_2 (OCFCF_2)_m OCF = CF_2 + CF_2 = CF_2 \longrightarrow$$
$$\qquad\qquad\qquad\quad\underset{CF_3}{|}$$

$$(CF_2 CF_2)_n\text{-}CFO(CF_2\text{-}CFO)_m CF_2 CF_2 SO_2 F \text{ (XR resin)}.$$
$$\qquad\quad\underset{\underset{|}{CF_2}}{|} \qquad \underset{CF_3}{|}$$

This high molecular weight polymer may be fabricated by melting and can be processed into various forms, such as sheets or tubes, by using standard methods. On base hydrolysis, this resin is converted into Nafion^R perfluorosulfonate polymer:

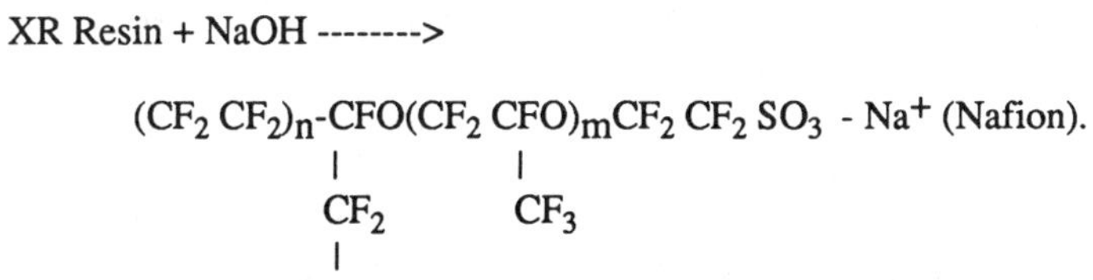

XR Resin + NaOH $\longrightarrow$

$$(CF_2 CF_2)_n\text{-}CFO(CF_2 CFO)_m CF_2 CF_2 SO_3 \text{ - } Na^+ \text{ (Nafion)}.$$
$$\qquad\quad\underset{\underset{|}{CF_2}}{|} \qquad \underset{CF_3}{|}$$

Figure 10-2: The preparation and structure of DuPont NAFION® perfluorosulfonic acid polymer.

digits are the equivalent weight divided by 100, and the last two the thickness in mils multiplied by 10.

Mechanical and dimensional stability of the polymer can be provided either by embedding PTFE mesh into the membrane or by supporting the membrane externally on a backing structure. Such supported Nafion membranes can withstand pressure differentials of greater than 70 atm and can operate at high pressure (200 atm). In both the wet and dry forms, Nafion displays excellent oxidative stability over the temperature range from 25~150°C. Its transport properties have been well characterized for membranes with equivalent weights in the range of 1100 to 1500. The degradation rate for recent Nafion 120 membranes can be expressed by:[1557]

$$r = 1.575 \times 10^{13} \exp[-11,731/T]$$

where r is the percent degradation after 10,000 h at a temperature of T (K).

Table 10-1 shows a comparison of the properties of Nafion with some other SPE membranes.[1557] General Electric Co. began to use Nafion membranes for fuel cell purposes in 1966 and 1968 for the Biosatellite mission, and used them exclusively after 1969. Nafion membranes have continuously undergone improvement, and lifetimes of greater than 40,000 h now have been achieved at 80°C.

A great deal of effort has been directed towards obtaining membranes with the highest possible conductivity, since this results in lower internal resistance. Because ionic conductivity should be approximately proportional to ion exchange capacity, high capacity membranes would appear to be desirable. However, a high density of ionizing groups in typical membranes results in poor mechanical strength and a poor ability to withstand thermal and physical stress.[588, 1015]

The conductivity of an ion exchange membrane drops sharply with the extent of hydration.[823, 880, 1038, 1372, 1712, 2317, 2564] Membrane dehydration also results in shrinkage and in a reduction in membrane-catalyst contiguity, giving higher contact resistance between the membrane and the electrode. Furthermore, cracking or pinholing may cause a chemical short circuit, giving local gas recombination, hot-spots, and the possibility of explosion. Decreasing the water content of an ion exchange polymer renders it more resistant to the effects of dehydration [2379] and water transport,[1015, 2471] but will decrease ionic diffusion and electromigration.

Usually the gases used in an SPE fuel cell are saturated with water to reduce membrane dehydration. Since water is produced at the cathode, and is above all necessary for the effective transport of H⁺ across the membrane as hydrated H_3O^+ ion, hydration of the anode gas is absolutely necessary for correct cell operation. Unfortunately, there appears to be

TABLE 10-1
Properties of Ion Exchange Membranes[1557]

Property	Desired Value	Phenol Sulfonic	Polystyrene Sulfonic	Poly(trifluoro-styrene sulfonic)	Per-fluorinated Sulfonic (Nafion)
Ionic Resistivity at 25°C (Ω cm)	≤ 20	100	≤ 20	≤ 20	≤ 20
Ionic Resistance at 25°C (Ω cm^2)	≤ 0.7	10	≤ 0.7	≤ 0.7	≤ 0.7
Hydrogen Permeability at 25°C (cm^3 m^{-2} h^{-1} atm^{-1})	≤ 21.5	≤ 21.5	≤ 21.5	≤ 21.5	≤ 21.5
Mullin Burst Strength at 25°C (atm)	> 3.4	*	10.2	2.0~2.7	5.1~6.8
Tensile Strength at Break at 25°C (kg cm^{-2})	> 140	*	210	63	140~210
% Elongation at 25°C	> 100	*	100	40	> 100
Maximum Temperature of Thermal–Hydrolytic Stability (°C)	> 100	50	> 100	> 150	> 150
% Oxidative Degradation After 18 h Exposure at 80°C to 30% H$_2$O$_2$ + 20 ppm Fe^{++}	< 1	100	100	< 1	< 1
Crease-Crack Resistance (ability to withstand 5 folds after drying at 100°C)	pass	fail	pass	fail	pass

*cloth reinforced

considerably more gas permeation through the membrane when the gases are humidified. Since this results in both parasitic reactant consumption (especially at open-circuit) together with excessive heat generation, low rates of gas diffusion through the membrane are essential for effective fuel cell applications.

Table 10–2 shows that as expected, hydrogen (molecular diameter = 0.235 nm) has a greater permeation rate than oxygen (molecular diameter- 0.295 nm). It also shows that phenolic membranes have high gas permeabilities, and are therefore unacceptable for use in fuel cells. Thinner membranes give a decreased IR-drop, but show more gas crossover via permeation, and have lower physical strength than thick membranes. If the cell is operated above ambient temperature, certain membranes soften, and electrode catalyst particles may then contact through the membrane under the pressure of the backing plates, short-circuiting the cell.

According to some reports, the conductivity of ion exchange membranes increases exponentially with temperature, the effect being greater for cation membranes than for anion membranes.[1712] However, it also has been shown that the conductivity may reach a maximum around 50°C.[2317, 2740] At higher temperatures (80°C) some membranes may exhibit chemical instability and undergo dehydration. These results may be rationalized on the basis of the method of measurement. Above all, conductivity depends on water content of the membrane. Hence, results obtained at constant water vapor pressure will be very different from those obtained in an autoclave.

In most of the older ion exchange membrane fuel cell work, membranes were soaked in acid, followed by simple blotting, rather than being used in leached* form. It was usually considered that this procedure reduced cell internal resistance, presumably because the free or diffusable electrolyte film acted as a current-carrying bridge between the membrane and the electrode. However, if free electrolyte is used, water will accumulate at the cathode as a result of the cell reaction and by electroosmosis accompanying the liquid phase transport of hydrogen ions. As water accumulates on the membrane surface, any free acid resulting from membrane degradation diffuses into it, so as the water escapes, acid is gradually lost, giving an increase in internal cell resistance. In addition, the molecular water gradient may dehydrate the anode side of the membrane, giving lower H^+ ion hydration numbers, hence still greater cell internal resistance.

---

*Whether a truly leached form (i.e., one having no unbound free electrolyte) really exists is open to question, since even the slightest amount of chemical degradation of the polyelectrolyte may form water-soluble, low–molecular weight sulfonic acids.[1666] Also, it has been found that repeated washings in distilled water fail to remove a residual low level of anionic impurity from membranes.[1152]

TABLE 10-2
Gas Transmission Through Cationic Ion Exchange Membranes

Membrane Type	Temp. (°C)	Water Content (wt. %)	Transmission Rate (cm^3-cm/s-atm-cm^2)×10^9	
			H_2	O_2
Non-crosslinked PSSA*	22–24	31 (saturated)	5.5	3.7
	24	25 (nonsaturated)	4.1	—
PSSA grafted to poly-(trifluorochloroethylene)	22	saturated	0.16	0.013
	57	saturated	0.30	—
	70	saturated	0.42	0.13
Poly (phenolformaldehyde sulfonic acid)	23	—	524,000	246,000

*polystyrenesulfonic acid

It is often argued that ion exchange membrane fuel cells employing an acidic electrolyte are really no different from cells using a matrix-plus-electrolyte configuration, as in the PAFC. However, the ion exchange membrane cells can tolerate much higher pressure gradients, and ionic concentration gradients are minimal. In addition, the matrix itself is the electrolyte, so no neutral material is needed that might increase IR-drop. Indeed, the matrix itself is the electrolyte, so that the application of the Nafion membrane to the fuel cell can be truly said to have been a breakthrough. Above all, the major advantage of the system compared with the matrix PAFC is its ability to operate effectively at low temperatures. This results from the properties of the fluorinated sulfonic acid groups in the polymer on the oxygen reduction reaction. A further important advantage is the total immobilization of the acid groups in the electrolyte, so that contact with cell components is minimal. The only liquid in the system is pure water (required as potable water on the Gemini missions with earlier SPE cells), hence corrosion problems are minimized.

Since the membranes rely on hydrogen ion transport, the self-diffusion of this ion through these membranes has been investigated.[626, 2472, 2473] The prevailing theories of electromigration of hydrogen ions through polystyrenesulfonic acid have been developed in some detail.[368, 755, 823] As previously stated, it appears that the conduction process cannot operate in the absence of water in the membrane. In fact, water content has long been known to be the most important parameter in membrane conductivity. Indeed, increasing the number of moles of water per mole of sulfonate group from one to six, has been shown to increase membrane conductivity by as much as six orders of magnitude.[823, 1352] A more detailed discussion of the theories and mechanisms of membrane conduction processes can be found in Refs. 879 and 881. In view of the importance of water in the conduction process, for which ideally about three water molecules per proton are required, water must be recycled from the cathode to the anode. In fuel cells rejecting liquid water at the cathode, this can be carried out by wicking, even under zero-gravity conditions.[176, 1613] The necessity for providing water for proton transport through the membrane sets an upper limit to cell operating temperature, even under pressurized conditions. At atmospheric pressure, the upper temperature limit is about 90°C, beyond which drying out of the membrane and loss of conductivity occurs. The use of higher gas pressures allows the use of higher temperatures (up to about 150°C), but beyond operating pressures of a few atmospheres, this procedure shows diminishing returns, since increasing water vapor pressure with temperature reduces the effective partial pressure of the reagent gases. Consequently (see following), at a constant total pressure of 7.6 atm, cell performance is about the same at 104°C and 150°C.

The state of the art (ca. 1985) of the SPE fuel cell has been reviewed in

Ref. 176. Up to the present, a major feature of this system has been the type of interface between the anode and cathode catalyst and the polymeric electrolyte itself. This three-phase boundary interface can be regarded as two-dimensional, rather than three-dimensional as in the case of a liquid electrolyte, since the degree of interpenetration between the solid phases (electrolyte and catalyst) has been rather small. The normal method of fabrication of the interface has been a pressure-bonding operation at a temperature approaching 200°C between the membrane and a catalyst film containing Teflon, under conditions such that membrane dehydration and cracking could not occur.[176] The use of Teflon (at least at the cathode) was determined to be necessary in early work so that the catalyst would not flood from the presence of accumulated product water. In general, catalyst utilization has been rather poor because of the nature of the interface, and rather high loadings (5 mg/cm^2) of pure platinum have been traditional. Work is clearly necessary to improve the electrolyte-catalyst interface, and to reduce the platinum loading (for non-aerospace applications) by the use of supported catalysts and an improved technique for electrolyte-catalyst bonding.[176] This is discussed in more detail, as follows.

Many studies have been made on transport mechanisms and membrane properties for a wide range of SPE membranes. These include the transport of water, sodium ions, and hydroxyl ions through cationic membranes;[2133] the determination of the mobilities of water, cations, and anions in perfluorinated [2283] and Neosepta*[2163] ion exchange membranes; the flow of sulfuric acid through ion exchange membranes;[1853] measurement of the specific resistivity, permselectivity, and limiting current densities in commercial SPE membranes at 25~75°C;[301] a study of transient current effects in membranes,[1069] and the time required to attain limiting current densities at a membrane-electrolyte boundary.[981] They also include evaluation of the fixed charge density,[298] the non-uniformity of ionogenic groups,[2721] and the potential distribution[55, 306] in ion exchange membranes; determination of the electrical properties of ion exchange membranes having mobile ion exchange sites;[2365] the effect of membrane polarization on membrane conductivity;[871] the determination of the thermal conductivity of polystyrene sulfonic acid ion exchange membranes;[2109] the thermal conductivity and specific heat capacity of cationic membranes;[2110] and strong electric field effects in anionic membranes.[2279]

A large number of similar studies have been made specifically on

---

*A perfluorinated carboxylic acid membrane developed by Tokuyama Soda Company, similar to the Flemion® membrane developed by Asahi Glass.[176] These membranes are intended for the highest H$^+$/Na$^+$ selectivity for the chlorine-caustic industry.

Nafion membranes. These include electron microscopic[580] and X-ray scattering[2772] studies of the membrane morphology; effects of ion-clustering[279, 2768] and contact ion-pair formation;[1472] studies of the effects of moisture content and hydration on the density, expansion coefficient,[2409], and counter-ion behavior;[1642] ionic mobility and contact-pair effects;[1472] theoretical descriptions of the transport processes in Nafion using irreversible thermodynamics,[1398,1984] and simplified flux equations.[2671] In addition, measurements have been made of the sodium ion,[1638, 2768, 2769] cesium ion,[1638, 2768] and iodide ion[1638] diffusion coefficients in Nafion in a variety of solvents (water, methanol, acetonitrile, and propylene carbonate); of the selectivity coefficients of various cations in Nafion,[2337,2767] of the exchange rates of cations in Nafion in different aprotic solvents;[1637] of sorption phenomena occurring during the neutralization of Nafion;[2410] of chemical treatment of Nafion to increase the ion exchange capacity, selectivity, and other properties;[2074] of chemical treatment to increase the cationic transport number;[199, 1425] and of determination of the cohesive energy density in Nafion.[2771] The regeneration of Nafion membranes by treatment with acids,[1549] with bases,[2415] with acids and bases,[200,201] and by thermal and pressure treatments[2744] has also been discussed. Electrical methods to determine the degree of fouling of the membrane;[1727] metalizing the membrane to improve its conductivity;[1495,2366] reinforcing Nafion with fabric;[751,2065] and methods of bonding together sheets of Nafion [717,2314] also have received consideration. For use in brine electrolysis,[2315] Nafion has been treated with titanium phosphate[1832] and coated with PTFE.[1738] Electrodes have been coated with Nafion for use in electrolysis[1465] and to reject albumin.[2467] The rates and mechanisms of the transmission of gases through Nafion also have been investigated.

Many methods have been published or patented for the preparation of ion exchange membranes that may have fuel cell implications, including cationic sulfonated fluorocarbon membranes,[697, 698, 1162, 1223, 1426, 1833, 1834, 1949, 2112, 2193, 2239-2242, 2411, 2481, 2525, 2563] reinforced sulfonated fluorocarbon membranes,[210,751] and miscellaneous cationic membranes.[1607] The preparation of hydrophilic perfluoro [828] and methacrylate-based[1527] ion exchange membranes also has been reported, as have methods for producing membranes of complex shape.[1673]

Anionic membrane preparation methods include styrene-divinylbenzene-methacrylate polymers in PTFE-asbestos matrixes,[211,1092] quaternized heterogeneous anion exchange membranes,[1396] grafted polymer anionic membranes,[1091] and others.[2241, 2428] The different kinds of membranes and their areas of applicability have been reviewed in Ref. 1161.

Recently, a new membrane has been developed by the Dow Chemical

Electrolytes

Company, which is similar to Nafion, although it is of lower equivalent weight, therefore of somewhat higher conductivity (rather better than 0.05 $\Omega^{-1}\text{cm}^{-1}$; see Table 10–1). The material is prepared by a different chemical sequence to Nafion, and its formula is such that $m = 0$ in Fig. 10–2. It appears to be a very promising material for use as a fuel cell electrolyte, especially if some recent improvements in creating an optimal catalyst-electrolyte interface are used. This is illustrated by the data shown in Fig. 10–3, which shows some of the best selected results for a GE SPE cell using Nafion 1170 (175 micron thick) operating on H_2–O_2 at about 7.6 atm and at four different temperatures (cf. results in Ref. 176), and the latest data reported by Ballard in Canada [2332] for a cell operating with electrodes of about the same loading (~ 5 mg Pt/cm²) using the thinner (125 micron) Dow membrane at 82°C on pure H_2–O_2 at 5 atm pressure. The results obtained not only are about 30 mV better than the best GE data in the fractional A/cm² range, but also the remarkable results of 1.75 A/cm² at 0.7 V and 3.9 A/cm² at 0.5 V were obtained without any evidence of diffusion limitation, since the slope of the V-i curve corresponds only to IR-drop. This implies a proprietary improvement in the catalyst-electrolyte interface.

Other improvements in the catalyst-electrolyte interface are aimed at using low-loading carbon-supported platinum electrodes, so that a low-cost SPE system will result. In particular, Martin et al. [1693,1804] have devised a technique for dissolving either Nafion or the Dow membrane material by using a cosolvent technique, in which the fluoropolymer is first dissolved

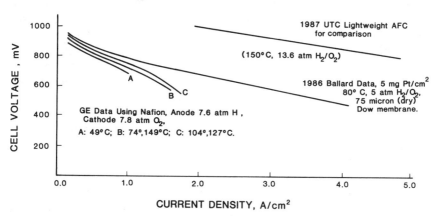

Figure 10–3: Fuel cell performance results (H_2/O_2) on solid polymer electrolyte cells. A–C: GE cells with DuPont NAFION® membrane. D: Dow perfluorinated sulfonic acid membrane. Anode and cathode Pt loading 4 mg/cm². For UTC data, see p. 175.

by autoclaving in an ethanol-water mixture, the resulting solution then being treated with a higher-boiling solvent miscible with water (e.g., ethylene glycol or trimethylphosphate) at a temperature higher than the boiling point of ethanol. The mixture then can be cast into a heated Teflon-bonded electrode or dry Teflon-catalyst mixture, and the cosolvent removed. The result is a well-dispersed fluoropolymer electrolyte-catalyst mixture that gives a good electrolyte-catalyst interface even with low-loading catalysts. Such electrodes may show oxygen reduction properties superior to those in direct internal contact with phosphoric acid, and may give superior cell performance in a contact with phosphoric acid electrolyte in a normal matrix, since the latter will provide a high-conductivity proton pathway from the anode at high temperature.[179] Martin's technique also can be used to cast very thin, low-resistance films of electrolyte.

C. OTHER AQUEOUS ELECTROLYTES

For the reasons made clear earlier, it has been generally concluded that only strong acids and bases are suitable electrolytes for low temperature fuel cells,[2699] although buffer electrolytes have been considered for certain applications.[314] As already discussed, buffers are generally inferior fuel cell electrolytes. Since their major conducting ion concentrations are low, a large polarization in the form of a pH gradient will exist across a fuel cell containing a typical buffer electrolyte. This may amount to an 800 mV loss even at low current density.[312, 1649] As expected, in buffer electrolytes limiting anodic and cathodic current densities due to ionic migration within the porous electrode structure are much lower than the gas diffusion limiting currents in strong acids and bases. Since buffer electrolytes generally do not have high conductivities, ohmic losses are higher than in strong acids or bases. However, where electrode structure is unimportant, as in the case for flow-by electrodes using electrolyte-soluble fuels, buffers may be fairly comparable to strong acids and bases under some conditions. A possible justification for the use of a buffer electrolyte under these circumstances is that cheaper non-noble metal catalysts that are unstable in acid electrolytes might then be used.

A special case of the use of a highly soluble buffer electrolyte was that of cesium carbonate for hydrocarbon and impure hydrogen oxidation. Cairns [534, 535, 542, 1017] claimed that this concentrated CO_2-rejecting electrolyte had the advantages of alkaline solutions from the viewpoint of catalysis. However, the use of cesium salts as electrolytes is impractical for most purposes for reasons of cost. Finally, weak acids and bases also give rise to unacceptable concentration polarization when they are used as fuel cell electrolytes.[245]

D. ORGANIC ELECTROLYTES

The solubility of oxygen in lithium perchlorate solutions in gamma-butyrolactone, propylene carbonate, dimethyl sulfoxide, and N-nitrosodimethylamine has been shown to be an order of magnitude higher than in aqueous KOH solutions.[872] From the standpoint of oxygen solubility, it might therefore be concluded that a useful air electrode in organic electrolytes may be possible. Tentative kinetic data, however, indicate that oxygen reduction is far from reversible in such media, with current densities only in the microampere range. These observations indicate a lack of feasibility for a power-producing air electrode in an organic system. However, voltammetric studies of oxygen reduction in dimethylnitrosoamine solutions have shown reduction currents in the 1 mA/cm^2 range.[2476] Whether this can be further improved is at present unknown.

E. MOLTEN CARBONATE ELECTROLYTES

Ionic conduction in fuel cells utilizing molten carbonate electrolyte is achieved by transport of carbonate ion, which is consumed at the anode and regenerated at the cathode. Representative cell reactions are

$$\text{Anode: } H_2 + CO_3^= \rightarrow H_2O + CO_2 + 2e^-$$
$$\text{Cathode: } \tfrac{1}{2}O_2 + CO_2 + 2e^- \rightarrow CO_3^=$$
$$\overline{\text{Cell: } H_2 + \tfrac{1}{2}O_2 \rightarrow H_2O}$$

MCFCs are usually most effectively operated at about 650°C to give best performance consistant with materials stability. The major advantage of molten carbonate electrolyte is that the molten carbonate fuel cell (MCFC) can readily consume common CO-CO_2–containing fuels, such as gasified coal or natural gas reformate, both carbon monoxide and hydrogen being directly utilized as fuel. Removal of carbon dioxide from the gas streams is thus unnecessary; indeed, addition of carbon dioxide to the cathode stream is required for correct operation of the cell. In addition, cell construction materials are relatively inexpensive, and the system produces high grade waste heat, enabling incorporation of the fuel cell into a *total energy system*, producing both electrical and thermal energy or increasing electrical output using a bottoming cycle. Finally, provided CO_2 is correctly supplied to the cathode, the electrolyte composition is invariant. The state of the art of the MCFC is discussed in Chapter 17.

The high operating temperature of the MCFC does not allow the use of very high surface area electrodes that are available for aqueous electrolyte fuel cells, since these would sinter rapidly at 650°C. However, this high operating temperature allows the use of nickel-based electrodes of modest area (0.3–1.0 m^2/g) on which reaction kinetics are high enough to allow typical PAFC performance to be exceeded. Hence, cell efficiency under atmospheric pressure operation is 12~15% higher than that of the PAFC under comparable conditions, resulting in considerably greater system efficiency. Improved efficiency, as well as the lack of requirement for scarce materials (platinum), are the major reasons for present interest in the continued development of the MCFC.

Various binary and ternary eutectics of sodium-lithium carbonates and sodium-lithium-potassium carbonates have been examined for use as electrolytes.[471,472,1513,2183] However, the preferred electrolyte since 1968 has been a binary lithium-potassium carbonate consisting of 62 mole percent lithium carbonate and 38 mole percent potassium carbonate, with a liquidus temperature of 490°C. Although it is in principle, possible to use the electrolyte as a free liquid between appropriate electrode structures, it has always been the practice to mix it with an inert powder to form a paste at temperatures above the melting point of the carbonate mixture or to contain it in a presintered inert porous matrix.[224,229] This matrix has a mean pore radius less than that of the electrodes, so that flooding is prevented and gas-sealing is assured. To overcome the technological problems associated with sealing, stability, electrolyte-electrode contact, and fabrication, most of the experimental MCFCs up to the late 1960s employed paste electrolytes containing magnesia as the inert filler. These pastes were prepared by cold-pressing, hot-pressing, hot injection, or extrusion, as well as by other special techniques.[2207] For the most part, cell resistances were found to depend on electrolyte paste composition and not on the electrode structure employed. Comparisons of measured electrolyte resistances with calculated values indicated typical porosity and tortuosity factors of about three.[2484]

Molten carbonates were originally thought to wet the porous nickel electrodes employed at the anode, so that the latter was considered to operate as a mensicus system covered by a film of electrolyte. Indeed, with magnesia as the electrolyte filler, anode flooding was often observed. More recently, it has been determined that nickel under reducing conditions is apparently not wetted by the electrolyte,[1207] and that the cause of electrode flooding was attributable to wetting and surface-area changes in the magnesia electrolyte filler, rather than to the nature of the anode material. Long MCFC lifetimes could be demonstrated only after lithium aluminate became available for use as the inert filler. While this compound was used in experiments by Broers and Ketelaar quite early in the development of the MCFC, its use in the United States at IGT resulted from attempts to use alumina binder and the ternary Li-Na-K carbonate eutectic. The result was a reaction between lithium ion in the eutectic and the alumina binder to give lithium aluminate, giving an electrolyte approximating to the Na-K binary mixture. This seemed to show inferior electrochemical properties compared with the ternary melt, which stimulated a range of studies in the late 1960s to determine the significance of the melt cation in determining the electrochemistry of the anode and cathode reactions. In previous work, melts had been chosen only on the basis of melting point. By 1970, it had been realized that certain melts, particularly those with high lithium ion composition, could strongly favor the reversibility of the cathode process. This led to the choice of the 62/38* Li/K eutectic as a preferred electrolyte. This composition is not necessarily that giving the best MCFC performance, as will be discussed in the following.

In all state-of-the-art MCFCs, the electrolyte is contained by capillary forces in the matrix of a porous ceramic *tile*. This is so called since it was originally made from hot-pressed particles of lithium aluminate ($LiAlO_2$) combined with the powdered electrolyte. When cold, it literally resembled a thick (2-3 mm) ceramic tile. In order to reduce IR-drop to a minimum, it is now made in the form of a thin (0.4 mm or less) layer, usually prepared by tape-casting.[457, 2243] The combined binder material plus electrolyte will hence be referred to as the *electrolyte matrix*.

This matrix typically consists of 40 weight percent $LiAlO_2$ and 60 weight percent carbonate, the total structure having a porosity of about 70%. The thickness and pore-size distribution should match that of the porous electrodes so that electrolyte is appropriately wicked into each electrode to establish a stable large surface-area interface for the electrochemical reactions. Electrolyte retention must also be sufficient to provide structural strength and prevent cross-diffusion of reactants. The ability of the electrolyte matrix to maintain its shape and size is critical to cell life. For instance, it has surprising rigidity at 650°C: Under a compressive load of 100 kPa, it might typically deform by about 1% in the first ten hours and less than 2% thereafter.

The stiffness, strength, and electrolyte retention capabilities of the electrolyte matrix are a function of the particle size and shape of the lithium aluminate particles, and the relative amounts of carbonate and lithium aluminate. There are at least three allotropic forms (alpha, beta, and gamma) of lithium aluminate, with the alpha-form having the highest density. The chemistry and the effects of temperature and pressure on the stability regimes of these allotropes are still under investigation. The most effective shape for the lithium aluminate matrix particles seems to be in the

---

*62 mol % Li_2CO_3 + 38 mol % K_2CO_3.

form of long rods or fibers of submicron diameter, when combined with carbonate melt giving maximum capillary strength and minimum pore size with a geometrical form giving minimal electrical resistance. The manufacturing method must yield the most stable allotropic form. The most common method previously used was the hot-pressing process discussed previously, in which finely divided alumina was reacted with the appropriate carbonate melt at about 630°C. The resulting product was cooled and ground, after which further carbonate was added to give the correct composition, the process being repeated several times. The final product was then hot-pressed at about 50 MPa (7250 psi) at a temperature of 10~20° below the carbonate liquidus (i.e., 450~490°C) to produce the so called tile.

For minimum electrolyte resistance, these were as thin as possible, but they were also designed to be sufficiently strong and impermeable to prevent anode and cathode gases from cross-mixing. Until 1980, for hot-pressed tile sizes up to 30 cm X 30 cm, a thickness of 1~2 mm was usually employed. It was found that unreinforced tiles had difficulty in withstanding the volume changes occurring during electrolyte freezing on thermal cycling; accordingly, starting about 1972 they were reinforced. The reinforcement consisted of stainless steel alloy fibers or screens, usually of the high-aluminum alloy Kanthal (of the Kanthal Corporation in Danbury, Connecticut), which produces a passive surface coating preventing short-circuiting of the anode and cathode. Such reinforced tiles were a temporary solution towards thermal cyclability.

A joint EPRI/DOE workshop on MCFC technology was held at Oak Ridge, Tennessee in November 1978 to assess the state of the art and to determine how best to scale up experimental laboratory cells to produce practical hardware. The general conclusions were that laboratory cell construction based on the use of hot-pressed tiles was unacceptable, particularly from the viewpoint of IR-drop; in addition, it would be impossible to scale up to the sizes of 1 m² that would be required for commercial cells. Electrolyte matrixes with thicknesses of 0.25~0.5 mm would be required, and these could not be reinforced with Kanthal fibers. It was hoped that these thin layers would be intrinsically more thermally cyclable than the thick tiles, especially if their thermal expansion coefficients were matched with those of the electrodes and other metal parts, (e.g., by the use of graded structures of cermet in the anode), extending from pure metal next to the gas spaces to pure ceramic in contact with the electrolyte.[239] Any minor nonhealing fissures could be prevented from inducing major gas crossover, leading to burning of electrodes, by the use of a protective fine-pore bubble-pressure barrier built into the anode, if necessary.[18, 176]

An alternative method of producing thermally-cyclable electrolyte matrixes was developed around 1978~1979 by UTC,[176] and arose from the observation that Kanthal wires or fibers did not mechanically strengthen the electrolyte layer as was first thought, but acted as crack arrestors, giving stress relief and preventing microcracks from completely traversing the electrolyte. In principle, any strong particles mixed into the matrix that are wetted by carbonate can do this, provided that their diameter is large compared with the crack width (i.e., compared with the particles of lithium aluminate). After early attempts to use large particles of other ceramics (e.g., strontium titanate, ceria), all developers now add large-diameter lithium aluminate particles. As a result, cells and stacks are now thermally cyclable. Further details are given in Chapter 17.

The 1978 workshop focused attention on two new methods of fabrication of the electrolyte matrix: tape-casting[176, 239] or electrophoretic deposition[174, 486] of the lithium aluminate as a separate thin layer, followed by impregnation with carbonate electrolyte. Both these methods are susceptible to simple scale-up and give high-quality products of uniform thickness; tape-casting being a somewhat slower technique, but having the advantage of also being able to fabricate electrode structures.

A number of reports on electrolyte layer fabrication have been published, including work by the General Electric Company,[189] United Technologies Corporation,[2600] the Institute of Gas Technology,[556, 1683, 1684] and Argonne National Laboratory.[236, 1413, 2021] Workers at Oak Ridge National Laboratory studied the transport properties of electrolyte matrixes.[602, 1963] Publications from Argonne National Laboratory [2020] and from the Institute of Gas Technology [629] have described the characteristics and development of matrixes containing lithium aluminate. Under certain conditions, for example when lithium aluminate reacts with components leached from nickel oxide substitute cathode materials such as p-type perovskites (e.g., lanthanum cobaltate), UTC has examined alternative matrix materials, particularly lanthanum aluminate.[457] Strontium titanate also has been claimed by GE to be advantageous.[117] However, the present trend is to use lithium aluminate because of its low cost and adequate stability.

In tape-casting, lithium aluminate powder of suitable particle size is made into a slip-like slurry with a suitable organic binder and solvent, according to proprietary formulations given in the references in the preceding paragraph. The viscous mixture is allowed to flow out of a slot in a hopper on to an endless moving belt, usually made from stainless steel strip. Belt speeds are typically about 1 cm/s. The mixture passes under a knife-edge a short distance from the hopper, the blade (*doctor-blade*) being adjustable to give the layer thickness required. Fabrication is done under a controlled atmosphere with recovery of the volatile solvent, and drying is frequently by infra-red lamps. The total length of the flat belt surface is frequently 10 m or more. The finished layer is peeled from the belt and cut to size. The cell is assembled with the green layer (as yet containing the binder, with no electrolyte), and electrolyte is added in a suitable form. In most early work, it was pre-impregnated into the cell anode components so that anode-cathode distance remained constant during the heating operation, which involved burnout of the organic binder and wicking of carbonate into the matrix layer. A more recent and very intriguing approach has been to use a built-up matrix consisting of alternate layers of tape-cast lithium aluminate and carbonate mixture. Shrinkage during initial heat-up has been shown to have no effect on cell performance.[1255]

In electrophoretic deposition, lithium aluminate is dispersed in a suitable dielectric liquid, such as 2-pentanol. The application of an electric field across two electrodes allows it to be deposited in the form of a uniform layer on one of these electrodes, preferably the anode itself. The back of the anode of course requires protection to avoid deposition of a two-sided layer. The resulting deposit builds up rapidly (about 10 seconds being required for a layer 0.25 mm thick at a typical field strength of about 100 V/cm), is very uniform and has particles that are highly oriented due to viscous flow effects through the medium. This has the effect of lowering tortuosity and improving conductivity compared with mechanically-prepared layers (see following). Electrophoretic deposition has been described in an EPRI patent,[174] and has been examined by ERC[125] and by GE[486] under EPRI funding. A further technique for the fabrication of the electrolyte is calendering,[117] which in principle could be used to form the complete electrolyte matrix in one operation. In this technique, lithium aluminate (or other matrix support) is milled with binders to give a suitable paste, which then is rolled between parallel rollers to give a flexible film.[117] General Electric was the only developer to examine this concept, and they found it advantageous to use binder particles precoated with electrolyte. Since the technique is rapid (although the film produced is less uniform than that made by the preceding methods) and in principle can produce films made from the correct electrolyte-support mixture, it may be applicable for mass production.

As an example of the properties of electrolyte matrixes based on post-impregnated structures using electrophoretically deposited lithium aluminate, GE's experience may be quoted.[980] Typical matrix layers consisted of 45% by weight lithium aluminate and 55% by weight of 62/38 Li/K carbonate with a net porosity of 51% and a mean pore size of 0.25 μm. At 650°C the conductivity of the electrolyte-saturated matrix layer was about 0.35 Ω^{-1} cm^{-1}, compared with a value of about 0.31 Ω^{-1} cm^{-1} for comparable hot-pressed tiles measured under the same conditions. The activation energies for the conductivities of the electrophoretic matrixes and hot-pressed tiles were 22.6 ± 2.1 and 33.1 ± 2.1 kJ/mol, respectively, the activation energy for the conductivity of the pure carbonate electrolyte being 22.1 kJ/mol. The difference in the activation energies for the two types of tiles has been attributed to carbonate expansion, since the aluminate particles in the hot-pressed tiles were loose and free to move as the carbonate expanded, whereas the aluminate particles in the electrophoretic tiles were joined into a rigid network as a result of a heat treatment (partial sintering) procedure used to produce improved rigidity. The IR-drops in experimental 2 mm-thick electrophoretic matrixes used for comparison with hot-pressed structures were somewhat greater than 85 mV at a current density of 160 mA/cm^2, which is at least four times greater than a tolerable value in a practical cost-effective cell in which a thin electrolyte layer is essential (in a PAFC, IR-drop would be about 15 mV under comparable current density conditions). However, GE's work showed that the conductivity of electrophoretic matrix layers was 12~15% better than that of hot-pressed formulations; and more recently, ERC has shown that the same is true in comparison with tape-cast layers.[125]

Nothing so far has been said about optimization of the electrolyte composition to obtain maximum cell performance, although it has been remarked that early experiments showed that the binary Na-K electrolyte gave rather poor cell performance. Various theoretical and experimental studies have been made of the behavior and characteristics of molten salt electrolytes. The theoretical concepts for predicting the physical properties of molten salt systems, such as conformal ionic solution theory which describes the salt from an electrostatic viewpoint,[2103] have been reviewed and their limitations discussed.[365] Work at Oak Ridge National Laboratory has dealt with theoretical studies of mass transport in molten salts[437, 441, 1901, 2551, 2554] and the determination of steady state composition profiles in molten carbonate electrolytes.[438, 439, 442, 2550, 2552, 2553] The ionic conductivity of molten salts has been reviewed[2466] and studied theoretically.[2380] A theoretical study of ion mobility in mixtures of molten salts also has been made. [1656]

Although all of the above are significant from the viewpoint of MCFC operation, the effect of the cation on electrochemical properties (i.e., on anodic and cathodic polarizations) is of the greatest importance. Researchers at the Institute of Gas Tecchnology have reviewed and analyzed the properties of molten lithium-sodium-potassium carbonate electrolyte in order to identify and optimize electrolyte composition regimes.[18, 176, 1938] Collected data for a large number of molten salt systems applicable to molten carbonate fuel cells can be obtained from the Molten Salts Data Center at Rensselaer Polytechnical Institute in Troy, New York.[1299, 1300]

F. SOLID ELECTROLYTES

As early as the turn of the century it was recognized that certain compounds in the solid state could function as electrolytes; the best-known of these being the Nernst Glower, a zirconia rod used as a high-temperature light source, which Nernst showed to be an oxide ion conductor, atmospheric oxygen being reduced at its negative pole, and gaseous oxygen being evolved at the positive. For all practical uses, a solid electrolyte must be as close as possible to a pure ionic conductor, with little or no electronic conductivity; that is, the transport number for electrons should be close to zero to prevent internal short circuiting of the solid-state device. It is also desirable that only one conducting ion should be involved. This means that only one ionic species, a reactant or product in either the anodic or cathodic process, should have a transport number of unity, the transport numbers of the other ions as well as that for electrons being zero. The condition of having only one mobile ion is necessary for the maintenance of a stable solid structure.

Ideally, the nature of the electrode reactions dictate that the conducting ions should either be H^+ or $O^=$. Solid oxides and ion exchange polymers are the two principal types of solid electrolyte that have been used in fuel cells, although other solid electrolytes continue to be investigated. Since all practical ion exchange polymers that have been used as fuel cell electrolytes are H^+-conducting acidic gels that (like all acids) rely on the presence of water molecules for ion transport, they have been discussed earlier in the section on acid electrolytes. The following section considers solid ceramic electrolytes, particularly solid oxides. A review of the various uses of electrically conductive ceramics has been published.[1066] Today, one of the most important applications of these materials is for oxygen sensors in automobile engines, which measure the oxygen partial pressure compared with that of a reference source using the reversible (Nernst) potential difference between two electrodes connected to a high impedance transistor voltmeter.

Solid Oxide Electrolytes

These materials are impervious ceramics that have the ability to conduct a current by the passage of oxygen ions ($O^=$) through the crystal lattice at sufficiently high temperatures. Requirements for a solid oxide fuel cell (SOFC) electrolyte include: high oxygen ion conductivity, low electronic conductivity, phase stability, mechanical strength, gas tightness, thermal shock resistance, chemical resistance to fuel cell reaction gases (i.e., to oxidizing and reducing atmospheres), and compatibility with electrode and interconnection materials.

The only serious choice of material has been zirconia (ZrO_2), stabilized in its cubic form by doping with small amounts of calcia (CaO), yttria (Y_2O_3), ytterbia (Yb_2O_3), or a mixture of heavy rare earths ranging from dysprosia (Dy_2O_3) to lutetia (Lu_2O_3).[812] Doping with scandia (Sc_2O_3) also gives good results, although it is costly.[2385] Electrolytes based on ceria

(CeO_2),[599,883,2148,2786] thoria (ThO_2),[2218,2427] and yttria[344,767,1550,1715] also have been considered for use as solid oxide electrolytes, but have proven not to be competitive with zirconia because of their generally lower ionic conductivity. In addition, they usually show some electronic conductivity, and have poor chemical stability over certain of fuel cell oxygen partial pressure ranges. Ceria has been studied as a potential fuel cell electrolyte, but its relatively high electronic conductivity in reducing atmospheres is undesirable. Rare-earth-doped bismuth oxide (Bi_2O_3) also has been considered,[2405] since its conductivity at 850°C is about the same as that of doped zirconia at 1000°C; however, it does show some volatility at higher temperatures and is again unstable at low oxygen partial pressures such as would be encountered at the fuel cell hydrogen electrode. However, there have been recent claims that suitable dopants can result in stable operation.[1679]

Although the conduction mechanisms in solid oxide electrolytes still are not fully understood, the ionic conductivity of stabilized zirconia results from the presence of oxygen vacancies in the lattice produced by doping. From the atomic viewpoint, the movement of ions through the material must be regarded as successive jumps between individual lattice sites or interstitial sites. The introduction of defects into the crystal structure enhances ionic mobility by introducing lattice distortion and lattice defects, thereby reducing the activation energy for intersite jumps. Defects include: extra interstitial ions wedged into the ideal lattice; charged vacancies, as from ions missing from the ideal lattice; and foreign ions, which may be at interstitial sites or substituted in normal lattice sites. The cubic solid-solution lattice of doped zirconia has all the electrical properties required for an oxygen ion conductor, with negligible electronic conductivity at 1000°C. In addition, the metal ions in the cubic-fluorite crystal lattice are tightly locked into place so that cation mobility is very low,[2114] and there are oxygen vacancies in the anion sublattice, giving high oxide ion mobility.

It is well established that maximum oxygen ion conductivity is obtained in the zirconia system if the concentration of the doping oxide is near the lower limit for the stability of the cubic phase. At temperatures greater than about 300°C, fuel cells employing solid zirconia electrolytes generate the theoretical open circuit voltage of about 1 V, depending on temperature and reactant partial pressures. However, in order to use doped zirconia as a practical electrolyte, temperatures of at least 800°C are required in order to overcome the activation energy of oxygen ion transport and increase conductivity to sufficient levels. At 1000°C, the conductivity of doped cubic zirconia becomes about one tenth of that of molten carbonate melts at 650°C, so this is generally considered to be the minimum temperature for fuel cell operation using thin electrolyte films. Table 10–3 shows the resistivities of some of the more usual formulations. These compositions

TABLE 10-3

Resistivities of Various Doped Zirconia Electrolytes

Formula	Resistivity, Ω cm	
	800°C	1000°C
$(ZrO_2)_{0.85}(CaO)_{0.15}$	200	40~50
$(ZrO_2)_{0.90}(Y_2O_3)_{0.10}$	40~50	10
$(ZrO_2)_{0.92}(Y_2O_3)_{0.04}(Yb_2O_3)_{0.04}$	20~25	5~6

have been shown to exhibit long-term stability in fuel cell operating environments at 1000°C, over 40,000 h being demonstated by Brown Boveri,[2128] and have oxygen ion transport numbers very close to unity.

While calcia-stabilized zirconia has a lower conductivity than yttria- or ytterbia-stabilized zirconia, and also is susceptible to an ageing process that decreases its conductivity still further,[274] there is continued interest in its possible use owing to its low cost and improved sinterability compared with that of rare-earth-stabilized material. When formed as a thin film, resistivity-thickness products of 0.04 Ω cm^2 can be approached in calcia-stabilized zirconia films of sufficient thickness (10~200 µm) to be still gastight. In general, conductivity can be increased in three ways: by increasing the temperature, by further reducing thickness, or by the development of better materials. Most effort has been devoted to the latter; that is, on the development of improved compositions.

Numerous reviews on solid oxide electrolytes have been published,[194, 480, 738, 1074, 1075, 1724, 1777, 2568-2570, 2686, 2729] as well as papers on the theory and status of ionic and mixed conductors that are applicable to energy conversion systems,[2684] on defect interactions in solid oxide electrolytes,[1897] and on the effects of the type and concentration of dopants on the electrolyte domain boundaries.[2510] Methods have been given for the design of electrodes[2424] and of electrolyte sample holders[436] to measure the electrical properties of solid oxide electrolytes. The theory of, and methods of measuring diffusion coefficients and the partial conductivities of the mobile species in mixed ionic-electronic conductors such as solid oxides also have been discussed.[768]

Doped zirconia has been the subject of the majority of work. Soviet workers in particular have been especially active in this field. Various methods of fabricating gastight zirconia films have been published,[1247, 1529, 2407, 2521] including the use of sputtering,[367] thermal decomposition of alkoxide vapors,[1246] sintering,[1062, 1218, 1553, 2385] and electrochemical vapor deposition.[1247, 1251] A comparison of crystal structure, porosity, electronic conductivity, and oxygen permeability of various types of doped zirconia fabricated by different methods (ceramic, plasma spraying, hot pressing) has been published.[484, 1966] The large number of studies made on the ionic conductivity of doped zirconia include: the effects of oxide dopants on the conductivity of yttria-doped zirconia;[951] the ionic conductivity of calcia-, yttria-, and scandia-doped zirconia;[1229, 2468] the effects of temperature and sintering time on the resistivity of scandia-doped zirconia;[1228] the effects of temperature and oxygen partial pressure on the conductivity of calcia-doped zirconia;[1632, 1707] the conductivity of calcia- and yttria-doped zirconia in the temperature range of 800~2000°C;[61] the effect of gas flow on the conductivity of zirconia;[1376] bulk and grain boundary resistance of yttria-doped zirconia;[1230] the effect of grain size on the conductivity of yttria-doped zirconia;[1235] the effect of dopants on the grain boundary resistivity of scandia-doped zirconia;[1994] and the effects of calcia and tantalum oxide on the conductivity of the zirconia-ceria system.[1003] Studies also have been conducted on the real and imaginary impedance components at the interface between platinum and yttria-doped zirconia,[2287] and on the effects of alumina and silicon dioxide on the impedance and structure of calcia-stabilized zirconia.[296] The contact resistance between two samples of yttria-doped zirconia pressed in contact with each other has been examined,[822] and a method for measuring the contact resistance at the interface between platinum or silver and scandia-doped zirconia has been discussed.[1996]

The effect of temperature on the polarization resistance of the electrode-zirconia interface has been studied for copper, palladium, platinum, silver, and nickel electrodes.[1995] The dielectric properties of doped zirconia also have been measured,[850, 2180] as has its thermal conductivity[850, 918, 2180, 2603] and thermal diffusivity.[1790] Measurements of the coefficient of thermal expansion of zirconia also have been made,[850, 2180, 2603] since this parameter must be compatible with the thermal expansion coefficients of other components (e.g., electrodes, cell-to-cell interconnections) to ensure thermal cycability. Other studies have included the oxygen permeability of calcia-doped zirconia as a function of the oxygen partial pressure,[1262] the enthalpy and heat capacity of zirconia at high temperatures,[587] and the phase diagram for the calcia-zirconia and yttria-zirconia systems.[2088] The ion transport numbers in yttria-stabilized zirconia also have been determined.[484]

As has been remarked already, other solid oxides besides yttria- or calcia-doped zirconia have been considered. These include perovskite oxides, bismuth trioxide-doped zirconia, and bismuth trioxide-doped tungsten trioxide;[77] layered structured oxides such as $K_{0.72}(In_{0.72}Sn_{0.28})O_2$;[1425] alkali metal aluminates;[2029, 2778] and certain superionic conducting oxides.[656]

Beta-Alumina

Beta-alumina is a solid sodium-ion–conducting ceramic that has received considerable attention in recent years, since it is the electrolyte used in the sodium-sulfur battery. In it, the sodium ion has a transport number of unity, and electronic conduction is negligible. As already stated, sodium ion conduction is not appropriate for use in fuel cells consuming hydrogen and oxygen, but the very high conductivity of the material at 300–350°C (as much as 0.3 Ω^{-1} cm^{-1}) has inspired a search for derivatives that would possess highly mobile ions (notably H$^+$) that could be produced and consumed in fuel cell electrode reactions. While all attempts to do this have failed so far, the material is reviewed below as a representative of a class of superionic conductors that may ultimately be useful as fuel cell electrolytes.

Beta-alumina is a class of nonstoichiometric crystalline compounds having the general structure $Na_2O \cdot xAl_2O_3$, in which x ranges from 5 (β''-alumina) to 11 (β-alumina). The material consists of layers of loosely packed sodium and oxygen ions sandwiched between close-packed layers of oxygen atoms in the spinel blocks. The bridging oxygen atoms in the loose layers have the effect of providing a spacing that permits rapid interstitial migration of sodium ions, giving high sodium-ion conductivity at temperatures above 250°C. The β'' form has higher sodium ion conductivity than does the β form. Below about 1550°C there is always some β''-alumina present, which appears to exist in epitaxy with the more stable β-alumina phase. Both strength and resistivity decrease as grain size increases, the latter being controlled largely by the sintering conditions. However, the electrical properties of most polycrystalline sintered beta-alumina ceramics appear to be determined more by the characteristics of the grain boundaries than by the interior of the grains. Doping with magnesium oxide or lithium oxide increases the sodium ion conductivity and results in the stabilization of the β'' phase. Sodium ion diffusivity in beta-alumina at 300°C is about 1×10^{-5} cm^2 s^{-1}, which is about half that for the sodium ion in molten sodium nitrate at the same temperature. The specific resistivity of beta-alumina is about 11 ± 5 Ω cm at 300°C, and 9 ± 4 Ω cm at 350°C. As just stated, that of β''-alumina can be as low as 3 Ω cm in this temperature range.

Numerous papers have been published which deal with methods of preparation,[2066] crystal structure and morphology,[411, 430, 729, 1355, 1739, 2150] and transport and physical properties[129, 258, 412, 413, 506, 649, 654, 730, 1006, 1158, 1204, 1278, 1953, 1954, 2045, 2149, 2164, 2244, 2370, 2422] of beta-alumina. Many reviews also have been written about the material.[66, 655, 658, 882, 1197, 1199, 1382, 1383, 2046, 2488]

As has already been emphasized, sodium-ion–conducting beta-alumina would seem to be of little relevance as a fuel cell electrolyte, not only because of its conduction properties, but also because of its instability in the presence of water or water vapor. However, it was soon discovered that whereas the oxide ion in the beta-alumina lattice is quite stable, the sodium ion readily exchanges with hydronium ion upon exposure to water vapor.[771, 833, 2697] The attraction was the possibility of producing a stable proton conductor that could operate at 300~500°C, which might be of value as a hydrogen-air fuel cell electrolyte, since at these temperatures noble metal catalysts would not be required. Assuming an electrolyte thickness of 0.1 cm, an effective proton conductor should have a maximum resistivity of about 2 Ω cm in this temperature range, and must be stable at water vapor partial pressures up to about 0.33 atm.[834] Some hope existed that hydrogen-substituted beta-alumina (*hydrogen beta-alumina*) might be the desired electrolyte.

Hydrogen beta-alumina was first prepared in 1968. It is generally made by boiling sodium beta-alumina in concentrated sulfuric acid at 295°C.[461, 834] Although hydrogen beta-alumina is stable up to about 700°C, it has been found that its composition varies widely over the temperature range between 20~550°C.[834] Up to about 225°C the form corresponds to H_3O^+ beta-alumina, in which the ratio of H^+ to H_2O is unity. At 225°C the material undergoes a reversible partial dehydration, accompanied by a drop in ionic conductivity, to give the H^+–H_3O^+ form with an H^+/H_2O ratio greater than unity. At about 800°C, this is followed by irreversible decomposition to α-alumina plus water. Over the temperature range 20~200°C, the specific conductivity κ, expressed in Ω^{-1} cm^{-1}, of single crystal H_3O^+ beta-alumina at a temperature of T K is given by the expression [834] $k = 280 \exp[-9058/T]$. This corresponds to 10^{-10} Ω^{-1} cm^{-1} at 25°C, compared with a value of about 10^{-2} Ω^{-1}cm^{-1} for single crystal sodium beta-alumina at the same temperature. Thus, the conductivity of H_3O^+ beta-alumina is certainly much lower than that of sodium beta-alumina, making it unsuitable as a fuel cell electrolyte. Conversely, it has been reported that the conductivity for H_3O^+ ß″-alumina is about 5×10^{-3} Ω^{-1} cm^{-1} at 25°C,[835] a value approaching that of sodium beta-alumina, so that it would appear that an H_3O^+ superionic ceramic conductor for use as a fuel cell electrolyte may be a future possiblity if a stable ceramic framework (i.e, not β''-alumina) can be synthesized which is resistant to attack by water vapor.

Some recent reports relating to hydrogen beta-alumina include a review of ion-ion and ion-lattice interactions in β- and β''-alumina;[837] a discussion of the formation, decomposition, and conduction mechanisms in H_3O^+ β''-alumina;[835] the thermal stability of hydrogen beta-alumina;[1313, 1314] a study of the structure and properties of hydrogen beta-alumina;[307] and conduction mechanisms in hydrogen beta-alumina and in hydrated sodium β- and β″-alumina.[836]

Other Solid Ionic Conductors

A number of new potential solid electrolytes have been examined in recent years, some of which might be applicable to fuel cells. Among these are materials based on solid polyethers such as polyethylene oxide and polypropylene oxide.[193] These form cation-conducting adducts with selected alkali metal salts, and show conductivity (although at a low level) in the absence of water at moderately high temperatures, so they must be distinguished from aqueous gel polymers of the SPE type.

Polyurethane films containing particles of beta-alumina have been used as cation-conductors by General Electric Co.[838] Solid solutions of phosphate-containing materials such as $Na_{1+x}Zr_2Si_xP_{3-x}O_{12}$,[2033] or $Na_{1+x}Y_xZr_{2-x}(PO_4)_3$ and $Na_xZr_{1+x}Nb_{1-x}(PO_4)_3$,[1844] have been reported to be good sodium ion conductors. Solid sulfate-based electrolytes have been reported to have cationic conductivities at 300~800°C comparable to those of sodium beta-alumina or molten salts, but not adversely affected by moisture.[1116] One such substance, triethylene diamine sulfate, reportedly has a conductivity at 87°C of 10^{-3} Ω^{-1} cm^{-1}.[2408] Other sulfate-based solid electrolytes include lithium sulfate (575~860°C) [197] and $Li_{1.28}Zn_{0.36}SO_4$, $Na_{1.6}Mg_{0.2}SO_4$, and $Li_{1.72}Mg_{0.14}SO_4$ (22~745°C).[1117] Dehydrated sodium thiosulfate, developed by Deutsche Automobilgesellschaft, is reported to have a conductivity of 10.4 Ω^{-1} cm^{-1} at 364°C.[2628]

Sodium-conducting glasses also have been employed as solid electrolytes; for example, by Dow Chemical Co. in its sodium-sulfur secondary cells.[1943] Glass electrolytes theoretically possess a number of advantages over crystalline solids. For example, they have no grain boundaries, can easily have a wide composition range, and their physical properties can be modified by thermal treatment. In general, they have only a weak electronic contribution to conduction, can be made in thin films, and possess good plasticity. Oxide, doped oxide, sulfide, and fluoride glasses have received attention.[2087] Glasses such as B_2O_3-Li_2O-LiCl display lithium ion mobility.[1060, 1598] Some other solid lithium-ion–conducting solids are Li_xNiO_2 and Li_xCoO_2;[999] and Li_2MgCl_4, Li_2MnCl_4 and Li_2CdCl_4 (200~800°C).[1652]

All of the above materials are cation conductors, and so far only the smaller alkali metals have shown even moderate mobility. Whether they can be replaced by the proton for use as fuel cell electrolytes is open to some doubt. The proton is much too small and has too high a peripheral field to have significant mobility, and must be associated with a carrier molecule to give adequate conductivity. In aqueous media, the carrier molecule is

H_2O, giving the hydronium ion H_3O^+, or one of its hydrated forms such as $H_9O_4^+$. The H_3O^+ ion has about the same diameter as the potassium ion, but it is well known that the enhanced conductivity of protons compared with K^+ ions is due to the ability of the proton to *hop* via a Grotthus chain mechanism from carrier molecule to carrier molecule. If the Grotthus chain cannot be set up (e.g., in the absence of water), then proton conductivity will be poor or non-existent. A good example is the conductivity of the monohydrate of trifluoromethane sulfonic acid referred to earlier in this chapter. When pure, it is apparently fully ionized into H_3O^+ ions and $CF_3SO_3^-$ ions, but its conductivity is poor; it essentially behaves as a low-melting potassium salt. In contrast, the anhydrous acid is a nonconductor, the proton being completely immobile. Aiming to make solid hydronium ion conductors for use at temperatures over 150–200°C appears to be pointless, since the stability of the ion is not great. However, the use of carrier molecules for protons other than water is certainly possible: Good examples include NH_3, giving ammonium ion, NH_4^+; and, more important for the temperature range above 200°C, the phosphoric acid molecule, H_3PO_4, giving the ion $H_4PO_4^+$, which is responsible for the anomalously high conductivity of concentrated phosphoric acid at high temperatures. More information must be developed on the appropriate groups that should be included in solid proton conductors to give high conductivity by complexing protons with forces in the correct range (i.e., neither too strong nor too weak). At the present time, this subject is largely unexplored.

Among the new proton-conducting solid electrolytes, $H_3PMo_{12}O_{40} \cdot 29H_2O$[1849, 1850, 2777] and $H_3W_{12}PO_{40} \cdot 29H_2O$[1851, 1922] appear to show some promise. The latter has a reported conductivity at 25°C of about $0.2 \, \Omega^{-1} \, cm^{-1}$, combined with a low activation energy and a good resistance to attack by water vapor. The Japanese Agency of Industrial Science and Technology and Hitachi have used these electrolytes in hydrogen-oxygen fuel cells generating 100 mA/cm² at 0.5 V. These electrolytes seem to rely on water as the molecule allowing high proton mobility, since the materials lose ionic conductivity at temperatures much above the boiling point of water. Other similar proton-conducting solid electrolytes are $H_4SiW_{12}O_{40} \cdot xH_2O$ and $H_4GeW_{12}O_{40} \cdot xH_2O$.[1923]

Programs to develop practical solid proton conductors have been outlined by the Center for Materials Research at Stanford University,[1198] and also in Denmark.[1315] The object of the latter work is to attempt to develop a direct-methanol fuel cell under European (EEC) cooperation. This would operate at temperatures higher than those in the PAFC, and would (if possible) need no noble-metal catalysts, so that it could be a possible future replacement for the IC engine in vehicles. The main difficulty involved in achieving such a goal will be corrosion, the intrinsic problem associated with all materials exposed to air cathode conditions in acid media, especially at high temperatures. In particular, the search for a non-noble insoluble air cathode material, probably an oxide, will prove difficult.

In view of the above remarks, an anionic conducting system, perhaps involving some solid electrolyte allowing either hydroxide or carbonate conduction, might be a more appropriate future goal. In anhydrous media, OH^- ions are not subjected to the same constraints as protons from the viewpoint of conductivity and mobility. This is very well illustrated by the difference in conductivity between molten potassium hydroxide and pure trifluoromethane sulfonic acid. If a solid structure contains the correct combination of cations, it may have the properties of a CO_2-rejecting OH^--ion conductor. A number of new solid anionic conductors (principally fluoride ion conductors) also have been described in the literature. These include $Pb_{0.75}Bi_{0.25}F_{2.25}$, $Pb_{0.875}Th_{0.125}F_{2.25}$, $KBiF_4$, $RbBiF_4$, α-$TlBiF_4$, α- and β-$PbSnF_4$,[1060, 1610, 2090] and KF-doped PbF_2.[1610] It remains to be seen if a continuation of this work can lead to the development of an efficient, CO_2-rejecting electrolyte exhibiting the required OH^- or $CO_3^=$ conduction.

CHAPTER 11

Anodic Electrocatalysis

A. PRINCIPLES OF ELECTROCATALYSIS

The term *electrocatalysis* refers to the heterogeneous catalysis of a charge transfer reaction across an electrode-electrolyte interface.[1187] In the case of anodic dissolution or cathodic deposition of a metal, the charge is transferred to the electrolyte by the passage of metal ions across the electrode-solution interface, electrons being correspondingly passed between the electrodes via the external electronic circuit. For a redox reaction, where the electrode is invariant and a species in solution is either oxidized or reduced, charge is transferred across the interface between the electrode and the electrolyte by the passage of electrons. If the reactant is adsorbed on the electrode surface, part of the transfer is electronic (from the bulk electrode to the adsorbed species), and part ionic (the adsorption–desorption of the reacting species before or after electron transfer). The reactions normally associated with electrocatalysis in fuel cells are those of the redox type, but involving catalytic adsorption.

In electrocatalysis, as in chemical catalysis, the reaction rate is a function of the catalytic activity of the catalyst, which in turn depends on the effectiveness of adsorption in reducing the reaction activation energy and hence increasing reaction rate. However, in electrocatalytic reactions, the rate also is significantly influenced both by the presence of an electric field across the electrode-electrolyte interface and by the nature of the electrolyte. The importance of these latter factors was first investigated seriously in the late 1950s and early 1960s when work was initiated on the influence of electrode potential on electrosorption and on the mechanisms of electrooxidation reactions.

The word electrocatalysis seems to have been coined by W. T. Grubb in 1963,[1040] although it appeared in the German-language Soviet literature as early as 1929.[1442-1444] By the early 1960s, the mechanisms of the hydrogen and oxygen electrodes were beginning to be understood, at least in aqueous media, but little knowledge was available concerning the mechanisms of electrochemical organic oxidation, and direct anodic oxidation of hydrocarbon fuels had not yet been practically demonstrated. Earlier work in the 1920s and 1930s provided extensive knowledge of the nature of the electrified interface between liquid mercury electrodes and aqueous electrolyte solutions,[1005] but little was known about the electrical double layer (i.e., the charged electrolyte-electrode interface maintaining the electrode potential difference) at solid electrodes. The nature and microscopic properties of an electrocatalyst surface, or the behavior of non-aqueous electrolytes, essentially were totally unknown. As a result of the refinement and development of the techniques of electrode kinetics, solid-state physics, surface science, and quantum chemistry, we are now in a much better position to understand the fundamental nature of electrocatalysis. Significant progress has taken place in this subject over the past ten years, and this can be expected to continue over the next decade.

Function of an Electrocatalyst

The function of any catalyst is to increase the rate of reaction by providing an alternate reaction path with a lower activational energy barrier compared with that of the uncatalyzed path. In conventional chemical catalysis, once the catalyst has been chosen it is possible to further increase the reaction rate constant only by raising the temperature. In electrocatalysis, however, the activation energy barrier also can be lowered by increasing the electric field strength across the electrode-electrolyte interface, as is implied in the Tafel equation discussed in Chapter 2. As this equation shows, it is often possible to increase the reaction rate in the forward direction (and correspondingly reduce it in the backward direction) by more than an order of magnitude by appropriately changing the electric potential difference by as little as 100 mV or less.

Although the potential differences between the electrode and the bulk solution are quite small, they occur across a region that extends out from the electrode surface by only 1 nm in aqueous solvent, giving an extremely high local electric field strength ($\sim 10^8$ V/cm) supported by a layer of solvent (water) dipoles at the electrode surface (the Helmholtz double layer). At thermodynamic equilibrium for each half-cell reaction (each of which must include electrons of mean energy—i.e., at the Fermi level—in the electrode as reactants and products at the anode and cathode, respectively), the electrode Fermi level-solution potential difference will be at a fixed (and unmeasurable) level. However, the difference between the Fermi level potentials between the two electrodes under thermodynamic equilibrium conditions is equal to the free energy of the overall cell reaction expressed in V (see Chapter 2). This is the theoretical potential difference that would be measured at no net reaction rate, or electron flow; that is, with an instrument of infinite resistance. As soon as a net current is allowed to flow by lowering the value of the resistance in the external circuit, the Fermi levels in the two electrodes shift to provide a driving force for the transfer of electrons to and from the electrodes, the changes at the individual electrodes automatically adjusting such that the anode and cathode reaction rates are equal and opposite. This is the situation in a primary battery or fuel cell: In an electrolyzer or in the charging of a secondary battery, an external d.c. power source is connected in the external circuit to push the cell reactions in the opposite direction from the spontaneous process.

It is not the act of electron transfer between oxidant and reductant (fuel) molecules that distinguishes electrocatalysis from conventional heterogeneous chemical catalysis. It is rather the separation of this transfer into two half-reactions one oxidative, the other reductive, each at a different location and often on a different type of (optimized) catalyst surface. Thus, instead of a direct electron transfer between reactants in solution or those adsorbed on a common catalyst, the electrons are made to transfer through an external electrical circuit under a free energy (voltage) difference in which they can produce useful electrical work. It now also becomes possible to use two different catalysts: one specific to the anode reaction, the other to the cathode reaction. This latter feature of electrocatalysis enables an optimization both of catalyst efficiency and of cost.

It is beyond the scope of the present work to provide a detailed review of the theoretical aspects of electrocatalysis and charge transfer reactions; many such reviews can be found in the literature.[168, 857, 1424, 1812, 1910, 2245, 2353, 2631, 2760, 2761] Instead, some of the more important features of electrocatalysts for specific use in fuel cells will be discussed.

Requirements for a Fuel Cell Electrocatalyst

A fuel cell electrocatalyst should possess electronic conductivity, should be stable in the fuel cell environment, particularly under the difficult conditions at the cathode, and should display suitable adsorption characteristics for reactants and/or reaction intermediates. Chemisorption on the electrocatalyst surface must take place for effective reaction activation, compared with that in a simple redox reaction on an inert electrode. Since essentially all reactions of interest for fuel cells, with the possible exception of certain processes occurring at high temperatures, involve adsorbed intermediates, it is vitally important to understand the nature of the adsorption processes involved.[1237]

Chemisorption can be associative or dissociative.[1187] Typically, in the former, the π-orbitals in a double bond of the adsorbate form two single bonds to the catalyst surface; whereas in the latter, part of the adsorbed molecule splits off, both fragments chemisorbing separately. The anodic surface coverage, particularly on high surface-area platinum electrocatalysts, has been determined for a large number of organic molecules such as methanol, ethanol, formic acid, formaldehyde, ethylene glycol, methane, acetaldehyde, propanol, butanol, and carbon monoxide.[2269] It has been found that many of these compounds undergo dissociative chemisorption involving the splitting off of one or more adsorbed hydrogen atoms.[168] Hydrocarbons that chemisorb in acid electrolytes in this manner oxidize by reaction with adsorbed water at much higher rates and lower overpotentials (i.e., with lower activation energies) than those required for the same reactions on uncatalyzed (i.e., non-adsorbing) surfaces. The important point that the adsorbate must react with water to be oxidized in aqueous solutions should be noted, since water is the only phase containing available oxygen to oxidize C- to CO_2.

Theoretical studies using statistical probability theory have been made of the effects of electro-adsorption of organic substances on electron transfer and the kinetics of electrode reactions.[1416] Accurate methods of calculating the effects of adsorbed organic substances on the potential difference across the metal-solution interface also are available.[869] These methods take into account electrostatic dipole, quadrupole, octopole, and higher moments of the materials. The adsorption of surfactants that may be present in the electrolyte also can affect the kinetics of the charge transfer reaction; for example, by blocking suitable associated adsorption sites, as well as by affecting activation energies or bond strengths through electrostatic effects.[50]

As would be expected, the strength of the chemisorption bond is of great importance.[1187] A well-known example is the heterogeneous catalytic oxidation by an oxide catalyst, the activity of which is a function of the bond strength between the cation of the catalyst and the oxide ion.[424, 2160] However, when the chemisorptive bond strength is too high, a simplified analysis shows that it is too difficult for the products of the reaction to escape from the catalyst surface, thereby preventing the adsorption of further reactant by site blockage.[168] Conversely, when the bond strength is too low, insufficient reactant will be held onto the catalyst surface to permit a sufficiently high rate of electron transfer. At intermediate bond strengths the maximum rates of catalytic oxidation tend to be found. While this simplistic approach has been further elaborated,[140, 168] particularly to take the irreversibility of adsorbed intermediates into account, it still represents the general rule. However, for future work that might involve the tailoring of catalysts to particular processes, including changing reaction sequences to alter the rate-determining step, and thus lower the Tafel slope to reduce overpotential, a much more sophisticated approach must be used in future. The aim should be to eventually approach the degree of selectivity available in biological enzymes.

The most active electrocatalysts belong to the transition metals and certain of their compounds, as determined by their stabilities in the media and at the temperatures and potentials in question. The transition metals all have vacant d-orbitals in their atomic structures. By interacting with the electrons of adsorbable molecular groups, the vacant d-orbitals of these electrocatalysts produce chemisorption bonds of variable character. According to Pauling's theory of metals,[1986] the percentage of d-character of a metallic bond can be taken as a measure of the availability of vacant atomic d-orbitals for chemisorptive bonding. This commonly is referred to as the *electronic factor* in electrocatalysis, and has been widely used as a guide to the interpretation of the catalytic activity of metals.[428]

In addition to the electronic factor, the so-called *geometric factor* also is commonly cited as playing an important role in electrocatalysis. This refers to the observation that different types of sites, or groups of sites, on the surface of the same electrocatalyst, may display different electrocatalytic activity for the same reaction. The different crystal faces of a metallic electrocatalyst such as platinum suggest one obvious method of classifying different types of sites which may show different activity.[1725] To a reasonable approximation, the very small particles of supported dispersed platinum that commonly are used as fuel cell catalysts can be considered as groupings of individual single crystals, so it follows that the study of single crystal materials (which are fairly reproducible and amenable to various modes of classification) may yield a better understanding of the behavior of dispersed catalysts, which may show different proportions of individual crystal faces as a function of crystallite size.

A considerable amount of work has therefore been undertaken to study the relationships between the properties of small (~2 nm) supported particles of metallic electrocatalysts and those of smooth, bulk metal surfaces. For example, it has been shown that anion adsorption, which is known to affect the adsorption behavior of both hydrogen and oxygen on platinum, is influenced strongly by crystal orientation.[206, 1298] Similarly, the electrocatalytic activity of platinum deposited on pyrolytic graphite is a function of the degree of disorientation in the crystal structure.[1782]

In addition to crystal orientation, other site variations caused by crystal defects or microscopic areas of included or adsorbed impurity also can affect electrocatalytic activity. For example, crystalline steps often can serve as highly energetic sites where chemisorption may occur on a greater number of emerging orbitals than will be the case on a crystal plane. The surface morphology of the electrocatalyst is particularly important for adsorbates that are not very reactive. An increase in the rate of a reaction of the type (A + B → AB) may (at least in principle) be brought about by the provision of two different types of adsorption sites on the surface of the same electrocatalyst; one specific for A, the other for B.[1816] When associatively or dissociatively adsorbed molecules occupy many adjacent sites, these must be present in the correct number and with the correct steric geometry. The probability of finding suitable sites will obviously decrease markedly with the complexity of the adsorbate. Biological enzymes have evolved to allow the correct geometry for optimum adsorption, giving the least steric hindrance in the reaction sequence. A long-term goal in electrocatalytic research will be the design of catalysts that can mimic these natural structures for specific half-cell reactions.

Since the bond between a chemisorbed molecule and an atom at the surface of an electrocatalyst can be considered as an initial approximation to be covalent, the interaction between the electrons of these two species may be amenable to analysis using molecular orbital theory. In the final analysis, the geometric factor in electrocatalysis will be determined by the electronic factor and its three-dimensional geometry, since it is a consequence of the electronic structure of the materials involved.[419]

Evaluation of Electrocatalysts

Stability

A primary requirement of any electrocatalyst is stability in the operating environment of the cell. For this reason, corrosion testing almost always comprises the first step in the screening process for a candidate electrocatalyst. Corrosion testing usually begins with simple immersion of the material in the electrolyte at the fuel cell operating temperature in the presence of reactant, followed by potentiostatic testing in the absence of reactants, and finally by long-term testing under actual fuel cell operating conditions.

Electrocatalytic Activity

The next step in the evaluation process is the determination of the electrocatalytic activity of the candidate material. This is usually done by evaluating the polarization curve (electrode polarization vs. the logarithm of the current density to give a Tafel plot) for the reaction in the electrolyte of interest on a crystal of electrocatalyst of well-defined composition and surface area. If the reactant is a gas (e.g., hydrogen or oxygen), for convenience the electrolyte is saturated with it at atmospheric pressure. When possible, results also are obtained as a function of temperature. The value of the current density extrapolated back to the theoretical reversible potential (i.e., to zero polarization) is an estimate of the exchange current density, i_o, which is a measure of the maximum current that can be extracted at negligible polarization. The larger the value of i_o, the higher the intrinsic activity of the electrocatalyst. (The significance of the Tafel slope has been given earlier, in Chapter 10, Section B.) Electrocatalytic activity also can be evaluated by measuring the steady state current density that can be extracted at a given degree of polarization, or, alternatively, by measuring the extent of polarization at a fixed current density.

The activation energy for the reaction on a given electrocatalyst can be determined by measuring the current density, i, in the Tafel region at a constant reference potential at different temperatures and plotting $\ln i$ vs. $1/T$ (K). A linear plot indicates a reaction under the control of a single rate process. The activation energy is given by $-R[\partial \ln i / \partial(1/T)]$, where R is the gas constant and the expression in brackets is the slope of the $\ln i$ vs. $1/T$ plot. Low activation energies in the Tafel (activation controlled) region are usually characteristic of high electrocatalytic activity, at least according to the simple theory.[168]

Electronic Conductivity

To function correctly, an electrocatalytic particle in a porous electrode structure must be located at the site of electron exchange between the reactants and products of the reaction, and must serve as a conduction pathway for electrons to the current collector. The electrocatalyst on its support must therefore be an electronic conductor. If the electrocatalyst is a poor conductor or a semi-conductor it must be supported on or dispersed within a conducting material such as graphite. When using dispersed catalysts, it is important that there are sufficient reaction sites per unit weight and volume of the reaction zone of the electrode to ensure that the reaction can proceed at the required rate.

Precautions

To ensure valid measurements of the intrinsic electrocatalytic properties of materials, it is important to pay close attention to experimental procedures. First, care must be taken to ensure that the intrinsic activity of the catalyst is being evaluated, and not that of part of the electrode structure containing the catalyst.[970,1384,2503] In one instance known to the authors, the reported activity of an organic electrocatalyst was in fact due to electrocatalysis by the platinum mesh supporting the catalyst. When the platinum mesh alone was used, experimental results were exactly the same as those with the "catalyst" present. Similar difficulties may be encountered if care is not taken to exclude all traces of platinum from other components of the test cell, as well as from the test electrode itself. For example, if platinum counter electrodes are used in certain media, traces of platinum may dissolve and migrate through the test cell, ultimately becoming incorporated into the surface of the electrocatalyst under test, giving erroneous results.[877] Since platinum is often several orders of magnitude more active than the catalysts being studied, considerable care is required from the viewpoint of the design of the experiments.

When plotting polarization data to obtain Tafel slopes it is necessary to use both IR-free and diffusion-free voltage data. IR-free data can be obtained using oscilloscopes or IR-measuring circuits to determine resistive voltage drops, which then can be used to correct the measured data. Automatic IR-compensating circuits (e.g., IR-compensated potentiostats) are now available for this purpose. When making polarization measurements, it is also important that the test electrode be operated under conditions where the reaction does not depend on the diffusion of reactants or products to or from the reaction sites. This may particularly be the case if a high surface-area electrode is used in a *drowned* condition in a solution containing a dilute reactant; for example, oxygen at 1 atm pressure in an aqueous electrolyte. A useful technique in such cases is the use of an ultrathin porous diffusion electrode having one side in contact with the gas phase and the other in contact with the solution phase. Provided the thickness for which activity at constant potential is proportional to catalyst loading has been established, such electrodes are not diffusion-limited unless the current densities are very high. Alternatively, mass transfer effects (concentration polarization) can be corrected for by the use of planar electrodes using well-characterized and reproducible techniques, the best known being the rotating disk electrode.[2610]

Finally, although a fuel cell must be able to operate under practical conditions on relatively dirty reactants, meticulous care must be taken to ensure that the gases, electrolytes, and water used are of high purity to obtain significant, reproducible information on intrinsic electrocatalytic activity in aqueous media. In many cases it is sufficient to use double-distilled water, followed by pre-electrolysis of the electrolyte. However ultra-pure water occasionally must be employed. One recommended preparation involves desalination using reverse osmosis, followed by irradiation with ultraviolet light in the presence of H_2O_2 and HCl to oxidize organic compounds, followed by double de-ionization with a mixed bed exchange resin.[1908] However, such methods may result in introduction of organics and traces of chloride in the electrolyte. Solution pre-electrolysis [2610] is often used for final cleanup. Semiconductor-grade gaseous reactants with special-quality grease-free regulators now are readily available, and are usually satisfactory for normal work.

Experimental Evaluation Techniques

Adsorption

In the past, the principal methods employed for the qualitative and quantitative examination of electrosorbed intermediates have been linear sweep and cyclic voltammetry. Other electrochemical methods have included: rotating disk voltammetry,[31] computer simulation,[847, 1577, 2047, 2688] rotating ring-disk voltammetry,[1783] thin layer voltammetry,[31, 734, 1111, 1182, 1183] chronocoulometry,[127, 128, 612, 613, 734, 941, 1502, 1841, 1907, 1955] multipulse potentiodynamic techniques[1879] in conjunction with gas chromatography [1041, 1877] or mass spectrometry,[254] and numerous relaxation techniques.[251, 2125] Although these methods are sensitive and can detect small fractions of a monolayer of adsorbates, they are not very specific from the chemical viewpoint. However, voltammetric techniques have clearly shown that the adsorption energies of intermediates are different on different crystal orientations of an electrocatalyst.[2136] Radiotracer methods also have been used with varying degrees of success to investigate adsorption phenomena on electrocatalysts.[369, 532, 1163, 2024, 2292]

Surface Study Methods

As understanding of the phenomena underlying electrocatalysis becomes more sophisticated, it is increasingly important to develop *in situ* techniques under real-time operating conditions to investigate processes at the electrode-electrolyte interface. Surface characterization is especially necessary for alloy and nonmetallic electrocatalysts, because the surface properties of these materials are often quite different from those of the bulk material. Surface science methods promise detailed characterization of electrocatalyst surfaces as well as information on the structure and chemistry of adsorbed reactive intermediates.[379, 391, 672, 1911, 2136, 2228] These new methods provide information on the electronic and vibrational states of the catalyst surface, which are especially valuable in indicating the interactions between the catalyst and its substrate, so that more stable catalysts can ultimately be developed. Unfortunately, the electron spectroscopy methods cannot be applied *in situ*, and *in situ* optical methods have not been very successful to date, although Fourier-transform infra-red spectrometry (FTIR) is beginning to show promise.[339, 2040]

Many surface science techniques are currently undergoing rapid refinement.[1911, 2136] Methods currently being used for the characterization of electrocatalyst surfaces include: low energy electron diffraction, (LEED),[809, 1317, 1566, 2268] reflected high-energy electron diffraction (RHEED),[1466] interferometry,[909, 1120, 1209, 1434, 1615, 1823, 1904, 2177] ellipsometry,[264,

[311, 1288] internal reflection spectroscopy,[291, 1554] electron spectroscopy for chemical analysis (ESCA),[1319] infra-red reflectance Raman spectroscopy and the Raman microprobe,[1911] surface acoustic spectroscopy,[1911] X-ray photoemission spectroscopy (XPS),[1911] Auger electron spectroscopy (AES),[1310, 2444] secondary ion mass spectroscopy (SIMS),[2136] low-energy ion scattering spectroscopy (ISS),[2136] and linear a.c. electrode polarization impedance techniques.[743] Techniques that can be used to shed light on the structure and chemistry of adsorbed reactive intermediates include: ultraviolet (UV) and infra-red (IR) spectrophotometry, ultraviolet photoemission spectroscopy (UPS), thermal (flash) desorption (TDS), LEED, AES, vibration spectroscopy via low energy electron loss (EES), and extended X-ray absorption fine structure (EXAFS).[1911, 2136] The above give an indication of the wide scope of the modern methods now being developed and applied to the study of electrocatalysts and the probability that their use will lead to important future discoveries.

B. ELECTROCATALYSIS OF HYDROGEN

While most government agencies and academic institutions involved in fuel cell catalyst research and development have usually been generous in publishing their results, private corporations in this field generally consider it to be in their interest to keep some details proprietary, particularly in regard to preparation techniques. The discussion that follows is therefore somewhat less complete than might be desirable. On the other hand, sufficient information is available to give the reader a good grasp of the scope of work and of the state of the art. In addition, we should note that some newer catalysts discussed in the following have not undergone extensive life testing under realistic fuel cell operating conditions. It can therefore be expected that many showing initially high activity may not be stable as a function of time in a practical fuel cell.

The anodic oxidation of pure hydrogen can be considered to be a *facile* reaction, since it is readily effected by a fairly wide range of electrocatalysts in a variety of electrolytes. Hydrogen electrode current densities in excess of 200 mA/cm^2 are achievable at anodic polarizations of less than 30 mV under mild operating conditions with typical good catalysts. The process is the archetype of facility, in contrast, polarizations for the much more difficult oxygen reduction reaction might be 300~350 mV at the same current density. The reaction kinetics, mechanisms, and modes of electrocatalysis for the reaction are known in considerable detail.[168, 169, 671, 2136] It suffices to say that the hydrogen oxidation reaction is rapid compared with that for oxygen reduction because it involves only the breaking of one single bond, along with two electron transfer steps per molecule. In contrast, oxygen reduction requires the breaking of a double bond and four combined electron-atom transfer steps. In oxygen reduction, there is evidence that reaction intermediates are irreversibly adsorbed because of the affinity of the substrate catalytic species to form a very stable covalent product. In contrast, the hydrogen intermediates are much more reversibly adsorbed, making the desorption reaction much easier. Hydrogen oxidation in fact resembles chlorine reduction, both reactions involving breaking of a single bond with a total transfer of two electrons and participation of reversibly adsorbed species. A review of the preceding is given in Ref. 168.

As discussed earlier, the electrocatalytic oxidation of impure hydrogen, (i.e., the hydrogen-rich product from a hydrocarbon reformer containing varying amounts of CO, CO_2, CH_4, N_2, and H_2O,) can be carried out in fuel cell systems other than those using alkaline electrolyte. However, it is somewhat more difficult than that for pure hydrogen, due to dilution effects as the reactant hydrogen is consumed, and from the effect of catalyst poisoning by CO and other compounds present. Some further implications of these are discussed in the following.

Noble Metal Electrocatalysts

Platinum traditionally has been the hydrogen electrocatalyst of choice for most aqueous acidic electrolyte systems. In work since the late 1960s, various forms of the metal have been put in thin layers of from 0.01 to 1.5 mg/cm² loading on substrates such as carbon, metalized Teflon, and a variety of other stable metals. Of metals that are able to withstand the corrosive conditions in acid electrolytes at the appropriate hydrogen potentials, platinum is generally considered to be the most effective electrocatalyst among the stable noble metals. These include the platinum group metals (gold, silver and mercury), all of which are thermodynamically stable to practical hydrogen electrode potentials.[2044] In addition, carbon and certain carbides exhibit kinetic stability.

The relative activities for the hydrogen reaction in acid electrolyte on clean catalyst surfaces descend in the order: platinum, rhodium $\geq$ palladium > ruthenium >> gold, WC_x.[2136] The gross exchange current density for the hydrogen oxidation reaction on supported platinum catalyst in phosphoric acid at 160°C is about 140 mA/cm².[614] Based on the true surface area of the platinum, this corresponds to a value of 30 ± 15 mA/cm².[2136] Under these conditions, the limiting current density for these electrodes using pure hydrogen is about 10^4 mA/cm².[614] Consequently, the hydrogen electrode polarization on supported platinum is only about

5 mV per 100 mA/cm^2 in the normal range of current density in a practical fuel cell. This polarization value is remarkable, and is about as close to reversibility as one could hope for.

In view of its outstanding stability and activity, platinum is the prime choice as an anode catalyst. Although not entirely unique in its properties as an electrocatalyst, it does have a fortunate combination of properties that promote electrocatalytic activity. Its vacant d-orbitals are such that the metal-reactant bond strength is neither too strong (in which case reaction products would be unable to desorb) nor too weak (which would saturate the surface with intermediates, preventing further reactant adsorption). Also, under anode reaction conditions, platinum does not form oxide layers, which would reduce rates for hydrogen activation by blocking sites[2130, 2232] as they do for oxygen reduction.[705] Platinum has both a good affinity for the adsorption of hydrogen atoms, as well as a high diffusion coefficient for surface diffusion of adsorbed hydrogen atoms (> 5x10^{-9} cm^2/s);[817] indeed, the two properties are certainly connected. This affinity also forms the basis for its ability to dissociatively chemisorb saturated hydrocarbons, and is probably associated with why platinum is one of the few metals that can catalyze hydrocarbon isomerization.[76, 429, 1714]

The extent of hydrogen adsorption on platinum is dependent on the electrode potential.[1993] The molar entropies and enthalpies of adsorbed hydrogen atoms on the (111), (100), and (110) crystal planes of platinum have been calculated and the values compared with experimental values obtained in acid solution.[1895] As would be expected, platinum surface orientation has a pronounced effect on hydrogen adsorption.[1909] A detailed review of the hydrogen fuel cell anode reaction on noble metals in phosphoric acid has been published.[949] It is precisely because the individual electrocatalytic modalities of platinum are not unique, but are rather fortuitously combined in a single material, that the search for cheaper materials that can function as effectively as platinum is being actively pursued.

When used as a hydrogen oxidation electrocatalyst in acid, platinum customarily has been used alone,[2051, 2162, 2227] but occasionally it has been alloyed with other noble metals such as iridium or ruthenium. For use as a hydrogen oxidation anode catalyst, platinum-ruthenium alloy was reported to be less expensive and to give better performance than platinum-iridium alloy.[1558]

Platinum and platinum alloy electrocatalysts can be prepared using Adam's method, which involves fusion of the metal chloride or chlorides with sodium nitrate, followed by washing and reduction with hydrogen.[29] The metal also can be prepared by the direct reduction of the chloride with NaBH$_4$,[491] HCHO,[2709] or other reducing agents, as well as by the thermal decomposition and reduction of platinum salts.[931] A method has been published for the preparation of small (3~120 nm) monodispersed platinum electrocatalyst particles.[2512] Platinum-ruthenium alloy catalyst can be prepared by the reduction of a mixture of mixed oxides of the metals.[1558]

Electrodeposited platinum black has often been used for kinetic studies. A good review dealing with various aspects of the nature and preparation of platinized-platinum electrodes has been published.[848] Electrodeposited platinum occasionally has been used as a fuel cell catalyst. The porosity and adherence of electrodeposited platinum of this type has been reported to be increased if the process is carried out in a bath containing diaminedinitroplatinum, ammonium nitrate, sodium nitrate, and ammonium hydroxide at industrial a.c. frequency and 50~120 mA/cm^2 at 70~80°C.[1506]

Although other noble metals have sometimes been alloyed with platinum for use as hydrogen electrocatalysts in acidic media, in general the other noble metals have not been examined alone for this purpose. Palladium, for example, has not been widely used as a hydrogen catalyst in acid media on account of its excessive affinity for hydrogen. Palladium and some of its alloys equilibrate rapidly with molecular hydrogen, which diffuses into the metal structure as hydrogen atoms to form an alloy.[169] This affinity forms the basis for the use of palladium as a selective diffusion membrane to separate hydrogen, such as from reformer product gases (Chapter 8). It also allows the α-phase palladium-hydrogen alloy to be used as a reliable reference electrode,[1258] whose electrode potential and physical properties can be correlated accurately with the chemical potential of the hydrogen. Because of the extensive dissolution of hydrogen in the metal, palladium hydrogen electrodes exhibit characteristic polarization transients when current is switched on or off.[799]

This miniaturized reference electrode has been used to establish the relationship between polarization and the concentration of dissolved hydrogen adjacent to the electrode.[1601] Studies of the adsorption and dissolution of hydrogen in α-palladium have shown that an upper limit exists for the polarization resistance of the reaction between adsorbed hydrogen atoms and atoms dissolved directly below the surface.[462] The adsorption of hydrogen on palladium can be suppressed by the presence of species such as Bu_4N^+ and As_2O_3 and increased by species such as thiourea.[1680] Interestingly, layers of adsorbed palladium on the surface of platinum have been shown to decrease the uptake of adsorbed hydrogen in acid (HCl) media.[2393] Hydrogen diffusion polarization at palladium and palladium alloy electrodes and its implications for the theories of polarization have been reviewed.[1602]

Like palladium, ruthenium also is known to absorb hydrogen.[1055] The electrochemical surface oxidation and reduction processes on ruthenium have been found to take place in or close to the same potential domain as hydrogen deposition and ionization.[1056] A study of the adsorption–desorption kinetics of hydrogen on the clean and sulfur-covered ruthenium (001) crystal face has shown that the hydrogen adsorption is dissociative and described by $(1-f)^2$ kinetics, where f is the fractional coverage with adsorbed hydrogen.[2235] The hydrogen desorption energy decreases linearly as f increases, ranging from about 110 kJ/mol at low coverage to about 46 kJ/mol at 80% coverage. The effect of adsorbed sulfur is to block the dissociative sites for hydrogen adsorption and recombination sites for hydrogen desorption. The influence of anodic film formation on the behavior of reduced ruthenium surfaces in strong acid has also been studied.[514] Ruthenium dioxide is a well-known anode electrocatalyst in industrial electrolysis processes. The diffusion coefficient of H^+ ion in ruthenium dioxide electrodes[2675] and the adsorption of $SO_4^=$ ion on mixed RuO_2-TiO_2 electrocatalysts[79] have been determined. In the latter study it was found that in acid medium, sulfate adsorption increased with increasing ruthenium dioxide content.

The kinetics of the hydrogen oxidation reaction also have been studied on rhenium electrodes in 1M sulfuric acid.[920] In the case of rhenium, a two-electron mechanism is indicated.

Poisoning of Noble Metal Electrocatalysts

At operating temperatures below about 125°C in acid electrolyte, carbon monoxide seriously poisons noble metal electrocatalysts, causing a large degradation in performance. At the present time (see Chapter 10), the only acid electrolyte capable of operating beyond this temperature is phosphoric acid. For supported platinum, the nature of the support may be an important factor in determining the severity of poisoning. For example, acid fuel cell electrodes consisting of platinum supported on activated carbon have been found to be less prone to carbon monoxide poisoning than those made from platinum supported on graphite.[2399] Studies of the CO-tolerance of carbon-supported platinum fuel cell electrodes have shown that the initial carbon monoxide tolerance is not significantly affected by the nature of the carbon support (i.e., by carbon crystallite size or by the method of catalyst preparation), but that the age of the electrode may have a determining effect on its tolerance towards carbon monoxide.[2641]

The adsorption characteristics of carbon monoxide and hydrogen on platinum in acid electrolytes have been studied.[963] At saturation coverage with adsorbed carbon monoxide, only about 3% of the platinum surface remained active under the conditions of the experiments. In addition, it has been found that an adsorbed CO molecule has no more effect in reducing the heat of adsorption of hydrogen on neighboring sites than does an adsorbed hydrogen atom.[2136] The poisoning effect of carbon monoxide therefore seems to be a simple competition with hydrogen for sites, leading to a surface blockage deriving from the relatively strong chemisorptive bond of the carbon monoxide molecule.

Carbon monoxide can be anodically oxidized on noble metals in acid solutions at electrode potentials that are sufficiently high to weaken its chemisorptive bond.[376] As expected, the adsorption properties of the surface of the catalyst also are of major importance for oxidation to occur.[1817] On polycrystalline platinum electrodes in acid solutions there is evidence that the adsorbed carbon monoxide is oxidized via different reaction paths, depending on the electrode potential.[345] Similarly, at appropriate anodic potentials carbon monoxide also can be oxidized on palladium in acid solution. In sulfuric acid, for example, the surface coverage of CO on palladium decreases to close to zero over a narrow potential range at greater than 0.65 V/SHE.[464, 465] At potentials below 0.9 V/SHE, adsorbed carbon monoxide oxidizes in the absence of significant quantities of adsorbed oxygen; whereas at potentials above 0.9 V, the oxidation is accompanied by the formation of the oxide layer.[464] It also has been reported that carbon monoxide can be oxidized on a freshly prepared tungsten carbide surface; however, the reaction appears to be much faster in alkaline solution than in acid solution.[2187] Tungsten sulfide and molybdenum sulfide[405] as well as a number of the so-called bronzes of this group of metals[2578] also have been examined, but they appear to exhibit less activity towards carbon monoxide oxidation than tungsten carbide. Unfortunately, the anode potentials required to oxidize carbon monoxide on any of these materials are higher than any that are feasible in an acid fuel cell, and they would in any case result in irreversible anode oxidation.

The mechanism of carbon monoxide poisoning on platinum hydrogen oxidation electrocatalysts has been studied. Adsorption of carbon monoxide on freshly prepared platinized-platinum has been shown to be similar to that on bright platinum, being mainly of one-site adsorption. However, as the platinized-platinum electrode ages, an increase in two-site adsorption coverage takes place, in which a CO molecule bridges two hydrogen adsorption sites. This results in a decrease in the rate of the electrochemical oxidation of strongly bonded species on the electrode,[2399] and accounts for the observation that carbon monoxide poisoning is much more pronounced on aged catalysts than on those that are freshly prepared. Hence, although under some circumstances there is some evidence for a one-site poisoning mechanism, a two-site blockage mechanism is more likely to be encountered in a working fuel cell. As expected for a two-site mechanism, it is observed that the decrease in the activity of a platinum electrocatalyst for hydrogen oxidation is proportional to $(1-f_{co})^2$, where f_{co} is the fractional coverage of the catalyst surface with adsorbed carbon monoxide.[614, 2136]

In addition to poisoning platinum, carbon monoxide also adversely affects palladium, and lowers the dissolution rate of hydrogen into the metal.[464, 465] A 60% coverage of a palladium surface with adsorbed carbon monoxide lowers the hydrogen dissolution rate by a factor of about two. At higher coverages the dissolution rate is further reduced, although some hydrogen dissolution occurs even at saturation coverage by carbon monoxide. Similar effects have been observed on gold-palladium alloys.[1001] Thus, for 20 atomic percent gold in palladium, adsorbed hydrogen can be oxidatively stripped only after first oxidizing any adsorbed carbon monoxide. On ruthenium, it has been shown that the amount of carbon monoxide uptake can be strongly influenced by the presence of preadsorbed sulfur on the catalyst surface, an inverse relationship existing between the coverage of preadsorbed sulfur and that of adsorbed carbon monoxide.[2234]

Several different approaches have been employed to solve the problem of carbon monoxide poisoning in acid fuel cells operating on reformate-type fuels (i.e., the PAFC). One common solution has been simply to raise the operating temperature to above 135°C, above which carbon monoxide poisoning becomes less serious. Thus at 190°C, up to 2% by volume carbon monoxide can be tolerated without posing any serious difficulties for the hydrogen oxidation reaction.[614] The exact percentage before poisoning effects are observed depends on the H_2/CO ratio at the anode exit, and also on the amount of sulfur (see following) present in the gas stream. Typically a utility fuel cell will tolerate an H_2/CO anode exit ratio of about 8 without poisoning, at S contents (expressed as H_2S plus COS) of up to about 80 ppm. A study of the temperature vs. CO tolerance relationships for platinum fuel cell anodes has been made that gives the temperature required to obtain an acceptable level of polarization for a given platinum loading and carbon monoxide content of the fuel gas.[856] Increasing the temperature, however, represents a compromise between lower poisoning but higher rates of corrosion and component degradation, particularly at the cathode.[394]

A second solution has been the development of alloy or doped catalysts that eliminate or reduce the adverse effects of carbon monoxide. The incorporation of small amounts of nickel during the preparation of platinum black, for example, produced an electrocatalyst that demonstrated excellent short-term tolerance towards carbon monoxide.[934] Shell Research, Ltd. (England) found that by alloying platinum with ruthenium, impure hydrogen from a reformer could be used without the usual performance degradation encountered with pure platinum.[2702] This improved performance was attributed to direct oxidation of carbon monoxide on the alloy surface, leaving vacant sites to allow oxidation of hydrogen. Although the catalyst loading was high (12 mg/cm^2), a 17-cell battery was operated at only 30°C for 15,000 h with no drop in performance. The U.S. National Bureau of Standards developed a hydrogen oxidation catalyst consisting of ternary modifications of tungsten carbide of the form Mo$_{1-x}$W$_x$C, for use in phosphoric acid fuel cells.[1729, 1730] It was claimed that these electrocatalysts can tolerate up to 3% carbon monoxide, even at room temperature. Other CO-resistant electrocatalysts include platinum- and rhodium-blacks on tungsten oxide, developed by American Cyanamid.[1881, 1883] Inclusion of sulfur in platinum anode electrocatalysts also has been claimed to result in improved tolerance towards carbon monoxide in acid electrolytes.[349, 2186] In addition, it has been reported that platinum covered with a monolayer of sulfur, selenium, tellurium, or lead exhibits better electrocatalytic properties than platinum alone.[2186] Finally, tin ad-atoms also have been reported to be effective for the oxidation of carbon monoxide on platinum or platinum-covered gold electrodes.[1817]

In the extreme cases it is possible to avoid carbon monoxide poisoning altogether by separating the hydrogen from the other components of the reformate fuel gas using a palladium or palladium-silver diffusion membrane or by using selective oxidizers to lower the carbon monoxide content of the fuel gas stream fed to the electrodes.[826] This approach is useful if fuel cell start-up is required at very low temperatures in small, man-portable units for military communications equipment; for example, using SPE cells of simple design (cf. Johnson-Matthey, Chapter 5).

As indicated above, small amounts of sulfur in or on the electrode can increase the resistance of platinum towards carbon monoxide, however platinum catalysts also are poisoned by sulfur compounds. Hydrogen sulfide, for example, may react with CO to form COS. Significant hydrogen sulfide poisoning has been observed in phosphoric acid cells using hydrogen fuel gas containing 2% carbon monoxide;[614] however, it seems to be now established that the PAFC with supported low-loading platinum anodes will operate satisfactorily at 200°C with 1.5% CO and up to 50~75 ppm of (H$_2$S + COS).[2146] Hydrogen sulfide also may react with the carbon substrate of the electrode to form free sulfur.[614]

The adsorption and electro-oxidation of sulfur and hydrogen sulfide on platinum in acid electrolyte have been studied.[2662] In the poisoning of platinized-platinum electrocatalysts by thiourea [SC(NH$_2$)$_2$], it has been shown that 65~70% of the sulfur is adsorbed in elemental form on the platinum surface, the remainder adsorbing as thiourea or as thiocyanate ion.[2382] The catalyst surface can be regenerated by anodic-cathodic cycling. Rhodium hydrogen anodes in sulfuric acid also have been reported to be poisoned by thiourea.[2383] In this case, when thiourea adsorbs, adsorbed sulfur forms on the rhodium surface and reacts to produce Rh$_2$S$_3$. The sulfur poisoning could be removed at 0.6~1.2 V, but the regenerated rhodium had only 30–50% of its initial capacity for hydrogen adsorption. In respect to sulfur dioxide as a poisoning impurity, there is evidence that phthalocyanines may be able to catalytically oxidize adsorbed SO_2 in acid solution.[2070]

The above discussion makes it clear that for complex fuel compositions, it may be expected that deleterious synergistic interactions may take place among various impurities present. The nature of these interactions must be carefully evaluated to establish their long-term effects on fuel cell performance. This is an area that has only fairly recently become the subject of a significant research effort.[2146]

Nickel Electrocatalysts

Because of its low cost, nickel has been the preferred hydrogen oxidation electrocatalyst for use in alkaline solutions. While the pure metal is about three orders of magnitude less active than platinum for the hydrogen electrode process, activity can approach that of noble metals in practical electrodes by the use of the highest surface area (Raney nickel, which is pyrophoric when dry), and a heavy catalyst loading (e.g., 20 mg/cm^2 or greater, compared with 0.25 mg/cm^2 for platinum supported on carbon). However, heavily loaded nickel gas-diffusion electrodes, such as the 80% Pt–20% Pd 10 mg/cm^2 pure black anodes used in the space shuttle fuel cell, certainly are not optimized for maximum catalyst utilization.

Nickel is thermodynamically stable as bulk metal up to +110 mV/SHE at pH 14 and at 25°C, whereas at pH 0 its domain of stability is negative of –550 mV/SHE,[2044] hence it cannot be used as a hydrogen electrode in acid solution. In alkaline media, a reduced nickel oxide layer, probably consisting of β-Ni(OH)$_2$, forms on the metal surface and is capable of absorbing hydrogen.[2605] Similarly, in alkaline solution surface-oxidized nickel has been found to be electrocatalytically more active than bare nickel for the hydrogen evolution reaction. However, this may simply be due to an increase in surface roughness on the reduction of the oxidized surface. Discharge of the electrolyte cation has been proposed as having a role in the mechanism of the hydrogen evolution reaction,[1891] but this may be disputed.[141] It also has been found that water adsorbs dissociatively on a nickel surface close to the hydrogen electrode potential, possibly forming complexes that take part in the hydrogen oxidation reaction.[2364]

Sintered nickel alone (Chapter 9) has been used as an electrocatalyst at 200°C in concentrated KOH in the Bacon Cell[212] and in molten carbonates at 650°C.[2451] The polarization of the anodic hydrogen reaction on nickel in alkaline solution is dependent on the form of the nickel employed. For example, nickel that has been sputtered onto vitreous carbon or epoxy

Anodic Electrocatalysis 331

resin has been reported to be a durable fuel cell anode material resisting irreversible oxidation.[422] Hydrogen evolution studies have shown that the electrocatalytic activity of electrodeposited nickel black anode catalysts depends on the presence of sulfur-containing species in the electroplating bath.[1723] The addition to the plating bath of small amounts of SCN^-, thiocarbonate, $S_2O_3^=$, or thiourea, for example, results in nickel electrocatalysts with lower hydrogen polarization. Whether such activity is maintained over time is open to question, however.

Raney nickel often has been used for fuel cell hydrogen anodes. Studies of the activity of Raney nickel indicate that when hydrogen chemisorbs on this material to less than a monolayer, the adsorbed hydrogen and nickel surface atoms alternately undergo place change.[2363] Additional layers of adsorbed hydrogen dissociate but do not affect the surface in the same manner. Other adsorbed species also probably contribute to the activity of Raney nickel. For example, in hydrogen evolution in KOH solution activity appears to be greatly enhanced by the specific adsorption of metal ions such as Cd^{++}, Pb^{++}, Ag^+, and Mg^{++}.[1498] Nitrates also have been shown to exert a beneficial effect, probably through the formation of catalytically active adsorbed complexes with the metal. Porous nickel plate coated with Raney nickel has been found to be more active than expanded nickel screen for hydrogen evolution, but this is almost certainly due only to surface area effects.[897]

The effects that carbon monoxide and hydrogen sulfide impurities in the hydrogen fuel gas exert on the performance of Raney nickel anodes have been studied.[2399] Methods have been reported for the preparation of porous Raney nickel electrodes containing Teflon,[2400] as well as for the electrodeposition of Raney nickel onto metal substrates.[1373] Methods for the preparation of porous nickel fuel cell electrodes for use in alkaline electrolyte are also available.[757]

The activity or stability of nickel hydrogen anode catalysts is sometimes improved by doping with small quantities of noble metals such as 0.1% platinum or palladium.[1593] Similarly, the incorporation of cadmium or lead atoms into the nickel lattice has been reported to impart a significant increase in electrode activity, resulting from the formation of new active sites of a more optimal metal-hydrogen bond energy.[1500] The addition of small amounts of copper or vanadium to nickel hydrogen evolution electrocatalysts also has been reported to decrease hydrogen overvoltage,[814] as has the treatment of Raney nickel catalysts with lead nitrate solution.[1499] Electroplated nickel-based molybdenum-vanadium alloy is very corrosion resistant, highly conductive, and has been reported to exhibit low hydrogen polarization.[1548] The addition of transition metal salts of titanium, manganese, zirconium, vanadium, chromium, tungsten, cobalt, and copper to porous nickel hydrogen anodes also is reputed to improve performance.[2067] The addition of cobalt as molybdate has been considered to be especially effective.[1593] Similarly, the activity of Raney nickel can be increased significantly by alloying with transition metals such as iron, titanium, or molybdenum.[817] The addition of these metals apparently increases the surface diffusion coefficient of chemisorbed hydrogen atoms on the catalyst surface. Nickel-aluminum-titanium alloy has been used as a hydrogen anode catalyst in concentrated KOH at temperatures up to 92°C.[521] The addition of alloying elements such as aluminum and titanium to Raney nickel also helps to prevent area loss by sintering in gas-diffusion electrodes.

Nickel Boride

Nickel boride (Ni_2B) has been shown to catalyze the anodic hydrogen reaction in alkaline solution.[1307, 1310, 1311, 1764, 1766, 2452] In KOH at 80°C, it has been reported to be as effective as palladium black for the anodic oxidation of hydrogen.[1310] Nickel boride also has been used in conjunction with sintered nickel in concentrated KOH at 80°C.[1593] The gradual deterioration of the activity of nickel boride electrocatalysts as a function of time may be the result of changes in structure or wetting. However, it is more likely to be a result of surface reduction to nickel, since the polarizations of degraded nickel boride and pure nickel are similar.[1008] Heating the catalyst in oxygen at 70~110°C for 0.75~2 hours has been found to reduce the rate of performance degradation in alkaline electrolytes.[1767]

Mixed Metal and Spinel Electrocatalysts

Mixed metal catalysts of the composition $LnM_{4.5-5.5}$, where Ln is a lanthanide (La, Sm) and M is Ni and/or Co, have been found to make good hydrogen absorption anodes that were claimed to eliminate the need for high pressure fuel cell operation to obtain good performance.[527] Since they adsorb and store up to one atom of hydrogen per M atom, they have also found use in a low-pressure version of the nickel-hydrogen secondary battery. A pressurized version of this battery, using a nickel oxide positive and a rechargeable platinum negative hydrogen electrode, storing hydrogen in a pressure vessel, is used as a long-lived power source in communications satellites.[773, 774, 2353]

Catalytic fuel cell anodes of composition $M_xMn_{3-x}O_4$ (where M is Cu, Ni, or Ag; and x =1~5) or $Cu_zNi_{1-z}Mn_2O_4$ (where $z \le 0.70$) containing 12~16% graphite, have been reported.[466] Spinels of the structure XY_2O_4 (where X is Fe, Zn, Mn, Ni, Co, Mg, Cd, or Cu; and Y is Fe, Cr, Mn, Ni, or Co [e.g.,

NiCo$_2$O$_4$]) are used as hydrogen evolution catalysts, and may find use as hydrogen anode catalysts.[1874] Siemens A.-G. has reported that mixed metal electrocatalysts for use as fuel cell anodes in acid electrolytes can be made by reacting a powdered metal of the iron group at high temperature with two of phosphorus, arsenic, antimony, bromine, sulfur, selenium, or tellurium.[1827]

Sodium Tungsten Bronzes

Tungsten-based electrocatalysts have shown some promise both as fuel cell anode and cathode electrocatalysts. Good reserves of tungsten are available in the United States, primarily from Union Carbide's mines in Nevada and California.[109] Sodium tungsten bronzes, Na$_x$WO$_3$, which originally were developed as potential electrocatalysts for oxygen reduction, also have been used to catalyze the anodic hydrogen reaction,[346, 584, 663, 707, 893, 998, 1227, 1659, 2076, 2248, 2273, 2578] both in alkaline [2248] and acid [2076] electrolytes. In the latter case, the material was doped with 1 ppm platinum. The tungsten bronzes will be discussed in more detail in the section on oxygen reduction electrocatalysts.

Tungsten Trioxide

Tungsten trioxide (WO$_3$) can be considered to be a limiting case of sodium tungsten bronze, Na$_x$WO$_3$, with $x = 0$. Tungsten trioxide, which exhibits good stability in acid electrolytes, has been found to catalyze the anodic hydrogen reaction in acid solution.[1150] When WO$_3$ is associated with platinum in a hydrogen anode, there appears to be a synergistic effect in respect to activity, the mixed catalyst being more effective than either catalyst alone.[2, 1150, 1151] The behavior of WO$_3$-Pt electrocatalysts for hydrogen oxidation has been reviewed.[2503]

Tungsten Carbide

Another promising non-noble material that can be used in acid electrolytes to catalyze the anodic hydrogen reaction is tungsten carbide, WC,[267, 355, 395, 934, 1798, 1826, 2187] supported on either carbon[355] or charcoal.[267] Tungsten carbide is of great interest since it is a stable conducting solid that is isoelectronic with platinum, and displays many of the adsorption characteristics of the latter.[1758] Tungsten carbide has been used in sulfuric acid at 60~70°C[395, 1798, 1826, 2187] and in concentrated phosphoric acid at temperatures up to 150°C.[2187] Material with a specific surface area of about 30 m^2/g yielded a hydrogen ionization current density of 170 mA/cm^2 at 200 mV polarization in acid at 90°C.[1967] The current generated on tungsten carbide hydrogen anodes has been observed to increase after polarization at > 300 mV.[1967] The material has been suggested as a replacement for noble metal electro-oxidation catalysts in acid media,[2632] although its polarization is considerably greater than that of optimized noble-metal electrodes. As an example, the 170 mA/cm^2 result just quoted would result in negligible polarization using a low-loading platinum on carbon electrode, although some of the difference may be explained by structural factors. The great catalytic interest of WC for hydrogen oxidation is its ability to give high CO-tolerance.[1597, 1729]

The kinetics of the hydrogen ionization reaction on tungsten carbide have been shown to be the same on both ultra-dispersed particles of the material and solid tungsten carbide electrodes, the rate determining step being anodic dissociative adsorption or cathodic recombination and desorption of molecular hydrogen.[1726] Other workers also have studied the kinetics of hydrogen evolution on finely divided tungsten carbide,[410] and dissociation of adsorbed hydrogen on tungsten carbide has been confirmed, involving slow surface steps.[2363] Neutron scattering has been used to study the adsorption of both hydrogen and water on tungsten carbide surfaces.[2364] At room temperature, the surface diffusion coefficient for water was found to be ~10^{-5} cm^2/s.

It has been observed that the addition of 10-20% silver to tungsten carbide particles increases the activity for hydrogen oxidation in sulfuric acid by 30% over that for tungsten carbide alone.[2271] It has been shown that platinum can be dispersed easily on tungsten carbide to give catalysts of high area and low platinum loading, resulting in a much more efficient use of the noble metal.[1597, 1729] The CO-tolerance of such electrodes already has been noted.

Tungsten carbide fuel cell catalysts are produced by carburization of a tungsten compound (e.g., WO$_3$) at high temperature (590–680°C), using a gaseous mixture of carbon monoxide and carbon dioxide with a CO/CO$_2$ ratio of 10~20.[402, 403] Tungsten carbide also can be prepared by heating WO$_3$ in a mixture of hydrogen and carbon monoxide at 700–850°C for 20 hours.[1872] Depending on the starting material (W, W$_4$O$_{11}$, WO$_3$, or H$_2$WO$_4$), the activity of tungsten carbide can vary by half an order of magnitude as a result of differences in surface properties (adsorption behavior) of the final product.[1885, 1886] No effect on activity is observed when the carburization atmosphere is changed from carbon monoxide to a mixture of carbon monoxide and hydrogen, or carbon monoxide and argon.[1885] Tungsten sulfide (WS) also has been suggested as an anodic hydrogen oxidation catalyst.[405] Pure tungsten alone has been evaluated as a hydrogen evolution electrocatalyst in sulfuric acid[263] and in hydrochloric acid [1691] electrolytes, but it is much less effective than platinum.[169]

Miscellaneous Hydrogen Electrocatalysts

In 1973, E. I. duPont de Nemours Co. announced the development of anode catalysts based on the Mo-O-S semiconductor system, that were able to oxidize hydrogen, carbon monoxide, unpurified reformer gas, and formaldehyde in acid electrolytes.[2724-2726] In 1.25 M sulfuric acid at 85°C these catalysts were able to sustain hydrogen ionization current densities of only about 35 mA/cm^2 with 480 mV polarization; however, in 50% phosphoric acid at 100°C current densities of 100 mA/cm^2 were obtainable at 100 mV IR-free polarization.[2725] Doping to modify their properties may allow a further improvement. The performance of these materials in alkaline solution was inferior to that in acid: In 8 M alkaline electrolyte at 65°C, only 35 mA/cm^2 was attainable at 120 mV polarization.[2725] Molybdenum sulfide also has been examined as a possible hydrogen anode catalyst. [2578]

Electrodeposited cobalt has been claimed to be a suitable catalyst for fuel cell hydrogen electrodes.[1294] Cobalt carbide also has been reported to be an active hydrogen anode catalyst,[1825] as has cobalt sulfide.[2053] Mixed cobalt-molybdenum oxides have been used to coat nickel anode catalysts for use in alkaline electrolytes in order to increase electrocatalyst activity.[489] Transition metal molybdates,[854] including cobalt molybdate,[153] are also of interest as high-performance catalytic materials for hydrogen oxidation and evolution in alkaline electrolytes. However, like many non-noble catalysts for hydrogen oxidation, care must be taken not to overpolarize them in the anodic direction, otherwise instability and/or dissolution may result.

Platinized boron carbide has been evaluated as a hydrogen anode catalyst in 2.5 M sulfuric acid at 25°C.[1151] Chelates such as copper phthalocyanine and palladium acetylacetonate, which have been developed primarily as oxygen reduction catalysts, also have been suggested for use as hydrogen oxidation catalysts in acid fuel cells.[1802] Transition metal silicides have been evaluated as catalysts for the anodic oxidation of hydrogen in acid fuel cells.[877] Although these materials are very stable in acid solution, their electrocatalytic activity has been found to be low. Niobium has been shown to be electrocatalytically active for hydrogen evolution in concentrated alkaline solutions, but it appears to be of little value for hydrogen oxidation.[1930]

C. ELECTROCATALYSIS OF HYDRAZINE

Hydrazine, like hydrogen, is readily oxidized using a number of electrocatalysts.

Nickel

Nickel, in a variety of forms,[1942] (e.g., porous nickel,[665, 1493, 1671, 1677] nickel black,[2108] and Raney nickel[198, 1046, 1453, 1473, 1535, 2198]) often has been used as an electrocatalytic anode material for the oxidation of dissolved hydrazine in low-temperature alkaline fuel cells. However, a difficulty encountered with nickel is its tendency to promote hydrazine decomposition to ammonia, particularly at open circuit. To avoid this problem, porous nickel hydrazine anodes have been replaced by porous iron electrodes.[2002]

Nickel Boride

Nickel boride (Ni_2B), usually on a nickel substrate, also has been used successfully as an anode catalyst in low-temperature alkaline hydrazine fuel cells.[1053, 1306, 1307, 1311, 1342, 1766, 1768] General Motors developed an electrode of this type that was capable of operating at 200 mA/cm^2.[1766, 1768] The electrode was formed by first co-depositing a porous layer of Mond nickel powder with nickel on an electroformed nickel sheet substrate, followed by catalyzing by chemically precipitating nickel boride into the porous nickel layer. Studies carried out by General Motors on this material showed that a gradual loss of boron occurred as a function of time, giving a slow voltage decay at the operating current density. This is similar to the observation noted earlier when this compound is used as a hydrogen oxidation catalyst. Heat treating the material in hydrogen has been found to help stabilize it against boron loss.[1768, 1769]

Noble Metals

Although noble metals such as platinum[1492, 2606] and ruthenium[1065] have been used as catalysts for hydrazine fuel electrodes in alkaline cells, palladium has had the most success among the noble metals.[1490, 1942] Palladium deposited on a nickel substrate has been commonly used.[1478, 1803] Palladium-doped Raney nickel has been shown to be significantly better than Raney nickel alone [2198], and work carried out by the U.S. Army demonstrated that palladium black in porous iron is even more desirable, since it produces no gassing at open circuit and only a very small amount of gassing during cell operation.[2001] Matsushita Electric has used palladium-nickel-tin catalysts precipitated in sintered metal structures as hydrazine fuel cell anodes.[1722]

The adsorption characteristics of hydrazine on platinum[1492] and on palladium black[1494] in KOH solutions have been studied. Platinum also has been used to oxidize hydrazine-borane (HB) (i.e., $N_2H_4 \cdot BH_3$) in both acid and alkaline solution.[802, 803] The use of various noble metal catalysts for the oxidation of hydrazine in industrial batteries has been discussed by a number of authors.[630, 955, 2167, 2449, 2474]

Miscellaneous Electrocatalysts for Hydrazine Oxidation

Silver can be used as an electrocatalytic anode material for the oxidation of hydrazine, but is unable to withstand significant polarization, and therefore has not often been used in practical hydrazine fuel cells where hydrazine concentration gradients may generate sizable diffusion overpotentials. Cobalt has been used with some success for the oxidation of hydrazine.[2655] A mixture of nickel-cobalt black was found to be more active than nickel black, and also less soluble than cobalt black.[1473, 1535] This combined catalyst also has been reported to give higher hydrazine conversion efficiency than pure cobalt black. Iron-nickel-cobalt catalysts have been used for the oxidation of 1 M hydrazine in 6 M KOH at temperatures up to 51°C.[2082] Tungsten carbide has been tried as a hydrazine oxidation electrocatalyst in both alkaline (6.5 M KOH) and acid (1M H_2SO_4) solutions at 70°C.[2187] In the alkaline solutions, current densities of up to 150 mA/cm² were reported. Although high polarization was encountered in the acid electrolyte, no open circuit decomposition of hydrazine was observed. Mo-O-S catalysts were examined in 2 M caustic solutions at 50°C containing a six volume percent solution of hydrazine monohydrate:[2725] They gave a current density of 35 mA/cm² at a polarization of 430 mV. Chelates, such as graphite-supported iron-tetrasulfonophthalocyanine (Fe-TSP) and Co-TSP, also have shown activity for hydrazine oxidation in alkaline solution; Fe-TSP being more effective than Co-TSP.[2790]

D. ELECTROCATALYSIS OF HYDROCARBONS

Direct electro-oxidation of hydrocarbons was the first goal of acid fuel cell researchers in the late 1950s and 1960s, because of the system simplicity and intellectual satisfaction of the direct hydrocarbon energy converter compared with the internal combustion engine. Unfortunately, simple hydrocarbon gases were shown to be anodically oxidized at unacceptably low rates, while higher molecular weight compounds generally proved to be even less reactive. Olefins, aromatics, and six-membered ring naphthenes have been shown to sometimes act as catalyst poisons. The practice up to the late 1950s (and to date) was to use massive loadings of platinum to electrocatalyze the direct oxidation of hydrocarbons, thereby obtaining at least measurable rates.

Various investigations have been conducted on reaction mechanisms for the direct electro-oxidation of hydrocarbons. Such studies include: the general problems of hydrocarbon oxidation;[668, 2581] hydrocarbon oxidation on platinum in 85% phosphoric acid at 150°C;[967] oxidation of the derivatives of methane, ethane, and propane;[348, 351, 1614] oxidation of olefins and paraffins;[2209] oxidation of normal alkanes on platinum;[1614] oxidation of multi-component hydrocarbon fuels on platinized-platinum in 95% phosphoric acid at 175°C;[1651] mechanisms of the oxidation of saturated hydrocarbons on platinum;[498, 1186] hydrocarbon oxidation mechanisms in low temperature fuel cells;[760] the effect of electrode surface roughness;[2519] and the effects of adsorption phenomena on hydrocarbon oxidation.[218] A good review of the anodic oxidation of hydrocarbons has been published in Ref. 546. In the subsequent sections, some of the anode catalysts that have been employed for the direct electrochemical oxidation of hydrocarbon fuels will be reviewed.

Methane

Methane has been directly electro-oxidized in sulfuric acid with 100% conversion to CO_2 and H_2O on Raney platinum at 100°C[353] and on platinized-platinum at 80°C.[1179] In the former study, current densities of 100 mA/cm^2 at 0.50 V/SHE were reported. The results of a study of methane oxidation on platinized-platinum in 0.5 M sulfuric acid at 60°C indicated that two electrosorbed species seem to be involved in the oxidation, namely a COH-type species and a CO-type species.[2381] The addition of nickel ions to 80-95% phosphoric acid electrolyte was said to enhance the ability of platinum to oxidize methane directly.[499, 2518] The presence of Ni^{++} at 130°C was found to considerably improve the resistance of platinized-platinum to self-poisoning, thereby extending its useful life and giving higher performance. It was concluded that the effect might be attributed to an increase in hydrocarbon solubility resulting from a water-structural effect.[499]

Ethane

Ethane has been reported to be oxidized completely at 150 mA/cm^2 and 0.50 V to carbon dioxide and water on Raney platinum in sulfuric acid at 100°C.[353] Ethane also has been electrooxidized on platinum in 0.5 M sulfuric acid at 90°C.[2519]

Propane

Propane has been electro-oxidized on platinum in sulfuric acid solutions with varying degrees of success. On Raney platinum at 100°C in 1.5 M sulfuric acid, 100% conversion was reported at 150 mA/cm^2 and 0.50V.[353] In a different study, current densities on platinum in 1.5 M sulfuric acid at 95°C were reported to be in the range of 10~30 mA/cm^2,[2633] while in yet another study on platinized porous carbon in 2.5 M sulfuric acid at 80°C, current densities were less than 1 mA/cm^2.[2209] The reaction in 2 M sulfuric acid at 65°C also was examined.[2027]

Propane oxidation was reported to take place on platinum in 85% phosphoric acid at 140°C at a current density of about 20 mA/cm^2.[2633] Oxidation was also shown to occur on platinum in 2 M phosphoric acid at 65°C.[2027] The kinetics of electro-oxidation of propane on platinum in concentrated phosphoric acid over the temperature range of 108~232°C have been investigated.[2587] As in the case of methane, the addition of nickel ions to the electrolyte was found to enhance the rate of the reaction in 80~95% phosphoric acid.[499]

More recently, studies have been made of propane electro-oxidation in a number of fluorinated organic acid electrolytes synthesized at the American University in Washington, D.C.[53] In parallel work, it was found that propane was oxidized on platinized-platinum in tetrafluoroethane-disulfonic acid at about the same rate as in phosphoric acid.[233] Platinum also has been used as the electrocatalyst for propane oxidation at 65°C in 2 M CF$_3$COOH.[2026, 2027]

Powdered nickel subsulfide (Ni$_3$S$_2$) has been reported to catalyze propane oxidation on stainless steel in mildly alkaline (pH 9) buffered electrolyte, apparently with no readily oxidizable intermediates, at a polarization of less than 180 mV and with approximately 100% current efficiency.[1655] No corrosion of the stainless steel was observed.

Butane

Butane has been oxidized on Raney platinum in 1.5 M sulfuric acid at 100°C with 100% conversion to carbon dioxide and water, at 90 mA/cm^2 and 0.50 V.[353] However, in a different study on platinized porous carbon in 2.5 M sulfuric acid at 80°C, butane could only be directly oxidized at less than 1 mA/cm^2.[2209] High initial activity for the direct oxidation of butane was reported on 50-50 cobalt-platinum alloy prepared by adsorption of cobaltous chloroplatinate on a carbon support followed by reduction, although activity could not be sustained.[805]

Octane

Octane has been reported to be oxidized on platinum in 0.1 M sulfuric acid at 80°C.[715]

Ethylene

Most studies of the oxidation of ethylene have been made using platinum catalyst and sulfuric acid electrolyte at temperatures up to about 90°C.[2519] It has been found that fairly high current densities are obtainable on Raney platinum at temperatures greater than 50°C.[353] However, in another study, current densities of only 1~5 mA/cm^2 could be achieved on platinized porous carbon in 2.5 M sulfuric acid at 80°C.[2209] The kinetics and mechanism of ethylene oxidation on platinized platinum in 0.5 M sulfuric acid have been studied.[2489, 2490]

Propylene

Propylene has been oxidized on platinized porous carbon in 2.5 M sulfuric acid at 80°C, giving current densities of 1~5 mA/cm^2.[2209] A study also has been made of the heterogeneous solution oxidation of propylene with molecular oxygen in aqueous suspensions of palladium or palladium-gold catalysts at 50~96°C, and the mechanism discussed with respect to fuel cell processes.[1415]

Methanol

Like hydrazine, methanol is considered to be a liquid fuel of relatively high activity. Until about a decade ago, there was greater general interest in using hydrazine in fuel cell systems than methanol, since hydrazine is more reactive. More recently, however, interest in hydrazine has fallen, partly on account of its toxicity and partly because no low-cost energy-efficient hydrazine synthesis process has been developed to the commercial stage (see Chapter 8). Consequently, interest in methanol as a direct fuel has increased, despite the fact that it is electrochemically much less reactive than hydrogen or hydrazine.

Methanol is expected to find wide commercial use in the future, first as a gasoline extender (octane rating enhancer) for internal combustion engines, and finally as a substitute for gasoline. Methanol has been seriously evaluated for direct use in low temperature fuel cell systems, particularly by Shell and Exxon/Alsthom in the 1970s. Owing to the solubility of methanol in aqueous electrolytes and its tendency to oxidize chemically at the cathode, the direct methanol route, while simple and compact in principle, presents difficult system challenges for the fuel cell designer. Methanol differs from hydrazine in that in both flow-through and flow-by anode systems, it is impossible to consume all the fuel and prevent migration of the remainder to the cathode area in a cell of typical design with an acidic (CO_2-rejecting) electrolyte. The higher reactivity of hydrazine in KOH, together with its tendency to decompose to give gaseous products, makes this problem less likely with hydrazine. The fuel cell system designer has of course the option of an indirect (e.g., reformed methanol) system, but the penalty is greatly increased weight and volume. The technical problem is to try to obtain fuel cell power densities using methanol directly that approach those with hydrogen electrodes.

The major application for a low temperature direct methanol-air fuel cell would be for vehicular transportation. To be practical as an electric vehicle power source, a methanol fuel cell anode must be capable of generating a power density of greater than 200 mW/cm^2, must have an operating life of at least 3000 h, and must not require a noble metal catalyst loading in excess of 1 mg Pt/cm^2.[501, 504] The catalyst loading criterion appears to be considerably lower than that achievable for the state of technology, and even it will limit the total market because of the problem of platinum availablity. Although the development of a high-performance direct methanol fuel cell is a high risk endeavor, the rewards in the transportation sector also may be very high. Several reviews of the electrocatalysis and development aspects of the direct methanol system have been published.[501, 1545, 1911, 2512] The most effective practical catalysts appears to be alloys, such as Pt-Sn.[83] Hitachi's work on the methanol acid-electrolyte fuel cell has been referred to in Chapter 8.[1753, 1755, 1758, 1759]

Methanol Electrocatalysis in Acid Solution

In aqueous acid solution, the overall anodic oxidation of methanol results in the production of carbon dioxide:

$$CH_3OH + H_2O \rightarrow CO_2 + 6H^+ + 6e^-$$

This is in contrast to the path in alkaline solution, which will lead to the production of carbonate, with subsequent carbonation of the electrolyte:

$$CH_3OH + 8OH^- \rightarrow CO_3^= + 6H_2O + 6e^-$$

To prevent irreversible carbonate formation, it is customary to carry out the reaction in a CO_2-rejecting acid electrolyte. The disadvantages of an acid electrolyte are that noble metal catalysts must be used in order to ensure catalyst stability and that high catalyst loadings are required to avoid self-poisoning.[504] The reaction customarily has been carried out on platinum or on a platinum-based electrocatalyst, using sulfuric acid as the electrolyte. The overall reaction for the oxidation of methanol in acid shows that the electrocatalyst must play a bifunctional role, in that it must simultaneously adsorb both methanol and water,[219, 1076] just as in the case of direct hydrocarbon oxidation.[168] Because water is a reactant, if a high water activity is not maintained in the vicinity of the anode the reaction rate will be decreased. This explains why increasing the temperature (which tends to concentrate the electrolyte, thereby lowering the activity of water) often results in only a small increase (or in even a decrease) in the apparent activity of the electrocatalyst.[504]

The rate of the methanol oxidation reaction is found to be slower in acid than in alkaline solution, however the reaction in the former media tends to yield a greater conversion to the final fully oxidized product (CO_2) than is the case in alkali.[504, 1545, 1639, 1911] In the latter, there is often considerable partial oxidation to formaldehyde and formate. In contrast, the reaction in acid electroyte results in essentially complete conversion, only minor amounts of formaldehyde and formic acid ultimately being produced.[1029, 1758] In acid, the main reaction proceeds in series via an adsorbed intermediate, formaldehyde, and formic acid:[1029]

$$CH_3OH \rightarrow CH_3OH_{ads} \rightarrow \text{(intermediate)} \rightarrow CH_2O_{ads} \rightarrow HCOOH_{ads} \rightarrow CO_2$$

It is known that in acid solution methanol dehydrogenates on platinum at potentials as low as 0.2 V, which suggests that the intermediate is the species COH,[504, 1077] which occupies one site:[1758]

$$CH_3OH_{ads} \rightarrow COH_{ads} + 3H^+ + 3e^-$$

The hydrogen atoms resulting from the dissociation first adsorb, then discharge rapidly with electron transfer. There also is some possibility that the intermediate species might be C_2O_2H (which would occupy two sites),[1758] or $CH(OH)_2$.[459] Further oxidation of the adsorbed COH intermediate to CO_2 takes place in acid at potentials greater than 0.5 V.[504] Further oxidation of the COH_{ads} produces only CO_2, whereas the oxidation of soluble methanol always produces some formaldehyde and formic acid.[1758]

The experimental evidence suggests that a second adsorbed species, $C_xH_yO_z$, forms in parallel with the adsorbed intermediate COH_{ads}, and that this second intermediate acts as a poison. It is strongly held on the elec-

Anodic Electrocatalysis

trode surface, in effect blocking reaction sites and causing a decrease in current density as a function of time.$^{(504, 1758, 1911)}$ Typically, the major portion of the drop in current density occurs within the first few seconds of closing the circuit, the steady-state current density after several hours being lower than the initial value by a factor of 10^4–10^5.$^{(1758)}$

The gradual long term decay often is recoverable in part by periodically holding the cell at open circuit.$^{(501)}$ To recover an approximation to the initial activity, the poisoning intermediate must be removed by reaction at higher potentials with adsorbed water or oxygen,$^{(1758, 1911)}$ in particular by pulsing to high anodic potentials, such as by short-circuiting the cell.$^{(384)}$ Efforts have been made to improve the performance of platinum, rhodium, and platinum-rhodium alloy electrodes in sulfuric acid by rotating the electrodes to continuously renew the electrode surface, but such a technique is not feasible for fuel cell applications.$^{(830, 1890)}$ The build-up of the poisoning intermediate occurs in both acid and alkaline media, and is clearly the major obstacle to be overcome in the electrocatalysis of methanol oxidation.$^{(1758)}$ The reaction also has been found to be very sensitive to poisoning by the ammonium ion.$^{(2653)}$

The precise reaction mechanism for methanol oxidation may be sensitive to both the method of preparation of the electrocatalyst and to the purity of the reactants employed.$^{(1758)}$ For example, the surface preparation of the platinum catalyst affects the composition of the methanol adsorption products.$^{(2687)}$ Similarly, the electrocatalytic activity of supported platinum seems very dependent on how it has been prepared. Although the surface area of a catalyst prepared in a hydrogen atmosphere is greater than that of one activated in air, the intrinsic activity of the former is lower.$^{(206)}$ Studies by Shell Research, Ltd. showed that platinum and platinum-ruthenium catalysts with much higher steady-state activities than that of ordinary platinum could be made by supporting the catalysts on specially treated carbon-fiber paper.$^{(205)}$ The high platinum activity and stability of these preparations may be a result of interaction with the carbon catalyst support. Cyclic voltammetry showed that this form of platinum exhibited different hydrogen adsorption–desorption, and adsorbed oxygen reduction characteristics compared with unsupported and conventionally supported platinum.

Studies using a variety of different acid electrolytes have shown that methanol oxidation on platinum is sensitive to the acid chosen.$^{(84, 1200, 1753)}$ The acid medium affects the reaction by influencing the activity of the water required as a reactant and by blocking reaction sites on the catalyst surface by adsorption of undissociated acid molecules or anions.$^{(1758)}$ The two acids frequently considered as alternatives to sulfuric acid have been phosphoric acid and trifluoromethylsulfonic acid (TFMSA). Phosphoric acid can be used at temperatures up to about 200°C (utilizing gaseous methanol reactant), whereas sulfuric acid can be used only up to only about 90°C, beyond which it starts to decompose.[1758] Hence, the use of phosphoric acid offers the possibility of faster reaction kinetics. However, typical platinum catalysts are not as active in phosphoric acid as they are in sulfuric acid, so that very high catalyst loadings were required to obtain reasonable performance in H_3PO_4.[84, 1572, 1753, 1911] The reason for the reduction in activity of platinum catalysts in phosphoric acid is partly a result of adsorptive poisoning by undissociated H_3PO_4 molecules and $H_2PO_4^-$ ions, both of which tend to be more difficult to desorb from the platinum surface than the corresponding species in sulfuric acid,[84, 1295] and partly due to the low water activity in H_3PO_4, as shown by its low water vapor pressure. Adsorption of other organic molecules on the platinum surface also strongly affects the electrochemical oxidation of methanol.[218]

Sulfuric acid and phosphoric acid have been compared as electrolytes for methanol-air fuel cells using standard methanol catalysts such as platinum black, 70 Pt-30 Ru, Pt-Sn, and platinum gray.[84] As expected, for both acids, as the acid concentration increases, catalyst activities decrease. As would be expected, the reaction rate on the self-poisoned catalyst surfaces has been found to be proportional to the activity of the water in the acid, which decreases as the acid concentration increases. Although these results indicate that phosphoric acid is inferior to sulfuric acid as an electrolyte for the anodic oxidation of methanol, it has been suggested that new developments such as bimetallic cluster catalysts and higher temperature operation (at which rapid internal reforming of methanol may be feasible) may make a re-evaluation of phosphoric acid worthwhile.[501]

Also, TFMSA monohydrate has been compared to sulfuric acid as a medium for methanol oxidation. Most noble metal electrocatalysts have a very low activity for methanol oxidation in the pure monohydrate because of the low water activity in this material.[1758] Aqueous solutions of TFMSA therefore give better performance than the pure acid.[1200] Up to 60°C, at TFMSA concentrations between 10~50 % by weight, activity in TFMSA is typically somewhat better than in 3 M sulfuric acid, and the self-poisoning effect also is less than in sulfuric acid. However, at temperatures greater than 60~80°C, aqueous $CF_3SO_3^-$ anion was reported to decompose to produce sulfur and hydrogen sulfide, both of which poison noble metal catalysts.[1200, 1758] In addition, at concentrations greater than 50%, TFMSA was reported to decompose at 0.0V to produce sulfur species that may poison the catalyst. However, recent work has shown that the $CF_3SO_3^-$ ion is indeed stable,[2146] so that the observed effects must have been due to the presence of impurities (incompletely fluorinated material) in the sample of acid used. The latter has turned out to be a frequent problem with synthetic fluorinated sulfonic acids. At the time of these studies (1977–1978), it was concluded that the slight improvement in performance at temperatures below 60°C did not justify the greater expense of TFMSA over sulfuric acid; however, this should be modified in the light of more recent work showing that fluorinated sulfonic acid polymers (including Nafion) are stable to much higher temperatures. Finally, we should re-emphasize the fact that methanol oxidation in a liquid electrolyte acid fuel cell requires careful design, to prevent methanol passing over to the cathode and causing depolarizion. The system should contain a conducting membrane, which may profitably be Nafion or a similar material. This also can serve as the electroyte, even at temperatures above 100°C, provided that methanol is supplied to the anode in the vapor phase along with steam to give the necessary high water activity for oxidation and to render the Nafion sufficiently conductive. The comments in Chapter 8 concerning in-cell methanol fuel processing should also be considered as an alternative method of direct methanol use in the fuel cell.

Electrocatalysts Used in Acid Solution

Essentially all the useful electrocatalysts that have been used for methanol oxidation in acid solution contain noble metals (notably, platinum) either alone (e.g., as Raney platinum[351]) or in alloy form. As discussed earlier, platinum is very effective for the adsorptive dehydrogenation of methanol to produce the reaction intermediate, COH_{ads}. However, platinum at low potentials has the disadvantage of not being a very effective adsorption surface for the water fragments required to further oxidatize the adsorbed intermediate to carbon dioxide.[504] Platinized-platinum has sometimes been heat-treated in an oxygen atmosphere to improve its adsorption characteristics, and thereby increase its activity for methanol oxidation in sulfuric acid.[2007, 2008] For the same reasons, heat-treatment in hydrogen also has been recommended.[2010] For the pure noble metal catalysts, the descending order of methanol oxidation activity in 2.5 M sulfuric acid has been determined to be as follows:[1568] Os > Ir, Ru > Pt > Rh > Pd. It should be noted that water adsorption increases on going from Pt to Ir, to Ru, and to Os.

The use of noble metal alloys has been shown to give considerably better performance than pure noble metals. Platinum, for example, has been alloyed with a number of metals, including: titanium,[1296,1572] tantalum,[1296] ruthenium,[354, 798, 804, 1114, 1296, 1568, 1572, 2700] palladium,[1568, 2389] rhenium,[1296, 1354] osmium,[1296] mercury,[1354] tin,[83, 570, 1296, 1298, 1354, 1572, 1753, 1754] arsenic,[1354] antimony,[1354] and bismuth.[1354] The improved activity obtained by the presence of the second metal seems to be related to the ease with which the second metal can adsorb oxygen.[1354] This increases the rate of methanol oxidation by providing sites for dissociative adsorption of water.[504] It also seems likely that alloying metals such as ruthenium, antimony, rhenium, and molybdenum can act as co-catalysts that allow oxidization of the poisoning intermediate as well as that of the adsorbed intermediate COH_{ads}. The higher oxygen affinity component has been suggested to act independently of the platinum, allowing oxidation of the poisoning residues via a surface redox process; alternatively, in a solid solution it may alter the electronic configuration of the platinum to enable stronger adsorption of oxygen.[1758] In any event, the operation of alloy platinum electrocatalysts is associated with adsorption of H_2O or rather OH_{ads} species at lower potentials. In addition, the presence of the alloying component may weaken the bond between the poisoning intermediate and the platinum surface. It is also possible that atoms of the second component may physically isolate surface platinum atoms from each other, thereby preventing bridging of poisoning residues across the platinum surface.[1758]

The most successful alloy catalyst has been platinum-ruthenium.[354, 798, 804] Platinum-ruthenium alloy catalysts have been supported on carbon in small stable clusters of 1.0~2.5 nm diameter, about the same size as dispersed platinum clusters.[1758] Fifty atomic percent platinum-ruthenium alloy was found to give outstanding performance in sulfuric acid.[1568] Although very high loadings (in excess of 200 mg/cm²) were required, current densities greater than 5000 mA/cm² at 0.55 V could be sustained with minimal self-poisoning.[1114, 1568, 2700] Platinum-ruthenium alloy catalysts also give good performance in phosphoric acid. United Technologies investigated methanol oxidation in 99.5% phosphoric acid at 160~204°C on platinum, platinum-tin, platinum-titanium, and platinum-ruthenium catalysts at loadings of 0.5 mg/cm² on Vulcan XC-T2R carbon at methanol partial pressures of 0.65 atm.[1572] All the mixed catalysts performed better than pure platinum, platinum-ruthenium being the most effective (< 300 mV polarization at 100 mA/cm²). Ruthenium dioxide-coated electrodes also have been found to be very effective for methanol oxidation.[515]

Other combinations that have been examined include 50 atom % alloys of various noble metals with palladium and with ruthenium.[1568] In sulfuric acid, the decreasing order of activity for 50 atom % palladium alloys was shown to be:

$$Pd\text{-}Ru > Pd\text{-}Os, Pd\text{-}Ir > Pd\text{-}Pt > Pd$$

The corresponding order for the ruthenium alloys was:

$$Ru\text{-}Pt \gg Ru\text{-}Rh, Ru\text{-}Pd, Ru\text{-}Ir > Ru\text{-}Os > Ru\text{-}Au > Ru$$

Platinum-palladium alloy at loadings of about 2 mg/cm² in a nickel matrix also were reported to be more active than ruthenium or palladium alone.[2389] In addition, platinum has been used alone in sintered porous nickel.[665]

Platinum-tin alloy electrocatalysts are also known to be very active for methanol oxidation. However, while they are more resistant to self-poisoning than pure platinum, they are easily poisoned by phosphate or silicate ions.[83, 570, 1753, 1754] Platinum containing tin appears to enhance methanol oxidation via a ligand effect.[1298] Research carried out by Shell has shown that the incorporated tin is present in a zero-valent state, ruling out operation via a redox couple. Instead, tin appears to affect platinum adsorption properties. This conclusion contrasts with results obtained by Soviet workers, who have reported that the doping of platinum with tin exerts no effect on promoting methanol oxidation.[1781] However, work in Japan has demonstrated that platinum-tin oxide catalysts show better activity than platinum alone for methanol oxidation in acid.[1358] Although the addition of sulfur to platinum inhibits the oxidation of methanol in sulfuric acid, it has been claimed that platinum containing MoS_2 is much more active than platinum alone.[350, 1758, 2186]

In addition to the use of noble metal alloys, it has been found that sub-monolayers of certain metals, when adsorbed on the surface of platinum, usually by underpotential deposition (UPD),[942, 943] also show increased activity for methanol oxidation. Thus, when metals such as lead, bismuth, cadmium, or tin (which have high hydrogen overvoltages) or metals which strongly adsorb oxygen or OH are co-reduced or co-adsorbed on the platinum surface, the effective life of the catalyst is increased.[1911] It has been suggested that the adlayer of the second element functions by suppressing hydrogen adsorption, thereby inhibiting the formation of strongly bonded poisoning intermediate.[42] It also has been speculated that the promoting effect of adsorbed ruthenium, tin, germanium, and osmium may result from the enhancement of -OH adsorption by the second component.[1815, 2656, 2657] Adsorbed elements that have been observed to have such a promoting effect on platinum in sulfuric acid solution include tin, titanium, rhenium, ruthenium;[570, 1297, 1298, 1354, 1754, 2657, 2796] lead, bismuth, cadmium;[42, 2796] and silver, iodine, and sulfur.[2796] Work at Brookhaven National Laboratory has shown that similar effects are observed on platinum that contains adlayers of lead, bismuth, thallium, and cadmium in both 85% phosphoric acid[47] and in 1 M perchloric acid.[1914] Soviet workers have reported that adlayers of copper on palladium also promote methanol oxidation in sulfuric acid.[2009]

It would appear that in acid solution, only electrocatalysts containing noble metals exhibit appreciable activity for methanol oxidation. Although it has been reported that tungsten carbide[252] and sodium tungsten bronzes (e.g., $Ni_{0.24}WO_3$,[497]) both show some activity, it is several orders of magnitude lower than that of platinum. Other reports state that high surface area tungsten carbide exhibits zero activity for methanol oxidation; although tungsten carbide is able to adsorb water strongly, it is unable to simultaneously adsorb methanol.[1758] Furthermore, tungsten carbide has been reported to be subject to corrosion in sulfuric acid.[1782] Various bronzes have been shown to exhibit negligible activity (<10 μA/cm^2) for methanol oxidation in 1 M perchloric acid.[2578] Work at Hitachi, Ltd., has indicated that molybdenum borides (MoB, MoB$_2$, and Mo$_2$B$_5$) show promise as methanol oxidation catalysts in both sulfuric acid and phosphoric acid.[1530]

Methanol Electrocatalysis in Alkaline Solution

As discussed earlier, the two major incentives for attempting to oxidize methanol in alkaline solution are the less corrosive nature of the electrolyte, permitting the potential use of inexpensive non-noble metal electrocatalysts, and the higher reaction rate compared with acid electrolyte. Conversely, the disadvantages are that carbonation of the electrolyte occurs and that the reaction produces significant amounts of formaldehyde and formate.[2590]

While it is true that in alkaline solution the reaction proceeds considerably faster than in acid, especially at lower temperatures, this may be due merely to a slower rate of adsorptive self-poisoning rather than the result of inherently faster kinetics.[2592] It is known, for example, that OH$^-$ ions are more readily desorbed from platinum than are the anions of acids such as sulfuric acid and phosphoric acid, and this may account for the higher steady state conversion rates observed in alkaline electrolytes.[1758] In alkaline solution, as in acid, an adsorbed intermediate is produced, which must be further oxidized to yield carbonate. In addition, in alkaline solution the second oxidative step does not take place at potentials much below about 0.4 V/NHE.[504]

Electrocatalysts Used in Alkaline Solution

Although some non-noble metal electrocatalysts exhibit activity for methanol oxidation in alkaline solution, the noble metals in general, display the highest activities, as is the case in acid media.[504] Thus, Raney nickel doped with 2 mg/cm^2 of platinum performs better than Raney nickel alone, and is able to sustain 100 mA/cm^2 at 25°C with much better resistance to self-poisoning than undoped catalyst.[408, 1593] Noble-metal methanol oxidation electrocatalysts that have been used in alkaline solution include platinum,[2301] Raney platinum,[351] Raney palladium,[353] and gold.[2567]

As in acid solutions, noble metal alloys show higher activities than those of the pure noble metals. Platinum-palladium, for example, gives a considerably higher conversion of methanol to carbonate ion than does pure platinum.[1758, 1838, 1903, 2723] The addition of bismuth, or less effectively lead, to platinum-palladium alloy catalysts further improves their long-term stability.[641] Similarly, platinum-gold alloys (in particular, 40% Pt–60% Au) have been found to be much more active than platinum alone for the electro-oxidation of methanol (and other aliphatic alcohols) in alkaline media.[311] This improved performance has been attributed to the lowering of the adsorption energy of the adsorbed organic and oxygen-containing species at the alloy surface. In 6 M KOH, the following descending order of activity for methanol oxidation on noble metal alloy catalysts (expressed in atomic %) has been observed:[2593] 75% Pt–25% Ir > 60% Pt–40% Ru > 67% Pt–33% Pd.

Other catalysts that have been examined for use in alkaline solution with varying degrees of success include palladium-silver,[353] palladium oxide–silver,[2737] Raney nickel,[504] nickel,[2123] nickel oxide (NiO),[2123] Raney nickel–titanium,[2068] and nickel–silver.[1721, 2390] Tungsten carbide has been found to have negligible activity in 6.5 M KOH solution at 70°C.[2187]

Use of Buffer Solutions

There also has been some interest in the use of buffer solutions as electrolytes for the oxidation of methanol, since aqueous buffer systems such as the $CO_3^=/HCO_3^-$ system are not corrosive yet are able to reject carbon dioxide.[504, 2653] The Exxon-Alsthom methanol cell, for example, employed a carbonate-bicarbonate buffer electrolyte, which allowed only modest current densities due to the onset of concentration polarization. As described in Chapter 10, concentrated carbonate solutions with very soluble cations (e.g., cesium carbonate or rubidium carbonate[534, 535, 542, 1017]) are of theoretical interest because they too are able to reject carbon dioxide via $CO_2/HCO_3^-/CO_3^=$ equilibria, are sufficiently concentrated to avoid marked concentration polarization, and also allow complete conversion of methanol to carbon dioxide.[504, 534] Cesium carbonate solutions can be used up to temperatures of about 140°C.[1758] Raney platinum-palladium alloy catalysts have been used with bicarbonate systems and have been found to be more active than platinum for the oxidation of methanol.[1568]

Summary

In summary, the ideal electrocatalyst for methanol oxidation must be bifunctiona (i.e., it must be able to adsorb and dehydrogenate methanol), and it must simultaneously be able to co-adsorb the oxidation products of water (i.e., $-OH_{ads}$ or $-O_{ads}$). Furthermore, for use in inexpensive systems for broad application, noble metal loadings should be no greater than about 1 mg/cm^2. For low temperature conversion, sulfuric acid may still be the best electrolyte, while at higher temperature, phosphoric acid (incorporating a methanol diffusion barrier) or Nafion-type materials are advantageous. For the latter classes of electrolyte above 100°C, methanol is best supplied in vapor phase along with steam.

To date, the most effective electrocatalysts for all aqueous electrolytes appear to be platinum alloys, particularly platinum-ruthenium. The activity of the catalyst appears to depend on the supporting material, since surface groups such as COOH and OH on carbon supports, for example, have been found to enhance the activity of dispersed platinum.[1758] This is an area that merits further investigation.

Ethanol

As with methanol, the majority of studies on ethanol oxidation have used noble metal electrocatalysts in sulfuric acid electrolyte. Ethanol at 2 M concentration has been oxidized to carbon dioxide on Raney platinum in 2.5 M sulfuric acid at 25°C and 80°C.[351] For diagnostic purposes, bright platinum also has been employed in sulfuric acid.[363, 2035]

A series of platinum alloy catalysts has been studied for the oxidation of ethanol in phosphoric acid electrolyte under the current advanced fuel cell operating conditions.[1572] As has been the case for methanol oxidation, all the alloy catalysts were superior to pure platinum, platinum-ruthenium again giving the best performance. Indeed, platinum-ruthenium yielded higher oxidation currents for ethanol than for methanol.

Ethanol also has been oxidized on mixed metal oxides such as $LaCo_{(1-x)}Ni_xO_2$ and $LaCo_{(1-x)}Fe_xO_2$. Low-spin transition metal ions should favor the transfer of the surface oxygen ions that are considered to be responsible for the oxidation process. In nickel-rich compounds, catalytic activity is related to the ability of nickel to lose oxygen at the temperature at which the cell operates.

Propanol

Isopropanol and n-propanol, each in 2 M solution in 2.5 M sulfuric acid, have been electro-oxidized at 25° and 80°C on Raney platinum to yield carbon dioxide. Gold-platinum alloy catalysts also have been found to be active towards propanol oxidation in sulfuric acid.[311] Studies of the oxidation of propanol on platinum in 1 M perchloric acid have indicated that the hysteresis commonly observed in ascending and descending voltage-current curves for the reaction may be caused by a reduction of inhibition effects when adsorbate layers are oxidized.[1892]

The oxidation of isopropanol in alkaline solution is exceptional among alcohol oxidation processes in that the usual self-poisoning effects are apparently absent.[504] The reason is that acetone is produced as a reaction product, enabling long-term operation without requiring oxidation of the carbon nucleus to CO_2. Unfortunately, the reaction is not very energy-efficient, since it involves only two electrons per three carbon atoms, compared with 18 for the complete oxidation of three methanol molecules.

Formaldehyde

It has been suggested that the oxidation of formaldehyde (HCHO) may follow two parallel reaction paths, leading either to the formation of carbon dioxide or to that of a chemisorbed self-poisoning intermediate.[1758] As in the case of methanol, the adsorbed intermediate is removed by reaction with chemisorbed -OH derived from water adsorption. A Soviet study has shown that in sulfuric acid, COH and CO are formed as chemisorption products.[80]

Formaldehyde has been oxidized at room temperature on platinized platinum in sulfuric acid.[1447] On platinum, platinum surface oxides seem to play a role in the rate-determining step of the oxidation process.[1534] Formaldehyde also has been oxidized successfully in sulfuric acid using a number of binary platinum co-deposited electrocatalysts (platinum-antimony, platinum-arsenic, platinum-bismuth, platinum-mercury, platinum-rhenium, platinum-tellurium, platinum-tin) and a range of homogeneous platinum-rhodium alloys of different surface composition.[1354] In all cases the activities of the alloys were better than that of pure platinum. The presence of adlayers of sulfur on the platinum surface, which inhibit the oxidation of methanol, appear to promote that of formaldehyde.[2186] Formaldehyde oxidation on tungsten carbide in 1 M sulfuric acid at 70°C has been found almost to essentially stop with the formation of formic acid as a product, which then is only very slowly further oxidized.[2187,2800] On Mo-O-S catalysts in 1.75 M hydrochloric acid, 21% by weight formaldehyde has given an oxidation rate of 40 mA/cm^2 at 85°C.[2725, 2726]

Formaldehyde has been oxidized on Raney platinum in 5 M KOH at 25°C and 80°C.[351] Studies of the various aldehyde oxidation mechanisms in alkaline solution on a number of electrocatalysts such as glassy carbon, platinum, mercury, copper, silver, nickel,[2565, 2566] and gold[2566, 2567] also have been made. A study of the oxidation of formaldehyde on silver in KOH solution showed that small additions of tellurium favorably affect the adsorption behavior of formaldehyde on silver.[2678] On tungsten carbide, the oxidation of formaldehyde at 70°C in 6.5 M KOH exhibited a high initial rate, but was soon inhibited by poisoning.[2187]

Acetaldehyde

Acetaldehyde, CH_3CHO, has been reported to undergo slow electrochemical oxidation at 70°C in 1 M sulfuric acid on tungsten carbide.[2187] However, the reaction was shown to stop at the acetic acid stage.

Formic Acid and Formate

The electrochemical oxidation of formic acid, like its precursor, formaldehyde, proceeds via a reaction scheme that involves the adsorption of both normal and self-poisoning intermediates prior to charge transfer.[48, 1758] The nature of the adsorption interaction in sulfuric acid between the acid and the oxidized platinum surface has been studied.[765] The effects of adsorbed organic molecules on the electro-oxidation of formic acid on platinum also have been investigated.[218]

The reaction has been studied in a number of acids, usually using platinum catalysts. Studies with single crystal platinum electrodes in 1 M perchloric acid have indicated that the intermediate -COOH is formed, which then reacts either with adsorbed hydrogen atoms to yield strongly bonded COH, or reacts further with HCOOH to yield $=C(OH)_2$.[48] The oxidation in perchloric acid is strongly dependent on the crystal orientation of the platinum surface, the highest and most reversible activity being on the (111) crystal plane, followed by the (110) and (100) planes.[48, 49] The high activity on the (111) crystal plane has been attributed to its lower hydrogen coverage, resulting in lower surface concentrations of the self-poisoning intermediate. Work at Brookhaven National Laboratory has shown that the underpotential deposition (UPD) of adatoms of metals such as bismuth, lead, cadmium, and thallium on the platinum surface appears to catalyze the oxidation of formic acid in acid solution. In 1 M perchloric acid at 25°C, for example, formic acid oxidation currents were 5~10 times greater on the underpotentially deposited surfaces compared with bare platinum; bismuth and lead being especially effective.[1914] Presumably the presence of the evenly dispersed metal adatoms on the platinum surface helps to prevent the adsorption of the poisoning organic intermediates, which requires two or more adjacent bare platinum sites.[1914]

Japanese workers have found that platinum-bismuth catalysts supported on carbon show higher activity for formic acid oxidation than pure platinum.[1814] Similar beneficial effects of UPD metal adatoms, especially lead, also have been observed in 85% phosphoric acid.[47, 2329] Similarly, tellurium adlayers on the platinum surface have been reported to greatly decrease electrode self-poisoning by adsorbed formic acid decomposition products in sulfuric acid.[2676] The effects of monolayers of sulfur, selenium, tellurium, and lead on platinum have been investigated in sulfuric acid at 30°C and 70°C;[2186] and pre-adsorbed sulfur layers on platinized-platinum have been found to exhibit a beneficial effect on the oxidation of formic acid.[666]

Extensive voltammetric studies of the effects of foreign metal adatoms on the surface of rhenium electrodes have shown behavior similar to that observed on platinum.[43] Conversely, the adsorption of silver adatoms on palladium has been found to reduce the rate of formic acid oxidation.[1468] In addition to noble metal electrocatalysts, tungsten carbide,[2187] cobalt sulfide,[2053] tungsten sulfide,[405] and molybdenum sulfide[405] have been examined as electrocatalysts for the oxidation of formic acid in acid electrolyte, although without much success. Some activity has been claimed for the cobalt complex of 5,14-dihydrodibenzo-5,9:14,18-tetra-aza[14]annulene.[2800]

In neutral electrolytes, platinum, palladium, gold, and rhodium are all reasonably active electrocatalysts for formate oxidation, whereas silver shows low activity.[294] The high activity of platinum has been shown to be related to local changes in pH in a working electrode and corresponding shifts in the conditions for the formation of platinum oxide on the catalyst surface.[45] However, as the pH is increased to a value above about 12.6, there is an abrupt decrease in the rate of formate oxidation on platinum in KOH.[2677] This appears to be caused by a change in the nature of the oxidizing species on the electrode surface. As in acidic media, the addition of small amounts of lead and thallium to platinum result in a considerable improvement in the rate of formate oxidation, presumably by preventing the formation of strongly bound intermediates through the suppression of hydrogen adsorption.[44]

In addition to platinum, palladium has also been a favored catalyst for the oxidation of formate in alkaline solution, and the reaction mechanism on this catalyst has been studied in some detail.[285, 600] Formate ion was oxidized in 6.5 M KOH in 2 M concentration solution on Raney palladium at 25° and 80°C.[353] The incorporation of 50 atomic percent gold into the palladium was found to increase the exchange current density for formate oxidation by a factor of ten over that for pure palladium.[293] Palladium-platinum alloys on nickel substrates also have been used for formate oxidation in KOH and NaOH, somewhat higher activity being found in KOH.[2396] The activity of this catalyst was shown to be apparently enhanced by the addition of carbamate ion.

Ethylene Glycol

In alkaline solution, ethylene glycol ($HOCH_2CH_2OH$) is oxidized anodically on noble metal electrocatalysts to form oxalate as the main product; some glycol aldehyde, glyoxal, glycolate, and glyoxalate being formed as intermediates or byproducts.[1105, 1106] On platinum, the process occurs via a one-electron reaction, desorption of an intermediate being the rate determining step. Platinum, gold, and 40 atom % Pt–60% Au all will catalyze the reaction at ambient temperature.[1105, 1106] The use of 50 atom % Ru-Pt alloy [2596] and 43 atom % Pt–34% Pd–23% Bi[642] also has given good results in alkaline media. A lead–platinum intermetallic, $PbPt_3$, coated on a nickel substrate has been found to catalyze the oxidation of ethylene glycol in KOH, yielding current densities a factor of ten greater than those on pure platinum.[1463] Pure Raney nickel will oxidize ethylene glycol to oxalate in alkaline solution at high current density at temperatures above 60°C with little long-term degradation in performance.[819]

A number of studies also have been carried out on the oxidation of ethylene glycol on platinum in acid solution.[1233, 1233a, 1234] On platinized platinum in 1 M perchloric acid solution, oxidation proceeds through glycoaldehyde to form glyoxal, glyoxalic acid, and oxalic acid as products. A linear sweep voltammetric study of the anodic adsorption and electrode behavior of ethylene glycol on platinized platinum has also been carried out in 1 M sulfuric acid electrolyte.[1056] A similar study has been conducted on the oxidation of glycolic acid, $CH_2(OH)COOH$, in alkaline solution.[1951]

E. MISCELLANEOUS FUELS AND ELECTROCATALYSTS

Cyclohexane and benzene have been electro-oxidized at low rates on smooth platinum in 85% phosphoric acid at 130°C.[2199] Ammonia, NH_3, has been electrolytically oxidized in 0.1 M NaOH solution on platinized platinum, rhenium-coated platinum, and palladium-coated platinum electrodes.[590, 591]

Organic solution catalysts (dissolved noble metal complexes), such as $(bpy)_2pyRu(OH)_2^{++}/(bpy)_2pyRuO^{++}$ (where bpy is 2,2'-bipyridine) and $(trpy)(bpy)Ru(OH)_2^{++}/(trpy)(bpy)RuO^{++}$ (where trpy is 2,2',2''-terpyridine), have been found to catalyze the oxidation of diverse compounds such as 2-propanol, ethanol, acetaldehyde, and unsaturated hydrocarbons at 25~50°C at applied potentials greater than about 0.84 V/NHE.[1818] At these potentials Ru(II) can be oxidized to Ru(IV). Organic dyes have also been considered as fuel cell catalysts.[1275] Solution catalysts may find application in fuel cells using dissolved fuels, however the long-term stability of low-

Anodic Electrocatalysis

cost compounds would have to be demonstrated and overpotentials must be drastically reduced from the values reported to date to be of practical value.

Another development that might eventually be applied to fuel cell catalysis is the chemically modified electrode, in which functional groups can be attached to the surface of a substrate electrode.[1842,2297] For example, vinyl ferrocene and ruthenium-pyridine complexes can be immobilized on the surface of specially abraded and etched glassy carbon electrodes; alternatively, electroactive ferrocene polymers can be deposited onto glassy carbon–platinum surfaces directly via radio-frequency plasma discharge.[1896] Similarly, manganese, iron, cobalt, nickel, copper, and zinc metalloporphyrins (e.g., tetra(aminophenyl)porphyrin) also can be immobilized onto glassy carbon surfaces,[2124] and ruthenium nitro complexes (e.g., Ru(III)NO$_2$) can be immobilized on a platinum electrode using an alkylaminesilane.[10] This type of chemical modification of an electrode surface could be especially useful for the oxidation reactions of organic fuels involving adsorbed intermediates, or for oxygen reduction reactions, which are considered in the next chapter. Formation methods for various reactive groups on glassy carbon and their surface evaluation have been discussed in the literature.[895] Again, chemical modification can only be considered if stability of the attached groups is high.

CHAPTER 12

Cathodic Electrocatalysis

A. ELECTROCATALYSIS OF OXYGEN REDUCTION

Availability and cost clearly make atmospheric oxygen the most desirable oxidant. In addition, the use of ambient air eliminates the need for storage tanks and reduces the weight and volume of the fuel cell system.

A fuel cell air electrode system is essentially only a classical oxygen electrode, for which the ideal reactions are:

In acid, $O_2 + 4H^+ + 4e^- \rightarrow 2H_2O$ $\Delta E° = +1.229$ V

In base, $O_2 + 2H_2O + 4e^- \rightarrow 4OH^-$ $\Delta E° = +0.401$ V

Depending on the electrolyte and on the system temperature, the reduction of oxygen is routinely catalyzed by a variety of materials, including noble metals and transition metals and their alloys, carbon materials, and organic and other compounds.[467, 1059, 2517] These will be discussed in greater detail in the following.

In a hydrogen-oxygen fuel cell under standard conditions, the theoretical reversible open circuit voltage in both acidic and alkaline media is 1.23 V. As is well known, the oxygen electrode does not behave ideally, and experimental open circuit voltages typically range from 0.98 V to about 1.10 V.[908, 1137, 2116, 2286, 2708] Since it is well established that the hydrogen electrode behaves nearly ideally over a wide range of operating conditions, especially in acid solution,[313, 385, 585, 929] these deviations almost always can be attributed to the non-ideal behavior of the oxygen electrode. Although the exchange current density for the hydrogen oxidation reaction in acid on platinum typically ranges from 0.1 to 100 mA/cm^2,[381, 585] for the oxygen reduction reaction under the same conditions the exchange current density has been measured at only about 1.3 µA/cm^2,[1142] which is a maximum value under the conditions normally encountered in aqueous fuel cells. At room temperature, the value is closer to 10^{-10} A/cm^2.[168, 705]

In alkaline electrolytes this irreversibility has often been related to the formation of peroxide via the reaction [2735]

$$O_2 + H_2O + 2e^- \rightarrow HO_2^- + OH^- \qquad \Delta E° = -0.076 \text{ V}$$

for which $\Delta E°$ is equivalent to +0.752 V v. hydrogren of pH 14. Peroxyl ions may reach high concentration near the electrode surface during oxygen reduction, thereby limiting the electrode current density due to large concentration polarization effects. In the traditional view of the process on inactive (non-adsorbing) electrocatalysts such as carbons and mercury, further electrochemical reduction of peroxyl ions according to

$$HO_2^- + H_2O + 2e^- \rightarrow 3OH^- \qquad \Delta E° = +0.877 \text{ V}$$

does not occur at moderate polarization levels (200~500 mV). However, the ideal four-electron transfer reduction to hydroxyl can be approached fairly closely at high current densities if the peroxyl ions formed at the electrode surface can be decomposed catalytically via the reaction

$$HO_2^- \rightarrow OH^- + 1/2 O_2$$

This reaction gives an overall 4-electron transfer by recycling oxygen. If this is the mechanism followed, the peroxide-decomposing ability of an electrocatalyst is one of the keys to the successful operation of high-rate air cathodes in alkaline systems. However, peroxide ion is not necessarily the route by which all electrocatalysts operate in alkaline solution, since the O=O bond in molecular dioxygen can be broken early on in the reaction sequence in the so-called four-electron pathway on certain surfaces. A major difference between the two-electron pathway and the four-electron pathway is in the pH dependence of overpotential. In the former, it increases strongly, generally by about 60 mV per pH unit, as pH is reduced. This is the case on carbon electrodes. In the latter, it is independent of pH, as on phthalocyanine-coated carbons and on platinum electrodes. This topic is controversial and has been reviewed.[132, 149, 255, 168] In the final analysis, the large deviation of open circuit voltage from the reversible oxygen electrode potential on practical fuel cell electrodes in both acid and alkaline electrolytes usually is attributable to mixed potentials resulting from interactions between the catalyst and oxygen. The nature of these interactions will be examined in the next section.

The incentive for the development of a truly reversible oxygen electrode is readily appreciated. If the open circuit voltage for the hydrogen-oxygen system could be increased from the customary 1 V to the reversible value of 1.23 V, it would be possible to increase the energy conversion efficiency of hydrogen-oxygen fuel cells by as much as 20%. Such a breakthrough would also be of great significance in the development of high energy metal-air secondary batteries such as the zinc-air cell.[144, 1301]

The Use of Platinum in the Electrocatalysis of Oxygen Reduction

As described in previous sections, one of the main objectives in fuel cell research over the past 25 years has been the development of a viable system using an acid electrolyte. Platinum has traditionally been employed as the oxygen reduction catalyst in such systems on account of its ability to meet the three criteria of: electrocatalytic activity, stability, and electronic conductivity. Platinum also has been used as an oxygen evolution electrocatalyst.[709, 2211, 2230]

The platinum loading of a typical PAFC oxygen cathode is in the order of 0.5 mg of Pt per geometric square centimeter of electrode area. Doubling the platinum loading reduces the polarization by about 25~30 mV at fuel cell operating conditions. (For the Westinghouse pressurized system, these consisted of an initial cell voltage of 0.7 V at 325 mA/cm^2 current density, 190°C, 4.8 atm pressure.[851]) However, doubling loading beyond about 1 mg/cm^2 reaches the point of diminishing returns, since catalyst utilization is markedly reduced as electrode thickness increases.

Platinum blacks of high surface area and activity for use as oxygen reduction catalysts can be prepared using a number of methods, including the reduction of chloroplatinic acid with formaldehyde [965, 970] or borohydride,[2206] spray techniques,[2611] deposition onto ion-exchange membranes[312], or reduction from an aqueous saturated solution of diaminetrinitronitratoplatinum using hydrogen and heating.[220] A good review of platinized electrodes, including the theory and methods of preparation, is given in Ref. 848. Methods also have been published for the preparation of mono-dispersed high surface area platinum crystallites of colloidal dimensions for use with oxygen cathodes.[358, 396, 1118, 1381, 2012, 2354, 2512, 2762] This has been the favored preparation technique for the past decade.

Numerous studies have been conducted on the kinetics and mechanisms of oxygen reduction on platinum electrodes. These have included such diverse topics as the role of proton transfer in the reduction mechanism,[156] the effect of electrode wettability on reaction rate (reflecting the fact that for a given catalyst loading, porous hydrophobic electrodes have about fifty times the active area of hydrophilic electrodes [990]), and the effects of various factors on transient electrode behavior.[992]

In general, it is found that over the pH range of 0.4 to 13, the kinetics of oxygen reduction on platinum are controlled by the rate of charge transfer involving adsorption of molecular oxygen, a reaction which tends to be inhibited by intermediates involved in the overall oxygen reduction process.[2266] It also has been found that the behavior of single crystal platinum is different from that of polycrystalline platinum, largely on account of the different chemisorptive bond energies and anion adsorption phenomena on the different crystal planes.[2135] This provides an explanation for the fact that the activity of electrodeposited platinum is a function of the current density used in electrodeposition, since crystal structure orientation is known to be a strong function of current density.[1808] Oxygen adsorption studies [1387, 1856, 1894] have shown that when oxygen adsorbs on platinum, both weakly bound and tightly bound oxygen are formed on the platinum surface. In acid media, the total quantity of weakly and strongly bound species is constant and is not dependent on temperature, but the temperature does affect the proportions of the two species.[1894]

Adsorption of acid anions has been proposed as strongly influencing the oxygen reduction reaction.[633, 634, 867] Oxygen reduction experiments carried out on platinum black in concentrated acid, for example, show that the oxygen reduction rate increases as the activity of the water increases, which may be related to the effects of adsorbed anions on the orientation of adsorbed water dipoles in the double layer.[1750] Similarly, adsorption of organic impurities can affect oxygen reduction reaction rate by influencing the orientation of adsorbate molecules.[1526] Refs. 168 and 2136 give an overview of the understanding of the kinetics of the oxygen reduction reaction on platinum up to the late 1970s.

The Oxide Film on Platinum

In spite of the almost universal choice of platinum as the oxygen reduction electrocatalyst in acid systems, its use is not without problems. One difficulty is a gradual loss of surface area with time, another is its less than ideal performance as an oxygen reduction catalyst (even though it is the best available), as reflected by both its oxygen exchange current density and non-ideal open circuit voltage. The reason that the ideal open circuit voltage of 1.23 V (at 25°C) is not achieved in practice is related first to the low exchange current density for oxygen reduction; and second, to the fact that platinum electrodes are not truly inert, but react with oxygen to form an electronically conducting partial monolayer of Pt-O.

Many aspects of this submonolayer *oxygen film* still are controversial, and have impeded progress in understanding the oxygen electrode. For example, there is not complete agreement on whether or not the film is a real phase oxide of platinum, and, if it is, which oxide or oxides.[2520] If it is not a true phase oxide, it is not certain whether it is a chemisorbed oxygen film, a *surface alloy* between the platinum and the oxygen, or oxygen dissolved in the surface layers of the platinum.[1145, 2226] In addition, depending on the circumstances (particularly on the anodic potential), the coverage can be partial, mono-, or multi-layered.[88, 343, 669, 962, 1144, 2459] There is no general agreement on the parameters involved in film formation, and on the conditions for a steady state, if they exist.

There are a number of reasons for these difficulties. In the first place, electrode prehistory strongly influences behavior, hysteresis often being observed. Factors such as crystal structure,[970, 2648] trace impurities, [2227, 2787] surface roughness, [1796, 2258] and higher oxide formation all can affect electrode behavior. Furthermore, B.E.T. surface area measurements, thermodynamic data on platinum oxides, and other studies that are made under carefully controlled conditions, are not necessarily valid in interpreting a reacting electrochemical system. Finally, the interpretation of classical voltammetric measurements[416, 417, 2694, 2695] is not always clearly defined. For example, the commonly used hydrogen capacity method of determining the true electrode surface area[2006, 2162] is complicated by incomplete knowledge of the structure of the hydrogen adlayer.[962]

Many studies of the adsorbed oxide layer on platinum have been made using a variety of electrolytes and conditions.[261, 343, 359, 386, 670, 708, 711, 962, 1144, 1331, 1388, 1670, 2432, 2601, 2602, 2647, 2660] The oxide film is known to affect the kinetics of both oxygen reduction[1648] and oxygen evolution.[709] For example, the hydrophilic properties of a platinum surface are determined by the degree of oxidation of its surface, and the adsorption and orientation both of adsorbed water molecules [1648] and of species such as H_2S,[2662] which often are present as impurities, are different on reduced and oxidized platinum. However, as in many other cases, these differences represent only short-term variations on the catalyst surface, since they will be removed as the catalyst ages with time.

In acid solutions, the oxide film is believed to be responsible for the open circuit voltage deviations of the oxygen electrode on platinum. The *polyelectrode theory* holds that there are two types of reaction sites for the oxygen reaction: bare platinum and Pt-O.[1139, 1140, 1142-1144] Thus, it is postulated that the anodic "corrosion" reaction

$$Pt + H_2O \rightarrow Pt\text{-}O + 2H^+ + 2e^- \quad \Delta E° = 0.88 \text{ V}$$

takes place on the bare platinum sites; whereas the cathodic oxygen reduction reaction

$$1/2 O_2 + 2H^+ + 2e^- \rightarrow H_2O \quad \Delta E° = 1.23 \text{ V}$$

occurs on the oxide-covered sites. Thus, an open-circuited oxygen electrode is in a steady state rather than at a true thermodynamic equilibrium, and the observed open circuit voltage under standard conditions falls somewhere between the two standard electrode potentials of the two constituent reactions.

When the electrode is pretreated so that all the sites consist of Pt-O, only the ideal, reversible reaction should occur, displaying an open circuit voltage of 1.23 V under standard conditions. This has been verified experimentally, in acid solutions[382,469,522,1140,1143,1147-1149,1591,2453,2658] as well as in alkaline solutions.[185] Close to the theoretical oxygen electrode potential, such oxide-covered platinum electrodes exhibit less polarization at the same current density than that extrapolated from those with bare platinum sites. Unfortunately, reversibility is destroyed as soon as the electrode is polarized beyond 50~60 $\mu A/cm^2$, which results in destruction of the oxide film. Thus, although more reversible oxygen electrode potentials can be obtained on Pt-O electrodes, this is true only at impractically low current densities.

In acid solution, it has been shown that a bare platinum surface is more active than a Pt-O–covered surface by about two orders of magnitude at current densities greater than about 100 $\mu A/cm^2$.[704,710] The reaction rate in alkaline solution also depends on the degree of oxidation of the platinum surface.[1647] The kinetics of the oxide film formation on platinum in KOH solution have been studied using galvanostatic charging techniques.[712,713]

Use of Platinum in Acid Systems

Phosphoric Acid

Phosphoric acid is the electrolyte of choice for the first generation of commercial fuel cells, and is therefore the electrolyte for which the most immediate need exists for understanding of the oxygen reduction process. Because of the corrosive nature of hot phosphoric acid at oxygen potentials, platinum appears to be the only practical oxygen electrode catalyst giving acceptable performance.[134-136] For use in phosphoric acid, platinum usually is deposited on a carbon substrate at loadings ranging from 0.25 to 1 mg/cm^2.[563, 1590, 2206, 2611]

Workers at Brookhaven National Laboratory have studied the mechanisms and kinetics of oxygen reduction on supported and unsupported platinum in phosphoric acid.[2443] In particular, attention was directed towards the evaluation of a number of Cabot carbon blacks for use as platinum catalyst supports.[2329] The electrodes studied consisted of a 12% PTFE-bonded carbon layer on a hydrophobic carbon paper with catalyst loadings of 1 mg/cm^2. Experiments were conducted in 85% phosphoric acid at 25°C and 135°C. Regal 660R carbon showed the most promise as a support allowing the highest development of platinum activity among the five Cabot carbons tested.[2329] At 135°C, these electrodes exhibited a voltage of 0.91 V at 0.1 mA/cm^2 (real surface area) and an oxygen reduction exchange current density of 0.1 μA/cm^2. Studies also have been made using platinum loadings of 0.35 mg/cm^2.[1935]

In spite of the amount of work expended in studying the mechanisms of the oxygen reduction reaction on platinum in phosphoric acid, the subject has yet to be completely resolved, owing to the sensitivity of the reaction to impurities and electrode pretreatment.[1545,1734] In addition, the characteristics of the reaction are strongly dependent on the phosphoric acid concentration used, which affects factors such as oxygen solubility, proton activity, and double layer structure.[2330] For example, a progressive decrease in the exchange current density for the oxygen reduction reaction on platinum occurs as the phosphoric acid concentration is increased from 7 to 56% by weight.[2006A] Changes in phosphoric acid solvent structure that occur at 56% by weight have pronounced effects on oxygen reduction kinetics. References 152 and 2762 provide overviews of the mechanisms and kinetics of oxygen reduction on platinum in phosphoric acid.

Under the sponsorship of the U.S. Army, United Oil Products Co. (UOP) and Energy Research Corp. developed a platinum-doped wetproofed cathode for use in phosphoric acid fuel cells that seemed to show better long-term endurance than carbon substrate cathodes prepared by conventional methods.[1485, 1606, 2669] This material, named *Kocite*, was prepared by gas phase deposition of carbon into platinum-containing porous alumina by the thermal-catalytic decomposition of an organic compound such as cyclohexane. The deposited carbon therefore had the structure of the original porous alumina, which was leached out with acid, leaving a platinum-containing porous carbon material. Platinum loadings between 0.5 and 3 mg/cm^2 were examined.[2669] Recent work on supported platinum electrodes on stable supports (particularly graphitized Cabot carbon blacks) is reviewed in Ref. 394.

Sulfuric Acid

Oxygen reduction at 25°C on platinum in sulfuric acid at various concentrations has been examined, including 0.05 M,[404, 1390, 2432, 2433, 2435] 0.5 M,[1665, 2787] 1 M,[1142, 1143] 2.5 M,[1068, 1570, 2440A] and 5 M.[1665] The reaction also has been studied at 70°C in 2.5 M sulfuric acid.[1068, 1570] In addition to bright platinum, platinum black,[1664] platinum on activated carbon,[2440A] and platinum on graphite[1068, 1570] have also been examined. The effect of arsenic and cyanide contamination of the sulfuric acid electrolyte has been investigated.[1390]

Trifluoromethyl Sulfonic Acid (TFMSA)

The oxygen reduction reaction has been found to be more rapid in TFMSA (CF_3SO_3H) than in phosphoric acid. Work at Brookhaven National Laboratory has shown that at 25°C, oxygen reduction is 50 times faster in TFMSA than in 85% phosphoric acid.[1915, 1917, 1918, 2329] Rotating ring-disk studies of the kinetics of the oxygen reduction reaction on platinum in aqueous TFMSA have shown that between 1.0 and 0.8 V, the Tafel slope is –60 mV/decade; while below 0.8 V, it is –120 mV/decade. The –120 mV/decade slope is the value commonly observed for the oxygen reduction reaction in most acid electrolytes, and indicates that the first step is a rate-controlling electron transfer reaction.[1918]

The -60 mV/decade Tafel slope observed above 0.8 V extends over three orders of magnitude at 80°C. In this region, the mechanism may involve an initial fast charge transfer step followed by a rate-limiting chemical step, since the pH dependence of the oxygen reduction reaction in this region is 1, and not 3/2 as has been observed in dilute perchloric acid.[705] In the –60 mV/decade region, the activation energy was measured to be about 85 kJ/mol, and about 23 kJ/mol in the –120 mV/decade region. While no peroxide was detected in the upper region, between 0.65 and 0.5 V, the reduction proceeds via a parallel mechanism in which both water and peroxide are produced; at more cathodic potentials the oxygen reduction essentially stops at the peroxide stage. The reduction is first order with respect to oxygen, which has a somewhat higher solubility in TFMSA than in phosphoric acid.[2006A] It is necessary to purify TFMSA very carefully since impurities tend to inhibit the four-electron reaction to water.[1915, 1918]

One advantage of TFMSA over phosphoric acid, at least in dilute solution, is that TFMSA anions do not strongly adsorb on platinum as do the anions of phosphoric acid.[1915, 2005A, 2330] This factor may explain the improved kinetics of oxygen reduction in TFMSA. For example, it has been found that the addition of phosphoric acid to TFMSA in amounts as low as 10^{-5} M lowers the rate of oxygen reduction.[2005A] Conversely, addition of small amounts of TFMSA to phosphoric acid *increases* the rate of oxygen reduction.[1916, 2329] For example, additions of TFMSA to give a 0.1 M solution in 85% phosphoric acid at room temperature have been found to increase the oxygen reduction rate by as much as 35%. This effect may be caused by enhanced proton activity or by an effective increase in the oxygen solubility.

In spite of its apparent advantages, TFMSA is not a practical fuel cell electrolyte. It requires careful purification, is relatively volatile, and wets Teflon, which rules it out as a phosphoric acid replacement. In addition, there is evidence that the high oxygen reduction activity observed on platinum rotating-disk electrodes in 0.1 M aqueous TFMSA is not observed

Cathodic Electrocatalysis

when attempts are made to use TFMSA under practical fuel cell operating conditions.[2005A] In concentrated acid, the liquid structure may be sufficiently different from that in 0.1 M solution to markedly influence oxygen reduction.

Other Acids

Although TFMSA itself is unlikely to be a practical fuel cell electrolyte, the fact that small additions of TFMSA to phosphoric acid promote the rate of oxygen reduction on platinum indicates that if protonic superacids* of low volatility can be developed, they might be attractive additives to phosphoric acid to enhance oxygen reduction at realistic fuel cell operating temperatures in the vicinity of 200°C.

To be a useful candidate, a new acid must have high conductivity, low vapor pressure, high stability, and high oxygen solubility. It should exhibit low specific adsorption on platinum; and should not wet Teflon, all at temperatures preferably in the neighborhood of 200°C. Indications were that some of the higher homologues of TFMSA could meet these criteria,[171, 2146] and could lower the polarization of the oxygen electrode by as much as 100~150 mV.[2005A] This provided a powerful driving force to synthesize members of this group of acids, although this has proved to be high-risk, difficult, and time-consuming.

For example, workers at the Lawrence Berkeley Laboratory studied the dimeric form of TFMSA, tetrafluoroethane-1,2-disulfonic acid, $(CF_2SO_3H)_2$ (TFEDSA[165, 2140]), synthesized under EPRI funding by Eco, Inc.[688] This showed very promising performance at temperatures below 100~110°C, but at higher temperatures forms a dihydrate which has a rather poor conductivity, with the result that the observed 40 mV gain in oxygen reduction kinetics would be more than offset by the increase in ohmic resistance in a typical fuel cell. Thus the initial promise shown for TFEDSA did not materialize.

Other superacids that may have some promise include difluoromethanediphosphonic acid (DFMDPA), difluoromethanedisulfonic acid (DFMDSA), trifluoroacetic acid (TFAc),[1916] soluble four-basic fluorosulfonic acids,[165] and soluble half-salts of a fluorosulfonic acid that contain an alkali metal cation to lower the vapor pressure at high temperatures.[164]

---

*The term *superacid* is commonly used in the fuel cell community to describe the family of fluorinated organic sulfonic acids that have a high degree of dissociation and proton activity. For example, 0.1 M TFMSA, with pK_a = -2, has the same pH as 0.7 M phosphoric acid, with pK_a = +3.[2005A] They should not be confused with the true superacids of the inorganic chemists.[1931]

The most recent data appear to show that the vic-diacids are not stable, and that perhaps all phosphonic acids hydrolyze at temperatures approaching 200°C.[2146, 2322] To date, no potential electrolyte materials with sufficient conductivity in this temperature range have been identified; however, other promising materials are based on perfluorodisulfone imide acids ($-CF_2SO_2NHSO_2CF_2-$).[2766]

In addition to the studies employing phosphoric acid, sulfuric acid, and, more recently the superacids, reduction of oxygen on platinum has also been studied in mixtures of nitric acid–sulfuric acid[785] and in perchloric acid.[2131]

Corrosion of Carbon Catalyst Support and Platinum Surface Area Loss

To provide maximum activity per unit mass of electrocatalyst, the platinum crystallites are normally dispersed on a high surface-area carbon support material. The performance of PAFC cathodes employing such supported platinum catalysts gradually decreases with time.[1672] On both supported and unsupported platinum electrocatalysts the decay, which is related to a decline in effective surface area, initially proceeds quite rapidly and then continues at a slower rate.[1934, 2358]

To improve PAFC efficiency and output, there has been a recent trend towards operating these cells at higher temperatures and pressures than was previously the case. For example, in order to achieve a desired heat rate of about 8400 BTU/kWh, Westinghouse phosphoric acid fuel cells now must operate at close to 0.71 V/cell at 250 mA/cm^2, which requires operation at 190°C and 5 atm pressure.[165] Similarly, to achieve a desired end-of-life heat rate of 8300 BTU/kWh, United Technologies Corp. must guarantee 0.73 V/cell at end-of-life at a current density of 216 mA/cm^2, requiring operation at 205~210°C and 8.2 atm pressure.[165]

Dispersed Platinum Catalysts

Standard PAFC oxygen catalysts, which are now universally prepared by a colloidal technique, consist of about 10% by weight platinum on a carbon support.[165] The platinum is in the form of small clusters of about 1 nm diameter, and has unusually high catalytic activity. Studies at Stanford University have shown that these clusters have a normal face-centered cubic structure with a Pt-Pt spacing that is typical of that found in bulk platinum metal.[701] When the crystallite size is sufficiently reduced on suitable supports to give electrodes of optimized structure, it becomes possible to achieve virtually complete utilization of the platinum electrocatalyst surface. This condition now is being approached, the surface area utilization of present platinum electrocatalysts being on the order of 40~50%. With such high utilization, the kinetic processes that take place in the electrode structure are no longer controlled by diffusion.[2355]

Increasing the platinum loading on the support does affect the specific surface area of the platinum, since the possibilities of recrystallization increase. For example, a 1% platinum loading on Shawinigan acetylene black has been shown to have an initial specific surface area of about 130 m^2 per gram of Pt, whereas increasing the loading to 20% platinum lowered the specific surface area to about 60 m^2 per gram of Pt.[2358]

Catalyst Supports

In addition to being electronically conductive and stable, the material used to support the platinum must be of sufficiently high surface area to allow effective dispersion of the platinum crystallites, thereby inhibiting eventual recrystallization. This is necessary in order to maintain as high a catalyst surface area as possible and to reduce mass transfer limitations within the electrode structure. The ideal specific surface area range appears to be about $4\text{~}6 \times 10^8$ m^2 per cubic meter bulk volume of support material.[161] For high surface area carbons, this usually translates to a specific surface area in the range of 200~250 m^2/g; whereas for a metal oxide support (such as doped Ta_2O_5), it might correspond to about 45~60 m^2/g. If the specific surface area of the support is greater than the above ranges it may be difficult to achieve a uniform catalyst dispersion, and diffusion losses inside the electrode also may occur. In any event, it appears that the platinum crystallite specific surface area increases with increasing specific surface area of the support material, but reaches limiting values in the support area ranges given above.[161, 2357]

Work in the early 1970s established that only carbon materials had the correct combination of corrosion resistance, electronic conductivity, and cost to be feasible for use as the catalyst support material for mass-produced fuel cell components.[165] Until about 1978, the most popular catalyst support for use in phosphoric acid fuel cells was Cabot Corporation's Vulcan XC-72R.[164] This material is a conductive furnace black, and was selected because of its high electronic conductivity, and because its high specific surface area of about 250 m^2/g seemed to be about the optimum value needed to avoid excessive crystallite coalescence during preliminary heat treatment of the catalyzed material. Even more important, it was found to accept colloidal platinum readily and was easily fabricated into Teflon-bonded electrodes of uniform and consistently high quality. Unfortunately, its corrosion rate was shown to be excessive at temperatures above about 180°C in phosphoric acid at fuel cell cathode potentials (~0.7 V), especially under pressurized conditions, when water vapor pressures are high.[394]

The target life of a PAFC should be about 40,000 h, at the end of which the platinum end-of-life specific surface area ideally should be at least 80 m^2/g of Pt.[2358] However, in concentrated phosphoric acid at 200°C, an oxygen cathode running at a typical potential of 0.7 V can degrade from an initial platinum active surface area of about 100 m^2/g, to an active area of about only 20 m^2/g after 40,000 h of operation.[161] This gradual loss in platinum electrocatalyst surface area has been one of the most serious technical problems encountered in the PAFC.

Platinum Surface Area Loss

There are three modes by which the active surface area of a platinum electrocatalyst can decrease: surface blockage, dissolution followed by precipitation, and migration and coalescence. The first mode consists mainly of a simple physical blockage of active sites; the second and third modes, which are loosely referred to as sintering, involve an increase in the average size of the platinum crystallites as a function of time.

Surface Blockage: The gradual decay in the performance of oxygen cathodes can be partially explained by the adsorption of impurities on the active sites of the platinum.[458, 706, 916, 1292, 1672, 2648] Numerous impurities are known to adsorb on platinum electrodes, and the degree of adsorption of many of these impurities is a function of potential. Cations,[1822] anions,[960, 962, 1144, 2227, 2787] and organic species[706, 961, 962, 2545] may be involved. Even under the best conditions, it is very difficult to remove all impurities from the electrolyte, even by the use of multiple recrystallizations, treatment with oxidizing agents, and pre-electrolysis.[962] In a technical electrolyte such as phosphoric acid, which must be used in a large-scale practical system supplied by technical grade gaseous fuels and air, the problem of purification becomes academic, especially when it is noted that some organic impurities in concentrations as low as 10^{-7} M can block and poison the cathodic reduction of oxygen.[706]

Adsorbed impurities on the platinum surface not only tend to lower the active surface area by simple physical blockage, but also tend to promote the two-electron reduction of oxygen to hydrogen peroxide instead of the desired four-electron reduction to water.[358] This bias towards the peroxide pathway comes about because the four-electron reduction involves kinetics that will be second order in the surface concentration of sites, whereas those for the two-electron kinetic process should only be first order.

As discussed earlier, the adsorption of phosphoric acid anions, or indeed of molecules of the acid itself, must be considered in the blockage process. In 85% phosphoric acid, it has been claimed that raising the temperature from 25°C to 150°C results in changes in the anion adsorption characteristics of the acid that result in a 5% decrease in the active surface area of the platinum in Teflon-bonded fuel cell cathodes. Similarly, at 150°C, increasing the phosphoric acid concentration from 85% to 105% decreases the active area by 5~10%.[1913]

Dissolution–Precipitation: The second mode of platinum surface area loss involves the corrosion (dissolution) of the platinum in the strong phosphoric acid electrolyte and its subsequent recrystallization (precipitation) in the form of larger crystallites of smaller surface area. Studies using an ion beam microprobe have shown that after 4500 h of operation, a significant amount of platinum can dissolve and migrate from the cathode of an operating phosphoric acid fuel cell and redeposit at the anode.[2330] Workers at Case Western Reserve University have determined the solubility of bulk platinum in phosphoric acid as a function of potential and temperature.[161] Rather high solubility values were obtained, and recent work has shown that the values do indeed reflect those for finely dispersed platinum crystallites employed in fuel cell cathodes.[175, 357, 2144] Platinum corrosion and recrystallization is therefore becoming a more serious problem as PAFC temperatures, pressures, and cathode potentials increase. For platinum black, the dissolution and precipitation mechanism has been shown to be important.[2358] In addition, impurities such as ferric chloride, arsenic, and possibly copper, can promote platinum dissolution, and should therefore be avoided.[358]

Migration and Coalescence: It has been known since the early 1970s that, with time, high surface-area platinum electrocatalysts sinter, that is, they undergo a structural relaxation that is accompanied by a decrease in active area.[342, 1141, 1672] Although work at Brookhaven National Laboratory using 85% phosphoric acid at 150°C showed that the rate of sintering of supported fuel cell catalysts is five times lower than that of unsupported catalysts, sintering of supported platinum should nevertheless be considered to be a serious problem.[1734]

When small platinum catalyst crystallites are deposited on high surface-area carbon, electronic interactions occur between the platinum and the support. This explains why the electronic properties of the support appear to play a definite role in determining the catalytic activity of the supported platinum. It seems probable that the origin of these interactions depends on the relative Fermi levels of the electrons in the metal and in the substrate.[1412] Because the interactions and adhesive forces between the platinum and the carbon are rather weak, under certain conditions it is possible for platinum particles to migrate over the carbon surface, forming larger particles of smaller active surface area via two-dimensional surface Brownian motion.[2355]

Sintering can therefore occur via one of two mechanisms, the driving force in both cases being a decrease in particle surface energy.[1412] The first mechanism (*particle migration*) involves the migration of discrete platinum crystallites across the carbon support and their subsequent coalescence to form larger particles. Depending on the conditions, the rate-determining step for particle migration can be either surface diffusion or particle coalescence. In any case, both processes will be potential-independent. The second mechanism (*Ostwald Ripening*) involves the migration of individual atoms of platinum from smaller particles to larger ones. On supported catalysts, the metal atoms can diffuse either on the surface of the support as adatoms or via the liquid electrolyte medium. The rate-determining step in this second mechanism may be either the surface diffusion of adatoms, or the rate of transfer of the metal atoms between the support surface and the growing metal crystallite. If reaction occurs via electrolyte transfer, the rate will be strongly potential-dependent.

Inevitably, sintering becomes more rapid as the temperature is increased. For example, Engelhard Minerals and Chemicals Corp. showed that 10% platinum on carbon cathodes operating at 0.7 V in concentrated phosphoric acid gave a 33% platinum surface area loss after 500 h of operation at 163°C, while operation at 204°C resulted in a loss of 68%.[1364]

Corrosion of Catalyst Support

As we have already remarked in Chapter 9, electrodes made on carbon support materials have the advantages of being the simplest materials to fabricate and they do possess unexpectedly high stability in the acid cathode environment. However, over long operating periods the carbon support does corrode to a greater or lesser extent, depending on the conditions and on the nature of the carbon. The problem of platinum sintering is closely linked with support corrosion, since the latter will allow platinum microcrystallites to become detached or dissolve, recrystallizing as lower surface area material, often attached to basal plane dislocations on the carbon surface. Platinum may even be a local catalyst for carbon corrosion, causing undercutting and crystallite detachment. Finally, platinum that has dissolved in the electrolyte can be lost completely from the cathode area, since it may diffuse to the anode and reprecipitate there.[1667] As carbon corrosion increases, the surface area of the carbon support decreases,[2358] accompanied by a corresponding reduction in platinum surface area.[161, 2357]

The corrosion rate of the carbon support, and hence the extent of surface area loss of the supported platinum, is influenced by several factors. As would be expected, increasing the temperature increases the rate of carbon corrosion.[2358] The corrosion rate also is affected by potential, the normal Tafel relationship applying. In general, the observed Tafel slope for carbon corrosion in phosphoric acid under utility fuel cell conditions has been observed to vary between 100 and 140 mV/decade, the more resistant varieties of carbon having the lowest slopes.[177]

At the normal cathode operating potential of about 0.7~0.8 V/NHE (depending on operating pressure), conditions and materials must be chosen so that both the rate of carbon corrosion and catalyst dissolution is minimal. However, if the cathode is operated at potentials above about 0.8~0.9 V/NHE (i.e., at part load), the rate of corrosion of the carbon catalyst support can increase tenfold.[2358] As a guide, the cathode potential on the NHE scale is about 50 mV above the measured cell voltage. It is therefore very important to maintain fuel cell operating parameters at design conditions and avoid exposure of the cathode to open circuit or part load, corresponding to cell potentials above 0.8 V. This can be done by oxygen depletion or nitrogen dilution in the cathode gas supply loop.

The corrosion rate of the carbon support also is affected by the concentration of the phosphoric acid, higher rates of corrosion being observed at lower acid concentrations.[2002A] Thus, over a range of temperatures, carbon supports were reported to corrode about five times faster in 103% phosphoric acid than in 106% phosphoric acid.[2357] However, later work using acetylene black support showed that the change in corrosion rate within this range of acid concentration was much less, being closer to a factor of two.[177,2361] As the concentration of phosphoric acid decreases below about 96% by weight, the acid changes from the *para* (condensed polymer) form to the normal *ortho* form. It is well established that carbon substrates corrode significantly faster as a function of acid concentration below about 96% by weight,[177, 2361] in spite of the fact that the activation energy for corrosion is lower in the lower concentration range.[2357, 2361]

The faster corrosion rates at lower acid concentrations at constant temperature correlate with increasing water activity, and for Vulcan XC-72, acetylene black and massive glassy carbon, it has been observed that the corrosion rate of the carbon support increases linearly with the logarithm of the vapor pressure of the water in equilibrium with the acid.[177, 2145, 2357, 2002A] This indicates that in this acid concentration range, water is a first-order reactant, hence the rate-determining corrosion reaction must involve attack by a primary water molecule on the carbon surface. However, in the range where condensed acids are stable, the corrosion rate is almost independent of water vapor pressure, indicating that the condensed phosphoric acid itself must supply the oxygen for reaction with carbon.[177, 2145, 2361] This may suggest some chemical instability of the condensed acids.

It is interesting that the limiting water vapor pressure at which the corrosion reaction mechanism changes has been observed to be 100 mm Hg for corrosion resistant carbon materials, and it appears to be independent of temperature in the range 150~180°C.[177]

The move towards higher temperature operation to allow improved performance led to the discovery that concentrated phosphoric acid starts to show significant evaporation losses at temperatures approaching 200°C. The evaporating species is P_2O_5, whose vapor pressure increases as water vapor pressure decreases. Accordingly, UTC has used a more dilute acid at 205°~210°C and 8.2 atm to suppress evaporation and ensure that the PAFC contains sufficient acid inventory for a hot lifetime of 40,000~60,000 h (about 0.8 kg/kW under UTC operating conditions; see Ref. 177). The acid concentration used has the disadvantage of being more corrosive towards the carbon catalyst support.[165] The change in operating conditions from 190°C to 210°C and system pressure changes of from 3.4 to 8.2 atm correspond to changes in phosphoric acid concentration from about 98% to 93%. Under these conditions, the carbon support corrosion rate will be several times higher at 210°C than at 190°C, taking into account the three factors of temperature, water vapor pressure, and increased cathode potential.[165] Thus, a particularly corrosion-resistant support is required for use under these conditions.

The time-dependent loss of performance of carbon-supported platinum fuel cell electrodes is a consequence of both the corrosion of carbon catalyst support materials and the sintering of dispersed platinum crystallites, both of which limit operation for practical lifetimes to temperatures just above 200°C. Improvements in electrode performance that can be obtained by operation at yet higher temperatures and pressures will depend on the development of improved materials. This is discussed in the following two sections.

Efforts to Reduce Support Corrosion

Considerable progress has been made since 1976 to develop more corrosion-resistant supports, and so to reduce the cathode materials limitations resulting from support corrosion. Starting in 1978, Exxon Research and Engineering Co. indicated that when activated carbon catalyst supports are pretreated with phosphoric acid, phosphate ester groups become attached to the carbon chains, and the resulting phosphonated carbons are both more corrosion-resistant and sinter-resistant than the untreated carbons.[625, 1969] It was suggested that in addition to promoting wetting of the carbon surface, the phosphonic acid groups act as anchor sites to immobilize platinum crystallites, slowing down initial crystallite migration and subsequent recrystallization.[161] In initial work, Exxon used the wood-based North American P-100 activated carbon, as well as a number of other similar materials such as American Norit Polycarbon C, Westvaco Nuchar, and North American P-108. After the treatment with phosphoric acid, the phosphorus content of these carbons ranged from 2.3~3.4 weight percent. However, after 100~150 h under the most severe conditions (0.9 V at 190°C), a considerable loss of phosphorus occurred due to irreversible hydrolysis of the phosphate ester groups in the treated carbons. Later, under EPRI sponsorship, a similar, though more complex, treatment was applied to Vulcan carbons by Eco, Inc. to produce fluorinated, phosphonated Vulcans exhibiting increased corrosion resistance.[161] While this approach showed some promise, it involved a complex and costly synthesis process and the products were difficult to wet and catalyze.

Some of the efforts to develop a replacement for Vulcan XC-72 at Brookhaven National Laboratory have already been mentioned. Five Cabot carbons for use as a support material for platinum catalysts in oxygen electrodes in 85% phosphoric acid at 25°C and 135 °C were examined.[2329] The carbons included Monarch 1300, CSZ 98, Mogul L, Vulcan XC-72R, and Regal 660R. Of these carbons, Regal 660R gave the best performance; that is, it was easiest to catalyze and fabricate into acceptable electrodes. This work indicated that the electrode-forming properties of Vulcan XC-72 were not unique, and that other viable support materials could probably be found.

Under EPRI support, Stonehart Associates obtained very promising results with steam-treated (900~950°C) Shawinigan acetylene black as a catalyst support material.[161, 164, 165, 2358, 2361] Because the steam-activated acetylene black, called *Consel* by the developer, has a B.E.T. surface area approximately four times that of the as-received material, it is able to support and retain platinum crystallite dispersions at much higher platinum loadings than the latter.[2358, 2361] Like Vulcan XC-72, it has a specific surface area greater than 200 m^2/g and readily accepts colloidal platinum. More important, its corrosion rate has been shown to be only about 1/100 that of Vulcan at 0.7 V/NHE in 200°C phosphoric acid under fuel cell operating conditions.[161, 164, 177]

Steam treatment at elevated temperatures removes the most corrodible parts of a carbon and essentially activates the material, resulting in a larger surface area. Treatment by CO_2 at high temperature, or oxidative treatment at lower temperatures, could result in similar morphological changes.

An alternative treatment is to heat the support material to high temperture in an inert atmosphere to give increased long-range order. As discussed in Ref. 177, this type of treatment can fully graphitize certain carbon materials on heating to temperatures of about 2700°C, while other carbon materials are scarcely affected. An example of the latter is acetylene black, which undergoes neither crystallographic nor corrosion-resistance changes when subjected to this heat-treatment. In contrast, Vulcan XC-72 is a good example of a graphitizable carbon black. After heat-treatment at increasing temperatures between 1500°C and 2700°C, its lattice parameters approach those of pure graphite and under the transmission electron microscope it starts to show definite graphite planes, rather than the disordered amorphous structure of the original material.

Many other furnace blacks show similar behavior, although the effects of heat treatment do depend on the structure of the original carbon. In general, it is found that as the heat treatment temperature is increased, the specific surface area of the heat-treated carbon decreases.[1657, 2357] For example, the specific surface area of Vulcan XC-72 falls from an as-received value of about 250 m^2/g to 80 m^2/g. The lower specific surface area of the resulting graphitic carbon yields a somewhat lower colloidal platinum specific surface after catalyzation. However, this small under-utilization of platinum is considered to be more than compensated by the improvement in the corrosion resistance of the support.[177]

During the late 1970s UTC conducted extensive proprietary examinations of corrosion-resistant graphitized carbon-black (furnace-black) supports. The Stonehart Associates' work starting in 1978[2361] helped to clarify many of the corrosion properties of these materials, particularly thermally-treated Vulcan XC-72. First, the heat-treatment temperature had little effect on the rate of corrosion at 1.0 V/NHE under fuel cell conditions in phosphoric acid. However, the Tafel slope changed dramatically from about 180 mV/decade for the as-received material to close to 100 mV/decade for fully graphitized material, most of the improvement taking place in the heat-treatment temperature range 1500°C–2200°C. This yielded an eightfold improvement in corrosion rate at 0.8 V/NHE (pressurized operating conditions) and a more than twentyfold improvement at 0.7 V/NHE (atmospheric pressure operating conditions). The advantage of graphitization is therefore clear.

The corrosion resistance of some common support materials has been found to increase in the following order (the latter three materials varying by a factor of about three[2002A]): as-received Cabot Vulcan XC-72 < Vulcan XC-72 heat-treated at 1200°C–2500°C < Vulcan XC-72 graphitized at 3000°C < as-received Shawinigan acetylene black < steam-treated Shawinigan acetylene black.

Steam-treated acetylene black is more difficult to make into electrodes with good structure than untreated Vulcan XC-72, and since mid-1986 the Shawinigan black starting material has been unavailable. However, an excellent substitute appears to be graphitized Carbon Black Pearls 2000 furnace black.[177] The latter is a substitute for the high-area, high-color channel blacks that were discontinued in the late 1960s. The starting

Cathodic Electrocatalysis

material has a specific surface area approaching 2000 m^2/g as a result of a preoxidative treatment that confers high corrosion resistance. After graphitizing, it has an area of 214 m^2/g and retains excellent corrosion resistance which may be five times better than that of acetylene black,[177] although this requires further confirmation.

A maximum carbon support corrosion rate of about 10^{-2} mA/mg measured at 0.9 V/NHE is considered to be acceptable for a 40,000 h PAFC cathode operating at 0.7V/NHE.[2002A] For use under pressurized conditions, the value should be an order of magnitude lower. With a 100 mV/decade Tafel slope, this will indicate complete corrosion of a 5 mg/cm^2 support in 90,000 h. While this corrosion rate sounds excessive, one should bear in mind that corrosion rates fall with time, typically by an order of magnitude over 100–200 h as the most corrodible material is consumed, and more resistant material remains.[177] All things being equal, the fall will be initially exponential, eventually tending toward a straight-line relationship; however, there is recent evidence that some further rise in corrosion rate occurs at long times as the corroded carbon becomes porous.[2147]

Care must therefore be taken in comparing corrosion results taken from different sources, since those that are obtained after 100 h of precorrosion at 0.9 V will give an order of magnitude lower corrosion currents at 0.8 V than those for specimens exposed only to 0.8 V for the same length of time. These and many other aspects of electrode corrosion have been reviewed.[177] This reference shows that these low rates of carbon support corrosion are now achievable in pressurized 95% phosphoric acid fuel cell systems at 200°C. Substrate corrosion can now be said to be under reasonable control, and should no longer be a limiting factor in the performance of the PAFC cathode under present operating conditions if the correct materials are chosen. If future operating conditions require greater catalyst support corrosion resistance, it may be necessary to employ special treatments such as pre-etching[164] or the incorporation of boron,[161, 2361] or to proceed with the use of a different, but more costly, catalyst support material such as a doped oxide,[161] or certain doped carbides such as TaC or doped SiC.[1341] However, new carbon blacks continue to become available as specialized fabrication methods and new precursors are used. Graphitic materials are attacked first at edge-planes. Recent work at LBL has developed a high-surface graphitic black that appears to possess no apparent edge-planes, each particle consisting of a convoluted tangle of graphitic microchains. It is claimed that this proprietary material has a corrosion resistance that is ten times better in alkaline solution than that of normal graphite or of the best graphitized carbon blacks. If this is also true in acid electrolytes, it may promise a breakthrough in PAFC cathode supports and structural components.[625] Corrosion data on a wide range of carbon materials under PAFC operating conditions are shown in Fig. 12–1.

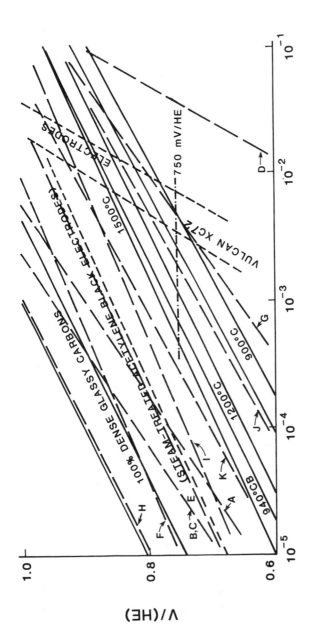

Figure 12.1: Composite Tafel plots for corrosion of various carbon materials in 200°C phosphoric acid. Plots generally taken after 10,000 minutes. *Dashed line:* Corrosion behavior of 5 mg/cm² Vulcan XC72 supports and steam-treated acetlyene black supports (right and left, respectively). *Lines with indicated C° temperature:* Plots for typical 67 wt% graphite powder-33 wt% phenolformaldehyde resins heat-treated at indicated temperatures. *D:* Non-heat-treated material. *G-J:* Range of 900°C heat treatments. *940°C, B:* Boron-doped material heat-treated at 940°C (see text). *K:* 5% Boron-doped resin treated at 900°C (see text). *I:* 1200°C heat-treated resin.

Efforts to Reduce Sintering

Several approaches have resulted in varying degrees of success in solving the problem of platinum sintering. It is well known that the loss of active platinum sites can be changed by impurity adsorption, which suggested one of the earliest approaches to preventing surface area loss. This showed that platinum black deposits containing cadmium or lead were more resistant than pure deposits to surface area loss in hot (80°C) phosphoric acid.[2454] The difficulty with this approach is that the adsorbate or additive employed must not adversely affect the oxygen reduction reaction and must be stable at oxygen electrode potentials.

The method of platinum preparation could exert some influence on its stability. For example, it was observed that Raney platinum was more stable than normal platinum black in acid electrolytes at elevated temperatures.[353, 354] Similarly, Teflon-bonded platinum cathodes prepared from PtO_2 were reported to be more stable in hot acid than those prepared from platinum black,[664] and platinum black prepared at higher temperatures had greater stability than that prepared at room temperature.[2034] Thus, alloying of platinum, the use of Raney platinum, or other variations in the method of formation of the electrocatalyst may all impart some extra stability to the catalyst.

Of more recent interest is the method of immobilizing platinum crystallites to prevent their migration along the surface of the carbon substrate. United Technologies, for example, has claimed that crystallite migration can be significantly lowered by depositing porous carbon on and around the crystallites, the porous carbon being laid down by heating the platinum catalyst in carbon monoxide at 260~649°C.[2539]

Another possible method for solving the platinum sintering problem is the concept of *in situ* electrochemical regeneration, investigated at Brookhaven National Laboratory.[1734, 1912, 1932-1934] When unsupported PTFE-bonded platinum black oxygen electrodes containing 12 mg Pt/cm² were examined at 150°C in 85% phosphoric acid at 0.50 V, initial current densities were 120~160 mA/cm². After operation for 1140 h, both the platinum surface area and the output current density had decreased by about 40%. However, it was found that by cycling the electrodes at 100 V/s between 0 and 1.3 V, about 95% of the original current density and platinum surface area could be restored, after which the performance decay more or less again followed the original pattern.[1912, 1932] Tests also were carried out on standard Vulcan-supported fuel cell oxygen cathodes containing 0.35 mg Pt/cm² manufactured by ERC.[1734, 1933, 1934] When these electrodes were aged at 150°C in 85% phosphoric acid, 30~50% of the original platinum surface area was lost after 5112 h. Cycling between 0 and 1.4 V at 200 V/s for about 2 h restored 77~95% of the original surface area.

So long as the voltage sweep rate was greater than about 100 V/s, cycling treatment did not damage the carbon support. However, at low sweep rates some carbon oxidation occurred.

Perhaps the most encouraging progress to date has come from the work at the Lawrence Berkeley Laboratory, which developed Engel-Brewer intermetallic platinum catalysts of improved stability and oxygen reduction activity.[161, 164, 2137, 2138] During the same period, details on similar catalysts were published in the patent literature by UTC.[1283-1285] These catalysts contain platinum and Group IVB and VB elements, and are made by a simple chemical treatment involving precipitation of the chloride salt of the base metal from solution onto a standard carbon-supported platinum catalyst. In the exploratory work, this consisted of 10% by weight platinum on Vulcan XC-72R, fabricated by Prototech using a colloidal procedure in which platinum clusters are adsorbed onto the carbon surface from a colloidal suspension.[1280, 1281, 2013-2017] After deposition of the appropriate alloying element, the catalysts were heat-treated in a helium-purged induction furnace to form the intermetallic alloy.[2137]

The resulting catalysts consist of single-phase bimetallic clusters of 3~5 nm diameter on the Vulcan XC-72 surface, having platinum to base metal ratios of 3:1 and 5:1. The intermetallic compounds have a face-centered cubic structure of the Cu_3Au type, and are all significantly more active towards oxygen reduction than is platinum alone, typically by a factor of two to three at constant potential. Thus, intermetallic catalyst loadings of 0.5 mg/cm^2 in PTFE-bonded fuel cell electrodes have yielded performance gains of 20~40 mV over conventional supported platinum at 215 mA/cm^2 in 97% phosphoric acid at 177°C. In air at 1 atm pressure, the following cathode potentials (in mV) were obtained at 215 mA/cm^2 for the following electrocatalysts: heat-treated colloidal platinum, 702; untreated colloidal platinum, 688; VPt_3, 725; $HfPt_4$ and $ZrPt_4$, 722; $TaPt_5$ and $NbPt_5$, 715. Although some ultimate dissolution of the base metal from the surface of the clusters was considered probable, stability tests in concentrated phosphoric acid at 177°C showed that these catalysts have good resistance to sintering and did not appear to decompose during a 10^2 h time-scale.[2137] The higher activity of the heat-treated platinum sample is interesting, since its surface area fell from an initial value of about 130 m^2/g after preparation to 70~80 m^2/g after the heat-treatment operation at 800°~900°C. Most of this can be attributed to the lower intrinsic activity of very small platinum microcrystallites for oxygen reduction, which may fall by about a factor of two in this specific surface-area range due to crystal face effects.[456] This has been often discussed in the literature, and is reviewed in detail in Ref. 177. Impurity removal, and perhaps the higher resistance of alloys to impurities,[135, 137] may also be important in determining final activity.

Work was conducted under EPRI funding by Stonehart Associates to determine the loss of surface area of the Pt_3V intermetallic as a function of time compared with pure Pt on the same Vulcan substrate.[177] Since the alloy had been heat-treated during preparation, its specific surface area as measured by CO adsorption was lower than that of the pure Pt precursor (about 53 m^2/g vs. 70 m^2/g). However, measurements out to 5000 h appeared to indicate that the alloy maintained its surface area and activity better than that of the pure Pt control. For example, after this time, the pure Pt sample had a specific area of about 35 m^2/g, whereas the alloy was still over 40 m^2/g.[177] Even so, parallel experiments using Rutherford Backscattering Spectroscopy at Los Alamos National Laboratory[1667] detected vanadium at the cell anodes that had been leached from the cell cathodes. There was also no doubt that the activity of the vanadium alloy cathodes tended rapidly towards that of their pure Pt precursors. At first, this problem was attributed to a lack of correct formation of the Pt_3V intermetallic in these electrode compositions.

However, in 1982 and 1983, two U.S. patents were granted to UTC, no doubt after a great deal of in-house experimentation, indicating that indeed, Pt-V compositions of all types, while more active than pure platinum, lost vanadium, along with their activity, as a function of time.[1573,1574] Indeed, at 177°C in phosphoric acid, 65.5% by weight of the vanadium was reported to be lost within 48 h. Under the same conditions, only 37.5% of the chromium alloying element in the proposed improvement was lost. Chromium addition certainly appeared to give increased stability over vanadium, along with the same advantage of improved activity. This immediately opened up a series of questions regarding the mechanism of action of these alloys. For example, chromium should not form Engel-Brewer intermetallics, yet it certainly showed much higher initial activity than pure platinum, as was demonstrated by work at Giner, Inc. in 1982.[177,1289] At this point, it seemed that the improved activity of virtually any platinum alloy was a general phenomenon resulting from lattice order-disorder, which might change the bond-strength of reaction intermediates and therefore the rate of dioxygen reduction. Indeed, attempts were made to correlate activity with interatomic distance in different alloys.[177,1288]

However, the active chromium alloys still showed a higher rate of decay than that of heat-treated Pt, and after 2000 h showed approximately the same performance as the latter[177] (Fig. 12-2). Starting in 1982, Giner, Inc. started to develop a new series of proprietary ternary alloys under DOE-NASA funding.[1290] These were reported to contain transition metals other than chromium, because of UTC's patent position. Giner's alloys, of unspecified composition, showed 30 mV higher activity than pure Pt (i.e., about the same initial activity as the chromium alloy, but with a lower

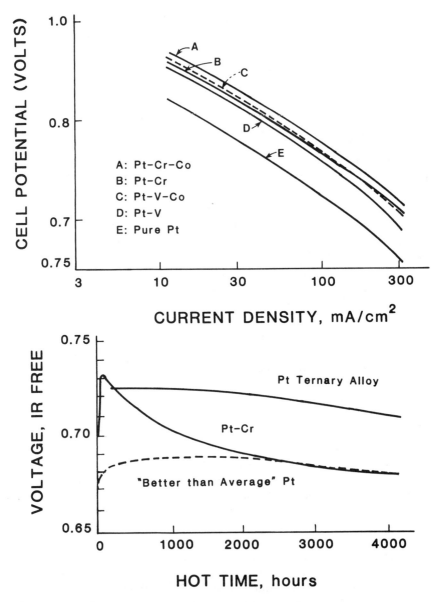

Figure 12–2: (Above): Cell performance with platinum and binary and ternary alloy cathodes (H_2–air, 190°C, phosphoric acid, Ref. 1625). (Below): Degradation of proprietary Pt ternary alloy (Griner, Inc.), Pt-Cr, and pure Pt cathodes (190°C, phosphoric acid, H_2–air, utilizations 80% H_2, 50% O_2) as a function of time at 150 mA/cm².[177]

decay rate; Fig. 11-2). Finally, in 1984, another UTC patent was granted, claiming higher activity and stability for Pt-Cr-Co ternary alloy.[1645] These have been recently reviewed.[177]

The above alloy developments are of great interest, but the true long-term stability of all transition-metal alloys is now open to doubt. Their higher activity may indeed be intrinsic, but it could also be the result of the formation of some type of Raney platinum at the surface of each catalyst particle by leaching of the alloying element from the electrode surface. It has also been suggested that it is extrinsic, and results from changes in electrode surface chemistry after heat-treatment of the platinum in the presence of transition metal salts to form the alloys. These may result in changes of electrode structure that simply allow higher utilization of the total platinum catalyst area available.[177] Yet there is no doubt that the alloys are formed, as can be seen from X-ray diffraction results. Recently, it has been suggested that Pt-Cr alloys in a certain composition range are simply poisoned less by the ever-present impurities under real conditions than is pure platinum, so that they appear to show higher activity.[790, 975] This theory receives some support from previous work on Pt-Ru alloys in phosphoric acid, which showed that indeed those containing 4~10 atomic % Ru were more active than pure platinum in the presence of impurities.[135, 137]

Whatever the mechanism, noble metal alloys (e.g., those containing small amounts of ruthenium or other hard stable Group VIII elements) may give higher initial activity than that of pure platinum, together with greater stability, since they harden the microcrystallites and should not be leached out under PAFC cathode conditions. This concept is being actively pursued.[1341, 2362]

Use of Platinum in Alkaline Systems

Many studies have been conducted on the reduction of oxygen on platinum in alkaline media. Aqueous potassium hydroxide has almost universally been chosen as the electrolyte because its conductivity and carbonate ion solubility are both somewhat higher than for sodium hydroxide. For laboratory kinetic studies, 0.1 M KOH at room temperature has usually been selected,[404, 1590, 2258, 2432, 2433, 2435] although in some work, concentrated electrolytes closer to the concentrations employed in fuel cells have been used.[1514] Oxygen reduction also has been studied in molten KOH electrolyte,[1919] in which O_2^- ion is produced as a first reaction product. In aqueous KOH solution, both tightly bound and weakly bound oxygen (from water oxidation) appear to be adsorbed on platinum. Although at room temperature the total quantity of bound oxygen (–OH or –O) is greater in aqueous KOH than it is in aqueous sulfuric acid, in KOH the total adsorbed oxygen at a given potential decreases as the temperature increases.[1894] Oxygen reduction on carbon-supported platinum has been studied employing 8.7 M KOH at 25°C,[340] and 5 M KOH at 30°C and 70°C.[1068] The reduction on graphite-supported platinum in 6 M KOH has been investigated at 25°C and at 70°C.[1570]

Alloys of Noble Metals and Platinum

The oxygen reduction reaction has been studied on a number of platinum noble metal alloys in both acid and alkaline media. A number of alloys of platinum, palladium, ruthenium, iridium, osmium, and gold have been evaluated as oxygen reduction catalysts in sulfuric acid at 25°C.[648] Platinum-iridium was examined in 30% sulfuric acid at 82°C.[2457] Platinum-ruthenium alloy has been used at 25°C as an oxygen reduction electrocatalyst both in 0.5 M sulfuric acid and in 0.1 M KOH.[1391] Dense platinum-rhodium alloys have been examined in dilute acid solution,[481, 1138, 1146] and as a result, platinum-ruthenium alloys were studied in phosphoric acid over a wide temperature range in work referred to previously.[135, 137] A conclusion of this work was that alloy catalysts may give better results than platinum for oxygen reduction under impure (i.e., ordinary) conditions; whereas they are inferior under conditions of high purity, where pure smooth platinum is without doubt the most effective catalyst. However, as has been discussed earlier, such a conclusion may require modification for high surface area materials.

Platinum-ruthenium alloys have also been used as oxygen evolution catalysts.[1559] United Technologies Corporation has patented oxygen reduction catalysts comprising alloys of platinum with titanium, silicon, strontium, and/or cerium base metals.[1279] These alloys were deposited on carbon support particles to yield finely divided alloy electrocatalysts with specific surface areas of ≥ 50 m^2/g of noble metal. Research at Shell International indicated that dipping a platinum alloy electrode (e.g., a piece of platinum-gold foil) into a molten tin salt (e.g., $SnCl_4$), followed by washing it with water, produced an electrocatalyst with an activity 50 times greater than that of pure platinum.[1756]

Other Noble Metals

The whole range of noble metals in the form of bright solid electrodes has been studied for oxygen reduction activity in phosphoric acid over a wide temperature range. A review of this work, which defines some of the factors controlling activity under these conditions, is given in Ref. 140.

Palladium has been used as an oxygen reduction catalyst in 0.05 M sulfuric acid at 25°C.[2433] However, it is not stable under PAFC cathode conditions, besides being about 50 mV less active than platinum.[135] It has also been examined in aqueous 0.1 M KOH at 25°C,[2433] and in 2 M KOH at 25°C and 75°C.[966] In aqueous KOH solution, the oxygen reduction activity of palladium-cobalt alloy appears to be superior to that of a number of other electrocatalysts in the following order of decreasing activity: palladium-cobalt > cobalt-palladium > silver > nickel > iron.[821] Palladium-gold alloy has been used in fuel cell electrodes, but it was abandoned in the late 1960s since the palladium was shown to be too soluble in concentrated KOH at 70°C.

Ruthenium and iridium electrodes have been examined for oxygen reduction, particularly in sulfuric acid.[648] However, for both metals, activity in acid electrolyte is much inferior to that of platinum.[140] On a bare ruthenium cathode in sulfuric acid, it has been reported that about 90% of the oxygen reduces directly to water, and 10% to hydrogen peroxide.[2773] On a ruthenium dioxide surface, however, all the oxygen reduces directly to water, with no peroxide formation. Studies of the adsorbed oxygen and oxide layers show that ruthenium and iridium behave differently from the other noble metals in that their adsorbed oxygen or oxide layers are not readily reduced at low potentials.[319, 2075]

Work conducted at United Technologies Corp. has, however, shown that the resistance of RuO_2 to reduction depends on the method of formation of the oxide on the ruthenium surface.[1409] Thus, ruthenium dioxide deposited on graphitized carbon black and formed by heating in air is very difficult to reduce in sulfuric acid by electrochemical treatment, whereas that formed anodically (to 1.3 V) can be reduced to bare ruthenium by cathodically sweeping to 0.0 V. The electrical conductivity of ruthenium dioxide has been shown to be affected by its morphology, which also may affect the mechanism of the oxygen reduction reaction on its surface.[1633]

There are indications that at potentials greater than about 0.7 V/NHE organic compounds such as 1,3-dienes adsorb strongly on ruthenium dioxide,[1788, 1789] which may affect its oxygen electrode behavior. Cyclic voltammetry on ruthenium in alkaline solution[516] and the formation and reduction of the anodic film on ruthenium have been studied.[513] Since about 1970, ruthenium dioxide has been a well known and favored electrocatalyst (especially on a titanium substrate) for the oxygen evolution reaction.[260, 1641, 2037, 2251, 2326] Iridium dioxide has also been used for this purpose.[884, 2037, 2326] The use of ruthenium dioxide as a medium for the kinetics of electrode processes has been reviewed.[1569]

Gold has been examined as an oxygen reduction catalyst in both acid[648] and alkaline[1023] media. Oxygen reduction on pure gold in acid is relatively irreversible,[140] owing to the low d-band character of gold and the consequent low heat of adsorption of oxygen.[46,140] Both Au and Au_2O_3 have been studied as oxygen evolution catalysts.[2230]

In alkaline solution, gold has been found to be an active and electrochemically stable catalyst.[1023] Its use in alkaline fuel cells has been reported to eliminate the problem of catalytic metal migration from the cathode, which may be both a source of performance degradation and of eventual short-circuiting. The oxygen cathodes in the UTC Space Shuttle Orbiter fuel cell consist of Teflon-ated 20 mg/cm² gold 10 wt % platinum black on gold-plated nickel screens.[394] The platinum does not change electrocatalytic activity, but simply acts as a sintering inhibitor. These electrodes have excellent activity and stability, and normally operate at about 80°C in concentrated KOH, although they have been successfully used at temperatures higher than 120°C for special applications.[1706]

It has been claimed that alloying gold with platinum, rhenium, nickel, or copper can give performance levels in 35% KOH at 67°C, or better than 800 mA/cm², which is equivalent to or even better than that obtained with platinum.[1023] Oxygen reduction on gold-platinum alloys in 2 M KOH has been studied at 25°C and at 75°C.[966] Workers at Brookhaven National Laboratory have shown that the adsorption of foreign metal adatoms on gold catalyzes the oxygen reduction reaction in both acid and alkaline solutions.[46] Rotating ring-disk experiments have shown that bismuth adatoms, for example, can effectively lower the oxygen reduction overvoltage in 85% phosphoric acid at 25°C by about 300 mV. Unfortunately, bismuth adatoms promote the formation of large amounts of peroxide. Adsorbed lead appears to act in the same manner. Tin adatoms seem to promote complete oxygen reduction to water, although the reaction takes place at high overpotentials, while adsorbed submonolayer silver appears to hinder oxygen reduction. The Brookhaven study concluded that in phosphoric acid pure platinum appeared to be the most effective catalyst for oxygen reduction.[46]

Silver

Silver has long been recognized as a reasonably effective oxygen reduction catalyst in concentrated aqueous KOH,[966, 1103, 1532, 2221, 2431] although the remarks at the end of this section concerning its cost-effectiveness should be noted.

In particular, finely divided Raney silver has been favored for fuel cell cathodes. However, it may be attacked under cathode conditions at elevated temperatures and high KOH concentrations. Thus in 7 M KOH at

96°C, calcium-doped Raney silver cathode catalyst undergoes some dissolution, which increases as KOH concentration increases.[991] Operating the cathode at low potentials significantly reduces the rate of attack, at the expense of fuel cell efficiency. Doping Raney silver with gold has been reported to improve its oxygen reduction performance.[818] Similarly, additions of zinc have been stated to improve the characteristics of Raney silver electrodes.[576A] The addition of ≥10% mercury to silver has been found to stabilize performance.[1153, 1552, 2523] Nickel has commonly been alloyed with silver.[1532] Methods have been reported for the preparation both of active silver[2416] and of active silver on pyrolytic graphite for use in fuel cell oxygen cathodes in KOH solution.[1807]

Numerous investigations of the kinetics of the oxygen reduction reaction on silver in KOH solution have been made. The rotating ring-disk electrode has been used to determine the relationship between the mechanism of oxygen reduction on silver and adsorbed oxygen.[866] The behavior was observed to depend on the physical nature of the silver used. For example, it has been reported that amorphous silver is more active than crystalline silver.[1924] In porous electrodes, activity variations can be attributed to differences in surface areas, porosities, and electrical resistivities.[1828] On silver, as on platinum, in very pure electrolyte the four-electron reduction to water appears to predominate, with the production of only very small amounts of peroxide.[865] As the electrolyte becomes increasingly contaminated, the production of peroxide increases. The mechanism of the peroxide reaction is dependent on pH, temperature, and peroxide concentration.[1884] Whether oxygen reduces via the two-electron process to peroxide or via the four-electron process to water depends on the catalyst, but not on the electrode substrate, provided that the substrate is sufficiently electronically conductive.[2622] For a given catalyst, simple methods are available for evaluating whether the peroxide is decomposed chemically or electrochemically.[2555] Silver has been determined to be a good peroxide decomposition catalyst, being comparable in this respect with palladium and nickel.[1628] An investigation of peroxide decomposition on silver in 1 M KOH showed that various chemisorbed surface compounds and non-stoichiometric surface oxides are involved in the process, depending on the concentration of hydrogen peroxide.[2254] A detailed investigation of the oxygen reduction reaction kinetics on silver on porous activated carbon in 0.1~12 M KOH solution at 25°C and 70°C showed that oxygen reduction is first order with respect to oxygen.[1533] Studies of the reduction of atmospheric oxygen on silver in 0.1~3.0 M KOH have shown that the oxygen reduction rate decreases as the concentration of KOH increases.[1395] This was explained by double layer changes, but more probable explanations are mechanistic changes or changes in oxygen solubility. Silver has been used as an electrode material for the anodic evolution of oxygen in alkaline solution.[842, 1751]

Several studies have been made to optimize the fabrication and performance of practical porous oxygen fuel cell cathodes using silver on activated carbon substrates. Voltage-current density curves have been reported,[86] and for DSK-type porous silver cathodes there are useful correlations between the current density values and the relative contributions of the different types of polarization.[2302] An investigation into the effects of an interruption of oxygen supply to porous silver diffusion electrodes was made by subjecting them to potentials in the hydrogen evolution region.[2295] The results of this study showed that prolonged polarization at these negative voltages had no detrimental effect on the cathodes, so any reduction of oxides or oxide films present under fuel cell cathode conditions is reversible. For porous hydrophobic oxygen cathodes containing high loadings (~50 mg /cm^2) of Teflon-ated silver catalyst pressed onto a nickel screen substrate, an optimization study showed that in 7 M KOH the optimum PTFE composition was only 7.5~10 wt %.[414] Generally, 30~40 wt % is considered optimal, so these electrodes are unlikely to be of uniform structure. As the temperature of these electrodes was raised above 160°C, where silver becomes soluble, oxygen reduction activity decreased as expected, as catalyst surface area decreased. Similar optimization studies have been made in 6 M KOH of the effects of Teflon content, the method of carbon activation, and the effect of oxygen pressure for porous activated carbon oxygen cathodes containing ≤ 5% silver.[1946, 1947] Methods for the fabrication of wet-proofed porous oxygen cathodes consisting of silver-nickel catalysts on carbon black substrates pressed onto nickel screens[1906] and for the preparation of porous sintered nickel oxygen cathodes containing silver catalyst[2582] have been reported.

In general, silver cathodes have lost favor because of their relatively poor stability. Also, the requirement of high silver loadings to obtain the same performance as that on low-loading platinum on carbon electrodes of the type used in phosphoric acid electrolyte tends to be self-defeating. For example, since the oxygen reduction exchange current on silver is at least an order of magnitude lower than that on platinum, much higher loadings (say, 20 mg/cm^2) are required to approach the activity of platinum. However, at such loading, catalyst utilization may only be 10%, compared with 80% for platinum at 0.5 mg/cm^2. As loading is further increased, effective utilization falls further. The result is that platinum is cheaper than silver as a catalyst in terms of dollars per kW at the same cell efficiency. Very low silver loadings will allow all the silver to be used, but will give low-performance electrodes with short lifetimes.

Nickel

Nickel is of course unusable as an oxygen cathode in acid solution. Although its electrocatalytic activity in alkaline solution (where it is present as a more or less hydrated oxide) is not particularly great,[1009] it has often been used as a catalyst support material in alkaline media; and sometimes as an oxygen catalyst, since it will resist anodic corrosion,[195,1515] is a good electronic conductor even when oxidized, and is relatively inexpensive. At high temperature (over 200°C) in molten aqueous KOH it starts to show high activity (as in the Bacon Cell, see Chapters 9 and 10), but this is due to a change in the rate constant and mechanism of the oxygen reduction process, rather than to an intrinsic catalytic property of nickel oxide, for example in its lithium-doped form. This has been discussed in more detail in Chapter 10.

When the metal has been examined as an oxygen reduction catalyst in KOH at temperatures of less than 100°C, it has usually been in the form of Raney nickel, whose preparation is given in reference 1044. It has been reported that a nickel-lanthanum alloy consisting of 90% nickel and 10% lanthanum deposited on carbon is more active for oxygen reduction in 0.1 M KOH at 25°C than is pure nickel.[1392, 1393] Nickel-cobalt alloys have also been reported to show good activity.[1215] In addition to its use in oxygen reduction, oxidized nickel has been commonly used industrially as an oxygen evolution anode material in hot concentrated KOH solution.[478, 721, 1515, 1643, 2326]

Several studies have been published investigating the parameters involved in the fabrication and performance of nickel oxygen cathodes for fuel cell use. These include a study of the effects of sintering conditions on the structure and strength of sintered porous nickel.[2482] In addition, the activity of activated carbon-supported nickel electrocatalysts for oxygen reduction in alkaline solution has been examined as a function of the activation method, texture, porosity, and surface area.[2285] Japanese workers claim to have developed a nickel fuel cell air electrode for use in alkaline solution that has almost the same activity as silver, with a service life in excess of 1200 h at 50 mA/cm^2.[1217]

Carbon

Carbon in its many forms is inexpensive, and it has sufficient electrical conductivity and kinetic chemical stability as an oxygen cathode up to about 200°C in acid electrolytes and to about 80°C in alkaline solutions.

However, in acid media the electrocatalytic activity of all forms of pure carbon for oxygen reduction is poor, even though it has been examined as an oxygen reduction catalyst in sulfuric acid[1122] and in a mixed sulfuric and nitric acid electrolyte.[2265] In the latter, it may only serve as a reduction surface for nitrate ion.

In alkaline electrolytes, however, the situation is significantly better, and finely divided active carbons[132,149,155,1002,1532,2526] and graphites[132,149,155,754] have been used alone as oxygen reduction catalysts in concentrated KOH solutions. For example, pure carbon high surface-area carbon black and activated carbon electrodes can be made with initial activities close to those of silver-containing electrodes.[155] A detailed study was conducted by Exxon Research and Engineering Co. using the ring-disk electrode technique to evaluate the use of carbon as an oxygen reduction catalyst in a variety of fuel cell electrolytes.[1968] Rate constants and polarization curve analyses have been reported for the oxygen reduction mechanism on soot-coated pyrographite disk electrodes over the pH range of 0.3~14.[85]

An investigation of the oxygen reduction kinetics in KOH on a range of carbon blacks, active carbons, and graphites examined the effects of pH, electrolyte concentration, oxygen partial pressure, and temperature.[155] At constant current density, the reaction was shown to be pH independent, and a reversible O_2/HO_2^- couple appeared not to be involved in the mechanism under the HO_2^- concentrations existing under normal reaction conditions. Under near open circuit conditions, however, there is evidence that the ionization of oxygen in alkaline solutions operates reversibly through the two-electron mechanism to produce HO_2^-,[2032] as earlier work by Berl had shown.[327] Thus, near equilibrium, the oxygen reduction reaction on carbons is like that reported on mercury,[222] whereas at high current densities it is truly under kinetic control. Under these conditions, the reaction rate constant is rather independent of the carbon type, and involves a second rate-determining electron transfer step, the first step being reduction of O_2 to O_2^-.[2032]

It has been found that surface species that are formed on the carbon surface under oxygen cathode conditions can exert important effects on carbon activity.[2479] Such surface groups may be adsorbed oxygen and the quinone/hydroquinone system. Similar quinone-like and hydroquinone-like surface oxides that affect the electrical resistance of the electrode also have been found to form during oxygen evolution on graphite in sulfuric acid solution,[1541] and it has been claimed that hydroxide-coated carbons are effective oxygen reduction catalysts in hydrazine fuel cells with 30% NaOH electrolyte.[1704] The presence of such surface species can be used to explain the observation that some types of active carbon have lowered oxygen reduction activities after anodic oxidation, which are attributed to changes in resistivity and structure, while other types appear unaffected.[2624]

Shell International reported a catalyst preparation method that utilizes such surface species.[1757] The method involved oxidation of a polycrystalline graphite-containing carbonaceous support to form acid or basic oxide surface groups. The oxidized carbon support was then impregnated with a solution containing cations or anions of an electrocatalytically active metal which will respectively form salts with the acid or basic oxide surface groups. This process is followed by a subsequent reduction to the metal.

Several studies have addressed the various parameters that influence the performance of practical oxygen electrodes made from carbon. An investigation of the method of deposition of Teflon wetproofing agent on active carbon for oxygen reduction in acid showed that PTFE exists in two different states, related to its particle size, which affects both electrode hydrophobicity and performance.[1319] The existence of two types of PTFE may help to explain the initial dependence of the electrochemical characteristics of wetproofed active carbon oxygen cathodes on their Teflon content.[249] This of course may depend on the type of PTFE used, which varies from manufacturer to manufacturer. The effect of Teflon loading and sintering on electrode performance has been studied.[1432, 2208, 2613, 2766]

The performance of carbon oxygen cathodes in alkaline fuel cells, like that of carbon-supported platinum cathodes in PAFCs, gradually shows an irreversible decline in activity as a function of time. However in the case of carbon-based electrodes in alkaline solution, the loss in performance seems to be more a problem of mechanical degradation of the carbon material than one of changes in its intrinsic electrocatalytic characteristics.[2626] It is known, for example, that the apparent oxygen reduction activity of some carbons is merely a direct function of the available carbon surface area,[155] suggesting that differences in wetting due to changing surface oxidation with time may affect the characteristics of the three-phase boundary. It also has been observed that the slow deterioration of PTFE-bonded two-layer carbon air electrodes in concentrated KOH occurs only when carbon dioxide is present in the air, and that the rate of decline in electrode activity increases as the KOH concentration increases.[2132] Furthermore, the rate of electrode ageing also is strongly affected by the relative humidity of the air and by the porosity of the electrode. This suggests a carbonate formation effect on the three-phase boundary.

A study of the optimization of the macrostructural parameters of two-layer wetproofed carbon fuel cell oxygen electrodes showed that polypropylene was the best matrix material to contain activated carbon catalyst.[913, 1220-1222] The original electrodes were made by hot-pressing the active layer (consisting of 18 mg active carbon/cm^2 plus 2 mg Teflon/cm^2) and the supporting layer (consisting of carbon black wetproofed with Teflon) into a Teflon matrix. These electrodes were able to operate in 7 M KOH on oxygen from the air at a current density of about 100 mA/cm^2 and on pure oxygen at about 300 mA/cm^2. At 50 mA/cm^2, the operating life was increased from around 700 h to greater than 3000 h by replacing the Teflon matrix with a polypropylene matrix.

Methods of preparing active carbon with 85% active surface area requiring a minimum use of binders, and of fabricating porous active carbon-coated Teflon oxygen cathodes are given in references 1563 and 2196, respectively. Two reviews of the use of carbon in fuel cell electrodes have been given by Kordesch.[1482,1485] Reference 1485, prepared for the U.S. Department of Energy, is especially comprehensive.

Pure carbon electrodes are unlikely to be much used in alkaline solutions in future, because low platinum loadings on carbon supports, and more recently pyrolyzed transition metal macrocyclics on carbon supports,[394,2259,2260] now show greater activity than the carbons alone by about 100 mV in concentrated KOH, and up to 250 mV in dilute KOH. However, past studies on carbons as catalysts, particularly from the viewpoint of electrode stability, are particularly important in determining carbon support preparations and surface treatments together with future electrode structures.

Perovskites

Much of the electrochemical irreversibility of the oxygen reduction reaction on conventional catalysts may be a consequence of their inability to adsorb oxygen in a single dissociative step. Such a step would involve two electrons, which would make it improbable unless it could conveniently take place as two successive rapid reactions.[168] It has been recognized since the late 1960s that a common feature of many of the best catalysts in alkaline electrolyte may be a high paramagnetic susceptibility, which can be related to the number of apparent unpaired d-band electrons per atom alluded to earlier in the discussion of the catalytic behavior of the noble metals. In one view of the effective catalysis of the oxygen reduction reaction, the paramagnetic electrons in the oxygen molecule should couple with those in the electrocatalyst to facilitate side-on adsorption.[816, 988] When oxygen molecules adsorb in this manner, the O-O bond may easily split on a surface with neighboring -OH and -O groups via a mechanism involving a chain of hydrogen bonds. This is the so-called pseudo-splitting mechanism suggested by U. R. Evans.[816]

Although the oxygen reduction reaction on noble metal electrocatalysts such as platinum usually occurs on the bare metal metal surface, it can also take place (though at lower rates) on an oxide-covered surface.[140,168,382,705,2281] It has been found that a number of mixed oxides are more stable in strongly alkaline solutions (e.g., 50% KOH) than simple oxides in the voltage region where oxygen reduction takes place.[2281] It has been shown that among these mixed oxides, perovskite electrocatalysts promote the desired side-on adsorption of the paramagnetic oxygen molecule in alkaline media, enabling dissociative adsorption to occur.[1761] Perovskite is the mineral $CaTiO_3$, which has given its name to the family of oxides with the same structure. This can be represented by the formula RMX_3, where R is an alkaline earth metal, M is a transition metal cation, and X is usually an oxygen ion. A material exhibiting the ideal perovskite structure consists of a simple cubic structure in which R forms the cubic lattice, M is at the body-centered position, and X is at the face-centered positions. In this configuration six X's surround the M in the lattice, and twelve X's surround the R. Conversely, each X is surrounded by two M's and four R's in a distorted octagonal structure. Perovskites having the ideal cubic structure are rare, and most exist as distorted cubic, monoclinic, rhombic, or tetragonal systems. Typical perovskites include $BaTiO_3$, $KNbO_3$, $BaSnO_3$, $CaZrO_3$, and $SrCeO_3$.

Of the various perovskite oxides investigated as oxygen reduction catalysts, strontium-doped lanthanum cobaltites such as $La_{0.5}Sr_{0.5}CoO_3$ seem to have exceptional properties. These materials are very stable in alkaline solution, even under anodic conditions, $La_{0.6}Sr_{0.4}CoO_3$ being more stable than $La_{0.8}Sr_{0.2}CoO_3$.[1719] Furthermore, studies have confirmed that oxygen appears to be dissociatively adsorbed side-on on these materials.[2504] In 45% KOH solution at room temperature, strontium-doped lanthanum cobaltites have been found to yield a steady and reversible open circuit voltage of 1.22 V that varies with the oxygen partial pressure according to the Nernst equation for a four-electron process.[2501, 2502, 2504] Work using lanthanum nickelate ($LaNiO_3$) has shown that on this material, oxygen reduces completely to hydroxyl with no peroxide formation.[1718] Studies of oxygen reduction on $Nd_{0.5}Sr_{0.5}CoO_3$ in 45% KOH indicate that the rate determining step of the reduction reaction is the adsorption of oxygen.[2773] The extent of oxygen coverage in 45% KOH has been found to be about 1% and to be independent of temperature over the temperature range of 25°C to 80°C for both strontium-doped lanthanum and niobium cobaltites.[2773] In addition to the above mixed oxides, $BaFe_{1-x}Co_xO_{3-a}$ (where $x = 0.1$~0.5, and $a \ll 1$) also has been reported to be useful as an oxygen reduction catalyst in alkaline fuel cells.[2404]

The mechanism of the oxygen reduction reaction on perovskite oxides seems to be dependent on a suitable combination of electronic and ionic conductivity, which is affected by factors such as oxygen partial pressure and transition metal cation. For example, studies employing Ni-Co-O catalysts have shown that a change in the oxygen ionization mechanism takes place during the transition from nickel oxides to nickel cobaltites, and that this mechanism change is brought about by the presence of the Co^{+3} ion.[1593,2498] In oxygen-deficient perovskite oxides, it has been found that p-type electronic conductivity is predominant at high oxygen partial pressures, whereas oxide ion conductivity predominates at low oxygen partial pressures.[483] Ionic conductivity is related to oxygen vacancy concentrations and the size of the dopant ions[483, 1718] as well as to the particle size distribution of the catalyst.[1445] At 76°C, the oxygen diffusion coefficient in $La_{0.5}Sr_{0.5}CoO_{3-y}$ was determined to be about 1.4×10^{-13} cm^2/s,[1445] while that in $Nd_{0.8}Sr_{0.2}CoO_3$ at 25°C is 1.4×10^{-11} cm^2/s.[1903] For $NdCoO_3$, it has been found that the resistivity at 25°C can be lowered by several orders of magnitude to minimum values in the 10^{-4}~10^{-1} Ω cm range by doping with strontium, calcium, or barium.[1903] For ferrites such as $La_{1-x}Sr_xFeO_{3-y}$, it has been found that the catalytic activity for oxygen reduction in alkaline media is related to the fraction of vacancies, and that the ferrites themselves undergo reduction concurrently with the reduction of oxygen.[479] The physicochemical and electrochemical properties of perovskite oxide catalysts have been reviewed. [2418]

Methods have been disclosed for the preparation of fuel cell electrodes using $Nd_{0.8}Sr_{0.2}CoO_3$, $Nd_{0.5}Sr_{0.5}CoO_3$, $Sm_{0.7}Ba_{0.1}Ca_{0.1}CoO_3$, and other perovskite oxide catalysts.[1528, 2524] In addition to being used as oxygen reduction catalysts, perovskites such as $Ba_2Mn RuO_6$,[2326] $La_{1-x}Sr_xCoO_3$,[1446, 1719] $La_{1-x}Sr_xMnO_3$,[1716] $La_{0.2}Sr_{0.8}Fe_{0.2}Co_{0.8}O_3$,[1720] $SrFeO_3$,[1717] and $LaNiO_3$ also been used as catalysts for oxygen evolution.

Although the low temperature performance of perovskite oxygen reduction catalysts such as strontium-doped lanthanum cobaltite is far lower than that obtainable with platinum catalysts, at temperatures above 170°C their performance starts to approach that of platinum. For this reason, mixed oxides of the perovskite family may ultimately provide a cheaper substitute for platinum oxygen reduction catalysts for higher temperature alkaline fuel cell systems. However, in acid electrolytes, perovskite oxide catalysts exhibit high polarization for oxygen reduction at current densities as low as only 0.1 µA/cm^2[2578] and are in any case unstable, rendering them unsuitable for use in acid fuel cells.

Spinels and Mixed Metal Oxides

In addition to the perovskites, another group of metal oxides, namely the spinel family, has also shown some promise as oxygen reduction catalysts in alkaline solutions. The spinels comprise a class of hard, variously colored minerals, consisting chiefly of mixed oxides of metals such as aluminum, magnesium, zinc, or iron. Some translucent varieties, such as the so-called ruby spinel, are used as gemstones. The formula for a so-called typical spinel can be represented by XY_2O_4. The spinel unit lattice contains eight XY_2O_4 units arranged in a fairly complex configuration in which the oxygen ions form face-centered cubic structures containing tetragonal oxygen cages that are partly occupied by the metals X and Y. Normal spinel-type compounds can be classified by their ionic packing factor, defined as the ratio between the cation radius to the anion interstitial radius. The electronic conductivity and the magnetic ordering of a spinel can be correlated with this parameter.[2195]

Examples of normal spinels are $MgAl_2O_4$, $FeAl_2O_4$, $CoAl_2O_4$, Na_2MoO_4, and Ag_2MoO_4. In addition to these normal spinels, another type of spinel structure exists, known as an inverse spinel. This can be represented by the formula $X(YX)O_4$, in which the X-positions in the normal spinel structure are occupied by half the Y's of the inverse structure, with the remainder of the Y's and all of the X's of the inverse structure occupying the Y positions in the normal structure. Examples of inverse spinels are $Fe^{III}(Fe^{II}Fe^{III})O_4$, $Fe^{II}(TiFe^{II})O_4$, $Cr(Fe^{II}Cr)O_4$, and $Zn(SnZn)O_4$. Other materials, here referred to as mixed oxides, fall into neither the spinel nor perovskite classification, although both spinels and perovskites can of course be classified as mixed oxides.

As with the perovskites, crystals having spinel structures usually display special conductivity and magnetic characteristics that account for their ability to act as oxygen reduction catalysts. The electronic conductivity, work function, and surface semiconductivity of a spinel or mixed oxide catalyst can be varied to yield a material displaying optimum oxygen reduction activity.[2496] The electronic conductivity, for example, of an active spinel-type mixed oxide fuel cell catalyst such as $Co_xFe_{3-x}O_n$ or $Co_xNi_{3-x}O_n$ ($x = 0.05$~3; $n = 3$~4), can be increased by doping it with boron.[1431] It has been reported that the electrochemical reduction of oxygen on cobalt, magnesium, cerium, and manganese cobaltite spinels in KOH solution operates via the intermediate formation of peroxide before final four-electron reduction to water.[2599] Reviews of the properties of spinel-type oxide electrodes,[2437] and of the role of conductive transition metal oxides,[2487] and of oxides in general[2696] in the oxygen reduction reaction, are available. Both simple and mixed oxide electrodes have been experimentally investigated for use as oxygen electrodes.[719]

Some of the spinels and mixed oxides that have been evaluated for use with oxygen reduction electrodes in concentrated KOH solution include those based on cobalt, magnesium, iron, copper, and aluminum,[2735] together with nickel-iron compounds.[720] In addition, Exxon Research and Engineering Co. reported that $Bi_2(Ru_{2-x}Bi_x)O_{7-y}$ is an effective oxygen reduction electrocatalyst in 3M KOH at 75°C.[1168] Similarly, E.I. DuPont de

Nemours and Co. reported that delafossite-type oxides such as $PtCoO_2$, $PdCoO_2$, $PdRhO_2$, and $PdCrO_2$ are good catalysts for oxygen reduction in NaOH, and that the oxygen activity of these oxides correlates with that of the noble metal cation in the metallic state, platinum being more active than palladium.[566]

In addition to the work done with these materials in alkaline solutions, some investigations also have been made of their utility for acid systems. For example, the U. S. National Bureau of Standards carried out a number of potentiodynamic and galvanostatic studies on some of the more promising mixed oxides to determine if they could be used as oxygen reduction electrocatalysts in phosphoric acid fuel cells.[329] The results showed that barium ruthenate and the systems Ti-Ta-O, V-Nb-O, V-Ta-O, and Ce-Ta-O were stable in hot phosphoric acid at temperatures up to about 150°C. Soviet workers have investigated a number of simple and complex oxides of cobalt and/or nickel for oxygen reduction in acid media, as well as the effects of adsorbed oxygen on the reduction rate, and have reported that the electrodes containing cobalt possessed the highest activity.[1394]

Several authors have reported methods of preparing spinel and metal oxide electrocatalysts and electrodes. Japanese workers have described the preparation of sintered spinel-type electrodes using $Cu_xFe_{3-x}O_4$ or $Zn_xFe_{3-x}O_4$ ($x = 0.005~0.4$) with resistivities of about 0.044 Ω cm.[2634] Methods also have been reported for the preparation of oxygen cathodes on metallic screens with compressed mixtures of 85~95% metal oxide and 5~15% of conducting powdered material such as silver, nickel, cobalt, or acetylene black, plus a binder.[1711] Westinghouse Electric Corporation has patented an air electrode formulation that makes use of $CuWO_4$, $NiWO_4$, and/or $CoWO_4$, in addition to other materials, as oxygen reduction catalysts.[528] The effects of catalyst preparation parameters on the performance of simple and complex oxide oxygen reduction electrocatalysts have also been investigated,[1394] and the preparation of oxygen reduction catalysts of the mixed oxide type for use in alkaline solutions has been reviewed.[1009]

As in the case of the perovskite oxides, the spinels and other mixed oxide catalysts have also been used as oxygen evolution catalysts. Those which have been used in alkaline solutions for this purpose include $NiCo_2O_4$,[842, 1304, 1305, 1515, 2507, 2326] Co_3O_4,[1515] transition metal oxides in general,[1225] and mixed nickel-lanthanide group metal oxides.[864] Those used in acid solutions (usually sulfuric acid) have included transition metal oxides;[1224] mixtures of tin, antimony, and manganese oxides;[1600] and mixed oxides of ruthenium (especially Ru-Ir-Ta).[2328]

As is the case for many other oxygen reduction electrocatalysts, the performance of spinels and mixed oxides tends to decrease with time. Studies of Co_3O_4 spinels in 1~12 M KOH indicate that the decrease in oxygen reduction activity can be correlated with changes in the crystal structure of the surface layer of the electrodes.[780] Identical observations have been made for other cobaltites, such as $CdCo_2O_4$, $MgCo_2O_4$, $MnCo_2O_4$, and $NiCo_2O_4$.[2436] Similarly, long-term (5000 h) studies of the performance of cathodically polarized hydrophobic air electrodes using nickel-cobalt spinel catalysts have shown that the reduction in electrode activity over the first 1000 h can be correlated with an increase in the porosity of the active surface layer towards liquid, and that after an initial decline, a period of stable performance usually is observed.[2497] Hydration products of the spinel oxides do not appear to exert any noticeable effects on electrode performance stability. In addition, part of the deterioration can also be attributed to excursions outside of the optimum voltage domain. For example, $NiCo_2O_4$ is stable in 5M KOH at potentials greater than 0.75 V, but at 0.50 V the spinel structure is completely destroyed, forming compounds such as Co_2O_3, NiO, CoO, $CoO.H_2O$, and NiOOH, none of which is active for oxygen reduction.[2505] An extensive review of a large number of candidate mixed oxide compounds has indicated that, at the present time, none of these catalysts simultaneously displays sufficient electrical conductivity, resistance to anodic dissolution, and catalytic activity to be a viable oxygen reduction catalyst for use in aqueous electrolyte fuel cells.[1009]

Tungsten Carbide

The U.S. Bureau of Mines evaluated a large number of materials for use as oxygen reduction catalysts in 0.5 M sulfuric acid.[568] The purpose of the investigation was to search for an abundant, low cost substitute for platinum or at least to increase the activity of platinum already in use. The materials studied included carbides, silicides, phosphides, borides, nitrides, oxides, and various metals. The results of the study showed that the oxygen reduction activity of tungsten carbide (WC), although not as high as that of platinum, could be improved by doping with platinum, by varying the stoichiometry, or by sputtering a thin layer of platinum on the surface of the tungsten carbide. Tungsten carbide also has been used by Westinghouse Electric Corporation in conjunction with cobalt, silver, and acetylene black as an active material for porous air electrodes in nickel fiber plaque substrates.[611] The primary objective of this work was the development of a rechargable (bifunctional) air electrode for use in iron-air batteries.

Tungsten Bronzes

The so-called bronzes comprise a series of homogeneous, non-stoichiometric electronically-conducting compounds of the general formula M_xTO_3, where M is an alkali or alkaline earth metal, T is a transition metal, and x varies from 0 to 1.[569] The tungsten bronzes, especially sodium tungsten bronze (Na_xWO_3), have been reported to be effective electrocatalysts for the oxygen reduction reaction. Numerous reports have described various methods of preparing tungsten bronzes[246, 346, 584, 663, 1227, 2077, 2273, 2319] as well as their physical properties.[2073, 2685] Alkali metal tungsten bronzes can be synthesized by the electrolysis of fused isopolytungstates at temperatures of 600~900°C[2319] or by the cathodic reduction of melts of salts such as Na_2WO_4–Li_2WO_4–WO_3.[246] Sodium vanadium bronzes can be made by reaction in evacuated ampules at 600~640°C for about 200 h via the reaction

$$xNaVO_3 + xVO_2 + (1-x)V_2O_5 \rightarrow Na_xV_2O_5 \qquad (0.1 \leq x \leq 0.5)^{[246]}$$

If sodium tungsten bronze is treated anodically, it is possible to form one of a range of hydrogen tungsten bronzes, H_xWO_3.[598, 1136, 2664] Depending on the potentials employed, either semiconducting or essentially metallic phases of the hydrogen tungsten bronze can be produced.[2664] Hydrogen tungsten bronzes can also be prepared from oxidized tungsten electrodes or from WO_3 by cathodic treatment in acid electrolyte at potentials outside the range of hydrogen evolution.[598, 1136] The chemical potential of hydrogen (μ_H) in H_xWO_3 exposed to sulfuric acid solutions has been determined as a function of x, from $x = 0.002$ to $x = 0.5$.[681]

The tungsten bronze family has excellent stability in acid solutions, even at oxygen evolution potentials, but is rather less stable in alkaline media.[1945, 2320, 2321] Tungsten bronzes doped with cerium, such as $Na_xCs_yWO_3$, which can be made from ternary Na_2WO_4–Cs_2WO_4–WO_3 melts, are also resistant to corrosive environments.[2318]

The tungsten bronzes have been claimed to have activities comparable to that of platinum under certain conditions in acid media.[1945] Values of 10^{-14} A/cm^2 and 4×10^{-5} A/cm^2 have been reported for the exchange and limiting current densities, respectively, of the oxygen reduction reaction on crystals of pure $Na_{0.7}WO_3$.[2663] Sodium tungsten bronzes[707, 1749, 1749A, 2079, 2246] and cobalt tungsten bronzes[2079] have been examined as oxygen reduction electrocatalysts in sulfuric acid. Platinum-doped sodium tungsten bronze,[1749] cerium tungsten bronze,[2578] and ytterbium tungsten bronze[2578] have been similarly examined in perchloric acid. Sodium tungsten bronze also has been tested as a potential oxygen reduction catalyst in 0.01~5 M NaOH solution.[2248]

The catalytic properties of the tungsten bronzes may be attributable to their surface properties. Although bulk bronzes have a metallic nature, studies have indicated that their surfaces act as semiconductors in the presence of electrolyte, probably as a result of the leaching of metal ions from the surface.[1749A] Such selective leaching has been observed to take place both in acid[1748] and alkaline[1064, 2248] solutions. For example, Na_xWO_3 (0.4 < x < 1), has been shown to exhibit Na-deficient epitaxial surface films that markedly influence its catalytic properties.[204] These surface films can be formed in acid, alkaline, and neutral solutions by anodic oxidation of a tungsten bronze electrode.[204, 2320, 2321] It would appear that the resulting surface is essentially pure WO_3.[391, 885, 1752, 2079]

Work undertaken at Keio University in Japan has shown that the anodic oxidation of sodium tungsten bronze in acid or neutral solutions causes the formation of an n-type semiconducting surface layer having a specific resistivity of 10^{-5}~10^{-4} Ω cm. This results from preferential dissolution of the sodium from the bronze surface, during which the bronze surface is transformed from a cubic to a tetragonal crystal structure.[2320, 2321] Cesium-doped sodium tungsten bronzes also display electron n-type semiconductive properties.[2318] For the sodium tungsten bronzes studied at Keio University, photochemical measurements showed that at the surface the band gap energy was 2.8 eV, and that the flatband potential varied with pH, being equal to 0.73–0.06 pH (V/SHE). The conduction band edge was situated between two energy positions, corresponding to 0.77 V and 1.06 V. Although the basic properties of the semiconducting surface layer were similar to those of the bulk oxide, the stability of the surface layer in alkaline solutions was less. The Japanese workers reported that single crystals of tetragonal $Na_{0.07}WO_3$ had an electrical conductivity of 0.11 $Ω^{-1}$ cm^{-1} and an activation energy for conduction equal to 0.025 eV.

Partial molar thermodynamic data have been obtained for sodium tungsten bronzes, and the thermodynamic and transport properties of these materials have been discussed in terms of available structural data.[2339] Electronic conductivity values as high as 2.5×10^4 $Ω^{-1}$ cm^{-1} have been reported for some tungsten bronzes.[2578] A proposed model for electron transport in sodium tungsten bronze is based on a two-phase model consisting of clusters of metallic and nonmetallic regions,[2665] but this has been disputed elsewhere.[569] Other workers have used nuclear magnetic resonance measurements to explain the conduction process in the low-mobility region of sodium tungsten bronze (0.22 < x < 0.57), where a metal-insulator transition may occur.[2511] The chemical diffusion coefficient for sodium in sodium tungsten bronze has been reported to be less than 10^{-12} cm^2/s, so that ionic conductivity is much less than electronic.[2339]

In the late 1960s, there were glowing reports on the efficacy of the tungsten bronzes for the oxygen reduction reaction.[2247] However, a great deal of conflicting evidence regarding their usefulness exists. For example, some work has indicated that various pure tungsten and niobium bronzes exhibit negligible oxygen reducing activity in acidic media ($HClO_4$), and only minor activity in alkaline media.[2578] Other reports indicate that the oxygen reduction activity of these materials can be attributed largely to inclusions of platinum impurities at levels in the order of 200–400 ppm,[868, 1748, 1749, 1749A, 2248] which may have resulted from the use of platinum counterelectrodes in the molten salt techniques used for growing single crystals. It is known that sodium tungsten bronzes are readily contaminated with platinum through the use of platinum counterelectrodes or by pre-electrolyzing solutions using platinum electrodes.[868, 2663] The amount of platinum inclusion that is required to produce platinum-like behavior is quite small, and it has been shown that doping pure sodium tungsten bronze with about 800 ppm of platinum will increase its activity to about the same level as that obtained with pure platinum.[2663] However, there is some evidence to show that platinum-doped tungsten bronzes are not especially effective oxygen reduction catalysts,[2080] particularly in concentrated phosphoric acid, which may form phosphotungstates giving surface deactivation.[146]

The adsorption of platinum on the surface of sodium tungsten bronze has been found to change the Tafel slope for the oxygen reduction reaction from a value close to $-4RT/F$ to $-2RT/F$.[886] It is possible that the adsorbed platinum forms surface states that pin the Fermi level close to the surface, which becomes completely metallized. In such cases, the changes in potential difference will occur entirely across the Helmholz layer, which will explain the $-2RT/F$ Tafel slope then observed. Even so, this explanation will not account for the $-4RT/F$ seen on undoped material; slopes as great as this are not accounted for by the classical electron transfer theory either on metallic or semiconducting substrates.[168]

An additional problem with the tungsten bronzes is that the catalytic effects seem to be present only in the monocrystalline material, and are lost when the crystals are ground into powder.[1945] This introduces many practical limitations with regard to their use in porous fuel cell electrodes.

Metal Chelates

The fact that certain metal chelates (transition metal macrocyclics) are able to catalyze the oxidation of organic compounds by oxygen has been recognized for many years.[1312, 1976] A possible mechanism is as follows:

$$\text{chelate} + O_2 \rightarrow \text{chelate-}O_2$$
$$\text{chelate-}O_2 + RH \rightarrow \text{chelate} + R^* + HO_2^*$$

where RH is a proton source and R^*, HO_2^* are radicals. In an electrode, the last step is more likely to involve electron transfer rather than radical formation. In the first step, electron transfer from the metal chelate to the oxygen molecule takes place, resulting in a weakening of the O-O bond, thereby increasing the rate of the following steps, and possibly allowing

Cathodic Electrocatalysis 399

complete O-O bond dissociation early in the reaction sequence. Overall, however, the mechanism of oxygen reduction on chelates is somewhat ill-defined, and it is generally accepted that the first step, preceding dioxygen adsorption, is a reduction of the central ion in the chelate to give a lower valency state. This is then followed by bonding of oxygen, and subsequent electron transfer to give the original valence state of the central ion and superoxide (or peroxide after involvement of a proton) as products.[168] As would be expected for a mechanism of this type, there certainly appears to be a relationship between the redox potential of the central ion and catalytic activity.[145, 394] However, for the most effective catalysts (see following) the initial redox process is not rate-determining and the rate-limiting step occurs rather late in the reaction sequence and may involve O-O bond-breaking.[145]

Because many of the chelates are relatively stable in acids, there has been some interest in exploring their use as inexpensive non-noble substitutes for platinum oxygen reduction catalysts in acid fuel cells, although, as the following shows, more success has been achieved in alkaline systems.[394] The activities of various types of chelates for the oxygen reduction reaction in fuel cells have been reviewed,[699, 2556] and a model based on molecular orbital theory has been proposed to explain the electrocatalysis of oxygen reduction by transition metal chelates.[1274]

While many chelates have been examined,[394] attention has naturally been focused on those showing the highest stability under oxygen electrode conditions. A common feature of these compounds is the MN_4 structure, where M is the central metal atom, which is coordinated in a planar configuration to four nitrogen atoms, which are part of a stable cyclic aromatic structure. The structures of the compounds discussed in the following are given in Fig. 12-3.

One of the most thoroughly examined series of macrocyclics has been the phthalocyanines. Those studied have included complexes of iron, cobalt, nickel, copper, magnesium, and hydrogen.[470, 841, 2078] Methods for their synthesis and testing have been reported.[1772, 2750] Iron, nickel, cobalt, copper, and manganese phthalocyanine complexes have been examined as catalysts in air electrodes.[2776] In addition to their use as oxygen reduction catalysts, phthalocyanines also have been used as anode catalysts in redox fuel cells.[331]

The decreasing order of oxygen reduction activity of high surface-area carbon-supported metal phthalocyanines in both 4 M sulfuric acid and in 6 M KOH has been reported to be Fe > Co ≈ Ru > Mn > Pd ≈ Pt > Zn.[2575] Of the various phthalocyanines evaluated for the reduction of oxygen, the complexes with cobalt and copper appear to be the most stable, while those with iron and cobalt seem to have the best combination of activity and stability.[1064, 1308]

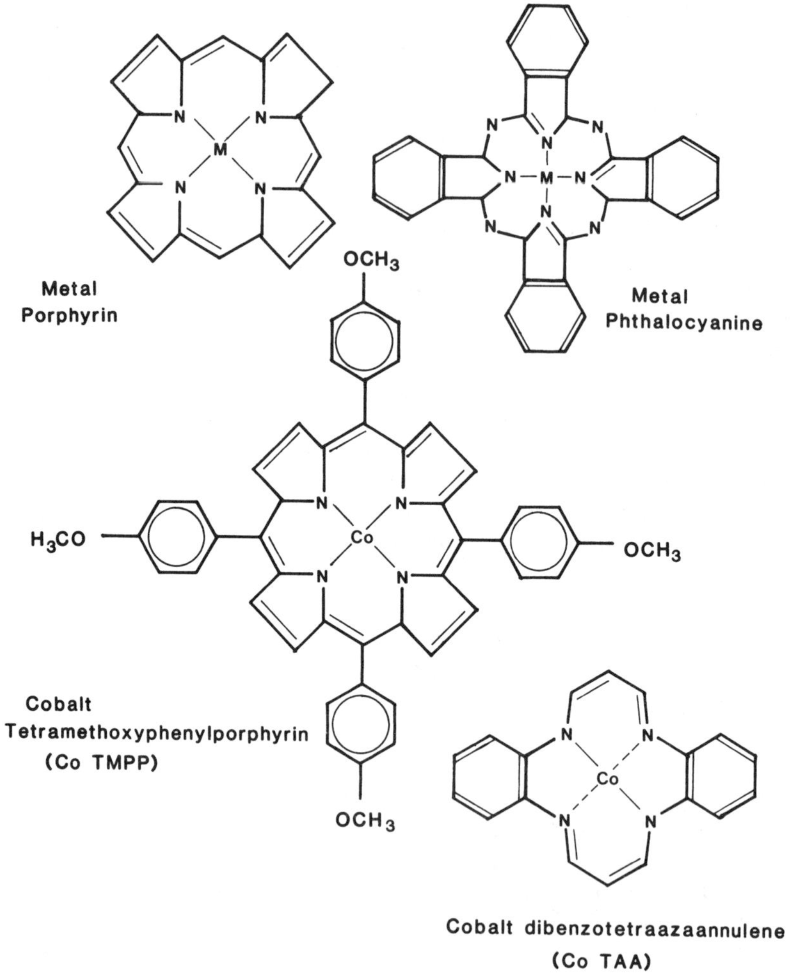

Metal Porphyrin

Metal Phthalocyanine

Cobalt Tetramethoxyphenylporphyrin (Co TMPP)

Cobalt dibenzotetraazaannulene (Co TAA)

Figure 12–3: Some macrocyclic catalyst structures. M = transition metal.

The iron phthalocyanines have generally been the favored macrocyclics for the catalysis of the oxygen reduction reaction. It has been reported that the room temperature performance of carbon-Teflon oxygen reduction electrodes containing polymeric iron phthalocyanine in 3 M sulfuric acid is 20~80 mA/cm^2 at 0.85~0.65 V, with less than 100 mV degradation over 1000 h at 20 mA/cm^2.[1772] In acid solutions, the initial step of the oxygen reduction reaction on iron phthalocyanine complexes is believed to be the chemisorption of the oxygen molecule by the central transition metal atom.[700]

Iron phthalocyaninetetrasulfonate (Fe-PTS) and the corresponding cobalt compound (Co-PTS) have received particular study. Both are water soluble transition metal macrocyclics that strongly adsorb on noble metals and on graphite in both acidic and alkaline solutions.[2763] When adsorbed at monolayer levels on graphite, both complexes have a pronounced catalytic effect on the reduction of oxygen in both acidic and alkaline media.[2788–2790] The iron complex apparently promotes oxygen reduction to water via the four-electron pathway, whereas the cobalt complex operates through the two-electron mechanism, giving hydrogen peroxide. The solubility behavior of iron phthalocyanine polymers and their interactions with carbon substrates have been investigated using isotope technical methods and Mossbauer spectroscopy.[1771]

In addition to applications in sulfuric acid, wetproofed carbon electrodes containing 5~20% iron phthalocyanine monomer also have been investigated as oxygen cathodes in phosphoric acid solutions, with emphasis on the effects of changing the phosphoric acid concentration, the iron phthalocyanine concentration, and the partial pressure of the oxygen.[1948]

Another phthalocyanine complex that has shown some potential for use in phosphoric acid is the cobalt chelate of 5,14-dihydro-5,9,14,18-dibenzotetraaza[14]annulene (Co-TAA).[687,2643] A method for the preparation and activation of Co-TAA on acetylene black or active carbon for use as a fuel cell oxygen reduction catalyst has been reported.[2799] Eco, Inc. (Cambridge, Massachussetts) has demonstrated that Co-TAA can be covalently linked to the surface of Vulcan XC-72 carbon to yield a stable and active oxygen reduction catalyst.[2643] A Teflon oxygen cathode containing Vulcan-supported Co-TAA matrix was tested at 135°C in 85% phosphoric acid at 0.65 V for over four weeks with negligible loss in catalytic activity. The observed activity of this cathode was claimed to be equivalent to that of a similarly prepared 10% platinum-catalyzed cathode. Eco has claimed that not only is this catalyst considerably cheaper than platinum, but also is less prone to catalyst poisoning. The mechanism and kinetics of the oxygen reduction reaction in both acid and alkaline solution on platinum, gold, silver, and carbon compared with the same materials coated with cobalt dibenzotetraazaannulene have been examined.[1274]

In alkaline solutions a number of phthalocyanine complexes have shown promise. Dow Chemical reported that 10% metal phthalocyanines incorporated into a porous nickel substrate show good performance as oxygen reduction electrodes in NaOH solution.[732] Japanese workers have reported excellent performance in KOH solutions for sintered wetproofed porous nickel gas diffusion electrodes containing a dual catalyst consisting of a mixture of palladium and cobalt phthalocyanine.[44] It has been claimed that cobalt phthalocyanine on glassy carbon exhibits a much longer life for oxygen reduction in 6 M KOH when it is co-deposited with polystyrene.[1397] It has been reported that iron naphthalocyanine supported on carbon black is a good oxygen reduction catalyst in alkaline solution and also appears to have good stability.[1668] Polymeric iron phthalocyanine deposited on acetylene black by precipitation from a concentrated sulfuric acid solution has been reported to have good activity for oxygen reduction in neutral solutions (50 mA/cm^2 at –196 mV polarization).[1347]

A number of studies have dealt with the mechanism of the oxygen reduction reaction on phthalocyanine complexes, including the mechanisms on carbon-black and graphite-supported materials in acid media.[1276] For cobalt phthalocyanine in sulfuric acid, the first electron transfer seems to be rate determining, whereas in KOH the second electron transfer seems to be the rate-determining step.[2575] For polymeric iron phthalocyanines in alkaline media, however, the rate-determining step appears to be a chemical dissociation of the O-O bond.[145] The rotating disk technique has been used to study the effects of the nature of the central metal ion (iron, manganese, cobalt) of the phthalocyanine complex on the mechanism and kinetics of the reduction of oxygen to peroxide over the pH range 0.3~14.2,[2069] and voltammetry has been employed to investigate the ability of various metal phthalocyanines to decompose H_2O_2 to H_2O in sulfuric acid solution.[1437]

The electrocatalytic activity and stability of the metal phthalocyanines depend on two requirements: electronic conductivity in the solid phase, and reversibility of surface oxygen adsorption.[151,157] These conditions are fulfilled by means of a particular ligand field structure involving a multi-spin state configuration of the central metal ion. Studies of the oxygen reduction reaction in 2 M sulfuric acid using a number of iron phthalocyanine complexes deposited on carbon blacks and Cabot Vulcan-6 showed that the electrocatalytic activity of the complexes appears to increase as their electronic conductivities decrease.[700,1509] Similar studies have given identical results.[1509] Studies conducted in France on monomeric, dimeric, and polymeric forms of iron phthalocyanine have shown that the spin and oxidation state of the central iron atom (determined by Mossbauer spectroscopy), the electron delocalization in the organic ligand, and the character of the Fe-O_2 bond are all important factors in the catalytic process of oxygen reduction on these materials.[1690] The organic ligand structure may be regarded as having a fine-tuning effect on these properties.

Similar work carried out by workers at the U.S. National Bureau of Standards using monomeric and polymeric iron phthalocyanines has confirmed that both the degree of polymerization of the phthalocyanine and the spin configuration of the central iron atom play an important role in catalytic activity.[145] Where charge transfer is the rate-controlling step in the oxygen reduction reaction, electrocatalytic activity of the phthalocyanines appears to increase with the ease of electron availablity from the material under reductive conditions. This is certainly associated with their electronic conductivity characteristics, which seem to be related to charge-transfer complex formation. For example, reaction with molecular dioxygen can produce charge transfer complexes if the material is in the correct spin state.[145] As a further example, the electronic conductivity of phthalocyanine complexes can increase by 7~8 orders of magnitude when they are exposed to halogen vapors.[2743] Similar changes occur upon exposure to a variety of gases such as nitrogen dioxide, nitrogen monoxide, sulfur dioxide, oxygen, nitrogen, and carbon monoxide.[2161] The presence of a metal greatly increases the sensitivity of the compounds to these effects, which can be explained in terms of a conduction model in which gas absorption produces acceptor levels in p-type phthalocyanines, giving rise to increases in the carrier density.[2161] Conductivity anisotropy observed in copper and magnesium phthalocyanines may be attributed to the different mobilities of the charge carriers in the different layers, and appears to be strongly influenced by sample purity and substrate temperature.[841]

Electrochemical studies of thin layer deposits of hydrogen (i.e., metal-free), zinc, and nickel phthalocyanine indicate that the electrochemical behavior of these compounds can be explained by considering them as relatively well-behaved p-type semiconductor electrodes.[825] Basic studies of the electronic properties of ten phthalocyanine complexes of copper, chromium, and manganese have been reported.[1669]

The electrocatalytic activity of transition metal phthalocyanine complexes for oxygen reduction has also been found to be light sensitive. A study of the effects of photo-stimulation showed that photo-induced growth of the electrode surface takes place, which might be explained by the passage of photo-stimulated injection currents through the porous electrode surface.[2267] Investigation of the activity of thin film deposits of phthalocyanine showed that the dark current increases and the photo current decreases as the work function of the substrate decreases.[1219] Reflectance spectroscopy has been used to examine the optical properties of iron and cobalt phthalocyaninetetrasulfonate (Fe and CoPTS),[2763] and the activation energy of the dark electronic conductivity of osmium phthalocyanine has been determined.[1464]

The porphyrins are a second major group of transition metal macrocyclics that show promise as oxygen reduction catalysts. They have been shown to be stable and catalytically active in acid electrolytes.[2184, 2189] For example, it has been reported that Co(II) tetra(p-methoxyphenyl) porphyrin (CoTMPP) is more active than iron phthalocyanine, one of the most effective oxygen reduction catalysts of this group.[2189] Other porphyrins that have shown oxygen reduction activity, particularly in acid solution, belong to the group of the tetraarylporphyrins.[299] Dimeric porphyrins such as dicobalt porphyrin dimers on pyrolytic graphite also show activity for oxygen reduction, although the reaction proceeds with significant amounts of peroxide formation in parallel with the four-electron process to water.[653] In the latter work, the original aim was to be able to prepare dicobalt structures in which dioxygen would simultaneously bond to both central ions, leading to automatic bond-breaking early in the reaction sequence and a reversible oxygen electrode.

Studies at Tohoku University in Japan have indicated that Fe(III) and Co(III) tetra-o-aminophenylporphyrins catalyze oxygen reduction in acid media, the iron complex being more effective than the cobalt complex.[1438] Shell International Research has shown that iron complexes of tetra(p-substituted)phenylporphyrin can be made more active by heating in an inert atmosphere at 300~1000°C for $\geq$ 20 minutes.[2576] For example, the activities of iron tetra(p-isopropyl)phenylporphyrin and iron tetra(p-methoxy)phenylporphyrin (FeTMPP) were improved by heating for 1 h in nitrogen at 700°C and 800°C, respectively. Several reviews of the electrochemistry of the porphyrins and metallo-porphyrins have been published.[510, 722, 849, 2438]

Like that of the phthalocyanines, the electrocatalytic activity of the porphyrins depends on their ability to adsorb oxygen reversibly on the surface and on their electronic conductivity in the solid phase.[151, 157] The ionic state of a metalloporphyrin can affect its behavior as an oxygen reduction catalyst.[1374] In alkaline electrolyte, for example, the anionic form of the chelate predominates, which should favor electron localization on the adsorbed molecule and O-O bond rupture with reduction to water. However, in acidic media, the neutral form of the chelate is the favored species.

Thus, in sulfuric acid, the HSO_4^- ion is adsorbed and coordinates to Co(II), Fe(II), and Mn(II), resulting in molecular oxygen adsorption in the side position to form a σ-bond with subsequent reduction via the two-electron path to hydrogen peroxide. This mechanism has been confirmed by work at Ohio State University, which showed that in acid solution, oxygen reduction on both Co(III) tetrapyridylporphyrin[337] and on iron tetra(N-methylpyridyl)porphyrin[336] yields hydrogen peroxide. Related research using Fe(III) tetrakis(o-aminophenyl)porphyrin and Fe(III) tetrakis[N-(2-hydroxyethyl)pyridiniumyl]porphyrin on glassy carbon indicated that the reduction of oxygen to hydrogen peroxide or to water can be controlled by changing the surface coverage of the porphyrin.[338] Based on the results obtained for dissolved, adsorbed, and polymer-coated porphyrins, the above authors have proposed a general mechanism for the electrocatalysis of oxygen on porphyrins.[338] The Battelle Institute in Geneva has conducted a study of the oxygen reduction reaction on iron, cobalt, nickel, manganese, copper, and zinc tetraphenylporphyrin (TPP) on pyrographite, gold, platinum, and glassy carbon surfaces in both KOH and sulfuric acid.[302]

Although phthalocyanine macrocyclics are stable in acid media at low temperatures, they tend to decompose when the temperature is raised.[687, 2643] For example, in sulfuric acid, iron phthalocyanine shows signs of decomposition at temperatures in excess of about 50°C.[1510] CoTAA cannot be used at temperatures greater than 135°C, which prevents its use in utility PAFCs. In any case, it is about 25 mV less active than a typical supported platinum electrode at 10 mA/cm^2, and becomes appreciably less active at higher current densities (see following). Porphyrins also have a tendency to dissolve in the concentrated phosphoric acid electrolyte required for utility fuel cells.[687] It would therefore appear that the prognosis for the use of pure chelate macrocyclics in acid fuel cells is not very good at the present time,[177, 1911] though they may find application as oxygen reduction catalysts in alkaline fuel cells, where their activity is higher and the environment is less severe.

A major problem in making practical fuel cell electrodes using pure macrocyclics deposited on carbon has been the fact that their activity tends to fall off rather rapidly at current densities on the order of 100 mA/cm^2 in what appears to be a premature limiting current. This has been alluded to in the case of CoTAA in acid solution, and it is also shown in the case of the more active FeTMPP (tetramethoxyphenylporphyrin).[177] In alkaline solution, iron phthalocyanine electrodes may show excellent activity at low current density, but are decidedly inferior to even pure carbon electrodes at the current densities of 200~250 mA/cm^2 expected in practical fuel cells.[145] Whether the problem is a result of electrode structural effects, or can be attributed to electronic resistance (due for example to poor particle-to-particle contact) is not known.

Some attempts to improve electrode structure led to the discovery that cobalt-containing macrocyclics, in particular CoTMPP, gave improved activity after heating the carbon-supported material to high temperatures, even to 800°C.[1432, 2208, 2613, 2766] After such pyrolysis, the organic structure of the macrocyclic can be expected to have disappeared, and the cobalt should be present in the form of oxide. Whether or not the N_4 group is still present is controversial: It has been claimed that no trace of the typical Mossbauer characteristics of FeN_4 exist in pyrolyzed iron macrocyclics.[2259, 2260] However, recent work appears to imply that the presence of nitrogen is important for catalytic activity, since carbon blacks containing nitrogen (e.g., obtained by pyrolysis of nitriles or pyrrole) can be treated with a cobalt salt to produce a material whose catalytic activity is similar to that of pyrolyzed carbon-supported CoTMPP.[2303, 2765]

While the precise nature of this new family of catalysts derived from the transition metal macrocyclics is unknown at present, their somewhat serendipitous development has led to a genuine breakthrough in the electrocatalysis of oxygen reduction in alkaline solution. The activity of such electrodes is higher than that of platinum, and they can be made so as to show no fall-off in activity at high current density. Indeed, in their latest version, they can operate at almost 1 A/cm^2 on air at 1 atm pressure in 6M KOH at 50°C, or at 3~4 A/cm^2 on pure oxygen, as shown in Fig. 12-4. Their stability in alkaline solution is good, and it appears to be improved by the fact that demetallation is reversible, so that a reservoir of metal oxide in the electrode will maintain activity via a dissolution-precipitation mechanism in the electrolyte.[2303,2765] A life of several thousand hours in alkaline media seems now assured using a low-cost technology and readily available materials. It is not impossible to foresee such electrodes in future automobile alkaline fuel cells operating on pure hydrogen or ammonia, without the restraint of the problem of platinum supply.

Even more remarkable are the tentative results that have been obtained in acid media. While the stability of what appears to be a carbon-dispersed transition metal oxide in alkaline electrolyte is thermodynamically possible, it is totally unexpected in acid media at relatively high temperatures (e.g., 100°C in phosphoric acid). However, pyrolyzed carbon-supported FeTMPP electrodes do have acceptable stability up to this temperature, although unfortunately their stability, which seems to be potential-independent, diminishes rapidly as the temperature is further raised.[335] Their activity, which is compared with that of typical platinum and CoTAA electrodes[177] in Fig. 12-5 in H$_3$PO$_4$ at 100°C, is almost 100 mV higher than

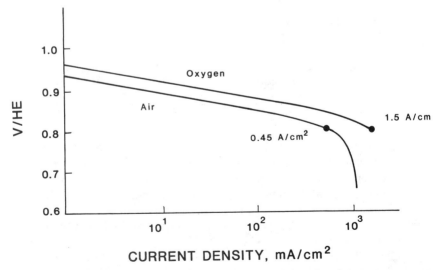

Figure 12-4: Oxygen/air cathode voltage as a function of log current density for pyrolyzed Cobalt TAA electrodes on high-surface-area carbon in 6N alkaline electrolyte at 60°C.

Cathodic Electrocatalysis

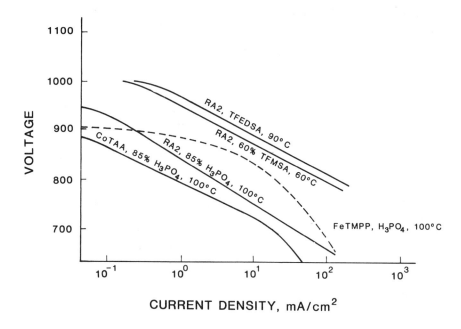

Figure 12–5: Comparison between commercial low-loading Pt electrodes (0.3 mg/cm² Pt on Vulcan XC-72, Prototech, Inc., RA2) and CoTAA, FeTMPP electrodes for oxygen (1 atm) reduction in 85% H_3PO_4 at 100°C. The upper curves show the same electrodes in tetrafluoroethane disulfonic acid (TFEDSA) at 90°C, and in 60% trifluoromethane sulfonic acid (TFMSA) at 60°C.

that of a low-loading platinum electrode at 10 mA/cm². However, they do show the same rapid fall-off in performance at higher current densities. In recent work, UTC claims to have shown that this is not due to an effect of electrode structure, but to "a resistance effect in the catalyst layer on the carbon."[335] This, however, remains controversial. While no promises in regard to the future progress of this new technology can be made, further knowledge of the type of pyrolyzed macrocyclic compounds formed on carbon surfaces may allow great progress in oxygen electrode electrocatalysis in acid electrolyte.

Miscellaneous Electrocatalysts

In addition to the above, numerous other materials have been examined as oxygen reduction catalysts. Although none of these materials has been particularly successful, it is instructive to briefly list some of them to indicate the general breadth of the search for the ideal fuel cell oxygen reduction catalyst.

The reduction of oxygen on single crystal titanium dioxide (rutile, TiO_2) has been investigated in a number of electrolytes using both cyclic voltammetry and the rotating ring-disk electrode.[1978] The reaction mechanism was found to proceed via a surface species at potentials positive to the flatband potential in the electrolytes used. Most of the oxygen reacted via a four-electron pathway to water, with a small percentage proceeding via a two-electron pathway to hydrogen peroxide. Titanium dioxide is widely used as an oxygen evolution catalyst in both acid and alkaline solutions.[517, 1474, 2194, 2230] Its use as an electrocatalyst has been reviewed.[1569]

Lead has been evaluated as an oxygen reduction cathode material in 4.25 M sulfuric acid in an investigation into its possible use in sealed lead-acid storage batteries.[1389] Potentiostatic switch-on curves for a smooth lead-disk electrode were plotted in the oxygen reduction potential region at different electrode rotation rates in air, oxygen, and argon atmospheres. The use of copper and brass as oxygen reduction electrodes for use in fuel cells was studied using the ring-disk electrode technique to generate forward and reverse polarization curves in neutral electrolytes containing chloride, sulfate, and nitrate ions.[241, 242]

In a British patent, Exxon Research and Engineering Co. reported a fuel cell cathode for use in mildly alkaline buffered electrolytes in the pH 7 to 10 range consisting of 45~55% manganese sesquioxide (Mn_2O_3) supported on high surface-area carbon.[2780] When this cathode was tested in a hydrogen-oxygen fuel cell operated at 75°C in a 1 M solution of potassium carbonate–bicarbonate as electrolyte, it was able to deliver a current density of 500 mA/cm² at 0.6 V. Polish workers have reported a method of preparation of a similar catalyst for use in oxygen diffusion electrodes.[1556]

Soviet researchers have studied the rate of oxygen reduction on magnetite (Fe_3O_4) electrodes containing 20~30% PTFE emusion on a nickel grid in 0.1 M KOH solution.[2089] The rate of oxygen reduction depended on the extent of oxidation of the magnetite. On reduced magnetite, for example, the oxygen reduction exchange current at 25°C was 1.6×10^{-8} mA/cm², while on nonreduced magnetite it was $(2\sim6) \times 10^{-9}$ mA/cm². The Tafel slope in each case was -70 to -80 mV/decade. The reaction was found to be first order with respect to oxygen, and the slope of the log-log plot of the oxygen adsorption rate vs. the electrode surface coverage had a value of 1.2, which was said to correspond to that expected for nondissociative oxygen adsorption on magnetite. Oxygen evolution has also been studied on iron oxide films in concentrated KOH solution.[1007, 1924]

Diamond Shamrock Technologies S.A. has patented an oxygen electrode containing a germanium sulfide (GeS_2) catalyst in a sintered porous titanium matrix which is claimed to be useful for either oxygen reduction or oxygen evolution in acid solution.[1875] After 240 h of operation at 60 mA/cm^2 as an oxygen evolution anode in 10% sulfuric at 60°C, this electrode was reported to have undergone only negligible wear and no catalytic ageing.

Robert Bosch G.m.b.H. in Germany investigated the use of transition metal chalcogenides as oxygen reduction catalysts for fuel cell use in sulfuric acid medium.[253] In 2 M sulfuric acid at 70°C, the Co-S and Co-NiS systems showed the highest catalytic activity (0.225 mA/cm^2 at 0.6 V). The substitution of sulfur with selenium or tellurium decreased activity in the order S > Se > Te. Unfortunately, the stability of any of the compounds was insufficient for use at current densities greater than 5 mA/cm^2 in acid electrolytes.

A solution catalyst consisting of cupric chloride ($CuCl_2$) dissolved in hydrochloric acid electrolyte (43% $CuCl_2$ + 3% HCl + 54% H_2O) was reported to effectively catalyze oxygen reduction on a graphite substrate at current densities of 300 mA/cm^2 at 65°C.[693, 694]

B. ELECTROCATALYSIS OF OXIDANTS OTHER THAN OXYGEN

For certain applications where cost is not an important factor, and where the oxidant must be stored, hydrogen peroxide has been used as an oxidant. The advantages of hydrogen peroxide over oxygen are that it is a liquid, it is infinitely soluble in cold water, and is chemically more reactive than oxygen at ambient temperatures. Hydrogen peroxide has been regularly used as an oxidant in fuel cells for small submarines since it provides greater chemical energy per unit volume[1677] and gives fewer buoyancy and pressure differential problems[1000] than a gaseous reactant. In such cases, the hydrogen peroxide is usually catalytically decomposed at the fuel cell cathode.

As was previously discussed, peroxide decomposition catalysts may also play an important role at the oxygen cathode in alkaline systems. In some cases, carbon may be a primary oxygen reduction catalyst to peroxide, but the further reduction of the peroxide to water requires catalysis, normally via the disproportionation mechanism giving water and dioxygen, the latter being thus recycled. Some materials that have been shown to decompose peroxide to water and dioxygen in alkaline media include silver, platinum,[1956] various bronzes,[2578] some spinels and mixed oxides,[2735] porous carbon,[1677] and iron phthalocyanine.[2189]

In acid solutions, in addition to platinum,[2434] it has been found that some of the bronzes appear to be good hydrogen peroxide decomposition catalysts. For example, in 1 M perchloric acid, some bronzes have been found to be superior to silver and gold in their ability to decompose hydrogen peroxide over the temperature range of 35~65°C.[2578] Similar results have been demonstrated in 0.5 M sulfuric acid with $Na_{0.65}WO_3$, $Co_{0.03}WO_3$, and WO_3,[2079] as well as with platinum-doped Na_xWO_3.[2080] If indeed these or similar relatively cheap materials prove to be stable, long-lived peroxide decomposition catalysts, it may then be possible to produce an effective non-noble metal oxygen cathode for acid fuel cells by including an inexpensive electrocatalyst that need only reduce oxygen to hydrogen peroxide.[2578] However, the object of such a search may prove to be very elusive.

Finally, it has been found that carbon dioxide may be electrochemically reduced to oxalate anions in nonaqueous solvents.[912] This suggests that it might be eventually feasible to use this abundant combustion product of fossil fuels (an environmental problem because of the greenhouse effect) in some kind of driven fuel cell that can produce useful products. A solar energy–driven system is a future possibility.

CHAPTER 13

Fuel Cell Stack Materials Selection and Design

A. MATERIALS SELECTION

The development of high power density fuel cells has necessarily resulted in the use of reactive electrode–electrolyte systems operating at elevated temperatures. Although prototypes of these devices show great promise, the corrosive cell environments, particularly at the cathode, have resulted in considerable problems concerning the choice of construction materials. Ideally, cell components should be inexpensive, easy to fabricate, should have stability in the cell environment, and should not adversely affect the functioning of the electrodes, such as poisoning from traces of dissolved construction materials. Materials problems exist in all fuel cell types, including the all-solid high temperature SOFC. Several reviews on the materials of construction of electrochemical cells[1195, 1661] and on materials problems in fuel cells[303, 447, 1102, 2447] have been published. Readers are also referred to the general reviews in Ref. 1989.

From a thermodynamic viewpoint, the least corrosive of the electrolytes commonly employed is aqueous potassium hydroxide, followed by molten carbonate and then by hot phosphoric acid, which must be considered to be the most corrosive. However, as we have repeatedly stressed, thermodynamics are only part of the story in respect to the stability of materials at the cathode of the different types of fuel cell. Only certain oxides and noble metals should be thermodynamically stable at the fuel cell cathode in basic systems, which include aqueous KOH (in which the oxides may be hydrated) and molten carbonate. In hot phosphoric acid, with the exception of the passivating films on niobium and tantalum, oxides should dissolve to give salts, and only certain noble metals (Au, Pt, Rh, Ir and possibly Ru) should at least be on the verge of stability. These stability zones for aqueous systems are given in Ref. 2044.

However, in determining stability, we must take into account the kinetics of dissolution, as well as the thermodynamic properties of the materials. In the case of molten carbonate media, the high temperature makes the rate of attack on susceptible materials rather great, so that thermodynamics are a good guide to stability. In the aqueous media (KOH and phosphoric acid), many materials that are calculated to be thermodynamically unstable have such slow dissolution rates that the stability is assured over many thousands of hours. As is clear from Chapters 9 and 12, this is particularly true for carbon at the oxygen cathode in both acid and alkaline solutions.

Without this property of carbon, the construction of a cost-effective PAFC would be impossible. However, for kinetic reasons, carbon is more readily attacked by aqueous KOH than by phosphoric acid. Carbon can be used under cathode conditions up to 210°C in the latter system, whereas in alkaline cells its use is limited to about 75~80°C. In consequence, with regard to conducting materials required for electrode structures, KOH can be considered to be more corrosive in practice than phosphoric acid. Fortunately, for structural components where electronic conductivity is not required, many plastics are suitable for use in contact with KOH at say 70°C, whereas only PTFE derivatives are stable in 210°C phosphoric acid. In all media, the reducing conditions at the cell anode represent a less severe environment than those at the oxidizing cathode.

Aqueous Alkaline Systems

As indicated above, the principal danger of corrosion is on the oxygen or air side of the cell, where strongly oxidizing conditions exist. At temperatures near 200°C in alkaline media, nickel is the most durable metal for use as electrodes and cell parts, since its oxide is very insoluble in the electrolyte. At temperature below 100°C, nickel and silver, either solid or electroplated, are the two metals most commonly used for cell parts, especially where electronic conduction is required. Silver-plated nickel anode screens are used in the Space Shuttle Orbiter cell.[394]

Certain alloys containing metals such as nickel, chromium, and molybdenum are stable up to 80°C, as is Monel, an alloy of nickel and copper. Electrodeposited magnesium has been used successfully for the bipolar plates between cells, and is attractive on account of its low density. In the Space Shuttle Orbiter fuel cell, these are gold-plated to improve conductivity.[394] With regard to nonmetallic electrical conductors, carbon has been used extensively up to 65°C. Many different kinds of plastics have been used, (e.g. as binders for carbon particles) but most are limited to temperatures below 100°C.

It has been found that lithiated, high surface-area nickel oxide is a fairly corrosion-resistant oxygen catalyst at 200°C. Bacon found that if a newly sintered porous nickel electrode was first soaked in an aqueous solution of lithium hydroxide, dried, and then heated in air for a few minutes at 700°C, an electrode with good performance and corrosion resistance for over 1500 hours at 200°C was obtained. Later, it was found more convenient to replace lithium hydroxide with lithium nitrate solution, which improved performance by increasing electrode surface area.[28, 394] At ambient temperatures, platinum is more effective for oxygen electrode use than any other metal. However, if cell temperatures exceed about 80°C, the slight solubility of platinum oxides in concentrated KOH may lead to what is known as platinum transfer, that is, slow dissolution from the cathode, followed by transfer and re-deposition either on the anode or in the matrix, if present. In the space shuttle fuel cell, gold black with 10% platinum as a sintering inhibitor is used,[394] gold is certainly more stable than platinum. As indicated in Chapter 12, corrosion-resistant spinels such as $CoAl_2O_4$ have sometimes been incorporated into the cathode, since they are excellent peroxide decomposition catalysts, tending to make the oxygen reduction process go towards a complete four-electron transfer.

With regard to gasket materials, at temperatures of 200°C or above, it is probable that PTFE (Teflon) is the only insulating material that will resist the corrosive action of strong KOH and oxygen for long periods. For the Apollo fuel cell, operating at about 260°C, a special grade of this material was developed, since normal grades, which may not be 100% fluorinated, could react to produce nickel fluoride in contact with nickel.[1171] In addition to its corrosion resistance, PTFE retains sufficient strength at 200°C to resist the tendency to creep. The sealing arrangement in the Apollo fuel cell is illustrated in Refs. 394 and 689. At temperatures below 100°C, various other plastics can be used successfully for cell parts, including gaskets, and in some designs the whole cell stack has been encapsulated in a resin such as epoxy.[138, 453, 454, 1267]

A special gasketing problem to contend with in alkaline systems is associated with so-called creep of the alkaline electrolyte. One driving force for this phenomenon in cathodes can be illustrated: The production of hydroxyl ions from the electrochemical reduction of oxygen occurs in a thin reaction zone above the electrolyte meniscus at the three-phase boundary. The movement of water into this zone causes it to increase in both conductivity and thickness, leading to a movement of the reaction zone in the electrode.[1202] Alkaline creep is a function of the cation present, its concentration, the relative humidity, and on any additives that may be present.[724] Where seals are involved, the rate of water transmission through sealant materials, and seal compression are important.[724] Creep also is related to surface tension (the Marangoni effect), vapor phase transport, and electroosmosis.[273] Materials that have been recommended for preventing alkaline creep include polyaminoamides[2528] and special polymeric sealants.[2659]

The long-term (8000 h) compatibility characteristics of various structural materials for use in hydrogen-oxygen alkaline cells have been evaluated.[1023] Exposure of candidate materials to 35% KOH at 93°C showed that the most compatible materials were polypropylene, polysulfone, and 50% asbestos-filled polyphenylene sulfide. In regard to resistance to saturated water vapor at 110°C, the most effective materials were polyarylether, polyarylsulfone, polysulfone, polypropylene, and 50% asbestos-filled polyphenylene sulfide. A sintered polysulfone structure, made wettable by electroless nickel plating, could be substituted for sintered nickel as the electrolyte reservoir plate (ERP), corresponding to the ribbed substrate in the PAFC (see Chapter 16 and Ref. 177). This material had only one sixth the weight of the nickel sinters used in the Space Shuttle Orbiter cell, which themselves accounted for 50% of the total cell weight.[394]

Matrixes

An important component of many aqueous alkaline fuel cell systems with noncirculating electrolyte, and in some cases with circulating electrolyte, has been a matrix or diaphragm inserted in the electrolyte to separate the anode electrode structure from that of the cathode. Traditionally, this diaphragm has been made from porous chrysolite, a fibrous type of asbestos having the composition $Mg_3Si_2O_5(OH)_4$. However, the trend toward cell operation at higher temperatures has shown this material to be unstable in 40~60% KOH at 150~200°C, primarily because of dissolution of SiO_2 from the matrix.[1706] A review of the use of asbestos membranes in fuel cells[2352] and a discussion of their modes of failure[1806] can be found in the literature. A general review of separators and membranes in electrochemical power sources also has been published.[1587]

It has been found that the stability of asbestos diaphragms can be improved by pre-leaching SiO_2 from the asbestos,[682] by incorporating magnesium silicate into the membrane,[1345] or by the addition of silicate ions to the electrolyte itself.[682, 1011, 1090] In addition, various treatment procedures have been devised to improve the stability of asbestos membranes. These include treatment with chloroprene rubber,[21, 22] polymer emulsions,[731, 778] and PTFE;[1164, 1650, 2571, 2720] and the addition of polyesters and inorganic fillers[2255] or organic titanates.[1121] More recently, matrixes made from potassium titanate, especially potassium octatitanate, have been shown to have good resistance to chemical attack in 45% KOH at 150°C for about 10,000 hours.[660, 1080, 1706, 2041] Since the titanates are typically in the form of 10 μm needles, diaphragms using them tend to be more fragile than those made from long chrysotile fibers. Diaphragms also have been made from combinations of potassium titantate and PTFE[2478] as well as from potassium titanate containing 10~15% asbestos.

There also is interest in the research and development of thermoplastic polymers for use as matrixes.[1585, 1840, 1859] Examples include diaphragms composed of PTFE-polypropylene[1107] and of PTFE.[781, 1264, 1694] A review of polymer matrixes for use in alkaline media has been published.[1839] Finally, various novel membrane materials have been studied, such as silica gel-polyethylene,[782] polybenzimidazole,[589, 1695] zirconia gel,[1731] synthetic asbestos,[1346] and porous sintered nickel diaphragms.[1999]

For cells using dissolved fuels such as hydrazine and methanol, and/or dissolved oxidants (hydrogen peroxide), the use of a diaphragm is necessary to prevent diffusion of the fuel (or oxidant) through the electrolyte to the opposite electrode, where direct chemical reaction may occur. Membranes that have been used in alkaline hydrazine systems include asbestos[2000, 2740] and plastics such as polyethylene-acrylic acid.[2527] When this approach is used, there is a trade-off between improved reactant utilization and increased cell internal resistance due to the presence of the diffusion barrier. Methanol causes especially severe problems if it migrates to the cathode, where it wets the Teflon, causing the electrode to flood, with a corresponding decrease in performance. Attempts[2736] have been made to protect the cathode by inserting a grid electrode in front of the cathode to oxidize the methanol; mechanical barriers also have been used.[804]

Aqueous Acid Systems

The range of materials that can be used in hot phosphoric and sulfuric acid is much less than in alkaline electrolytes. As already remarked, the thermodynamics, (i.e., the solubility of metal ions and the higher absolute electrical potentials encountered in the fuel cell stack) result in the corrosion of most metals in acidic media. The choice of materials has therefore always been very restricted with acid electrolytes; and the greatest barriers to reducing cost and increasing reliability in phosphoric acid fuel cells have been the problems associated with materials.[2059, 2060] While sulfuric acid cannot be readily used at temperatures above 100°C because of water vapor pressure, heat balance, and stability problems, it is still occasionally recommended for use as a pure hydrogen fuel cell electrolyte, largely on account of its lower corrosivity (at 70~80°C) than that of phosphoric acid at higher temperatures.

In the past it was usual to construct many of the parts of experimental acid fuel cells from metals such as platinum, gold, or tantalum, sometimes as electroplates; for example, gold on tantalum to reduce electrical resistance. Platinized titanium was also successfully used. A search for alternate, less expensive materials showed that the binary alloys 20 Mo-Ni and 70 Ni-Ti, and the quaternary alloy 23 Mo-30 Nb-35 Ti-12 V had corrosion rates of less than 25 μm per year in 95% phosphoric acid at 165°C.[2204, 2499] The addition of up to 5 weight percent Fe to the Mo-Ni alloy did not affect its corrosion resistance appreciably. Hastelloy B, a commercial alloy of similar composition to the binary Mo-Ni alloy, has been found to be equally resistant to 95% phosphoric acid at 165°C. Earlier studies[2202, 2203] showed that 48 Fe- Mo, 1.8 Cr-6 Nb-78 Ta-13.7 V, and 4.6 Cr -60.5 Ta-34.9 V exhibited corrosion resistance to 95% phosphoric acid at 200°C; and that 67.4 Mo-Ti, 0.5 Mo-7.95 Ti-91.5 W, 27 Ta-Ti, and 6.8 Mo-29.3 Nb-52.3 Ti-11.6 V exhibited resistance to 95% phosphoric acid at 165°C.

As indicated in Chapter 16, the problem of cell structural materials has been eliminated by the use of graphite. In the 1970s, some attempts were made to use plastic bonded carbon structural components, for example, for bipolar plates.[177] Phenolic resins were shown to be effective up to about 165°C,[177] as discussed in the following section on matrixes. H-resin, a Hercules product, was shown to be stable under PAFC cathode conditions, but its developer discontinued manufacture because of its high cost, complex fabrication, and the lack of a sufficiently large market. Polyphenylene sulfide was used in the UTC 1 MW demonstrator in 1976, but tests showed that decomposition of this polymer was occurring during operation.[177] In consequence, all developers now use pure carbons (for cells operating under less extreme conditions) or graphite, as discussed in the following section.

Reactant manifolds are invariably of the external manifold type, with the exception of small stacks operated by Fuji Electric in the early 1980s (see Chapter 16). The manifolds must provide gas flow distribution, maintain pressure-tight seals against the fuel cell stack, and must be compatible with hot phosphoric acid, fuel, and air. Early PAFC stacks used hot-molded phenolic-impregnated fiber glass as the manifold material, but this material was found to be chemically and structurally unsound for use in the larger stacks and more extreme environments now in use. United Technologies has used reactant manifolds with PTFE frames that are made from stamped low-carbon steel, coated with perfluoralkoxy (PFA) protective films.

For pressurized water-cooled stacks, it has been found that standard heat transfer materials such as copper or stainless steel must be Teflon-sheathed for service in the hot phosphoric acid cooling plate environment. The use of Teflon-sheathed aluminum is not possible, except perhaps for dielectric liquid cooling systems. Graphite and certain graphite-resin materials are stable, but present problems in sealing and making electrical connections. United Technologies used copper cooling coils coated with thin protective films of PTFE and PFA.[2060] They then examined copper tubes imbedded in graphite, and have experimented with all-graphite coolers. Because of dissimilar metal corrosion in the steam circuit if copper coolant tubes and stainless steel tubes elsewhere are used, future cooling plates will have serpentine PTFE/PFA coated stainless steel coolant tubes.[448]

Sealants

Numerous proprietary sealing materials have been developed for use in the PAFC, all of which seem to show some problems. United Technologies claims to have developed a sealant usable in hot phosphoric acid at temperatures up to 204°C. This contains PTFE, a fluorocarbon oil, and a filler such as silicon carbide (SiC) or graphite.[2515] The company also has developed a thermally conducting caulk containing fluorocarbons and graphite that is said to resist phosphoric acid absorption.[1696] Other sealants often used in phosphoric acid fuel cells, particularly for sealing graphite parts, include Viton or Teflon. Viton, which is the trademark for a series of fluoroelastomers based on the copolymer of vinylidene fluoride and hexafluoropropylene, can be used in a cured form or an uncured (thermosetting) form. These sealants probably contain graphite, since they are normally black. Teflon sealants are of two types: a softer variety containing TFE (tetrafluorinated ethylene), and a stiffer material based on EPC (ethylene propylene copolymer). For utility fuel cell use under pressurized conditions, it is generally conceded that the stability of Viton is insufficient.

Electrode Assemblies

As discussed in Chapters 12 and 16, the electrode parts used in the PAFC almost always consist of mixtures of catalyst and PTFE dispersed on electrode substrates made from graphitized carbon or graphite resin composites in the form of a graphitic paper or felt material. This can be flat or ribbed, the latter being particularly used if an electrolyte reservoir is incorporated into the electrode assembly.[177] Fibrous carbon of high graphite content also has been used as an electrode substrate material, either as chopped fibers added to resins that are then pyrolyzed to give ribbed substrates,[177] or in the form of graphite cloth when the cell design requires flat substrates.[2025, 2061]

The catalysts used in the phosphoric acid fuel cell are invariably pure platinum (at the anode) or platinum alloys (at the cathode), dispersed on appropriate carbon-based supports (Chapter 12). However, in sulfuric acid and probably also fluorinated sulfonic acids, tungsten carbide (WC), silver-doped tungsten carbide, or cobalt phosphide (CoP_3) catalysts may have sufficient activity to be of value at the anode; while metal chelates and thiospinels of transition metals may sometimes be acceptable at the oxygen cathode.[303] These catalysts have been discussed in greater detail in Chapters 11 and 12.

The current collectors present one of the more serious materials problems in acid fuel cells. Only gold, platinum, tantalum, and the more resistant forms of carbon are sufficiently resistant to attack in hot phosphoric acid, although as discussed earlier, specialty alloy current collectors have been examined for the cathode, though with little success. However, all these materials have lower electrical conductivity than silver or copper, and, with the exception of graphite, all are prohibitively expensive. Even so, the electrode support screen in Teflon-bonded electrodes, which may equally serve as the current collector, was often made from gold-plated tantalum for early 1970s laboratory testing.

A succession of different current collector materials has been examined in past work, including dual porosity highly wetproofed structured electrodes that contain relatively little electrolyte on the gas side. Some ideas along these lines have been given in Ref. 177. This may allow the possibility of cheaper cathode-side current collectors; for example, carbons or carbon-plastic composite materials instead of pure graphite may be used as current collectors and gas distribution plates in cells of appropriate design.[177] The disadvantage of this approach is that it prevents the use of an electrode reservoir on the cathode side.

Separators and Matrixes

Bipolar separator plates for the PAFC are commonly made from high density compression-molded graphite, small-pore carbon foams, or molded graphite, or from graphite-glassy carbon made by converting graphite-phenolic resin composites by heat treatment;[617] or from materials made from reticulated vitreous carbon or carbon cloth (Engelhard).[2060] Methods of selecting and evaluating graphite-resin composites for molding bipolar plates have been discussed.[650] Some of the approaches used will be discussed further in Chapter 16, and some alternatives have been reviewed in Ref. 177.

PAFC systems invariably use a matrix to retain the electrolyte in the cell. The matrix must have a high porosity of small pore-size so as to remain always wetted, and be able to absorb large quantities of electrolyte. In addition, it must be chemically, mechanically, and dimensionally stable in the reaction environment; should be electronically nonconductive; should

Fuel Cell Stack Materials Selection and Design 419

have a bubble breakthrough resistance greater than 35 kPa; and should be cheap and amenable to mass production.

Commercially available matrixes[784] and other matrix materials potentially suitable for phosphoric acid service[71] have been investigated. One material which showed some early promise consisted of Kynol (a crosslinked amorphous phenolic polymer) fibers bonded with 10% Resinox (a type of phenolic resin) binder, and containing 50% by weight 325-mesh tantalum oxide (Ta_2O_5) to improve wettability.[562] This material was produced by a paper-making technique that gave matrixes of high uniformity and as thin as 0.013 cm, giving a resistivity of 7.87 Ω cm when wetted with 105% phosphoric acid at 140°C. The use of matrixes made from resin-bonded polymer fiber sheets in concentrated phosphoric acid at 150°C has been reported by AEG-Telefunken.[870] However, since that time the temperature of the PAFC has risen to 190~210°C and the use of resins is no longer feasible. The three major U.S. PAFC developers (UTC, ERC, and Engelhard Industries) now all use matrixes made from Teflon-bonded silicon carbide,[2060] typically 90 wt% SiC in <10 µm particle size together with 10 wt% colloidal PTFE.[444]

Use of Plastics for Aqueous Fuel Cell Components

Because of their low cost and ease of fabrication, plastic materials would be very desirable for fuel cell construction. The use of plastics may possibly minimize the need for fasteners and adhesives, since plastic components can be snapped together to give rigid structures. Where this is impractical, adhesives are generally easily applicable. In some cases no further structural components may be required, except perhaps for a carrying frame or cradle. For example, the Goodyear Tire and Rubber Co. has developed methods of building high temperature fluorocarbon elastomeric fuel cell containers that reduce the problems of voids and air bubbles that may occur in the seam areas of such components.[661] A workshop was held in 1978 to recommend future directions in energy-related polymeric research.[1660]

For use in aqueous fuel cell construction, the ideal plastic material must meet the following requirements:

1. Chemical resistance to the electrolyte at both ambient and operating temperatures.
2. No reaction with cell reactants and products.
3. No contamination of the electrolyte.
4. Favorable long-term ageing properties.

5. Low creep.
6. Ease in joining and sealing.
7. No embrittlement at -40°C, at least for alkaline applications.
8. Low cost.
9. Ease of fabrication at relatively low temperatures, preferably with reusable scrap and kerf material.

Since no real plastic material will simutaneously satisfy all the above criteria, in practice a compromise must be made. Selection of the best plastics is further hindered by the fact that it is difficult to predict whether or not a material for which data look promising will actually perform satisfactorily, since many of the properties are quoted at room temperature and in the absence of chemical and electrochemical gradients.

Attention should be paid to the following five potential problem areas when selecting a plastic material for fuel cell applications:

Mechanical Stability

The product selected must have sufficient structural integrity to meet the demands of load and temperature that will be required of it. Retention of physical properties over the desired working range is more important than the values of the properties at the most extreme conditions. It is also essential to consider the ways in which the fabricated part might fail. For example, tensile strength for these materials is often the only quoted parameter, yet for fuel cell applications compression resistance, creep resistance, and impact strength may be more important. Because of the difficulty of making a selection that is based only on the conventional physical properties of plastics, empirical methods of predicting service life of fuel cell materials (such as elastomeric baffles) have been formulated.[328] Table 13-1 gives some of the physical properties of various candidate plastics that might be used in fuel cells. Tables such as this should be used only as a general guide for performance under normal conditions.

Chemical Stability

Chemical environments attack plastics via two basic mechanisms: chemical attack and physical attack. Chemical attack includes oxidation of the plastic, reaction of functional groups in or on the polymeric chain, and depolymerization, resulting in structural weakness and eventual disintegration. Physical attack includes solvent absorption (which may result in

TABLE 13–1
Physical Properties of Selected Plastics*

Code:
- CPE = conventional polyethylene
- LPE = linear (high density) polyethylene
- PP = polypropylene
- PMP = polymethylpentene
- FEP = Teflon® fluorinated ethylene propylene copolymer
- ETFE = ethylene-tetrafluoroethylene copolymer
- PC = polycarbonate
- PVC = rigid polyvinyl chloride

Property	CPE	LPE	PP	PMP	FEP/ETFE	PC	PVC
Temperature limit, °C	80	120	135	175	205	235	70
Specific gravity	0.92	0.95	0.90	0.83	2.15	1.20	1.34
Tensile strength, MPa	14	28	35	28	21	55	45
Brittleness limit, °C	–100	–100	0	—	–270	–135	–30
Flexibility	excellent	rigid	rigid	rigid	excellent	rigid	rigid
Transparency	translucent	translucent	translucent	clear	translucent	clear	clear
Autoclavability	no	caution	yes	yes	yes	yes	no

*Source: The Nalge Co.

swelling, softening, or solvent permeation); dissolution in a solvent; leaching of plasticizers (resulting in embrittlement); and interaction of potential stress cracking agents with residual fabrication defects in the material (leading to cracking).

In general, increased exposure time, temperature, and concentration will increase the rate of such attack. Plastic materials containing functional groups such as esters, which may hydrolyze in fuel cell operating environments, must be rejected *a priori*. Other polymers such as polyethylene and polypropylene may swell in the presence of hydrocarbons. The most stable polymer materials, and the most acceptable on all grounds other than cost, are the fluorocarbons, such as polytetrafluorethylene (PTFE) fluorinated ethylene propylene (FEP-Teflon®), and partially fluorinated materials such as ethylene-tetrafluoroethylene copolymer (ETFE-Tefzel®). The latter, whose structural formula is $(CH_2\text{-}CH_2\text{-}CF_2\text{-}CF_2)_n$, is related to Kynar®, $(CH_2\text{-}CF_2)_n$; but has chemical properties that are not identical. KEL-F® (CF_2-CFCl) should also be mentioned, as should Saran® (polyvinylidene chloride, the chloro-equivalent of Kynar®).

While the utility of the fluoro-based materials, particularly various grades of PTFE, has been proven in porous three-phase boundary electrodes, their hydrophobic character may be a disadvantage for some applications. The properties of PTFE have been reviewed elsewhere.[2403] Plastics such as PTFE and polyethylene often are used for inlet and outlet tubes, and for channels for electrolyte flow in alkaline fuel cells.

The choice of one plastic material over another is usually a compromise between cost and durability. For example, polyethylene is inexpensive but is only stable to around 80°C, whereas PTFE is stable to temperatures above 200°C, even in strong alkalies, but is expensive and may show a tendency to creep. United Technologies (IFC) uses PFA (perfluoroalkoxy) Teflon films on manifold hardware; and the same material has been recently examined by Engelhard for a gas-tight film in their bipolar plate, to replace polyether sulfone (PES) and polyetheretherketone (PEEK) films (see Chapter 16).

Table 13–2, based on data from the Nalge Co., presents a guideline to the chemical resistance of selected polymer materials.

Heat Build-up and Thermal Expansion

Since carbon-based polymers are good thermal insulators, provision in the cell design must be made to avoid excessive internal heat accumulation that may adversely affect fuel cell performance. The use of polymers in contact with dissimilar materials (e.g. metals) may give problems because of the differing coefficients of thermal expansion.

TABLE 13-2

Chemical Resistance of Selected Plastics*

Code:
- CPE = conventional polyethylene
- LPE = linear (high density) polyethylene
- PP = polypropylene
- PMP = polymethylpentene
- FEP = Teflon® fluorinated ethylene propylene copolymer
- ETFE = ethylene-tetrafluoroethylene copolymer
- PC = polycarbonate
- PVC = rigid polyvinyl chloride

Resistance Key:
- 1st letter = chemical resistance at 20°C
- 2nd letter = chemical resistance at 50°C
- E = no damage after 30-day constant exposure; may tolerate years of exposure
- G = little or no damage after 30-day constant exposure
- F = some attack after 7-day constant exposure
- N = noticeable attack within minutes to hours after exposure (failure, however, may take years)

	CPE	LPE	PP	PMP	FEP/ETFE	PC	PVC
Phosphoric acid, 1~5%	EE	EE	EE	EE	EE	EE	EE
Phosphoric acid, 85%	EE	EE	EG	EG	EE	EG	EG
Sulfuric acid, 1~6%	EE	EE	EE	EE	EE	EE	EG
Sulfuric acid, 20%	EE	EE	EG	EG	EE	EG	EG
Sulfuric acid, 60%	EG	EE	EG	EG	EE	GF	EG
Sulfuric acid, 98%	GG	GG	GG	GG	EE	NN	NN
Formic acid, 3%	EG	EE	EG	EG	EE	EG	GF
Formic acid, 50%	EG	EE	EG	EG	EE	EG	GF
Formic acid, 98~100%	EG	EE	EG	EF	EE	EF	FN
Nitric acid, 1~10%	EE	EE	EE	EE	EE	EG	EG
Nitric acid, 50%	GG	GN	FN	GN	EE	GF	GF
Nitric acid, 70%	FN	GN	FN	GF	EE	NN	FN

Table 13-2 (continued)	CPE	LPE	PP	PMP	FEP/ETFE	PC	PVC
Perchloric acid	GN	GN	GN	GN	GF	NN	GN
Hydrochloric acid, 1–5%	EE	EE	EE	EG	EE	EE	EE
Hydrochloric acid, 20%	EE	EE	EE	EG	EE	GF	EG
Hydrochloric acid, 35%	EE	EE	EG	EG	EE	NN	GF
Hydrofluoric acid, 4%	EG	EE	EG	EE	EE	GF	GF
Hydrofluoric acid, 48%	EE	EE	EE	EE	EE	NN	GF
Potassium hydroxide, 1%	EE	EE	EE	EE	EE	FN	EE
Potassium hydroxide, conc.	EE	EE	EE	EE	EE	NN	EG
Sodium hydroxide, 1%	EE	EE	EE	EE	EE	FN	EE
Sodium hydroxide, 50% to sat.	EE	EE	EE	EE	EE	NN	EG
Ammonium hydroxide, 5%	EE	EE	EE	EE	EE	FN	EE
Ammonium hydroxide, conc.	EG	EE	EG	EG	EE	NN	EG
Methanol, 100%	EE	EE	EE	EE	EE	GF	EF
Ethanol, 40%	EG	EE	EG	EG	EE	EG	EE
Ethanol, 100%	EG	EE	EG	EG	EE	EG	EG
Isopropyl alcohol, 100%	EE	EE	EE	EE	EE	EE	EG
sec-Butyl alcohol, 100%	EG	EE	EG	EG	EE	GF	GG
tert-Butyl alcohol, 100%	EG	EE	EG	EG	EE	GF	EG
n-Butyl alcohol, 100%	EE	EE	EE	EG	EE	GF	GF
Isobutyl alcohol, 100%	EE	EE	EE	EG	EE	EG	EG
Propane gas	NN	FN	NN	NN	EE	FN	EG
Hexane	NN	GF	GF	FN	EE	FN	GN
n-Heptane	FN	GF	FF	FF	EE	EG	GF
n-Octane	EE	EE	EE	EE	EE	GF	FN

*Source: The Nalge Co.

Electrical Conductivity

In general, the electronically insulating properties of polymers is advantageous for use as cell container materials, frames, gaskets, and so on. However, stable organic polymers that have good electronic conduction would be desirable for use as lightweight current collectors or electrode substrates. The battery industry has made a great effort to develop electronically-conductive plastics for use as lightweight grids in storage batteries. Usually such plastics contain a conductive material such as graphite or metallic particles. Thus far, attempts to either overcome the inherent nonconductivity of plastics, or to make conducting composites, have not been very successful. For this reason, metals or dense carbon are still the preferred materials for current-carrying fuel cell components.

Gas Permeability

Another problem is the possible permeation of gases and vapors through plastic fuel cell components.[2062] The gases and vapors usually considered are hydrogen, oxygen, carbon dioxide, water vapor, and gaseous catalyst poisons such as carbon monoxide. Although losses caused by the permeation of reactant gases through the cell walls may theoretically result in energy losses, in practice, these are negligibly small. For example, the maximum loss of hydrogen through a 100 cm^2 sheet of 10-mil thick polystyrene is only 1.2 Ah/day per atmosphere of pressure differential, and the loss through rigid PVC is 500 times smaller still. However, a practical problem may be the hazard resulting from the accumulation of an explosive gas such as hydrogen that has permeated into a nonvented area through the walls of a fuel cell under continuous operation.

Although the partial pressure of carbon dioxide in air is only about 0.0003 atm, it is virtually zero above concentrated alkaline solutions because of chemisorption through reaction with hydroxyl ions:

$$2KOH + CO_2 \rightarrow K_2CO_3 + H_2O$$
$$K_2CO_3 + CO_2 + H_2O \rightarrow 2KHCO_3$$

This will result in a very small permeation flux of carbon dioxide through a plastic fuel cell wall into the electrolyte. This phenomenon should be given some consideration during the design process of an alkaline fuel cell stack for use under normal terrestrial conditions. However, this phenomenon will not of course occur in acid electrolyte systems.

Very small quantities of gases such as hydrogen sulfide or carbon monoxide are known to be strong catalyst poisons. For fuel cells working in areas polluted with these gases the choice of low permeability plastics will be imperative. For example, the vapor pressure of hydrogen sulfide above strong alkali solutions is low, so even at low ambient atmosphere concentrations of hydrogen sulfide, driving forces for permeation will exist. Similar problems may be caused by organic vapors, particularly those with a relatively high critical temperature, since these vapors often show high permeabilities in organic polymers despite their high molecular weights.

Many factors affect the permeability of a given polymeric material to a gas. The permeability generally increases with increasing water content, especially if the water content is above 20%. The permeability increases exponentially with temperature in an Arrhenius fashion, but is usually independent of pressure (up to several atmospheres) and thickness of the plastic material.

As a general rule, gas permeability decreases with increasing polymer molecular weight, density, and degree of crosslinking, since segmental chain mobility and the ease of hole formation are decreased by these changes. The addition of a plasticizer decreases the cohesive forces between the chains, producing an increase in segmental mobility and thereby increasing gas permeability. Since permeability with respect to a particular gas is determined by the solubility and diffusion coefficient of the gas in the given polymer, permeability does not necessarily increase with decreasing gas molecular weight. For example, in silicone rubber, the permeability of a molecule as large as $n\text{-}C_8H_{18}$ is twelve times greater than that of hydrogen. Table 13-3 lists various plastic polymer materials in order of increasing permeabilities to hydrogen, oxygen, carbon dioxide, and water vapor.[1577]

Molten Carbonate Systems

The materials of construction (i.e., for the bipolar separator plate) in the molten carbonate fuel cell must be impermeable to fuel and oxidant gases, and stable in the electrolyte at all potentials encountered in the cell. Until the early 1970s, the only metals that were known to exhibit long-term resistance to the melt under all conditions were certain noble metals and their alloys. The best of the alternative materials appeared to be type 347 stainless steel, but practical evaluation showed that this material was not really suited to long-term operation over several years. A number of other alloys (e.g., Hoskins 815, Kanthal A, and Kanthal A1) were found to have superior properties. From the viewpoint of corrosion resistance, the most effective alloys were those that contained aluminum. This property has been attributed to the formation of aluminate surface films. Kanthal A, for example, required a period of about five years to corrode to a depth of 25

TABLE 13-3

Permeability Ranking* of Various Plastics [1577]

Plastic	H_2	O_2	CO_2	H_2O
Polyvinyl fluoride	1	—	—	4
Polyvinyl chloride (rigid)	2	—	2	9
Polyester-terephthalate	3	—	—	5
Polyvinyl chloride (non-rigid)	4	1	—	14
Polyimide	5	2	1	8
Nylon 66	5	3	3	7
Nylon 6	5	—	4	12
Polytrifluorochloroethylene	6	3	—	1
Polyethylene (high density)	7	4	5	3
Cellulose triacetate	8	6	—	15
Cellulose acetate	8	6	—	17
Chloroprene (pure)	8	—	—	—
Butadiene polystyrene	9	—	—	—
Butadiene acrylonitrile	9	—	—	—
Polycarbonate	10	6	8	13
Polypropylene	11	5	6	2
Polyethylene (medium density)	12	7	10	2
Polyethylene (low density)	13	7	11	2
Neoprene	14	—	—	14
Fluorinated ethylene-propylene	14	8	9	2
Ethylene vinyl-acetate-butyrate	16	9	—	16
Acrylonitrile-butadiene-styrene copolymer	17	11	8	11
Butadiene methyl-methacrylate	18	—	—	—
Polystyrene	19	10	7	10

*The lower the number, the lower the permeability

µm in a ternary carbonate eutectic paste containing 53 wt % lithium aluminate at 700°C.

The corrosion rate of these partially immersed alloys was found to be greater under an inert nitrogen atmosphere than under an oxidizing atmosphere.[723] This may have been due to nitrogen impeding passive film formation; or, as is more likely, it may have been a result of the loss of carbon dioxide from the melt, so that the oxide content of the melt was higher than it would have been under equilibrium conditions. Aluminum nitride, AlN, has been shown to offer some promise as a corrosion-resistant material in high temperature molten lithium salt electrolytes.[548] Passive film formation on zirconium in molten alkali metal carbonates at 600~800°C also has been studied.[1962] Lithium zirconate (Li_2ZrO_3) also appears to merit attention as a protective coating. This material has been reported to be stable for 1000 hours in molten mixtures of lithium carbonate–sodium carbonate at 700°C in the presence of reformed naphtha fuel gas.[1597]

However, materials that form highly-passivating films (e.g., the aluminum-containing stainless steels) are of little value for major cell components since their oxide films are nonconducting. Hence, they cannot be used as bipolar plates, and can only find use for special purposes, such as nonconducting fibers for electrolyte layer reinforcement. However, at the cell edge, where rapid corrosion can take place as a result of the contact of metallic components with both oxidizing and reducing atmospheres, the nonconducting properties of aluminate films can be put to good use by the technique of aluminizing (see Chapter 17).

For laboratory-scale standard single-cell components (cell housings and gas ducts), dense alumina or type 316 stainless steel is used.[188] It has been claimed that in the presence of the electrolyte and matrix, a coating of lithium aluminate ($LiAlO_2$) forms on the surface of type 316 stainless steel, serving to protect it.[2386] However, as reviewed in Chapter 17, the more likely protective film component is lithium ferrite, $LiFeO_2$. Type 316 or related stainless steel cannot be used under anode conditions except in dry gas atmospheres (i.e., at low fuel conversions) since their corrosion rate is then too great. On the other hand, they show good corrosion resistance under cathode conditions. This is examined in more detail in Chapter 17 and in the following section.

Separators (Bipolar Plates)

Gas separators and current collectors should be combined in practical stacks in order to eliminate contact resistances. In addition to having corrosion resistance and thermal compatibility with other cell components, the bipolar plate must also provide electrical contact between adjacent anodes and cathodes. As has just been remarked, the presence of an electrolyte film on a continuous metallic conductor at the cell edge can cause enhanced corrosion because of contact with both oxidizing and reducing atmospheres, giving high electrode potential difference across the sample. The anode wet seal area, which typically will see the largest oxygen partial pressure difference (that between anode gas and depleted cathode gas on the outside of the stack), and hence the greatest potential difference, is particularly vulnerable. Protection is afforded by putting a large electronic resistance in the corrosion circuit by aluminizing or aluminiding the seal area to form passive films on the bipolar separator plate. Similarly, it has been shown that the incorporation of 2~70% Al in the surface of stainless steel gives protection at the anode wet seal.[555] The use of a high-temperature aluminum paint also has been examined.[117]

Although type 316 stainless steel has usually been employed for the bipolar plate, most recently by the Japanese developers under low (25~35%) fuel conversion conditions, there is no question that a thin (i.e., inexpensive) bipolar plate made from this material will not stand up under real cell conditions. Studies by the General Electric Co. under contract to the U.S. DOE showed that although type 316 or 316L stainless steel separators were able to meet DOE's cost goal of \$14/kW (1980), they were unable to meet the life goal of 40,000 hours.[485] The corrosion life of type 316 stainless steel at the anode is certainly a function of fuel utilization, fuel composition, and the thickness of the electrolyte film covering the steel. It now is known that types 310 and 446 stainless steels are superior to type 316 for use as bipolar plate.s[485] Under typical cell operating conditions, for example, type 310 stainless steel forms cathode corrosion scale at rates only 30~50% of that for type 316, when exposed in the form of a thin sheet to anode gas on one side and to cathode gas on the other.[2282] While aluminum-rich diffusion-bonded protective coatings may meet a passive lifetime goal of 40,000 h, they will result in too high a film resistance for efficient current transport between cells.

The best suggestion is that the anodic side of the stainless steel separator sheet should be coated or clad with an anodically inert metal such as nickel or copper.[485, 556] Such clad steels are, however, expensive at present, and their long-term stability may be uncertain on account of, for example, the effects of internal carbide formation.[485] Internal carbide formation at the stainless steel–nickel interface may increase the likelihood of embrittlement failure, especially during thermal cycling. This subject is discussed in greater detail in Chapter 17.

An alternative may be conducting ceramic, or at least ceramic-clad, cell bipolar plates. Pure ceramic materials will be difficult and costly to fabricate and it may be impossible to accomplish good electrical contact between cells, at the same time maintaining gas-tight anode and cathode compartments. The difficulty of mechanical strength throughout the battery is also clearly a problem if ceramic components are used.

Electrolyte Matrixes

Electrolyte matrix materials for molten carbonate fuel cells have been discussed in Chapter 10.

Anodes

The anodes in molten carbonate fuel cells have presented no real materials problems from the corrosion standpoint, provided that purified fuel gases are used. So long as the anode potential is maintained below that for nickel oxidation, porous sintered nickel is both a satisfactory and long-lived electrode material that has been used since the early molten carbonate fuel cell work of Broers and Ketelaar.[471, 474] The electrodes are made without a support, or on a nickel exmet (expanded metal) or screen. However, woven ceramic cloth also has been used as an electrode substrate material.[937] An alternative anode material to nickel is copper.[559] This has the merit of a higher oxidation potential, but is more difficult to stabilize against in-cell sintering because of its lower melting point. Work on copper electrodes has not gone beyond the demonstation stage.

The only major problem of nickel anodes under pure fuel gas conditions is purely mechanical: that of continuous, slow sintering, resulting in non-optimum pore size, and creep. In 1976–1977, sintering was perceived as the major problem; however, it was effectively prevented by the use of additives to the electrode.

Since the late 1970s porous sintered nickel anodes have usually been alloyed with 2~10 wt % alloyed chromium.[1686] Cobalt also has been examined as an alloying agent. More recently, proprietary additives have been included in electrodes to stabilize them from further sintering, pore growth, shrinkage, and loss of surface area.[747] These have been identified as containing chemically precipitated lithium aluminate.[239]

The addition of stable ceramic particles to nickel and copper surfaces is known to inhibit sintering via a diffusion-blocking mechanism.[2205] The beneficial action of chromium additions on anode sintering can be largely attributed to the insolubility of Cr_2O_3, or more accurately $LiCrO_2$, in the electrolyte. Particles of the latter ceramic form on the nickel sinter surface, and act as a sintering inhibitor. Furthermore, chromium oxide scale may act as a diffusion barrier against the entry of carbon into the nickel, thereby helping to prevent carburization.[485] The method by which cobalt may reduce sintering is disputed, but vacancy-blocking has been suggested.[2085] In practice, both cobalt and chromium are expensive, and neither is completely satisfactory. The problem of creep, leading to stack deformation, is still not certainly cured: some recent successful approaches to both sintering and creep resistance are discussed in Chapter 17.

It has been found that nickel anode current collectors and electrode substrates undergo relatively rapid corrosion in the presence of sulfur-containing gases (>10 ppm S), such as those that are produced by coal gasification or the high-temperature steam reforming of hydrocarbons. Sulfur can enter the anode as hydrogen sulfide in the fuel or the cathode as sulfur dioxide. In the latter case, it would be derived from the anode exhaust combustion products used to add CO_2 to the air supply. At the cathode, the sulfur dioxide is absorbed into the electrolyte as sulfate, which then is transported through the electrolyte matrix and regenerated at the anode as either H_2S or COS.[627] Corrosion of nickel parts then takes place via the formation of a low-melting Ni-NiS eutectic.[628]

Nickel anodes containing 0.5~20% of surface-stabilizing additives such as Cr, Zr, or Al have been reported to be reasonably resistant to corrosion in the presence of sulfur-containing gases.[1685, 1703] Cobalt-based anodes have also been claimed to show promise with respect to sulfur tolerance.[627] Copper may also be better than nickel in this respect.[559] Energy Research Corp. (ERC) developed a co-precipitation spray-drying process for fabricating high surface-area nickel powders containing Co, Cr, Cu, Mo, or ceramic oxides to produce porous sintered nickel anodes with improved sinter resistance and sulfur tolerance.[1703] Titanium carbide (TiC) and $Mg_{0.05}La_{0.95}CrO_3$ also have been claimed to show promising sulfur-resistant anode behavior.[2179] A discussion of potentially more sulfur-tolerant anode substrates can be found in Ref. 628.

Cathodes

Silver was almost exclusively used as the cathode material in most of the pioneering molten carbonate fuel cell work, either as pure silver or as silver-plated stainless steel.[723] CuO and NiO were also examined as cathode materials, as were $Ag-Cu_2O-ZnO$,[259] Pd, and Pt.[224] It has been reported that alloying silver cathodes with small amounts of aluminum (~4%) or zinc (~5.5%) improves their performance in molten Li-K-Na carbonates.[620] Similarly, it has been reported that CuO cathodes also can be improved by the addition of small amounts ($\leq 5.4\%$) of aluminum.[621] This presumably leads to doping of the structure with lithium aluminate, stabilizing it against sintering and porosity changes, as in the case of the anodes discussed in the previous section.

Corrosion problems were, however, soon detected with CuO;[475] consequently, because of the cost of Pt or Pd, silver became the cathode material of choice in the 1960s. Although silver performs remarkably well as an oxygen electrode, via a mechanism involving molecular oxygen dissolution,[162] the long-term stability of silver proved to be poor. At 700°C and 750°C the vapor pressure of silver is about 0.27 mPa and 0.93 mPa, respectively. At 700°C a cathode thinning rate of about 1 mm per 10,000 hours can be expected as a result of volatilization of the metal,[723] although not all the silver will be irrevocably lost, since most will condense out on colder parts of the system. At temperatures even of the order of 650°C, where its vapor pressure would be 1/3 of that at 700°C, pure silver (m.p. 962°C) cannot be used for systems with long lifetimes. Silver loss via evaporation can to a certain extent be controlled by alloying with less volatile constituents or by plating with a protective metal such as nickel, which in the form of nickel oxide produces a coating that is fairly effective in its ability to lower the vapor pressure of the silver.

A further serious limitation is the tendency of silver to dissolve and migrate through the electrolyte. Although its solubility in the melt is low, preferential dissolution and precipitation result in the removal of silver from the cathode and deposition as bulk silver in areas near the anodes, apparently through reaction with diffusing hydrogen. Dendrites may even form between these areas of preferential deposition and the anode, leading to a decrease in cell output and short-circuiting. Silver cathode leads used in laboratory test cells often are plated with palladium or nickel or are encapsulated in oxidation-resistant tubing to reduce losses.

On account of the difficulties experienced with silver cathodes in the 1960s, it has been usual to employ porous sintered nickel oxide (NiO) as the cathode material. When NiO comes into contact with molten lithium carbonate, some Li^+ is incorporated to form a solid solution, giving a reasonable electronic conductivity (~10 Ω^{-1} cm^{-1} at 650°C). To achieve the desired conductivity and to prevent the depletion of lithium from the electrolyte, cathodes usually have been doped with 2~3% lithium in a pre-oxidative step before use, although this is not strictly necessary as *in-situ* doping also can occur. Pre-lithiated, pre-oxidized porous nickel cathodes may, however, have better resistance to changes in pore size and surface area loss than cathodes that are oxidized and lithiated *in situ*.[1703] The chromium content of the cathode should be low, since it has been found that Li_2CrO_4 and other chromium-containing species are about 100 times more soluble in the catholyte than in the anolyte.[485] In consequence, the same plaques should not be used for anodes and cathodes.

Because the IR-drop of the nickel oxide cathode can be a significant portion of total cell IR-drop, cathodes should be as thin as possible (0.10~0.15 mm). Thin cathodes also show improved performance.[457, 1018, 2291] To give *in-situ* oxidized nickel plaques better structural integrity, stainless steel screens have often been incorporated.[1703] Since thin pre-oxidized cathodes for large cells (up to 1 m²) are virtually impossible to handle, the recent trend has been to go back to the use of the *in-situ* oxidation technique.

Since 1980, it has been known that nickel oxide has sufficient solubilty in the melt to reprecipitate near the anode, as described for silver. However, its solubility is much lower than that of the latter. Even so, it is not expected to attain a 40,000 h lifetime except in atmospheric pressure cells under normal operating conditions.[182] This problem is discussed in further detail in Chapter 17.

Electrocatalysts and Current Collectors

Electrocatalysts, in the normal sense of the term (high-surface area material with specific catalytic properties for the electrode reaction in question), are not necessary in molten carbonate fuel cells. At the normal cell operating temperature (~650°C), the relatively low-surface area electrode substrate materials are themselves sufficiently active electrochemically, and rates do not depend much on the nature of the material used. However, electrocatalytic effects have been studied in the literature, for example, on Ni, Au,[158, 937, 2614] and Pt or Pt-Rh alloy.[2614] The electrocatalytic properties of Ni,[2178] Co,[87, 2178] and Cu at the anode have been examined. Silver also has been studied as a cathode catalyst.[159] Perovskites of the type used in the SOFC (see following) initially were reported in the patent literature to be stable alternative cathode materials to NiO,[455] but this was later shown to be inaccurate.[457]

Anode and cathode current collectors in experimental cells usually are made from type 316 stainless steel,[188] although types 310 and 410 have also found favor.[2179] At the anode under high-conversion conditions, nickel or Inconel current collectors are most satisfactory. Comments on practical combined current collector–bipolar plate materials already have been discussed, and are considered in more detail in Chapter 17. Details of screening experiments for materials compatibility in molten carbonate fuel cells,[440] and a review of general materials problems in high temperature electrolysis and fuel cells[1902] are available.

A serious problem in practical MCFC stacks that has not been considered in the preceding is the difficulty of bonding and edge-sealing of parts under thermal stresses; for example, oil-canning effects with thin sheet-metal hardware,[1255] and stack distortion resulting from anode creep. Unfortunately the solutions proposed so far are of a proprietary nature.

Solid Oxide Systems

Electrolytes

The most successful electrolyte material that has been used in high temperature SOFCs is yttria-doped zirconia. To give sufficient electrolyte stability in the 1000°C operating environment, attention has had to be given to the properties and behavior of stabilized zirconias, particularly the interdependence of composition, crystallography, temperature, and conductivity. At elevated temperatures, ageing phenomena associated with decreasing conductivity with time are attributable to subtle crystallographic changes. These can be suppressed to levels that do not seriously affect long-term stability by employing zirconia of sufficiently fine-grain size. The reader should refer to Chapter 10 for a more complete discussion of this and other solid oxide electrolytes. Chapter 18 contains a summary of some recent work.

Electrode Materials

Electrode material for the SOFC must meet the following requirements:[810, 2129]

1. Moderate cost.
2. Good adherence to other cell components.
3. Thermodynamic stability.
4. Chemical inertness.
5. High electrochemical activity.
6. Minimal interdiffusion with adjacent materials.
7. Low volatility.
8. Compatible thermal expansion with other cell components.
9. A superficial resistivity $\leq 0.2\ \Omega\ cm^2$.
10. Ease of fabrication into thin (~30 μm) porous layers that resist excessive sintering.
11. Good mechanical strength.

Cathodes

On account of the highly oxidizing cathode environment at 1000°C, only three classes of possible cathode materials exist: noble metals, semiconducting oxides, and conducting metal oxides.

Among the metals, only silver, palladium, and platinum have been given serious consideration. Silver may be attractive because of its cost, high conductivity, and high permeability to oxygen; however, its vaporization rate is too great at 1000°C.[686] In any case, the pure metal would be liquid at the normal cell operating temperature (m.p. 962°C), so stable alloys would be required.

Only platinum comes close to satisfying the chemical stability criterion, but even it reacts to some extent with oxygen and zirconia,[810, 2128] and is in any case too costly for practical use in the loadings that would be required. In spite of these drawbacks, some serious studies have been conducted using platinum cathodes. For example, Matsushita Electric developed a method for forming a thin platinum electrode substrate onto a zirconia surface. The method involves applying a platinum powder, firing, applying a platinum salt solution over the fired platinum powder, and refiring to pyrolyze the platinum salt to form metallic platinum.[2746] During the fabrication of the zirconia substrate tube a coating containing SiO_2 is first applied. This is then fired, and etched with hydrofluoric acid to improve platinum adhesion.[2745]

Fuel Cell Stack Materials Selection and Design

The second class of cathode materials examined consists of oxides that are semi-conducting or only moderately conducting. Because of their low conductivities, it has been customary to use embedded current collectors to ensure adequate current and potential distributions. Unfortunately, the requirements of these current collectors are similar to those of metal cathodes, so that platinum is the only choice, making this type of electrode structure uneconomical.

The third, and most successful, class comprises cathodes made from oxides with high electronic conductivity, including those in the p-type perovskite group. Some of the most promising materials examined to date have included:[811] Sn-doped In_2O_3, Ca-doped $LaCoO_3$, Sr-doped $LaCoO_3$, $LaCoO_3$, $PrCoO_3$, Bi-doped $LaNiO_3$, Sr-doped $LaMnO_3$, and Sb-doped SnO_2. Doped indium oxide appeared to be particularly promising, since it is a purely electronic conductor with a very low $O^=$ ion mobility; and a study of oxygen transport through indium oxide and its stability was carried out at the Brookhaven National Laboratory.[2718] However, a drawback of this material is its relatively high vapor pressure. Lithium-doped nickel oxides were considered in early work, but were found to lose their conductivity at elevated temperatures. Lanthanum cobaltite has been found to react somewhat with zirconia; however, praseodymium cobaltite has shown considerable promise for long-term operation. Its main disadvantage is its coefficient of thermal expansion, which differs greatly from that of zirconia, giving rise to spalling effects when the cell is thermally cycled. However, it might be possible to control these effects by the use of composite cathodes.

Cathode leads for use in laboratory cells have also been a problem. Since they must conduct electricity from a low temperature environment outside the cell to a high temperature environment inside, only metals can be considered. Again, the choice is limited to the noble metals. As was remarked above, pure silver cannot be used at the cell operating temperature. Some success has been achieved with silver-palladium alloys and with coatings of nickel (m.p.1455°C) and other materials on the surface of the silver.

Anodes

The reducing environment of the fuel gas permits the use of a range of metals as anode materials. In particular, porous nickel has been favored, although other materials also have been examined. As would be expected, nickel has a tendency to exhibit grain growth and densification at the operating temperature of 1000°C. This can to a large extent be controlled by incorporating zirconia into the porous nickel sinter to yield a cermet, or ceramic-metal composite. The difference in expansion coefficient between the nickel and the zirconia can be accommodated by the porous nickel structure, which can deform and stretch under various conditions of thermal stress. Hence spalling of the nickel anode away from the electrolyte layer has not been a problem in the SOFC.

Thin anode films of 30~50 μm thickness are produced by plasma spraying of fine metallic powders or by the sintering of fine-grained metal-zirconia cermets, although other techniques also exist (see Chapter 18). Both processes produce porous films with good adherence. Occasionally the metal layer has been laid down on the zirconia electrolyte by electrodeposition. The thin anode layer should have a superficial resistivity (specific resistivity-to-thickness ratio) of 0.04~0.10 Ω cm^2 at 1000°C.[2129] Table 13-4 presents typical values for thin film anodes on 0.005 to 0.01 cm-thick doped zirconia at 1000°C.[1104, 2128]

In addition to nickel, other materials such as Co; Ni-stabilized and Co-stabilized zirconia cermets; and CeO_2 doped with Y_2O_3 or with La_2O_3 also have been considered as anode materials.[810] Battelle has patented a method of making μm-thick layers of Yb_2O_3-doped zirconia anode coatings using cathodic sputtering techniques.[326] At Westinghouse, the nickel anode (see Chapter 18) is stabilized by electrochemical vapor deposition of yttria-doped zirconia into part of its open porosity.[496]

Interconnect Materials

The interconnect material (ICM) in a solid oxide electrolyte fuel cell is used to electrically interconnect the anodes and cathodes of the thin layer stacked cells in series. The requirements for an ICM are:[2128, 2153, 2157]

1. High electronic conductivity ($\leq$ 50 Ω cm in the cell-operating environment).
2. Low ionic conductivity (especially towards cations).

TABLE 13-4

Resistivity-to-Thickness Ratios of Thin Film Anodes on Doped Zirconia at 1000°C [1104, 2128]

	Resistivity/Thickness (Ω)
Plasma-sprayed nickel	< 0.04
Nickel or cobalt zirconia cermet	< 0.01
Electrodeposited nickel	~ 0.2

3. Chemical and mechanical stability (including low volatility and nonreactivity with adjoining cell components) in both air and fuel gas at 1000°C.
4. Mechanical compatibility (adherence and thermal expansion) with the electrodes and electrolyte.
5. Fabricability into ultra-thin (<40 µm) gas-impervious layers.
6. Absence of mass-transport effects in the presence of chemical gradients that might lead to the formation of voids or high contact resistances.
7. No time-dependent phase changes or recrystallizations between 25°C and 1000°C.
8. Low materials and fabrication costs.

Interconnect materials usually are made from perovskite oxides with the structure $LnMO_3$, where Ln is a rare earth, an alkaline earth, or Pb; and M is one of the group Cr, Co, Mn, Fe, or Ni. These materials are good electronic conductors, similar to the sodium tungsten bronzes, Na_xWO_3 ($0 < x \leq 1$). Partially filled conductive electron bands occur in these materials, which impart high electronic conductivity.

The ICM materials that have been most studied include Sr-, Mg-, or Ni-doped $LaCrO_3$, $CoCrO_3$, and Sr-doped $LaMnO_3$; however, it now appears that only the rare-earth chromites will remain essentially invariant in the fuel cell operating environment.[2718] It is generally agreed that doped $LaCrO_3$ is the most promising ICM material for practical applications. Lanthanum is the preferred rare-earth element because of its cost, together with the conductivity and thermal expansion characteristics it imparts to the mixed oxide. Chromium is unique in that its trivalent state is stable over the 18 orders-of-magnitude difference in oxygen fugacity encountered in the cell. The use of dopants such as Sr, Mg, or Al permits a selective control of electronic conductivity and thermal expansion.

$LaCrO_3$ undergoes an undesirable orthorhombic-to-rhombohedral phase transition at 270°C, involving two forms of the same basic perovskite structure having different degrees of crystal distortion. Fortunately, this transformation can be eliminated between 25°C and 1000°C by substituting Sr or Al for part of the original La or Cr.[2153] The National Bureau of Standards has investigated Sr-doped $LaCrO_3$–and Sr-doped $YCrO_3$– ICMs on Mg-stabilized zirconia and Y-doped ceria.[330] The results showed that the resistance of the interface between the doped $LaCrO_3$ interconnect, and the metal anode is very sensitive to the partial pressure of oxygen in the surrounding atmosphere. For example, if the oxygen partial pressure is high, the resistivity of the ICM is low, even at room temperature. It is believed that this behavior involves the diffusion of oxygen in and out of

Schottky-like barriers in the conductor, changing the number of charge carriers through partial compensation of the Sr^{++} dopant with oxygen vacancies. Doping $LaCrO_3$ with Mg, increases its permeability to oxygen, thereby allowing a sufficient access to oxygen to maintain high electrical conductivity.[2155] With regard to chemical stability of these materials in the cell environment, it has been found that $LaCrO_3$ (and $PrCrO_3$) do not react with yttria-stabilized zirconia at temperatures below 1300°C, whereas $LaCoO_3$ (and $PrCoO_3$) are only stable up to about 1200°C.[1109] Useful operating lifetimes in excess of five years would seem to be readily achievable with doped $LaCrO_3$ interconnect materials.[2716]

Workers at BNL have studied the use of materials with plastic properties as ICMs to allow stress-relief under operating conditions. These include Nb-doped rutile, further doped with soda-lime glass, together with TiO_2-doped iron oxide glass composites.[2157, 2158] Because of their physical properties, these materials result in less cracking under thermal stress than doped perovskites. In addition, they exhibit attractively low resistivities: For example, that of Nb-doped TiO_2 is about 10 Ω cm at cell operating temperatures. In spite of this, none of these materials appears likely to surpass doped $LaCrO_3$ as a practical ICM.[2327]

Table 13–5 lists some of the properties of a number of ICMs. In addition to the materials listed in the table, $La_{0.8}Sr_{0.2}Cr_{0.8}Ni_{0.2}O_3$ also has been

TABLE 13–5

Selected Physical Properties of Various Interconnection Materials [1109, 1809]

	Resistivity, Ω.cm		Coefficient of Thermal Expansion at 1000°C, K^{-1}	Melting Point, °C
	25°C	1000°C		
$La0._{0.84}Sr_{0.16}CrO_3$	0.2~1.0	< 0.03	—	—
Mg- or Ni-doped LaCrO3	—	< 0.5	—	—
$LaCrO_3$	—	10	~9 × 10⁻⁶	2430
$CoCrO_3$	—	< 50	—	—
PrCrO3	—	1	~9 × 10⁻⁶	—
$La_{0.5}Sr_{0.5}CoO_3$	10^{-4}	~6 × 10⁻⁴ *	—	—
$LaCoO_3$	—	0.1	~21 × 10⁻⁶	~1400
$PrCoO_3$	—	0.1	~21 × 10⁻⁶	—
$LaNiO_3$	—	—	—	~1400
$La_{0.7}Sr_{0.3}MnO_3$	~9 × 10⁻²	0.01~5	—	—
$LaMnO_3$	—	—	—	~1880
$La_{0.5}Se_{0.5}FeO_3$	~7 × 10⁻²	5.5 × 10⁻³**	—	—
$LaFeO_3$	—	—	—	~1880

*1200°C
**900°C

Fuel Cell Stack Materials Selection and Design

identified by the Battelle Institute as a promising candidate for this application.[276]

A number of theoretical and practical studies have been made to understand and improve the characteristics of ICMs. For example, the University of Illinois has developed a conduction model for rare earth perovskite oxides.[2717] The diffusion coefficients and solubilities of oxygen in interconnection materials have been studied at BNL.[2751,2752] The University of Missouri-Rolla, Westinghouse, and ANL have investigated the dependence of the thermal expansion, lattice parameters, rates of volatilization, and electrical conductivities of $LaCrO_3$-based oxides as a function of composition.[74] Westinghouse has published discussions of different methods of preparation of modified lanthanum chromite ICMs, including r.f. sputtering, chemical vapor deposition, and electrochemical vapor deposition.[2153, 2674] The developers particularly favor the latter method to produce material doped with Mg to improve electronic conductivity, and with Al to improve thermal expansion properties. Westinghouse has reported that this formulation has remained leaktight and uncracked over the duration of a 700 h operational test that included three thermal cycles.[2154] This so-called electrochemical vapor deposition technique is discussed in more detail in Chapter 18.

Dornier System G.m.b.H. has developed methods for making doped $LaMnO_3$, $LaCoO_3$, and $LaNiO_3$ components and for bonding conductors in high temperature fuel cells or electrolysis cells.[2212] The company has filed for patents for the production of mixed oxides such as $La_{0.8}Pb_{0.2}Cr_{0.3}Mn_{0.7}O_3$, whose electronic conductivity is claimed to be independent of oxygen partial pressure at 1000°C.[2214] In another patent they prepare these materials by spray-atomizing a solution of the required cations with sufficient oxygen into a reaction tube at 500~1000 K, and by precipitating the resulting oxide in a cyclone, to yield products such as $La_{0.8}Ba_{0.2}Cr_{0.3}Mn_{0.7}O_3$.[2213]

Japanese workers have published details for fabricating high conductivity (< 10 Ω cm) metal oxide structures that undergo low evaporation losses.[1470, 1471] A typical composition is $La_{1-x}Ca_xCr_{1-y}Ni_yO_{3-z}$, where $x = 0$~0.1, $0 < y < 0.6$, and $z < 0.2$. The rate of evaporative loss during 200 h at 1800°C was minimal when $y = 0.33$, and was relatively unaffected by the value of x.[1470] The Battelle-Frankfurt has reviewed the selection of interconnect materials that form gastight and mechanically stable connections between solid ceramic electrolytes, and electrical connections between electrodes.[275]

Miscellaneous Cell Components and Sealants

The structural requirements for solid oxide fuel cells are provided to a large extent by the rigid electrolyte, hence the question of structural materials for the cell body is not as important as it is for the MCFC. A review of the materials problems encountered in SOFCs[1196] and a survey of the use of ceramics in battery and fuel cell applications[482] have been published.

Of greater concern is the problem of high-temperature bonding materials and sealants. A general discussion of ceramic-metal battery seals for aerospace batteries has been published by Ceramaseal, Inc.[449] Sandia Laboratories has developed a technique for sealing platinum to zirconia that involves deformation of the platinum against the zirconia.[2298] Kent, George, Ltd. in the United Kingdom has developed a method of bonding a metal such as platinum to a ceramic such as yttria-stabilized zirconia that involves coating the zirconia surface with a thin layer of paste containing small metal flakes. This coating is bonded to the substrate by sintering in air.[2152] A second method involves bonding the metal by painting, spraying, or plasma coating, followed by sintering in a hydrogen atmosphere.[2091] The British National Research Development Corporation has developed a heat-resistant ceramic-to-metal sealant containing a microcrystalline material composed of CaO 5~40, SiO_2 35~70, MgO 9~30, and Al_2O_3 3~30 percent. This sealant material is claimed to be stable to 1050°C without softening, and to have a leak rate of less than 10^{-7} Lusec.[1581] The U.S. Government has developed a high-strength hermetic ceramic-to-metal sealing method that involves heating a chemically cleaned wire-like metal gasket and a ceramic member (Al_2O_3 or ZrO_2) while simultaneously deforming the metal gasket against the ceramic member at a temperature equal to 30~75% of the melting point of the metal.[741] Toshiba Electric Co., Ltd. has developed an oxide composite for sealing metal to metal, ceramic to ceramic, or metal to ceramic. The sealant contains Al_2O_3 35~55, CaO 35~55, Y_2O_3 0.3~30, SiO_2 <4.9, MgO <15, SrO <10, BaO <10, ZrO_2 <10, B_2O_3 <10, Na_2O <15, K_2O <15, Nb_2O_5 <20, and Ta_2O_5 10 weight percent.[2412-2414] When gas-tightness is especially important, gold washers can be used in place of ceramic powder pastes to seal cells.[2802] The unique Westinghouse cell design, which avoids tight sealing of components, is described in Chapter 18.

Some single cell solid oxide fuel cell materials life data at 1000°C for periods up to about four years are available, but further long-term data are still required to give confidence in system viability.[810]

B. FUEL CELL STACK DESIGN

A fuel cell system consists of an assembly of a large number of unit cells into a *stack* or fuel cell battery. The design of the stack must provide for mechanical integrity; maintenance of pressure seals; intercell electrical connections; supply of reactants from a common manifold to each cell, together with the recycling of reactants as necessary; the removal of reaction products; and the removal of heat. In addition, sufficient electrolyte must always be present for correct operation and to prevent gas leaks between electrodes. In alkaline systems with a flowing electrolyte design, the flow loop always contains excess electrolyte and acts as a reservoir. However, in matrix cells, internal reservoirs must be incorportated or means of electrolyte replenishment must be used. In this regard, aerospace alkaline cells incorporate electrolyte reservoir plates in each cell, and the PAFC generally requires sufficient internal reservoir capacity to compensate for electrolyte evaporation over the lifetime of the cell. The particular problems posed by the reservoir of the PAFC are considered in detail in Chapter 16.

Design Parameters and Correlations

The energy conversion efficiency in the stack is perhaps the most important parameter in a fuel cell system, and is usually the starting point for system design. Equations have been developed that give the fuel cell energy conversion efficiency as a function of the maximum theoretical cell voltage and of the current efficiency.[320] These equations have been generalized to include fuel gas mixtures containing quantities of conventional fuels such as methane, which cannot participate directly in the power-producing electrochemical reactions. Theoretical efficiencies have been determined for a large number of fuel cell systems such as hydrogen-oxygen, hydrogen-chlorine, hydrazine-oxygen, hydrazine–hydrogen peroxide, zinc-oxygen, carbon-oxygen, carbon monoxide–oxygen, methane-oxygen, propane-oxygen, ammonia-oxygen, methanol-oxygen, formaldehyde-oxygen, and formic acid–oxygen.[1862] In addition, the efficiency of converting hydrogen via a number of different types of fuel cells, including acid, alkaline, molten carbonate, and solid oxide systems, has been evaluated.[2325] The efficiencies, probable costs, and environmental emission levels of fuel cells have been compared with those of other energy conversion systems such as conventional internal combustion engines, gas turbines, and closed cycle (Stirling) engines.[2172]

Numerous studies on the optimization and correlation of stack parameters have been made. Such studies help the designer to understand the interrelationships that determine the performance of a given design. For example, the voltage-current curves and performance of a 1 kW 40-cell hydrogen-air stack have been explained on the basis of a thermodynamic development centering around the supply of fuel.[2113] In another study, the functional relationships that exist between electrochemical properties, cell structure, and electrolyte volume have been used to determine cell voltage as a function of electrolyte content.[2620] Other publications include: a general method of fuel cell optimization based on the mass of the fuel cell;[59] the use of equations for selecting the optimum structural and working parameters of fuel cells based on the current distribution throughout the stack, and that of current-producing and side reactions;[58] modeling the relationships between operating conditions, capital cost, and fuel consumption to determine the optimum operating current density as a function of energy consumption and investment;[1378] the effects of the current collector arrangement on the current density distribution in cells with thin monopolar electrodes, dissolved reactants, and gaseous reaction products;[1504] optimization of the operating conditions for hydrogen-air fuel cells;[2253] and methods for obtaining the optimum operating voltage or current density for a hydrogen-air fuel cell operating at a fixed power output at a given air-to-fuel ratio.[1867]

Electrical Configuration

Since each unit cell in most systems is advantageously constructed in the form of a thin sandwich or wafer, all stack configurations must first take this into account. For cells with gaseous reactants, each electrode in such an arrangement is backed by a reactant flow chamber or plenum permitting a uniform reactant supply to the total surface of the electrode. The structure is kept as thin as possible to maximize energy density and to minimize mass-transport losses and IR-drop, the lower limit being set by the pressure drop for the reactant across the plane of the electrode. Except in some alkaline systems that are still of 1960s design, the electrolyte is not circulated and is retained as a thin layer between the electrodes, usually via surface tension in a stable porous non-electronically conducting matrix material. This arrangement reduces ionic transport and ohmic resistance losses in the electrolyte to a minimum, serves as a barrier between the reactant gases, and prevents short-circuiting by accidental contact of electrodes subjected to mechanical pressure. The electrodes themselves either must be good electronic conductors, or if their conductivity is insufficient, must be intimately bonded to a conducting current collector, so that the ohmic losses in the electron flow circuit can be minimized. Because of the wafer-like shape of each cell, a large number can be stacked efficiently in a small space to create a compact stack capable of producing a relatively large d.c. power density.

The preceding type of stack configuration is the commonest employed, but it does possess certain disadvantages. For example, because of the small thickness of the unit cells, it may be difficult to retain and replenish the electrolyte, if addition to replace loss or irreversible change is necessary. Construction of thin electrodes of sufficient strength, rigidity, and

Fuel Cell Stack Materials Selection and Design 443

conductivity may also prove to be difficult. In addition, since the close spacing of the cells restricts access to the electrodes, the efficient supply of reactants and removal of reaction products, heat, and electricity can result in difficulties.

Stacked cells can be connected electrically by bipolar or monopolar arrangements. In medium-to-large cells, the bipolar arrangement is most commonly used. In it, cells are simply stacked upon one another so that the plane of the cathode of one cell contacts that of the anode of the adjacent cell via an impervious electronically conducting sheet or plate. The bipolar arrangement has the advantage of connecting the cells electrically in series without the need for external wiring between each cell, thereby eliminating ohmic interconnection losses. More important, in the bipolar arrangement, the current is collected perpendicular to the plane of the electrodes; therefore the conduction distances are short, leading to very low IR-drop between individual cells. In contrast, in the monopolar arrangement, the current must be carried along the plane of the electrodes to convenient anode and cathode collection points or edge terminals, as in a lead-acid battery. This results in IR losses along the plane of the electrode as the current for the whole cell is accumulated from the points remotest to the terminals to the latter themselves. These IR losses distort the natural current distribution in the bipolar cell, and lead to lower average current densities and lower cell voltages, hence lower system efficiency.

However, a constructional disadvantage of the bipolar system is the fact that a conducting separator in intimate electronic contact with both cells must be placed between each cell to prevent mixing of the fuel and oxidant supplied to adjacent anodes and cathodes. This separator may be in contact with a continuous film of electrolyte on both sides and therefore may be subjected to the electrochemical potential difference between individual anodes and cathodes, resulting in corrosion. In addition, the whole series of cells must be conceived as one block or pile (i.e., stack) and suitable mechanical arrangements must be made to ensure continuous electronic contact without any partial breaks or interruptions at any point between the surfaces of individual cells or groups of cells. Finally, a minor disadvantage of the bipolar arrangement is that it can only be series connected, and is therefore a high-voltage d.c. device, which is usually desirable for power production with electronic conversion of d.c. to a.c. The monopolar system can also be parallel-connected to give a low-voltage, high-current system, which might be more useful for certain applications; for example, for direct connection to electrochemical synthesis cells.

In the bipolar arrangement, each cell must be individually supplied by anodic and cathodic reactants, whereas in a monopolar system, alternate cells may be reversed so that they are placed anode to anode, and cathode to cathode in the so-called bicell configuration.[1488, 1735] In this monopolar configuration, the unit cells usually share common gas chambers, thus simplifying the reactant feed system. Conversely, it now becomes necessary to make the electrical interconnections externally via a complex alternating pathway, if series connection of cells is desired (cf. Fig. 13–1,

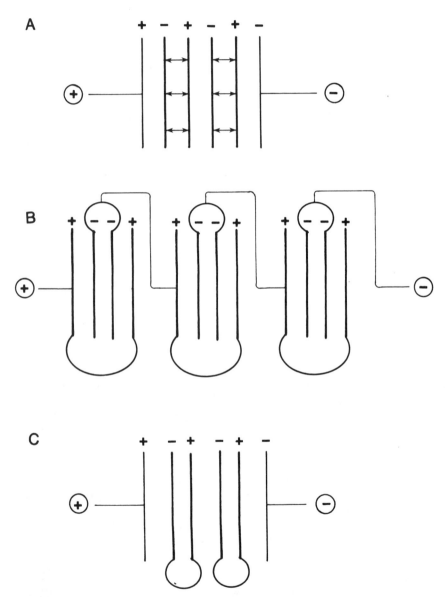

Figure 13–1: Different intercell connection arrangements. A: Bipolar (internal back-to-back connections with lowest internal resistance); B: Edge-collected bicell;[177] C: Edge-collected monopolar.

which summarizes the connection alternatives for both arrangements). This usually results not only in a further increase in ohmic losses, but also in sealing and corrosion problems.

Although it is customary to connect the cells within a given stack exclusively in either a bipolar or monopolar arrangement, occasionally both bipolar and monopolar electrode groupings may be found within the same stack.[1461]

For applications that require mobile fuel cell power plants, it is important to minimize stack weight and volume. As an example, Hitachi has devised cell-stacking structures designed to improve the specific weight and volume power densities of low-temperature fuel cells.[1371,1925] To avoid the problems of external electrical connection that are encountered with conventional monopolar stacked structures, Hitachi has developed a bipolar stack design that makes use of a U-shaped electrode frame and an I-shaped reagent chamber frame. It is claimed that this U-shaped cell-stacking structure reduces internal ohmic connection losses to levels lower than those encountered in conventional bipolar designs. In two further developments, use is made of three- and four-chamber spiral stacking structures that consist of only one planar component. In these spiral structures each cell component is stacked in series, rotated from the previous component by 120°, resulting in a spiral flow of fuel, oxidant, electrolyte, and electricity in the stack. Table 13-6 shows the improved performance of these new designs. Alsthom in France also developed a spiral-type fuel cell that was claimed to have a reduced electrical resistance.[529]

Manifolding Arrangements

The monopolar bicell system described in the previous section can have one of the simplest manifolding arrangements, since the gap between

TABLE 13-6

Performance of Hitachi Stack Designs for Use in Alkaline Hydrogen-Oxygen Fuel Cells[1902]

	Power Output (W)	Current Density (mA/cm2)	Electrode Area (cm2)	Temp. (°C)	Stack Power Density	
					W/kg	W/L
Conventional monopolar	—	—	—	—	35	45
U-shaped bipolar	500	80	256	30	64	93
3-chamber spiral	150	80	256	65	77	98
4-chamber spiral	150	70	56.3	25	86	139

electrodes of adjacent cells can be made relatively wide compared with the length of the cells. When air (scrubbed air for alkaline systems in most cases) is used as the oxidant, natural convection can be used to supply this reactant if the cells are relatively small. Since the transverse IR-drop in the monopolar components also puts an upper limit on the total depth of each cell measured from the terminal area, this approach is feasible in cells having a depth of about 15 cm, with an upper limit of perhaps 30 cm, depending on the conductivity of the current collector used in the cathode and on the current density employed.

Systems of this type, requiring at the most, an external air blower, have been generally favored for metal-air systems, which are fuel batteries in which the anode (in this case in a bicell arrangement) is a reactive metal plate shared between adjacent cells. Such systems have included zinc-air,[2698] iron-air,[2698] and more recently aluminum-air,[2174, 2175] all with alkaline electrolyte. The latter has been of particular importance to alkaline fuel cell development, which has otherwise remained rather stagnant, because it has allowed the introduction of the advanced pyrolyzed macro-cyclic cathodes described in Chapter 12.

Small alkaline systems using the bicell arrangement[394, 1488, 1735] can be simply manifolded; for example, by using natural convection on the cathode side. However, the anode side requires more care, since the product water in a H_2-O_2 cell must then be either removed from the anode side, or from the electrolyte itself. In the first case, a condensing system is required in an anode feedback loop (active water rejection); or a passive water rejection system via the electrolyte is required as an integral part of each cell if circulating electrolyte is not used (as in the Allis-Chalmers aerospace system, Ref. 176). Otherwise, water rejection can be in specialized evaporators in series with a circulating electrolyte.[394, 1792, 2368, 2369] In contrast, acid systems (the PAFC and the SPE) reject water on the cathode side, so that oxygen systems generally require a cathode feedback loop, whereas air systems with low cathode reactant utilization can use simple active evaporative product removal under the right cell temperature conditions. If pure hydrogen is the fuel, the anode side can be dead-ended. This approach has often been used, but it has its particular design and operational problems.

A terrestrial fuel cell system using both diluted fuel (e.g., reformed methane) and diluted oxidant (e.g., air, or air with CO_2 in the case of the MCFC) requires both inlet and outlet manifolds for both the fuel and oxidant, both of which may require feedback loops. In general, evaporated product water can be removed using the appropriate reactant loop, depending on the electrolyte employed, at least in systems with stationary electrolyte. The cells will also require a cooling system. If the design uses circulating electrolyte (which we consider to be an obsolete approach), then cooling can be conducted via the electrolyte circulation system, which can also serve for water removal, for example using special evaporation cells as in the Siemens alkaline system.[394, 1792, 2368, 2369] If stationary (matrix) electrolyte is used, a separate cooling circuit may be necessary if heat removal cannot be effectively carried out using, for example, the process air-circulation system. Thus, in the most general case, up to six different manifolds may be necessary for each cell; for example, anode inlet-outlet, cathode inlet-outlet, and cooling flow (or electrolyte circulation) inlet-outlet.

Supply and exit systems requiring six manifolds can therefore be a considerable problem for the designer of a typical four-sided stack, although what are essentially flattened six-sided arrangements (particularly the Westinghouse approach to the PAFC stack discussed in detail in Chapter 16) are certainly possible. For liquid cooled stacks, as used by UTC and Engelhard for the PAFC, the cooling manifolding in the form of small-diameter cooling tubes into special cooling plates (e.g., every five active cells), can be incorporated into the reactant gas manifold hardware because of its small overall volume. If gas (i.e., air) cooling is used, the manifolding will depend on whether process air can be used in one form or another, in which case a common supply and exit manifold for both reactant and cooling air can be used, making only four manifolds in all. This is the case in systems of very varied conception and design; for example, in the original DiGas™ air-cooled ERC PAFC (Chapter 16), and in the usual MCFC (Chapter 17) and SOFC (Chapter 18) arrangements. However, if process air and cooling air are completely separated, which results in improved heat-transfer design in a high-performance PAFC, then six-sided manifolding (cf., Westinghouse, preceding) is necessary. Finally, a circulating electrolyte alkaline system requires small-diameter manifold tubing that occupies a much smaller volume than that of the gas manifolds. However, it must have a parallel entrance and exit from all cells in the stack; it is, therefore, more complex than the liquid-cooled PAFC case.

If flows of reactants, coolants, or electrolytes are such that they can be accommodated in channels of modest diameter, then they can be accommodated in the form of holes through the stack itself perpendicular to the plane of the cells, with supply to individual cells by narrow parallel take-offs and exit channels. This arrangement, referred to as internal manifolding, was originally developed for the electrolyte flow system in some alkaline stacks of the 1960s. For this application, the reduction of the supply channels to give the smallest diameter consistent with acceptable circulation power requirements is necessary to reduce the coulombic current losses that result from the ionic intercell connection via the electrolyte. This important topic, which may involve serious corrosion problems, has been discussed.[2791]

Any internally-manifolded system requires careful sealing between components of separate cells to prevent leaks. In principle, the liquid-cooled PAFC stacks could have their cooling systems manifolded in this way, but in practice the all-carbon construction of the stack makes sealing and thermal stresses difficult to handle, so the cooling liquid is supplied via external tubes rather than through internal holes. Internal manifolding could perhaps be used for the reactant gas circulation systems in PAFC stacks, but since each cell must be electrically isolated from the next, dielectric seals between each pair of cells must be provided in the internal manifold channels. However, it was decided at an early stage that the development of reliable seals would be difficult, so all current PAFC stacks (with the exception of a Fuji electric design; Chapter 16) use external gas manifolds, in which the dielectric seal to prevent shorting of cells is a sheet (or gasket) applied perpendicular to the cell plane along the side of each stack. The manifold is therefore a *box* supplying all the cells in parallel. We should note that this concept would be totally impractical for the manifold in an electrolyte circulation system, since hermetic sealing is virtually impossible and the volume of electrolyte in the external manifold would then be large enough to short out the entire cell stack.

In systems of the simplest type, the stacks will be rectangular with parallel flow channels for both anodic and cathodic reactants arranged along each side and at right angles. In such cases, the flow channels for both reactants and products also cross each other at right angles, so the system is termed *cross-flow*. Because of its simplicity of manifolding, this has been generally used by most recent fuel cell stack developers where circumstances have permitted. This includes the PAFC and MCFC stacks developed at UTC, as well as the liquid-cooled Engelhard and air-cooled ERC PAFC systems.[177] More details concerning the manifolding arrangement of these systems are given in Chapters 16 and 17.

While the geometric design of the flow channels of the anodic and cathodic reactants can be made as complicated as desired, only two practical arrangements have been demonstrated that can advantageously be used as alternatives to the cross-flow system. From the practical viewpoint, any alternatives must include the simplicity of the mechanical flow patterns of the channels inside the stacks, as well as the design of the manifolds feeding them.

As a result, the only alternative patterns that have emerged involve approximately parallel reactant flow, in which both fuel and oxidant enter at one face of the stack in a relatively complicated manifold arrangement,

Fuel Cell Stack Materials Selection and Design

and exit at the opposite face. The flow patterns can be arranged so that both flow together, going from *rich* to *lean* compositions (coflow), or the contrary (counterflow). As would be expected, the former gives the highest current density (and temperature) at the reactant inlet, where the thermodynamic driving force is greatest. However, in the counterflow case, the local thermodynamic reversible potential stays approximately constant as the reactants cross the cell, since the chemical potential change that results from the consumption of the fuel roughly compensates for that of the consumption of the oxidant. If rectangular cells are used, the manifolds can be applied to two parallel faces of the cell (Westinghouse PAFC), leaving the other two parallel faces for gas cooling manifolds. In the case of the MCFC, either internal or external parallel manifolding (or a combination of both) may be possible (Chapter 17).

These alternative flow patterns are found in the Westinghouse PAFC stack (roughly counterflow geometry), discussed in Chapter 16, and in some of the high-temperature fuel cells. Among the latter, Westinghouse SOFC design only allows the coflow configuration (Chapter 18), whereas the MCFC can advantageously be designed around either co- or counter-flow configurations (Chapter 17).

Some examples of different manifolding arrangements (Fuji Electric internally-manifolded 30 kW PAFC stack, externally-manifolded cross-flow MCFC, and the Ishikawajima-Harima Heavy Industries internally-manifolded co-flow MCFC) are shown in Fig. 13-2.

Pressure Seals and Pressure Control

Some of the earlier alkaline fuel cell stacks were sealed by complete encapsulation of the stack in an epoxy compound. (For examples, see Refs. 138, 453, 454, 1267.) This method, of course, renders it impractical to dismantle the stack for inspection, adjustment, replacement, or repair; and in addition makes the system very susceptible to cracking under thermal cycling conditions. A special silicone rubber adhesive (RTV-S691) that could find use in low-temperature alkaline fuel cells has been developed for the European Space Agency by Wacker-Chemie GmbH.[1115] This material, which was developed for use with solar collectors, is said to remain flexible at both very high and very low temperatures.

Low pressure fuel cells may be pressure-sealed by mounting the electrodes in each cell between a picture-frame gasket and clamping the stack together, using either bolts passing through the gaskets or an external clamping assembly. The same method is used in the high-pressure utility PAFC.[177] An alternative sealing method, which seems to be the only

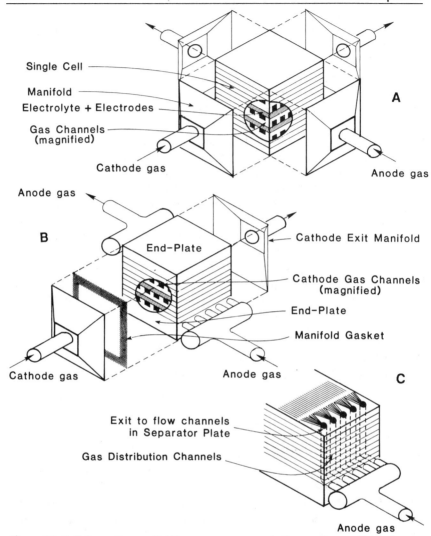

Figure 13–2: Schematic manifolding arrangements. A: Externally-manifolded cross-flow (phosphoric acid systems; Energy Research Corp., International Fuel Cells and Mitsubishi Electric molten carbonate systems); B: Schematic of Fuji Electric hybrid manifold cross-flow molten carbonate system (external cathode, internal anode); C: Flow pattern from gas entry port to bipolar separator plate and flow channels in B; D: Exploded view of Ishikawajima-Harima Heavy Industries co-flow molten carbonate system. Gas entry ports are as in C; E: Bi-polar separator plate for D (active area, 3600 cm^2).

Fuel Cell Stack Materials Selection and Design

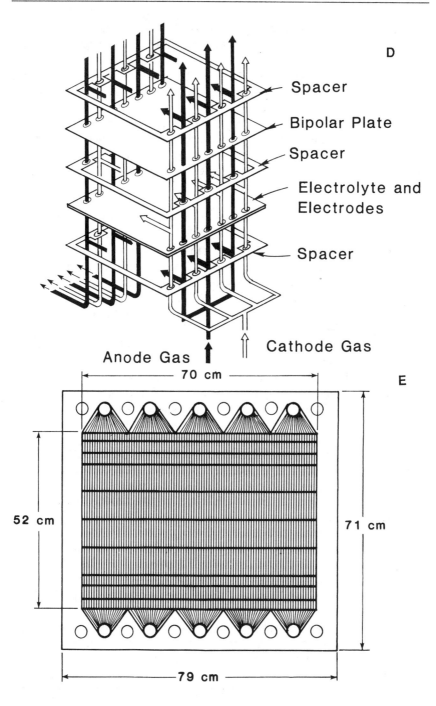

practical one for use in the MCFC, involves the use of the liquid electrolyte itself to form wet capillary seals. This is achieved by sandwiching together an electrolyte-saturated matrix, the electrode, and a separator plate.[2223,2492]

Gas seals are a particular problem as cell operating pressure increases. If the seals are not effective, gas can leak from the stack to the external environment. In addition to reactant loss, this may result in an explosive or environmental hazard. Unless they are intended for use with cryogenic reactants for space applications, system considerations dictate that alkaline fuel cell systems will operate at ambient pressure. However, the present trend in the case of the PAFC is to operate stacks at higher operating pressures (and temperatures) than formerly, so as to increase efficiency and power output per unit of electrode area, as well as to reduce the costs of piping. This subject is discussed in Chapters 12 and 16. For MCFCs, atmospheric pressure operation is favored in simple natural gas systems, but large generators operating in conjunction with an integrated coal gasifier will do so at high pressures (Chapter 17). The present generation of Westinghouse SOFCs are not sealed, and presently can only operate at atmospheric pressure (Chapter 18).

In general, operating pressures of more than about three atmospheres are considered too high for stack and manifold seals of low-pressure type. For stacks operating at higher pressures it is conventional to enclose the whole cell stack in an external steel pressure vessel under a blanket of pressurized nitrogen or air. By keeping the pressure of this external gas blanket equal to or greater than that of the internal stack pressure, stack seal integrity can be maintained even if low-pressure individual cell seals are used. A well-designed pressure vessel should be simple, cheap, and light; should not require a massive amount of precision machining or complex hardware development; and should be readily mass produced. Since cost is related to wall thickness per unit area, which is again directly related to design pressure, cost optimization of this part of the fuel cell system will depend on operating pressure, which is again related to fuel cell performance and cost

To prevent internal gas crossover through the electrolyte matrix from one reactant compartment behind one electrode over to the other, the pressure differential between the two compartments should be less than about 0.2 atm. To achieve sealing of this effectiveness in the pressurized MCFC, a bubble-pressure barrier (BPB) is necessary (see Chapter 17).

Gas cross-over can also occur across the bipolar plate, if separator plates made from porous materials such as graphite are used. The latter can equally absorb electrolyte from the electrolyte matrix. This problem of electrolyte extraction from the electrolyte matrix by absorption into separator plates has been encountered in both phosphoric acid fuel cells and in molten carbonate fuel cells. If electrolyte finds its way through to the next

Fuel Cell Stack Materials Selection and Design

cell, the usual result is catastrophic corrosion as a result of the creation of shunt currents throughout the stack.[177] It can only be solved by making the separator plate from a nonporous material, such as by filling with a suitable polymer. In this connection, PTFE is not very effective, since its permeability to gases is high: It protects the graphite bipolar plates from corrosion, but results in coulombic losses.

To maintain the required low pressure differentials across the cell, pressure control systems can be used that employ membrane valve control loops[2387] or ion-permeable partitions between anode and cathode compartments, in conjunction with special current regulating devices.[1291] By keeping one electrode compartment at a slightly higher pressure than the other electrode compartment, it is possible to prevent the passage of gaseous contaminants such as hydrogen sulfide from one half-cell to the other.[2801] When the stacks are not operating, the pressures in the reactant chambers are often kept above some predetermined value in order to prevent ingress of unwanted contaminants.[1019] Pressure control valves also can be used to connect the fuel cell stacks with the fuel processing subsystem.[910]

Stack Design Concepts

Many early design concepts are given in Ref. 1613, which include both monopolar and bipolar types, including the alkaline bicell system whose development has continued for specialized purposes.[1488, 1735] Many AFCs, including the Siemens system[1792, 2368, 2369] as well as the NASA-UTC advanced lightweight fuel cell,[1706] still use the monopolar approach. Both are described in detail in Ref 394. However, future AFC developments seem to be headed in the direction of bipolar technology. Some of the more advanced types of bipolar AFCs are discussed in Ref. 394, and those of SPEs in Ref. 176. PAFC and MCFC stack designs are discussed in detail in Chapters 16 and 17. Apart from the monopolar (e.g., Westinghouse) and bipolar (monolithic) SOFCs, which have unique construction problems from the viewpoint of stacking (Chapter 18), all mono- and bipolar fuel cells currently use a tight layered structure.

The patent literature for fuel cell design shows that a filter-press type of layered structure is used in the stack design of the majority of fuel cells. This consists of two end-plates, stressed by long bolts sometimes equipped with springs, which sandwich the whole group of stack components together in an arrangement applying as uniform a load as possible. Organizations that have used this type of design include Alsthom,[423, 603, 2094, 2654] Hitachi,[1133-1135, 1469, 2417] Electrochemische Energieconversie N.V.,[761, 1605, 2573] the French Institute of Petroleum,[578, 594, 596, 1020, 1021] Stamicarbon B.V.,[63] Siemens

A.G.,[1462, 2270, 2367] United Technologies Corporation,[1542] and others.[1623] A variation on the filter-press structure is to be seen in the Siemens alkaline system, which uses an inflated metallic bellows as a current collector between each cell.[177, 1792, 2368, 2369] This has the advantage of giving a much more uniform load than that in a simple filter-press arrangement, but at the expense of complexity and difficulty of current collection. Since each bipolar plate is inflated, current is effectively collected around its periphery, so the arrangement would have some of the current distribution problems of a monopolar system if developed in larger sizes.

Occasionally a novel configuration such as a rotating assembly that involves the alternate immersion and removal of the electrodes in the electrolyte[801] is proposed, but such nonconventional stack designs are very rare and would seem to be generally impractical.

CHAPTER 14

Reactant Processing and Clean-up Systems

A. FUEL PROCESSING

Although pure hydrogen energy carrier systems (*the Hydrogen Economy*) have received a great deal of attention over the past few years,[393] with few exceptions,[394] large-scale fuel cell systems currently under development are designed to operate on feedstocks such as hydrocarbons or coal, which must be upgraded to hydrogen-rich mixtures for use in the fuel cell stacks. Petroleum-derived fuels are processed by steam reforming, partial oxidation, or pyrolysis, while coal will be gasified. The methods and equipment that are (or will be) used to effect these upgrading processes have been discussed in detail in Chapter 8.

Hydrogen Enrichment

If required, hydrogen-containing fuels exiting the fuel processor can be made even richer in hydrogen using a number of techniques.

The rapid transport of hydrogen through palladium,[1089, 1634] already briefly mentioned in Chapter 10, has occasionally been used to purify reformer gases by passing the gases through diffuser membranes made from palladium-silver alloys.[793, 2461] However, in addition to being expensive, these alloys often have been unable to withstand vibration, thermal cycling, and deterioration from some of the reformed hydrocarbon products.[2134] The diffusion of hydrogen in thin palladium films can be significantly slower than in bulk palladium if the grain size is small, since the large numbers of grain boundaries present high resistance to hydrogen atom diffusion.[1166] This problem can be corrected to some extent by annealing, which increases grain size. Thin palladium membranes can be made by vapor depositing palladium or palladium alloys on thin sintered porous alumina plates having pore diameters ≤ 20 μm, provided that the alloy layer is thicker than the pore diameter.[1260] It also is possible to vapor deposit 0.5~17 μm-thick palladium or palladium alloy films on glass plates, from which the thin membranes can subsequently be peeled.[1261] Since the palladium diffusion method produces pure hydrogen, it may be especially valuable for small alkaline fuel cell applications. However, it can also be used with portable acid fuel cells that must start up from low temperature (see Chapter 6).

A second method of hydrogen enrichment that may have some value in fuel cell applications is the use of selective permeation through gas-permeable membranes. Of particular note are the Prism™ separators developed by the Monsanto Company.[1800] Monsanto Prism™ separators consist of special gas-permeable membranes fabricated into hollow fiber bundles of large surface area. Hydrogen-containing feed gas at a relatively high pressure is passed over the outside of the bundles, and hydrogen selectively permeates the bundles, to the lower pressure side inside. A pressure differential of ≥ 7 atm is usually required to achieve effective separation. The nature of the membrane material is such that so-called fast gases such as hydrogen, helium, carbon dioxide, and water vapor dissolve in and pass through the membranes much faster than so-called slow gases such as oxygen, methane, carbon monoxide, nitrogen, and C_2~C_6 aliphatic hydrocarbons, thereby enabling significant concentration of fuel streams containing more than about 30% hydrogen. Depending on the feed purity and pressure differential available, Monsanto claims that the hydrogen purity of the permeate gas typically is 86~96%, and can be as high as 99%. With further scrubbing, this may make the process of use even for alkaline systems, as well as for acid systems that must operate (or start up) at temperatures at which anode CO tolerance is low.

In addition, the process is reported to be tolerant to significant levels of impurities such as hydrogen sulfide, ammonia, water vapor, hydrocarbons (ethane through hexane), aromatics (such as benzene and toluene), carbon disulfide, and diamines. As the temperature increases, however, this tolerance may decrease. In addition, the operating temperature range for the membranes is rather restricted, being only 0°C to 55°C.

The Prism™ system has been in operation since 1977, and a lifetime in the order of at least five years is claimed, with no decrease in performance. However, to achieve this operating life, it may be necessary to pretreat the feed stream to remove solid particles or chemical contaminants.

Continental Oil Company has developed a novel process in which the gas from a coal gasifier, containing hydrogen, carbon monoxide, and carbon dioxide, is enriched in hydrogen via formate synthesis – decomposition.[2798] In the formate synthesis zone of the reactor, the gas contacts an aqueous solution containing alkali metal carbonate or bicarbonate to produce a solution containing formate and an effluent gas. The formate is subsequently catalytically decomposed in a formate decomposition zone to yield hydrogen-rich gas and a solution of carbonate or bicarbonate:

Formate synthesis $\quad CO + NaHCO_3 \rightarrow HCOONa + CO_2$
Formate decomposition $\quad HCOONa + H_2O \rightarrow H_2 + NaHCO_3$

In general, the only reason why hydrogen enrichment should be necessary in a fuel gas intended for use in a fuel cell system would be if the enrichment process itself could remove a contaminant, for example CO, that could result in electrode poisoning under normal operating conditions. This would be the case for a low temperature acid fuel cell, such as an SPE fuel cell. Purification is of course required to give pure hydrogen if an AFC is used. Beyond this, hydrogen enrichment, from the energetic viewpoint, will generally follow the law of diminishing returns. Separation of hydrogen will always require a reversible amount of thermodynamic work, which will indeed always be greatly exceeded in any practical device. Since all fuel cells have an essentially reversible hydrogen electrode, the gain in the fuel cell from using the enriched hydrogen (even at 100% conversion) will always be less than the work input required to produce the pure hydrogen used. In other words, and all other things being equal, it would be better from the overall energy viewpoint to consume the impure and dilute hydrogen directly. The only exception to this rule will be the presence of catalytic poisons (CO) or electrolyte poisons (e.g., CO_2 in alkaline systems) that can be readily removed via the separation process.

B. OXIDANT PROCESSING

For alkaline electrolyte fuel cell systems employing ambient air as oxidant, a major processing task is to remove carbon dioxide. Methods for carbon dioxide removal have been discussed in detail in Chapter 10.

Oxygen Enrichment

Pure oxygen produced by water electrolysis has usually been considered to be too expensive for fuel cell use. However, electrolytic hydrogen will be used as the primary second and third stage propellant for the European Space Agency's Ariane satellite launcher, starting about 1990.[982] Thus, it is likely the liquid oxygen from electrolytic sources may be directly used in aerospace fuel cells in the not too distant future.

The usual method of providing oxygen for aerospace fuel cells is by fractional distillation of liquid air. This is important for future use in MCFC systems, since studies have shown that an oxygen-blown coal gasifier system will result in greater overall efficiency and lower overall electricity costs than that of an air-blown system, taking into account the capital costs and energy requirements of the liquid oxygen plant.[1990, 2276] However, for the same reasons as those discussed above for hydrogen, a plant to produce either pure oxygen for the fuel cell cathode, or to enrich the oxygen content of the air, will not be cost-effective.

A possibly less costly method than fractional distillation may be to separate oxygen from air by using the method developed by the Sun Oil Company.[610] Pressurized air is fed to one side of a solid oxygen ion transporting membrane made from a material such as doped zirconia or the molybdates of bismuth, silver, or zinc. The pressure on the other (pure oxygen) side of the membrane is maintained at a lower value such that a pressure differential of 0.1~12 atm exists across the membrane. By short-circuiting porous electrodes on each side of the membrane, oxygen ionizes at the high pressure side to $O^=$ ions, which migrate through the membrane to the low pressure side, where they recombine to produce oxygen. This device is essentially a short-circuited oxygen concentration cell in which the partial pressure of the oxygen in the air is greater than that in the pure oxygen on account of the pressurization of the air. The cell, which operates at 400~650°C, also may be viewed as a short-circuited solid oxide electrolyte fuel cell (see Chapter 18). The use of urania-doped zirconia, which conducts both oxide ions and electrons, also has been suggested for similar applications.[216, 935]

C. GAS CLEAN-UP SYSTEMS

The primary fuel reactor produces processed gas streams that will be used in the cell stacks. These will require the removal of any impurities that may degrade fuel cell system performance. The primary contaminants are sulfur compounds, which also may include halides and other materials. Their removal is discussed in the following.

Sulfur Contaminants

Sulfur is the major contaminant encountered in processed fuels such as steam reformed fuel oil or gasified coal. For example, No. 2 fuel oil contains 2200~2600 ppm sulfur.[378, 1402] Sulfur compounds, notably hydrogen sulfide, adversely affect the performance of both PAFC and MCFC systems, although to different degrees, as has been discussed in Chapters 3 and 8.

Reactant Processing and Clean-up Systems

In PAFC systems, hydrocarbon fuels are generally steam reformed over nickel- or iron-containing catalysts, which are sensitive to sulfur. It has been found, for example, that some reforming catalysts that have useful lifetimes in excess of 1200 hours at sulfur levels of 20 ppm have lifetimes in the order of only about 8 hours at sulfur levels of 1500 ppm.[2702] In phosphoric acid systems, therefore, sulfur impurities impose more of a threat to the reformer than to the fuel cells themselves. The fuel processing chain for the processing of naphtha to be fed to a large-scale phosphoric acid fuel cell power plant, typically will consist of the following: a hydrodesulfurizer, a halogen guard, a zinc oxide sulfur absorber, a catalytic steam reformer, a high-temperature shift converter, a second halogen guard, and a low-temperature shift converter.[1081]

One way of detecting sulfur breakthrough from a desulfurizing bed (e.g., ZnO) is to pass a sample of the desulfurized gas through a sulfur-sensitive indicator to give a visual indication of when to recharge the bed.[792] More sophisticated methods make use of on-stream monitoring of sulfur levels using various instrumental methods of analysis coupled with microprocessors.

For MCFC systems, sulfur impurities affect the performance of the fuel cell itself, as can be seen from Table 14–1.[378] Even at levels of only 10 ppm it appears that sulfur causes a significant decrease in cell performance. Furthermore, the adverse effects become worse as the sulfur level and the operating pressure increase and the hydrogen content of the fuel decreases. It has been established that the major sulfur fuel contaminants are H_2S and COS (carbonyl sulfide).[427] Work carried out at the IGT and GE using bench scale-molten carbonate fuel cells, using a baseline fuel gas containing 36.5% hydrogen, 20.5% carbon dioxide, 11.6% carbon monoxide, and 31.4% water vapor, and a nominal 2 ppm hydrogen sulfide, indicated a small (15 mV) loss in cell voltage due to sulfur, but little

TABLE 14–1

Effect of Sulfur on Molten Carbonate Fuel Cell Performance[378]

Fuel Gas Composition	ppm S	Cell Voltage at 200 mA/cm2
52.7% H_2 + 10.7% CO	0	0.747
	10	0.710
	50	0.650
	100	0.599
	200	0.590
18.6% H_2 + 14.0% CO	0	0.685
	10	0.610

evidence of shift reaction poisoning.[1687] The cells contained Cr-doped nickel anodes and Li-doped nickel oxide cathodes, and were operated at atmospheric pressure in the temperature range of 650~675°C and at fuel utilizations of 50~75%. An analysis of the inlet and outlet compositions indicated that approximately 850 hours were required for the sulfur uptake in the cell to level off. After 100 hours of operation on clean gas, the level of sulfur in the outlet gas was lowered to 0.3 ppm H_2S and the cell performance had essentially returned to the intitial level. This indicates that the effects of sulfur are to some extent reversible, at least at low concentrations.

At levels of less than 1 ppm, hydrogen sulfide and carbonyl sulfide pass through the anode compartment without affecting cell operation. However, as already mentioned in Chapter 13, they pass on to the cathode as SO_x along with the carbon dioxide anode exhaust stream. At the cathode the SO_x is converted to sulfate, which migrates through the electrolyte to the anode, where it is reconverted to H_2S and released into the anode compartment.[427] The net result is a gradual build-up of sulfide in the anode region. At levels greater than 10 ppm, hydrogen sulfide reacts with the nickel of the anode and other metallic cell components, resulting in a deterioration in cell performance. Publications relating to the H_2S tolerance of the MCFC are given in Chapter 3.

The addition of cobalt to sintered nickel anodes has been found to improve the tolerance to hydrogen sulfide, but it is unacceptable on cost grounds.[1703] Thus, for the successful development of MCFCs integrated with coal gasifiers, H_2S and COS must be removed from the fuel gas down to levels of less than about 1 ppm.[1282] The build-up of sulfur in the cell can be controlled by limited anode venting and anode gas recycling, as well as by placing sulfur guards in the CO_2 loop. However, it will probably be more economical to employ two-step low-temperature sulfur removal upstream of the fuel cell stack.[427] Although the hydrogen sulfide concentration in the feed must be reduced to low levels for MCFCs, it is easier and cheaper to remove H_2S from a processed fuel gas than it is to scrub SO_2 from combustion stack gases. Again, since the fuel cell has a considerably higher efficiency than a conventional combustion plant, the fuel cell system will require a smaller and cheaper fuel processing train for a power plant of the same electrical output.

The alternative to removing sulfur from the processed fuel stream would be to develop more sulfur-resistant materials for use in the fuel cell stack itself. Because of the high cost and uncertainty of the ultimate benefits of changing stack construction materials, current thinking favors the use of sulfur removal subsystems, which are already to a large extent a commercially available and proven technology.

High-Temperature Vs. Low-Temperature Sulfur Removal for Molten Carbonate Systems

From the standpoint of thermodynamic efficiency and ease of integration into the fuel cell system, high-temperature sulfur removal would be preferable to low-temperature removal. Considerable success has been achieved by dry scrubbing coal synthesis gas with iron oxide–silica sorbents at about 650°C. Unfortunately, these sorbents are able only to lower the sulfur levels from about 6000 ppm (by volume) to about 200 ppm, especially if large quantities of carbon dioxide are present.[1282, 1286] The urgent need, therefore, is to find a second-stage hot regenerable sorbent system that is capable of further lowering the sulfur level to about 1 ppm.

Attractive high-temperature sorbent candidates that have been investigated include V_2O_5/V_2O_3, CuO/Cu, and WO_3/WO_2.[1286] For reasons of simplicity of operation it is likely that these systems, if proven successful, will be used in fixed bed reactors.[234] The reduced forms of these sorbents take up hydrogen sulfide and are converted to sulfides:

$$V_2O_3 + 3H_2S \rightarrow V_2S_3 + 3H_2O$$
$$Cu + 2H_2S \rightarrow CuS_2 + 2H_2$$
$$WO_2 + 2H_2S \rightarrow WS_2 + 2H_2O$$

After sulfur sorption they are regenerated by air oxidation at 650°C to displace the sorbed sulfur:

$$V_2S_3 + 11/2O_2 \rightarrow V_2O_5 + 3SO_2$$
$$CuS_2 + 5/2O_2 \rightarrow CuO + 2SO_2$$
$$WS_2 + 7/2O_2 \rightarrow WO_3 + 2SO_2$$

This is followed by conversion to the reduced forms, usually by reaction with hydrogen at 650°C:

$$V_2O_5 + 2H_2 \rightarrow V_2O_3 + 2H_2O$$
$$CuO + H_2 \rightarrow Cu + H_2O$$
$$WO_3 + H_2 \rightarrow WO_2 + H_2O$$

With the copper and tungsten compounds there tends to be some formation and accumulation of sulfates during the oxygen regeneration step, leading to a reduction in their subsequent ability to take up sulfur.

Vanadium trioxide (V_2O_3) also shows a reduction of its sulfur absorption capacity, apparently because of a reduction in specific surface area, since commercial supported V_2O_5 sorbents have been found to give much better (twentyfold) sulfur loadings and regenerability than unsupported V_2O_5, CuO, and WO_3. In any event, at the present stage of development, all three attain unacceptably low sulfur loadings of only 2~4% of the theoretical values.[1286]

In contrast to high-temperature removal, a variety of low-temperature sulfur removal methods, such as the Selexol process,[1794, 1868] are already commercially available. Low-temperature sulfur removal will however be more costly than high-temperature removal, because of the extra plant and heat-exchangers involved. Furthermore, it has been shown that low-temperature cleanup will result in overall fuel cell power plant efficiencies that are about 2% lower than those obtainable with high-temperature cleanup.[421, 427, 1282] Conversely, low-temperature removal already is well developed commercially, and is capable of lowering sulfur levels down to about 10 ppm. Since it is by now well known that even 10 ppm sulfur can cause problems in the MCFC (though not in the PAFC; Ref. 2146), it is probable that a low-temperature removal process will be followed by a ZnO bed to reduce the sulfur levels to less than 1 ppm. Fortunately, the additional economic penalty to get down to 1 ppm is relatively small.[427] It might be worthwhile to use a combination system incorporating a conventional high-temperature method (such as FeO_x) to lower the sulfur level down to about 200 ppm, followed by a low-temperature stage incorporating a zinc oxide bed to achieve levels of less than 1 ppm.

Although the choice of the cleanup system for a coal-fired MCFC power plant will largely depend on the coal gasifier selected, it appears that low-temperature cleanup, in spite of slightly unfavorable thermodynamic efficiency and expense, will be adopted. A recent study to evaluate the best commercially available hardware for use with 10~150 MW coal-fired molten carbonate systems operating on American coals and lignite, recommended a sulfur-removing scheme consisting of a diethanolamine process for removing sulfur compounds from the raw gas, followed by a conventional Stretford plant for recovering sulfur from acid gas.[1836] This selection was dictated by capital cost, operational experience, and performance.

Hydrodesulfurization

Hydrodesulfurization is a common industrial process in which dry vaporized fuel is reacted over a cobalt-molybdenum catalyst with excess hydrogen to convert organic sulfur compounds to H_2S and hydrocarbons. Thus, hydrodesulfurization commonly is used to remove sulfur from fuels such as naphtha prior to catalytic reforming. In the conventional hydrodesulfurization process, naphtha and relatively pure hydrogen are passed over the hydrogenation catalyst at 10~40 atm pressure. In many fuel cell applications, however, hydrodesulfurizers must operate at near ambient pressures and must be capable of utilizing relatively impure hydrogen from the fuel processor. Fortunately, it has been shown that hydrodesulfurization catalysts are capable of long-term operation under the conditions appropriate for integration into a fuel cell system.[1402]

Typical hydrogenolysis reactions of organic sulfur compounds in naphtha are:[1081]

$$C_6H_5SH + H_2 \rightarrow C_6H_6 + H_2S$$
$$C_4H_4S + 4H_2 \rightarrow C_4H_{10} + H_2S$$

In the next stage of the process the H_2S produced is removed by passing it over a zinc oxide bed:

$$ZnO + H_2S \rightarrow ZnS \downarrow + H_2O$$

Zinc oxide has been shown to be capable of continuously lowering H_2S levels from 250 ppm to less than 1 ppm in hot coal gases at 650°C, with no evidence of zinc sublimation or ZnO reduction.[1897] Complete regeneration of the bed (resulting in the release of SO_2) is accomplished by air oxidation at 650°C, with no formation of $ZnSO_4$. With naphtha, hydrodesulfurization can reduce sulfur levels from 20~500 ppm to less than 0.05 ppm.[1081]

In addition to cleaning the fuel, leaving the zinc oxide bed to an essentially negligible sulfur level, hydrodesulfurization also can saturate unsaturated hydrocarbons, thereby reducing carbon deposition problems.[2461] However, conventional steam reforming and hydrodesulfurization are limited to fuels with end points below about 205°C. Thus, the conventional process can be used for the naphtha cut in petroleum (which comprises about 30% of the crude barrel) but not for kerosene and distillate fuel oil.[1402] Also, when ZnO beds are used to process hot (650°C) coal gas, it appears that only very shallow sulfur penetration into the ZnO pellets occurs, so that the sulfur removal capacity of the bed decreases with repeated use at temperatures above about 430°C, probably as a result of pore plugging during sulfur removal and sintering during regeneration.[1282] Another deterrent to the process is the need for hydrogen recirculation.[2461] In spite of these drawbacks, hydrodesulfurization is a favored method for sulfur removal from the processed fuel cell feed gases currently used.

Other Methods of Sulfur Removal

In addition to the use of zinc oxide, Raney nickel also has been used at 450~600°C to remove hydrogen sulfide, the nickel being converted to NiS.[2702] Activated carbon[1762] and molecular sieves[1595] also have been used to remove sulfur from reformer gases.

Other Contaminants

Halides in fuels such as naphtha have deleterious effects on steam reforming and low-temperature shift catalysts, hence it is advisable that they be lowered to levels of less than 0.05 ppm using halogen guard catalysts before the fuel enters the reformer and low-temperature shift converter.[1081] Similarly, HCl contained in coal gasifier product gas would be expected to cause problems in the MCFC (formation of stable chlorides, corrosion) and should be removed from the feed gas.

Since different coals contain different impurities, the potentially adverse effects of a large number of possible contaminants on the performance of molten carbonate fuel cells must be investigated, and methods of detection and removal developed. For example, researchers at GE and IGT have studied the effects of potentially harmful hydrocarbon contaminants (methane, ethane, ethylene, and propylene) on molten carbonate fuel cell anode materials.[1687] At concentrations ranging from a few ppm to 10 volume percent of the fuel, it was found that none of these substances adversely affected either the catalytic activity of the anode or cell performance during single cell tests lasting over 100 hours at 160 mA/cm^2 and 0.75 V. The only contaminant that appeared to react was ethylene, which reduced to ethane when passed over Ni-10 Cr alloy. Methanol and a mixture containing 90% ethanol, 5% methanol, and 5% isopropanol also exerted no adverse effects on cell performance. A fuel cell was operated for over 300 hours with a continuous influx of 1.1% of the ethanol-methanol-isopropanol mixture with no apparent effects. Cresol also appeared to be innocuous. The effects of cyanides were still under investigation.

Oxygen and water can be removed from humid hydrogen streams by first passing the mixture through a porous bed of platinum, palladium, or nickel catalyst to remove the oxygen; followed by water adsorption at 10~70 atm pressure in a second porous bed containing a molecular sieve, charcoal, or silica gel,[2278] however, unless the amounts contained in the fuel stream are high enough to affect cell efficiency, such a separative treatment seems pointless.

CHAPTER 15

Heat and Mass Transfer: System Issues

If a fuel cell is to operate for more than a short time, a steady supply of reactants must be supplied to the electrodes. It also is essential that fuel cell waste products in the form of heat and chemical reaction products (usually water) should continuously be removed from the stack. In addition to these mass and heat flows crossing the system-surroundings interface, there also may be flows internal to the system; for example, of recirculating electrolyte, reactants, and coolants. In all but the simplest fuel cells, the design and engineering of the heat and mass transfer subsystems play a crucial part in achieving optimum performance. The most crucial of these is the mass transfer system, described in the following section.

A. REACTANT SUPPLY AND DISTRIBUTION

In early fuel cell systems such as those employed in the Gemini[176] and Apollo[394] missions, the reactants were fed individually to each cell through small tubes. Large numbers of these tubes were then connected to external manifolds in an expensive and time-consuming fabrication operation. For a large fuel cell system, which may contain thousands of individual cells, a more efficient, reliable, and economical approach is to include the supply manifolds as part of the electrode frames or separators of the internal manifold type introduced in Chapter 13. In this approach, the holes and channels in the frames of the stacked cells align to form continuous passages and networks of side passages through which reactants are fed and distributed to the appropriate electrodes. In an alternative approach, pressurized PAFC fuel cell stacks of UTC type are completely enclosed within a large cylindrical pressure vessel. The space between each side of the square stack and the pressure vessel then contains a reactant gas entry or exit manifold in a cross-flow arrangement. This type of external manifolding must be designed to make pressure-tight dielectric seals with the sides of the stack; be compatible with hot electrolyte, fuel, and air; and be relatively inexpensive. United Technologies' early manifolds were made from hot-molded phenolic-impregnated fiberglass; however, this material was found to be structurally and chemically unsound for larger stacks operating under more extreme environments. Later manifolds were made from stamped low-carbon steel, incorporating PTFE frames and coated with perfluoroalkoxy (PFA) films, external steel straps, and tightening devices being used to clamp the two halves of the manifolds together.

Feed Modes

Fundamentally, there are three methods of supplying gaseous reactants to the cells, namely the dead-ended, circulating, and flow-through modes.

In the dead-ended mode, reactant gas of the highest purity possible is fed through a pressure regulating valve from a pressure vessel containing liquid or gaseous reactant to the sealed electrode gas compartment. Although extremely simple, this method does not provide for the removal of heat, reaction products, or accumulated impurities or trace quantities of inerts from the cell. In addition, under fluctuating power conditions the system may occasionally be unable to adjust adequately to the required output because the fuel supply cannot be adjusted to match the demand. Provision must be made to remove the product (water) via a subsystem, which again may have a transfer rate that limits output. This feed mode was employed in the low-temperature Gemini mission SPE fuel cells, whose nonfluorinated electrolytes operated at very low current density.[176, 689] Water was produced on the oxygen cathode side in this design, and was removed as liquid via a simple wicking system. In alkaline systems, water is produced at the anode in the form of diluted electrolyte or as vapor, depending on the cell temperature. Hence, the dead-ended mode can only be universally used on the cathode side, and then only if pure oxygen is the reagent. However, the anode side can be used dead-ended with pure hydrogen fuel if product water (and waste heat) is removed via a circulating electrolyte system. If product water is produced in the vapor phase in the gas stream appropriate to the technology, a dead-ended system will not be applicable. Removal of inerts in the dead-ended mode is discussed in the following.

In the circulating mode, each cell is fitted with both inlet and exhaust ports and channels to give as close to uniform gas distribution as possible, and external pumps or blowers are used to circulate the reactants in feedback loops. Although the system is more complex than the dead-ended mode, the reactants themselves can be used to remove heat and water from the cell using heat exchangers and condensers in the feedback subsystems, and it is easier to match fuel supply with power demand. The Apollo fuel cell system was one of the earliest systems that employed a circulating feed mode.

In a system employing the flow-through mode, the reactant gas is fed to the cell on a once-through basis, after which it is exhausted from the stack, either to the ambient atmosphere or to some other subsystem such as a combustion preheater. This feed mode can also eliminate heat, reaction products, and inerts. It is less complex than a circulating system, since no circulating pumps or complex flow networks are required. Again, since there is always an excess of reactant in the supply stream, fluctuations in cell output demand can be reasonably well met. The major drawback is that if both the fuel and oxidant are pure (e.g., H_2 and O_2 in an aerospace system), then some of each must be sacrificed. Accordingly, this feed mode would only be selected in such cases when the requirements for light weight, simplicity, and compactness are more important than fuel economy. However, if the gases used are mixed with inerts (e.g., if the fuel is reformate and the oxidant is air), this method can be used advantageously. For example, for air electrodes, air can simply be blown through the cell and then vented to the atmosphere or used elsewhere. In acid systems, this stream can be advantageously used to remove product water and for cooling. The impure hydrogen fuel stream will be allowed to flow more slowly through the cell to consume the required amount of hydrogen (for example, 80%), the exit gas being used for process heat in the reforming subsystem. In contrast, in alkaline systems using pure hydrogen fuel, a feedback loop to remove water will be more advantageous on the anode side.

Reactant Conditioning and Control

In theory, the reactants for a fuel cell need only be supplied in direct proportion to the current drawn. In practice, however, allowance must be made for parasitic consumption of reactants, for example via chemical processes, internal short circuiting currents (shunt currents), and other losses. As these are generally difficult to quantify and may change with such variables as electrode ageing, it is common practice to maintain the fuel and oxidant concentrations at preset optimal levels regardless of the demands made by the fuel cell stack. This is achieved by the use of reactant sensors, which activate the reactant feed system. For a gaseous reactant such as pure hydrogen, the sensor simply may consist of a valve that maintains constant pressure, whereas other sensors may be more complex. For example, UTC developed a sensor-controller that determines anode hydrogen feed rate using a balancing e.m.f. derived from the IR-drop in a shunt, in series with the cell load.[2639] The balancing e.m.f. acts through a transducer to increase or decrease the gas pressure in a fixed-volume gas electrode.

For reformer-generated hydrogen fuels, UTC has described a fuel control system in which the fuel processor is isolated from the fuel cells by a demand valve.[827] The fuel processor operates at elevated pressure and generates a reservoir of hydrogen, which is supplied to the fuel cells on demand, based on the pressure downstream of the demand valve and on the temperature in the hydrogen reformer. The French Institute of Petroleum has described an automatic device that uses pumps with monitoring of the electrolyte circulation rate to control the feed rate of oxygen or air to the fuel cell stack.[593]

For dissolved reactants such as methanol or hydrazine, more complex sensors that generate a signal proportional to dissolved reactant concentration may be required. Dissolved hydrazine, for example, can be controlled at levels of 0.4 ± 0.1 % by means of electrical concentration sensors that control a feed valve into the electrolyte.[1375] Some sensors make use of electrochemical devices that apply either voltage or current sweeps, the maximum coulombic current response in such sweeps being a function of the concentration of dissolved reactant. Such electrochemical sensors must be either insensitive to, or compensable for, variations in temperature or electrolyte concentration, and should have a fairly fast response time, preferably on the order of a few seconds or less. Liquid hydrocarbon fuels that can be dispersed into aqueous electrolytes can be kept at a constant concentration by recirculating the depleted fuel through membranes that are permeable to water but impermeable to the hydrocarbon, thereby concentrating the fuel for recycle.[2579]

The French Institute of Petroleum developed a method of supplying reactant fluid to a fuel cell at constant pressure in a closed loop system, in which the fluid is replenished with fresh fluid and then partly removed on an intermittent basis, as determined by fuel cell output.[1022] In addition, if each chamber of the fuel cell stack is fitted with its own means of controlled fluid supply and removal, steady fuel cell operation can be enhanced.[597] During cell operation, each chamber undergoes at least one rinsing stage. However, whether such an increase in complexity is practical remains doubtful.

Workers in the Soviet Union have developed equations that predict the optimum circulation rate of reactant solutions, based on fuel cell current density and gas evolution rates.[1496] Methods also have been developed that enable calculation of the reactant solution feed rates required to eliminate the formation of stagnation zones in fuel cells with corrugated electrode structures.[1497]

Heat and Mass Transfer: System Issues

In addition to controlling the concentration and feed rate of a reactant, a number of additional aspects of reactant conditioning must be taken into consideration. For example, for gaseous reactants it is important to control moisture content. In a simple case, if the gas is too humid, water may condense at cool points in a stack operating at less than the boiling point of water, resulting in electrode flooding, conduit blockage, or electrolyte leaching or dilution. Conversely, if the gas is too dry, electrodes, membranes, or electrolyte matrixes may dehydrate, resulting in decreased performance or even in cell failure. The above observations are particularly appropriate in the case of SPE systems, which require humidified fuel to give a highly conducting electrolyte membrane. If the latter dries out, not only does performance suffer, but the membrane may shrink completely and crack, resulting in destruction of the cell.

Moisture control of a gaseous reactant is seldom a problem in a fuel cell system using a flow-through feed mode, but it may be troublesome in systems that employ dead-ended or circulating gas supply. In one UTC circulating feed design, the dew point of the reactant gas is regulated via an injector in an integral heat exchange relationship with the stack.[1012] In this design, waste heat from the cells is used to continuously heat recirculating reactant gas to prevent condensation of water and to maintain a constant dew point from the time the reactant gas leaves the cells until it is mixed with preheated fresh reactant gas in the ejector. The dew point is controlled by controlling the recirculation rate. The problem of moisture elimination is treated in more detail in the upcoming section on heat and water removal.

A problem that occurs in systems employing dead-ended electrodes is the build-up of inerts such as nitrogen, which cause a drop in overall performance. In the earliest systems, the cathode gas compartments were periodically purged with air to remove accumulated inerts. However, periodic purging results in fluctuating cell performance as a function of time, and the system as a whole is in reality no longer dead-ended, but operates in a discontinuous flow-through mode. For systems using technically pure reactants, a more efficient approach uses a controlled bleed at the dead ends of the gas compartments of the electrodes. This amounts to a low-rate flow-through system which permits continuous removal of noncondensable inerts and impurities, and produces a more uniform cell output. This method has been recommended by the Institute of Gas Technology for alkaline hydrogen-oxygen fuel cells.[947]

As discussed in Chapters 3, 8, 12 and 14, the removal of catalyst poisons such as carbon monoxide or hydrogen sulfide is also necessary for certain types of fuel cells. For alkaline fuel cell systems using air electrodes, it is mandatory that carbon dioxide be removed from the air to prevent carbonation of the electrolyte. It has been suggested that the removal of carbon dioxide from reactant air may also improve the performance of cathodes in cells not using aqueous alkaline electrolytes.[2509]

Examples of Feed Distribution Designs

In this section we shall mention only a few of the many designs that have been disclosed for supplying reactants to fuel cells.

Siemans A.-G. has described plate-type distributors for use with filter press-type fuel cells using gaseous reactants and circulating liquid electrolytes.[1460, 2272] The distributor plates, which supply both reactants and electrolyte, are located in the center of the stack, parallel to the end plates. Each plate contains integral channels parallel to the plane of their surfaces, opening into branch channels passing through the distributor. These correspond to main electrolyte and reactant arteries passing through each cell of the battery. The use of such combination separator-distributor plates reduces electrolyte leakage losses. The plates, which are usually about the same thickness as a unit cell, are readily manufactured from plastics such as polymethacrylates for use with low-temperature alkaline electrolytes.

Varta Batteries has disclosed a feed distribution system that uses a vertical arrangement of in-series connected fuel cells and electrolyte-fuel mixture tanks.[2313] The latter are connected through overflow lines, with a funnel-shaped suction header in each tank leading to the main electrolyte-fuel transport line of each cell.

Alsthom has described hexagonal[2093] and cylindrical[2095] thin-plate reactant feed distributors that can be employed in high-power filter press-type fuel cells, particularly those using circulating alkaline electrolytes; for example, with dissolved hydrazine and hydrogen peroxide reactants.

The French Institute of Petroleum gives a method for circulating hydrogen and air through a fuel cell by using centrifugal pumps provided with liquid electrolyte seals.[2056] The electrolyte entrained in the recirculating gases is separated in the pumps by centrifugal forces, and becomes part of the liquid seals.

B. ELECTROLYTE FEED AND DISTRIBUTION

In fuel cell systems that employ circulating electrolyte, it is important that sufficient electrolyte of suitably uniform quality be distributed and confined to the narrow spaces between the electrodes. Both the movement and distribution of electrolyte in fuel cells have been studied.[1560] In matrix-type fuel cells employing immobilized liquid electrolyte, it is possible to

Heat and Mass Transfer: System Issues 471

determine the electrolyte concentration during cell operation by monitoring cell operating parameters such as the temperature of the hydrogen-steam mixture at the condenser and fuel cell outlets, the hydrogen circulation rate, and the pressures of the operating gases.[994]

For long-term operation, most circulating alkaline electrolyte hydrogen-air systems have required an in-loop electrolyte conditioning system to remove carbonates formed in the electrolyte as a result of contact of the electrolyte with carbon dioxide in the reactant gases. As outlined in Chapter 10, one method of decarbonating alkaline electrolytes involves passing the carbonated electrolyte through a regenerator cell where the carbonate is converted to hydroxyl ions and vented carbon dioxide by providing a substantial OH^- ion gradient between the electrodes of the regenerator cell. Both Alsthom-Exxon[788] and UTC[1361] have studied this method of electrolyte conditioning.

A novel electrolyte conditioning process, claimed to increase the efficiency of a hydrogen-oxygen fuel cell using free sulphuric acid electrolyte, involved adding 0.02~0.05 weight percent of a foaming agent such as Carboxane TW-100, Aromax T/12, or Galfac RE-610 to the electrolyte.[1852] The reactant gases were blown into the electrolyte to create a foam in which the electrodes operated. This was claimed to greatly improve gas-phase mass transport.

A number of electrolyte distributor designs have been described in the literature. General Electric developed a special plate for use in solid polymer electrolyte (SPE) cells that contains a number of parallel conductive ribs contacting the electrode, which in turn is bonded to the SPE membrane.[1579] These ribs define a network of fluid distribution channels through which a secondary acid electrolyte and reactant gas can be brought to the electrode and then removed. Siemans A.-G. designed an overflow chamber to control the electrolyte level in a fuel cell stack.[1268] The French Institute of Petroleum has disclosed a fuel cell containing special arrangements in the electrolyte supply circuit for improving the recovery and recirculation of the electrolyte.[595] Modular frames fitted with drainage and supply orifices, and containing provision for internal flow distribution also have been disclosed.[2508]

C. HEAT AND WATER REMOVAL

Because of the importance of heat and water removal to the optimization of the steady state performance of a fuel cell system, numerous studies have been made that attempt to analyze the transport phenomena involved. For example, the advantages and disadvantages of different heat removal schemes have been examined, and approaches have been outlined for providing simple but cost-effective and reliable means of fuel cell heat removal.[1697] Non-equilibrium thermodynamics has been used to analyze the spatial distribution of temperature and heat release mechanisms at individual electrodes in fuel cells operating under load in non-isothermal steady states.[1256, 2691]

It appears to be possible to extend cell service life and better maintain high current densities when the electrodes are cooled where product water is not formed.[2420] Thus, in a hydrogen-oxygen fuel cell stack using 30% potassium hydroxide electrolyte, it appears to be preferable to remove heat at the oxygen electrodes, whereas in acid cells it is better to cool the hydrogen electrodes. Similarly, UTC has found that for molten carbonate fuel cells at a given power density, cooling the oxidant gives somewhat better efficiency than cooling the fuel,[1113] since (as in the alkaline cell) the cathode represents the site of the heat of reaction in this system. This observation is particularly important if process gas cooling is chosen. However, for liquid cooling or gas cooling using a cooling plate for every five or so cells, it is academic.

Methods of determining the best choice of inlet and condenser temperatures in fuel cells have been discussed.[509] The operating temperature of a fuel cell battery can be regulated by the use of appropriately connected temperature sensors, transducers, and valves to control the heating of the intake air to the cathode chamber reaction temperature.[401] Using the general laws of continuity, diffusion, and heat transfer, models have been developed describing the combined heat and water transport processes in hydrogen-oxygen fuel cells that make use of circulating excess reactant gas to remove heat and product water.[283] Convection processes in electrochemical systems have been modeled in terms of the Nusselt number.[714] Methods have been developed that permit calculation of the required water by-pass ratios in condensers to remove heat and reaction water in fuel cell stacks.[1467] An analysis has been made of the effects on fuel cell behavior of the moisture transport processes that occur when water is removed from one of the electrodes.[2309] Volume average velocities have been used to analyze transient electrolyte water evaporation and heat rejection in hydrogen-oxygen fuel cells where circulating excess reactant gas is used to remove heat and water.[282, 284]

In addition, a mathematical transport model has been derived to study the cathode electrolyte drying transients that occur immediately following an increase of the load current.[295] This model was tested using data from a circulating reactant-type high power density hydrogen-oxygen matrix fuel cell. Cell geometrical parameters, such as the thickness of the porous nickel electrolyte reservoir plate, were found to exert a significant effect on the transient response of the cell. Finally, it has been observed that electromagnetic waves and rapidly reversing magnetic fields may increase the rate of the transport processes in fuel cells, thereby improving the electrical output and efficiency.[1320] Again, the practicality of this observation is open to doubt.

Water Removal

The by-product of the reaction in hydrogen-oxygen fuel cell systems is water, which must be removed from a stack operating at steady state at temperatures lower than its boiling point to prevent electrode flooding, electrolyte dilution, or the blocking of gas flow. Even under these operating conditions, the removal of water of reaction poses only minor problems if the reactant gases are circulated through the cell. For example, the hydrogen fuel gas stream can be used to carry product water in the vapor phase away from the electrodes so that it can be removed downstream in a condenser or other suitable separator. This type of water removal, in which vapor is eliminated by a circulating fluid, becomes more efficient the higher the temperature of the cell. It is normally termed *active* or *dynamic* water removal, and may be essentially viewed as a forced convection process. Conversely, methods that remove water by only diffusion or natural convection may be termed *passive* water removal processes. Active water removal is the preferred mode for large fuel cell systems because it is capable of handling much larger mass flows.

Many active water removal concepts have been developed. Exxon Research and Engineering Company has described methods that use part of the fuel cell waste heat to evaporate water from the electrolyte. A portion of this evaporated water can be later condensed and recycled to the stack for control of the electrolyte concentration.[787] Exxon also has disclosed the design of a plate that uses heat transfer and capillary action to remove the heat and water generated in a fuel cell.[202]

Siemans A.-G. has developed a method of removing water and heat from the hot electrolyte of a hydrogen-oxygen fuel cell by which the hot electrolyte is contacted with one of the reactant gases in a cooler. The reactant gas takes up water vapor and simultaneously removes heat via the heat of evaporation. The water-containing gas is then led through a compressor and water separator.[2629, 2630]

In other examples, UTC has developed a system design that both removes and de-aerates reaction water and delivers it to a steam generator where it can be used for steam reforming.[2754] The U.S. Department of the Navy has disclosed a product water management system that removes the steam generated in the fuel cells with a circulating stream of hydrogen gas.[1565] Fuji Electric Company claims a method of preventing the accumulation of reaction water in the electrode chamber. Their technique makes use of the gas pressure of either the hydrogen or oxygen stream to remove water from the chamber and add it to the electrolyte, eliminating the need for pumps.[1793] A device for measuring and adjusting the electrolyte concentration with water also is disclosed in the same patent. Another method of removing water from the electrolyte involves leading the electrolyte to an evaporation chamber held at a reduced pressure, from which the evaporated reaction water is then led to a condensation chamber at an elevated pressure and condensed.[468]

When the reactant gases are not circulated, an independent, usually passive, method of water removal is employed. As has already been mentioned, GE used a passive wicking method of water removal in its Gemini SPE fuel cells. After studying various wicking materials they found that Refrasil, a high silica content (99.3% SiO_2) woven glass material supplied by Hitco Corporation, is an excellent candidate for long-term wicking in solid polymer systems.[608] This type of material is ideal for use with SPE membranes because it does not contain any exchangeable cations, particularly Na^+, that might exchange with the active H^+ ions in the electrolyte. Tests showed that the wicking action of Refrasil wicks actually improved after 5000 hours of use.

Another method of removing water (and heat) with a minimum of mechanical moving parts is to evaporate the product water through a capillary membrane adjacent to the electrodes.[647] The vapor space behind the membrane is preferably evacuated to a pressure equal to the vapor pressure of the electrolyte at its operating temperature. With this method, temperature is not a critical factor and control is simple, and the method is particularly appropriate for use in space, as in the Allis-Chalmers alkaline system.[176] The performance of a hydrogen-oxygen fuel cell that employs such a self-regulating capillary membrane has been modeled, taking into account the various transfer processes occurring in the membrane, porous electrodes, and gas compartments.[7, 9, 606, 2618, 2619] Similarly, Siemans A.-G. use such water-permeable diaphragms to transport reaction water from the fuel cell electrolyte to an adjacent chamber where it is condensed.[1014, 1052]

A third method of passive water removal is to employ an electrolyte-filled porous water transport plate that retains reactants in the cell, but permits water to diffuse through the plate and evaporate into a water cavity.[1023] In the hydrophilic configuration the electrolyte is retained by a wetted porous plate, whereas in the hydrophobic configuration the electrolyte is prevented from entering the water vapor cavity by a non-wetting porous layer. Varta Battery has developed a fuel cell water removal subsystem based on these concepts.[2713] In the Varta design, diluted electrolyte is kept at a constant concentration by feeding it to one side of a porous evaporation wall. Concentrated liquid electrolyte is removed from the wall while the evaporated water fraction is drawn off from the other, wettable, side of the wall and condensed. A gas overpressure on the evaporation side inhibits the passage of liquid through the wall.

A fourth, novel method of passive water removal involves operating the electrodes submerged in a dielectric liquid saturated with reactant gas.[2580] A dielectric liquid that is immiscible with water is chosen, so that the water formed by reaction rapidly is rejected from the electrode interface, resulting in improved cell performance.

The relative merits of the removal of reaction water by using condensation from circulating gas, diffusion slit evaporators, and electrodialysis cells have been discussed in a review paper on process engineering in fuel cell plants.[2414]

Cooling

The total temperature differential between any two points in a fuel cell stack may be thought of as consisting of a ΔT in the stacking direction (i.e., axially through the stack) and a ΔT in the reactant flow direction (i.e., across the face of an individual cell). If the cell temperature becomes too high, excess dehydration may occur or the stack materials may fail. On the other hand, if it becomes too low, cell performance will drop and the cell may fail to give any effective output if, for example, carbon monoxide poisoning of noble metal anode catalysts takes place. A knowledge of, and the control of, the temperature gradients in the stack are important from the standpoints of cell performance, life, corrosion, and electrolyte loss. In phosphoric acid systems, for example, an acceptable temperature range was reported to be about 175~205°C in 1979.[2060] Currently, it would be more like 190~210°C, depending on pressure conditions, the upper limit being dictated by materials considerations.[177] Since the overwhelming majority of fuel cell reactions are exothermic, cooling is required to maintain the stack within the fairly narrow optimum temperature range.

Passive Cooling

The simplest way to remove heat from a fuel cell is to use a purely passive technique, in which heat is conducted away from the electrodes through the cell frames and is radiated to the surroundings. This technique is only appropriate for small systems. In alkaline systems where plastic separator plates are used, passive heat transfer can be improved by incorporating metallic heat conducting pins into the plates.[2263] Such pins not only conduct heat, but also can serve as electrical contacts between cells. Similarly, exterior cooling fins sometimes can be built into the external edges of the cell plates, and external blowers can be used to pass atmospheric air over the stack to aid cooling.

Passive evaporative cooling systems that make use of capillary forces in a hydrophilic porous structure (such as a wick) to pump cooling liquid to the area between adjacent cells also have been employed, using the principle of the heat-pipe. A disadvantage of this method is that if any vapor is formed within the wick or liquid feed manifold it can block the liquid from the wick, resulting in high local temperatures. Blockage of the liquid also can result from the flash-off of any noncondensable gases that may have been dissolved in the liquid coolant, a problem that is aggravated under low gravity conditions.

An improvement is to use a non-wettable wick structure.[1023, 2338] In this approach a coolant, such as water, is fed under pressure to a cavity on one side of a hydrophobic separator located in the fuel cell so as to efficiently remove heat. The coolant vaporizes upon absorbing waste heat from the fuel cell, and then passes through the non-wetting separator to the surroundings.

Active Cooling

As fuel cell systems become larger in power and volume, and are fitted into more compactly designed packages, passive heat removal becomes inadequate, and active cooling techniques involving the passage of a cooling fluid through the stack become necessary. Cooling only the surface of a long battery of fuel cells with a large cross-sectional area will not prevent overheating, especially if the current density increases with temperature, as is usually the case. To achieve acceptable temperature gradients, the usual practice is to position cooling ducts between blocks of cells (typically about every fifth cell) thereby causing most of the heat to flow axially through the system, rather than along the plane of the cells.[1156]

Basically, two types of active cooling systems can be employed: those using gases and those using liquids as cooling fluids. Gas cooling can be further subdivided into what has come to be known as process gas cooling; DIGAS cooling; and finally the Westinghouse-Sanyo method, which uses completely separate process gas and cooling streams and manifolding. The latter, referred to earlier, is described in detail in Chapter 16.

Process Gas Cooling

Process gas cooling is the simplest of all methods of active cooling. One of the gaseous reactants, usually cool air, is supplied in several times the stoichiometric quantity to absorb heat and act as the stack coolant. After

Heat and Mass Transfer: System Issues

exiting the stack, the hot excess air can be either vented, purged, or recycled after passing through a heat exchanger. It is also possible to use the fuel gas as the coolant. This approach was employed in the Apollo fuel cells, where the hydrogen was circulated through the cell stack and then through a cooler in a recycle loop. The major advantage of process gas cooling is that no special cooling plates or stacking arrangements are required. This method cannot of course be used if the reactant gases are supplied in the dead-ended mode.

DIGAS Cooling

DIGAS cooling, which is an abbreviation of *distributed gas* cooling, is a stack cooling concept developed by ERC.[177,2060] The air stream, once inside the air manifold, is split into two separate streams: one a reactant stream, and the other a coolant stream (see Fig. 15–1). When this method is used in the PAFC, the reactant air is distributed to the cathode via channeled bipolar plates made from a pyrolyzed graphite-resin composite material that has a thermal conductivity of about 5.2 W m^{-1} K^{-1} in the through-plane direction.[617] The coolant part of the air flows through special channeled cooling plates interspersed in the stack after every fifth cell. The flow split

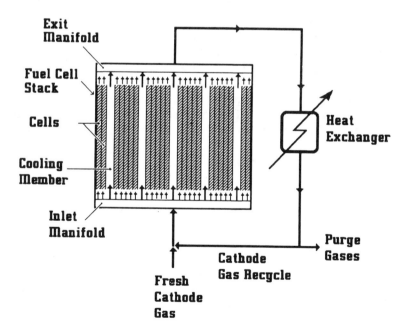

Figure 15–1: Energy Research Corporation's DIGAS cooling concept.[177, 2060]

is determined by the relative cross-sectional areas of the two flow paths. Typically there is a 10:1 split, with about 30 stoichiometric quantities of air used for cooling and 3 stoichiometric quantities of air used to supply the cathodes. By choosing short flow paths and carefully-designed channels for the air, the pressure drop and any electrolyte loss across the cells can be minimized.[1701]

For cells with an active area of 0.12 m^2, ERC has opted for rectangular cells (about 30 cm × 43 cm) rather than square cells of the same area.[617] Each air stream flows independently through the stack and is recombined at the exit manifold, after which it flows through an external loop where heat is removed, oxygen-depleted air is purged, and fresh air is added to the system.

Liquid Cooling

For systems employing circulating liquid electrolytes, it is possible to pass the electrolyte through a heat exchanger so that the electrolyte also serves as a coolant. In such cases, special corrugated separator plates are sometimes located adjacent to the cell electrodes. These plates create alternate flow channels for the electrolyte and process gas, and aid in cell cooling.[237]

For cells using immobilized matrix containment, it is not possible to use the electrolyte as a cooling liquid. For this type of system, the usual method of liquid cooling makes use of a separate liquid coolant that is passed through specially designed cooling plates in the stack. Liquid cooling may be effected by the coolant remaining completely in the liquid phase (i.e., using only the sensible heat of the coolant to cool the stack) or by partial vaporization of the cooling liquid, whereby use also is made of the heat of vaporization of the coolant. The liquid coolant can be either separately manifolded to the stack or led to and from each cooling plate via individual supply lines.

United Technologies Corporation has used two-phase water cooling for its phosphoric acid fuel cells,[2060] which has also been taken up by the Japanese developers Toshiba, Hitachi, Fuji Electric, and Mitsubishi Electric (see Chapter 16). Up to the present at UTC, coolers have consisted of a number of parallel thin-walled two-pass copper U-tubes passing through the graphite separator plates in the stacks, as indicated in Fig. 15–2. The headers supplying the copper tubes are made from stainless steel. To protect the copper tubes from the phosphoric acid environment and to isolate them electrically from the cells, they are coated with a thin film of Teflon or PFA, which somewhat lowers heat transfer efficiency. In alternative designs, the coolant tubes are connected to the stack coolant supply conduits by dielectric hoses of high length-to-diameter ratios to provide an

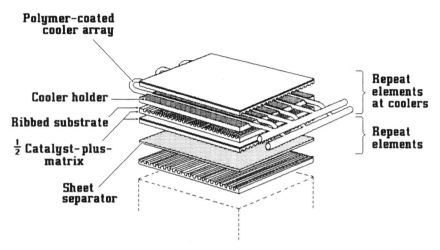

Figure 15-2: United Technologies Corporation's liquid cooling concept.[2060]

impedance path of several hundred thousand ohms to prevent short circuiting within the stack in the event of flaws in the protective coating.[1024] In order to inhibit corrosion of fuel cell components, should any shunt currents flow through the coolant fluid, UTC has proposed dissolving a reducible material, either in a gaseous form, such as hydrogen,[1873] or as an iron-containing salt, such as $Fe(NH_4)_2(SO)_2 \cdot H_2O$,[1360] in the coolant. Traces of hydrazine could also be used. When the salt is used, the iron goes into solution in the ferrous state and is sacrificially oxidized to the ferric state at the positive end of the stack. The ferric iron is subsequently reduced back to the ferrous state at the negative end of the stack. In order to reduce the corrosion risk, UTC at one point attempted to develop an all-graphite cooler, but this proved to be not economically viable at that time.[2541] United Technologies also has developed a thermally conductive grease-like sealing material for use in phosphoric acid cells. This material, which is claimed to be stable in hot phosphoric acid environments, consists of a phosphoric acid base with fillers such as carbon black, graphite, and silicon carbide.[2514] As indicated in Chapter 3, some unforeseen corrosion effects have occurred in the pressurized water-cooled UTC system, which have required extremely careful control of impurities and residual oxygen levels to avoid appreciable build-up of corrosion products in the system manifolding and supply tubing. To avoid this problem, the oxygen concentration in the cooling system of the 4.5 MW TEPCO demonstrator had to be maintained below 10 ppb, and the 40 kW demonstrators required chemical cleaning of their cooling system about every 1500 h. In future, UTC will not use a bimetallic cooling system, but one which will be of stainless steel throughout, with Teflon-coating inside the pressure vessel. Similarly, to prevent blockage, no headers will be used, and the cooling plates will contain serpentine stainless steel tubing.[448] Finally, we should point out that active cooling by evaporation of injected water can be carried out in the cells themselves *via* the cathode or anode reactant supply channels in some systems.

Engelhard fuel cell stacks also use liquid cooling. Engelhard claims that a cooling plate made from conventional materials such as aluminum that can be fabricated readily using conventional brazing techniques can be used in the PAFC.[2060] Their system uses a dielectric liquid such as Monsanto Therminol 44 as coolant, and employs baffles to promote good heat transfer control. It is claimed to offer a life of five years in a phosphoric acid environment, and is said to be potentially cheaper than other single- and two-phase liquid coolant systems.

Comparison of Active Cooling Techniques

The major advantage of process gas cooling is its extreme simplicity and low cost. Furthermore, since each cell is cooled equally, there is virtually no ΔT in the stacking direction. Conversely, because the reactant accumulates heat as it flows across the face of the cell, an appreciable ΔT is built up in the flow direction, giving uneven performance within individual cells. In addition, the large volumes of gas flowing through channels of small cross-sectional area, generate high pressure drops across individual cells, giving process gas cooling the highest power requirement of all the various cooling technologies. The large quantities of gas passing over the cathodes (in the typical arrangement) also result in high rates of evaporative electrolyte loss, leading to short life or the need for frequent electrolyte replenishment.

DIGAS cooling is relatively simple and reliable. Since the cooling plates are similar in design to the bipolar cell plates, no special manifolds, seals, or connections are necessary to integrate them into the stack. Its reliability is a result of its simplicity, since interruption of the cooling air supply, (e.g., by blockage) will also result in stack shutdown. Small air leaks usually do not cause problems, which is certainly not the case for leaks in a liquid-cooled system (see following). Although DIGAS cooling requires more power than liquid cooling, its power requirement is still low and is in any case much less than that required for process gas cooling, since the majority of the gas supplied passes through the relatively large channels of the cooling plate. Finally, the temperature distribution within the cells is more uniform than when process gas cooling is used, although the latter gives a smaller temperature gradient in the stacking direction.

ERC has carried out an interesting comparative study of the temperature distribution in process gas–cooled cells and DIGAS-cooled cells.[617] To evaluate process gas cooling, a three-cell stack employing 13 cm × 38 cm cells was assembled with thermocouples located both in the bipolar plates and in the adjacent gas channels of the middle cell. Although the gas temperature is usually assumed to represent the cell temperature, the real cell temperature is more closely approximated by that inside the bipolar plates. The results, presented in Fig. 15–3, show that although there is a considerable ΔT across a process gas–cooled fuel cell, the actual cell temperature distribution is more uniform than indicated by the gas temperature alone, due to in-plane conduction effects. However, it also is readily seen that by midway through the cell there is very little temperature difference between the bipolar plates and the gas, indicating that most of the heat transfer occurs only near the gas inlet.

To evaluate the efficacy of DIGAS cooling, a ten-cell stack comprising 28 cm × 36 cm cells was assembled with a graphite cooling plate sandwiched in the center. This size and configuration closely represents the design of a full scale system. Figure 15–4 shows the air temperature profiles obtained

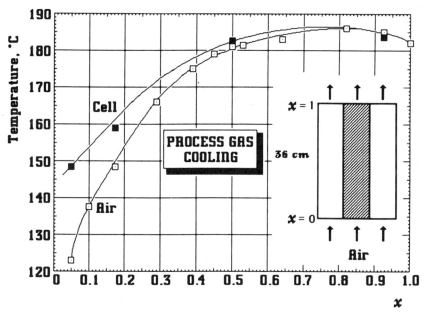

Figure 15–3: Temperature distribution in three-cell process gas-cooled assembly.[617]

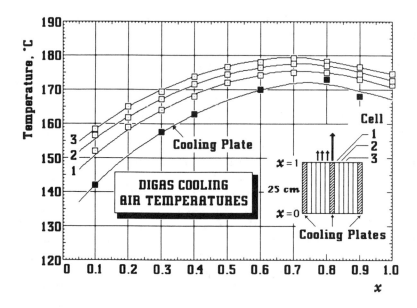

Figure 15-4: Temperature distribution in ten-cell DIGAS-cooled assembly.[617]

for this system under steady operating conditions. The temperature profile is much more evenly distributed in DIGAS cooling than in process gas cooling. The maximum air ΔT for the DIGAS case was only about 23°C compared with 62°C for process gas cooling. Since the temperature distribution in the bipolar plates has been shown to be more uniform than that in the air stream, it is reasonable to expect a temperature variation of only 10~15°C across a typical DIGAS-cooled PAFC. Since the temperature differential in the stacking direction was shown to be about 2~3°C, then with cooling plates at every fifth cell, the maximum air temperature differential in the stack will be about 27°C, with a corresponding cell temperature differential of perhaps about 15~20°C. It also can be seen from Fig. 15-4 that a temperature difference between the cell air and the cooling plate air is maintained throughout the length of the cell, so that heat is transferred to the gas flow throughout the length of the cell, rather than in just the first half of the cell when process gas cooling is used.

When the process air and the cooling air are completely separated, we have the system preferred for a pressurized air-cooled PAFC system by Westinghouse and Sanyo, the licensees of ERC. The main reason for choosing this approach is that of overall system efficiency, at the expense of increased complexity, since it requires more manifolding arrangements in the cell stack together with air circulation systems of increased size. Its major advantage is to reduce the size of the cathode feedback loop, and its consequent pumping power requirement.

The main advantage of liquid cooling is the fact that it offers the most efficient heat removal and recovery, giving by far the flattest cell temperature profile, since external heat recovery is most efficient with liquid heat exchangers. For example, in a typical liquid-cooled system, the temperature differential within individual cells can be kept to less than 5°C.

A liquid-cooled system also requires the lowest auxiliary power for a given heat exchange capacity. The total parasitic power consumption for liquid cooling is typically only about 1% of the total fuel cell system output, compared with about 2% and 5% for DIGAS and process gas cooling, respectively. Parasitic power requirements are clearly very important when multimegawatt power plants are involved. Finally, liquid cooling will result in the lowest electrolyte loss, since the coolant does not contact the electrolyte in the cell. Therefore the latter does not evaporate in association with the cooling process.

The disadvantages to be weighed against the above advantages of liquid cooling include high cost, complexity, and relatively low reliability. Liquid cooling is by far the most expensive of the common cooling procedures, typically being projected to be on the order of 25~50% of stack cost, compared with less than 5% for either DIGAS or process gas cooling. The complexity of liquid cooling is necessarily high because of the need for specially designed coolant plates and their associated hardware.

Since standard heat transfer materials such as copper or aluminum corrode in hot phosphoric acid, these materials must be coated or sheathed in protective films, and even then their performance is unreliable in the long term. Furthermore, it appears that the use of common PAFC fuel cell materials such as graphite or graphite-resin composites cannot be used for liquid cooling because of sealing and connection problems. Looked at from the viewpoint of the cell environment, if the protective films on the cooler are penetrated by phosphoric acid, there is a very high risk of corrosion failure and possible stack shutdown. Similarly, corrosion from the environment inside the cooling system, no matter how high its purity, cannot be discounted, as has been previously discussed for the UTC TEPCO unit.

If the coolant is a conductor, care must be taken to avoid shunt currents and short circuits, which can result in parasitic power consumption or even stack failure. The latter is particularly important when pressurized water is used as the coolant, since serious leaks could lead to dramatic and catastrophic stack failure by dissolution and leaching of the electrolyte, followed by attack on the pressure vessel and all associated components. If the coolant is an organic dielectric liquid, leakage could be catastrophic to electrode catalysts. Table 15–1 presents a summary made by NASA of the merits and deficiencies of the various cooling schemes considered for

TABLE 15-1

Comparison of Cooling Methods for Phosphoric Acid Fuel Cells [2060]

System Feature	Process Gas Cooling	DIGAS Cooling	Liquid Cooling
Construction simplicity	simple	simple	complex
Electrolyte loss	high	low	low
Reliability	high	high	low
External heat exchange	fair	fair	good
Cost of cooling system, % of stack	5	5	25~50
Total differential temperature, °C	35	25	10
ΔT_{cell} in stacking direction, °C	0	8	8
ΔT_{cell} in flow direction, °C	33	17	3
Total auxiliary power requirement, %	5	2	1
Stack auxiliary power requirement, %	3	0.5	0.5
Pressure drop across cell, kPa	0.87	0.10	0.10

use with PAFC systems.[2060] A final choice made for a given system will depend on the user's priorities and the manufacturer's expertise and experience.

D. START-UP AND SHUTDOWN

Starting a fuel cell system can never be simply a matter of opening the reactant feed valves and then throwing a switch. Depending on the characteristics of the system, a routine of flushing, purging, and activation procedures must be followed to ensure that all cells are functioning according to design, and that no mishaps such as the development of back e.m.f.s or eventually hydrogen explosions take place. All subsystems must be functioning properly before the stack itself is put on stream.

Similar procedures must be followed when the system is shut down or left inactive for appreciable periods. The effects of shutting down a fuel cell system depend on the types of electrodes, electrolyte, concentration, and shutdown mode. For example, when PAFCs are shut down and allowed to cool to low temperatures (-55°C) damage to the electrodes can result, which will significantly lower cell performance.[23] Dilution of the electrolyte prior to shutdown and freezing, causes more severe damage than when the electrolyte is concentrated; and electrodes with low hydrophobicity sustain more damage than those having higher hydrophobic characteristics. As a consequence, the general feeling is that utility PAFC fuel cell

Heat and Mass Transfer: System Issues

stacks should always be maintained at temperatures higher than the freezing point of the acid in the cell. Care should also be taken to avoid unduly large pressure differentials across electrolyte matrixes and membranes. In the case of high temperature systems, careful heating and cooling operations must be followed to prevent thermal stress and damage to the stacks.

Each system has its own peculiarities, as worked out by the designers, manufacturers, and operators. Although little information is available in the literature regarding these critical operations, the following have been published:

One way of starting up hydrogen-oxygen alkaline cells using porous electrodes containing platinum-group catalysts, involves first filling the electrolyte and gas chambers with electrolyte (25~35% KOH solution) at a temperature of 80~105°C over a period of 2~6 hours from the gas chambers, to the electrolyte chambers via the porous electrodes.[286] To start up the stack, the electrolyte should be displaced from the gas chambers with the reactant gases, the spent solution disposed of, and the stack sealed. This method is claimed to prevent metallization of components in the electrolyte loop, the appearance of short circuits, and blockage of electrolyte passages. To enhance the activity of platinum-group metal oxide catalysts in circulating electrolyte alkaline stacks, the electrodes can be treated with hydrogen at 80~105°C for 3~6 hours while submerged in circulating KOH solution.[287]

A start-up procedure claimed to minimize explosions in hydrogen-oxygen fuel cells consists of supplying hydrogen to the hydrogen gas compartments for 1~10 minutes, followed by blowing an inert gas into the hydrogen compartments, and then supplying oxygen to the oxygen gas compartments.[636] When the correct open circuit voltage is attained, hydrogen should again be fed to the hydrogen chambers, and stack operation then begun.

With regard to shutdown procedures, it has been claimed that the reliability and stability of low temperature hydrogen-oxygen fuel cell batteries may be improved if cells are shut down by first lowering the current load down to about 30 mA/cm^2 before cutting off the reactant supply.[1819] Another method of shutting down hydrogen-oxygen fuel cells is to connect a shorting circuit in parallel to the eternal load, followed by bleeding off the residual fuel and air, then filling the cell with nitrogen gas to retain the activity of the electrolyte (in this case, phosphoric acid).[2401]

A study also has been made concerning the start-up and shutdown periods required for platinum anodes and cathodes in both alkaline and acid fuel cells when the reactions are governed by surface reaction control, diffusion control, and convection control.[60]

E. MECHANICAL COMPONENTS

Once the type of fuel cell system has been chosen and its dimensions determined for a particular application, consideration must then be given to selecting mechanical components such as blowers, pumps, and the motors that are required to provide the necessary fluxes of reactants, coolant, and (where necessary) electrolyte. Typically, fuel cell ancillaries consume 4~15% of the gross d.c. electrical output of the stack. Ideally, these ancillary components should have a longer life than the fuel cell stack to give maximum reliabilty; should be compact, lightweight, and efficient; and should operate at low noise levels.

As in most cases, a compromise among these criteria usually must be made. Simple examples show that air blowers may operate at high efficiencies, but their actual power consumption may be rather high due to the volume of gas they must pump, especially for air delivery at high pressure. On the other hand, liquid coolant pumps have low efficiencies, but consume relatively little power, due to the small volume of material that must be circulated to achieve the correct temperature distribution.

Earlier devices were designed for aircraft and aerospace applications, and were intended to be highly reliable and compact; but they were in contrast costly, noisy, and often inefficient. While the latter factors were perhaps of secondary importance to these special applications of fuel cells, they assume much higher importance for successful large-scale commercial fuel cell systems. The general requirements for auxilaries are therefore high efficiency, low power consumption, low cost, and acceptable reliability. In addition to the above, factors such as availablity and delivery time are also important for the efficient planning of a production line. It is often difficult to decide whether to design, fabricate, and test auxiliaries in-house or whether to purchase them from outside suppliers. In the latter case, the decision may have to be made between buying an off-the-shelf item that may require modification, or buying a high-cost, custom-made single item that may have a long delivery time.

The following discussion of blowers, pumps, and motors has been taken from Refs. 1327 and 2134.

Blowers

Air blowers are usually the largest consumers of ancillary power required for ambient pressure fuel cell stacks. Along with the corresponding turbocompressors for pressurized systems, they are the noisiest components of most fuel cell systems and are probably the least straightforward to specify and design. Two general classes of air blowers are used: the aerodynamic type and the positive displacement type.

Rotary aerodynamic-type blowers impart kinetic energy to the air stream via rotating axial or centrifugal fans or impellors, and then allow the air stream to slow down, thereby converting the kinetic energy to energy in the form of excess pressure. The pressure rise can be increased by multi-staging, but adverse aerodynamic effects (e.g., surging and stalling) then occur more frequently, and some efficiency also is lost. The mechanical efficiency of such machines is relatively high.

Positive displacement blowers, which effect pressurization by physical compression, usually have better adiabatic efficiencies but worse mechanical efficiencies than aerodynamic blowers, because of higher frictional losses. Since the inefficiencies of blowers result in heating of the gas stream, some of the losses are recoverable as an increase in pressure. The two types of positive displacement blowers commonly used are the reciprocating bellows, or diaphragm type, and the sliding vane type. A double-acting bellows, either single- or multi-stroke, is needed to provide a reasonably steady air flow. This type can be very quiet but the mechanical and volumetric efficiencies and bellows life can be low. Vane blowers are even less efficient but have the advantage of being much more compact than diaphragm blowers; for example, vane blowers sometimes occupy only about one tenth the volume of a diaphragm blower of equivalent capacity.

Sliding vane blowers have a large stable range of delivery rates, whereas aerodynamic blowers are severely limited by the onset of surging at low flow rates and high pressures. Thus the positive displacement blower has an advantage over the aerodynamic blower in its tolerance to variations in air flow or pressure requirements. For ambient pressure fuel cells with power outputs above about 10 kW, however, flow rates are so high that centrifugal blowers become more efficient than positive displacement blowers. For mechanical simplicity, air blowers and electrolyte pumps in circulating electrolyte alkaline systems have often been driven by a common shaft.

Pumps

Seal-less pumps for circulating electrolyte alkaine systems are commonly employed to eliminate the leakage and power losses associated with rubbing seals. For pumping corrosive electrolyte, magnetically-coupled rotary pumps, which have perfect seals, are often employed. Lifetimes of thousands of hours without appreciable wear are frequently reported for magnetically coupled, plastic-bodied pumps driven by d.c. motors. These pumps, however, have limited torque capacities and are rather bulky. They may be either hydrodynamic or of the positive displacement type. The latter variety is generally preferred because it can readily pump a gas-liquid mixture. Reciprocating bellows pumps have also occasionally been used, but they may suffer from fatigue of the bellows and valves, and the reciprocating mechanism may be noisy and inefficient.

The materials of construction for pumps have the same requirements as discussed in the earlier sections on materials in Chapter 13. An additional aspect of circulating electrolyte system design is that the pump must handle liquids of widely varying viscosity; for example, the viscosity of 7 M KOH solution undergoes a tenfold increase as its temperature falls from 70°C to -30°C. Prewarming of the electrolyte at very low ambient temperatures by the use of chemical or electric heating will mitigate this problem and the associated high electrolyte pressure-drops in the circulating system. Many of the same design considerations also apply to coolant pumps and fuel pumps for reformers, except that the conditions encountered are then usually less corrosive. Peristaltic pumps sometimes are used for the fuel supply in small systems.

Motors

As with blowers and pumps, small electric motors have lower efficiencies than larger ones. In a small d.c. motor, a high rotation speed is essential to obtain a good power-to-weight ratio. Unfortunately, high speed means high windage and bearing losses, a higher rate of brush wear, and increased commutation losses. Static as well as rotary commutation may be used, but, in both, must be designed to minimize radio frequency interference.

Another problem that must be considered with a stand-alone fuel cell system having a variable load is the fact that fluid flow rates must increase as current load increases. In practice this means that the voltage supply to the motors should increase while the fuel cell voltage decreases. This necessitates suitable control circuits.

Turbocompressors

Turbocompressors are important to the efficiency of large fuel cell units, since their use increases stack voltage at a given current density. They are normally powered by hot-air turbines using waste compressed air from the stack, some fuel being burned as necessary to provide any excess power requirements. In the proposed 7.5 MW Westinghouse PAFC system (Chapter 16) a steam-condensing cycle is used, but this has the disadvantage that no excess for cogeneration purposes is available from the fuel cell unit. Since compressor and turbine efficiencies depend so strongly on their size, the use of turbocompressors can have diminishing returns in small units, since more and more fuel is required to operate them, thus compen-

Heat and Mass Transfer: System Issues 489

sating for the increase in stack output. Since an improvement of 1% efficiency in the turbocompressor can add about 0.5% to total plant efficiency, the tendency has been to increase efficiency as much as possible.

As an example, small off-the-shelf turbocompressors (e.g., the 2500 Nm^3/h Shimazu unit used in the Japanese LTLP 1 MW demonstrator; see Chapter 5) now have efficiencies of 70$^+$ %, whereas a larger off-the-shelf machine that would have been available in 1976 for the New York 4.5 MW demonstrator would only have been about 67% efficient. As a result, a special air-bearing turbocompressor was developed for the New York and TEPCO units, with an efficiency of 71.5%. According to Japanese sources,[72] it is now expected that magnetic bearing machines in this size range can now be made with 73% efficiency, and Kobe steel has developed an 8 atm, air-bearing machine with 74% efficiency. While it has normally been said that the lower size limit for a pressurized PAFC unit would be a minimum of 1 MW, and more favorably 10 MW, Fuji Electric is designing a small 50 kW natural gas or LPG unit operating at 2 atm, under Tohoku Electric Co support. It is intended for use on remote islands. The turbocompressor is a tiny (157 Nm^3/h) unit with an efficiency of about 60%, being developed by Ishikawajima-Harima Heavy Industries.[72]

F. ELECTRICAL DESIGN AND POWER CONDITIONING

The electrical arrangement normally employed in small fuel cell systems, which may consist only of one small stack or even of a single cell, is to place the cells in series, using monopolar or bipolar configurations, and external or internal electrical connections. Some of the details of these configurations have been discussed in Chapter 13. In very large fuel cell systems, which may contain many hundreds or thousands of cells, the common procedure is to arrange the cells in series-connected stacks, each consisting of several hundred cells, and then to connect these stacks electrically in parallel. Blocks of series-connected stacks also may be connected in series, with groups of series-connected blocks then connected in parallel to develop the desired system voltage and current output. For large systems such as these, which may generate hundreds of megawatts in the future, great care must be taken to ensure that seemingly mundane components such as electrical connections, switches, and circuit protection are appropriately designed. In addition, intercell shunt or leakage currents, which are particularly important in flowing electrolyte systems, but which still can appear in matrix cells (cf. Chapter 17) must be minimized.

In addition to these electrical design factors relating to the power source itself, consideration must also be given to the end use of the power generated by the system. Some of the output will be required as d.c. power for auxilaries, and the majority as a.c. power for the main system output. Furthermore, various auxilary devices will be operated at different voltages. Thus power conditioning also is an important aspect in the design and operation of a given fuel cell system.

Aspects of Electrical Design

As has been briefly discussed in Chapter 13, it is important that electrical connections should have minimal electrical resistance and maximum corrosion resistance. A review of the types of electrical connections used in energy conversion technology has been published.[2222] In this regard, it is often true that seemingly simple problems such as making suitable electrical connections are difficult to solve in practice. The following examples will serve as an indication of the nature of some of the problems of connector design.

To increase output voltage and reduce electrical losses in fuel cells employing carbon electrodes, good electrical contact throughout the electrode must be provided. In cells of the present generation, such contact is provided by pressure alone (Chapter 16). In the past, however, many devices to aid electrical contact were described. For example, McGraw Edison Co. claimed a carbon cathode design with one or more cavities leading through the top face to a lower portion of the electrode.[2215] Each cavity was to be provided with a wire coil inserted under tension so that when released it would spring out to engage the cavity wall under pressure and provide a low resistance connection continuously throughout the cathode. Preferably, the coil also would have an upper tail extension connected to the cathode terminal for providing a direct connection of the terminal with the lower portion of the cathode. Similarly, the Monsanto Co. developed a method of making a mechanically tight electrical connection between an electrode cable and an electrode that involved compressing the electrode lead and a slug of electrically conductive material, such as solder, within a hole drilled axially in the electrode, followed by the addition of an insulating material such as Scotchcast resin to fill the remainder of the hole.[366] A second insulating material was then added over the connection, making it resistant to chemical attack. In some types of fuel cells, especially those employing solid electrolytes, high current pulses and voltage surges can damage the electrolyte film. These voltage and current excursions can be reduced by connecting condensers between the anodes and cathodes of the cells.[1609]

A problem that is troublesome in fuel cell systems employing circulating liquid electrolytes is that of minimizing the so-called leakage or shunt currents that flow through the electrolyte channels, which can thus effectively short circuit the stack.[2791] Workers at UTC developed mathematical models that allow the calculation of shunt currents and the effective operating potentials in multicell stacks containing electrolyte interconnecting paths between individual cells.[1362,1363] Numerical results were also presented for stack designs employing acid electrolytes. Similarly, work reported in the Soviet literature has described an experimental-theoretical optimization model for the design of the supply channels of fuel cells so that leakage currents and power losses are kept to a minimum.[2623] In this model, power loss is given as a function of the leakage current. The electrical energy losses through liquid distribution lines can be kept at tolerable levels in fuel cell stacks that are subdivided into groups and further subdivided into blocks by providing elongated electrolyte passages between the blocks.[2707] Another method of lowering leakage currents is to electrically isolate the cells by the introduction of air bubbles at regular intervals into the electrolyte inlet or outlet manifolds, thereby preventing the formation of a continuous electrolyte path between the cells.[321]

Power Conditioning

The volt-ampere curve characteristic of a given fuel cell system depends on the properties of the fuel cell itself, and the value of the voltage at a given current will vary with the external electrical load. The curve falls from a high voltage at no load (which will actually consist of the relatively small parasitic load required to operate the blowers, pumps, motors, solenoid valves, etc. that operate the system) to a much lower voltage at full load. This performance curve presents a challenge to the design of the power conditioner, which must function to maintain a constant a.c. output independent of the d.c. stack voltage variation as a function of load. In addition, the relatively low voltages generated by the series-connected fuel cell stacks may have to be increased by the power conditioner to produce the high voltage utility-quality a.c. output voltages required for most applications.

A number of techniques have been employed for power conditioning that incorporate semiconductor and integrated circuit technology. Some of the simpler techniques, requiring minimal voltage regulation, produce square waves, while more sophisticated approaches, obligatory for connection to the utility grid, are capable of producing precision sine waves and exact power. Depending on the design, some power conditioners are able to accept fuel cell power from any point on the fuel cell operating curve, while others require that the fuel cell be maintained at a fixed operating point (see following).[853] The smallest power conditioners produce about 140 VA,[565] whereas the large units for utility applications must be capable of dealing with hundreds of megawatts in the future. With the rapid advances that now are taking place in solid state electronics, power conditioners rapidly are becoming more precise, reliable, compact, and cost effective. Already, GTO thyristors capable of handling 4.5 kV, 3.5 kA have been developed (for example, by Fuji Electric, Ref. 72) and the latest projection for the capital cost of large utility-quality inverters is about $100/kW(1986).

Although fuel cells are able to operate over a significant portion of the volt-ampere curve, large systems are designed to operate most efficiently when the voltage and current values of the individual cells are kept close to a fixed, predetermined design value, known as the operating point. For this reason it is important to use some form of control device to maintain the cell voltages close to the constant output values. In addition to the problems of power regulation given above, the cell voltage must not be permitted to rise to values much higher than those at the operating point for any length of time under low-load conditions. This results from the fact that if the cathode potential rises, catalyst degradation and corrosion problems will almost certainly occur, at least in the case of the aqueous systems (PAFC, SPE, AFC).

Several electrochemical methods can be used to control the output voltage of a fuel cell; that is, the potential of the cathode. One consists in varying the partial pressure of oxygen in the cathode feedback system. Thus the output from a hydrogen-air fuel cell can be lowered by recirculating part of the cathode exhaust gas back to the cathodes.[2100] Control also can be achieved by varying the total reactant pressure. For example, NASA has tested the latter method on hydrogen-oxygen fuel cells designed for the Space Shuttle Orbiter, and has determined that it is a better method of voltage control than the use of reactant partial pressures. It was found to be simpler to apply, to provide a faster response to voltage changes, to give better control, and to require less complex instrumentation.[2058]

For fuel cells in which the reactants and products are dissolved in the electrolyte, the cell voltage is a function of the electrolyte composition. Based on this relationship, equations have been derived for different modes of voltage stabilization.[1049, 1050] The results show that the optimum method of voltage stabilization depends on the net energy loss incurred in implementing the stabilization procedure. The equations give the required electrolyte flow rates and the effective stabilization times. One method of voltage control for cells employing dissolved reactants is to make use of a feedback loop wherein the reactant pump increases its pumping rate when the cell output voltage falls below a certain level.[1377]

Heat and Mass Transfer: System Issues

G. BOTTOMING CYCLE SYSTEMS

The high-grade waste heat from a large high-temperature fuel cell system may be used in a bottoming cycle (or cycles) to generate more electricity. In conceptual designs for central station baseload plants, 30~40% of the electrical power will be generated by the bottoming systems using waste heat from the fuel cells and the coal gasifier, the remainder being from the fuel cell stacks. The principal types of bottoming systems that may be employed are gas or steam turbines. In general, small plants (<10 MW) will usually not be large enough to justify the added expense of a bottoming cycle. Plants in the 50~100 MW power range will probably utilize gas turbines, while large plants of greater than about 500 MW capacity will advantageously employ large steam turbines in addition to gas turbines.

Gas Turbines

The turbine is the almost universal means of extracting energy from a high pressure gas. In a gas turbine, air is compressed and heat is then added by the combustion of fuel in the compressed air, resulting in a power-producing expansion process. Part of the power produced by the turbine (generally about two thirds) is used to supply power for the compressor, the remainder constituting the net power output. Unlike a reciprocating engine, which operates on one mass of the working fluid at a time in a piston and cylinder arrangement, the working fluid (the hot gas) in a gas turbine flows without interruption through the turbine, passing continuously from one single purpose device (compressor, combustor, turbine) to the next. Gas turbines only deliver useful power in the upper 10~15% of their speed range, the maximum speed being controlled by some operating limit. Changes of turbine speed, and hence power output, are controlled by increasing or reducing the fuel supply at a constant rate.

Some of the advantages of gas turbines include:[2201]

1. Fast response to load changes.
2. Operation on a variety of fuels.
3. Low initial cost.
4. Short delivery time.
5. Relative ease of silencing.
6. Independence of a cooling water supply.
7. Low particulate (smoke) emissions.
8. Availablity of large engines (ca 150 MW).

9. High reliability.
10. Easy starting ability.
11. Small area required for plant site.
12. Rapid repair time.

The ratio of the power of a gas turbine to its cost increases as the turbine inlet temperature, hence the efficiency, increases and as the efficiencies of the other components are increased with the introduction of higher complexity (e.g., multi-shaft designs with a greater number of stages). These are necessary, since higher temperature operation requires a correspondingly higher compression ratio. The temperature and pressure limits are dictated by the best available materials of construction.[1093] For example, the temperature limit is set by turbine blade creep, with a loss of strength at elevated temperature; whereas the pressure ratio is limited by rotor tip velocity, which determines material stress. To overcome some of these limitations, it is common to artificially cool the blades (e.g., by air-cooling holes) and to operate at transonic and supersonic flow at the compressor inlet. The actual efficiency of a gas turbine is very dependent on the aerodynamic design and on the quality of the sealing to minimize leakage flows both in the compressor and at the ends of the turbine blades.

On account of the high temperature remaining in the large quantities of exhaust gas from a gas turbine, in a large power plant it is economical to further use this heat to raise steam by feeding the engine exhaust gas into a boiler. This is used to drive a steam turbine, thereby generating more power. Typical simple-cycle gas turbine efficiencies are in the order of 20~30%, with values as high as 35% being obtainable from higher pressure, aeroderivative engines. Information is available for all commercial stationary gas turbines on an annual basis.[123]

Steam Turbines

There are a number of types of steam turbines available.[2176, 2336] The most simple type is the *straight noncondensing turbine,* which is designed to expand steam from the boiler pressure down to the lower pressure required for process steam or other uses. The distinguishing characteristic of this type of steam turbine is that its exhaust is higher than atmospheric pressure. The power output of a noncondensing steam turbine is a function of the initial steam conditions, the turbine exhaust pressure, and the process steam demand, and usually is limited by the process steam demand. A straight noncondensing turbine governor varies the turbine inlet steam flow to control the turbine speed as the electrical generator output increases or decreases in response to any changes in the connected system load. It may operate as an isolated unit or synchronized in parallel with other generating units. Noncondensing turbines only extract a portion of the energy available in the steam, the remainder being available for other purposes, such as preheating or for chemical processes.

Systems with steam demands at two or more pressure levels can often benefit by the use of *automatic extraction noncondensing turbines* that provide the flexibility to respond automatically to variations in steam demands at two or more pressure levels. Automatic extraction turbines are designed for extraction of considerable quantities of steam at some stage during the expansion, the pressure at this stage being a maintained constant at some predetermined value over a large part of the extraction and load range by automatic valving. A single automatic extraction turbine is essentially the equivalent of a high-pressure noncondensing turbine placed ahead of a low-pressure turbine. Thus, the high pressure section exhausts part of its steam to the low-pressure section, and the rest is bled off for use elsewhere, such as for industrial cogeneration. Because of the flexibility and greater design complexity of this type of turbine, it generally costs more than an equivalent condensing turbine, which is discussed in the next section. It also may be somewhat less efficient than an equivalent straight condensing turbine on account of its extra valving and design complexity. Turbines of this type typically are rated at 10~40 MW with inlet steam conditions of 42 atm/400°C to 100 atm/510°C. Extraction pressures of 11~45 atm for the high pressure and 3~15 atm for the low pressure systems are typical.

The most common type of steam turbine used in central station applications is the *condensing turbine*, consisting of a steam turbine to which main steam is supplied at any economical pressure and temperature. Some steam normally is extracted for feedwater heating, but most of the steam goes entirely through the turbine to the condenser. Unlike a noncondensing steam turbine, which also produces process steam, the main function of a condensing turbine is to produce power, usually in a cycle designed for the highest steam economy amd hence overall efficiency. Automatic extraction condensing types of steam turbine-generators usually are used for applications in which only electrical output power is required, and are available individually in sizes up to about 150 MW. Steam turbine plants used for electricity generation achieve fuel HHV-a.c. power electrical efficiencies in the order of 35~38% in very large plants (> 200 MW), while smaller units (< 20 MW) achieve efficiencies of about 30%.

Choice of Bottoming System

A number of factors must be carefully evaluated before the optimum turbine design for a given system can be determined. Important factors in the choice of steam turbines, for example, include the required load range over which full extraction must be available and the division of operating time between various electrical loads and various process flows. Table 15–2 lists some comparative data for gas turbines and steam turbine plants.[1093]

Since fuel cell modules are relatively small power producers compared with turbines, the overall bottoming cycle fuel cell power plant size will in many cases be dictated by the efficiency-size characteristics of the bottoming subsystem and the plant reliability requirements (i.e., the number of independent power trains required). In other systems, the minimum size may be determined by the fuel processor, particularly for a coal gasifier. All the bottoming systems just mentioned are available commercially, and the choice will depend on the trade-off between efficiency (i.e., fuel cost) vs. capital and maintenance costs.

Fuel Cell System Examples

A conceptual coal-based 100 MW molten carbonate fuel cell power plant designed by General Electric, and Pacific Gas and Electric serves to indicate the relative contribution of bottoming cycles to overall plant output.[1520,2097] This plant would incorporate a pressurized entrained bed oxygen-blown Texaco coal gasifier with a low-temperature Selexol process to clean up the fuel gas. With the exception of the fuel cells, all the components of this system are now commercially available.

TABLE 15–2

Comparison of Gas Turbines and Steam Turbines[1093]

Parameter	Gas Turbines	Steam Turbines
Usual power range	100 kW~100 MW	1 MW~1000 MW
Approximate capital cost[1]	$80~250/kW	$140~240/kW [2]
Approximate installed cost[1]	$150~300/kW	$800~1200/kW
Size and weight	small	large
Typical plant life	15~30 years	25~35 years
Lubrication needs	oil topping up	negligible
Maintenance periods	monthly and annual	daily
Inspection/overhaul period	several years	annual
Typical electrical efficiency	20~40%	30~40%
Fuel grade	high to medium	medium to low
Fuel type	liquid or gas	gas, liquid, or solid
Staffing requirements	none extra to plant	extra required

[1] 1979 dollars
[2] turbine only

The fuel cell sections of the plant would be modular, each containing sixteen 1-MW fuel cell stacks sharing one gas turbine. Four such modules located in parallel provide superheated steam for one steam turbine. The superheated steam provided by each module is obtained by passing recycled cathode exhaust (CO_2-air mixture) through a steam generator. Excess CO_2-air passes through a gas turbine before final venting to the atmosphere. The energy flow for this plant is summarized in Table 15-3, which shows that 41% of the gross electrical generation is provided by the bottoming system. For this power plant, which has an overall electrical efficiency of 48.9% (HHV of the coal to a.c. power), the fuel cells, gas turbines, and steam turbine produce 59%, 25%, and 16%, respectively, of the gross electrical power.

A more advanced coal-fueled molten carbonate system, designed by UTC, also has been proposed.[1520] In this system, coal would be gasified in an air-blown catalytic gasifier operating at a much lower temperature than the previous system. By the use of a catalyst and a lower gasifier temperature, the main product from the gasifier is nitrogen-diluted methane. After high temperature desulfurization by a process that is still under development, the methane-containing fuel would be supplied to an internal reforming molten carbonate fuel cell containing a special reforming catalyst. As in the previous system, the hot gases leaving the fuel cell are routed through a turbocompressor before being vented to the atmosphere. However, no steam turbine is used in the design.

TABLE 15-3

Energy Balance for Coal-Fueled 100 MW Baseload Plant[1520]

Electric Power Component	% of Coal HHV	MW Power
Fuel cells	33.3	68.1
Turbocompressor	12.1	24.7
Steam turbine	9.2	18.8
Expansion turbine	1.9	3.9
Auxiliary power requirements	−7.6	−15.5
	48.9	100.0
Thermal Energy Components		
Stack loss	23.0	
Fuel processing	8.6	
Heat rejection in cooling tower	13.0	
Inverter and generator losses	4.5	
Unaccounted losses	2.0	
	51.1	

TABLE 15-4

Energy Balance for Advanced Coal-Fueled UTC Direct Molten Carbonate Fuel Cell System[1520]

Electric Power Component	Power Generation, % HHV
Fuel cells, a.c.	49.9
Turbocompressor, a.c.	9.5
Recycle turbocompressor, a.c.	1.8
Auxiliary power requirements	−1.4
Net power plant efficiency, %	59.8

Table 15-4 shows the energy flow for this system. It is clear that significant gains in efficiency can be made, compared with the more conventional GE design, without the need for a steam turbine bottoming cycle by careful thermal integration of advanced components. However, substantial further development will be needed before such a system can become a reality. Other schemes are described in Refs. 421, 492, 493, 1521, 1787, and 1977.

The systems just described would both be baseload plants. For small on-site generating systems of 1~2 MW output, it would be uneconomical to provide bottoming cycles, and such systems could not operate with integrated coal gasifiers. If they operate on coal gas from a remote gasifier, their overall coal-a.c. efficiency will by necessity be lower than that possible with large advanced systems that employ integrated gasifiers and bottoming cycles. However, if methane is available instead of coal, the higher efficiency of fuel processing will result in high fuel-a.c. efficiencies in small units. For example, small internal reforming molten carbonate fuel cell systems operating on natural gas may be capable of overall plant electrical efficiencies greater than 50%, without bottoming systems. These are discussed in more detail in Chapter 17. In the 5 MW class, a higher efficiency may be obtained by the addition of a simple bottoming cycle. For example, the GE-PG&E design described in Ref. 1207 includes a hot-air turbine.

All these conceptual bottoming-cycle plants are designed around the use of high temperature MCFCs, which are still relatively far from commercialization. An interesting study was made by Kinetics Technology International Corporation which was more realistic in that it proposed a coal-fueled 50 MW power plant containing the cell stacks of four UTC prototype 11 MW FCG-1 PAFC systems.[1836] This plant design incorporated a small commercial fixed bed air-blown gasifier to process North Dakota lignite, producing a low BTU coal gas as the source of hydrogen, as well as 11% of the BTU content of the lignite as product tar and oil. A diethanolamine process was used to remove the sulfur compounds from the coal gas, followed by a conventional Stretford plant[1794, 1868] to recover the sulfur from the acid gas.

Overall, five cases were studied. The bottoming cycle for the baseline case (Case A) consisted first of a catalytic combustor for the fuel cell anode exhaust gases, which would contain 15% of the original hydrogen content. The resulting gas stream was then mixed with the fuel cell cathode exhaust gas and passed through a turboexpander to produce additional power to run the other compressors that provide the air and fuel flow to the fuel cell stacks.

Case A (*Baseline*) took no credit for the economic or thermal value of the tar-oil and the waste heat produced by the plant. Case B (*Auxiliary All-Steam Electric Plant*) would use all the available tar-oil and reject heat to generate steam at 42 atm and 482°C to run through an expansion steam generator. In Case C (*Export Steam Plant*), the steam would be sold across the fence instead of being used in a bottoming system for auxiliary power. In Case D (*Most Electrical Power Capacity Design*), electric power is generated from the fuel cells, from steam produced by the combustion of tar-oil, and from the use of a turboexpander generator running on the excess hot gases. In Case E (*Tar and Oil Sold Across the Fence*), the tar and oil would be completely removed from the system, but with some energy or economic value (i.e., the heating value of the tar-oil would be subtracted from the HHV of the lignite fuel to calculate the overall heat rate). In this case, the tar-oil is sold across the fence and the excess hot gases expanded in a turboexpander to generate an additional 4.0 MW of power.

The results of these five cases are shown in Table 15–5. It is evident from this table that bottoming systems are much less important for low temperature fuel cell systems such as the PAFC than they would be for high temperature systems. Even in Case D with the maximum amount of heat recovery, the bottoming subsystems account for only 15% of the total electrical power output, compared with values of 30~40% for the high temperature MCFC designs. From the capital cost figures it can be seen that, while Case D has the lowest plant capital cost per kilowatt of electrical output, it is only slightly more economical (by 9.3%) than the baseline case without a bottoming cycle.

From the preceding, it can be seen that bottoming cycle systems are needed to achieve high system efficiencies with practical fossil fuels, particularly coal; and that the higher the system temperature, the more important it becomes to include bottoming systems to increase efficiency.

TABLE 15-5

Assessment of Coal-Fueled 50 MW FCG-1 Phosphoric Acid Fuel Cell Power Plant[1836]

Case	Plant Output (MW)	Bottoming Power as % of Total Power	Overall Heat Rate (BTU/kWh)	Total Installed Capital Cost ($/kW)*
A: Baseline	43.2	0	12,170	1,424
B: Auxiliary All-Steam Electric Plant	48.3	10.6	10,880	1,348
C: Export Steam Plant	43.2	0	12,170	1,473
D: Most Electrical Power Capacity	50.9	15.1	10,330	1,291
E: Tar and Oil Sold Across the Fence	47.2	8.5	9,940	1,321

*Thirty-year levelized bus bar costs in constant 1981 dollars; lignite cost at $25/ton delivered in 1981; lignite cost escalated at 2.8% per year; 70% plant capacity factor.

Since the bottoming system may account for a significant part of the plant capital cost, the increase in efficiency that it contributes must be optimized to give the lowest overall cost of electricity.

PART IV

State of the Art

In Part I of this book we examined the basic principles and inherent scientific advantages of the fuel cell as an energy conversion device. In Part II some of the concepts and areas of application of fuel cell systems were considered, and the numerous past and present fuel cell programs undertaken in various parts of the world were reviewed. In Part III we considered in detail the design considerations that must be taken into account in order to engineer a full-scale operating fuel cell system.

In Part IV we will look at the current state of the art of the three most extensively researched types of fuel cell: the Phosphoric Acid fuel cell (PAFC, Chapter 16), the Molten Carbonate Fuel Cell (MCFC, Chapter 17), and the Solid Oxide Electrolyte Fuel Cell (SOEFC, Chapter 18). These three technologies are generally referred to as First, Second, and Third generation fuel cells systems, respectively, corresponding to their relative nearness to commercialization. For each type we describe the design concepts and hardware of the major developers, examine the present levels of performance, consider manufacturing cost, and discuss problem areas that need further investigation.

CHAPTER 16

Phosphoric Acid Fuel Cells

A. UNITED TECHNOLOGIES CORPORATION FUEL CELLS

Since January 1985, the Power Systems Division of UTC has been engaged in a joint venture with Toshiba, whose primary purpose is to develop and market utility PAFCs. The title of the venture is International Fuel Cells (IFC). However, for convenience, the old corporate name is used in the following, except for new commercial developments taking place after that date, particularly specific commercial utility fuel cell designs.

Phosphoric acid fuel cells developed by UTC up to about 1978 consisted of a sandwiched configuration as shown in Fig. 16–1, in which the ribs in the cross-flow configuration are rotated through 90° for clarity. The phosphoric acid electrolyte, maintained at about 200°C, was contained in a porous, hydrophilic matrix located between the catalyst layers of the anode and cathode. In 1975, U.S. Patent 3,859,139 described an early method for applying catalyst to each surface of the matrix. In this technique, the pores of the inert porous hydrophilic matrix are first completely filled with a volatile filler (e.g., water), which is followed by freezing the water inside the matrix pores. While the volatile filler is in the solidified form, a catalytic mixture is applied to each side of the matrix surface from an aqueous suspension containing PTFE and platinum black catalyst to lay down the desired amount of catalyst and wetproofing agent. The presence of the solidified filler in the pores of the matrix would ensure the production of a catalyst layer of uniform thickness and would also prevent the electrically-conductive catalyst particles from entering the matrix and possibly causing internal short circuits. After pressing the catalyst layers onto the matrix, the structure should be slowly heated to a temperature of about 280°C, at which the solidified water is completely removed while the polymer-catalyst layer is sintered or bonded to the matrix. This manufacturing technique was claimed to produce an electrode assembly with internal matrix pores essentially free of catalyst with a flat, uniform catalyst layer on the surface of the matrix. The fabrication of relatively thin matrixes

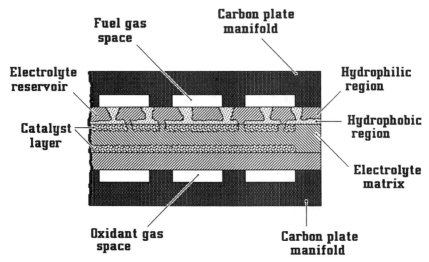

Figure 16–1: Sandwiched configuration of phosphoric acid fuel cells developed by United Technologies (*ca* 1978).

(250~500 µm) was thereby made possible. This patent was claimed to be of use with both acidic and alkaline fuel cells. Suitable matrix materials included asbestos mats, polymer membranes, and ceramic materials. This method is cumbersome from the viewpoint of fabrication since it requires a freezing operation, but it has been quoted for historic interest. It has since been supplanted by simpler and more rapid techniques of assembly of cell components, which allow the use of much thinner matrixes with lower IR drop (100~130 µm; see following).

A later UTC patent (U.S. Patent 4,017,664 in 1977) recommended the use of silicon carbide as the matrix material for the PAFC. Silicon carbide is virtually inert to hot phosphoric acid and provides all the necessary and desirable properties required for a matrix that should give a service life of at least 40,000 hours under normal fuel cell operating conditions. Silicon carbide is now universally used for this application. Mitsubishi Electric in Japan has shown that the optimum average size of the silicon carbide particles for use in electrolyte matrixes is about 1 µm, and that the optimum matrix thickness is 100~130 µm.[1423]

Another feature of the UTC design has been the use of an electrolyte reservoir to prevent electrolyte depletion in the silicon carbide matrix as a result of slow evaporation with time. One of UTC's earliest descriptions of a matrix electrolyte cell design incorporating an acid reservoir is described in U.S. Patent 3,905,832 (1975). Here, the compact cell comprises a pair of opposed electrodes, an electrolyte matrix containing an aqueous, ion-conductive electrolyte between these electrodes, together with an electrolyte reservoir positioned behind and partially defined by at least one of the electrodes. Furthermore, the electrode partly defining the reservoir has a continuous hydrophobic surface and selectively hydrophilic areas that are substantially uniformly distributed within the boundaries of the electrode surface. The electrolyte volume of the cell is controlled by the movement of electrolyte between the electrolyte matrix and the reservoir through the selectively hydrophilic areas of the electrode via capillary action, thereby stabilizing the electrochemical performance of the cell. By the time that the above patent application was made, it was realized that the structural components in acidic cells should be made of carbon, since all other inexpensive materials resulted in unacceptable corrosion rates, as has been discussed in Chapter 13.

The function of the electrolyte reservoir is described in more detail in a 1977 patent (U.S. Patent 4,035,551), in which excess electrolyte wicks into the porous, hydrophilic reservoir layer through the catalyst layer and fills the smaller pores within the reservoir. The larger pores remain empty and provide clear passageways for the reactant gas to reach the catalyst. The preferred configuration is one in which the reservoir layer is the electrode substrate itself; that is, on the gas side of the electrode. The catalyst layer is bonded onto the surface of this substrate. As described in U.S. Patents 4,038,463 (1977) and 4,064,322 (1977), the hydrophilic reservoir, which is behind and adjacent to one of the catalyst layers, can include impregnated hydrophobic material to provide reactant gas passages to the catalyst layer throughout a relatively small volume of the reservoir material. In addition, the reservoir layer also may include impregnations of material similar to those used for the electrolyte matrix to facilitate transfer of the electrolyte between the matrix and the reservoir by capillary action. These concepts are indicated in Fig. 16-1.

Figure 16-2 shows the earlier, so-called conventional UTC cell configuration, in which the repeating cell elements consisted of dense graphite gas manifold plates with ribbed flow fields on both sides in a cross-flow geometry. These also served as gas separator plates. Located between the two separator plates were the hydrophilic anode and cathode substrates, each on a carbon-felt or paper backing, having a semi-dry catalyst layer, together with half of the electrolyte-saturated matrix on the outer face. The multilayers of suitably partially wetproofed carbon-felt or paper backing represented the electrolyte reservoir areas. Assembly of the cell involved contacting the matrix layer of one cathode onto that of one anode. To provide efficient gas manifolding in the cross-flow geometry chosen, it was necessary that the ribs on the fuel side of each separator plate should run at right angles to those on the air side, resulting in a design that was not amenable to continuous fabrication; either by machining from solid in the earliest pure carbon or graphite concepts, or by molding in graphite-resin,

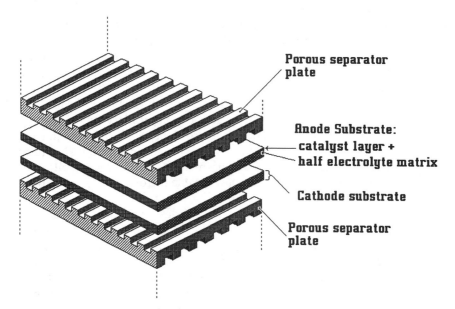

Figure 16-2: United Technologies Corporation phosphoric acid fuel cell conventional cell configuration.

followed by heat-treatment and surface finishing.[1485] This system was used in the stacks for the 1 MW prototype (with molded graphite-polyphenylene sulfide bipolar plates, and finally in the stacks for the New York 4.5 MW demonstrator, which had graphite bipolar plates; cf. Chapter 3).

In any case, the electrolyte reservoir capacity of this design using multiple carbon paper layers was very limited, and it involved a large number of carbon-to-carbon contacts that could result in a high ohmic resistance. Accordingly, it was supplanted by the so-called ribbed substrate design, shown in Fig. 16-3, in which the ribs are formed on the reactant gas side of the electrode backing or substrate itself. A catalyst layer and a half-thickness matrix is applied to the nonribbed (flat) side of the porous substrate to form each electrode. The original patent (No. 4,115,627 in 1978) proposed the use of semi-wetproofed ribbed substrates to create gas channels through the continuous surface on which the electrodes are applied, so that both reactant gas in the flow channels and electrolyte stored in the ribs of the reservoir could be made equally available to the electrode-matrix interface. However, experience later showed that this

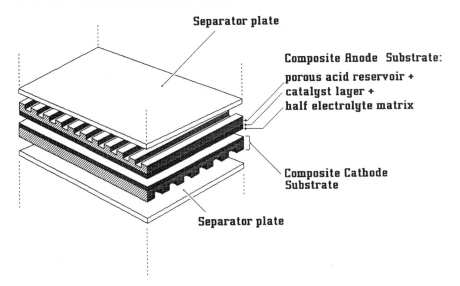

Figure 16-3: United Technologies Corporation phosphoric acid fuel cell ribbed substrate cell configuration.

was not necessary, since sufficient wetproofing occurred during the sintering of the Teflon-catalyst mixture in the electrode layer.[177]

Unlike the conventional UTC cell design, the ribbed substrate material is amenable to continuous processing since the ribs on each backing sheet run only in one direction. Separately manufactured impermeable flat graphite sheets, slightly more than 1 mm thick, serve as bipolar gas separator plates contacting the substrate ribs for the anode and cathode of adjoining cells in UTC's usual cross-flow arrangement. The ribbed substrates typically had a thickness of about 1.8 mm, giving a total cell weight (without cooling plates and electrolyte) of about 5.4 kg/m^2. As in the conventional stack arrangement, cooling plates were placed approximately one every five to eight cells, so that the pitch is five cells per inch (about 5 mm/cell).

The thick ribbed substrates allowed about a fivefold increase in effective electrolyte volume, compared with that of the conventional stack configuration when both anode and cathode substrates are used as reservoirs. Approximately 1.5 kg of electrolyte per m^2 could be stored, and it was estimated that this would be sufficient for more than 40,000 h of operation at 205~210°C and 8.2 atm pressure. This had already been shown to greatly increase the long-term stability of cell performance in laboratory cells and short stacks, as is reviewed in Ref. 177.

The ribbed substrate configuration was used in the TEPCO 4.5 MW demonstrator plant, for which fabrication started in 1980, and in all the UTC 40 kW on-site power plants. While it is also planned for the first commercial plants, new developments are taking place that may result in an elimination of the flat bipolar plate, which is not only difficult to manufacture and handle in large sizes, giving a high reject rate, but also results in a non-negligible contact resistance between cells. (This will be discussed in a later section.)

Table 16–1 lists some the UTC patents that relate to the development of graphite plates, matrixes, reservoirs, and cell configurations up to 1979.[1485]

TABLE 16–1

Selected U.S. Patents of United Technologies Corporation That Relate to Phosphoric Acid Fuel Cells[1485]

Patent No.	Year	Description
3,634,569	1972	Preparation from graphite powder of dense graphite plates for use in phosphoric acid fuel cells.
3,650,102	1972	Technique for the production of carbon paper for use as electrode substrates.
3,694,310	1972	Production of wettable organic fiber matrixes by heat treating phenolic resin fibers.
3,801,374	1974	Preparation of corrosion-resistant electrically- and thermally-conductive molded graphite coolant and support plates.
3,855,002	1974	Prevention of escape of reactant gases by sandwiching an electrolyte-saturated matrix between a separator plate and an electrode, utilizing the electrode for support and the electrolyte itself to provide a wet capillary seal against gas escape.
3,857,737	1974	Method of depositing fine platinum crystallites on conductive wetproofed carbon fiber substrates for use as electrodes in phosphoric acid fuel cells.
3,859,138	1974	Production of Pt-catalyzed infusible electrically conductive wetproofed electrode substrates from cured novolac fibers and silicon dioxide fibers.
3,859,139	1975	Method of fabricating a composite electrolyte matrix–electrode assembly and depositing catalyst particles on the outside faces thereof.
3,905,832	1975	A compact fuel cell containing a porous electrolyte-containing matrix between the two electrodes and an electrolyte reservoir positioned behind and partially defined by at least one of the electrodes.
3,912,538	1975	Catalyzed porous pyrolytic carbon-coated fiber substrates containing hydrophobic layers.

Phosphoric Acid Fuel Cells

Patent No.	Year	Description
3,932,197	1976	Method of applying platinum catalysts to hydrophilic regions of a wetproofed carbon substrate.
3,956,014	1976	Production of alternate layers of porous hydrophobic and hydrophilic catalyst-containing carbon that offer little resistance to the passage of reactants and products.
3,972,735	1976	Production of wetproofed catalyzed carbon paper electrodes.
3,944,686	1976	Vapor deposition of pyrolytic graphite on porous fibrous sheets of carbon.
3,979,227	1976	(Similar to 3,932,197.)
4,000,006	1976	Application, using a screen printing technique, of a thin continuous uniform electrolyte matrix layer on the surface of an electrode to retain the electrolyte.
4,017,664	1977	Silicon carbide electrolyte retaining matrix for phosphoric acid fuel cells.
4,035,551	1977	Use of an electrolyte reservoir matrix.
4,038,463	1977	Electrolyte reservoir for a fuel cell.
4,064,322	1977	Electrolyte reservoir for a fuel cell.
4,043,933	1977	Porous platinum-containing electrodes made from flocculates.
4,054,687	1977	Production of a partially graphitized carbon support material for platinum catalyst that is resistant to platinum migration losses.
4,064,207	1977	Porous carbon sheets produced from inexpensive carbonizable filaments.
4,058,482	1977	Production of platinum-containing porous PTFE sheet for bonding to carbon support plates.
4,078,119	1978	(Similar to 3,932,197.)
4,080,413	1978	Porous carbon sheets produced by the pyrolysis of acrylic filaments.
4,115,528	1978	Production of strong, corrosion-resistant, thermally- and electrically-conductive porous carbon substrates for fuel cells.
4,125,676	1978	Porous open-cell carbon foams made from vitreous carbon for use as a gas distribution layer in phosphoric acid fuel cells.
4,137,372	1979	Method of depositing porous carbon on and around supported platinum crystallites to reduce the rate of platinum recrystallization.
4,137,373	1979	(Similar to 4,137,373.)
4,150,076	1979	Production of catalyzed wetproofed carbon-paper sheets for use as fuel cell electrodes that are flood-resistant at open circuit air potentials.
4,163,811	1979	Electrodes made by deposition of colloidal solutions of PTFE and catalyst onto conductive substrates.

A further review of the UTC patent literature involving carbon or carbon-related fuel cell components was given in 1983.[167A]

Description of A Proposed 11 MW UTC Power Plant

As outlined in Chapter 3, the baseline prototype design for the generic commercial UTC power plant, code-named the FCG-1 (Fuel Cell Generator Number 1), was to be an upgraded version of the 4.5 MW demonstrators. Until 1986, the design was in a continuous state of flux as new concepts, optimizations and layouts were successively explored. At the time of writing (December 1986), the design of the final 11 MW power plant for initial commercial applications had been determined. This plant was to be known as the IFC PC-23. The PC-23 differed somewhat from the interim FCG-1 design, but since many more details are known about the latter, it will be discussed in detail here.

A comparison of the characteristics of the FCG-1 and 4.5 MW plants is given in Table 16–2.[646] Compared with the 4.5 MW demonstrators, the FCG-1 design was simplified to increase reliability and reduce cost. This

TABLE 16–2

Comparison of UTC FCG-1 Baseline Commercial Power Plant with 4.5 MW Demonstrator Power Plant[646]

	FCG-1 Baseline Plant	4.5 MW Demonstrator
Rated power, MW net a.c.	11	4.5
Minimum power, % of rated	0	25
Heat rate (HHV) at rated power, BTU/kWh	8300	9300
Rated power efficiencies (LHV), %:		
Reactant supply	83.5	86.5
Power section	57.5	48.7
Power distribution	99.6	99.6
Inverter	96.0	96.0
Mechanical	96.1	96.8
Overall	**44.1**	**39.0**
Standby fuel consumption, % of rated	4	15
Fuels, primary	naphtha	naphtha
Fuels, secondary	natural gas	natural gas
Fuels, optional	coal-derived	—

would result from the use of more conventional components; for example, commercial tube-and-shell heat exchangers rather than customized compact flat-plate units; and from the use of a smaller number of large-area cells per MW. The system was much less compact than the demonstrators, allowing greater accessibility for maintenance purposes. The improved heat rate of the FCG-1 (and of the PC-23) compared with that of the demonstrators resulted from the enhanced cell performance allowed by the use of improved materials in the cell stack operating at higher temperature and pressure.

The provisional FCG-1 was proposed to consist of a d.c. module, a power conditioner, a water treatment system, a heat rejection system, and a control system.[1086] A number of auxiliary subsystems would provide the fluids and electrical power required for the power plant operation. Part of the power plant system will be the *controller*, which coordinates the operation of the power plant and the auxiliary subsystems. The d.c. module was expected to consist of four integrated subsystems, namely the fuel processor, the d.c. power section, the air pressurization subsystem, and the thermal management system. As already indicated, it was intended to employ commercially available heat exchangers, contact coolers and turbocompressors, and it was redesigned for improved maintenance access with simplified interconnecting piping configurations. The cost was also lowered because the assemblies could be more easily fabricated and they would contain fewer parts than those of the demonstrators.

The power conditioner was planned to convert the output d.c. power from the fuel cell stack module to three-phase, 60-Hz a.c. power at utility line voltage. The unregulated d.c. output from the stacks was to be fed to inverter bridges, which would transform it into utility-quality three-phase a.c. This would then go through series reactors and be stepped up to the line voltage via an output transformer. The series reactors provide an inductive impedance between the inverter bridges and the utility line for control purposes. A.c. and d.c. switchgear and fuses were planned for isolation and fault protection.

The water treatment system cleans and returns water recovered from the d.c. module. The system receives untreated water from a storage tank, removes dissolved gases and suspended solid contaminants, and then heats the feedwater prior to pumping it to the d.c. module. The thermal energy used for heating the treated feedwater would be extracted from the d.c. module blowdown and condensate. Waste heat removal from the d.c. module would use air-cooled fin-fan coolers in the planned design.

The control system would include a power plant controller as well as controllers for each of the subsystems. These were planned to provide the logic and sequencing of all components and controls. While operation via the operator panel at the site was initially planned (as for the 4.5 MW demonstrators), there is also provision for power plant operation from a remote operator panel via a link to the power plant controller. In the mature commercial unit, remote dispatch, along with unattended operation, will be the operating mode so that labor costs can be minimized.

In the original FCG-1 concept, the major goal was to use the same fuel processing system as the demonstators, which would be upgraded in capacity by the use of higher pressure (8.2 atm compared with 3.4 atm in the demonstrators). This would allow the reformer to process a proportionately greater mass of gas, resulting in a greater total stack current. Since the steam requirements at the increased pressure also require a somewhat increased fuel cell operating temperature (205~210° vs. 190°C for the demonstrators), overall operating voltage would be increased, so that the fuel processor of the original 4.5 MW system would produce a total net output of 11 MW. The cell area would at the same time be increased from 0.34 m^2 (3.7 ft^2) to 0.93 m^2 (10 ft^2), resulting in a reduction in the total number of power section cell stacks from 20 to 18.

This philosophy has been largely adhered to. Aside from the use of thicker-walled pressure vessels to accomodate the increased diameter cells operating at higher pressures, many components were to be the same for the proposed FCG-1 and its successors as for the 4.5 MW demonstrator. These included the fuel processing beds in the hydrodesulfurizer, the zinc oxide absorber unit, the low-temperature shift converter, the high-temperature shift converter, and the piping and valve sizes. Although they used the same technology as that employed in the demonstrators, the dry cooling tower; the water treatment subsystem; pumps and blowers; the thermal management system, its auxiliary burners, and steam separator; and the condensers were resized in the FCG-1 design.

The key element in this philosophy, namely the use of the same reformer as that of the demonstrators, had to be abandoned. The New York experience showed that even a modified version of the original compact reformer would operate at a higher temperature than expected, and the temperature would inevitably increase further as the system pressure increased from 3.4 atm to 8.2 atm. The flame temperature would then be over 1650°C, requiring special materials, including zirconia insulation and sheathing for the reformer tubes. Inevitably, this led to a redesign so that operation could be carried out at lower temperatures with the use of less costly, conventional materials. In future, improved technology will also be employed for the power section, the flow sensors, and the power conditioner; this being the result of improvements in the state of the art, and of the experience acquired in New York and Goi. Fuel processor development is described in Ref. 1936.

Major simplifications incorporated into the design of the proposed FCG-1 power plant would result in a 37% reduction in the overall number of parts of the system compared with those in the 4.5 MW demonstrator designed in 1979. The distribution of the parts reduction was as follows:[646] major components, 33%; control valves, 50%; sensors, 21%; miscellaneous, 44%. The complete 11 MW FCG-1 system would consist of 17 truck-transportable pallets, all of which would be factory-assembled. There would be nine cell stack pallets each containing two stacks. Each of these power section pallets would weigh less than 30,000 kg and would be readily truck transportable. The proposed reformer would weigh about 29,500 kg and would be shipped as a separate pallet, 3 m square by 3.5 m high. The remaining seven pallets would all be 3.5 m high, with a maximum width of 3 m and a maximum length of 8.2 m. None of these would weigh more than 32,000 kg. Interconnecting piping and cabling between the pallets would be installed at the site. A three-unit 33 MW dispersed generating installation would have a footprint of only 72 m by 78 m, and would be capable of changing power output over the entire range (including standby at zero power) in 15 seconds to meet sudden load demands.

Cost reductions of about 47% relative even to the original (1980) FCG-1 configuration may be expected in future as a result of the use of larger cells, of improved cell stack manufacturing techniques, of commercial heat exchangers and turbocompressors in simpler system and pallet designs, and of advanced controller technology with a two-bridge inverter.[646] A further per kW cost reduction of 4.5% might be expected because of the improved performance that may be obtainable from the use of advanced catalyst formulations at lower catalyst loadings. It has been anticipated that improvements in performance may lower the heat rate from 8300 BTU/kWh to 7900 BTU/kWh on natural gas fuel by making further improvements to turbocompressor and inverter efficiencies, and to improved system chemical engineering. The commercial driving force at the present time is to lower cost, as well as to further improve efficiency, to the 7800 BTU/kWh~7200 BTU/kWh range, to allow the system to compete with future extrapolations of alternative utility generating technologies such as the combustion turbine combined cycle system. Marketing and competition for the fuel cell has been reviewed in Chapter 2 and in Refs. 177 and 432.

Ongoing Performance and Technology Improvements

The Consolidated Edison 4.5 MW demonstrator was planned to operate at 3.4 atm (50 psia), 190°C (375°F), and at a current density of 250 mA/cm^2 (232 ASF). As described in Chapter 3, operation of the system proved to be impossible as a result of the shelf life of the cells in the stacks. However, the improved ribbed substrate configuration used in the stacks for the TEPCO Goi 4.5 MW demonstrator did indeed operate at the design conditions.

As indicated earlier, the ribbed substrate configuration was upgraded to operate at 8.2 atm (120 psia) and 207°C (405°F) in the 11 MW FCG-1 design, which could be projected to allow a theoretical increase in operating current density to 432 mA/cm^2 (400 ASF) at constant voltage. Such an increase in stack operating pressure might be expected to increase rated power output by a factor of two without increasing the size of most stack components.[1588]

To demonstrate the improved performance possible from operation at higher temperature and pressure, a 3.7 ft^2, 20-cell stack with 8 cells per cooling plate, was endurance tested at 216 mA/cm^2 (200 ASF), starting in early 1983, with a planned termination in mid-January 1985 at 16,000 h.[177] Its performance was compared with that expected from the New York demonstrator.[1334] The test stack unit cell voltages for initial performance and after 5000 hours of operation were 0.745 V and 0.71 V, respectively, compared with the corresponding values of only 0.66 V and 0.62 V for cells similar to those that were to be used in the New York and Goi plants. In comparing the results, the 16% difference in current density should also of course be taken into account. In the current density range in question, this difference corresponds to about 15 mV on the Tafel plot, so that the total improvement in performance was about 75 mV over that of the demonstrator systems.

The voltage-time curve for this stack is shown in the upper plot in Fig. 16-4. The lower solid line on this plot was the performance goal for the 4.5 MW demonstrator, whereas the upper line is the projected performance decay curve for a stack operating under high-temperature, high-pressure conditions from data obtained from laboratory cells. This is UTC's so-called E-line, the approximately exponential voltage-time curve at a current density corresponding to an average unit cell voltage of 0.73 V at end-of-life (40,000 operating hours). This voltage will result in the guaranteed system end-of-life heat-rate of 8300 BTU/kWh under specific operating conditions, which are discussed in some detail in Ref. 177. One of the factors in achieving this higher operating voltage was the reduction of the 350 mA/cm^2 current density of the 4.5 MW demonstrators to 216 mA/cm^2 in the FCG-1 design.

It is clear that the performance decay of the short stack shown in Fig 16-4 was inadequate, although it did serve as a good demonstration of the potential lifetime of such stacks, which will always behave somewhat differently from laboratory cells because of their temperature and reactant distribution profiles. Since single cells have completed up to 100,000 h of life, there is now great confidence in the ability of components of appropriate design and made from the correct materials to give the type of service required.[177] To demonstrate the progress that has been made between late 1982 and 1985, the second plot in Fig. 16-4 shows the voltage-time performance of the second DOE/NASA/UTC full-scale 0.93 m^2 (10 ft^2) 28-cell short

Phosphoric Acid Fuel Cells

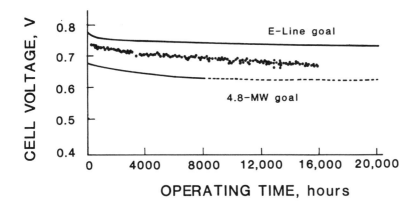

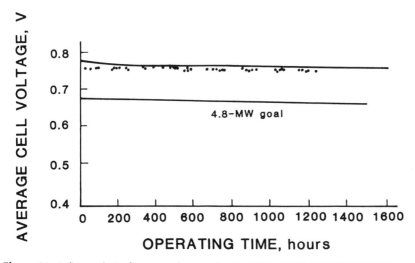

Figure 16–4: (Upper): Performance history for the 1982–1983 DOE/NASA/UTC 0.34 m² electric utility stack to 16,000 h. Cells 3.7 ft², 30 cells, 8 cells/cooler, 120 psia, 207°C average temperature, 216 mA/cm², 85%/70% reactant utilization. (Lower): Second DOE/NASA/UTC 0.98 m² short stack. Cells 10.3 ft², 28 cells, 7 cells/cooler, other conditions as above.[177]

stack, fitted with 7 cells per cooling plate because of the total number of cells used. It can be seen this time that performance is much closer to the E-line, and that the linearized decay rate over the first 1000 hours is this time only about 1.5 mV/1000 h, compared with about 5 mV/1000 h (decaying to 2~2.5 mV/1000 h after 5000 h) in the case of the 3.7 ft² stack.

This reflects the use of improved (and better matched and selected) materials. Even more remarkable is the performance being obtained in laboratory cells using special electrode formulations,[177] as shown in the upper plot of Fig. 16–5, where the E-line is now exceeded by some 20 mV.

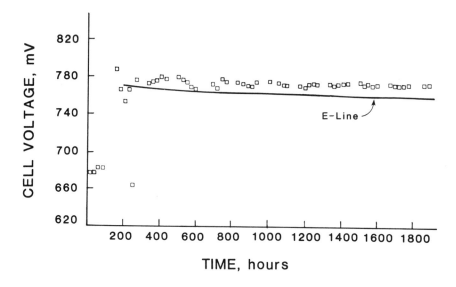

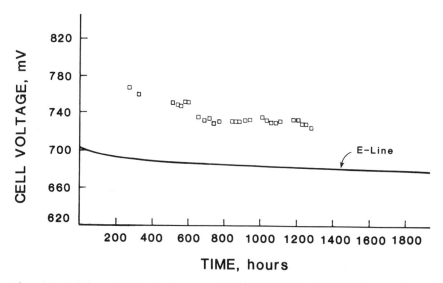

Figure 16–5: (Above): Output voltage vs. time for the DOE/NASA/UTC high-performance electric utility cell number 6319: Conditions as Fig. 16–4. (Below): Output voltage vs. time for a 0.9 g/cm² Pt cathode cell with advanced structure operating at 432 mA/cm²: Other conditions as above.[177]

Phosphoric Acid Fuel Cells

Finally, the lower plot in Fig 16-5 shows a specially-designed laboratory cell with lower IR-drop and 0.9 mg/cm^2 platinum equivalent cathode catalyst loading, rather than the standard value of 0.5 mg/cm^2. If all of the extra platinum is active, this should raise the cell potential by 25 mV. However, the cell in Fig. 16-5 was operated at twice the normal current density (432 mA/cm^2), which should lower the potential by 30 mV via the Tafel relationship, and add a further 25 mV in IR-drop. The total difference between the upper and lower plots is 35 mV after 1000 h, which is in good agreement (to within 5 mV) with the given estimations. Thus, if performance can be improved by about a further 15~20 mV at 432 mA/cm^2, it should be possible to operate at an end-of-life heat-rate of 8300 BTU/kWh with the total number of stacks, hence per kW stack cost, reduced by a factor of two. The possibilities for accomplishing this, and its implications as far as system cost is concerned, are discussed in the following.

The 8.2 atm, 3.7 ft^2, 30-cell short stack and subsequent stacks tested employed a simplified lower cost cooler configuration. This design simplified construction by eliminating many tube connections, significantly reducing the number of coolers per stack and increasing the number of allowable cells per cooler from four to eight. The new coolers employed a serpentine configuration which was estimated to be cheaper and more reliable than previous designs.[448] The plates had grooves that held the cooler tubes, and clamshell holders were used to contain the serpentine coolers. In addition, more conductive material was used in the plates to improve heat transfer through the assemblies. Initial 10 ft^2 coolers of this improved design were tested.

Automatic handling and more efficient heat treating processes were developed that significantly advanced the separator plate fabrication technology, resulting in improved materials quality with a lower level of impurities. This led to fewer pinholes, reduced material loss, and thinner molded separator plates. Now, 10 ft^2 separators are being routinely fabricated. Similarly, an improved 10 ft^2 substrate forming technology was put in place using a dual-belt forming machine that is able to manufacture substrates while incorporating several processes which previously were carried out independently. This equipment is able to form electrode substrate at the rate of 10 ft^2/minute, and represents a major step towards lower cost substrates. Work also is underway on the development of lower cost graphite fibers and alternative bonding resins for use in the green substrates before baking and graphitizing.

Finally, stack manufacture has been significantly improved by the use of automatic electrode processing technology for 10 ft^2 components. This incorporates the advanced alloy catalysts and supports discussed in Chapter 12, along with lower cost procedures for preparation of the catalyst layer. The use of an automatic electrode spray catalyzation machine and flow field machining has significantly increased the rate of the electrode fabrication process. The use of an improved matrix has resulted in better cross-pressure resistance and lower cost. The electrode substrate and matrix in the latest design are filled with acid automatically.

In conjunction with the DOE-sponsored higher pressure cell stack technology program, efforts are also being made to develop a higher pressure reformer tube and to demonstrate that commercially available contact coolers will meet the requirements for gas stream condensers. Several reformer tube configurations have been tested; and results have demonstrated that scale-up of the reformer tube height from 1.6 m (4.5 MW design) to 2.3 m, and increasing the reformer operating pressure from 5.4 atm to 10.9 atm will provide the 91% naphtha fuel conversion at better than the 85% energy efficiency required for the 11 MW power plant. Similarly, test results have shown that commercial contact coolers, which cost ten times less than those previously used, are able to meet the requirements of the 11 MW power plant exhaust gas condenser.

The 4.5 MW demonstration plant employed expensive custom-made 71.5% efficiency turbocompressors with air bearings to pressurize the reactant gases (Chapter 15). Under the sponsorship of Niagara Mohawk and Northeast Utilities, three commercial production-type turbocompressor systems were tested to show that the required pressure ratio of 8.4/1 can be obtained and to demonstrate that it is possible to obtain the 67% efficiency required for the 11 MW power plant using simple modifications. Both objectives were met, efficiencies of 66~68% at the 8.4/1 pressure ratio being obtained. These commercial turbocompressors are cheaper than the previous specially-designed units by a factor of five. Advanced Japanese turbocompressor designs, discussed in Chapter 15, potentially offer even higher efficiencies.

Improvements in the power conditioning subsystem also are being developed to increase its capacity at lower cost.[1588] Both DOE and EPRI are sponsoring work on larger, higher-voltage-rated thyristors, an improved commutation circuit, and advanced logic.

Many of the cited modifications (where appropriate) are being incorporated into the advanced on-site cell, which will follow on from the 40 kW unit described in Chapter 3. As was already indicated in that chapter, economies of scale point to a larger atmospheric-pressure on-site unit to attain the minimum cost goal of about \$2500/kW to achieve initial market penetration.[696] As the technology matures, costs will become lower and the market will broaden. The present aim is to use an atmospheric-pressure version of an updated TEPCO-type stack, producing about 200 kW instead of 275 kW, at 216 mA/cm^2 instead of 250 mA/cm^2. By using a novel proprietary stack construction concept, which may result in very considerable cost savings, it has been possible to reduce IR-drop. Combined with the use of advanced catalysts, the latest results (Fig. 16–6) now lie 20 mV

Phosphoric Acid Fuel Cells 519

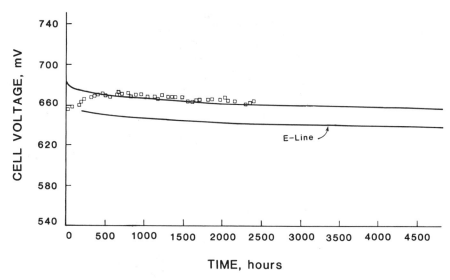

Figure 16–6: Performance of an on-site cell using "Configuration B" bipolar plate (see text). Atmospheric pressure, 190°C, 216 mA/cm², 80%/60% reactant utilizations.[177]

above the E-line for the on-site system, and indicate that 0.655 V cell potential is not an unreasonable end-of-life goal at 216 mA/cm², 190°C and atmospheric pressure (i.e., 75 mV below the FCG-1 pressurized system operating at 15~20°C higher temperature at the same current density). Such a system should have a very long lifetime and may well find a very broad market as mature production is achieved. Since the new construction method is related to some of the concepts used by Engelhard, it will be discussed later in the section of cost reduction and future system improvement.

B. WESTINGHOUSE–ENERGY RESEARCH CORPORATION FUEL CELLS

Mark II 7.5 MW Fuel Cell System

As was described in Chapter 3, the ultimate objective of the joint Westinghouse–ERC phosphoric acid fuel cell program was announced as the production and operation of two prototype Mark II 7.5 MW power plants by mid-1987. Figure 16–7 shows the design configuration for one of the 375 kW fuel cell modules that will comprise the power section of the Mark II system.[177, 2732] This system is air-cooled by a modification of the DIGAS cooling technique developed by Energy Research Corporation, which is discussed in Chapter 15, and in more detail in Ref. 177.

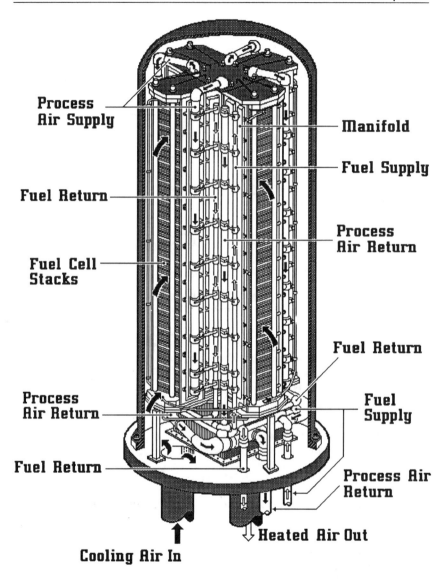

Figure 16-7: Design configuration of 375 kW fuel cell module of Westinghouse-Energy Research Corporation Mark II fuel cell system.[177, 2732]

Phosphoric Acid Fuel Cells

The final selection of 7.5 MW as the size of the system was based on careful design studies of systems ranging from 1.5 MW to 15 MW. The results of these studies indicated that a gradual reduction in the initial heat rate from 8770 to 8560 BTU/kWh at the projected cell voltage used, could be expected as a function of increasing plant size, but that there was no advantage to be gained in the cost of electricity by increasing the size of the power plant above about 4 MW. The actual net a.c. output from the 7.5 MW Mark II system is 6.9 MW. The complete power plant will consist of two parallel banks, each containing ten 375-kW modules.

The power modules are planned to have variable pressure and temperature control, the original design operating conditions being 4.8 atm (70 psia) and 190°C at a current density of 325 mA/cm^2 at a planned cell voltage of about 0.69~0.70 V. The design operating point was chosen as the optimum among a number of competing factors.[2732] Operation at higher pressures, for example, results in a tradeoff between an increase in fuel cell output and rotating group power losses. Based on the cell voltage parameters assumed, a minimum heat rate and cost of electricity would occur in the pressure range 3.4 to 5.1 atm. Similarly, operation at higher temperatures results in increased fuel cell performance, though at the cost of increased fuel cell degradation rates, especially at part power. For this reason, the acceptable operating temperature range was restricted to 190°C~204°C.

One of the first realizations of Westinghouse was that the original DIGAS cooling concept was inappropriate for a high-efficiency utility system. To obtain maximum efficiency in a pressurized unit, it would be necessary to separate the process air and cooling air streams. Accordingly, the original ERC DIGAS concept was abandoned. The Westinghouse system therefore uses six manifolds: entry and exit for each of the two reactants, plus two more, along the long side of the rectangular cell, for the cooling air stream, which is diverted to cooling plates, arranged one every five cells. A further description follows.

With the exception of the complex manifolding arrangement, the design of the individual Westinghouse-ERC cells is similar to that of UTC cells of "conventional" configuration, in that a sandwich configuration consisting of advanced matrices (MAT-1, a silicon carbide layer on the anode side and a Teflon-bonded active carbon bubble pressure barrier at the cathode), and graphite separator plates with reactant channels is used. The acid management concept employed in the Westinghouse-ERC cells is designed to accommodate volume changes during transients, in this case not via an in-cell reservoir, but by the use of intercell and cross-cell wicking arrangements. The arrangement therefore provides for on-line acid replenishment as required.[557] We should state that UTC has never found an assembly using this type of acid replenishment satisfactory for cells operating under pressurized conditions, although it may be appropriate for an atmospheric pressure system.

In 1982, test facilities were in place for the fabrication of pilot units, and 12-cell stacks of full scale hardware (30 cm × 43 cm active area) have been examined since that time.[851] By 1982, a production capacity of 2~3 MW/year already was available, requiring 7000~8000 ft^2 (of a total of 13,000~15,000 ft^2) of manufacturing area. Full-sized repeat components such as molded bipolar and cooling plates had been produced, and 23- and 80-cell stacks had been built and tested at atmospheric pressure; while 5, 9, 12-, and 50-cell stacks had been built and tested at operating pressures up to 7 atm. A full-scale Mark II design stack comprising 23 cells was constructed. This stack used two cooling plates with five cells between each cooling plate and two cells beyond each plate, this configuration being representative of the full size Mark II arrangement. The design incorporated a *zee* pattern bipolar plate and a *tree* pattern cooling plate, both of which are represented in Fig. 16–8. The *zee* pattern bipolar plates permit the separation of process and cooling gas streams without complicated manifolding, whereas the variable area *tree* cooling plate configuration achieves uniform cell temperatures with useful cooling gas temperature rises.[507] With the manufacturing facilities then available, the molding time for the Mark II bipolar *zee* plates was less than one minute. Heat treatment and fabrication methods for producing corrosion-resistant components have been developed starting from plates containing graphite flakes and about 30 wt % phenolic resin. This is discussed in more detail in Ref. 177.

In early 1983, cell voltages of 0.71~0.735V were attainable using reformed fuel at a current density of 100 mA/cm^2 at 190°C.[851] Under these conditions, the ohmic loss in the matrix was about 8 mV/cell, and the contact resistance loss was about 19 mV/cell. The latter could be reduced, among other ways, by changing the carbon backing paper. Operation for greater than 20,000 hours seemed to indicate a performance drop of only about 0.2 mV per thousand hours. Platinum cathode catalyst loadings of 0.5 mg Pt/cm^2 yielded IR-free cell voltages of about 0.76 V. The cell voltage was found to decrease by about 27 mV for each halving of the platinum catalyst loading, which is close to the value expected from the Tafel equation. As a result of these experiments, it was considered that a cell voltage of about 0.75 V could be obtained using proprietary alloy catalysts of the type described in Chapter 12.

Later work, based partly on UTC results, showed that such performance could not be sustained with the cathode components then available. In particular, the corrosion of both the Cabot Vulcan XC-72 cathode support and of the carbonized (but not graphitized) graphite powder-resin bipolar

Phosphoric Acid Fuel Cells

plate would be a serious problem, as outlined in Chapter 12. As a result, a target cell voltage of 0.68 V at 325 mA/cm^2 at 4.8 atm at 190°C was set. Since the technology used was not very different from that of the UTC 4.5 MW demonstrators, which could only offer about 0.65 V under roughly similar conditions, it is no surprise that a new current density of 267 mA/cm^2 was adopted in May 1985 to increase the cell potential to a more acceptable final value that would give a system heat rate of about 8300~8500 BTU/kWh.[177] Westinghouse short stacks under test at 4.8 atm during 1986 still show rather high degradaion rates after about 1000 h of operation compared with those of UTC. For example, around about the 1000 h point, Westinghouse has set a goal of 7~8 mV/1000 h of degradation, and practical results with pressurized 9-cell stacks were about twice this level. This should be compared with the 1.5 mV/1000 h registered by UTC, as shown in Fig. 16-5. This apparently required some fine tuning of cell materials and design. In 1985, Westinghouse lost its option to construct a 7.5 MW plant with Southern California Edison Corp. at Long Beach, California, and has since been concentrating on the development of a 1.5 MW demonstrator unit.

Westinghouse 7.5 MW System Description

The process flow diagram for the proposed Westinghouse-ERC Mark II phosphoric acid fuel cell system is shown in Fig. 16-8.[2732] The fuel processor is a self-contained subsystem. The only interfaces with the rest of the plant are the water feedlines, and the fuel cell anode supply and exhaust lines. The complete fuel processing subsystem, which has been thermally integrated, should achieve a process efficiency of 90%. Natural gas or naphtha fuel would be fed to the fuel processor, where the fuel will be preheated, vaporized, desulfurized, and mixed with superheated steam. This mixture enters the reformer at 427°C and is heated to 816°C for the reaction. Two alternate types of reformer are planned for use with this unit. The first is a monolithic unit of Westinghouse design, the second a tubular, modular system designed by Haldor-Topsøe under EPRI support. The latter consists of a series of annular tubes of heat exchanger type, the burner being at the bottom of each. Since the system is modular, it can easily be factory-built and field assembled.[119] In contrast, the Westinghouse single-unit reformer is a counterflow, bayonet-type forced convection heat exchanger, the thermal needs of which are provided by combusting preheated anode exhaust gases with a small amount of desulfurized fuel along with a 10% excess of preheated air.

The fuel mixture enters one annulus and is heated by the combustion gas directly, as well as regeneratively by the reformed gas as it returns in an adjacent annulus. The hydrogen-rich product gases exit from the reformer

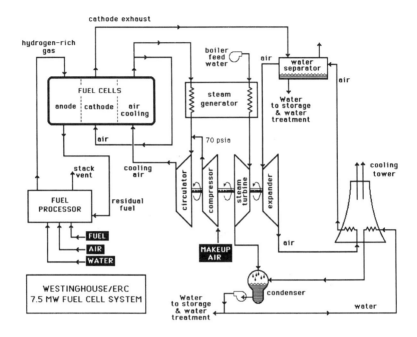

Figure 16-8: Simplified process flow diagram for proposed Westinghouse-Energy Research Corporation Mark II phosphoric acid fuel cell system.[2732]

at 593°C, after which they are cooled to 371°C before entering a high-temperature shift converter. This step is followed by further cooling to 204°C before they enter the low-temperature shift converter to reduce CO to acceptable levels. The final product leaving the low-temperature shift converter will contain about 70% hydrogen and 1.5% carbon monoxide.

The processed fuel will next be fed to the power section, which would consist of twenty 375-kW fuel cell modules arranged hydraulically in parallel, and electrically in a series-parallel arrangement. Each 375 kW module (see Fig. 16-7) would consist of four stacks, each of four hundred 30 cm × 43 cm cells, to give a total of 1600 individual cells per module.[851] A complete 375 kW module will be about 3.35 m high, 1.4 m in diameter, and will weigh 4500–5500 kg. Each cell in this proposed design embodies *zee* process flow plates, as described (Fig. 16-9), for both process flow streams in order to minimize thermal gradients across the cell. The *zee* plates have manifolds arranged on opposite faces of the shorted side of each cell in what appears to be a counterflow arrangement, so that a reactant (fuel or air) enters half of one side of the cell, and exits on the other half of the opposite side. This is apparent from Fig.16-8. We should note that in the latest design, the *zee* plates on the fuel side have been given a

Phosphoric Acid Fuel Cells

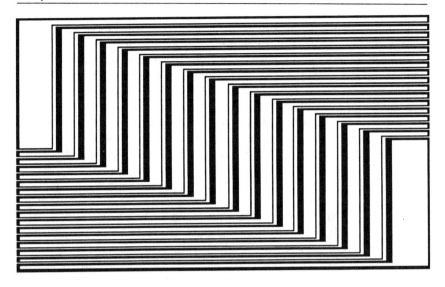

"Zee" PATTERN BIPOLAR PLATE

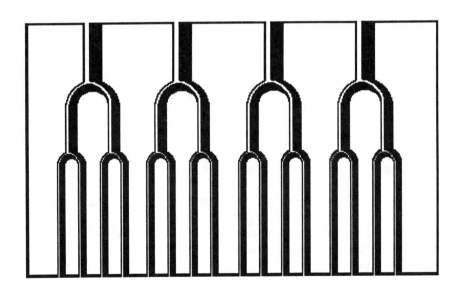

"Tree" PATTERN COOLING PLATE

Figure 16-9: Concept of Zee pattern bipolar plate and Tree pattern cooling plate used in Westinghouse-Energy Research Corporation Mark II fuel cell system.

different geometry to increase the channel tortuosity, allowing a better control of fuel flow (hence utilization) via the pressure drop.

The air coolant stream entry manifolds are on the outer long sides of each of the four stacks arranged crosswise in the module. The coolant air passes through *tree* cooling channels in a cooling plate located between each group of five cells. Circulated air is used as the fuel cell coolant and the *tree* coolant channels regulate the heat transfer coefficient in such a manner as to give a nearly uniform temperature distribution throughout the cells. The cooling air would undergo a 32°C rise in temperature between inlet and outlet. The air used to cool the stacks enters the bottom of the pressure vessel of each module, flows around the stacks, and enters from the outside edge, as indicated in Fig. 16–7. The cooling air flows radially through the cruciform stack arrangement to the center space of the module, and exits from the bottom of the module. By connecting the four stacks in the module in series, 375 kW of d.c. electricity at 1120 V will be produced at the original power density levels. For the complete system, two modules would be connected electrically in series and each two-module package would be connected electrically in parallel to produce 7.5 MW d.c. (at the original design current density) at 2240 V.

A portion of the cooling air leaving the fuel cells would be fed to the cathode compartments of the fuel cell, where it can be used as oxidant for the fuel cell reaction. The remaining cooling air would be used to provide heat for generating the low pressure steam required to drive the steam turbine. The cooling air exiting the steam generator would be circulated back to the fuel cells by means of a large gas circulation blower directly coupled to the steam turbine. The 70 psia (4.8 atm) makeup air needed to replenish the portion of the process air routed to the fuel cell cathodes is supplied by a compressor. An air expander driven by the cathode exhaust air leaving the fuel cells and a small motor would be used in conjunction with the steam turbine to drive the circulator and compressor. The complete unit (i.e., the circulator, compressor, steam turbine, expander, and motor) is planned to be a modular factory assembled package.

The process air exiting the expander would go through dry cooling towers to a direct contact cooler condenser, the condensed water passing to the water treatment and storage system. The latter is planned to deliver water to a steam drum heated by the fuel cell cooling gas. The saturated water from this drum will be supplied to the fuel processor, while the steam will be fed to the steam turbine.

The d.c. output from the fuel cell power modules will be consolidated and inverted by the power conditioning subsystem, providing electrical control and protection and interfaces with the power plant main transformer. The design of the inverter is based on a line commutated 3.4 MW unit originally designed and built at the MHD Component Development and Integration Facility in Butte, Montana.

Phosphoric Acid Fuel Cells

The Mark II fuel cell power plant would be operated at part load by varying fuel cell temperature and pressure according to a controlled relationship. By operating in this manner it is possible to keep the concentration of the phosphoric acid electrolyte approximately constant over the complete operating range, thereby enhancing the lifetime of the cells. For example, at 25% of rated power, the fuel cells would operate at 2.6 atm pressure, 170°C, and at a current density of 74 mA/cm^2.

Table 16-3 shows the 7.5-MW Mark II system design specifications that have been established to meet the needs of a broad range of utility and industrial cogeneration applications.[2732]

TABLE 16-3

Design Goals of Westinghouse-ERC 7.5 MW Phosphoric Acid Fuel Cell System[2732]

Performance*	Full Power	25% Power
Gross fuel cell power output, MW$_e$ d.c.	7.50	1.88
Net plant electrical power output, MW$_e$ a.c.	6.91	1.58
Net plant heat rate, BTU/kWh	8766	8723
Fuel cell operating temperature, °C	190	170
Pressure, atm	4.76	2.60
Unit cell voltage, V	0.700	0.777
Current density, mA/cm^2	325	74
Operational		
Application	Intermediate load following	
Real power response:		
Cold start to standby	< 4 hours	
Standby to maximum power	< 20 min	
Operation	Automatic dispatch from 0% power at standby to full power	
Normal operating range	25~100% of rated power	
General		
Fuel capability	Dual (natural gas–naphtha)	
Availability	90%	
Plant life	25 years	
Construction	modular	
Transportability	truck (normal permits)	
Installed capital cost:		
Prototype	$2600/kW	
Commercial	$700/kW	

*Beginning of use

C. ENGELHARD INDUSTRIES PHOSPHORIC ACID FUEL CELLS

As outlined in Chapter 4, Engelhard Industries Division of Engelhard Minerals and Chemicals Corp. has been actively engaged for about twenty years in the development of 50-kW dielectric-liquid-cooled phosphoric acid fuel cells for use as on-site energy systems using reformed methanol as fuel. The Engelhard system has been unique in using a novel approach to cell construction.

Fuel Cell Description

Each 25 kW Engelhard stack was planned to contain 176 cells (each of area 33 cm × 58 cm [13 in. × 23 in.]) operating at 180–204°C and at atmospheric pressure, giving 0.65 V/cell and 162 mA/cm^2. The 25 kW system as planned would have been about 8 ft high, and would be started by heating up the stack and fuel processor by combustion of some of the methanol fuel. The ultimate objectives for this on-site integrated energy system include a cost of \$200/kW (1977 dollars) and a lifetime of at least five years.[1364] Particular emphasis was placed on low-cost end-plate design with low IR-drop. Initial experiments involved the use of vitreous carbon-clad base-metal plates, but occasional flaws have made these unsatisfactory for a five-year lifetime. In 1984, a 37.5 μm gold sheet with grafoil (flexible graphite gasketing material) backing was used, and in 1985 a new gold-clad base-metal wire current collector in a fan-shaped arrangement was tested. The recoverable gold cladding is expected to add less than \$10/kW to stack cost.[1367]

Operating Conditions

Engelhard has investigated the use of both pressurized and nonpressurized operation in systems in this size range (25 kW). Pressurized operation has been shown to achieve higher hydrogen utilization rates, improve cathode kinetics, and thus permit smaller area hardware. Conversely, increasing the hydrogen utilization rate tends to lower the cell voltage because of the effect of the Nernst equation on the essentially reversible anode. For example, increasing the hydrogen utilization rate from 70% to 78% has been shown to lower the cell voltage by about 3.6%.

After an analysis along these lines, Engelhard found that the combined electrical-plus-thermal efficiency is almost the same for both small pressurized and nonpressurized systems. Key factors in determining the attractiveness of pressurized operation are the availability of suitably efficient, economical, and flexible turbocompressors, and the ratio of the thermal to the electrical load.

Electrodes and Catalysts

Engelhard electrodes have normally used 10% Pt-on-carbon catalysts for both anodes and cathodes at nominal loadings of about 0.5 mg Pt/cm^2.[1325] However, the company has spent a considerable effort on the development of advanced electrocatalysts for both the cathode and anode. Thus far, improvements have been made in improving the stability of both the carbon catalyst support as well as the surface area of the catalyst under typical cell operating conditions. Thus, in tests in 105% phosphoric acid at 204°C and at a cell voltage of 0.70 V, the surface area of a baseline 10% platinum catalyst levels off after several hundred hours from an initial value of about 156 m^2/gram to 30~40 m^2/gram,[1364] whereas that of an advanced catalyst may decline to only 60~80 m^2/gram.[1365] The performance goal for platinum surface area is generally considered to be 60~80 m^2 Pt/g Pt after 10,000 hours;[1364] however, it is hoped that advanced cathode alloy materials may improve upon this in the future.

Advanced bimetallic anode catalysts were shown to have improved resistance to ageing in the presence of 2% carbon monoxide. More recent efforts have focused on cathode catalyst formulations that provide greater activity for the oxygen reduction reaction and that exhibit greater carbon monoxide tolerance for the anodic process. By as early as 1981, the performance objectives just outlined had been exceeded in single cells operating at 191°C. Electrode performance with the stated catalyst loading goals have been given as 0.68 V (IR-free operation on hydrogen-air) at 161 mA/cm^2 (150 ASF), with a total of no more than 1 mg of platinum metal per square centimeter of total cell area.[1364] The present favored cathode catalyst is the E3 platinum binary alloy formulation on Shawinigan acetylene black support, although later formulations (e.g.: E7, a binary; and E8, a ternary) are under test. Present performance is typically 0.71 V per cell at the standard current density after 1000 h, with a decay rate at that time of about 2.5 mV/1000 h. Engelhard progress is summarized in the reports under Ref. 1367.

Electrolyte Matrix

An important part of the Engelhard program was the development of hydrophilic, thin laminated electrolyte matrix structures of low IR-drop that facilitate routine electrolyte replacement via a wicking mechanism to prevent long-term performance degradation resulting from electrolyte loss. Their approach therefore has been different from that of UTC, who has always favored the reservoir scheme. As already discussed, present indications are that whereas the latter is necessary in pressurized systems, it appears to be possible to incorporate electrolyte replenishment systems in atmospheric pressure cells. Engelhard's development of an electrolyte replenishment scheme has permitted fuel cell operation at higher temperatures than were possible previously. Engelhard's electrolyte management system is based on automatic replenishment, according to cell demand. The electrolyte reservoir is external to the cells but located inside the stack. This technique has been stack-tested successfully for thousands of hours, and a low-cost version having lower IR-drop (30 mV maximum at 150 mA/cm^2) now is being developed. The projected cost goal for the matrix was given as \$32/m^2 in 1977 dollars.[1364]

Acid transport to the inside of the stack is effected by wicking along a carbon paper material in contact with the electrodes. The favored material has been manufactured by Kureha in Japan, but alternatives must now be examined since Kureha no longer manufactures this product, and Stackpole abandoned the equivalent U.S. product (PC206) in 1986. Similarly, Shawinigan acetylene black also is no longer available, so a substitute support must also be found. For this purpose, Engelhard is examining Gulf acetylene black (GAB). Non-availability of materials is definitely becoming a problem for some PAFC developers. However, it is fortunate that other carbon manufacturers are interested in the potential of the PAFC market, particularly Great Lakes Research Corporation (P. O. Box 1031, Elizabethton, TN 37644-1031) in the United States, which has developed a low-cost, induction-graphitized corrosion-resistant support under EPRI funding,[1339] and Toray Industries (2, Nihon-Muromachi 2-chome, Chuo-Ku, Tokyo 103) in Japan, whose products TGP-090 and TGP-120 are being examined by Westinghouse.[852] Important parameters of papers are: porosity and pore size distribution, conductivity, mechanical strength and handleability, and thickness tolerances. The latter is important since the thin papers cannot be surface-finished, unlike the thick ribbed substrates used by UTC.

Bipolar Plates

Engelhard fuel cells make use of a multi-element bonded so-called ABA-type configuration for the bipolar plates, as indicated in Fig. 16–10. In this arrangement, the two halves of the bipolar plate are cemented together by the B element. The A elements are porous gas-distribution members, and both ribbed and nonribbed configurations have been evaluated.[1365] Unlike UTC, Engelhard did not elect to use porous ribbed elements for electrolyte storage, and in consequence the plates must be wetproofed. The B elements were originally resin-CVD (chemical vapor deposition)–treated sheets of woven fabric carbon paper. The most recent B elements used films of

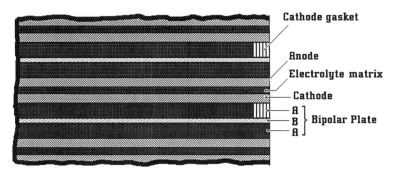

Figure 16–10: Engelhard phosphoric acid fuel cell construction.[1365]

polyethersulfone (PES), polyetheretherketone (PEEK) and PFA Teflon, which are bonded under uniform temperature conditions at about 300°C, 370°C, and 400°C, respectively to give a conducting (literally glued) interface between the two elements of the bipolar plate. Since the plate must be wetproofed with Teflon at a sintering temperature of 288°C, PES is no longer the preferred polymer since softening can occur during wetproofing.[1367]

The bipolar separator plate is a key component affecting the cost of a phosphoric acid fuel cell stack, and the development of multi-element bipolar plates based on commercial graphite (originally Pfizer porous vitreous carbon; Ref. 177) has now been put into practice in stacks.[505] The Engelhard approach to low-cost separator plates has been to use inexpensive precursor materials coupled with high throughput continuous processing.[1366] One candidate precursor material for the gas distribution A elements of the bipolar plates was needled rayon felt costing $1~$3 per square meter. The material is carbonized by continuous processing through a hot-zone at a rate of 8~15 cm/min, after which it is upgraded by a chemical vapor deposition process to pyrolytic carbon. During the heat treatment process at 1600~1900°C, its density increases from 0.1 g/cm^3 to 0.5 g/cm^3. The projected high-volume cost goal for bipolar plates has been given as $43/m^2 (1977 dollars).[1365] The IR-drop through the bipolar plate should be no more than 3 mV at 150 mA/cm^2.[1364] There is no question that Engelhard's approach to this problem is a major advance towards a practical atmospheric pressure PAFC system, and it could also be used for an advanced AFC. Some projections of this technology are discussed in the last section of this chapter.

Cooling Plates

Engelhard's cooling system uses either corrosion-protected metal or nonmetallic cooling plates, the latter being preferred. The mineral oil dielectric coolant is circulated externally through fiberglass braided Teflon distribution tubes, and internally through Viton hoses. The dielectric liquid flows via forced circulation through cooling plates, which are inserted among the cells in the stack. Since the system is not pressurized, the danger of rupture is small compared with a water-cooled system. The use of a dielectric liquid coolant permits efficient heat recovery at high heat transfer temperatures, which not only permits the use of small heat exchangers, but also requires minimal pumping power compared with a gas-cooled system. Headers located within the reactant gas manifolds of the fuel cells are used to distribute the coolant to the individual cooling plates. The headers and cooling plates are connected using short lengths of Viton tubing, which is flexible and electrically nonconductive. Cooling plates are inserted into the stack at four-cell intervals, with one also placed at each end of the stack, inboard of the current-collecting plates. These end cooling plates are used to control overall heat loss.[1325]

Engelhard has made considerable efforts to improve the resistance of their cooling plates to corrosion from hot phosphoric acid.[1366] One early, relatively successful approach was to coat the cooling plates with a specially developed, thermally conductive protective coating that prevented access of the phosphoric acid to the metallic copper-coated surface of the plates. A more recent advance has been to develop completely resistant nonmetallic cooling plates made from polymer-containing carbon. A cost goal for the cooling units of $32 (1977 dollars) per square meter of cell area has been suggested.[1364]

Fuel Processor

Engelhard's technology development program has concentrated on the development of a 50 kW isothermal methanol steam reformer, which would be coupled with two 25-kW series-connected fuel cell stacks and a grid-connected power conditioner. The fuel processor has a shell-and-tube configuration containing 120 1-inch diameter tubes containing the reforming catalyst. A number of steam reforming catalysts have been evaluated for use with this reactor. Complete methanol conversion has been obtained in integrated reformer-burner test units at a design space-velocity of 1000 scfh exit hydrogen per cubic foot of catalyst.[1365]

D. REQUIRED IMPROVEMENTS IN PHOSPHORIC ACID FUEL CELLS

At the present time, the phosphoric acid fuel cells of International Fuel Cells are technologically the most advanced from the viewpoint of performance and decay rate. The proposed 1986 systems of IFC would have had end-of-life heat rates of about 8300 BTU/kWh on natural gas fuel. The key to wide commercial use of phosphoric acid fuel cells will be to first reduce cost, then to decrease the heat rate; that is, to improve the cell energy conversion efficiency by increasing the cell voltage. For example, a 40 mV improvement in cell voltage would give the same overall change in electricity cost as a 7.5% reduction in overall system cost.$^{(167)}$ The simplest way to increase cell voltage will be to operate at even higher temperatures and pressures than the 205°C and 8.2 atm currently used at IFC. Unfortunately, catalyst substrates of acetylene black or graphitic type are not sufficiently corrosion resistant to permit these more extreme operating conditions. To this end, new substrates, such as carbides and silicides, are being investigated. In particular, high surface area (100 m^2/g) titanium carbide$^{(167)}$ showed some promise but reproducible corrosion resistance has not been achieved. Other carbides (e.g., tantalum carbide and doped silicon carbide) may permit an increase in cell voltage of as high as 60~70 mV above the present value of about 0.73 V, leading ultimately to an end-of-life HHV heat rate of as low as 7600 BTU/kWh. Some of these problems are reviewed in Chapters 12 and 13.

At these higher operating temperatures, it will be impossible to employ separator plates made from conventional graphite-resin mixtures carbonized at low temperatures of 900~1200°C, because of their relatively high porosity and low corrosion resistance. However, it may be possible to protect them from electrolyte penetration by extensive wetproofing, along the lines used by Engelhard in its ABA plate structure. Engelhard has also developed a method of sealing the surface porosity of carbon bipolar plates prepared by heat treatment at lower temperatures in the 1200~1600°C range using chemically-resistant nonporous coatings such as glassy carbon, chemical-vapor–deposited carbon coatings, or conductive CVD carbide coatings.$^{(1367)}$ This approach may be more economical than costly heat treatment at very high temperature. However, even if these alternatives are used, the use of fully graphitized materials and dense glassy carbons heat-treated at 2000~2900°C may be necessary to achieve the required stability at the cathode. Boron additives seem to impart an enhanced *pregraphitization*, and may permit heat treatment at somewhat lower temperatures than usually required for these materials.$^{(2361)}$

As is apparent from the previous discussion, different developers are taking different attitudes towards cell construction. International Fuel Cells uses the all-graphitized ribbed substrate stack, incorporating reservoirs at both the anode and the cathode. In this respect it has been followed by Toshiba, Hitachi, and Fuji Electric in the Japanese national development program (Chapter 5). Mitsubishi Electric, also in the national program, is reportedly taking a somewhat different approach with a conventional stack containing a reservoir. Energy Research Corporation, followed by its licensees Sanyo and Westinghouse, uses a solid graphite ribbed bipolar plate with thin flat electrode backing layers (substrates) in a stack containing no reservoir, into which electrolyte can be wicked along the backing paper from a series of holes in the stack, each provided with wicks from a common reservoir. Engelhard uses an arrangement not unlike ERC's, but has the unique built-up ABA bipolar plate.

However, other concepts also are emerging. The first of these was Toshiba's so-called hybrid stack,[177] which can be best described as a combination of the UTC ribbed substrate at the anode, with a conventional cathode, which contacts the flat side of a bipolar plate ribbed on the cathode side only. Consequently, the cathode is of conventional construction with the electrode on a ribbed backing paper. While the arrangement may have been originally devised only to avoid infringement of UTC's patents, it is nevertheless of considerable interest from the technical viewpoint. Before returning to this subject we should point out that Toshiba's association with UTC in the IFC joint venture only involves system components and design, not stack components. Thus stack development by the two companies is kept in two quite separate compartments, with no mutual exchange of information.

Toshiba's hybrid stack has an electrolyte reservoir on the anode side only. This makes sense, since corrosion is not a problem at the anode, so a carbon, rather than graphite, ribbed substrate could be used there if necessary. In addition, since hydrogen diffusion is much more efficient than that of oxygen, a ribbed substrate of the correct design at the anode side may be made as thick as necessary to give the correct reservoir capacity. Indeed, the ribbed porous reservoir does not even have to serve as the electrode substrate at the anode: It can be turned around, so that the ribs face outwards as in a conventional stack. This sounds like a retrogressive approach, but this is not necessarily true since the anode paper electrode backing need not be an expensive graphitized material, and this approach may make the anode reservoir concept more efficient. The better efficiency may result from the fact that the ribs (the gas channels) are in direct contact with the electrode substrate, so that the solid part of the porous substrate is remote from the ribs, hence can be made as thick as necessary to accommodate the electrolyte required without a complex selective wetproofing procedure. Such an approach may have many advantages for a low-cost stack, since the anode reservoir will consist of cheap carbons only, and the cathode side can use a minimum of relatively

Phosphoric Acid Fuel Cells

costly graphite parts that are thoroughly wetproofed to ensure that as little acid contact as possible, hence as little long-term corrosion, takes place.

Toshiba's approach to the reservoir stack has a major constructional disadvantage, since it uses a solid graphite bipolar plate ribbed on one side (the cathode) only. While such plates can be made by machining from the solid for demonstration purposes, solid pressed plates that are ribbed only on one side have a tendency to warp during bake-out and graphitization. In contrast, high porosity ribbed substrates are sufficiently flexible for this to be accommodated. The warping problem has already been observed with the tree-pattern cooling plate in the Westinghouse program. In contrast, pressed plates that are similarly ribbed on both sides, such as the Westinghouse bipolar plate, are substantially flat after bake-out. Toshiba's stack might therefore be advantageously built up using an ABA (or even an AB) approach on Engelhard lines, which would be technically feasible and would be substantially cheaper than any single-piece solid graphite structure.

Another interesting approach combining features of the UTC ribbed substrate stack and the Engelhard ABA system is Kureha's KES-1 structure, announced in May 1985.[177] This consists of two ribbed anode and cathode substrates of UTC type, that are collimated together using a carbonized, then graphitized, resin sheet in a manner that avoids warping and cracking, due to the use of "a new idea," according to Kureha publicity. Corrosion resistance and strength are also claimed to be exceptional. Thus, Kureha's concept allows the production of a ribbed substrate UTC stack without the need for an expensive, flat graphite bipolar insert plate which has a high reject rate and gives problems of contact resistance.

One or more of the aforementioned approaches, or a combination of some of the ideas, is certain to lead to a better, cheaper and more effective PAFC stack design.[177] Indeed, the proprietary alternative UTC stack configuration B whose performance under atmospheric pressure conditions is given in Fig. 16–5 already contains some of the above approaches towards a lower cost, lower IR stack. It is also possible that these stack approaches can be applied to low-cost, high-efficiency alkaline stacks operating at 70~80°C.

Depending on how much of the above research is successful, heat rates on the order of 7200~7800 BTU/kWh should be attainable in second generation PAFCs operating at higher temperature and pressure on light distillate or even biogas fuels at the current densities in general use at the present time. However, as is discussed in detail in Chapter 2, cost reduction will be the major emphasis at first, and the trade-off will be between heat rate and stack power density (i.e., current density). The stack components themselves in limited production will probably cost $1000 kW (1986) under U.S. or Japanese labor conditions, compared with $50/kW for stack materials and $100/kW (at 5 g Pt/kW) for the platinum-based catalyst at late-1986 prices. There will therefore be great pressure on the developers to reduce costs. The easiest way to do this will be to increase stack performance (e.g., by reducing IR-drop), but to use the improved performance by increasing current density at the same stack voltage (i.e., at the same system heat rate). Since absolute cathode voltage does not change, this conservative approach implies no materials improvements from the corrosion viewpoint. By this means, stack cost can be rapidly reduced, giving an increased production rate leading to further cost savings. When total system costs are reduced to levels such that the total cost of the stack (without catalyst) becomes a relatively small fraction of total system cost (perhaps 20%; see following), consideration to reducing heat rate (increasing voltage) can be given by reducing current density and introducing new materials and construction techniques, where necessary. Some of these scenarios have been reviewed in Ref. 177.

Under the best materials scenario, PAFC operating temperatures of approximately 240°C may ultimately be possible. These are high enough to permit the use of direct internal-reforming methanol cells with simplified external fuel processing hardware and with methanol-a.c. heat rates of perhaps 6800 BTU/kWh or less.[167] This would be extremely attractive for cheap PAFC units requiring little systems engineering in a coal-based energy economy with a methanol energy vector. The latter may be the precursor of a purely hydrogen energy economy, which may ultimately be necessary because of the greenhouse effect.[393, 789]

E. COST OF PHOSPHORIC ACID FUEL CELLS

As discussed in the previous section, there is still a need for technical improvements in PAFC systems. However, from a purely technological standpoint, present state-of-the-art systems can be considered to be at the point of operational viability, with already attainable lifetimes of 40,000 h and acceptable end-of-life performance. The question now becomes one of cost.

Fuel cell experience over the past thirty years has clearly shown that a large public sector financing has been required for the emergence of this new technology. What now has to be ascertained is whether or not fuel cell systems are commercially and economically viable. At the present time, it is very difficult, if not impossible, to obtain meaningful data on the cost of fuel cell systems. However, Table 16-4 indicates some cost figures that have been suggested in recent years. The figures given in this table should be taken only as very approximate guides. In 1970, total system costs in the order of $150~200/kW were quoted for the TARGET program, equivalent

TABLE 16-4

Estimated Capital Costs of Phosphoric Acid Fuel Cells

Item	Estimated Cost ($/kW)	Reference No.	Year
UTC 11 MW system[1]	670	1969	1982
UTC 11 MW system	1000	1980	1982
UTC 11 MW system[2]	< 1000	1988	1982
UTC 11 MW system[3]	< 1500	1989	1982
Westinghouse–ERC 7.5 MW system:[4]			
Stack only	326	2578	1981
Projected total fuel cell system[5]	558	2578	1981
Prototype integrated system[6]	1838	2578	1981
Prototype non-integrated system	1326	2578	1981
Westinghouse-ERC 7.5 MW system:[7]			
Prototype	2600	1987	1982
Commercial	700	1987	1982
Japanese 1 MW commercial system	421~460[8]	2511	1982
50 MW integrated cogeneration plant[9]	2200	2529	1983
UTC 40 kW OS/IES[10]	435	1332	1979
Engelhard 50 kW OS/IES	200	1979	1982
Japanese OS/IES[11]	650~766[8]	2558	1983
UTC fuel cells for vehicular propulsion[12]	150~250	1973	1982
Westinghouse mobile 60 kW military air controller radar power unit[13]	2740	1977	1982

(1) Includes site construction costs. (2) Installed cost of ultimate commercial power plant in volume production. (3) Operation on coal gas; cost includes complete gasification system. (4) 1980 dollars. (5) Includes assembly, instrumentation, installation, and engineering fees. (6) Operation on coal gas; complete optimal thermal integration of system for maximum efficiency (rotating equipment and auxiliaries in particular are more expensive than for non-integrated system). (7) Installed; 1981 dollars. (8) Based on exchange rate of ¥261/dollar. (9) Including integrated (North Dakota lignite) coal gasification plant; 1983 dollars. (10) Installed cost, 1977 dollars; operation on reformed natural gas. (11) Based on total production of 4000 MW. (12) Based on production of 100,000 units per year; 1981 dollars. (13) Based on production of 1000 units; complete self contained unit.

to $450~600/kW in 1986 dollars. However, as the nature and magnitude of the problems to be solved became more clearly understood, cost figures have gradually escalated. For large systems operating on reformed hydrocarbon fuels, net plant costs for units in mature production somewhat less than $1000/kW now appear to be acceptable. If the plant is to operate on coal gas, the figures rise to about $1400/kW. As an example, IFC appears to be prepared to sell the first three 11-MW units for about $3500/kW, followed by a further 18~30 units at $1800/kW; further units then becoming available at about $1000/kW as production increases, start-up costs are absorbed, and experience is acquired. However, most fuel cell workers are hopeful that, with volume production, realistic technology improvements, and successful operational experience, cost figures will ultimately be significantly lower. As indicated in Table 16–5, this certainly has been found to be the experience for the stack costs of aerospace fuel cell systems, and there is no reason to believe that the same will also happen for commercial and industrial systems.

In a relatively mature atmospheric pressure on-site fuel cell system the fuel cells themselves may correspond to about one third of the total system cost. In the more complex pressurized electric utility units, the stack cost (without catalyst) might descend to 20% of total system cost in mature production.[177] If an internal reforming methanol system could be manufactured, or if pure hydrogen were ultimately available, the efficiency of the plant would not only increase, but the plant cost would also be significantly decreased by the elimination of the fuel processing subsystem and simplification of many of the auxiliary subsystems such as those required for gas cleanup, leading to appreciably lower electricity costs and much improved competitiveness with alternative systems such as the advanced combined cycle gas turbine.

TABLE 16–5

Cost of Aerospace Fuel Cells[297]

Project	Year	Stack Cost ($/kW)
Apollo	1965	225
Shuttle demonstrator	1974	65
Orbiter	1976	25
Advanced lightweight fuel cell	1979	6

CHAPTER 17

Molten Carbonate Fuel Cells

A. INTRODUCTION

In Chapters 3 and 10 the potential system advantages of high-temperature fuel cells over the PAFC were discussed. These include higher efficiency due to lower polarization, the possibility of internal reforming, and the use of bottoming cycles to generate further electricity from high grade waste heat. Overall plant cost may be lower than that of the PAFC, since hydrocarbon fuel processing subsystems will be less complex, and noble metal catalysts will not be required. Finally, with an appropriate gasification plant, the high-temperature systems should be able to operate efficiently on coal.

MCFC and SOFC systems are not in so advanced a stage of development as the PAFC because of scale-up problems associated with high temperature operation. Thus, PAFC systems operating at about 190–210°C, have been built on the MW scale (demonstration stage), whereas MCFC stacks, which operate at temperatures of about 650°C, reached a maximum scale of 20 kW in 1986 (pilot plant stage). Similarly, solid oxide electrolyte fuel cells, which operate at temperatures of about 1000°C, have thus far been built only in units up to about 300 W (laboratory stage). A summary of work up to 1984 is given in Refs. 19 and 495. In this chapter, the state of the art of the MCFC will be examined, and some future projections of the technology considered.

B. OPERATING PRINCIPLES

Figure 17–1 shows a simplified schematic of a coal-fired molten carbonate fuel cell. Coal is gasified to produce a clean fuel mixture of hydrogen and carbon monoxide, which reacts at the anode with molten $CO_3^=$ ions to produce carbon dioxide and water vapor, releasing electrons to the external circuit. The anode reactions shown in Fig. 17–1 show CO as being

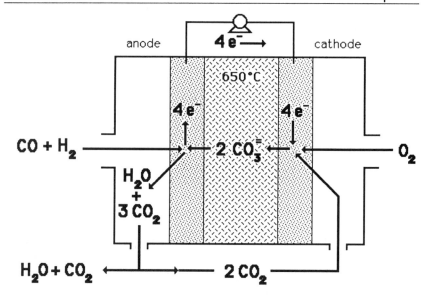

Figure 17-1: Simplified schematic of a coal-fired molten carbonate fuel cell.

electrochemically active. Although a pure CO/CO_2 mixture will react at low rates,[148, 2243] hydrogen is oxidized much more rapidly and will be consumed first. Consequently, the steam available in the anode compartment from hydrogen oxidation, especially near the anode outlet, allows a rapid water-gas shift reaction to take place:

Molten Carbonate Fuel Cells

$$CO + H_2O \rightarrow CO_2 + H_2$$

In the presence of metallic cell components in the anode compartment, at 650°C, carbon monoxide present in the fuel tends spontaneously to shift to hydrogen, thus improving anode kinetics. As hydrogen is depleted from the fuel, and water vapor partial pressure increases, shifting becomes increasingly favored. This direct chemical consumption of CO to produce hydrogen effectively results in the latter being the only electrochemically active species.[14, 2243] For this reason, the anode reaction in the MCFC is normally written:

$$H_2 + CO_3^= \rightarrow CO_2 + H_2O + 2e^-$$

In many cases, anode gas recycle schemes are planned that would prevent carbon (soot) formation via the normally irreversible (at 650°C) Boudouard reaction:

$$2CO \rightarrow C + CO_2$$

This reaction can be prevented by keeping the CO partial pressure low, with the CO_2 partial pressure sufficiently high to always be in a thermodynamically safe composition range. The simplest way to achieve this condition is to ensure that the steam content of the fuel gas entering the cell and within the cell anode is sufficiently high so that $(pCO)^2/pCO_2$ is always kept in the appropriate range via the water-gas shift equilibrium. Safe gas composition ranges are given in Ref. 457. Either steam injection or anode gas recycle can be used. In both cases, this results in some decrease in the hydrogen partial pressure of the fuel gas, giving a slightly lower fuel cell voltage (and efficiency). Anode gas recycle requires pumping power, whereas steam injection does not. An economic trade-off must therefore be made between cost and system efficiency, which requires careful computer modeling and attention to system detail.

Carbon dioxide produced at the anode is cycled to the cathode, where it reacts with incoming electrons and atmospheric oxygen to regenerate the carbonate ion consumed at the anode. Although the detailed mechanism of the cathode reaction is not completely understood, in electrolytes with high lithium ion contents (such as the eutectic mixture used in the molten carbonate fuel cell) it is believed that the reaction proceeds at least partially via a peroxide path:[139, 2022]

$$1/2 O_2 + CO_3^= \rightarrow CO_2 + O_2^= \quad \text{(Step 1)}$$
$$O_2^= + e^- \rightarrow O^= + O^- \quad \text{(Step 2)}$$
$$O^- + CO_2 + e^- \rightarrow CO_3^= \quad \text{(Step 3)}$$
$$O^= + CO_2 \rightarrow CO_3^= \quad \text{(Step 4)}$$

In this process, the second electron transfer step appears to be rate-determining close to oxygen electrode equilibrium. If this occurs, the exchange current of the reduction process should be proportional to $(pO_2)^{0.375}(pCO_2)^{-1.25}$ (see Ref. 150). In practice, the measured oxygen partial pressure dependence is indeed small and fractional, but that for CO_2 is also small and is positive, not negative. This seems to indicate that CO_2 molecules are involved in the rate-determining step in practical fuel cell electrodes; for example, in slow neutralization of $O^=$, corresponding to step (4) above. Analysis shows that the CO_2 reaction order for the exchange current would then be -0.25, which is again experimentally incorrect. If the rate-determining process was a chemical step (reaction of CO_2 with $O^=$) taking place after electron transfer, then a negative (-0.5 dependence) should again be observed. One way to rationalize the observed reaction orders is to push the rate-determining step further back down the reaction sequence, involving CO_2 as a reactant in step (2):

$$O_2^= + CO_2 + e^- \rightarrow CO_3^= + O^-$$
$$O^- + CO_2 + e^- \rightarrow CO_3^= \quad \text{(rate-determining step)}$$

The exchange current should now be proportional to $(P_{O2})^{0.125}(P_{CO2})^{0.25}$, which is close to the experimental values.[2243] It can be shown that the reaction orders at equilibrium will be the same if the superoxide equilibrium:

$$3O_2 + 2CO_3^= \rightarrow 2O_2^- + 2CO_2,$$

which is known to occur in high-potassium melts,[150,158] is observed before the first electron transfer step. The latter will then be preceded by the reaction

$$O_2^- + e^- \rightarrow O_2^=$$

However, if this is the case, the two processes could be distinguished by the reaction orders measured at constant potential, which would be proportional to $(pO_2)^{0.5}(pCO_2)^{1.0}$ and $(pO_2)^{0.75}(pCO_2)^{1.5}$, and by the difference in the measured values of the transfer coefficient. In the proposed peroxide process, the reaction

$$CO_2 + O^- + e^- \rightarrow CO_3^=$$

would be an electron transfer rate-determining step following one previous electron transfer, whereas in the superoxide mechanism it would have followed two previous rapid electron transfers. The transfer coefficient (see discussion in Chapter 2) for the processes then would be 1.5 and 2.5 respectively, the smaller value indicating the more probable mechanism. Further work is required to determine complete reaction parameters, which are important for the design of more effective cathode structures (cf. Ref. 2243).

Addition of the net anode and cathode reactions gives the overall cell reaction, which is simply the oxidation of carbon monoxide and hydrogen to produce carbon dioxide and steam. Each cell can generate electrical energy at 0.7~1.0 V, depending on gas composition and pressure and current density. The overall plant reaction in the example shown in Fig. 17–1 is the combustion of coal (simplified as being equivalent to carbon) to produce carbon dioxide.

Carbon dioxide can be transferred from the spent anode stream to the cathode in two ways. The first and simplest method is to burn the anode exhaust gas (which contains unreacted fuel) with excess air in a catalytic burner to generate carbon dioxide and steam. After condensing the steam to increase the CO_2 partial pressure, the resulting mixture containing excess nitrogen is added to the cathode gas supply. This method is inexpensive, but dilutes the oxidant with nitrogen, lowering performance. The alternative is to separate CO_2 from the anode exhaust via the use of a (conceptual) product exchange device, which can be regarded as a magic box using chemical or physical processes to separate relatively pure CO_2 at high energy efficiency.

The technical attraction of this method is that it would furnish a richer oxidant, resulting in a higher cell voltage. However, in any practical system the increase in cell efficiency (and improvement in the cost of power on the cell level) should be weighed against the cost and power requirements of the product exchange device. This will however allow the production of a richer fuel gas for recycling (i.e., the remaining hydrogen after CO_2 separation), thereby allowing better fuel utilization, hence higher system efficiency, at the expense of higher cost and complexity. In principle, CO_2 can be removed after condensing steam from the anode exhaust using pressure-swing adsorption, but the requirement for compression work makes this method energetically costly. A better method of separation would use a temperature-swing, and thus make use of fuel cell waste heat. For example, absorption cycles based on the use of carbonate formation from magnesium oxide, followed by thermal decomposition, have been examined. Perhaps the most intriguing method of all would separate as much as possible of the hydrogen from the anode exit by passing the gas through a subsidiary fuel cell anode operating at the highest possible utilization; for example, that of a PAFC. In this way some energy is recovered from the depleted stream, and a largely pure mixture of CO_2 and steam would result. This would then be added to the cathode stream after steam separation by condensation.

Internal Reforming Systems

As outlined in Chapter 8, an advanced form of the MCFC that shows promise for operation on natural gas fuel is the internal reforming molten carbonate fuel cell (IRMCFC or *direct fuel cell*), which catalytically converts methane directly to hydrogen in the anode compartment, eliminating the need for a separate reforming reactor.[1980] In theory, such a fuel cell is intrinsically efficient because the heat and the water vapor produced via the fuel cell reaction supply both the *in situ* energy and reactant requirements for reforming. Other advantages claimed for the IRMCFC compared with an external reforming unit, include greater system simplicity, heat exchangers of significantly reduced total area operating at lower temperatures, multifuel capacity (e.g., methanol), and possibly a minimum number of non-standard parts giving lower cost and increased reliability. Compared with the PAFC, the lack of noble metal catalysts and high quality waste heat for adsorption air-conditioning in small on-site units represent further advantages.

Laboratory-sized IRMCFC cells being developed by ERC have already demonstrated lifetimes of almost one year (see Chapter 8). Experiments have indicated that methane conversion increases as the cell current density increases, and reaches essentially 100% for hydrogen utilizations greater than 50%.[683] Calculations indicate that an internal reforming molten carbonate fuel cell system could operate on natural gas with an overall system efficiency of over 50%.[1207, 1982, 1983] This concept may be important for small (1~2 MW) on-site electrical generators, in which the additional cost of a bottoming cycle to further increase efficiency would be uneconomic, but which still produce some high-quality waste heat for heating and air conditioning.

In principle, IRMCFCs also could be profitably used in a coal-fueled system.[1520] For this application the coal could be gasified in an air- or oxygen-blown catalytic gasifier operating at lower temperatures than the majority of the coal gasifiers reviewed in Chapter 8. In the latter case, excess steam would be required to maintain the low operating temperature. Thermodynamic considerations would favor production of methane, rather than hydrogen, as a product at the lower operating temperature. After cleanup, the fuel gas could be fed to the internal reforming fuel cells. By careful system integration, thermal losses could be kept low, and the use of fuel cell waste heat to upgrade fuel quality could yield significant gains in overall system efficiency compared with that obtainable using a standard high-temperature coal gasifier. However, substantial technical process and further equipment development would be required before a system of this type could be commercialized.

The standard nickel anode has only sufficient activity to reform a small fraction of methane to hydrogen in a normal MCFC. Both ERC and the IGT have both been active in the development of catalysts (typically active nickel at loadings of 20~100 mg/cm^2) on inert catalyst supports (such as LiAlO$_2$) for the internal reforming reaction.[558, 1964] Preliminary results indicated that the methane conversion is hindered by the presence of carbonates at concentrations greater than about 5%. More detail on internal reforming systems and programs have been presented in Chapters 3 and 8.

C. CELL DESCRIPTION

In Chapter 3, several integrated molten carbonate fuel cell systems and their reactant flow sheets were discussed. In this section, emphasis will be on the cell and stack of the MCFC. Figure 17-2 shows a published design configuration for a proposed molten co- or counter-flow carbonate fuel cell stack.[20] Each unit cell would be planned to have a total thickness of about 5 mm and an area of about 1 m^2.[19] In the power plant, cell assemblies in the form of vertical 1 MW stacks were proposed, each comprising perhaps 500

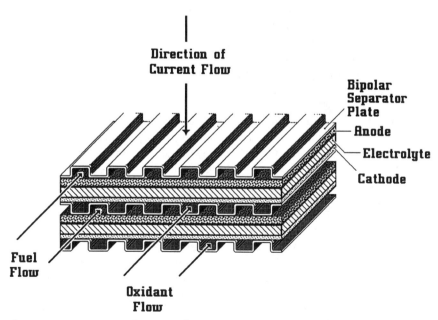

Figure 17-2: Design configuration for molten carbonate fuel cell stack.[20]

series-connected cells. Bipolar metal separator plates would serve both to carry the current from one cell to the next as well as to separate the cathode of any given cell from the fuel fed to the anode of the adjoining cell.

As in all proposed designs, the anode and cathode would be separated by an electrolyte structure that consists of a porous matrix made from finely divided submicrometer lithium aluminate ($LiAlO_2$) or strontium titanate ($SrTiO_3$) ceramic particles, as discussed in Chapter 10. This would contain about 45% by volume of the molten carbonate electrolyte, in most proposals a eutectic mixture of 62 mol % lithium carbonate and 38 mol % potassium carbonate. The carbonate-containing tile remains essentially rigid at the 650°C operating temperature of the stack, about 125~200°C above the melting point of the electrolyte. Electrolyte matrixes were typically about 1.5 mm thick in early designs, and less than 0.5 mm in more recent work.

Since the late 1960s, cathodes have almost always consisted of porous nickel oxide (NiO), generally formed by *in situ* oxidation of a nickel sinter. *In situ* oxidation results in the automatic accumulation of 2~3 cation % lithium, which makes the cathode structure an electronically conducting ceramic, $Li_xNi_{1-x}O$, where x is in the range of 0.022~0.04.[2022] At 650°C, the resistivity of the material is about 0.2 Ω cm. The thickness of the cathodes has normally been maintained at about 0.3 mm, lower than that of typical anodes in order to lower total electronic resistance and to obtain highest performance. The relationship between cathode performance as a function of thickness and pore size has been studied.[457]

The cathode material typically has (or rather develops) a porosity of about 55% and a mean pore size of about 10 μm. The cathodes usually develop a bimodal pore distribution, the small pores being flooded with electrolyte, thereby providing extended reaction surface and ionic conductivity; and the large pores remaining open, facilitating mass tranport in the gaseous phase. Lithiated nickel oxide cathodes can be fabricated in several ways, both by *in situ* oxidation and starting from nickel oxide powders. Reference 2022 provides an excellent review of these and other aspects of the development of cathodes for molten carbonate fuel cells.

Molten carbonate fuel cell anodes have almost consistently been made from porous sintered nickel (see Chapter 13), with a thickness of 0.5~0.8 mm and a porosity of 55~70%, the mean pore size being approximately 5 μm. Because anode kinetics are faster than those at the cathode, the effective surface area can be lower for the anodic processes. Partial flooding of the comparatively thick anode is therefore acceptable, and it can provide a useful means of accommodating variations in the total carbonate content of the cell.[2022] However, the solution to the problem of an electrolyte reservoir in the MCFC has yet to be found (see following), compared to the corresponding problem in the PAFC discussed in Chapter 16.

Molten Carbonate Fuel Cells

Ideally, MCFC anode pores should be somewhat smaller than the larger gas-filled pores of the cathode, so that the latter operate, in a sense, dry, but larger than the electrolyte-filled pores of the cathode. The matrix should have a pore size that is smaller than that of other components, so that it always remains filled with electrolyte (see Ref. 2243). Typical pore size distributions for the anode matrix and cathode are shown in Fig. 17-3.

Ceramic materials must be incorporated into the anode structure to stabilize it from further sintering, pore growth, shrinkage, and loss of surface area. Until recently, sintering control has been via the use of Ni-Cr alloy powder or Ni and Cr mixtures, usually in the 2~10 wt % Cr range.[1686] The formation of, first Cr_2O_3, followed by surface formation of $LiCrO_2$, has been shown to be very effective in preventing drastic surface area loss by sintering.[2243] However, surface formation of $LiCrO_2$ has two disadvantages: first, it is wetted by the electrolyte, changing anode surface area and three-phase boundary characteristics; and second, a long-term growth of wettable fine-pore structure occurs as Cr^{3+} ions diffused from the electrode to form $LiCrO_2$. Recent work has concentrated on electrode structures that contain more stable oxidizable sinter-resistant additives.

In work under EPRI sponsorship, IGT has investigated the use of copper

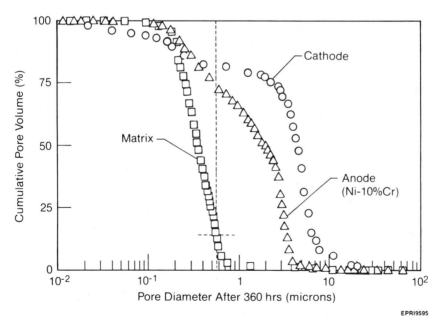

Figure 17-3: Pore size distribution requirements for well-matched anode, electrode matrix, and cathode components in the MCFC.[2243]

as a substitute anode material for nickel.[558] Copper has several advantages over nickel, such as higher electrical and thermal conductivity, a better matched coefficient of thermal expansion with other cell materials, and lower cost and better domestic availability in the United States. More particularly from the technical viewpoint, its superior resistance to anodic oxidation, enabling operation at higher current densities, is most important.

Reactant gases are fed in parallel to all cells in the stack via common manifolds. Designs using both external and internal manifolding have been proposed,[19] (see Chapter 13, Fig. 13-2). A common external manifolding scheme is one in which large fuel inlet and outlet manifolds are placed against opposite sides of the stack, with corresponding oxidant manifolds placed against the remaining two sides. Internal manifolding schemes make use of cell internal components to direct the reactant gas flows in parallel from cell to cell, each individual cell being uniformly supplied. A similar arrangement is used for the exiting gases. Cells can be supplied in cross-flow, counter-flow or co-flow configuration, where these refer to the anode-cathode gas-flow patterns. For example, in Fig. 17-2, a GE co-flow design is shown. A counter-flow design of the type used consistently by UTC since 1980 in a series of different stacks is shown diagramatically in Figs. 13-2 and 17-4, which also shows an alternative design with sheet metal gas channels. These should be compared with the co- or counter-flow structure in Fig. 17-2. The UTC stack design, with cross-flow manifolding, is reviewed in Ref. 1408.

Cross-flow manifolding has the advantage of simplicity, but it results in a complex current density and temperature pattern, since anode and cathode gases whose composition and temperature both vary from inlet to exit interact at right angles to produce the current density and temperature map. An example is shown in Fig. 17-5. Counter-flow and co-flow (Fig. 17-5) are necessarily simpler, since the flows are parallel. The former, in particular, results in the lowest temperature extremes, since the gases are in extreme concentration configurations (i.e., paired weak and rich) as they pass through the cell, so that the Nernst potential (i.e., the driving force for the reaction) and the current density remain moderately constant across the cell. In co-flow, the composition, Nernst potential, and temperature have highest values at the inlet side, and lowest values at the outlet side. Overall, counter-flow and co-flow may be most effective for internal reforming, where a large amount of heat is required for anode gas transformation in the early part of its flow zone.[1965] Recently it has been argued that co-flow cells using sheet metal may be more effective from the viewpoint of thermal stress, including the *oil-canning* effect.[121]

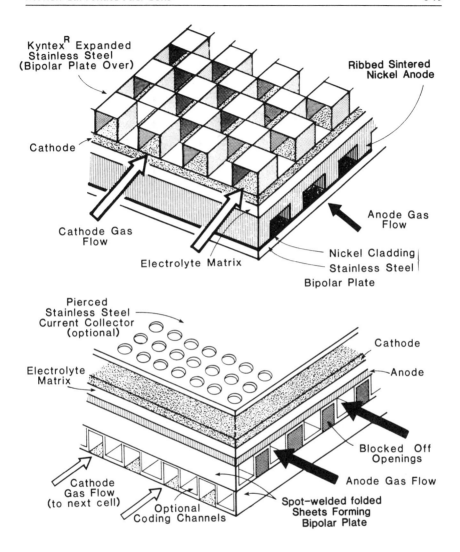

Figure 17-4: (Above): Schematic UTC crossflow stack design using ribbed anode, flat bipolar plate and gas channels and current collector made from Kyntex® three-dimensional expanded metal. (Below): Schematic ERC plane anode system with two spot-welded channeled sheet-metal parts to form combined current collector/bipolar plate. The pierced stainless steel current collector to add mechanical strength and to improve electrical contact area to the cathode is an undesirable part of many present designs, since it increases contact resistance and acts as a lithium ion sink.

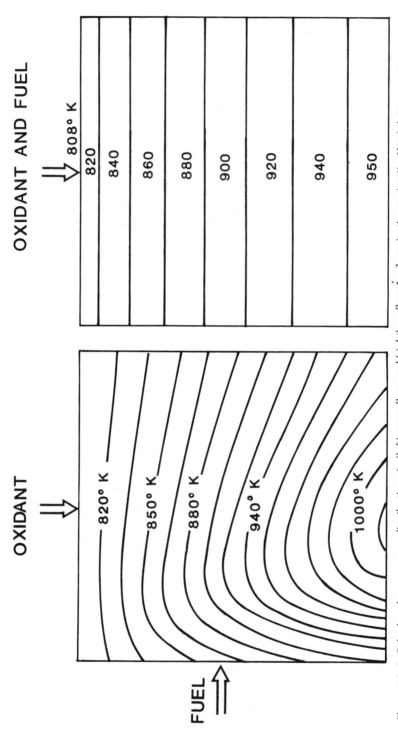

Figure 17-5: Calculated temperature distributions in (left) cross-flow and (right) co-flow 1 m² nonisothermal cells. Gas inlet temperatures 800 K. Low BTU gas assumed, atmospheric pressure, 75%/25% utilization, cell resistance 0.5 Ω-cm², Assumed i_{av} is 200 mA/cm².

Molten Carbonate Fuel Cells

Most developers have used more complex cell configurations than that shown in Fig. 17-2. For example, cell construction and geometry have been such that separate current collectors have been required (see Fig. 17-4). These have included cathode configurations such as a perforated stainless steel plate current collector in contact with a flat cathode on one side and with a ribbed bipolar separator plate on the other,[2092] or alternatively a cut and folded, ribbed current collector of expanded metal type in contact with a flat cathode and separator plate (Fig. 17-4). These are easy to make and use for the construction of laboratory cells, but the cost of this large area of stainless steel components would be prohibitive in cost-effective stacks. In addition, as we will see in a later section, the large area of stainless steel acts as a lithium sink by forming a corrosion scale, which gives poor contact resistance. The only cost-effective structure is likely to consist of a simple sheet-metal co- or counter-flow arrangement, as shown in Fig. 17-2.

United Technologies has consistently used ribbed anode structures with flat bipolar plates (Fig. 17-4), although the ribbed nickel anode itself is likely to prove too costly for a commercial unit. Since their system design, like their PAFC, utilizes externally manifolded cross-flow, flow on the cathode side is assured by a ribbed stainless steel exmet current collector (Kintex®, the Kintex Corp. in Danbury, Connecticut) in contact with the bipolar plate. Other developers have built small cross-flow stacks using machined ribbed bipolar plates (e.g., Toshiba, Fuji, Hitachi, Ref. 2092, often combined with pierced current collectors), and ERC has used inconel Kintex material on the anode side of flat bipolar plates, but these approaches will also have to be abandoned for cost-effective stacks. At the present time, IHI and IGT are unique in pursuing the co- or counter-flow approach shown in Fig. 17-2.

D. PERFORMANCE

Output

With a given set of components, the output of a cell depends on the hydrogen content of the fuel used, the percentage hydrogen utilization, and the operating pressure. Oxygen utilization is usually low enough to have little overall effect. These are illustrated in Table 17-1.[378] Percentage hydrogen utilization mainly influences the gas composition contribution to the equilibrium cell exit voltage, as determined by the exit Nernst equation:

$$\Delta E_{cell} = \Delta E^{\circ}_{cell} = \frac{RI}{4\mathcal{F}} \ln \left\{ \frac{(pCO_2)^2_{an}(pH_2O)^2_{an}(a_{CO_3^=})^2_{cat}}{(pCO_2)^2_{cat}(pO_2)_{cat}(pH_2)^2_{an}(a_{CO_3^=})^2_{an}} \right\}$$

where the partial pressures are exit values, and in which the anodic pH_2, pCO_2, and pH_2O values take into account the water-gas shift equilibrium. When the gases are shifted to equilibrium, the term in braces becomes simply $([CO_{eqm,anode}]^2[O_{2,cathode}])/([CO_{2,eqm,cathode}]/[CO_{2,cathode}])^2$, or $([H_{2,eqm,anode}]^2[O_{2,cathode}])/([H_2O_{eqm,anode}]/[H_2O_{cathode}])^2$, all other terms canceling. Taking this into account also allows us to express the open circuit potential (either at the entrance or exit, or indeed anywhere in the cell) in the alternative form

$$E_{cell} = E^{\circ} - (RT/2\mathcal{F}) \ln [(pCO)(pO_2)^{1/2}(CO_2)_{cat}]/(CO_2)_{an}$$

with a similar equation for the hydrogen-oxygen-water process with the appropriate equilibria inserted.

Unless there are gas cross-leaks, molten carbonate fuel cells always locally reproduce the theoretical open circuit voltage within 1~3 mV. [14] For the fairly representative fuel shown in Table 17-2, the open circuit voltage at 650°C is about 1.056 V at 1 atm and about 1.116 V at 10 atm.

Typical molten carbonate fuel cells will operate at a current density of about 200 mA/cm² and at a unit cell voltage of more than 0.8 V under pressurized conditions.[19] As outlined in Chapter 3, because of faster

TABLE 17-1

Effect of Pressure and Fuel Utilization on Molten Carbonate Fuel Cell Performance[378]

(at 50% oxidant utilization and 162 mA/cm² current density)

Fuel	% H_2 Utilization	Operating Pressure, atm	Cell Output Voltage, V	Power Density mW/cm$_2$
Reformed CH_4*	25	1	0.823	133
	50	1	0.780	126
	75	1	0.740	120
Coal gas†	75	1	0.670	108
	75	10	0.840	136

*60% hydrogen
†Low BTU coal gas, 16% (hydrogen + CO)

Molten Carbonate Fuel Cells

TABLE 17-2

Representative Fuel and Oxidant Mixture for Molten Carbonate Fuel Cell[14]

Fuel: 197 BTU/SCF		Oxidant	
H_2	51.2%	O_2	11.7%
CO	9.7%	N_2	64.1%
H_2O	28.5%	H_2O	7.1%
CO_2	10.5%	CO_2	17.1%

kinetics, molten carbonate fuel cells are able to operate at voltages which are about 100 mV higher than the voltages achievable in the PAFC under the same pressure conditions, despite the thermodynamically higher value of the reversible voltage in the latter. Similarly, both thermodynamic advantages and in some cases lower ohmic losses enable the molten carbonate fuel cell to attain an operating voltage up to 150 mV higher than that of the SOFC. For the molten carbonate fuel cell, these voltages translate into a system energy conversion efficiency that is 15~25 incremental % higher than is possible with the PAFC or SOFC. When the overall efficiency of the complete power plant is taken into consideration, it has been estimated that for operation at atmospheric pressure using pipeline-quality natural gas as fuel, an overall system efficiency of better than 50% can be expected for MCFC systems, compared with values in the low 40s for SOFCs, and a value of slightly over 40% for PAFCs.[19] Somewhat higher values may be possible if operation can be conducted at higher pressures.

Losses

The ohmic losses in molten carbonate fuel cells are reasonably well defined and understood, and can be minimized by a good choice of current collector materials, designs having a minimum number of contact resistances; and by reducing the thicknesses of the cathode and of the electrolyte matrix layer. However, all the preceding refers to initial resistance values; as cell components age, changes can take place that are not easily quantified.

For cathodes which are made by the *in situ* oxidation of nickel plaques,[20-22] in cells with thick (1.5 mm) electrolyte layers, about 40% of the total IR-drop occurs in the cathode and its contacts, first with the cathode current collector, then with the cathode side of the bipolar plate. Most of the remaining IR-drop occurs in the electrolyte matrix. As an example, a typical good 100 cm² laboratory cell of this type had an ohmic resistance of about 5 mΩ. This is about five times higher than that in a good PAFC, for which ohmic polarization is now only about 10 mV at a current density of 100 mA/cm². The use of thinner electrolyte layers (e.g., those feasible with the now-standard 0.5 mm tape-cast technology[2243]) can lower overall resistance by 40% to about 0.3 Ω cm², or about three times that of the PAFC. Further improvement will demand new cathode current collector structures, particularly from the viewpoint of the contact resistance between the cathode and the stainless steel bipolar plate. This must involve fewer contact surfaces between the cathode and the bipolar plate itself, particularly those involving the subsidiary stainless steel current collector that most developers have used in their cells. The use of these additional surfaces results in higher contact resistances, more poorly conducting films in the electronic circuit, and a high area sink for cell corrosion products. A further discussion of this problem follows.

Compared with the ohmic losses, the activation, diffusion, and mass transport voltage losses in MCFCs are much more difficult to model, since the effects of electrode kinetics and electrode structures are not completely understood, and it is difficult to determine detailed current density distributions. Half-cell studies are being made of electrode kinetics and of the effects of differential fuel conversion to clarify some of these non-ohmic polarization mechanisms. However, compared with the PAFC, these components are small; for example, total polarization below that of the exit Nernst potential may be 150 mV, including 50 mV of IR-drop. The corresponding figures for the PAFC might be 350 mV, including 15 mV of IR-drop.

Table 17–3 shows an estimated distribution of the component voltage losses in a cell operating at 200 mA/cm² on reformed methane at 1 atm (75%

TABLE 17–3

Estimated Distribution of Voltage Losses in a Molten Carbonate Fuel Cell[12]

Operating conditions:	200 mA/cm², 650°C, 1 atm
	Reformed methane
	75% hydrogen utilization
	50% oxidant conversion
Cell open circuit voltage	1.026 V
Ohmic drop	–0.068 V
Concentration polarization	–0.115 V
Activation and mass transfer polarization	–0.148 V
Cell Output Voltage	0.695 V

hydrogen utilization, 50% oxidant conversion).[12] Ohmic drop is 0.068 V, whereas total polarization is 0.146 V. The concentration polarization represents the difference between the open circuit potential with the gases at zero utilization and the exit Nernst potential with the gases at their real utilizations.

Performance and Lifetime Goals

For a 650 MW plant operating at 5~10 atm on purified gas from a coal gasifier, it would be desirable to achieve the following goals:

1. A cell output voltage of ≥ 0.80 V at 160~200 mA/cm^2.
2. A stack life in excess of 40,000 hours, corresponding to six years at a 75% capacity factor.
3. A 30-year life for the balance of plant.
4. Fuel utilization $\geq 75\%$ on low BTU coal gas using burner-type carbon dioxide recycle from anode exit to cathode inlet.
5. The cells must operate well under steady-state and transient conditions of power demand, gas flow, pressure differential, temperature differential, and thermal cycling (unless provision can be made to keep the stack hot during shutdown, which is not at present envisaged).
6. There still should be good system availability when some stacks are shut down for maintenance or replacement.
7. No open circuit failure or progressive cell failure should be allowable.
8. Short-circuit failure or lower-than-design performance should be limited to 2~3% per stack.

If the above performance goals of cell voltage and fuel utilization can be attained, calculations have shown that an overall plant efficiency of 50% (coal to a.c. power) is possible using a combination of steam and gas turbine bottoming cycles.[421] In addition, SO_x and NO_x emissions will be very low.[182] While these are very attractive, the most important economic criterion will be the cost of electricity, which must be at least competitive with that derived from other advanced energy conversion technologies which may be less efficient, but which have lower operating and maintenance (O&M) costs and capital charges. Foremost among these technologies is the coal-gasifier combined cycle (CGCC) plant, which may reach a coal-a.c. efficiency of 45%.[631,1155] This technology will differ from that of the MCFC in that all the gas will be burned in the gas turbine, rather than only the anode tail-gas as envisaged in early coal-burning MCFC schemes.[421, 1521, 1977] Waste heat is recovered from the gas turbine exhaust and from elsewhere in the plant to provide further power at higher total efficiency by the use of a steam cycle.

Apart from the coal gasifier and its associated gas cleanup system, the CGCC plant, whose prototype is at Cool Water, Southern California,[631,1155] has components whose technology represents a fairly low risk in terms of lifetime use. The reliability of gas and steam turbines over their respective lifetimes is well known, whereas that of the MCFC is not. Similarly, maintenance requirements, such as turbine reblading, are state of the art, and represent a relatively small O&M cost. In contrast, fuel cell stacks costing $300/kW that might have to be replaced four times over the operating lifetime of the plant (which has a total capital cost of $1200/kW) may far outweigh any small improvements in efficiency between two systems of comparable capital cost, one of which is a CGCC, the other a CGMCFC plant. At current coal costs of $2/MMBTU, a saving of five percentage points in efficiency is equivalent to only 0.15¢/kWh, or only $60/kW over the stack operating lifetime of 40,000 h. Hence, unless stack cost can be reduced, and/or lifetime increased, there will be little incentive to use MCFC technology. This observation has been made in recent reports.[1521, 1977]

Introduction of the MCFC into central power plants is therefore likely to be a gradual technology, taking place as confidence builds up in stack reliability as a result of initial use in on-site applications, where higher life-cycle costs can be accepted because of the higher price of alternative delivered energy for on-site use, compared with bus bar production costs for central power stations. These applications should allow cost reductions and improved lifetimes. The attractiveness of MCFC-based central station systems will in any case increase as coal prices increase, due to higher demand as fossil fuels become more scarce after the year 2000.

E. PROBLEM AREAS

Design and Fabrication

For commercial viability, mass production methods must be perfected for the manufacture of components that are currently still being made by hand machining and other preproduction techniques. These are now under active development, no major difficulties being foreseen. It also is important to design and develop effective methods for stacking hundreds of cells without encountering manifolding and pressure seal problems, and to scale up and integrate cost-effective hardware and subsystems. Eventually, and as a longer-term goal, work will have to be undertaken to design

Molten Carbonate Fuel Cells

improved carbon dioxide recycle systems and improved gasifiers, rather than relying on commercially available equipment that does not integrate perfectly into the most cost-effective power plant design.

The Electrolyte Layer

During thermal cycling, large volume changes in the matrix occur because of melting and freezing of the electrolyte. These changes create stresses which produce microcracks and morphology changes in the electrolyte matrix. In addition, the gross thermal expansion coefficient of the solidified electrolyte during a thermal cycle will generally differ from that of metallic parts of the cell, creating further stress. Under low pressure operating conditions, this causes little difficulty, but when cells are operated at 5~10 atm, pressure differentials of as little as 0.2 atm across the electrolyte matrix can cause significant gas cross-over between anode and cathode, destroying cell performance. The cracking problem was observed to be especially serious with thick hot-pressed tiles. To alleviate it, reinforcement materials such as metal and ceramic fibers, fiber mats, and woven screens were incorporated into the tile structures, as discussed in Chapter 10. Because of inadequate strength and corrosion of the reinforcement materials by molten carbonate, these attempts were only moderately successful in long-term operation.[20] Other approaches have included cell designs that avoid mechanically loading frozen electrolyte matrixes. Some of these were moderately successful, but tended to be costly.

The use of gas bubble pressure barriers (BPBs) at the electrode-electrolyte interface, particularly at the anode, and the successful addition of crack arrestors to prevent crack propagation in thin electrolyte layers has already been discussed in Chapter 10. This is likely to continue to be a favored technique in the future. Similarly, the transition from thick hot-pressed electrolyte tiles to thin electrophoretic, tape-cast or calendered layers that are easy to manipulate in large sizes in the green state has already been introduced.

Although the thin electrophoretic electrolyte matrixes developed by GE and ERC are much less prone to cracking on thermal cycling than hot-pressed tiles, under certain circumstances they still showed a tendency to crack. Part of the reason may well have involved their structure, as GE claimed,[278] but most of the work conducted by GE was not on thin, low-IR structures, but on electrophoretic reproductions of thick hot-pressed layers. Thin electrophoretic layers have low IR-drop compared with tape-cast matrix structures,[125] but have a disadvantage in that large crack-arresting particles cannot easily be incorporated into them in a controlled manner.

Currently, the potentially cost-effective tape-casting technique is favored by all developers. It was originally developed by UTC, among others,[1519] as a development from the electronics industry, where similar methods are used to form thin dielectric sheet materials.

The method involves making a viscous slurry of lithium aluminate particles in an organic solvent that also contains a deformer, a plasticizer, and an organic polymeric binding material.* This slurry is spread onto a substrate to a thickness of 1.0~1.3 mm, and the solvent then evaporated to produce a very pliable sheet of organic-containing material about 0.5 mm thick. This sheet, still containing the organic binders, is next placed between anodes and cathodes and assembled into a cell. In the original method of cell assembly, the anodes and cathodes were preloaded with the appropriate carbonate electrolyte, and the cell was heated up in the presence of air. When the temperature is sufficiently high, the organic materials burn out, leaving behind a porous lithium aluminate structure between the electrodes. The carbonate electrolyte preloaded into the electrodes or into a separate tape-cast layer then melts and wicks by surface tension into the porosity of the matrix structure. This type of tile structure is moderately dense and has a specific conductivity that is not significantly different from that of the original hot-pressed tiles. The tape-cast electrolyte matrix layers are, however, thinner (0.5 mm vs. 1.5~2 mm for hot-pressed tiles), and therefore give a lower cell internal resistance. As remarked above, such cells can operate at about 200 mV below the exit Nernst potential, or 0.85 V at 10 atm pressure and at 175 mA/cm^2.[1519]

Scale-up of this process does not appear to present problems, and green matrix sheets typically can be made in widths of about 90 cm or greater, and in any desired length with suitable equipment. Between 1981 and 1985, UTC used 0.1 m^2 matrix layers for experimental 10- and 20-cell stacks, but they are currently proceeding to a ~90 cm square unit. The UTC 0.1 m^2 stacks using tape-cast electrolyte matrixes have been subjected to 4~5 thermal cycles without any problems of gas crossover.

Before withdrawing from the DOE fuel cell program, General Electric developed a calendering technique that does not require the use of a solvent. Instead, a thermoplastic organic binder alone is mixed with the lithium aluminate particles and the dry mixture heated to 120~130°C, and calendered out between rollers to form a 0.5 mm-thick pliable sheet. This sheet appears to be similar to that formed by tape-casting, but due to the mechanical work in formation, it may have fewer defects, particularly air bubbles, than tape-cast sheet. This technique has already been introduced in Chapter 10.[117]

---

*To avoid manufacturing hazards and the need for solvent recovery, the trend is now towards the use of aqueous slurries.

Separator Plate Corrosion

Of all the components in a molten carbonate fuel cell, the separator plate dividing the anode of one cell from the cathode of the next is probably the most susceptible to corrosion, since it must be a metallic conductor, possibly in contact with thin continuous films of electrolyte on both sides. In addition to having to withstand attack from corrosive impurities in the anode and cathode reactant gases together in their own electrolytic environment, it must be resistant to the complete range of conditions found in a cell. This represents the whole range of chemical environment in the cell, varying from a highly reducing atmosphere at the anode inlet to a mildly reducing one at the anode outlet and finally to an oxygen-containing atmosphere in the cathode compartment. Type 316 or type 310 stainless steel sheet has been commonly examined as a separator plate, and has been shown to be reasonably satisfactory under laboratory conditions in the cathode environment, where a relatively stable and protective oxide scale forms in the presence of molten carbonate. However, as we shall describe, hydrogen diffusion through the thin separator plate itself may result in anomalous effects in the formation of scale on the cathode side.

Stainless steel separator plate corrosion is a problem in the reducing environment of the anode, since under the wet, mildly reducing atmosphere at the anode exit these materials are definitely unstable and may show catastrophic corrosion (see Chapter 13). For these reasons, the anode side of the stainless steel separator is clad with nickel or, in experimental IGT work, with copper.[559] Stainless steel of 310 or 316 compositions can, however, be used under dry anode conditions, such as at low hydrogen utilization. While this is out of the question for practical fuel cell use, it may be exploited in technology demonstrations, as in recent Japanese work.[2092] Both nickel and copper are thermodynamically totally inert to corrosion under any reasonable anode conditions, copper being particularly able to withstand large accidental or intentional polarization.

At the anode side of the separator plate, a problem may arise from the high gas-phase carbon activity, together with its high mobility in nickel, which can result in carbon transport from the fuel gas into the separator plate, giving carburization and embrittlement of the stainless steel and possible delamination at the nickel–stainless steel interface.[558, 2023] In contrast, copper is relatively resistant to carbon transport.

A number of studies have been performed by IGT and elsewhere on the rate of corrosion of stainless steels under typical anode and cathode gaseous atmosphere conditions. Some of these are illustrated in Fig. 17-6. It can be seen that SS 316 is apparently adequate at the cathode, but will show generally poor stability under many anode conditions, as just discussed. Such experiments have included a range of materials. Unfortu-

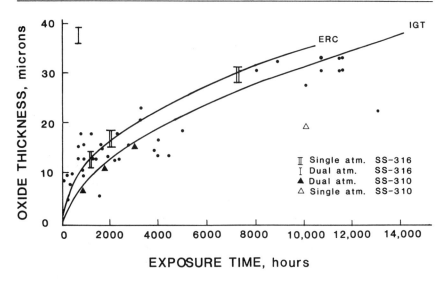

Figure 17-6: Corrosion of cathode side of SS 310 and 316 bipolar plates exposed to standard oxidant and Li/K carbonate eutectic at 650°C. Points are "single-atmosphere" corrosion measured on SS 316 by IGT. Other symbols are "single atmosphere" and "dual atmosphere" (hydrogen on reverse side) data obtained by ERC as indicated.

nately, the primary conclusion has been that no single inexpensive material is stable under all anode and cathode conditions.

The preceding tests were conducted on massive samples of the various materials completely surrounded by the test atmosphere. Work conducted at IGT in the late 1960s showed that thin metal parts exposed to anode and cathode atmospheres on each side could experience anomalous corrosion behavior due to hydrogen diffusion through the metal, which could lead to reducing conditions on the inside of a cathodic oxide scale, inducing spalling by the formation of water vapor under the oxide film. Recent tests have shown that nickel-clad thin sheet type 316 stainless steel under anode-cathode biatmospheric conditions can give corrosion rates on the cathode face that are up to four times higher than the typical values in Fig. 17-6. These result from oxide spalling on the cathode face, particularly on areas of the sheet metal where the oxide is subject to stress, such as bends or curves.[2282] Fortunately, the same study has shown that SS 310, a higher-chromium alloy, while having a slightly higher intrinsic rate of scale formation, does not readily suffer from spalling. This is presumably due to its lower hydrogen permeability, resulting from its Cr content.

In recent years there have been indications that for test periods on the order of 25,000 to 40,000 h, corrosion problems at the cathode side of an apparently effective separator plate may eventually appear. In general, increased contact resistance between the cathode and the current collector and bipolar plate was observed due to the build-up of a thickening Li-containing film as a function of time. Lithium ferrite ($LiFeO_2$) is the final stable cathode corrosion product on stainless steels, and it was being considered as a candidate for precoating the cathode side of the separator plate.[20] It is possible that electronically conducting protective ceramic coatings may eventually be developed for use on the separator plates.

The stainless steel at the cathode side of the cell is a major sink for lithium ion as it slowly and progressively forms this protective layer. This has been shown to reduce the lithium inventory of the cell, thereby changing electrolyte composition and performance. It also increases IR-drop by reducing electrolyte volume and by the formation of the $LiFeO_2$ film. The electrolyte loss effects can be controlled by the use of electrolyte reservoirs, either internal or external to the cell, but contact resistance effects depend on the number of intercomponent contacts involving a lithium ion sink that are present. Clearly, a reduction in the number of sandwiched sheet metal components would be desirable to better maintain performance. Some further observations on this topic follow, under the problems associated with the cathode.

Attack via corrosion currents may also be a problem at the outer edges of the separator plate where a wet seal is formed between the separator plate and the electrolyte matrix. The wet seal consists of thin continuous layer of carbonate electrolyte extending around the edge of the plate, and therefore electrolytically joining anodic and cathodic areas of adjacent cells. This can lead to a catastrophic corrosion situation in which the rate of attack on the cathode side of the metal plate is only limited by the resistance of the thin electrolyte film. To eliminate this type of attack, it is necessary to put a dielectric layer (ionic insulator) in the circuit between the metal plate and the electrolyte film. This can be effected by applying a stable nonconducting film. Techniques have been developed to protect the wet seal regions of the separator plate with aluminide or aluminized diffusion coatings. These coatings are typically applied to the stainless steel by flame or plasma spraying of metallic aluminum, followed by a 1000°C diffusion anneal.

Any reduction in corrosion rate will result in significant cost savings, since the separator plate probably will be the most expensive element (in terms of $/kg of material) in the cell. It is desirable to have separator plates of about 0.2 mm thickness with no additional current collector plates. This would correspond to about 1.25 kg/kW, or a maximum of $15/kW at present prices for the materials content of the clad sheet.

Seals

External manifolding for the reactant gases requires gaskets to make gas seals with the edges of stack sides. At present, at least in UTC practice, gaskets are made from a fibrous ceramic material, for example zirconia felt, filled with electrolyte. At ERC, zirconia felt gaskets filled with other ceramics are being considered to reduce their total carbonate uptake.[125] Finally, at Fuji in Japan, solid zirconia gaskets have been used in small stacks constructed to high mechanical precision, and Toshiba is examining zirconia felts filled with borate glasses.[2092]

Although no serious mechanical problems (as distinct from those involving electrolyte transfer through porous gaskets; see following) have been encountered with short stacks and short duration runs, when the stacks are scaled vertically upward (e.g., to the 100-cell level and beyond) it is anticipated that there may be sealing difficulties over long duration runs associated with accommodating the mechanical dimensional differences encountered. As discussed in the next section, seals of poor design can result in serious loss of electrolyte from the cell. In addition, the use of molten electrolyte in the gaskets to maintain a gas-tight seal results in inevitable shunt currents along the edges of the stacks, leading to coulombic losses. The internal manifolding first used in the GE designs, later in those at Hitachi and IHI in Japan, may present fewer sealing and shunt current problems than does the external manifolding used by UTC and ERC. In 1986, Fuji Electric had a unique cell design with internal anode manifolds and external cathode manifolds (see Fig. 13–2).[2092]

Electrolyte Loss

Electrolyte loss has been a problem in molten carbonate fuel cells and stacks and may result in being the major life-limiting factor. As discussed earlier, after cell start-up, the molten carbonate electrolyte forms lithium-containing passivating films on stainless steel cell components and cathode lithiation occurs. These reactions consume LiO, with liberation of CO_2, thus reducing the volume of the electrolyte in the cell and taking it to a non-optimized, potassium-rich composition. While this might be solved by pretreating the metal components and prelithiating the cathodes, the more recent tendency has been to initially fill with a non-eutectic lithium-rich electrolyte; for example, 70 atomic % Li, 30 % K.[1301] The reduction of the total amount of stainless steel in the system cathode area again cannot be overemphasized.

Until recently, the primary electrolyte loss mechanism was believed to involve hydrolysis of the electrolyte to produce hydroxides, which then vaporize into the fuel and air streams and are thus lost from the cell:

$$Li_2CO_3 + 2H_2O \rightarrow 2LiOH + H_2CO_3$$

The evaporation rates at both the cathode and at the anode have been examined in detail by IGT using a thermobalance method.[1941] The general conclusion was that if the gases passing through the cell did indeed become saturated with carbonate-derived vapors, losses over the lifetime of the cell might be severe, and, as in the case of the PAFC, an electrolyte reservoir would be required. However, since the losses would be proportional to the gas volume throughput, they would be inversely proportional to the operating pressure. This was encouraging, since it indicated that reservoir volumes would be smaller under pressurized operating conditions; indeed, at 6~8 atm, enough melt could probably be stored in the cell to allow 40,000 h lifetime.

In 1982, UTC observed severe carbonate losses from the 10-cell 0.1 m^2 stack then being examined under DOE support.[2243] Much of the carbonate loss could not be accounted for, and it amounted to almost 40% of the initial inventory. Careful analyses of the exiting gas streams, which were scrubbed free of carbonate by downstream hardware, demonstrated that evaporation failed to account for the observed loss. Indeed, it seemed that evaporation rates were only about 10% of those estimated from the IGT data, making this loss mechanism relatively minor.

Finally, by analysis of the electrolyte inventories of individual cells after operating later stacks, it was established that carbonate migration took place from the positive to the (upper) negative end of the stack, via the porous manifold gaskets. This is illustrated in Fig. 17-7. Mathematical modeling showed that the mechanism is electroosmotic. Transfer of electrolyte from the interior of the cells to the gasket areas is schematically explained by the same figure. If a film of electrolyte extends from the interior to the exterior of the cell along an electronically conducting component in contact with an anode, the gas atmosphere on the outside of the stack (normally cathode gas or cathode gas exhaust) will see a local anode potential on this component. This gas will accordingly be reduced to carbonate ion at the outside of the stack. To complete the electronic circuit, an equivalent amount of anode gas is oxidized in the anode compartment, hydrogen and carbonate ion giving water and CO_2. The ionic circuit is completed in the electrolyte film, through which cations must migrate to preserve electroneutrality. Hence, it appears that electrolyte has been transferred from the inside to the outside of the cell. This theory was confirmed by experiment, which showed that the rate of transfer from the inside to the outside of the cell would depend on both the electronic and ionic conductivities involved; that is, on the ionic film thickness and on the electronic resistance in the anode component extending to the outside of the stack, as well as on the external gas atmosphere.

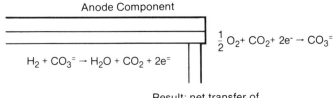

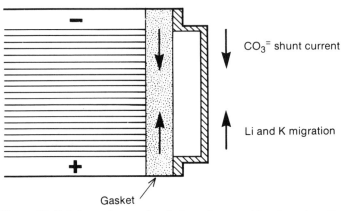

Figure 17-7: Schematic transfer of electrolyte outside of anode cell component and electroosmotic migration from the positive to the negative end of the cell stack via porous dielectric manifold gaskets.[2243]

Once outside of the cells and in the porous dielectric represented by the manifold gasket, the electrolyte would creep to the negative stack endplate under the influence of the electric field gradient present between the cells. Again, the rate of creep depends on the flow resistance inside the manifold, and also on the field strength. The latter (in V/m), depends on the cell potential and on the difference between cells, so is relatively fixed. Otherwise, the rate of transfer is fixed by the rate of carbonate formation outside the stack, on the rate of wicking of this carbonate into the gasket, and finally on the flow resistance of the gasket (i.e., on its thickness and porosity).

All the preceding can be controlled. In the 1981~1982 stack, which was intended for use as a prototype of a 100-cell 0.1 m² stack, randomly selected components were used with large dimensional tolerances that were intended to represent production line hardware. In consequence, to accommodate these tolerances, the manifold gasket was much thicker than in previous stacks. Accordingly, the electrolyte loss also was much greater than in previous stacks, in which this transfer mechanism was not noticed.

Molten Carbonate Fuel Cells

Later stacks have been built to better tolerances, with improved wet-seal areas to reduce the rate of transfer to the outside of the cells. Since transfer depends on manifold capacity, it should become less important as cell electrolyte inventory increases relative to the latter; that is, with cells of larger area. The use of tighter manifold seals will clearly reduce electroosmotic flow, but this is not always easy to achieve in a large stack which has not been precision machined, although it has been achieved in small stacks constructed by Fuji Electric. As already discussed, these are sufficiently well machined to be able to use a *window-frame* gasket made from solid zirconia.[2092] The flexible gas manifold seals can also be filled to reduce porosity; for example, with chemically precipitated zirconia or lithium aluminate. This approach is being taken by ERC (see preceding). Finally, the problem may be considerably reduced by using internal manifolding, where porous intercell gaskets may be avoided and where the pathways may be more tortuous. However, this yet remains to be demonstrated.

After transfer of the electrolyte to the negative pole of the stack, it was observed by UTC to disappear. Part of this disappearance may have been due to slow electrolysis under the combined potential of the cells in the stack, the current being limited by the ionic resistance along the manifold material. Such electrolysis could result in formation of alkali metal ion in the vapor phase at the negative end of the stack, resulting in an aerosol of oxide or hydroxide after reaction with the surrounding atmosphere. The resulting material could then have been absorbed, virtually without trace, in the large amounts of porous insulating material surrounding the stack. This mechanism was never established with certainty.

A certain degree of electrolyte fractionation was also observed by UTC, involving a relative enrichment of potassium ions in the cells towards the top of the stack via the manifold transfer mechanism.[2023] This phenomenon is caused by an unequal flux of lithium and potassium ions under the influence of the electrical field along the manifold. This observation was interesting, since normal thinking would have considered Li^+ to be the most mobile cation in the melt.[2243] As a result of this observation, UTC attempted to eliminate the problem by increasing the Li^+ content in the initial melt so that the mobilities of both cations were adjusted to approximate equality. United Technologies considered that 70 atomic % Li^+, 30% K^+ would give approximately the optimum ratio, enabling a constancy of composition throughout the stack no matter what quantity of electrolyte was transferred from cell to cell via the gasket electroosmosis mechanism. The experiment was carried out in the sixth 0.1 m^2 stack, in late 1983 (Fig. 17–8). The result was an unqualified success, because this melt was shown to give a higher cell performance than previous compositions after lithium loss by attack on stainless steel cathode components. The average cell

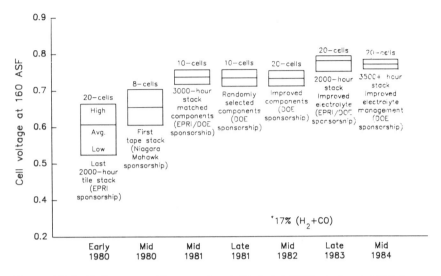

Figure 17–8: Performance of successive 0.93-dm² MCFC stacks at UTC (now International Fuel Cells., Inc; IFC) since 1980, at 650°C, 4.4 atma, with low BtU fuel (17% H_2 equivalent). Incremental improvements in overall voltage and in performance dispersion between cells at 172 mA/cm² are shown. Stacks 1–5 used 62:38 (atomic ratio) Li/K carbonate electrolyte, whereas the electrolyte composition of Stack 6 approximated to 30:70 Li/K (see Figure 7). Stack 7 operated under reduced temperature conditions for improved lifetime. The final stack (No. 8, not shown) operated for 5000 h.[2243]

potential was improved by some 60 mV, but this would have been anticipated from the results in Fig. 17–9[1939] concerning melt composition and anode, cathode and IR performance.

Since the electroosmotic pumping effect depends on the potential gradient in the stack, the electrolyte inventory in a given cell depends on its derivative; that is, on the change in the potential gradient with distance. In the center of a stack, this does not vary, so average inventory does not change with time. As Fig. 17–10 shows, inventory is only really affected in the first five and last five cells in the stack. These cells might be designed differently to accommodate these changes, for instance by the inclusion of end reservoirs, so as to prevent cell failure due to drying out (at the positive end of the stack) or flooding (at the negative). With a clever stack design, the manifold migration mechanism can indeed be used to advantage, since in principle it allows cell inventory and composition to be maintained as a function of time.

All the preceding must be borne in mind in the design of cost-effective stacks. However, a further effect must not be neglected. If a continuous

Molten Carbonate Fuel Cells

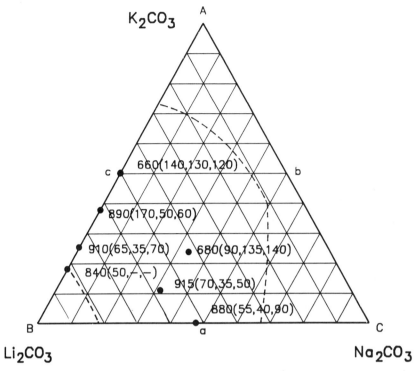

Figure 17-9: Effect of electrolyte composition on performance of cells with similar components.[1939] Figures (in mV) show cell potential and (in parentheses) IR drop and anode and cathode polarizations, respectively, at 650°C, 160 mA/cm².

electrolyte layer exists from the top to the bottom of the stack via the porous manifolds, then corrosion, as well as electrolysis of the electrolyte, can take place via the shunt current driven by the stack potential difference. United Technologies has examined this problem and has made modifications to their manifold seals and cell design on the lines proposed by Exxon in a different context.[2791] These have greatly reduced shunt currents.[1776] As a consequence of the above modifications, UTC's electrolyte-limited stack life without the use of specialized electrolyte reservoirs has risen from 2000 h to perhaps 15,000 h in the latest 20-cell 0.8 m² stack (mid-1986). This has already been discussed in Chapter 3.[1519]

Contaminants

The harmful effects of sulfur contaminants on molten carbonate fuel cells have been discussed in detail in Chapter 14, where it was stated that sulfur

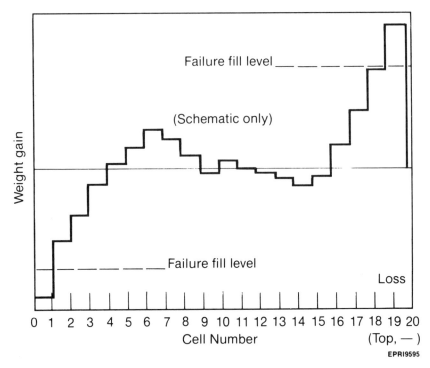

Figure 17-10: Schematic electrolyte distribution is UTC Stack No. 6 (Figure 17-8) after test termination. Electroosmotic migration (Figure 17-7) is shown, affecting mainly cells 1-3 and 17-20.

levels in the reactants should be brought down to levels below 1 ppm. Even the small amounts of sulfur-containing odorants that are added to natural gas to facilitate leak detection can be harmful and should be removed.[1518]

The effect of other possible fuel contaminants, such as ammonia, arsenic, and heavy metals that might be found in coal gas from a state of the art gasifier, is at present unknown. Work is clearly needed in this area.

Anode Creep

Anode creep (i.e., shrinkage of anode thickness that occurs when cells are under load) is a fairly recent problem that came into focus when compressive loads in stacks were considered in design evaluations. Early anodes were about 1.5 mm thick and were not stiff, so that after 500~1000 hours under a 4 atm load they were found to shrink 0.02~0.03 mm in thickness. The quoted load is typical to maintain at least a reasonable contact resistance: It should be remembered that the bottom cells of a 500-cell

Molten Carbonate Fuel Cells

MCFC stack will experience a load of 0.5 kg/cm² due to the weight of the stack alone. A non-uniformity of creep (e.g., as a result of temperature gradients across the stack) can open up gaps between the anode and the electrolyte matrix, giving a lower cell performance as a result of increased contact resistance.[1519]

Since this effect only becomes important in a cell stack, it was not observed in early single cell work. It was first recognized in about 1982, and one solution at that point was to chemically incorporate about 2~5% of microdispersed oxides (such as $LiAlO_2$) into the sintered nickel anode structure.[1964] The most recent approach is to use oxide dispersion hardened nickel sinters containing aluminum or titanium.[559, 2025, 2061] These appear to give less than 0.5% creep, which is considered to be acceptable even in 1 m² stacks having a greater than 50°C temperature gradient across the plane of the cells. This is illustrated in Fig. 17–11.

In 1980, GE recognized this problem and resolved it in a different way by developing ceramic-core nickel anode particles. They used small particles of ceramic materials such as strontium titanate or lithium aluminate, which were given a coating of nickel or copper by electroless deposition. An anode material consisting of about 75% ceramic with an outer metallic

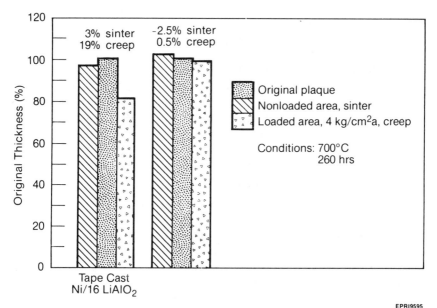

Figure 17–11: Comparison of creep of ERC tape-cast mixed Ni-16% $LiAlO_2$ anode, and dispersion-hardened Ni anode under accelerated conditions (700°C, 3.4 atma applied pressure).

surface resulted. When these particles were heated under pressure, a sintered porous anode material could be produced in which the metal particles have a ceramic core, resulting in a nonshrinking anode structure.[1207] The theory of ceramic inhibition of sintering and creep has been examined in detail.[1328]

Because of its lower melting point and surface energy compared with that of nickel, marked particle sintering and creep have been observed in the tape-cast sintered copper anodes investigated by the Institute of Gas Technology.[558] Indications are that even particle-stabilized copper anodes will require ceramic reinforcement to reduce local deformation and creep under practical stack conditions.

All-ceramic electronically conducting anodes should be completely creep-free. Some perovskite materials that have shown promise have been examined by Ceramatec in its (unsuccessful) program to produce a new cathode ceramic material under EPRI sponsorship.[1386] It will not be surprising if totally ceramic anodes are developed within the next few years.[20]

Cathode Dissolution

A serious problem with nickel oxide cathodes surfaced in mid-1981 when workers at United Technologies Corporation found small metallic nickel inclusions in the electrolyte after long duration runs of many thousands of hours in cells with thin, tape-cast electrolyte matrixes. They had not been previously observed when thick hot-pressed tile electrolytes were used. It was known that nickel oxide can react with gaseous carbon dioxide to form equilibrium solubilities of a few ppm of dissolved nickel carbonate in the electrolyte; for example, the solubility of NiO in the melt at 650°C is about 12 ppm.[2290] The dissolution probably occurs according to the reaction

$$NiO + CO_2 = Ni^{++} + CO_3^{=}$$

The equilibrium constant for this appears to be about 5.7×10^{-6} at 600°C.[2022] The Ni^{++} ion thereby formed in the vicinity of the cathode migrates through the electrolyte towards the anode under the influence of concentration and voltage gradients. Somewhere in the electrolyte between the anode and the cathode the electrical potential becomes sufficiently negative (i.e., the hydrogen partial pressure is sufficiently high) that Ni^{++} ion is reduced, eventually depositing sufficient bridged grains of metallic nickel across the cell to cause short circuiting. A variable, random front can exist across the cell in areas where hydrogen has managed to diffuse. This problem was confirmed by workers at GE, IGT, ERC, and ANL. The key factors involved in the problems of NiO dissolution are its proportionality to CO_2 partial pressure (i.e., to total system pressure), and to the reciprocal of electrolyte

Molten Carbonate Fuel Cells

thickness. In addition, the solubility of NiO may be increased by as much as a factor of ten when the cathode gas is humidified.[1369] Presumably this results from the local increase in OH^- ion content in the melt, which must influence the equilibrium concentration of Ni^{++} ion. Total disappearance of the cathode layer is never in question, since loss under any conditions would never be greater that 10% over 40,000 hours. Above all, the problem revolves around the precipitation of nickel in a thin matrix layer which itself may be thinner than the cathode. Present indications are that at atmospheric pressure, the present generation of thin cell structures may yield lifetimes of 40,000 hours, whereas lifetimes will be proportionately reduced as pressure is increased. Since the coal-gas MCFC is planned for operation at 6~10 atm pressure, the problem of the life of present generation cathodes is serious. Estimated lifetimes to shorting (4 wt% Ni in the electrolyte) are shown as a function of time in Fig. 17–12.

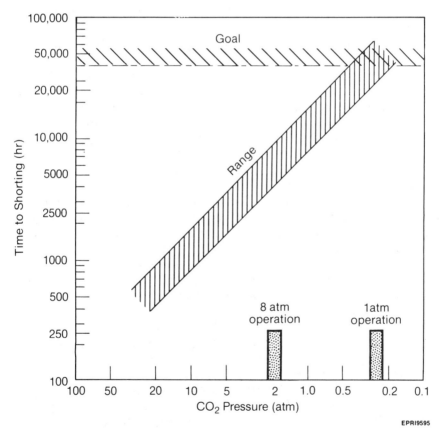

Figure 17–12: Probable life capability of Standard NiO cathode as a function of pCO_2 composition (UTC).

Several possible approaches are being studied to address this problem. The first involves developing a new range of stable cathode materials to replace nickel oxide. In addition to reasonable cost, the three main criteria for a new candidate cathode material are (a) adequate catalytic activity, (b) electronic conductivity in the range $0.1\sim 1$ $\Omega^{-1}cm^{-1}$, and (c) chemical and mechanical stability in the cell operating environment.

So far as is known, catalytic activity is not a problem at the cathode, since the peroxide reduction steps are essentially redox reactions whose rates are material independent.[139, 142, 150, 158, 2548] For conductivity and stability reasons, the material must obviously be a mixed oxide with majority and minority cations. An important factor in choosing both is that the minority cation should be able to induce electronic conductivity in the oxide by doping under typical cathode oxygen partial pression conditions. A mixed valency system is therefore required, in which one species must be a transition metal. It is important that the dopant should not differentially leach from the material, since this can cause undesirable changes in electronic conductivity.

Since dissolution and redeposition of the cathode material can cause long-term coarsening of the cathode microstructure, there is an upper limit to tolerable solubility. Close to this upper limit, both majority and minority cations should not reduce to metal in the anode environment, to avoid reductive transfer across the electrolyte matrix layer. If these thermodynamic constraints are not met, a material of negligible solubility is required. In general, lithium compounds of the type that are stable corrosion products at the fuel cell cathode are favored.

After the failure of materials such as p-doped perovskites (which were too soluble) and n-perovskites (which showed poor conductivity under cathode, though not anode, conditions[1386]), possible candidates now include electronically conducting ceramics such as Mg-doped lithium manganate (Li_2MnO_3) and Mn-doped lithium ferrate ($LiFeO_2$), both of which have resistivities of less than 5 Ω cm in air at 650°C and low solubilities in the cathode environment (respectively, ~1.3 and ~9 ppm solubility in the melt at 650°C).[1525, 1786, 2290, 2289] Cr- or Zr-doped zinc oxide also showed some promise from the standpoint of conductivity, but was found to have too high a solubility in the melt (about 200 ppm at 650°C) and tends to become incorporated into the lithium aluminate matrix.[1370] A further possibility is doped yttrium iron garnet, and possibly some magnetoplumbite structures.[1386] Some other lithium-based materials, such as doped mixed titanates-zirconates, may also be possible.[1386]

A second approach is to change the operating conditions of the cell in order to lower the solubility of nickel carbonate in the electrolyte. From the dissolution equilibrium, it is obvious that the higher the partial pressure of the carbon dioxide, the more Ni^{++} ion will be formed. Similarly, if the oxide ion content in the melt is changed by varying the cation composition, this may also reduce solubility.

Changing the standard electrolyte composition may therefore be one method to reduce nickel oxide cathode dissolution. It has been found, for example, that at 650°C, NiO is only half as soluble in a 38 mol % lithium carbonate eutectic mixture as it is in the standard 62% eutectic mixture.[1369] Alternatively, additives such as Mg^{++} also seem to be effective. Work is being actively pursued in this direction at the present time.

Another possible approach is to simply lower the partial pressure of carbon dioxide in the cathode gas from about 25% to about 10% and to use a greater degree of recycle. However, this will give some penalty in the cell voltage; though with careful design, UTC believes that the overall efficiency of a pressurized coal-based system will be only slightly affected, since more waste heat from the fuel cell will then be available for use in a bottoming cycle.[1173]

The final philosophy is to simply regard nickel oxide dissolution as a nuisance which must be tolerated, after making a few minimal adjustments to fuel cell operating conditions; for example, by using nonhumidified cathode gas, which is in any case advantageous from the viewpoint of reversible cell potential. Since the thickness changes that occur are such that the cathode will continue to function over 40,000 h, all that must be done within the cell itself is to prevent nickel granules from forming near the anode and thereby possibly causing a short-circuit. IGT has examined this approach in some detail. Post-mortem analyses of many cells have shown that nickel particles form in empty interstices in the matrix, i.e., those unfilled by carbonate. This is not surprising, since the melt itself does not wet pure nickel. In addition, the reaction is controlled by nucleation rate, and requires hydrogen supersaturation compared with the thermodynamic equilibrium value. Further, it was noted that little nickel nucleation was seen in old cells that did not incorporate an anode-side bubble pressure barrier, in which the melt would be expected to have a higher hydrogen partial pressure. The concept of IGT is therefore to add a seeded nickel nucleation region to the electrolyte matrix near the cathode, and to examine operation of the cell at an anode gas overpressure. Normally, cells operating with exhausted cathode gas on the outside use a small cathode gas overpressure, for safety reasons.

One of the preceding approaches will most probably be effective. In any case, nickel oxide will be effective as a cathode material in early small units operating at atmospheric pressure. Large pressurized units, which may require a substitute cathode material, will not be placed in production until the small units have proved themselves, which will take several years. This gives some breathing space for development of a nickel oxide substitute.

Other Cathode Problems

A further problem of NiO cathodes, and perhaps of any similar ceramic material, is that under the anticipated conditions of end-plate load the material may not be physically strong enough to resist cracking. Cathodes formed by the *in situ* oxidation of nickel plaques are structurally weak, since the sintered nickel precursor microstructure tends to break into clusters of smaller particles with no interconnections.[2022] Cathodes have often been reinforced with nickel or type 316 stainless steel mesh or screen, incorporated during the cathode fabrication. Better fracture strength has been obtained using 316 stainless steel for the reinforcement material as opposed to nickel, since the latter slowly oxidizes completely to NiO during cell operation.

Cathodes formed by *in situ* oxidation also tend to swell during formation, thereby disrupting the wet seals between the cell components and the electrolyte matrix. Reinforcement with stainless steel mesh greatly reduces such swelling. There is some evidence that the addition of 5 atomic % silver and 5 atomic % cobalt also somewhat improves the endurance and performance of nickel oxide cathodes, but these compositions are only being examined by one developer—Hitachi.[1449]

Tests conducted at ANL showed that when porous sintered nickel plaques are oxidized in the absence of lithium ions, the original pore structure is more or less maintained, no macroscopic expansion occurs, and the plaques remain flat. Furthermore, it was found that such pre-oxidized cathode structures were easily lithiated either in-cell or out-of-cell with no significant dimensional changes. As a result of these findings, Argonne now uses pre-oxidized and *in situ* lithiated cathodes. This technique has completely eliminated earlier problems of cathode expansion causing wet seal leakage.

F. COST OF MOLTEN CARBONATE FUEL CELLS

The ultimate cost of commercial high-temperature fuel cells (HTFCs) is even less certain than that of phosphoric acid fuel cells, because of the uncertainties resulting from their less advanced state of development. However, because they can consume carbon monoxide as a fuel, and are capable of directly converting natural gas to hydrogen within the cell, early HTFC systems will be significantly simpler than those using phosphoric acid electrolyte. This should result in relatively lower plant costs, possibly by as much as a factor of two. It appears that the cost of the actual fabricated carbonate fuel cell stacks will be comparable to, or even somewhat less than, that of phosphoric acid fuel cell stacks.[19]

Since there are no particularly expensive or exotic materials in a molten carbonate fuel cell, the stack cost is determined essentially by its specific performance (kW/m^2 and kW/kg) and manufacturing cost. If proposed performance goals can be met, the projected mature-production capital costs might be (in 1986 dollars) $120/kW for the fuel cell itself (10 atm, 40,000 h life); that is, cells and container only, without valves, pipes, controls, and so on. This is equivalent to $150/$m^2$, of which $80/$m^2$ or $8/kg would be materials cost. A total plant capital cost of perhaps $800/kW is not unreasonable.[12] An internal reforming unit, with its simpler chemical engineering system, might cost less than $500/kW.[182]

Table 17-4 shows the estimated cost (in 1982 dollars) of a 50 kW direct methane-fueled on-site IRMCFC fuel cell system designed by ERC.[1980] This atmospheric pressure system would have a stack containing about

TABLE 17-4

Estimated Cost of ERC 50 kW Methane-Fueled Direct Molten Carbonate On-Site Fuel Cell[1980]

		$/kW*
Fuel Cell Stack:		
anodes	55	
cathodes	18	
bipolar hardware	41	
electrolyte tiles	3	
auxiliaries	13	
assembly	50	
Total Stack Cost		180
Fuel Processor		30
Heat Exchangers		100
Inverter		100
Controls and Miscellaneous		140
Assembly		110
Total Manufacturing Cost		660
Total Installed Cost		1025

*1982 dollars

200 cells, each of active area about 0.18 m^2 operating at 200 mA/cm^2 and at a voltage somewhat ambitiously estimated at 0.75 V, giving an overall system HHV efficiency of 52~55%. The figure of $660/kW represents the estimated cost of the unit as manufactured in a limited production facility. Other costs, including installation, were estimated at $365/kW.

In a subsequent study, ERC investigated the feasibility of their internal reforming fuel cell for a larger sized plant of 2 MW, also to be used for dispersed generation, see Fig. 8–3 for flow diagram.[1981] This natural gas-fed baseline system would have an estimated heat rate of 6450 BTU (HHV)/kWh (53% overall efficiency) and operated at atmospheric pressure, at a current density of 160 mA/cm^2, at a unit cell voltage of 0.73 V, with 90% overall methane utilization (83% per pass), and 45% fuel recycle. The total plant investment was estimated to be $1230/kW (1983 dollars) on a limited (i.e., preproduction) basis. This results in capital charges equivalent to about 12 mills/kWh. Table 17–5 shows some costs and other characteristics for this plant, which could be commercially available by 1991. Many items used are commercial catalog components, except for the fuel cell stacks and the high-temperature blower. The system is by no

TABLE 17-5

Estimated Costs of ERC 2 MW Dispersed Generator Methane-Fueled Direct Molten Carbonate Fuel Cell System[1981]

Unit size, rated net electrical output (MW)		1.8
Heat rate (BTU/kWh; higher heating value basis)		6450
Preconstruction and licensing lead time (years)		1~2
Capital Costs ($/$kW_{a.c.}$)*	Total plant investment	1230
	Start-up, inventory, land	26
	Working capital	40
	Total capital requirement in service	1296
Operation and Maintenance Costs	Fixed ($/kW-yr)	14
	Incremental (mills/kWh):	
	variable, incl. fuel cell replacement	18.8
	consumables	0.6
First year cost of electricity† (mills/kWh)		83.7

*Constant January 1983 dollars
†At a fuel price of $7.76/$10^6$ BTU (HHV) in 1991

Molten Carbonate Fuel Cells

means optimized, and its multiplicity of small stacks requires an excessive amount of structure and piping. The purpose of this design was to see if a system of this simple type, sharing the same stack as that of a small on-site unit, could be a commercial proposition in limited production. It would appear to be so, and thus this approach may represent a potential market strategy for a developer. An optimized IRMCFC system in mature production can be expected to be much cheaper.

General Electric Company designed a 675 MW MCFC power plant operating on Illinois Basin #6 coal.[2097] This system was planned around a Texaco-type entrained bed gasifier, and would have an overall design electrical efficiency of 49.9%. GE's studies indicated a plant equipment investment cost of about $1200/kW (mid-1981 dollars), which includes materials, labor, engineering, sales taxes, and process and project contingencies, but not cell replacement expenses and capital charges. Using a plant capacity factor of 70%, calculations indicated a projected first-year cost of electricity (COE) of 60~65 mills/kWh. The analysis showed that if the replacement cells are capitalized, 75% of the COE value depends on fixed equipment costs while only 25% depends on variable operating costs, and that the cost of the coal used (which is related to plant efficiency) does not strongly affect the COE. Table 17-6 shows the calculated distribution of the fixed and operating costs. The costs just cited are all summarized in Table 17-7.

TABLE 17-6

Breakdown of Cost of Electricity from General Electric 675 MW Coal-Fired Molten Carbonate Fuel Cell System[2097]

		%	
Fixed Costs	Fuel processing	19.5	
	Fuel cell replacement	13.5	
	Bottoming system	10.5	
	Fuel cells	9.0	
	Stack–module piping, etc.	9.0	
	Land	5.2	
	Inverters	3.8	
	Miscellaneous	4.5	75.0
Operating Costs	Coal	16.8	
	Miscellaneous	8.2	25.0
			100.0

*Based on first-year costs.

TABLE 17-7

Estimated Capital Costs of Molten Carbonate Fuel Cells

Item	Estimated Cost ($/kW)	Reference No.	Year
ERC 50 kW on-site internal reforming system*	1025	1922	1982
ERC 50 kW internal reforming system, stack only	660	1922	1982
ERC 2 MW internal reforming dispersed generator	1296	2519	1983
General Electric 675 MW coal-fired system	1200	2001	1982

*Natural gas-fueled; total installed cost

CHAPTER 18

Solid Oxide Electrolyte Fuel Cells

A. INTRODUCTION

As the name implies, solid oxide electrolyte fuel cells (SOEFCs) utilize a solid oxide, usually doped zirconia, as the electrolyte. When this material is heated to about 1000°C, it becomes sufficiently conductive to oxide ions to serve as a solid state electrolyte. The first use of a solid oxide electrolyte was by Nernst in 1900 in his *glower* light source (see Chapter 10 and Ref. 1861), which used zirconia doped with 15% yttria. The first solid oxide electrolyte fuel cell was built by Baur and Preis in 1937.[269] This cell used magnesia- or yttria-stabilized zirconia at temperatures greater than 1000°C, but was able to sustain current densities in the order of only 1 mA/cm^2.

The real pioneering work on the practical SOEFC was begun in the early 1960s by workers at the Westinghouse Research and Development Center in Pittsburgh, and at Brown-Boveri and Cie in Heidelberg, Germany. The cells built by these workers were the first to achieve current densities of greater than 100 mA/cm^2 using hydrogen and oxygen. The basic components of the present high-temperature SOEFC were developed by Westinghouse Electric Corporation through a project supported by the U.S. Office of Coal Research, ending in 1972.

Principles of Operation

Figure 18-1 shows the basic principles of operation of a high temperature SOEFC operating on hydrogen fuel and oxygen oxidant. At the cathode, oxygen reacts with incoming electrons from the external circuit, ionizing to form oxide ions, which migrate to the anode through the zirconia electrolyte. At the anode, hydrogen reacts with these oxide ions to produce water vapor, which is accompanied by the liberation of electrons to the external circuit. The overall process is simply the reaction of hydrogen with

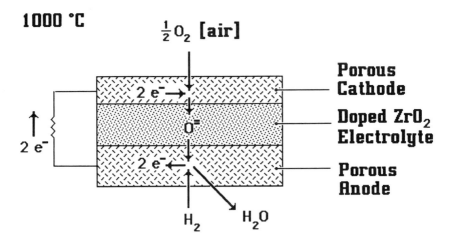

Figure 18–1: Principle of operation of a high temperature solid oxide electrolyte fuel cell.

oxygen to produce water, and it involves no secondary species that must be recycled, as is the case with CO_2 in the MCFC. The reversible voltage for the process is given by the Nernst equation:

$$\Delta E_{cell} = \Delta E°_{cell} - \frac{RT}{2\mathcal{F}} \ln \left\{ \frac{(pH_2O)_{an}}{(pH_2)_{an} \cdot (pO_2)^{1/2}_{cat}} \right\} \qquad [18\text{–}1]$$

Depending on the actual composition of the fuel, water vapor, and oxidant, the open circuit voltage of a SOEFC is usually 0.8~1.0 V.

Since oxide ion is the only conducting species, and the oxygen partial pressure on the fuel side is at the extremely low thermodynamic value representing that for the dissociation of water vapor, the SOEFC can be viewed as an oxygen concentration cell.[2129] Thus, the chemical equilibrium constant for the reaction

$$H_2 + 1/2\ O_2 \rightleftharpoons H_2O$$

is given by

$$K = \frac{pH_2O}{pH_2 \cdot pO_2^{1/2}} \quad [18\text{-}2]$$

However, since

$$\Delta E° = -\frac{\Delta G°}{n\mathcal{F}} \quad [18\text{-}3]$$

and

$$\Delta G° = -RT \ln K \quad [18\text{-}4]$$

then substitution of Eq. 18-2 into Eq. 18-4; of Eq. 18-4 into Eq. 18-3; and of Eq. 18-3 into Eq. 18-1 gives

$$\Delta E_{cell} = \frac{RT}{2\mathcal{F}} \ln\left\{\frac{K \cdot (pH_2)_{an} \cdot (pO_2)_{cat}^{1/2}}{(pH_2O)_{an}}\right\} \quad [18\text{-}5]$$

In the anode compartment, under reversible conditions, the dissociation equilibrium between water vapor, hydrogen, and oxygen will be given by

$$H_2O \rightleftharpoons H_2 + 1/2\ O_2$$

The equilibrium constant, K_D, for this dissociation is given as

$$K_D = \frac{(pH_2)_{an} \cdot (pO_2)_{an}^{1/2}}{(pH_2O)_{an}} = \frac{1}{K},$$

from which

$$\frac{K \cdot (pH_2)_{an}}{(pH_2O)_{an}} = \frac{1}{(pO_2)_{an}^{1/2}} . \qquad [18\text{-}6]$$

Substitution of Eq. 18-6 into Eq. 18-5 yields

$$\Delta E_{cell} = \frac{RT}{2\mathcal{F}} \left\{ \frac{(pO_2)_{cat}^{1/2}}{(pO_2)_{an}^{1/2}} \right\} \qquad [18\text{-}7]$$

which is the equation for a reversible oxygen concentration cell with an oxide ion transport number of unity, and which involves the half-cell reactions:

Cathode	$1/2 O_2 + 2e^- \to O^=$
Anode	$O^= \to 1/2 O_2 + 2e^-$
Cell	$1/2 O_{2\,(cathode)} \to 1/2 O_{2\,(anode)}$

Thus, from an academic viewpoint the driving force for the SOEFC is seen to be the difference in oxygen partial pressure on either side of the electrolyte. For water vapor dissociation at 1000°C, K_D has a value of ~4 × 10^{-10}. If the partial pressures of the hydrogen and water vapor in the anode compartment are equal, then from Eq. 18-7, the partial pressure of the oxygen in the anode compartment is of the order of ~1.6 × 10^{-19} atm. Taking air at 1 atm as the cathodic reactant, Eq. 18-7 gives an open circuit value of 1.14 V.

Solid oxide electrolyte fuel cells also can use carbon monoxide as well as hydrogen as a fuel. While there is some tendency for the carbon monoxide to force-shift, even at the high operating temperature, with direct consumption of H_2, that is,

$$CO + H_2O \to CO_2 + H_2,$$

in the presence of water vapor, it can also undergo the direct anodic reaction

$$CO + O^= \to CO_2 + 2\,e^-,$$

where the effective oxygen partial pressure in the anodic compartment is determined by the equilibrium

Solid Oxide Electrolyte Fuel Cells 583

$$CO_2 \rightleftharpoons CO + 1/2 O_2$$

Since catalytic effects are not important at the high operating temperature of the SOEFC, in principle, hydrocarbons and other organic compounds can also be oxidized, but there will always be a tendency for them to crack, giving carbon and hydrogen (Chapter 8). As in the MCFC, cracking can be prevented by steam injection, but the net result is essentially internal reforming, followed by rapid oxidation of the hydrogen and carbon monoxide produced.

B. FEATURES OF SOLID OXIDE ELECTROLYTE FUEL CELLS

In addition to those common to all fuel cells, there are several features of solid oxide electrolyte fuel cells that make them very attractive for utility and industrial applications.[843, 844, 1160, 1252]

First, the very high operating temperature (~1000°C) ensures that all fuel compositions, when combined with water vapor when necessary, will oxidize rapidly and spontaneously to thermodynamic completion if sufficient air is provided on the cathode side. The high temperature of reaction does not require expensive catalysts, and permits the direct processing of the fuel in the fuel cell itself, thereby adding to the simplicity of the fuel processing subsystem. Solid oxide systems are able to attain at least 96% of the theoretical voltage at open circuit, any losses being due to minor cross-leaks and to the effect of the small electronic conductivity of the electrolyte. In the absence of major IR-drop, they will operate at much higher current densities than molten carbonate fuel cells with high conversion of the fuel.

Since a solid oxide electrolyte fuel cell operates via transport of oxide ions rather than that of fuel-derived ions, in principle, it can be used to oxidize any gaseous fuel. Like the molten carbonate fuel cell, it can consume carbon monoxide as a fuel, but it does not require recycling of carbon dioxide from the anode to the cathode, leading to a further simplification of the system compared with that of the MCFC.

High-temperature operation and a tolerance to impure fuel streams (see following) may make solid oxide systems especially attractive when combined with coal gasification plants. The heat released in the fuel cell reactions can be efficiently transferred and used for coal gasification or hydrocarbon reforming. In consequence, no fuel need be burned to provide the heat of reforming, as would be the case in alkaline electrolyte or phosphoric acid systems.

Because the solid oxide electrolyte is normally very stable, no electrolyte migration problems under cell operating conditions exist as in the

MCFC.[843] In addition, the composition of the electrolyte is not a function of fuel or oxidant composition. Because no liquid phases are present, there are no three-phase interfaces to maintain, no problems with pore flooding, and no catalyst wetting problems. Solid oxide cells also have a good tolerance to overload, underload, and even to short-circuiting. The latter has been regularly demonstrated with 1984 technology Westinghouse cells, whose anodes can be totally oxidized after cutting of the fuel stream, then regenerated by contact with fuel gas without performance degradation.[496]

Another advantage inherent in an SOEFC is that cell components can at least in principle be fabricated into a variety of self-supporting shapes and configurations that may not be feasible with cells employing liquid electrolytes. Through the use of innovative designs, the self-supporting nature of the solid oxide electrolyte makes it potentially feasible to achieve very high power densities, since a large quantity of extra mass or volume may not then be required for use as cell containers and support structures. This is particularly true of the *monolithic* SOEFC cell discussed later. The potentially high power-to-weight and power-to-volume ratio for such SOEFC systems may eventually enable their use in mobile power sources as well as in those for stationary applications. Solid oxide cells also operate independently of gravity, which is desirable for traction and aerospace applications. Provided that the cells are well designed and component materials are carefully matched, for thermal expansion, solid oxide fuel cells should tolerate thermal cycling more easily than molten carbonate cells because no phase transitions occur between the ambient and operating temperature.

Since high-temperature water electrolysis is both thermodynamically and kinetically better than low-temperature electrolysis, solid oxide cells can be operated in reverse, as high-temperature water electrolyzers capable of operating at current densities in excess of 500 mA/cm^2 at cell voltages in the order of 1.3 V. This makes solid oxide systems attractive for use as energy storage devices and for peak power leveling applications.

Tolerance to Contaminants

Experiments have shown that solid oxide electrolyte fuel cells operate equally well on dry or humidified hydrogen or carbon monoxide fuel, or on their mixtures. In addition, the high operating temperature considerably reduces the possibility of catalyst poisoning. It is claimed by Westinghouse that the SOEFC nickel-cermet anode has a high tolerance towards sulfur; for example, the presence of 50 ppm by volume of hydrogen sulfide in the fuel lowers the cell operating voltage by only about 5%, apparently without causing any permanent damage to the cell, since the original performance can be restored when the impurity is removed from the fuel.[496]

Nickel anodes can react with sulfur-containing impurities in the fuel to form nickel sulfide, the minimum tolerance to sulfur occurring at local Nernst potentials of 0.7~0.8 V, where nickel is more readily oxidized. Furthermore, the resistance of nickel to sulfidization appears to be somewhat greater in hydrogen-steam fuel mixtures than in carbon monoxide–carbon dioxide mixtures. However, the tolerance increases with increasing temperature. Thus, in a typical fuel consisting of 25% H_2–H_2O and 75% CO–CO_2, at 700°C the tolerance of a nickel anode to sulfur is only ~5 ppm, whereas at 1000°C it is ~90 ppm. Therefore, at 1000°C, nickel or cobalt cermet anodes have been assigned sulfur tolerances of 90 and 200 ppm by volume, respectively.[843]

These sulfur tolerance levels are one to two orders of magnitude higher than those for other types of fuel cells, and permit the use of high-temperature sulfur removal methods, which are more energy efficient than the low-temperature methods that are required by the MCFC to lower the sulfur content of the gas to less than 1 ppm. This tolerance to fuel with a high impurity content may make the SOEFC very attractive for operation on heavier fuels such as diesel oils and, more importantly, coal gas. Moreover, for power plants operating on natural gas, it may completely eliminate the need for desulfurization.

Typical SOEFC System Configuration

Figure 18–2 shows a natural gas–fueled SOEFC system configuration being developed by the Argonne National Laboratory.[19] For thermodynamic reasons, the overall efficiency of a SOEFC system may be somewhat lower than that of a MCFC. Solid oxide systems may, however, be better able to follow load variations and have fewer system components, though they will often have a much larger number of small cells. In the system shown in Fig. 18–2, for example, it has been possible to omit the hydrodesulfurizer because of the high sulfur tolerance of the SOEFC; furthermore, since the basic chemistry of a SOEFC is different from that of a MCFC, it is also unnecessary to provide for the recycle of anode exit gas to the cathode. In the system shown, natural gas first is mixed with steam to stabilize against carbon formation, and then is preheated to about 700°C prior to entering the fuel cell stack. The residual fuel leaving the stack at 1000°C is burned. These hot exit gases are used first to preheat the incoming air to the cathode and then to generate steam, some of the latter being added to the incoming fuel and the remainder used to preheat the fuel. This

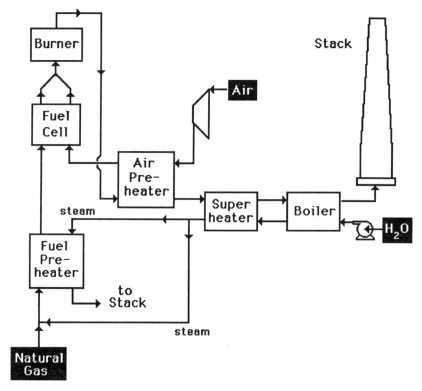

Figure 18–2: Natural gas-fired solid electrolyte fuel cell system configuration designed by Argonne National Laboratory.[19]

simplified chemical engineering system may possibly lead to a lower cost of electricity than that obtainable from a MCFC system, provided stack costs are comparable in both cases.

At a fuel utilization of about 90% and 0.65 V/cell, the overall electrical efficiency of the system is about 45%, based on the higher heating value of the fuel and with no bottoming cycles. Owing to its simplicity, such a system presumably would have a significantly lower capital cost than that of a corresponding MCFC system, again with the proviso that stack costs are comparable in both cases.

C. SOLID OXIDE FUEL CELL PROGRAMS

The energy crisis of 1973~1974 led to a renewed interest in the United States in fuel cells as the most efficient way to utilize coal. The Energy Conversion

Solid Oxide Electrolyte Fuel Cells

Alternatives Study, referred to in Chapter 3, indicated that high-temperature solid oxide electrolyte fuel cells could potentially yield the lowest cost of electricity of all fuel cells.[295, 490] In 1976, ERDA (now DOE) renewed its support of the SOEFC program. The goal of the program was to establish the technical feasibility, performance, and life of series-connected, multicell stacks of thin-film solid oxide electrolyte fuel cells. Westinghouse Electric Corporation was the main contractee to develop the cell technology, with various national laboratories and agencies being involved in more fundamental research on materials and electrode kinetics.

The original Westinghouse-DOE program was divided into four phases as follows:

Phase I (1976~1978): Demonstrate the feasibility of the thin cell concept and improve the durability of the interconnection material.

Phase II (1978~1980): Lay down the specifications of all cell components, fabrication techniques, and sequencing; demonstrate at least 1000 hours life at 1000°C and 400 mA/cm^2 of the interconnection material with less than 10% voltage degradation in a 10-watt laboratory stack; design, fabricate, and test a 30-watt 20-cell stack; fabricate and test ten stacks to demonstrate reproducibility.

Phase III (1981~1989): Design, build, and test 1-, 10-, and 100-kilowatt modules; demonstrate at least 10,000 hours life.

Phase IV (1990~2000): Design, build, and demonstrate a 5-megawatt coal gasifier–SOEFC power plant.

Reductions in the U.S. energy budget over the past several years have somewhat lowered the solid oxide program goals. Since about 1980, annual funding has been more or less constant at the two million dollar level, with a small level of supporting research being directed towards the investigation of fundamental cell phenomena and alternative concepts. It appears likely, however, that with the recent successes of the Westinghouse thin layer cell, the program will expand over the next few years.[19]

In Heidelberg, West Germany, workers at Brown-Boveri and Cie A.G. have been involved in the development of solid oxide electrolyte fuel cells since the early 1960s. The earlier Brown-Boveri work made use of self-supporting zirconia tubes, but now supported thin-film cells are being developed.[2128]

In Japan, researchers at the Electrotechnical Laboratory of the Agency of Industrial Science and Technology are developing SOEFC systems as part of the government-supported Moonlight Project. The goal of the Japanese research is to establish the technology for thin-film cells to be used in highly efficient generating systems for the effective use of limited energy resources.

D. SOLID OXIDE FUEL CELL COMPONENTS

The materials used in solid oxide electrolyte fuel cells and their required properties have been discussed in considerable detail in Chapters 10 and 13. Here, only brief mention will be made of the materials that are currently in favor:[20, 495, 843]

Electrolyte: Yttria-stabilized zirconia deposited on a calcia-stabilized zirconia support tube by electrochemical vapor deposition (see following) has been found to provide the best combination of high oxide ion conductivity and other physical, chemical, and electrical properties for use as the electrolyte. Yttria or calcia doping is required to stabilize the cubic crystal structure and to create the oxygen vacancy defects required for the enhancement of ionic transport. In the case of yttria-stabilized zirconia, it has been found that the maximum oxygen ion conductivity is obtained when the Y_2O_3 concentration is 16.9% by weight.

Cathode: Porous strontia-doped lanthanum manganite perovskite of composition $La_{1-x}Sr_xMnO_3$ (0.10 < x < 0.15). The strontium doping confers p-type electronic conductivity through the creation of electron holes. The porosity is required to facilitate mass transfer of oxygen to the cathode-electrolyte interface.

Anode: A cermet of metallic nickel or cobalt with yttria-stabilized zirconia. The functions of the zirconia phase are to support the nickel or cobalt catalyst, to inhibit coarsening of the catalyst particles, and to provide an anode thermal expansion coefficient compatible with those of the other cell materials. The final cermet contains 20~40% interconnected porosity to facilitate the transport of the reactant and product gases.

Interconnection material: The requirements for the interconnection material are perhaps the most stringent of those for all the components. In addition to the general requirements of any material for high-temperature operation, the interconnection material (ICM) must possess good electronic conductivity at 1000°C since it serves as the cell electrical current collector. Also, since it must be impervious to both fuel and oxidant gases, the ICM must be capable of being fabricated into a thin, dense layer able to withstand the stresses encountered during both fabrication and cell operation. Being exposed to both cathode and anode environments, the ICM must be chemically stable over an extremely wide range of oxygen partial pressures (10^{-18} ~ 1 atm) without forming any highly resistive phases as a result of chemical reaction or interdiffusion with the anode, cathode, or other materials. At the present time, magnesium- or strontium-doped lanthanum chromite

($LaCrO_3$), formed by electrochemical vapor deposition, appears to best satisfy these criteria.

The different components require different preparation conditions in terms of atmospheres and temperatures, such as for sintering. The cells must therefore be constructed in such a way that successive preparation steps for individual components do not affect the properties of those that have been previously prepared. This requires both cell designs and fabrication processes of considerable ingenuity.

Thin Film Technology

Various SOEFC configurations have been examined. The earliest designs used flat plate disks, but this configuration presented severe sealing problems. Next, batteries were formed by joining cells, each consisting of individual cylinders, together in bell-and-spigot configurations to give a system electrically connected in series. Because the current must be conducted along the plane of each cell, there is an upper limit to size with this configuration because of the electronic IR-drop in the current collector. Generally, a cell length of about 0.5 cm was used, so that with typical tube diameters each cell only produced about 0.1 W. This required a multiplicity of components. The configuration also was difficult to seal, as well as being expensive to manufacture. Self-supporting zirconia electrolyte tubes were also employed. These were made by special pressing and sintering techniques, had wall thicknesses greater than 0.3 cm, and hence high resistances. In addition, they were expensive and unsuitable for mass production techniques.

As a consequence of these difficulties, current designs make use of the so-called thin wall concept, in which thin films are deposited on porous support tubes. These methods result in total cell thicknesses on the order of only 100–200 μm. Such thin cells have lower internal ionic resistances, thereby giving higher efficiencies and opening the possibility for operation at somewhat lower temperatures than the 1000°C customarily employed. Thin wall designs also have low material and manufacturing costs, are suitable for automated production, make efficient use of raw materials, and are easily assembled into large multicell arrays. By analogy with lower temperature cells, these are usually called stacks, even though they are not stacks in the true sense (i.e., they do not consist of assemblies of flat elements). The aforementioned problem of current collection must always be carefully considered in these systems. Essentially all solid oxide electrolyte fuel cells now use the thin film concept.

It has already been remarked that unlike most other fuel cells, for which the various components can be fabricated separately (employing an unlimited range of processing conditions) and subsequently assembled to give complete cells, for thin wall SOEFCs it is not possible to select the processing conditions for each component independently. Since each thin component is layered sequentially over the previous component, the sintering temperature for each succeeding layer must be lower than that of the preceding layer to avoid altering the microstructure of the preceding layers.

The technology of materials formulation and of the thin film deposition techniques involved in the fabrication of SOEFC components is now in a fairly advanced stage, largely through the efforts of the research undertaken by the Westinghouse Electric Corporation. The analyses of the performance of SOEFC systems, however, still contain a large element of uncertainty, because long-term thermal and electrical performance of a full-size stack have yet to be demonstrated. For this reason, the funding and the state of the art for the SOEFC, which is usually regarded as third-generation technology, are not at as high a level as for the molten carbonate fuel cell (second-generation technology). However, the initial performance of small-scale stacks looks sufficiently promising, and the detailed design, engineering, and construction of generator-sized hardware are now starting to be undertaken.

Example of the Thin Wall Concept

Since the properties of the materials used in thin wall cells are sensitive to impurities, clean processing methods must be employed during their fabrication. Accordingly, precise control of the operating conditions is required to obtain proper compositions and structures. Important parameters in fabrication include surface preparation, the partial pressures of the reactant gases, the temperature, and the source and nature of the substrate.

The following example will clarify the principles of the thin wall concept by outlining the details of the fabrication of one of Westinghouse Electric Corporation's earlier solid oxide electrolyte fuel cell stacks, shown in Figure 18–3.[1247]

Support Tube

In this particular design, a zirconia tube stabilized with 15% calcia was employed as the support tube (Fig. 18–4A). This was formed by the

Solid Oxide Electrolyte Fuel Cells

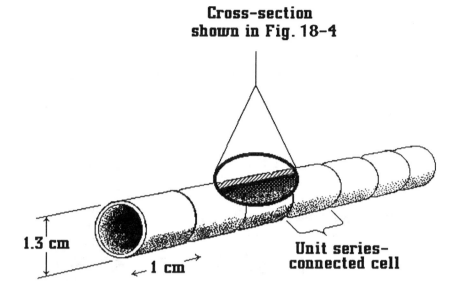

Figure 18–3: Seven-cell stack of Westinghouse tubular supported thin-wall solid oxide electrolyte fuel cells (conventional design).[1247]

extrusion of a plasticized material of controlled grain size after mixing in a pug mill, followed by firing at 1800~1900°C to give a strong, porous sintered body. The final support tube was 1~2 mm thick, 1.3 cm in diameter, and of about 30% porosity, containing pores of 2~10 μm dimensions.

Application of Anode

To apply the anode layer, the support tube was first surface-sanded and then dipped in a slurry of NiO and yttria-stabilized ZrO_2 (NiO/ ZrO_2 ratio ≥ 2/3). After evaporating the solvent, this layer was sintered at 1600~1700°C (Fig. 18–4B). The sintering time is critical to prevent diffusion of the NiO into the support tube, which would impart stress and induce

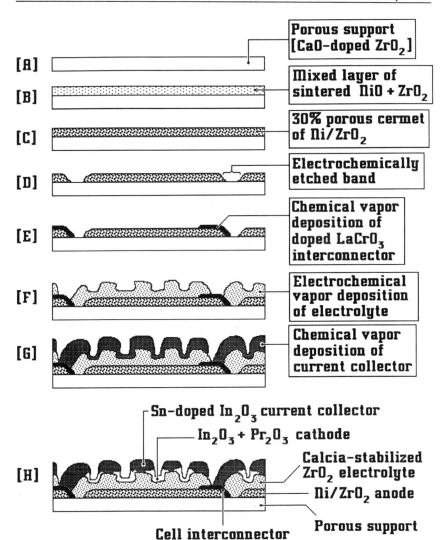

Figure 18-4: Steps in the fabrication of early Westinghouse thin-wall solid oxide electrolyte fuel cell.[1247]

electronic conductivity. Next the sintered material was reduced at a temperature lower than the melting point of the metal in a stream of hydrogen at about 1350°C to form a 20 μm-thick 30% porosity nickel-zirconia cermet (Fig. 18-4C). In order to obtain the proper cermet structure, water or carbon dioxide was added to the hydrogen stream to ensure that the reduction takes place at high oxygen activities. Sintering at 1350°C prevents subsequent sintering of the anode structure during cell operation at 1000°C.

Application of the Interconnect Material

The anode cermet layer was next appropriately masked and electrochemically etched to produce a band-ring structure that would form the basis for the final multicell structure (Fig. 18–4D). The etched bands were about 1 mm wide, and the cermet-covered areas about 9 mm wide.

After further masking, an approximately 4 mm-wide band of 20 µm-thick (Mg + Al)–doped lanthanum chromite ($LaCrO_3$) interconnector material was then formed by chemical vapor deposition (Fig. 18–4E). The layer is laid down by the following vapor phase reactions:

$$2CrCl_2 + 3H_2O \rightarrow Cr_2O_3 + 4HCl + H_2$$
$$2LaCl_3 + 3H_2O \rightarrow La_2O_3 + 6HCl$$
$$MgCl_2 + H_2O \rightarrow MgO + 2HCl$$
$$2AlCl_3 + 3H_2O \rightarrow Al_2O_3 + 6HCl$$

The laying down of the interconnection material is one of the critical steps in the thin cell formation process. To achieve good intercell connection, the seal must be gas-tight and must not overlap too much over active area of the cell.

Application of the Electrolyte

The electrolyte consists of a 20 µm-thick layer of doped zirconia, laid down using a combination of chemical vapor deposition (CVD) at 1150°C for pore closure and electrochemical vapor deposition (EVD) for film growth (Fig. 18–4F). The corresponding reactions are as follows:

Pore closure (CVD):
$$ZrCl_4 + 2H_2O \rightarrow ZrO_2 + 4HCl$$
$$2YCl_3 + 3H_2O \rightarrow Y_2O_3 + 6HCl$$
Film growth (EVD):
$$H_2O + 2e^- \rightarrow H_2 + O^=$$
$$ZrCl_4 + 2\,O^= \rightarrow ZrO_2 + 2Cl_2 + 4e^-$$
$$2YCl_3 + 3\,O^= \rightarrow Y_2O_3 + 3Cl_2 + 6e^-$$

The EVD step is a key element in Westinghouse's fabrication technology, since it allows formation of dense films of conducting oxides by relying on their ability to show a small amount of electronic transport along with majority oxide ion transport. A layer of zirconia particles was also incorporated during this step to make the upper layer of the electrolyte film porous.

During the pore closure reaction stage, care was required to prevent the zirconia support tube being attacked by CVD reactants or reaction products, which could leach CaO from the stabilized zirconia support tube, resulting in loss of mechnanical strength. Typical reactions that might be involved in this attack include:

$$CaO + 2HCl \rightarrow CaCl_2 + H_2O$$
$$2CaO + ZrCl_4 \rightarrow ZrO_2 + 2CaCl_2$$
$$3CaO + 2LaCl_3 \rightarrow La_2O_3 + 3CaCl_2$$
$$2CaO + 2CrCl_2 + H_2O \rightarrow Cr_2O_3 + 2CaCl_2 + H_2$$

Application of the Cathode Current Collector

The electrolyte film was then powder-masked and a porous layer of tin-doped indium oxide (In_2O_3) with 5~10 µm pore diameter was deposited by chemical vapor deposition at 600~800°C (Fig. 18–4G):

$$2InCl_3 + 3H_2O \rightarrow In_2O_3 + 6HCl$$
$$SnCl_2 + H_2O \rightarrow SnO + 2HCl$$

Formation of the Catalytic Cathode

The porous electrolyte layer under the indium oxide current collector was next impregnated with a solution of praseodymium nitrate, $Pr(NO_3)_3$, which is thermally decomposed to form praseodymium oxide, Pr_2O_3, when the tube is heated to operating temperature (Fig. 18–4H). Praseodymium oxide was used to reduce cathodic polarization.

The complete cell, shown in Fig. 18–4H had a thickness of about 100 µm (excluding the support tube).

Electrochemical Vapor Deposition

The briefly discussed electrochemical vapor deposition process, which is capable of forming a dense uniform film of zirconia electrolyte to the desired thickness, is an interesting thin film technique that incorporates both chemical and electrochemical reactions.[1247] To form the film, an

Solid Oxide Electrolyte Fuel Cells

equimolar mixture of hydrogen gas and water vapor is passed through the porous support tube at 1200~1300°C under low pressure conditions, while the appropriate metal chloride vapors surround the outside of the tube. Typically, the metal chloride vapor, $MCl_{2(g)}$, may be formed from the metal oxide by a reaction such as

$$MO_{2(s)} + 2C_{(s)} + Cl_{2(g)} \rightarrow MCl_{2(g)} + 2CO_{(g)}$$

The delivery rate of the metal chloride is controlled by regulating the flow of chlorine over a mixture of granular oxide-lampblack mixture. The final cation composition of the deposit is close to that in the source oxide materials.

As indicated in Fig. 18-5, zirconium chloride vapor reacts anodically with oxide ions from the zirconia substrate at the outer surface of the tube to deposit solid zirconia, with the release of gaseous chlorine:

$$ZrCl_{4(g)} + 2O^= \rightarrow ZrO_{2(s)} + 2Cl_{2(g)} + 4e^-$$

At the inner surface of the tube, oxide ions are regenerated cathodically by the reaction of water vapor:

$$2H_2O + 4e^- \rightarrow 2H_2 + 2O^=$$

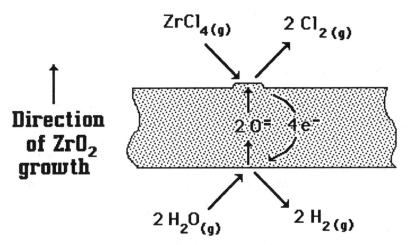

Figure 18–5: Principle of electrochemical vapor deposition of zirconia.[1247]

Other possible cathode reactions include the reduction of carbon dioxide or of oxygen:

$$CO_2 + 2e^- \rightarrow CO + O^=$$
$$O_2 + 4e^- \rightarrow 2\,O^=$$

The net result of this rather unique process is the formation of a dense film of zirconia oxide–ion-conducting electrolyte on the outer surface of the tube, made possible by the small, but not negligible, electronic conductivity of the material. Such films are stabilized in the face-centered cubic structure and are self-leveling. The zirconium chloride vapor is occasionally doped with 5 mol % cerium to increase (by a factor of about 3) the electronic conductivity of the ZrO_2 deposit in order to permit more rapid growth rates. The partial pressure of oxygen at the growing surface is kept very low (~10^{10} atm) by the use of a reducing atmosphere to drive the $O^=$ ions through the zirconia so that growth rates are as high as possible.

The preceding descriptions are given to show the method by which Westinghouse thin-film cells were constructed before 1980 in as much detail as possible. Many of the technical aspects are still proprietary, particularly those concerning the EVD operation. The steps used for the present generation of cells are described in Section F, though necessarily in less detail because of their even more proprietary nature. It should be apparent to all readers that the fabrication procedures involve many complex steps. So far, no costs are available, but it is apparent that they should be less than 4¢/cm² (including materials) if the system is to be competitive. This corresponds to about $300/kW for the fuel cell stack elements. Whether this can be achieved is unknown at present.

E. PERFORMANCE OF SOLID OXIDE FUEL CELLS

The electrode kinetics of solid oxide electrolyte fuel cells depend on component morphologies, which in turn are strongly affected by fabrication methods. Depending on the conditions, the kinetics can be controlled by the adsorption or desorption of reactants or products, by surface diffusion on the electrode, by diffusion in the electrolyte, by oxygen diffusion in the bulk of the electrode, or by electronic defect migration in the electrolyte.

Tables 18–1 and 18–2 show some single cell voltage-current characteristics for recent Westinghouse thin film solid oxide fuel cells.[1252, 2171] These cells differ in geometry from those outlined in the previous section, and are described in Section F. The data in Table 18–1 show that a cell operating at a constant 80% fuel utilization at 1020°C on a typical coal-derived fuel

TABLE 18-1

Single Cell Voltage-Current Characteristics of Westinghouse Thin Film Solid Oxide Electrolyte Fuel Cell: Effect of Air Vs. Oxygen[1252]

Temperature: 1020°C
Fuel: (67% H_2 + 22% CO + 11% H_2O), constant 80% utilization
Oxidant: Air, 4 × stoichiometric
Active area: 80 cm²

	Open Circuit		160 mA/cm²		320 mA/cm²	
	Air	Oxygen	Air	Oxygen	Air	Oxygen
Cell voltage, V	0.775	0.825	0.655	0.720	0.480	0.585
IR-drop, V	—	—	0.105	0.105	0.210	0.210
Polarization*, V	—	—	0.065	0.000	0.085	0.030

*(O.C.V.) - (Cell Voltage) - (IR-drop)

TABLE 18-2

Single Cell Voltage-Current Characteristics of Westinghouse Thin Film Solid Oxide Electrolyte Fuel Cell: Effect of Fuel Composition[2171]

1020°C
Air oxidant

	160 mA/cm²		320 mA/cm²	
	H_2 Fuel*	H_2-CO Fuel**	H_2 Fuel	H_2-CO Fuel
Cell Voltage, V	0.775	0.727	0.495	0.445
IR-drop, V	0.155	0.155	0.310	0.310
Polarization, V	0.120	0.140	0.250	0.275
	1.050	1.022	1.055	1.030

*(97% H_2 + 3% H_2O); O.C.V. = 1.055 V
**(38.8% H_2 + 58.2% CO + 3% H_2O); O.C.V. = 1.023 V

consisting of (67% H_2 + 22% CO + 11% H_2O), and on air at 25% utilization may achieve an output voltage of about 0.655 V at 160 mA/cm².[1252] This voltage corresponds to 120 mV below the theoretical exit Nernst voltage of 0.775 V, the inlet Nernst voltage for this fuel and oxidant being 0.97 V for a co-flow arrangement. More recent cells have achieved voltages of at least 0.67 V at 160 mA/cm² under the same conditions.[843] Cell performance has been stable over the thousands of hours of operation thus far achieved.

Because of the high value of the term $RT/\mathcal{F}$ in the Nernst equation at the cell operating temperature, the open circuit voltage of a SOEFC is strongly dependent on fuel composition. A comparison of the open circuit data presented in Tables 18–1 and 18–2 clearly illustrates this point. It should be remembered that since the electrodes are equipotential surfaces, the theoretical open circuit voltage in a practical cell operating in a co-flow configuration can be simply calculated using the outlet gas composition. Whereas inlet fuel compositions usually yield calculated open circuit (Nernst) voltages of about 1.0 V, outlet compositions usually yield values closer to 0.8 V. For this reason, the data presented in Table 18–1 are more representative of commercial operation than those presented in Table 18–2, since in the former, the percentage fuel utilization has been kept at a constant value of 80% by appropriate variation of the fuel flow as a function of current density. Thus, while the cell output voltages in Table 18–1 are lower than those in Table 18–2, performance data in Table 18–1 actually are superior to those in Table 18–2 because both the polarization and ohmic losses are lower.

Operation using carbon monoxide instead of hydrogen as fuel always appears to result in a reduction in cell performance. For example, the output from a thin film Westinghouse cell operating at 1020°C and 0.775 V dropped from 160 mA/cm^2 on pure hydrogen (97% H_2 + 3% H_2O) to 143 mA/cm^2 when operated on (38.8% H_2 + 58.2% CO + 3% H_2O), corresponding to a 10.6% drop in the current density.[2171] Alternatively, at a constant current density of 160 mA/cm^2, the cell voltage was reduced from 0.775 V to 0.727 V when the pure hydrogen fuel was replaced with the hydrogen–carbon monoxide fuel (see Table 18–2). This can of course largely be explained by the differences in the ΔG values between hydrogen and carbon monoxide. When CO-rich fuels produced by coal gasification processes are used, carbon deposition must be avoided, such as by the injection of steam or by the addition of small amounts of ammonia or sulfur compounds.[2171]

Although the rate of fuel oxidation at the anode depends on the electrode material and on the presence of surface oxide films, in general, the anodic reactions of hydrogen and carbon monoxide at 1000°C are considered to exhibit negligible activation or concentration polarization in state-of-the-art thin film cells. The performance of the cathode, however, presents more serious problems. Although activation polarization at the cathode is considered negligible, cathode performance can be hindered by mass transfer limitations if care is not taken. The data in Table 18–1, for example, show that when the cell is operated using pure oxygen, which is impractical in commercial cells, there is negligible cathodic polarization. However, operation on air results in a cathodic polarization of 65 mV at 160 mA/cm^2. This indicates some mass transfer limitation (concentration

Solid Oxide Electrolyte Fuel Cells

polarization) at the cathode, resulting from diffusion losses in the porous support tube–cathode combination. Continuous effort is being made to reduce this problem by seeking ways to decrease the thickness of the support tube (currently ~1 mm thick) and of the air electrode (currently ~0.7 mm thick), at the same time increasing their porosity. However, as the structure described in the next section will show, decreasing the air electrode thickness will be counterproductive in regard to overall IR-drop.

The data in Table 18–1 show that when the air electrode concentration polarization and the cell internal resistance drop is added to the cell voltage, the theoretical open circuit voltage is attained. This confirms that the anode exhibits negligible polarization, hence requires no further optimization in performance.

In state-of-the-art SOEFCs of no matter what technology, the major source of voltage loss is the internal electrical resistance within the cell. In earlier cell designs that utilized self-supporting zirconia tubes as electrolyte, the resistance of the electrolyte itself contributed a major portion of the ohmic drop. Table 18–3, for example, shows a breakdown of the various voltage losses in a Brown-Boveri and Cie 120-cell solid oxide fuel cell stack of this type.[2129] This design necessarily required a very thick electrolyte (producing almost half of the total ohmic drop), which led developers to

TABLE 18–3

Voltage Losses in 120-Cell Self-Supported Solid Oxide Electrolyte Fuel Cell Stack[2129]*

Operating temperature, °C:	1000		
Current density, mA/cm^2:	500		
Fuel/Oxidant:	H$_2$-CO/air		
			% of IR-drop
Ohmic drop, V:	electrolyte	0.148	46.3
	anode	0.025	7.8
	cathode	0.098	30.6
	ICM	0.049	15.3
		0.320	100.0
Electrode polarization:		0.347	
Single cell voltage:		0.375	
Open circuit voltage:		1.042	

*Self-supported tubes of 2.2 cm diameter. Unit cell length = 1.2 cm; thickness of zirconia electrolyte = 0.05 cm.

abandon self-supporting systems in favor of thin film cells. By way of contrast, data obtained from an early 7-cell stack of Westinghouse thin film cells operating at 0.64 V, 400 mA/cm^2, and 990°C indicated that in the thin film design, only about 12% of the total ohmic drop occurred in the electrolyte and anode combination.[2327]

The data of Table 18–3 also show that ohmic drop in the anode is small, whereas that in the cathode and interconnect are of major significance. In the Westinghouse thin film 7-cell stack,[2327] the cathode and interconnect, respectively, accounted for 31% and 61% of the total ohmic drop.

In state-of-the-art cells, the resistivity of the interconnection material has been reduced to an acceptable level, so that most of the ohmic drop now occurs at the cathode. In cells of the new Westinghouse configuration discussed in the following section, which are inverted compared with those previously described (i.e., the cathode is on the inside of the tube, not on the outside), great care must be taken to control the processing conditions during deposition of the zirconia electrolyte over the cathode. This is because the high-temperature halide vapor atmosphere can react with the doped lanthanum manganite cathode to form undesirable compounds that seriously increase the interfacial resistance at the cathode-electrolyte interface.[495] Fortunately, the formation of these compounds can be avoided or minimized by exercising careful control of the processing conditions during the initial phase of electrolyte nucleation.

F. STATE-OF-THE-ART SOLID OXIDE FUEL CELLS

Westinghouse Solid Oxide Fuel Cells

The conventional design for a SOEFC battery has been one in which a number of narrow cylindrical cells are joined end-on-end, and electrically connected in series to build up a stack of cells of a desired output, as shown in Figs. 18–3 and 18–4. Although this design has been useful in establishing thin-film technology and performance capability, fabrication and configurational problems do not make it optimal for practical systems. As well as the inherent IR loss involved in current collection along the surface of the multiplicity of narrow cells, a further major problem encountered was the inherent susceptibility of the ceramic-to-ceramic contacts to thermal stress.

In 1980, Westinghouse Electric Corporation developed a new design in which the electrical connections and power take-offs use low cost metallic components, since they are in the reducing atmosphere at the anode side, rather than at the cathode.[495, 843, 1252, 495] In this new design, each tube

Solid Oxide Electrolyte Fuel Cells

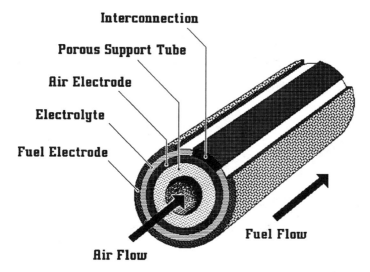

Figure 18–6: Westinghouse tubular solid oxide electrolyte fuel cell.[495]

consists of a single cell, as shown in Fig. 18–6. Unlike the earlier concepts, the air is supplied to the cathodes from on the inside of the tubes, the fuel being provided on the outside, allowing easy connection between cells in the reducing environment. It is interesting that this approach was not considered earlier on the grounds that it would require oxygen diffusion through the porous support tube to the cathode, which was thought to result in higher mass-transport polarization than that for hydrogen diffusion to the anode. This inverse cell arrangement has proved to be an unqualified success.

In this design concept, densely packed bundles of long single cells are connected in series or parallel using external metallic conductors in the external anode environment. These connections are ductile at the high cell operating temperature, and eliminate any ceramic-to-ceramic contacts between individual cells (see Fig. 18–7). Fig. 18–8 shows a side view of this configuration in a fuel cell power reactor.[1252] The fuel gas (i.e., hydrogen and carbon monoxide) surrounds the tubular cell and flows axially from one end of the reactor to the other. The cell tubes are closed-ended at the point where fuel gas enters. As the fuel flows over the active surface of the tubes, it of course becomes depleted, and the exiting fuel gas finally flows through a porous alumina felt diffusion barrier into a combustion com-

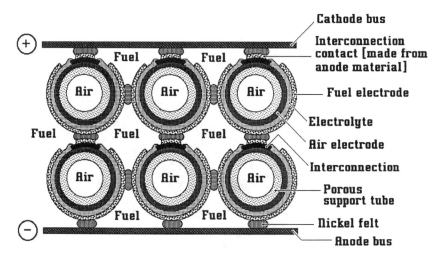

Figure 18–7: Schematic cross-section of Westinghouse tubular solid oxide electrolyte fuel cell bundle.[495]

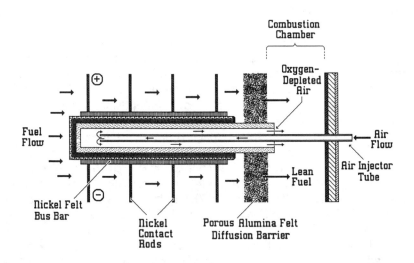

Figure 18–8: Side view of Westinghouse tubular solid oxide electrolyte fuel cell in a fuel cell power reactor.[1252]

partment, where it is burned with the spent air stream. Cell cooling is carried out using an excess flow of air, and the sensible heat contained in the effluent gases combined with combustion of exhaust fuel, is used to preheat the air entering the cell. In effect, this configuration integrates a high-temperature heat exchanger and the fuel cell bundle, without the need for any gas-tight seals. The use of a leaky porous seal to separate the combustion chamber from the fuel-rich zone minimizes thermal stresses on the ceramic tubes. Representative performance data from this type of cell are given in Table 18–1.

The closed-ended support tube is made in the same way as that described previously for the interior-anode system, sintering being conducted at 1650°C. The same materials as before are used for the interconnect, electrolyte, and fuel electrode, allowing a good match of thermal expansion characteristics. A new air electrode material, Sr-doped lanthanum manganite, was developed to replace tin indium oxide, which proved to be too volatile. The air electrode material is applied by filtration from a powder suspension via a partial vacuum maintained on the inside of the support tube. Sintering to form the electrode is then conducted in an oxidizing atmosphere at 1350°C. The tube is then axially masked, allowing deposition of the linear dense interconnect layer by CVD and EVD. The interconnect is then masked for deposition of the thin (~50 µm) electrolyte layer, which wraps around the edges of the interconnect as shown in Fig. 18–7. The critical phase in the fabrication of the cells is pore closure in the air electrode by the electrolyte during the process of deposition of the latter by successive CVD and EVD processes, following the method described earlier. The nickel anode layer is then applied as a slurry after masking of the interconnect and the borders of the electrolyte, and a further CVD-EVD process is then used to grow more doped zirconia electrolyte in the pores of the anode layer. This acts as a sintering inhibitor and maintains the correct three-phase boundary characteristics.

Each tube in experimental cells of this design is 30 cm long with an active zone of 21 cm; in the final commercial design it is anticipated that the tubes may be as much as 100 cm long. However, both mechanical aspect ratio in manufacture, and air flow and distribution down the center of the operating tube may make this goal difficult to achieve. Since current collection in the cells is radial, there is an upper limit to tube diameter; otherwise, IR-drop in the cells will become too great and will lead to uneven current distribution, most of the current coming from the active parts of the cell that are close to the interconnect. A diameter of 1.7 cm has been established as optimal, so that aspect ratio gives an upper limit to tube length. However, IR losses are still quite high, and consideration may in future be given to the use of thicker and more porous electrodes (to avoid any increase in mass-transport polarization), and possibly new materials of improved conductivity. An electronically conducting support tube may even be a possibility, although materials cost might then be a problem.

Westinghouse cells of this new seal-less design have been operated for hundreds of hours with essentially flat performance. Cells have been thermally cycled and have been shown to be able to tolerate abuse. For example, the anode will tolerate overload to the point of oxidation; after reduction it shows its original performance. In addition to continuing research on the development of component materials of improved characteristics, the major development thrusts now are to prove the design for long-term operation and to verify all heat and mass flows as well as electrical and mechanical performance. With DOE support, Westinghouse has now initiated detailed design, engineering, and construction of generator hardware, initially consisting of bundles of 24 cells, each with 21 cm active length. Typical performance on reformed methane at 90% utilization is 0.65 V at 160 mA/cm^2, although certain cells have delivered as much as 0.69 V under the same current density conditions. Each cell therefore produces about 12 W, or 290 W for the 24-cell bundle. Future large cells may be able to produce 50 W each.

Although it is too soon to be able to predict the cost of SOEFC tubes with any accuracy, the materials costs are reportedly of the order of \$50/kW (1982 dollars), and that of the support tubes about \$10/kW. The tubes will deliver a maximum of only 50 W each, and even this upper limit is doubtful. The tubes require a multiplicity of complex fabrication steps, involving many masking operations. Present cost of the preproduction system is about \$1000/W, and cost-reduction effort similar to that for photovoltaic cells, with robotic fabrication, will be required. Eventual costs similar to those predicted for similar ceramic items may be eventually attained. For example, sintered alumina tubes for sodium vapor lights may cost on the order of 10¢/cm^2, while the β-alumina electrolyte tube for the sodium-sulfur battery is predicted to be 1¢/cm^2.[2661] The latter is controlled by market requirements, and is equivalent to \$35/kW or \$100/kW for the electrolyte alone. If SOEFC tubes can be made for costs approaching the lower limit of this window (corresponding to fabrication added cost between \$100/kW and \$1000/kW), the future of the tubular SOEFC should be assured. However, the ß-alumina electrolyte is a much less complex object than the completed SOEFC cell, so much remains to be done.

Westinghouse has privately stated that its studies have shown that based on a production of 150,000 30-cm tubes/day (1.8 MW/day), the manufacturing cost can be reduced to 35¢ per tube, or \$30/kW, giving a total tube cost of (with starting materials) \$80/kW. If this can indeed be achieved, then the Westinghouse aim for a total mature production cost of about \$300/kW may be within reach. Whether large practical multimegawatt systems can be built up from a multiplicity of small cells from the chemical engineering viewpoint also remains to be seen; and their use in coal-fired systems depends on the possibility of pressurized operation, which has not as yet been demonstrated. If goals are met, the overall cost of a coal-fueled power plant may be on the order of $1500/kW.[2171]

Japanese Solid Oxide Fuel Cells

In Japan, solid oxide electrolyte fuel cells of conventional thin-film internal anode tubular configuration are being developed by the Electrotechnical Laboratory as part of the Moonlight Project.[1846] The Japanese originally tested bell-and-spigot systems, but owing to the difficulty of fabricating thin layers of electrolyte, they later switched to thin layer type cells on porous alumina support tubes, the thin layers being applied by a flame spray process.

For their cells, the workers at the Electrotechnical Laboratory have used 71 cm long by 21 mm O.D. and 16 mm I.D. porous alumina tubes of 20% porosity for supporting the cell components. The support material has a thermal expansion coefficient of $8.0 \times 10^{-6} K^{-1}$. Calcia-stabilized tubes have thermal expansion coefficient values of about $11.5 \times 10^{-6} K^{-1}$, which better matches the values of the other structural components of the cell, but are not practical because of their susceptibility to thermal shock during the flame spraying process. The active fuel cell portion of each assembly occupies the middle 32 cm of the tube.

In cell fabrication, a 100 µm layer of alumina powder is first sprayed onto the outer surface of the alumina tube using a d.c. arc plasma spray process. This layer is applied to ensure gas-tightness of the inter-connecting portions of the cells. Next, NiO powder is sprayed onto the surface using an acetylene-oxygen thermo-spray process to form a layer of about 100 µm thickness, the layer later being reduced to metallic nickel with fuel gas to form the anode. Yttria-stabilized zirconia $((ZrO_2)_{0.92}(Y_2O_3)_{0.08})$ powder of 15 µm mean particle size is then applied using a plasma spray process to produce a 150 µm-thick electrolyte layer. The interconnects consist of a plasma-sprayed Ni-Al underlayer and a $LaCrO_3$ upper layer, each of about 100 µm thickness. The upper layer is omitted for cells used for short duration (100 hours) tests. Next, a 150 µm-thick cathode layer is applied using a thermo-spray process, using either $LaCoO_3$ or strontium- or calcium-doped lanthanum cobaltite. This is followed by the application of Ni-Al powder electrical current leads by spraying. These are spray-coated with alumina powder to produce oxidation-resistant gas-tight layers. Soldered connections are made to the end caps at each end of the tube to conduct the electric current to the water-cooled metallic jackets, which hold the cell stack in place, guide the fuel gas to the inside of the stack, and conduct the exhaust gases (residual fuel and oxidation products) away from the stack.

The fuel cell stacks are assembled vertically in a furnace, the oxidant being supplied from the top and the hydrogen fuel supplied through a top jacket, while the water vapor and residual fuel gases are exhausted through a bottom jacket. Table 18–4 shows the hydrogen-oxygen and hydrogen-air performance of a demonstration battery consisting of eleven 12-cell stacks connected electrically in series, tested towards the end of 1982. Table 18–5 shows the output characteristics of three stacks of $LaCoO_3$–yttria-stabilized zirconia–Ni cells that were tested for 100 hours.

TABLE 18–4

Ouput Characteristics of Japanese 132-Cell Solid Oxide Electrolyte Fuel Cell Battery [1846]

	Operation on H_2-O_2		Operation on H_2-Air	
Maximum power output (W)	260		200	
Battery performance	Volts	Amperes	Volts	Amperes
	118	0	112	0
	92	2	79	2
	65	4	51	4
	42	6	24	6
	16	8	10	7

TABLE 18–5

Characteristics of Japanese Solid Electrolyte Fuel Cell Stacks Used for 100-Hour Testing [1846]

	12-Cell Stack	16-Cell Stack	20-Cell Stack
Effective stack length (mm)	324	320	320
Unit active cell length (mm)	11	8	5
Unit interconnector length (mm)	12	12	11
Operating temperature (°C)	1030	1050	1030
Total active area (cm²)	87.0	84.5	66.0
Performance:			
Maximum power output (W)	49	57	47
(unit cell: V @ ma/cm²)	0.458 @ 103	0.453 @ 92.4	0.490 @ 72.8
Cell open circuit voltage (V)	0.988	1.000	1.010
Cell voltage @ 50 mA/cm²	0.708	0.664	0.637
Cell voltage @ 100 mA/cm²	0.467	0.419	0.295

Monolithic Solid Oxide Fuel Cells

Certain flat-plate SOEFC designs, using plasma-sprayed electrolyte layers and for example, electrodes, arranged in a bipolar array with proprietary corrugated metal bipolar plates have been described,[1180, 1181] and other variants are under study.[1679] However, since 1983, Argonne National Laboratory (ANL) has been working on a flat-plate SOEFC of a radically new design, supported by DOE and the Defense Advanced Research Projects Agency (DARPA).[844] The Argonne cell, termed a *monolithic fuel cell*, employs the thin ceramic layer components of existing SOEFCs in a strong, lightweight honeycomb structure that promises to give much higher energy and power densities than conventional configurations by the reduction of IR-drop. Figures 18–9 and 18–10 illustrate the monolithic concept. In it, fuel and oxidant are conducted through alternating passages in the corrugated structure, which is composed of thin layers (25~100 μm) of active cell components fabricated into the unique honeycomb shape

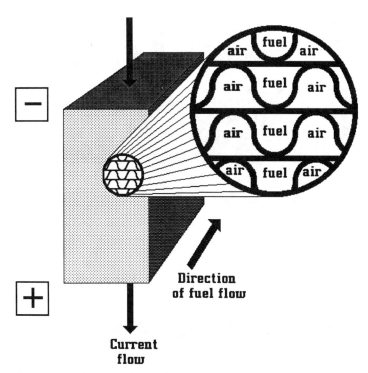

Figure 18–9: Argonne National Laboratory monolithic solid oxide electrolyte fuel cell concept.[844]

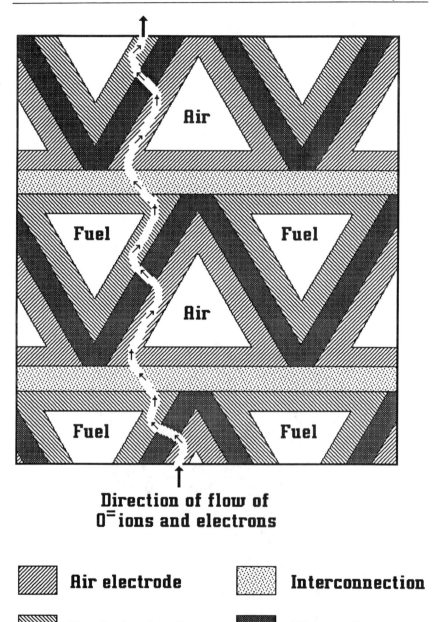

Figure 18–10: Schematic diagram illustrating flow of electrons and $O^=$ ions through Argonne National Laboratory monolithic solid oxide electrolyte fuel cell.[844]

possible only with a solid electrolyte. The corrugations also form the gas seals at the edges of the structure. The system can be co- or counter-flow.

Because individual cells in this design are only 1~2 mm in thickness, the active surface area per unit volume becomes very large, about 10 cm^2/cm^3, compared with values of about only 1~2 cm^2/cm^3 for conventional fuel cells. The small size of the unit cells not only increases the power density over conventional cells by a factor of at least five, but also decreases the voltage losses due to IR-drop by decreasing the current path length in the electrodes, as shown in Fig. 18-10.

Because of the low internal resistance and high Nernst driving force in the region of the fuel inlet in a co-flow design, the current density is much higher at the inlet region of the cell than it is near the outlet. As pointed out earlier, cell voltage losses in a solid oxide electrolyte fuel cell are primarily determined by the IR losses in the cell. Since the fuel electrode is an equipotential surface, the low current co-flow current density near the outlet of the cell will be desirable because only a small IR loss (1~5 mV) will occur near the cell outlet, so that the performance of the fuel electrode will indeed approach the maximum theoretical exit Nernst voltage. As a result of the variation in current density through the stack, each cell can operate within a few percent of maximum efficiency even at high power output.

The high potential power density obtainable from a monolithic cell, which results from the small cell pitch, the self-supporting compact container design, high specific active surface area, low internal resistance, and therefore high current density, is a very strong incentive towards its development. The principal challenge is to develop the technology required to fabricate the unique structures involved. Thus far, attempts have mainly used tape-casting techniques similar to those used for forming the matrix layer in the molten carbonate fuel cell, followed by sintering. Since sintering temperatures and atmospheric requirements for each component are different, a better approach might be to adapt some of the techniques used by Westinghouse or other methods, such as those used by Japanese researchers described in the last section. Some aspects of this fabrication technology have been reviewed.[182]

Potentially, the monolithic technology may be able to operate at an average current density four times higher than that of Westinghouse cells, and at 50 mV, greater cell potentials. Military applications are obvious, even those for one-shot applications with no thermal cycling requirement. This accounts for DARPA's present interest in the technology. However, civil applications are also of great potential interest. As an example, a cell stack with a volume of about 1 ft^3 (28.3 L), weighing less than 30 kg, could produce 30 kW (nominal) power output from a reformable methanol-water mixture. Calculations show that the sintered honeycomb should be capable of withstanding thermal cycling because of the relatively well-matched thermal expansion characteristics of the individual layers, and could be started up from cold by burning fuel in a few minutes. In this respect it is important to note that it has the structure of a ceramic heat exchanger, and can be manifolded in the same manner. Having the advantages of material costs of less than $10/kW, with very high compactness and power density, it could represent the twenty-first century replacement of the internal combustion automobile engine using methanol or even hydrogen as fuel, provided that its technical fabrication problems can be overcome.

G. PROBLEM AREAS AND FUTURE RESEARCH DIRECTIONS

Based on relatively short duration tests, the materials just described appear to be reasonably stable and compatible. Cells have been subjected to one-hour thermal cycling from room temperature to 1000°C many times without any apparent loss in performance. Whether or not they will endure the targeted life of 40,000 h of commercial operation has yet to be established. For example, some interaction has been observed to take place at the interface between the cathode and other cell components. The deleterious interdiffusion of other components during long-term operation at 1000°C also is of concern. On the more positive side, older technology Brown-Boveri cells showed good maintenance of performance using materials similar to those described over more than 35,000 h of operation.[2128]

The conductivity of the cathode, and also of the other cell components, could stand improvement, since ohmic losses account for the principal voltage loss mechanisms in solid oxide electrolyte fuel cells. This is especially true of the Westinghouse and other thin film systems, less so in regard to the monolithic concept. For the latter, the development of suitable fabrication techniques will be most important, followed by detailed design; for example, of suitable manifolding, conductive end-plate and bus configurations, and thermal start-up techniques. For the supported thin-film configurations, an important area of research will be to increase the strength and thermal shock resistance of the electrolyte support tube, at the same time reducing its resistance to gas phase transport to allow operation at higher current densities.

In addition to the necessity of continuing development of more stable and more highly conductive component materials with correct thermal properties, it will be eventually necessary to develop much less complex and less expensive fabrication techniques and processes for all cell designs. Some improvement in the oxidation resistance of the anode may be desirable to make it more resistant to fuel distribution fluctuations that occasionally may be encountered within cells.

Solid Oxide Electrolyte Fuel Cells

It is apparent that most of the problems being encountered with solid oxide electrolyte fuel cells are those particular to high-temperature materials, as opposed to those involving electrochemical kinetics or mass transport limitations. In particular, the development of more highly conductive materials permitting operation at lower temperatures would not only alleviate many of materials problems currently seen, but might also make solid oxide fuel cells attractive for transportation applications. Interdiffusion between components, particularly of impurities, and its effect on local electrical properties requires further study. Some of this has been recently reviewed.[495]

Rapidly expanding progress in ceramics technology shows great promise for application to solid oxide fuel cells.[20] For example, recent developments in the chemical processing of ceramic powders, such as the use of organometallic precursors, sol-gel synthesis methods, and evaporative decomposition of solutions, offer potential for improving the purity, particle distribution, morphology, homogeneity, and sinterability of the ceramic powders used in the manufacture of SOEFC components. Similarly, developments based on the use of transformation toughening and the use of second-phase particulates as crack arrestors are likely to find applications for increasing the fracture toughness and thermal shock resistance of SOEFC components. Furthermore, the extension of recent innovations in ceramics coating technology, such as the use of sol-gel processing, electron-beam evaporation, chemical vapor deposition, and sputtering, are likely to result in both fabrication advantages and improvements in materials. Thus, many of the thick-film, thin-film, and multilayer processing technologies that have been developed for use in the microelectronics industry are likely to result in important improvements in light-weight, high power density fuel cell systems in the near future. We should also finally point out that the materials for the high T_c superconductors first reported in late 1986 are very similar to the perovskites used in the SOEFC, although they are based on Cu, rather than Mn or Cr. This discovery promises rapid progress in thin film ceramic manufacturing technology.

CHAPTER 19

Conclusion

INTRODUCTION

It is customary to apologize for the fact that a book of this type can never be up-to-date. The main text was frozen in June 1987, and it has been since then impossible to make more than minor changes to both it and the references. However, in the final 15 months of preparation before publication, much has happened in regard to fuel cell power systems throughout the world. This has been very positive outside of the United States, and somewhat negative within the latter. This conclusion provides an opportunity to update the progress of technology development and fuel cell power system policy throughout the world up to September 1988. Some of this updated information was presented June 6 and 7, 1988 at the Fuel Cell Symposium, Tokyo '88 and at the International Energy Agency Fuel Cell Workshop that followed it. At the latter, it was decided that international working groups for research on the problems of the MCFC, SOEFC, and advanced AFC should be established.

AMERICAN ELECTRIC UTILITY POLICY

In the United States, the major disappointment has been the lack of a coherent electric utility program to order the International Fuel Cells (IFC, United Technologies–Toshiba) PC-23 11 MW natural gas–powered phosphoric acid fuel cell (PAFC) unit, which is envisaged as the entry-market workhorse electric utility fuel cell power system in the present text. In March 1988, IFC (working with Bechtel) published a report (E. W. Hall, L. M. Handley, and G. W. May, "Capital Cost Assessment of Phosphoric Acid Fuel Cell Power Plants for Electric Utility Applications," AP-5608, Electric Power Research Institute, Palo Alto, 1988) stating that at a manufacturing rate of 1500 MW per year (136 units), 11 MW PC-23 units should cost $1075/kW installed. The cost should descend, with increasing production, through a mass production value of $740/kW to a final figure of $600/kW in the late 1990s. The HHV heat rate of such units, reflecting the estimated requirements in Ref. 432 (p. 49) should approach 7000 Btu/kWh (48.75% efficiency), down from 8300 Btu/kWh (41.12% efficiency), the value estimated for the early models of the PC-23.

Since Ref. 432 has shown that the relatively small dispersed PAFC power plant must compete head-on in the intermediate-load market with medium-sized (100-150 MW) advanced combined-cycle units, it is important that both heat rate, and simultaneously, capital cost, should be brought down. In a change of emphasis from its previous program, the Electric Power Research Institute is actively pursuing these goals. Overall, it is clear that the usual advantages of the small, dispersible fuel cell power plant unit that have been frequently cited in the previous literature and in this text, will not on their own be sufficiently advantageous to sell the electric utility fuel cell. What is also now clear is that in the United States there is no longer a critical mass of utilities willing to purchase IFC PC-23 11 MW units because of their high initial cost and the lack of guarantees that they will operate as the developer has announced. Other factors, underlined by Ed Gillis of EPRI on April 28, 1988 in Toronto, have been: the present excess of utility baseload, de facto deregulation of the industry, the present low cost of fuel, and the excess of cheap Canadian power.

On March 30, 1988, in Atlanta, Georgia, the Fuel Cell Users Group (p. 57), after a certain number of previous delays, decided to let the sunset clause in its articles of incorporation take effect, and to disband. The general feeling was that it had presented the fuel cell concept to the electric utilities, as was required by its charter, and that it was now up to the utilities to go forward to the next step, that of commercialization. Its place will be at least in part taken over by the Industrial Fuel Cell Users Group, whose original charter was for concentration on the on-site market. The situation concerning the electric utility PAFC was summarized by several speakers at the Atlanta meeting. The original intention throughout 1986 and 1987 was that there would be an initial order of approximately five PC-23 11 MW units by U.S. utilities. Interested utilities were Puget Sound Power and Light; an East Coast consortium led by Northeast Utilities, Consolidated Edison (New York), Public Service Electricity and Gas (New Jersey), and Boston Edison; and finally an American Association of Public Utilities (APPA) group led by the city of Palo Alto, California. For various reasons, the different utilities decided not to purchase the 8300 Btu/kWh heat rate demonstration units at this time. High cost, as well as the problem of heat rate, was cited as one reason for this decision. In some cases, the non-availablity of a low-cost fuel was indicated. For example, Puget Sound Power and Light could not arrange a low-cost off-gas supply from a nearby refinery, and had no natural gas available. Since IFC said it could not use the marketing strategy outlined in Fig. 2–8 on p. 54, and insisted on selling the first units at price corresponding to real manufacturing cost, the demonstration was more than the utilities could afford.

The original cost estimate for each of the first three units was $40M (1986), or $3600/kW, for delivery in 1990–1991. This was to be followed by a cost of $20M per unit ($1800/kW) for the next 20 units. Ron Belval of the Palo Alto public utility summarized the situation as far as APPA was concerned: The cost of an 11 MW unit was quoted at $48M, with $1.2M for spares, an additional $7.5M for the alternative advanced *Configuration B* stack technology (see p. 519), $10M for operating costs, and $3.5M for the site (low for Palo Alto), giving a total of $70.2M, far beyond Palo Alto's commitment of $20M. While the existence of cheap landfill gas in Palo Alto was an advantage, one major problem was finding a site in a costly suburban environment, which would require public hearings. Certainly, the city of Palo Alto could not go it alone in effecting a demonstration, and even the APPA was too small to serve as the only potential utility PAFC market, even though there are 2100 utilities with capacities under 100 MW (95% of the total), and Palo Alto is in the upper 5% of these. Both Belval and Michael Bergman pointed out that such utilities really required smaller units with dual- or multi-fuel capability, in the capacity range 2.5-3 MW, rather than those of fixed 11 MW size. In particular, Belval remarked that the suppliers must understand and respond to the market, must offer guaranteed-quality technology, and he strongly criticized IFC for its marketing strategy and lack of understanding of the electric utility industry. He stressed the importance of the selection of the correct host utility, and stated frankly that the PAFC may *never* be cost-effective, in spite of the new Configuration B cell stack and the proposed eventual 7200 Btu/kWh heat rate. On the positive side, he added that EPRI staff had given strong support and assistance for the project, and the APPA would continue to work with EPRI to see that the technology evolves in the correct direction. For the time being, the program was being well served by the proposed demonstrations in Goi, Japan and in Italy, which are discussed later.

In complete contrast to the APPA requirements, Bill Messner of ConEd, representing a congested urban utility, stated that the 11 MW unit was too small for application in New York City, and that a 100 MW unit consisting of 25 MW modules would be closer to the ideal. This should have a much smaller power footprint (instead of 0.34 m^2/kW as in the PC-23, 0.09 m^2/kW would be preferred for a greenfield site, and 0.04-0.07 m^2/kW for a repowering site, respectively. The latter unit would be designed by ConEd and Bechtel Corporation, using IFC technology as a basis, and would be constructed in three stories with a total height of more than 20 meters. To be cost-effective, new Configuration B stacks with three times the area of those of the PC-23 (i.e., 2.8 m^2 or 30 ft^2) operating at increased pressure (13.6

atma instead of 8.2 atma) would be required in a plant costing a maximum of $985/kW (greenfield site) or $1005/kW (repowering site). Maximum operating and maintenance costs would have to be 0.68¢/kWh (greenfield site), and 0.76¢/kWh (repowering site), representing two around-the-clock operators. The plant capacity factor would be 75%. For the time being, a heat rate of 8300 Btu/kWh would be sufficient for the natural gas–powered plant.

ConEd would propose installing 2400 MW of such units in New York City, at the rate of 12 25-MW modules per year. The fuel cell power plant is considered by ConEd to be more capital intensive than combined cycles, and possibly more so than the atmospheric fluidized bed combustor (AFBC) steam cycle. However, the problem that will arise in the future is where the combined cycle and AFBC systems can be sited. In addition, the utility does not plan any further new or replacement capacity until the year 2005, so it does not require a demonstrator now.

The preceding show that the design of a large electric utility PAFC power plant should be flexible, and should be scaled for the customer's requirements. In particular, the key issue in fuel cell power plant design is now a question of engineering, and not of electrochemistry, materials, or stack technology. For example, it is expected by IFC that the desired HHV heat rate of 7200 Btu/kWh (47.4% efficiency) can be attained only using system improvements, and not electrochemical improvements allowing operation at higher cell voltage.

THE TOKYO ELECTRIC POWER COMPANY PAFC DEMONSTRATOR

The only utility that plans to go ahead at the present time with a 10-11 MW PAFC unit is Tokyo Electric Power (TEPCO), at the same Goi site used for the 4.5 MW demonstrator (Chapter 3). While IFC had previously declared that it would only supply complete PC-23 plants as a matter of policy, TEPCO has succeeded in negotiating an arrangement whereby IFC only supplies the complete cell stacks in its pressure vessels, Toshiba being responsible for the design and construction of the balance-of-plant. From their previous experience with the 4.5 MW unit, reliability and maintainability are the key factors in fuel cell operation. Accordingly, TEPCO has decided to make a unit that is less dense than the PC-23, and it will use a 100 m × 60 m (0.6 ha, 1.5 acre) site that is 25% larger than that proposed for the PC-23. An artist's rendering of the power section of this plant is shown on the jacket of this book. The unit, with 18 PC-23 stacks, is essentially designed by TEPCO, and will have the same heat rate as that proposed for the PC-23 (8300 Btu/kWh). Construction will start in October 1988, with the process and control (PAC) test planned for 1990. They emphasize that this will be a demonstrator, not a commercial unit, and it will be a test vehicle for reliability of major components, performance in different modes, the use of cogeneration heat, and environmental impact. It will be a 66 kV unit using natural gas fuel, producing less than 10 ppm of NO_x, and will have an acoustic signature less than 55 dBA at the site fence. It will be designed to operate between 30 and 100% of nominal power, and will have a cold start-up time of 6 h (2.5 h hot start-up time).

THE ITALIAN 1 MW PAFC DEMONSTRATOR

The Italian fuel cell program (Progetto Volta) is taking much the same approach as TEPCO in regard to PAFC development. The Italian Fuel Cell Working Group (chartered in March 1984: the utility AEM-Milan, Ansaldo, Belleli, Fiat, CISE, De Nora, ENEL, ENI, Gedol, Breda, ITAE-CNR, Italgas, Ministry of Defense, OTB Partecipazioni, Milan Polytechnic, Universities of Milan and Messina) planned a PAFC demonstration on about a 1 MW scale. The group made a decision to use a mature imported stack technology, and after evaluating Fuji Electric and IFC, it negotiated the purchase of two PC-23 stacks from the latter. These will be incorporated in a plant designed by KTI, using the Haldor-Topsøe reformer technology developed for Westinghouse under EPRI support (p. 523). In addition, on-site 25 kW Fuji units will be evaluated at Casaccia and Bologna. Other PAFC work in Italy is represented by the 2.5 kW unit purchased by CISE from ERC in 1982, which has been used for anode gas composition studies (e.g., sulfur tolerance) and transient response.

THE JAPANESE 1 MW PAFC DEMONSTRATORS

The latest developments in the Japanese Moonlight Program PAFC 1 MW demonstrators (p. 137-139) were clarified at the Symposium in Tokyo held on June 6 and 7, 1988. Both units have had their share of problems, similar to those in the UTC (IFC) 4.5 MW demonstrators. The Chubu Electric unit, located at the Chita power plant (two 700-MW units with a 600 m smokestack) occupies a 1800 m² site. It has four stacks, two by Toshiba and two by Hitachi, operating at 6.3 atma and 185°C. The system has a Hitachi turbocompressor, and the top-burner 19-tube Toshiba reformer is 6 m × 1.5 m. The system required reformer repair (replacement of liners, reformer tube burning, plugged burner) for six months to mid-1987. In addition, erosion of the five cooling tubes per cooling plate and perforation has occurred on the 1A and 1B stacks. Care is required in controlling differential pressure (10 cm of water in operation, 40 cm on shutdown). Both start-up and shutdown times are longer than specified (6.7 h and 1.5 h, versus 4.0 h and 1.0 h, respectively). The system generated 1 MW at 3:30 P.M. on December 16, 1987, and it will be tested to December 1988, and subsequently dismantled. Up to June 1988, there had been 51 incidents, 47% involving the reformer, 25% the turbocompressor, 14% other system components, and 14% the fuel cell stacks. The latter have therefore shown rather good reliability for an untested technology.

The Fuji-Mitsubishi Kansai unit consists of four 260-kW stacks (two Fuji Electric with ribbed substrates, two Mitsubishi Electric with ribbed bipolar plates). It is housed on a 3600 m² site in a 700 m² building. Up to June 5, 1988, the unit had had 137 start-ups, 30 of which were for electricity generation, which covered 483 hours (1935 reformer hours and 789 PAC test hours), with a cumulative 184,088 kWh of electricity generated. The NO_x and SO_x levels were adequate (under 20 ppm and 0.1 ppm, respectively), and the unit would go from 25 to 100% load in one minute. While it was designed for automatic operation, a one-man round-the-clock monitor was used. Start-up time was longer than expected (5.5 h rather than 4.0 h). Overall efficiency achieved so far is only 32% (compared with the design value of 40%) since a low-efficiency turbocompressor is used and the residual methane level from the 19-tube HK40 steel bottom-burner swirling-flow reformer, operating at 3.0/3.5 steam-to-carbon ratio, is high. Between January 20, 1987 and April 21, 1988 there were 78 incidents, mostly involving electrode differential pressure, cooling pipe clogging, and reformer pressure rise at high load, which resulted from reformer pipes touching. The ceramic reformer tube protection also cracked. Pressure rise in the reformer was cured by modifications to the turbocharger nozzle. However, problems with the electrochemical operation of the fuel cell stacks have been minimal, though cells had to be replaced because of defective separators, and care is required on start-up to avoid accumulation of anode gas at the air electrode. Plans are to continue testing through November 1988, and dismantle the unit in January 1989. Future plans are to have a 10 MW unit in 3-4 years, with solicitation of a joint study with other Japanese utilities and Government bodies such as NEDO.

OTHER JAPANESE PAFC UNITS

The 200 kW NEDO-funded Fuji on-site methanol PAFC system (p. 140) for use on remote islands will be intended to be a unit of the highest reliability, and will replace diesel generators (e.g., 1400 kW at Toro). An efficiency of 37.5% is expected, with cold start-up in 3 h and output change from 20% of rated load to 100% in 30 s. Any 20% load change should be virtually instantaneous. The corresponding Mitsubishi Electric 200 kW natural gas–powered unit will have a 36% electrical and 44% heat efficiency, and will be aimed at a 40,000 h service life. A maximum of 5 g/kW of platinum will be used, and NO_x emissions are planned to be under 20 ppm, with noise level at 55 dBA at 5 m. The system has a three-layer ribbed bipolar plate (somewhat similar to that developed by Engelhard, p. 530) consisting of a glassy carbon gas-impermeable central area and porous carbon-graphite reservoirs. Carbon paper electrodes (3600 cm^2) of Prototech type are used. The proprietary matrix sheet is a thin 55.8% porous layer with a contact angle of 70% for phosphoric acid. Mitsubishi has supplied some interesting data on acid loss by evaporation from this unit. The effective amount of acid retained by the system is 100 mg/cm^2 (about 600 g/kW), whereas loss at 205°C is 300 mg/cm^2 after 40,000 h. Hence, loss will be made up at 10,000 h intervals. Loss doubles on going from 205° to 220°C. The unit will require a high V, low A inverter, and will be made in two sections (10 m long total), so that it can be carried on two small trucks. A 100 kW unit (330 cells) of this type will be installed at the Hotel Plaza, Osaka, in a project in collaboration with Kansai Electric Power Company and Osaka Gas.

The final Japanese PAFC system is the joint venture between Sanyo and TEPCO to build an air-cooled Energy Research Corporation technology 220 V, 200 kW four-stack cogeneration unit at the site of the Shin-Tokyo power plant (Toyosu, Koto-ku), which started generating in September 1987. Toyo, Yogogawa Electric, and Meidensha were involved in system design and construction. The purpose of this plant is to demonstrate grid connection, cogenerated heat use, the reliability of the air-cooling, and the design of the control and maintenance subsystem. The atmospheric-pressure system will be rated at 35% HHV efficiency on natural gas, and operates at 0.64 V, 160 mA/cm^2. The system, which occupies 12 m x 19 m, is comparatively simple, with only five heat exchangers and a rectangular box 12-tube reformer with 15 bottom burners. Inverter efficiency is 90% at full load, and about 80% at 25% load. In June 1988, the plant was reported to have operated for 111 h, of which 96 were grid-connected, and to have produced 15,400 kWh up to the end of May 1988. Future demonstration plans include a lower system volume and cost (presently ¥2,500,000/kW, which must be reduced by a factor of ten), improvements in the distribution of reformer burners, lower heat losses, increased fuel processor reliability, reduction in cooling blower power, and reduction of the number of parts. In particular, the stacks will be reduced to two, and will incorporate a built-in, simplified inverter.

AMERICAN ON-SITE PAFC UNITS

At this point, it seems appropriate to review the on-site PAFC situation in the United States. As Karen Trimble stated at the March 30, 1988 final meeting of the Fuel Cell Users Group, the PC-25 IFC 200 kW on-site unit will use the IFC Configuration B stack design (p. 519, 535) and will operate at 36% HHV electrical efficiency, corresponding to a cell voltage of 0.62 V. It will produce this voltage at 348 mA/cm^2 (see Fig. 16–6, p. 519). In contrast, the PC-18 40 kW units operated at 0.665 V (about 40% electrical efficiency) and 115 mA/cm^2, so that the cell stack component capital cost in the PC-25 is reduced by a factor of 2.8. A 20 y lifetime is expected, with major servicing every five years. The stainless steel serpentine cooler tubes should alleviate any of the corrosion and blockage problems seen in the 40 kW unit, which were virtually eliminated after modification of the cooling system of the latter. The British Columbia Hydro 40 kW unit is still operating, and has not required service during a continuous run of 2206 operating hours (5253 total operating hours grid-connected and 2869 non–grid-connected) as of April 1988. It operated at an average HHV efficiency of 39.4% in the non–grid-connected mode over 1452 hours, which was greater than IFC's predictions. NO_x levels for the PC-25 are expected to be under 0.9 g/GJ (12 ppm), with SO_x at about 15 mg/GJ (virtually unmeasurable), and a noise signature of 60 dBA at 5 m. So far, about 30 orders for 200 kW PC-25 units have been received, promising a successful program as time progresses. System cost will be the key item: The 40 kW units cost about $8000/kW in 1984, whereas the present system should sell for $2500/kW, falling to $1000/kW in the future.

THE PAFC ELSEWHERE

Danish utilities and Sydkraft in Sweden are considering the purchase of 200 kW on-site units (Sweden has an initial $1.6M program to initiate new fuel cell research from 1987–1990). There is interest in Norway in the acquisition of 40 MW of PAFC stacks that would use the hydrogen produced by chlorine-caustic plants. After developing small alkaline units as a first step, HIDRILA in Spain is hoping to start a program on PAFC development, which will be followed by an MCFC program (discussed later). Italy is interested in the development of 1–5 kW PAFC units for military field use.

SPECIALIZED LOW-TEMPERATURE FUEL CELL SYSTEMS

These are receiving attention throughout the world. The most striking news has been the performance of the IFC alkaline lightweight fuel cell operating at 150°C and 13.6 atma on pure hydrogen and oxygen. It is capable of 0.8 V at 5 A/cm^2 and 1.0 V at 2 A/cm^2 (see footnote on p. 175 and Fig. 10–3, p. 295). It uses a similar gold-platinum alloy cathode to that in the Space Shuttle Orbiter fuel cell with a new ultra-thin matrix material (0.05 mm) giving very low IR drop (about 30 mΩ-cm^2, less than one third of the value in the PAFC). The anode-side electrolyte reservoir plate is presumably made from graphite, and the bipolar plate (co- or counter-flow configuration) is simply folded sheet metal, presumably gold-plated magnesium. One unique characteristic is the active cooling system, which uses direct injection of water into the reactant flow channels. All this results in an ultra-light unit, whose stack components weigh about 1.8 kg/m^2. Hence, the stack components alone have a specific power of 22 kW/kg. With all cooling, product water rejection, and control subsystems, the complete 90 kg system will produce 300 kW, not far from the values for a commercial jet engine. This system has the same weight as that of the original space shuttle unit, which produced 12 kW maximum power, limited by cooling and power conditioning requirements.

The Ballard data for a Dow-membrane solid polymer electrolyte cell achieved at only 80°C and at 5 atma H_2 and O_2, also shown on p. 295, are almost as spectacular. Maximum current density achieved was 4 A/cm^2 at 0.5 V, lower than that of the alkaline system due to the less severe operating conditions and the higher Tafel slope of the oxygen electrode reaction in alkaline media. Similar results (at 85°C and at 3.4 atma) were shown in a presentation at Toronto on April 4, 1988, which demonstrated a considerable difference between Nafion® 1170 and the Dow membrane. The former could only sustain 1.0 A/cm^2 at 0.5 V under the same conditions. The Ballard data with the Dow membrane have now been reproduced elsewhere (for example, at Dow Chemical). At Los Alamos National Laboratory and at Texas A&M University, a further advance has been made in reducing the platinum loading to 0.4 mg/cm^2 at the cathode and 0.1 mg/cm^2 at the anode (from 4 mg/cm^2 at both electrodes), yet still allowing at least 2 A/cm^2 at 0.5 V.

This has been effected by casting the solid polymer electrolyte inside conventional Prototech carbon-supported platinum electrodes using methods similar to those outlined on p. 295 (Martin et al.). We should point out that Ref. 176 stated that reduction in the platinum loading in the SPE system was one of the most important research problems to be solved to allow its more general application. At the present time, the self-supporting low-loading electrodes allow the use of more gel-like polymers, which may be much less expensive than the solid membranes that have been used previously. Already, Ergenics, Inc. of Wykoff, New Jersey is commercializing small SPE units (up to 500 W) that can be operated on rechargeable metal hydride cartridges, and can be used to supply power for such applications as professional videocameras, and Ballard is also examining other possible markets. These may include military backpack power packs, as well as fuel cell powered underwater vehicles and submarines. In Italy, Dow is collaborating with De Nora on a 10 kW unit. The use of total platinum loadings of 0.5 mg/cm^2 and a performance on ambient-pressure hydrogen and air of 700 mA/cm^2 at 0.5 V (S. Srinivasan, Texas A&M University) allows platinum cost less than \$30/kW. However, for the widest application (e.g., vehicles), the use of platinum must be further reduced, or eventually eliminated (see Fig. 12-5, p. 407), as is indicated by the arguments given on p. 54.

In Japan, a number of advances in small fuel cell technologies have been reported. A useful summary of all Japanese fuel cell activity (Fuel Cell R&D in Japan) was published in June 1988 by the new Japanese Fuel Cell Information Center (Institute for Applied Energy, SY Building, 1-14-2, Nishishinbashi, Minato-ku, Tokyo 105), which is an organization with 90 members. This summary included specialized fuel cells. The 7.5 kW, 380 kg Fuji Electric alkaline system has completed a field test of 18 months and has shown high reliability. Development of the direct methanol system continues at Hitachi, and Matshushita is developing an alkaline direct methanol system, presumably with an electrolyte that can be replaced, for applications such as lighthouses. Sanyo is developing a small hydrazine system, which has been used at 34 remote sites in Japan for applications such as power sources for pluviometers, water level meters, anemometers, microwave relay stations, salinity measurement, fish bait luring lights, and buoy beacons.

In Europe, the EEC is spending \$1.2M/y on novel materials for direct methanol and higher-temperature solid proton-conducting systems. Switzerland is developing methanol-air alkaline cells.

FUEL CELLS FOR FUTURE TRANSPORTATION NEEDS

The IFC lightweight space system uses literally gold-plated components and a 20 mg/cm^2 gold–10% platinum cathode similar to that in the Space Shuttle Orbiter PC-17C fuel cell (p. 164). It can therefore never be considered for widespread consumer applications; for example, as a power source for transportation when hydrogen fuel becomes available in quantity. The record U.S. summer drought of 1988, which followed a series of hot summers, was described by James Hansen of NASA-Goddard as the commencement of the *greenhouse effect*, at a U.S. Senate subcommittee hearing on June 23, 1988. Fossil fuel carbon dioxide may represent only about half of the total greenhouse effect, the rest resulting from methane, nitrous oxide and chlorofluorocarbon refrigerants in approximately equal

Conclusion

quantities (article in the EPRI Journal, Electric Power Research Institute, Palo Alto, June 1988). However, we should not fall into the trap of believing that fossil CO_2 is thereby less important than we had previously considered it to be, since we cannot deal directly with the emissions of CH_4 and N_2O which derive (apparently) from mostly agricultural sources, which are growing in the face of an expanding world population. For example, though CH_4 seems mostly to be historically derived from fermentation in swamps, it now comes from fermentation in the intestines of ruminants and from rice paddies. In general, it might be thought that emissions of this greenhouse gas should not have varied much over time, since agriculture might in the long run produce about the same amount per unit area as untamed nature. However, in a short time, CH_4 concentration in the atmosphere has almost tripled, going from a preindustrial level of 800 (or less) ppb to about 1800 ppb. Similarly, N_2O, for which figures are less clear, seems mostly to derive from either natural sources (leguminous plants) or artificial fertilizers, which tend to compensate each other on a global scale as land use changes.

While CH_4 and N_2O emissions seem to be out of our control, with a certain degree of good will, the use of chlorofluorocarbons can be banned. However, the Montreal Protocol on this subject, accepted by 31 nations, has only been ratified by the United States and Mexico. However, it does not really attempt to get its teeth into the problem of chlorofluorocarbon emissions, which it states must simply be reduced on a sliding scale until early in the twenty-first century. Dow Chemical has, however, stated that we *must* stop allowing the escape of these chemical compounds to the atmosphere as of *today*. While there may be some commercial intent in such pronouncements, we still must accept the inevitable: The fact is that these compounds are doing irreparable damage to the ozone layer, and are almost certainly worsening the situation closer to the earth's surface.

Assuming that emissions of the freon-type materials can be eliminated, the greenhouse effect can be controlled by the use of hydrogen fuel for transportation, other energy applications in major countries being served by electricity produced by methods that do not release CO_2 into the atmosphere. These would include nuclear and eventually solar sources, along with coal and other fossil fuels. In the latter case, after separation from the coal gas from an oxygen-blown gasifier, CO_2 would first be used as a useful saleable solvent for tertiary oil recovery (see W. G. Snyder and C. A. Depew, "Coproduction of Carbon Dioxide and Electricity," AP-4827; and T. A. Matchack, A. D. Rao, V. Ranathan, and M. T. Sander, "Cost and Performance for Commercial Applications of Combined Cycle Plants," AP-3486, Electric Power Research Institute, Palo Alto, 1986). When no longer required for this application it would be disposed of in used oil or gas wells, or in the deep oceans; and the hydrogen for transportation and electricity could be produced simultaneously, and perhaps both transmitted together via a high T_c superconducting cable, liquid hydrogen being the refrigerant (A. J. Appleby, testimony: before House Subcommittee on Science, Space and Technology, July 9, 1987, September 23, 1987, March 9, 1988; before the Senate Subcommitee on Energy Research and Development, September 23, 1987; before the House Subcommittees on Natural Resources, Agricultural Research and the Environment, and Science Research and Technology, June 29, 1988. See also EPRI Journal, June 1988).

Such hydrogen would be quite expensive compared to today's fuels, perhaps $13/GJ, equivalent to gasoline at the refinery at $1.67/gallon. The present cost of liquid hydrogen manufactured from natural gas purchased in large quantity (compared with its rather limited total production) is $16.30/GJ ($2.09/gallon gasoline equivalent). It would certainly be too expensive to burn in an internal combustion engine. However, it could be used at three times the efficiency in a vehicle fuel cell with inexpensive components based on the IFC lightweight alkaline technology operating at 60°C at 0.75 V, 0.45 A/cm² on CO_2-scrubbed air using the non-platinum cathode shown in Fig. 12–4, p. 406. Such a system could make large use of plastics, and could have a lightweight nickel-plated magnesium bipolar plate. Such a system, rated at 60 kW (80 hp), could weigh 32 kg for the stack parts, whose materials cost would be $123, based on an average materials cost of $3.85/kg. Even at today's liquid hydrogen cost, such a unit would be competitive on a ¢/mile basis to today's automobile, and it would be completely pollution-free.

MOLTEN CARBONATE FUEL CELLS: UNITED STATES

In the United States, total DOE fuel cell budget in 1987 was $28M, with $15.5 going to PAFC development, $7.6M to the MCFC, and $4.9M to advanced systems (largely solid oxide). In 1988, the corresponding figures were $13.2M, $11.1M, and $8.3M, with $1.4M for advanced research. Funding for FY89 is expected to be $9.5M, $7.0M, and $8.4M, with a further $1.6M going to advanced research. It should be noted that the spending for the PAFC is going down, whereas that for the SOEFC is ascending. However, these figures do not tell the whole story, since DOE has awarded three 24-month contracts of about $10M each, to ERC, IFC, and M-C Power, starting in early 1988. (M-C Power is a new company set up by IGT as a dedicated MCFC developer-manufacturer. It is collaborating with Combustion Engineering, Energy Center Netherlands, and IHI in Japan in this program.) These contracts are currently financed by a carry-over of money from previous years, and it is anticipated that there will be enough money available in the FY90 budget to complete them. In consequence,

Conclusion 625

U.S. Government spending on the MCFC is effectively $15M/y. Following completion of these contracts, and after successful demonstration of pressurized 5 kW–20 kW stacks with state-of-the-art components, a three-year $20M program is expected. It will involve 100 kW stack manufacture (1990), followed by field tests of natural gas and coal gas systems. General plans are to have small commercial systems available by 1995, and large central stations by 2000.

Each developer is taking a different approach to stack design. In the case of ERC and IFC, a cross-flow stack geometry similar to their previous designs (p. 450) will be used, whereas M-C power intends to use a formed sheet-metal structure with counter-flow geometry (see p. 451) of their IMHEX®(internally manifolded heat exchanger) design, resembling, as its name suggests, the plates of commercially available heat exchangers. This is sponsored jointly by EPRI and DOE. While this involves some element of technical risk, it has fewer parts and should give better electrical contact. In addition, it should reduce the electrolyte loss experienced with externally-manifolded cross-flow systems (p. 564).

The EPRI fuel cell program expects to spend $10.6M on the MCFC between 1988 and 1991, together with another $1.4M on exploratory research. Its total fuel cell budget is about $4M/y. The EPRI is involved in a joint program with DOE and PG&E for the development of the IRMCFC at ERC. Four stack tests were conducted by the latter during the period of June 1987 through June 1988. The first was an 11-cell, 1000 cm^2 unit containing two flat-plate reforming units (reforming plates) between cells three and four, and eight and nine. The latter are counter-flow with respect to the methane flow in the cell. This ran very satisfactorily for 1400 h at an average of about 0.78 V and 145 mA/cm^2. The second stack, of the same dimensions, used direct internal reforming in the anode chamber. It ran for 2700 h with a decay rate of 10 mV/1000 h (due to increasing IR drop) at 160 mA/cm^2 and 0.782 V early in life. Cells, however, showed wide dispersion (0.842 V to 0.677 V). The first 4000 cm^2 5-cell direct internal reforming stack operated in 1988 and was terminated in May. It operated at 0.76 V, 160 mA/cm^2, at 75% methane utilization, with an initially satisfactory decay rate. Plans were announced in June 1988 to operate a 4000 cm^2 5-cell indirect IRMCFC stack in July, followed by a 60-cell, 1000 cm^2 7.5 kW direct IRMCFC later in the year, followed by a 25 kW unit in 1989 and a 200 kW unit in 1990.

The EPRI support of new technology remains strong. At M-C Power, aluminum oxide–strengthened nickel anodes and nickel-coated ceramics (a continuation of the GE work) are being examined, as well as the endurance of tape-cast matrices and alternate cathode materials and ways of controlling NiO dissolution. For the latter, a chemical additive has been found that stabilizes and increases the performance of the cathode and reduces nickel dissolution by a factor of three. Niobium-doped ceria is being examined as a nickel oxide replacement at Case Western Reserve University, although conductivity at 650°C has not so far been encouraging. The work carried out at Ceramatec and reported in Chapter 18 on yttrium aluminum garnet and barium ferrite residue as alternative cathode materials continues.

The Gas Research Institute has not so far given high priority to the MCFC, as compared with the SOEFC. However, though their present budget for the former (1988) is $300K, this is expected to increase to $2.5M in 1993. In contrast, the SOEFC budget is expected to remain relatively constant at about $2M/y. They currently (September 1988) have a contract with M-C Power for on-site MCFC development. Their long-term objective is to develop a 1-2 MW compact on-site unit.

MOLTEN CARBONATE FUEL CELLS: JAPAN

In Japan, the MCFC Moonlight program has been reduced to three NEDO-supported companies—Hitachi, IHI, and Mitsubishi Electric. The 1988 objective for the first two is the development of cells with an area of 1.0 m^2 after the successful test of 10 kW class stacks in late 1986–1987, which was reported in the text (p. 136). The aim is a 100 kW system in 1991. Results are to be reviewed at end of that year, and, if successful, the program is to progress to a 1000 kW system, which it is hoped will be 45% efficient on natural gas (the 1 MW PAFC demonstrators only showed efficiencies slightly exceeding 30%).

In June 1988, Hitachi reported that work was in progress on the internally-manifolded 1.0 m^2 scale cell, which would be rectangular, made up of four 3600 m^2 parts. It would have the same NiO–Ag cathodes as were used previously, with new anodes consisting of Ni–Co–Cr alloy, rather than Ni–Co. The addition of 5-atom % of Mg, La, or Zr to the anode was stated as being advantageous in preventing shrinkage. Life-tests on these components were presently being conducted. The matrix would continue to be a lithium aluminate–alumina fiber sheet prepared by tape-casting, using the longest and thinnest alumina fibers available. The lithium aluminate particle size was 2×10^{-2} micron (maximum), 10^{-3} micron (mean), and 10^{-4} (minimum). The mixture is slurried and deaerated, reslurried, and mixed before tape-casting. Presently, batch production is used, with continuous manufacture planned for the future. Electrolyte loading in the free porosity is 100-110%, yielding a 1 Ω-cm^2 layer of unspecified thickness. The bipolar plate to be used would be 310 stainless steel (see p. 560). Hitachi reported that they were testing a 32-cell 10 kW stack under 1 atm and 3 atm conditions that had accumulated 2000 operating hours in June 1988.

After its very successful test of its 10 kW–class stack in late 1986, IHI reports operating a 1.1 kW, 1.0 m² single cell successfully for 1100 h up to May 1988. The 10 kW stack had a resistance of 0.5 Ω-cm², showed carbonate loss from one side of the cell with redistribution towards the exits, but unlike the work reported on p. 568, it showed no redistribution between cells. This tends to vindicate the use of the internally-manifold design over externally-manifolded concepts. It showed corrosion at the anode exit, but this is to be expected because of the use of non–nickel-clad 316 stainless steel hardware, which corrodes in a low-hydrogen, high–water vapor atmosphere, even that corresponding to 40% hydrogen utilization.

The later 1 m² cell was rectangular, measuring 1.8 m by 0.57 m, and was constructed in several parallel pieces. As in the case of Hitachi, lack of equipment of the appropriate size, in this case an 800°C electrode sintering furnace of sufficient length, has held up the development of one-piece cells. The co-flow bipolar plate consisted of two pressed metal plates with square channels welded to a flat plate so that a one-piece stainless steel and nickel structure can be built up to give a co-flow version of the lower scheme in Fig. 17-4, p. 549. Performance was 80 mV lower than that of small cells, since IR was 0.4 Ω-cm² higher, and current density distribution due to lack of uniform flow patterns limited fuel utilization to 60%, giving 0.72 V at an average current density of 150 mA/cm².

Presently, IHI is testing 100 cm² cells at 75% utilization, and has designed 2500 cm² sheet metal, 50 micron nickel-clad bipolar plates. These have laser-welded frames, and will be aluminum-clad at the edges using a newly-developed technique that can be applied to large areas. They have also designed and tested a flat-plate reformer using cell stack components, which consists successively of a reforming element with catalyst, a catalytic burner element, a gas preheater, a catalytic burner element, and a further reforming element. While this presently operates at 800°–850°C, it could eventually be used in an indirect IRMCFC (sensible heat reforming). In June 1988, plans were to have a 1 kW external reforming system for NEDO which would operate at 75% fuel utilization, which would be followed by a 10 kW system by April, 1989. In FY89 (starting April), a 50 kW system would be constructed, to be followed by a 100 kW system in 1991.

Good progress has been made in tape-cast components. Thickness can now be held to $\pm$ 25 microns, mean pore size to $\pm$ 0.5 microns, and porosity to within 1%. In the materials technology area, IHI has reported some interesting results. Observed carbonate losses with 35% Cr, 45%, Ni, 20% Fe (in mg/cm²) have been 5 (K) and 5 (Li) at 500 h, increasing to 50 (K) and 20 (Li) at 5000 h. The non-linear behavior of Li is presumably due to corrosion, but the high K loss seems unusual. Some interesting cathode nickel dissolution results have been noted. In the co-flow cell, much more dissolution takes place at the inlet than at the outlet. This is to be expected, since the cathode gas is richer in CO_2 at the inlet. However, the maximum density of redeposited nickel in the electrolyte lies closer to the cathode at the inlet, where the H_2 partial pressure is highest, showing that increasing the latter on the anode side (i.e., even to the extent to making the cell slightly leaky), may prevent nickel migration over to the anode. They report 25% of the cathode lost in 10,000 h, which is higher than rates reported elsewhere. While IFC has reported a nickel dissolution rate that is linear with time (Fig. 17–12, p. 571), IHI indicates a mg/cm² loss relationship at 0.3 atm CO_2 and 650°C equal to $0.26 t^{1/2}$, where t is the time in hours. Data on NiO dissolution are therefore still controversial.

Mitsubishi Electric is being supported by NEDO to develop a cross-flow internal reforming stack. This has a construction and performance (at 80% fuel utilization) similar to those of ERC reported earlier. Testing has been conducted on a 5-cell 961 cm² stack which showed excellent performance due to the use of a new Ni-Ru internal-reforming catalyst, since pure nickel catalysts were shown to degrade after 2500-3000 h. Average cell voltage at 40% methane utilization was 0.825 V at 150 mA/cm², with about 0.78 V at 60% utilization at the same current density. Decay, due to increasing IR drop, was still under 10 mV/1000 h.

Other companies and institutions performing MCFC R&D in Japan under the aegis of the Molten Carbonate Fuel Cell Power Generation System Technology Research Association (founded February 1988) are: the Government Industrial Research Institute, Osaka (GIRIO—materials development); the Central Research Institute of the Electric Power Industry (CRIEPI—pressurized test facility); Toshiba (improved cells and, together with Kobe Steel, the heat recovery system for the 1 MW MCFC demonstrator); Fuji Electric (improved cells); Tokyo Gas, Osaka Gas, Toho Gas, and Mitsubishi Electric (IRMCFC system design); Mitsubishi Metal Corporation (anodes); Nippon Kokan K.K. (NKK—new alloys, corrosion testing); Kobe Steel (materials, high-temperature desulfurization, turbocompressors); Nisshin Steel Corporation (alloys, corrosion); Kawasaki Heavy Industries (Ni-W anodes to replace Ni-Cr); Sumitomo Precision Product Company (stainless steel brazed-plate heat exchangers); the Japanese Fine Ceramics Center (materials); and finally Sanyo (ERC licensee). The number of developers, suppliers, sponsors, and so on, in the fuel cell area is now very large—at least 78 organizations can be identified. The MCFC Research Association coordinates activities with the goal of developing a 3 atma pressurized system to operate on coal gas stack in the 500 kW class at a cost of ¥50,000/kW. Plans are for baseload plants (coal) in the 600–1000 MW range, with 70–80% load factor, 15–24 h cold start time (4–5 h hot start) 5%/minute ramp rate, and an efficiency of over 50%; intermediate load plants (coal–natural gas) in the 200–500 MW range, with 50% load factor, 15–24 h (coal), 10–15 h (gas) cold start time (4–5 h and 2–4 h hot start, respectively), 5–10% ramp rate, efficiency over 50%; and peaking units (gas or methanol) of 20–30 MW, 4–6 h, 1–2 h cold and hot start, respectively.

Total spending proposed by NEDO over the 15-year periods from 1981 to 1995 is ¥57B, or $430M at the present exchange rate. However, total spending by all parties is likely to total $1.5B. Of the ¥57B total NEDO spending, ¥14B is for the PAFC, so the MCFC and SOEFC systems (particularly the former) are receiving the majority of funding. Spending by NEDO in the second phase, starting 1987, will be ¥34B. Spending on the PAFC is diminishing (1985–1988, ¥3830M, ¥2230M, ¥2540M and ¥1210M, respectively), whereas the MCFC shows the opposite trend (¥850M, ¥900M, ¥740M, and ¥2230M). The spending on the SOFC is quite small (¥90M, ¥60M, ¥80M, and ¥80M) for the same years.

MOLTEN CARBONATE FUEL CELLS: EUROPE

The Dutch and Italian programs are now well under way. The Netherlands is reaching the end of the first three-year phase of development ($5M from PEO, supplemented with $4M from ECN, and $0.6M from the European Community). Technology has been transferred from IGT (Chicago), a test facility constructed at Energy Center Netherlands (ECN), and research on component fabrication and new cathode materials carried out. The latter include $LiFeO_2$, Li_2MnO_3, ZnO, La perovskites, $LiFe_5O_8$ (which was unstable), and $Li_2Ti_{0.995}Ta_{0.005}O_3$. Tapes can be cast up to 45 cm wide, with scale-up to 60 cm in the near future. Cells up to 1000 cm^2 can be tested, and test times up to 5000 h have been achieved. A laser-welded sheet-metal bipolar plate with soft rails has been designed which is somewhat similar to that shown in Fig. 13–2E, p. 450. The first 10-cell 1 kW homemade stack is expected to operate in late 1988.

The next phase is expected to be for a period of 4.5 y, the government contributing $20–25M. The object of development will be a 3 × 100 kW pilot plant. The plan is to conduct market studies (1 y), component and stack fabrication technology (1.5 y), pilot plant construction (2 y), followed by plant operation, with commercialization in the latter half of the 1990s. In the meantime, the R&D program will be continued to ensure a competitive product. The major developer will be ESTN/Hoogovens.

The objective of the Italian program is the development of a 50 kW MCFC stack by the early 1990s. Present plans are for ENEA to spend about $40M (1987-1992), representing 80% of the development money, with 15% coming from Italian industry, and 5% from the EEC and other sources. Of this, over 50% is intended for molten carbonate. It is hoped that an international joint venture can be established, and discussions have taken place with the Dutch which promise that an agreement can be reached. The principal contractor is Ansaldo (Genoa), who has succeeded in constructing and testing five 20 cm² cells with Ni-Cr anodes and Ni-Ag cathodes made by cold-pressing and sintering. The cells had hot-pressed tiles, and gave 0.76 V at 160 mA/cm². Improvements in hot-pressing, prelithiated nickel oxide cathodes, and aluminized wet-seals gave cells that operated at 0.8 V, 160 mA/cm² initially, which fell to 80 mW/cm² after 1800 h. Tape-cast matrices, roll-bonded anodes, and prelithiated cathodes for 100 cm² cells are now being further developed.

ENEA-Casaccia is examining bimodal porous matrices, the characterization of new cathodes (e.g., Li_2MnO_3) by AC impedence techniques, and tape- and slip-casting in collaboration with CNR-Faenza. The Polytechnic of Milan is conducting materials research (corrosion, porous metal behavior, nickel oxide dissolution), together with fabrication of all-Italian cell components. At CISE (Milan), cathode preparation using different methods (*in-situ* lithiation, sputtering, evaporation), electrochemical half-cell testing, studies on nickel oxide dissolution under simulated conditions, and surface studies (Auger, ESCA, etc.). At CNR IRTEC (Faenza), tile manufacture and microstructural characterization is being studied. This includes optimal particle size distribution, forming technologies, binder removal after tape-casting, sinter cycles and microstructure. In Messina, CNR/ITAE obtained a complete set of state-of-the-art parts from IGT after signing a collaboration agreement, and the first cell was run in June 1984. One of the most interesting results obtained is the threefold reduction in decay rate obtained by operating at 600°C instead of the usual 650°C. This reduces voltage at constant current by 86%, but would provide an interesting way to introduce a reliable long-lived on-site internal-reforming unit operating at 0.65 V, rather than 0.73 V. While this would decrease electrical efficiency from about 53% to 46%, the electricity to reject heat ratio would have better adaption to the on-site application. The CNR/ITAE is also developing advanced internal reforming catalysts.

The EEC is spending about $1.3M for MCFC support studies, and finally, in Spain, HIDROLA proposes the development of a 100-200 kW MCFC unit, which would be followed by a commercial 100 MW plant in the late 1990s.

SOLID OXIDE ELECTROLYTE FUEL CELLS

Testing of a Westinghouse 24-cell, 400 W stack began at TVA in January 1987, and it completed 1760 h of testing connected to a television. Recently, a 324-cell has been put on test. Other Westinghouse stacks, which are currently available at an approximate cost of $250,000/kW, are on test at

Tokyo Gas and Osaka Gas (see following). A prepilot tube manufacturing plant, constructed on in-house funding, has come on-line during the summer of 1988. It is expected that this unit will allow a substantial reduction in tube costs. Westinghouse expects to have a 25 kW unit by 1989, systems of hundreds of kW by 1992–1993 for cogeneration, and MW-scale commercial plants by the year 2000.

In an extension of previous work in the United States, maufacturing methods for monolithic SOEFCs are being studied by Garrett, Combustion Engineering, and Argonne National Laboratory, as well as by UTC, Ceramatec, and IGT. Flat ceramic structures are being studied by the New Mexico School of Mines under EPRI funding. Texas A&M is also being supported by EPRI to examine an alternative method of manufacturing monolithic structures.

Tokyo Gas and Osaka Gas are testing 144-cell Westinghouse units, rated at 3 kW. The tubes are arranged in eight bundles, three in parallel, three in series. The fuel cell units, which contain about 25 kg of tubes and about 5 kg of active materials, consist of a box 2.4 m X 0.8 m X 2.1 m high, weighing 1300 kg. The Tokyo Gas unit had operated for 4503 h by the beginning of May 22, 1988, accumulating 9085 kWh. Its availability has been 99.6%, and it had had two start-ups. At Osaka Gas, unit number 1 had operated for 3013 h, yielding 6085 kWh. Availability was 97.9%, and four start-ups had been made. The corresponding figures for unit number 2 were 1268 h, 2809 kWh, 97.4%, with two start-ups. Performance of the tubes has improved substantially, now giving 300 mA/cm^2 at 0.6 V and 80% fuel (96% hydrogen) utilization, and 150 mA/cm^2 at 0.75 V under the same conditions. The unit voltage increased from 28 V to 32 V on going from 960°C to 1060°C, and at 1020°C the voltage dropped from 30 V to 25 V over 4000 h, due to increasing resistance. At Osaka Gas, this has been attributed to a poor nickel felt contact (see Fig. 18–7, p. 602). Noise level was 59 dBA 4 m from the unit, and NO_x was 1.3 ppm. Surface temperature of the unit was 42°C.

The Tokyo Electric Power Company is examining the design of coal-fired SOEFC power plants. Assumed fuel utilization is 80%, with cells operating at 0.7 V at a pressure of 8.15 atm. The proposed system is an oxygen-blown Texaco gasifier with Selexol cold gas cleanup. There is no anode gas recycle to avoid carbon formation, and the unit is claimed to be simpler and more efficient than a corresponding MCFC unit. Better sulfur tolerance is also claimed. In the 265 MW unit examined, coal feed is 533.1 MWth (68.6 metric tons/h), and the SOEFC produces 148.1 MW, the fuel-gas expander 8.8 MW, the gas turbine 57.7 MW, and the steam bottoming cycle 89.9 MW. Oxygen production requires 27.2 MW, the SOEFC 3.0 MW, and other parasitic losses are 9.3 MW. Overall plant efficiency is 49.7%.

In other Japanese work, the Electrotechnical Laboratory at Scuba, which has been involved in SOEFCs since 1972, is presently continuing work on a 500 W (at 0.76 V, 63 mA/cm^2) tubular traditional-configuration unit. While its other components are standard, it uses alumina seals and calcia-stabilized zirconia support tubes. Gas-tight films are plasma-sprayed, and porous films are flame-sprayed or slurry-painted. The unit has 15 cells in series, with 48 parallel bundles. The unit cell is 2.4 cm long, the interconnector 1.2 cm. Similar cells are being developed by the Engineering Division of TEPCO.

The National Chemical Laboratory for Industry is developing a flat-plate configuration unit using an Inconel bipolar plate, as well as internally-manifolded system of corrugated, pseudomonolithic type. Mitsui Engineering and Shipbuilding Company is exploring SOEFCs as power sources for LNG-powered ships. They claim a total weight reduction of 30%, and a volume reduction of 20%, compared with a conventional boiler and turbine system. Other organizations involved in SOEFC development are Toa Nenyo-Kogo (planar, 25 cm^2 cells), Onada Cement (plasma spray), Kyushu University (ceria electrolyte), and Nippon Kokan (tubular and planar units).

In Europe, the EEC is spending more than $4.4M over two years to support SOEFC activity, mostly at Dornier, with the latter providing a 50% cost share. Other work is being performed in six universities and at TNO (The Netherlands), British Gas, and GEC (United Kingdom). Two types of structures are being examined: the true monolithic honeycomb, and a multi-channel corrugated structure on a flat plate somewhat resembling that shown for the MCFC in Fig. 17-2, p. 545. The multi-channel plate would be fabricated by extrusion. Sol-gel powder processing, improved cathodes and electrophoretic deposition, and other preparation techniques are being examined. The ECN and the University of Delft are examining materials aspects of the aforementioned, and Dornier is performing the engineering analysis (mass and energy balances) and marketing studies.

The immediate aim is to fabricate a 50 W multi-channel structure. Evaluation of the Dornier 10 cm × 10 cm metallic bipolar plate, of a ceramic bipolar plate, of an ultra-thin low-IR ceramic structure capable of operation at 800°C, and of the structural alternatives themselves will be made in 1989, and a choice will be made on the design of a 3 kW unit.

Some SOEFC work is being conducted at CISE in Milan, outside of Progetto Volta. Denmark is studying the SOEFC at Risø University, and Norway has decided on a $500K program (in 1988) on SOEFC development in a collaborative program between the Center for Industrial Research, Oslo, the University of Oslo, and the University of Trondheim. Sweden has expressed the intention of purchasing a 25 kW Westinghouse SOEFC unit.

PART V

References

References

1. Aaronson, T., Alternative Technol. Power Prod., 2 (Energy Human Welfare–Crit. Anal.), 102, Macmillan, NY (1975).
2. Abbaro, S.A., A.C.C. Tseung, and D.B. Hibbert, J. Electrochem. Soc., **127**, 1106 (1980).
3. Abens, S., P. Marchetti, and M. Lambrech, National Fuel Cell Seminar Abstracts, 1978, p. 66.
4. Abens, S.G., J.A. Hofbauer, and P.G. Marcetti, National Fuel Cell Seminar Abstracts, 1981, p. 120.
5. Abens, S., R. Hayes, and W. Keil, National Fuel Cell Seminar Abstracts, 1982, p. 55.
6. Abens, S., M. Farooque, and T. Schneider, National Fuel Cell Seminar Abstracts, 1983, p. 111.
7. Abramzon, O.S., R.Kh. Burshtein, A.G. Pshenichnikov, S.F. Chernyshov, and Yu.G. Chirkov, Elektrokhimiya, **12**, 838 (1976).
8. Abramzon, O.S., R.Kh. Burshtein, A.G. Pshenichnikov, S.F. Chernyshov, and Yu.G. Chirkov, Elektrokhimiya, **13**, 334 (1977).
9. Abramzon, O.S., R.Kh. Burshtein, A.G. Pshenichnikov, S.F. Chernyshov, and Yu.G. Chirkov, Elektrokhimiya, **13**, 474 (1977).
10. Abruna, H.D., J.L. Walsh, T.J. Meyer, and R.W. Murray, J. Am. Chem. Soc., **102**, 3272 (1980).
11. Ackerman, J.P., and R.K. Steunenberg, Evaluation of Electrolytes for Direct Oxidation Hydrocarbon/Air Fuel Cells, U.S. NTIS AD-A Rep. No. 011277 (1975).
12. Ackerman, J.P., Proc. Symp. Electrode Mater. Processes Energy Convers. Storage, 634–643 (1977).
13. Ackerman, J.P., paper presented at 5th Energy Technology Conference, Washington, DC (1978).
14. Ackerman, J.P., personal communication, Dec. 1980.
15. Ackerman, J.P., and S.S. Borys, personal communication, Nov. 1980.

16. Ackerman, J.P., Argonne Nat. Lab., personal communication, Nov. 17, 1980.
17. Ackerman, J.P., paper presented at 1st National Seminar on Electrochemical Systems: Batteries and Fuel Cells, at Federal University of Ceara, Fortaleza, Ce, Brazil (1980).
18. Ackerman, J.P., and A.J. Appleby, "Proceedings of the DOE/EPRI Workshop on Molten Carbonate Fuel Cells," WS-78-135, EPRI, Palo Alto, CA (1979).
19. Ackerman, J.P., in "Progress in Batteries and Solar Cells," Vol. 5, 13 (1984).
20. Ackerman, J.P., and T.D. Claar, paper presented at A.S.M. Conference on Materials for Energy Systems, Washington, DC, May 1–3, 1984.
21. Adachi, T., T. Takeshita, and I. Matsuda, Japan, Kokai Tokkyo Koho, 80, 34,678 (1980).
22. Adachi, T., T. Takeshita, and I. Matsuda, Japan, Kokai Tokkyo Koho, 80, 44,564 (1980).
23. Adams, A.A., A.J. Coleman, and L.S. Joyce, National Fuel Cell Seminar Abstracts, 1983, p. 118.
24. Adams, A.A., and R.T. Foley, J. Electrochem. Soc., **126**, 775 (1979).
25. Adams, A.A., and R.T. Foley, Research on Electrochemical Energy Conversion Systems, U.S. NTIS AD Rep. AD-A023689 (1975).
26. Adams, A.A., and R.T. Foley, Research on Electrochemical Energy Conversion Systems, U.S. NTIS AD Rep. AD-A034454 (1976).
27. Adams, A.A., National Fuel Cell Seminar Abstracts, 1982, p. 145.
28. Adams, A.M., F.T. Bacon, and R.G.H. Watson, Fuel Cells, W. Mitchell, Jr. (ed.), Academic Press, NY (1963), pp. 129–192.
29. Adams, R. and R.L. Schiner, J. Am. Chem. Soc., **45**, 2171 (1923).
30. Adams, R.L., U.S. Patent No. 4,177,157 (1979).
31. Adams, R.N., Electrochemistry at Solid Electrodes, Marcel Dekker Inc., NY (1969).
32. Adlhart, O.J., Proc. 19th Power Sources Conf., 1 (1965).
33. Adlhart, O.J., and A.J. Hartner, Proc. 20th Power Sources Conf., 4 (1966).
34. Adlhart, O.J., and P.L. Terry, Engelhard Industries, Ammonia-Air Fuel Cell System, U.S. Govt. Contract No. F30602-71-C-0161 (1969).
35. Adlhart, O.J., and P.L. Terry, Phosphoric Acid Fuel Cell Stacks, Final Rep., Engelhard Industries Ltd., Newark, NJ, U.S. Army Contract DAAK02-68-C-0407, May 1969.
36. Adlhart, O.J., and P.L. Terry, Proc. 4th Intersoc. Energy Conv. Eng. Conf., 1048 (1969).
37. Adlhart, O.J., Proc. 24th Power Sources Conf., 182 (1970).
38. Adlhart, O.J. in "From Electrocatalysis to Fuel Cells," G. Sandstede, (ed.), Univ. of Washington Press, Seattle (1972), p. 181.

39. Adlhart, O.J., Proc. 7th Intersoc. Energy Conv. Eng. Conf., 1097 (1972).
40. Adlhart, O.J., M.F. Collins, R. Michalek, and P.L. Terry, Open Cycle Fuel Cell Power Plant Direct Currents, 1.5 KW, NTIS AD 764 285 (1973).
41. Adlhart, O.J., Energy Technology Handbook, McGraw-Hill, NY (1977), 5/49–4/73.
42. Adzic, R.R., D.N. Simic, A.R. Despic, and D.M. Drazic, J. Electroanal. Chem., **65**, 587 (1975).
43. Adzic, R.R., A.R. Despic, and A. Tripkovic, Ext. Abstr., Meet.–Int. Soc. Electrochem., 28th, **2**, 247 (1977).
44. Adzic, R.R., M.I. Hofman, D.M. Drazic, and A.R. Despic, Glas. Hem. Drus. Beograd, **44**, 417 (1979).
45. Adzic, R.R., M.I. Hofman, and D.M. Drazic, J. Electroanal. Chem. Interfac. Electrochem., **110**, 361 (1980).
46. Adzic, R.R., W.E. O'Grady, and S. Srinivasan, Brookhaven Nat. Lab. Rep. BNL 51198, UC-93, Energy Conversion TID-4500, FC-8, Mar. 1980, p. 1.
47. Adzic, R.R., W.E. O'Grady, and S. Srinivasan, Brookhaven Nat. Lab. Rep. BNL 51198, UC-93, Energy Conversion TID-4500, FC-8, Mar. 1980, p. 3.
48. Adzic, R.R., W.E. O'Grady, and S. Srinivasan, Brookhaven Nat. Lab. Rep. BNL 51198, UC-93, Energy Conversion TID-4500, FC-8, Mar. 1980, p. 8.
49. Adzic, R.R., W.E. O'Grady, and S. Srinivasan, Surf. Sci., **94**, L191 (1980).
50. Afanas'ev, B.N., Elektrokhimiya, **16**, 296 (1980).
51. Agency of Industrial Science and Technology (Ministry of International Trade and Industry), Govt. of Japan, Development of Electric Vehicles in Japan, Oct. (1973).
52. Ahlberg, E., and V.D. Parker, J. Electroanal. Chem. Interfac. Electrochem., **107**, 197 (1980).
53. Ahmad, J., Ph.D. dissertation, American Univ., Washington, DC (1980).
54. Aiken, M., Proc. 24th Power Sources Conf., 68 (1970).
55. Ait'yan, S.Kh., and M.L. Belaya, Elektrokhimiya, **16**, 695 (1980).
56. Akimaru, S., Y. Tsutsumi, K. Tamura, S. Ono, K. Ogimoto, K. Sakamoto, and N. Masunaga, National Fuel Cell Seminar Abstracts, 1983, p. 153.
57. Akimoto, Y., C. Iwakura, and H. Tamura, Denchi Toronkai, (Koen Yoshishu), 18th, 43 (1979).
58. Alashkin, V.M., B.P. Nesterov, and N.V. Korovin, Tr. Mosk. Energ. Inst., **248**, 33 (1975).
59. Alashkin, V.M., B.P. Nesterov, and N.V. Korovin, Elektrokhimiya, **13**, 311 (1977).
60. Albright, L.F., T.G. Rheofanous, and A.G. Roher, J. Electrochem. Soc., **123**, 445 (1976).
61. Alekseenko, L.S., G.Yu. Koifman, E.D. Posokhova, and D.M. Shakhtin, Izv. Akad. Nauk SSSR, Neorg. Mater., **10** (11), 2006 (1974).

62. Alfenaar, M., Ger. Offen. 2,610,253 (1976).
63. Alfenaar, M., and R.L.E. Van Gasse, Ger. Offen. 2,635,636 (1977).
64. Alfenaar, M., Ger. Offen. 2,827,971 (1979).
65. Ali, Z., R. Bauer, W. Schoen, and H. Wendt, J. Appl. Electrochem., **10**, 97 (1980).
66. Allen, S.J., L.C. Feldman, D.B. McWhan, J.P. Remeika, and R.E. Walstedt, Superionic Conduc., (Proc. Conf.), 279 (1976).
67. Allison, H.J., R. Ramakumar, and W.L. Hughes, Proc. 4th Intersoc. Energy Conv. Eng. Conf., 1042 (1969).
68. Altseimer, J.H., F. Roach, J.M. Anderson, and C.A. Mangeng, "The Potential Application of Fuel Cells in the Chloralkali Industry," DOE/MC/A146-1986 (LA-10516-MS), U.S. Dept. of Energy, Washington, DC (1985).
69. American Chemical Society, Fuel Cell Systems I, Adv. in Chem. Series 47, A.C.S., Washington, DC (1965).
70. American Chemical Society, Fuel Cell Systems II, Adv. in Chem. Series 90, Washington, DC (1969).
71. American Cyanamid, Hydrocarbon Fuel Cell Electrodes, Final Rep., U.S. Govt. Contract No. DA-44-009-AMC-897(T), Jan. 1969.
72. Anahara, R., Fuji Electric Co., personal communication, June 1986.
73. Anbar, M., U.S. Patent No. 3,741,809 (1973).
74. Anderson, H.U., R. Murphy, A.K. Fox, B. Rossing, and A. Aldred, Proc. Workshop High Temperature Solid Oxide Fuel Cells, Brookhaven Nat. Lab., 1977, p. 41.
75. Anderson, J.M., J.R. Joiner, and H.C. Maru, National Fuel Cell Seminar Abstracts, 1983, p. 91.
76. Anderson, J.R. and N.R. Avery, J. Catalysis, **5**, 446 (1966).
77. Anderson, M.P., Proc. Workshop High Temperature Solid Oxide Fuel Cells, Brookhaven Nat. Lab., 1977, p. 26.
78. Andersson, B.E.C., L.I. Carlsson, and R.C.R. Johnsson, Ger. Offen. 2,655,451 (1977).
79. Andreev, V.N., and D.V. Kokoulina, Elektrokhimiya, **15**, 1417 (1979).
80. Andreev, V.N., and S.A. Kuliev, Elektrokhimiya, **16**, 1451 (1980).
81. Andrew, M.R., W.J. Gressler, J.K. Johnson, R.T. Short, and K.R. Williams, Proc. 5th Intersoc. Energy Conv. Eng. Conf., 5–80 (1970).
82. Andrew, M.R., W.J. Gressler, J.K. Johnson, R.T. Short, and K.R. Williams, J. Appl. Electrochem., **2**, 327 (1972).
83. Andrew, M.R., J.S. Drury, B.D. McNicol, C. Pinnington, and R.T. Short, J. Appl. Electrochem., **6**, 99 (1976).
84. Andrew, M.R., B.D. McNicol, R.T. Short, and J.S. Drury, J. Appl. Electrochem., **7**, 153 (1977).

85. Andruseva, S.I., M.R. Tarasevich, and K.A. Radyushkina, Elektrokhimiya, **13**, 253 (1977).
86. Andrzejak, A., and T. Kuczynski, Chem. Stosow., **21**, 441 (1977).
87. Ang, P.G.P., and A.F. Sammells, Influence of Electrolyte Composition on Electrode Kinetics in the Molten Carbonate Fuel Cell, NTIS Rep. CONF-781063-2 (1978).
88. Angerstein-Kozlowska, H., B.E. Conway, and W.B.A. Sharp, J. Electroanal. Chem., **43**, 9 (1973).
89. Anon., Fuel Cells: Power Converter of the Future, Prod. Des. Eng., Part I, Sept.–Oct. 1963; Part II, Nov.–Dec. 1963; Part III, Mar.–Apr. 1964.
90. Anon., Prod. Eng., July 19, 1965, p. 52.
91. Anon., ASEA Cells Installed in Fork Lift, Frontiers in Fuel Cells, Vol. 4, 11, 1 (1966).
92. Anon., Chem. Ind., Oct. 29, 1966, p. 1815.
93. Anon., Fuel Cell Power Test, Automotive Industries, Dec. 1, 1966, p. 70.
94. Anon., Chem. Week, Dec. 10, 1966, p. 82.
95. Anon., G.M. Says Electric Car is 10 Years Away, Machine Design, Dec. 12–14, (1966).
96. Anon., Looking Ahead to the Fuel Cell, Gas World, **3**, June 1967.
97. Anon., Oceanology International, 12, July–Aug. (1968); also Undersea Technology, 32, Dec. (1968).
98. Anon., Canadian Western Natural Gas Co. pamphlet: Canada's First Fuel Cell Powerplant, (1972).
99. Anon., Chem. & Eng. News, June 26, 1972, p. 14.
100. Anon., Chem. & Eng. News, July 3, 1972, p. 16.
101. Anon., Natural Gas Service News, **78** (Fall), 1 (1972).
102. Anon., Chem. & Eng. News, Aug. 27, 1973, p. 15.
103. Anon., Chem. & Eng. News, Sept. 3, 1973, p. 32.
104. Anon., Chem. & Eng. News, Sept. 10, 1973, p. 19.
105. Anon., Chem. & Eng. News, Jan. 7, 1974, p. 31.
106. Anon., Business Week, Jan. 12, 1974, p. 28c.
107. Anon., Chem. & Eng. News, Apr. 8, 1974, p. 20.
108. Anon., Time Magazine, Apr. 15, 1974, p. 82.
109. Anon., Chem. & Eng. News, Oct. 21, 1974, p. 8.
110. Anon., The Catalystic Converter, Can. Motorist, **62**, 21 (1974).
111. Anon., Chemical Economy and Engineering Review, **12**, 41 (1980).
112. Anon., (UTC, Power Systems Div.), "On-Site Fuel Cell Power Plant Technology Development Program," 1982–1983 Annual Rep., GRI 82099, Gas Research Institute, Chicago (1983).

113. Anon., (UTC, Power Systems Div.), "4.5 MW Fuel Cell Development Program," EPRI Rep.. EM-3856-LD, Palo Alto, CA (1984).
114. Anon., (UTC, Power Systems Div.), "Onsite 40 kW Fuel Cell Power Plant Manufacturing and Field Test Program," DOE/NASA-0255-1 NASA CR-174988, U.S. Dept. of Energy, Washington, DC (1985).
115. Anon., (Onsite Fuel Cell Users Group), Market/Business Assessment Task Force, "The Gas Powercell National Market Report," Gas Research Institute, Chicago (1985).
116. Anon., Commercial Brochure: "The PC-23 Fuel Cell: A Strategic Alternative," International Fuel Cells, South Windsor, CT (1985).
117. Anon., (General Electric Co.), "Development of Molten Carbonate Fuel Cell Power Plant," Final Rep., Vols. 1 and 2, DOE/ET/17019-20, U.S. Dept. of Energy, Washington, DC (1985).
118. Anon., Occidental Petroleum Commercial Fuel Cell Brochure, 1985.
119. Anon., Haldor-Topsoe, Inc., Commercial Brochure, EPRI, Palo Alto, CA (1986).
120. Anon., Institute of Gas Technology/Illinois Institute of Technology, Ongoing work under EPRI Contract RP-1085-2 (1986).
121. Anon., Ishikawajima-Harima Heavy Industries, Fuel Cell Information Booklet (in Japanese), Tokyo, 1986.
122. Anon., (Temple, Barker, and Sloane, Inc.), "The Financial and Strategic Planning Benefits of Fuel Cell Power Plants," 2 Vols., EPRI EM-4511, Palo Alto, CA (1986).
123. Anon., (Gas Turbine World), Performance Specs, Pequot Publishing Co., Fairfield, CT (1986).
124. Anon., (Electric Power Research Institute), "Technical Assessment Guide," 1987 Edition, EPRI, Palo Alto, CA (1987).
125. Anon., (Electric Power Research Institute), EPRI Final Rep., RP-1085-3, Palo Alto, CA (1987).
126. Anon., Request for Proposal, Wright Patterson AFB, Commerce Business Daily, Feb. 3, 1987.
127. Anson, F.C., Anal. Chem., **38**, 54 (1966).
128. Anson, F.C., J.H. Christie, and R.A. Osteryoung, J. Electroanal. Chem., **13**, 343 (1967).
129. Anthony, P.J., and A.C. Anderson, Phys. Rev. B, **14**, 5198 (1976).
130. Antropov, L.I. and D.A. Tkalenko, Elektrokhimiya, **6**, 595 (1970).
131. Appel, M. and A.J. Appleby, C.R. Acad. Sc. Paris, Ser. C, **280**, 551 (1975).
132. Appel, M. and A.J. Appleby, Electrochim. Acta, **23**, 1243 (1978).
133. Appleby, A.J., D.Y.C. Ng, S.K. Wolfson, and H. Weinstein, Proc. 4th Intersoc. Energy Conv. Eng. Conf., 346 (1969).
134. Appleby, A.J., J. Electrochem. Soc., **117**, 328 (1970).

135. Appleby, A.J., J. Electrochem. Soc., **117**, 1159 (1970).
136. Appleby, A.J., J. Electrochem. Soc., **117**, 1373 (1970).
137. Appleby, A.J., J. Electroanal. Chem., **27**, 347 (1970).
138. Appleby, A.J., J. Jacquelin and J.P. Pompon, Extended Abstracts, 4th Inter. Symposium on Fuel Cells, Antwerp, Belgium (Oct. 1972).
139. Appleby, A.J., and S.B. Nicholson, J. Electroanal. Chem., **38**, App 13 (1972).
140. Appleby, A.J., in "Modern Aspects of Electrochemistry," Vol. 9, B.E. Conway and J.O'M Bockris, (eds.), Plenum, NY (1974), p. 369.
141. Appleby, A.J., J. Electroanal. Chem., **51**, 1 (1974).
142. Appleby, A.J., and S.B. Nicholson, J. Electroanal. Chem., **53**, 105 (1974).
143. Appleby, A.J., Proc. Cornell Int. Symp. Workshop Hydrogen Economy, 1973, 197–211 (1975).
144. Appleby, A.J. and M. Jacquier, J. Power Sources, **1**, 17 (1976).
145. Appleby, A.J., and M. Savy, U.S. National Bureau of Standards Spec. Publ. 455, 241 (1976).
146. Appleby, A.J., and C. Van Drunen, J. Electrochem. Soc., **123**, 200 (1976), and references quoted therein.
147. Appleby, A.J., Electrochemistry for a Better World, in "Electrochemistry: Past Thirty Next Thirty Years" (Proc. Int. Symp. 1975), Plenum, NY (1977), p. 373.
148. Appleby, A.J., and A. Borucka, J. Chem. Soc. Faraday Trans. I, **73**, 1420 (1977).
149. Appleby, A.J., G. Crepy and G. Feuillade, in "Power Sources 6," D.H. Collins, (ed.), Academic Press, London (1977), p. 549.
150. Appleby, A.J., and S.B. Nicholson, J. Electroanal. Chem., **83**, 309 (1977).
151. Appleby, A.J., and M. Savey, Proc. Symp. Electrode Mater. Processes Energy Convers. Storage, 321 (1977).
152. Appleby, A.J., Proc. Workshop Electrocatalysis Fuel Cell Reactions, Brookhaven Nat. Lab. 1978, p. 23.
153. Appleby, A.J., G. Crepy and J. Jacquelin, J. Hydrogen Energy, **3**, 21 (1978).
154. Appleby, A.J., and R.K. Sen, Proc. Workshop Electrocatalysis Fuel Cell Reactions, Brookhaven Nat. Lab. 1978, p. 84.
155. Appleby, A.J., and J. Marie, Electrochim. Acta, **24**, 195 (1979).
156. Appleby, A.J., Proc.–Electrochem. Soc. 1979, 80-3, 457 (1980).
157. Appleby, A.J., M. Savy, and P. Caro, J. Electroanal. Chem. Interfac. Electrochem., **111**, 91 (1980).
158. Appleby, A.J., and S.B. Nicholson, J. Electroanal. Chem. Interfacial Electrochem., **112**, 71 (1980).
159. Appleby, A.J., and S.B. Nicholson, J. Electrochem. Soc., **127**, 759 (1980).
160. Appleby, A.J., National Fuel Cell Seminar Abstracts, 1980, p. 45.

161. Appleby, A.J., "The Fuel Cell: A Practical Power Source for Automobile Propulsion," in Conf. Proc., Drive Electric 80, Wembley, London, Oct. 1980.
162. Appleby, A.J., and S.B. Nicholson, J. Electrochem. Soc., **127**, 759 (1980).
163. Appleby, A.J., and F.R. Kalhammer, in "New Energy Conservation Technologies and Their Commercialization," International Energy Agency, Springer, NY (1981), p. 3138.
164. Appleby, A.J., National Fuel Cell Seminar Abstracts, 1981, p. 46.
165. Appleby, A.J., National Fuel Cell Seminar Abstracts, 1982, p. 138.
166. Appleby, A.J., Proc. Renewable Fuels and Advanced Power Sources for Transp. Workshop, 55 (1982).
167. Appleby, A.J., National Fuel Cell Seminar Abstracts, 1983, p. 1.
167a. Appleby, A.J., Proc. Workshop Electrochem. Carbon, Aug. 17–19, 1983. Proc. Vol. 84–5, Electrochem. Soc., Pennington, NJ, p. 251.
168. Appleby, A.J., in "Comprehensive Treatise of Electrochemistry," Vol. 7, B.E. Conway, J. O'M Bockris, S.U.M. Khan, and R.E. White, (eds.), Plenum, NY (1983), p. 173.
169. Appleby, A.J., G. Bronoel, M. Chemla, and H. Kita, in "Encyclopedia of the Electrochemistry of the Elements," Vol. IX, Part A, A.J. Bard, (ed.), Dekker, NY (1983), p. 383.
170. Appleby, A.J., in "Proc. of the Workshop on the Electrochemistry of Carbon," S. Sarangapani, J.R. Akridge and B. Schumm, (eds.), PV 84-5, The Electrochemical Society, Pennington, NJ (1984), p. 251.
171. Appleby, A.J., in "Progress in Batteries and Solar Cells," **5**, 246 (1984).
172. Appleby, A.J., and P.N. Ross, in "Power Sources for Electric Vehicles," B.D. McNichol and D.A.J. Rand, (eds.), Amer. Elsevier, NY (1984), p. 887.
173. Appleby, A.J., "Fuel Cells: Trends in Research and Applications," Ext. Abs. UNESCO Meeting on Fuel Cells, Ravello, Italy, 1985, UNESCO, Paris (1985).
174. Appleby, A.J., U.S. Patent No. 4,555,453, Nov. 1985.
175. Appleby, A.J., in "Proceedings, Corrosion 86," National Association of Corrosion Engineers, Houston, TX, 1986.
176. Appleby, A.J., and E.B. Yeager, Energy, **11**, 137 (1986).
177. Appleby, A.J., Energy, **12**, 13 (1986).
178. Appleby, A.J., Proc. 10th IPMI Technical Conference, International Precious Metals Institute, Allentown, PA (1986), p. 379.
179. Appleby, A.J., U.S. Patent No. 4,610,938, Sept. 1986.
180. Appleby, A.J., (ed.), "Fuel Cells: Trends in Research and Applications," Hemisphere Publ., NY (1987).
181. Appleby, A.J., Energy, (in press, 1987).
182. Appleby, A.J., "Advanced Fuel Cells and Their Future Market," in "Annual Rev. of Energy," J. Hollander, (ed.), Annual Reviews, Inc., Palo Alto, CA, 1988, p. 267.

References

183. Applegren, G.H., and S.A. Vejtesa, "Technical Assessment Guide," EPRI P-2410-SP, Palo Alto, CA (1982).
184. Aramaki, I., and K. Izawa, Japan. Kokai Tokkyo Koho, **78**, 96,438 (1978).
185. Arapkin, N.S., Tr. Mosk. Energ. Inst., **248**, 95 (1975).
186. Archer, D.H., J.J. Alles, W.A. English, E.F. Sverdrup, and R.L. Zahradnik, in "Fuel Cell Systems I," Advances in Chem. Series **47**, A.C.S., Washington, DC (1965), p. 357.
187. Archer, D.H., L. Elikan, and R.L. Zahradnik, in "Hydrocarbon Fuel Cell Technology," B.S. Baker (ed.), Academic Press, NY (1965), p. 51.
188. Arendt, R.H., K.W. Browall, and N.S. Choudhury, Ext. Abstr., Meet.–Int. Soc. Electrochem., 30th, 83 (1979).
189. Arendt, R.H., and M.J. Curran, U.S. Patent No. 4,216,278 (1980).
190. Arendt, R.H., and M.J. Curran, J. Electrochem. Soc., **127**, 1660 (1980).
191. Arendt, R.H., and M.J. Curran, J. Electrochem. Soc., **127**, 1663 (1980).
192. Argade, S.D., and T.G. Coker, U.S. Patent No. 4,175,023 (1975).
193. Armand, M.B., J.M. Chabagno, and M.J. Duclot, Fast Ion Transp. Solids: Electrodes Electrolytes, Proc. Int. Conf., 131 (1979).
194. Armstrong, R.D., and T. Dickinson, Superionic Conduct., (Proc. Conf.), 65–80 (1976).
195. Armstrong, W.A., Charge-Discharge Cycling of DREO Air Electrodes, NTIS Rep. AD-A016721 (1975).
196. Aronson, R.B., Mach. Des., **49** (4), 20–22, 24 (1977).
197. Aronsson, R., B. Heed, B. Jansson, A. Lunden, L. Nilsson, K. Schroeder, C.A. Sjoeblom, J.O. Thomas, and B.C. Tofield, Fast Ion Transp. Solids: Electrodes Electrolytes, Proc. Int. Conf., 471 (1979).
198. Arshinov, A.N., Yu.L. Golin, N.V. Korovin, G.N. Maksimov, B.S. Pospelov, and A.A. Khakhalov, Elektrokhimiya, **13**, 1134 (1977).
199. Asami, S., T. Seita, and A. Shimizu, Ger. Offen. 2,718,307 (1977).
200. Asami, S., T. Kiyota, and A. Shimizu, Japan, Kokai Tokkyo Koho **79**, 155,995 (1979).
201. Asami, S., T. Kiyota, and A. Shimizu, Japan. Kokai Tokkyo Koho, **79**, 155,996 (1979).
202. Asher, W.J., U.S. Patent No. 3,957,535 (1976).
203. Ashworth R., and J.H. Williams, National Fuel Cell Seminar Abstracts, 1980, p. 100.
204. Atoji, M., Domain and Surface Structures of Sodium Tungsten Bronzes, Argonne Nat. Lab. Rep. ANL-78-63 (1978).
205. Attwood, P.A., B.D. McNicol, and R.T. Short, J. Appl. Electrochem., **10**, 213 (1980).
206. Attwood, P.A., B.D. McNicol, R.T. Short, and J.A. Van Amstel, J. Chem. Soc., Faraday Trans. **1**, 76, 2310 (1980).

207. Austin, L.G., R. Palasi, and R.R. Klimpel, in "Fuel Cell Systems I," Adv. in Chem. Series 47, A.C.S., Washington, DC (1965), p. 35.
208. Austin, L.G., Fuel Cells: A Review of Government-Sponsored Research 1950–1964, NASA (SP-120) (1967).
209. Austin, L.G., H.W. Pickering, B.G. Ateya, N. Rebert, and T. Poweigha, Evaluation of Fuel Cell Electrolytes for Direct and Indirect Oxidation of Carbonaceous Fuels, NTIS Rep. COO-2927-1 (1978).
210. Babinsky, A.D., and C.J. Hora, Belg. Patent No. 871,096 (1979).
211. Bachot, J., J.P. Quentin, and J.L. Bourgeois, Eur. Patent Appl. 4,237 (1979).
212. Bacon, F.T., Fuel Cell Symposium, A.C.S. meeting, Atlantic City, NJ (1959).
213. Bacon, F.T., Proc. Roy. Soc. Lond., **A334**, 427 (1973).
214. Bacon, F.T., in "Trends Electrochem.," Plenary Invited Contrib., Aust. Electrochem. Conf., 4th, **1976**, 27 (1977), Plenum, NY.
215. Bacon, F.T., J. Electrochem. Soc., **126**, 7C (1979).
216. Badwal, S.P.S., "Fluoride Related Electrode Materials for High Temperature Electrochemical Applications," Ph.D. dissertation, Flinders University, Adelaide, Australia, 1977.
217. Baehr, H.D., and E.F. Schmidt, Brennst.-Waerme-Kraft, **29** (10), 393 (1977).
218. Bagotskii, V.S., and Yu.B. Vasil'ev, in "Fuel Cells: Their Electrochemical Kinetics," V.S. Bagotsky and Yu.B. Vasil'ev, (eds.), Consultants Bureau, NY (1966), p. 99.
219. Bagotskii, V.S., and Yu.B. Vassilyev, Electrochim. Acta, **12**, 1323 (1967).
220. Bagotskii, V.S., G.C. Shteinberg, and N.A. Urisson, USSR Patent No. 506,266 (1977).
221. Bagotskii, V.S., Yu.M. Vol'fkovich, N.M. Danchenko, V.S. Karyakin, A.T. Ovchinnikov, and O.V. Chumakovskii, Elektrokhimiya, **13**, 525 (1977).
222. Bagotsky, V.S., and I.E. Yablokova, Zh. Fiz. Khim., **27**, 1663 (1953).
223. Bagotsky, V.S., and Yu.B. Vasil'ev (eds), Fuel Cells: Their Electrochemical Kinetics, Consultants Bureau, NY (1966).
224. Baker, B.S., L.G. Marionowski, J. Meek, and H.R. Linden, presentation at 145th National Meeting, A.C.S. (1963).
225. Baker, B.S., L.G. Marianowski, and J. Meek, Chem. Eng. Progr. **63**, 1 (1963).
226. Baker, B.S., and T.J. Joyce, Fuel Cells for Onsite Power Generation, paper presented at 26th Meeting of the American Power Conf., Chicago, (1964).
227. Baker, B.S. (ed.), Hydrocarbon Fuel Cell Technology, Academic Press, NY (1965).
228. Baker, B.S., L.G. Marianowski, J. Zimmer, and G. Price, in "Hydrocarbon Fuel Cell Technology," B.S. Baker, (ed.), Academic Press, NY (1965), p. 293.
229. Baker, B.S., L.G. Marionowski, J. Meek, and H.R. Linden, in "Fuel Cell Systems I," Advan. in Chem. Series 47, A.C.S., Washington, DC (1965), p. 247.

References

230. Baker, B.S., Proc. 21st Power Sources Conf. (1967).
231. Baker, B.S., Electrocatalysis with Trifluoromethane Sulfonic Acid as an Electrolyte, Electric Power Research Institute Spec. Rep. EPRI SR-13, Conf. Proc. Fuel Cell Catal. Workshop, pp. 128–130 (1975).
232. Baker, B.S., Spec. Rep.–Electric Power Research Inst., EPRI SR-13, Conf. Proc.: Fuel Cell Catal. Workshop, PB-245 115, pp. 128–129 (1975).
233. Baker, B.S., Electrolyte for Hydrocarbon-Air Fuel Cells, NTIS Rep. No. AD-A007220 (1975).
234. Baker, B.S., and R.N. Camp, Low Power Metal Hydride Fuel Cell/Battery Hybrid Systems, NTIS AD/A Rep. No. 005079/9GA (1975).
235. Baker, B.S., and M. Klein, U.S. Patent No. 3,935,029 (1976).
236. Baker, B.S., and L.G. Marianowski, U.S. Patent No. 4,009,321 (1977).
237. Baker, B.S., and D.J. Dharia, U.S. Patent No. 4,169,917 (1979).
238. Baker, B.S., and D.J. Dharia, U.S. Patent No. 4,182,795 (1980).
239. Baker, B.S., U.S. Patent No. 4,329,403, May 1982.
240. Balabrahman, P., and P.D. Sunavala, Indian J. Technol., **18**, 272 (1980).
241. Balakrishnan, K., and V.K. Venkatesan, Ext. Abstr., Meet.–Int. Soc. Electrochem., 28th, 2, 198 (1977).
242. Balakrishnan, K., and V.K. Venkatesan, Electrochim. Acta, **24**, 131 (1979).
243. Balasescu, G., Rom. Patent 66,376 (1977).
244. Banister, R.M., National Fuel Cell Seminar Abstracts, 1979, p. 37.
245. Bannochie, J.B., "Power Sources 3," D.H. Collins, (ed.), Oriel Press, Newcastle Upon Tyne (1971), p. 417.
246. Baraboshkin, A.N., K.A. Kaliev, and S.M. Zacharyash, Ext. Abstr., Meet.–Int. Soc. Electrochem., 30th, 202 (1979).
247. Barak, M., paper presented at H.E.W. Conf., Columbia University, Apr. 1967.
248. Baranov, V.I., N.V. Vdovichenko, V.M. Vlasov, A.M. Ivanov, G.F. Muchnik, I.B. Rubashov, and L.S. Tabakman, Elektrokhimiya, **8**, 694 (1972).
249. Baranov, A.P., G.M. Pokatova, L.N. Mokrousov, G.V. Shteinberg, A.I. Dribinskii, M.R. Tarasevich, V.S. Bagotskii, and Yu.G. Chirkov, Elektrokhimiya, **13**, 615 (1977).
250. Bard, A.J. (ed), Encyclopedia of Electrochemistry of the Elements, Vol. 1, Dekker, NY (1973).
251. Bard, A.J., and L.R. Faulkner, "Electrochemical Methods," Wiley, NY (1980).
252. Baresel, D., W. Gellert, J. Heidemeyer, and P. Scharner, Angew Chem. Int. Ed., **10**, 194 (1971).
253. Baresel, D., W. Sarholz, P. Scharner, and J. Schmitz, Ber. Bunsenges. Phys. Chem., **78**, 608 (1974).
254. Barger, H., and A. Coleman, J. Phys. Chem., **74**, 880 (1970).

255. Barger, H.J., Jr., and A.A. Adams, U.S. Patent No. 3,948,681 (1976).
256. Baris, J.M., C.D. Iacovang, and W.M. Vogel, U.S. Patent No. 4,058,482 (1977).
257. Barker, R.E., Jr., Pure Appl. Chem., 46, 157 (1976).
258. Barmatz, M., and R. Farrow, Ultrason. Symp. Proc., 662 (1976).
259. Barnochie, J.G., and C.G. Clow, in "Power Sources 2," D.H. Collins, (ed.), Pergamon Press, London (1970), p. 531.
260. Barral, G., J. Guitton, C. Montella, and F. Vergara, Surf. Technol., 10, 25 (1980).
261. Barrett, M.A., and R. Parsons, J. Electroanal. Chem. Interfac. Electrochem., 42, App. 1 (1973).
262. Barthelemy, R., National Fuel Cell Seminar Abstracts, 1981, p. 35.
263. Bas'ko, A.T., and F.N. Patsyuk, Ukr. Khim. Zh. (Russ. Edition), 46, 237 (1980).
264. Bashara, N.M., and D.W. Peterson, J. Am. Opt. Soc., 56, 1320 (1966).
265. Batzold, J.S., and W.J. Asher, Proc. 24th Power Sources Conf., 185 (1970).
266. Batzold, J.S., in "From Electrocatalysis to Fuel Cells," G. Sandstede, (ed.), Univ. of Washington Press, Seattle (1972), p. 223.
267. Baudendistel, L., H. Bohm, J. Heffler, G. Louis, and F.A. Pohl, Proc. 7th Intersoc. Energy Conv. Eng. Conf., 20 (1972).
268. Bauer, D., and P. Gaillochet, Chem. Non-Aqueous Solvents, 5A, 251 (1978).
269. Bauer, E., and H. Preis, Z. Elektrochem., 43, 727 (1937).
270. Bauer, H.H., Electrodics, Halsted Press (Wiley), NY (1967).
271. Bauer, R., and H. Wendt, A.I.Ch.E. Symp. Ser., 75 (185), 56 (1979).
272. Bauerle, J.E., National Fuel Cell Seminar Abstracts, 1979, p. 69.
273. Baugh, L.M., J.A. Cook, and F.L. Tye, Power Sources, 7, 519 (1978).
274. Baukal, W., in "From Electrocatalysis to Fuel Cells," G.Sandstede, (ed.), Univ. of Washington Press, Seattle (1972), p. 247.
275. Baukal, W., and W. Kuhn, J. Power Sources, 1, 91 (1976).
276. Baukal, W., W. Kuhn, H. Kleinschmager, and F.J. Rohr, J. Power Sources, 1, 203 (1976).
277. Baumgartner, C.E., R.H. Arendt, C.D. Iacovangelo, and B.R. Karas, National Fuel Cell Seminar Abstracts, 1982, p. 113.
278. Baumgartner, C.E., and B.R. Karas, "Investigation of Layered Structures for Carbonate Fuel Cells," EM-3090, EPRI, Palo Alto, CA, May 1983.
279. Bauminger, E.R., A. Levy, F. Labenski de Kanter, S. Ofer, and C. Heitner-Wirguin, J. Phys., Colloq. (Orsay, Fr.), (C-1), 329 (1980).
280. Baur, E., W.D. Treadwell, and G. Trumpler, Z. Elektrochem., 27, 199 (1921).
281. Baur, E. and J.Z. Tobler, Z. Elektrochem., 39, 169 (1933).
282. Bayazitoglu, Y., Ph.D. dissertation, University of Michigan, (1974).

References

283. Bayazitoglu, Y., G.E. Smith, and R.L. Curl, Hydrogen Energy (Proc. Hydrogen Econ., Miami Energy Conf.), 1974 B, 859 (1975).
284. Bayazitoglu, Y., and G.E. Smith, Int. J. Hydrogen Energy, **2**, 139 (1977).
285. Bazan, J.C., C.E. Mayer, and P.M. Taberner, Rev. Latinoam. Ing. Quim. Quim. Apl., **8**, 51 (1978).
286. Bazhina, O.S., Yu.L. Golin, A.I. Savchuk, S.Yu. Usmanova, V.E. Federovskii, and O.V.Chumakovskii, USSR Patent No. 564,669 (1977).
287. Bazhina, O.S., Yu.L. Golin, V.E. Federovskii, and O.V. Chumakovskii, USSR Patent No. 568,095 (1977).
288. Beard, R.B., S.E. Dublin, R.M. Korener, and A.S. Miller, Proc. 8th Intersoc. Energy Conv. Eng. Conf., 536 (1973).
289. Beck, T.R., Boeing Internal Report, Aug. 1964.
290. Beck, T.R., in "From Electrocatalysis to Fuel Cells," G. Sandstede, (ed.), Univ. of Washington Press, Seattle (1972), p. 379.
291. Beckman, K.H., and N.H. Harrick, in "Optical Properties of Dielectric Films," N.N. Axelrod (ed.), Electrochemical Society Inc., NY (1968), p. 123.
292. Becquerel, A.C., Traité d'Electricité, I, Paris (1855).
293. Beden, B., C. Lamy, and J.M. Leger, Electrochim. Acta, **24**, 1157 (1979).
294. Beden, B., C. Lamy, and J.M. Leger, J. Electroanal. Chem. Interfacial Electrochem., **101**, 127 (1979).
295. Beecher, D.T., Energy Conversion Alternatives Study (ECAS), Westinghouse Phase 1. Vol. 1: Introduction and Summary and General Assumptions, NTIS Rep. NASA-CR-134941 (1976).
296. Beekmans, N.M., and L. Heyne, Electrochim. Acta, **21**, 303 (1976).
297. Been, J.F., Proc. 14th Intersoc. Energy Conv. Eng. Conf., Vol. 1, 544 (1979).
298. Beg, M.N., F.A. Siddiqi, H. Arif, M.I. Khan, and A. Haq, Indian J. Technol., **17**, 419 (1979).
299. Behret, H., H. Binder, and G. Sandstede, Proc. Symp. Electrocatal. **1974**, 319 (1974).
300. Behret, H., H. Binder, W. Clauberg, and G. Sandstede, Proc. Symp. Electrode Mater. Processes Energy Convers. Storage, 519 (1977).
301. Behret, H., H. Binder, and A. Koehling, Proc. Int. Symp. Fresh Water Sea, **6** (**3**), 15 (1978).
302. Behret, H., W. Clauberg, and G. Sandstede, Ber. Bunsenges. Phys. Chem., **83**, 139 (1979).
303. Behret, H., H. Binder, and G. Sandstede, in "Materials Science in Energy Technology," G.G. Libowitz and M.S. Whittingham (eds.), Academic Press, NY (1979), Chap. 7, p. 381.
304. Bélanger, G., Electrochemistry Dept., I.R.E.Q., private communication (1974).
305. Bélanger, G., Fuel Cells: Their Development and Potential, ACS Symp. Ser., 90 (Chem. Energy), 303 (1979).

306. Belaya, M.L., and S.Kh. Ait'yan, Elektrokhimiya, 16, 691 (1980).
307. Bellamy, B.A., A. Hooper, A.E. Hughes, J.M. Newman, C.F. Sampson, and B.C. Tofield, Mater. Sci. Monograph, 6 (Energy Ceram.), 950 (1980).
308. Bellows, R.J., Ger. Offen. 2,725,995 (1977).
309. Bellows, R.J., and P.G. Grimes, in "Power Sources for Electric Vehicles," B.D. McNichol and D.A.J. Rand, (eds.), Amer. Elsevier, NY (1984), p. 621.
310. Belt, R.N., National Fuel Cell Seminar Abstracts, 1981, p. 117.
311. Beltowska-Brzezinska, M., Electrochim. Acta, 24, 247 (1979).
312. Belzer, M., C.A. Heath, and B.L. Tarmy, U.S. Patent No. 3,234,050 (1966).
313. Beltzer, M., and H.H. Horowitz, Electrochem. Tech., 1, 464 (1966).
314. Belzer, M., J. Electrochem. Soc., **114 (2)**, 1200 (1967); see also Nature, Jan. 1967, p. 63.
315. Belzer, M., and J.S. Batzold, Proc. 4th Intersoc. Energy Conv. Eng. Conf., 354 (1969).
316. Benczur-Urmossy, G., Ger. Offen. 2,418,742 (1975).
317. Benczur-Urmossy, G., Ger. Offen. 2,823,042 (1979).
318. Benczur-Urmossy, G., UK Patent Appl. 2,027,979 (1980).
319. Beni, G., L.M. Schiavone, J.L. Shay, W.C. Dautremont-Smith, and B.S. Schneider, Nature (London), **282** (5736), 281 (1979).
320. Benjamin, T.G., E.H. Camara, and J.R. Selman, Proc. 14th Intersoc. Energy Conv. Eng. Conf., 579 (1979).
321. Bennett, D.R., Brit. Patent No. 1,367,673 (1974).
322. Benoit, J., Forces, **20**, 49 (1972).
323. Berger, C., in "Fuel Cell Systems I," Advances in Chem. Series **47**, A.C.S., Washington, DC (1965), p. 188.
324. Berger, C. (ed), Handbook of Fuel Cell Technology, Prentice-Hall, Englewood Cliffs, NJ (1968).
325. Bergmann, E., and H. Tannenberger, Proc. Workshop High Temperature Solid Oxide Fuel Cells, Brookhaven Nat. Lab., 1977, p. 148.
326. Bergmann, E., and G. Horlaville, Brit. U.K. Patent Appl. 2,018,833 (1979).
327. Berl, W.G., Trans. Electrochem. Soc., **83**, 253 (1943).
328. Berner, W.E., Method to Predict the Service Life of Internal Fuel Cell Baffle Materials, NTIS AD Rep. No. 781260/5GA (1974).
329. Bertocci, U., M. Cohen, W.S. Horton, T. Negas, and A.R. Siedle, Mixed Oxides for Fuel Cell Electrodes, NTIS Rep. No. PB-248744 (1976).
330. Bethin, J.R., C.K. Chiang, A.L. Dragoo, and A.D. Franklin, National Fuel Cell Seminar Abstracts, 1980, p. 128.
331. Bett, J.A.S., and K.R. Lynch, in "Power Sources 3," D.H. Collins, (ed.), Oriel Press, Newcastle Upon Tyne (1971), p. 439.
332. Bett, J.A.S., R.R. Lesieur, D.R. McVay, and H.J. Setzer, National Fuel Cell Seminar Abstracts, 1978, p. 146.

333. Bett, J.A.S., R.F. Buswell, R.R. Lesieur, A.P. Meyer, and J.L. Preston, National Fuel Cell Seminar Abstracts, 1979, p. 144.
334. Bett, J.A.S., R.F. Buswell, R.R. Lesieur, and H.J. Seltzer, National Fuel Cell Seminar Abstracts, 1980, p. 57.
335. Bett, J.A.S., H.R. Kunz and S.W. Smith, "Investigation of Alloy Catalysts and Redox Catalysts for Phosphoric Acid Electrochemical Systems," FCR-6440, Los Alamos Nat. Lab., Los Alamos, NM (1984).
336. Bettelheim, A., and T. Kuwana, Anal. Chem., **51**, 2257 (1979).
337. Bettelheim, A., R.J.H. Chan, and T. Kuwana, J. Electroanal. Chem. Interfac. Electrochem., **99**, 391 (1979).
338. Bettelheim, A., R.J.H. Chan, and T. Kuwana, J. Electroanal. Chem. Interfac. Electrochem., **110**, 93 (1980).
339. Bewick, A., J. Electroanal. Chem., **150**, 495 (1983).
340. Biddick, R.E., W.J. Cummins, and R. Rubischko, Proc. 24th Power Sources Conf., 92 (1970).
341. Biddick, R.E., G.A. Mueller, R.R. Sayano, and H.P. Silverman, Proc. Symp. Batt. Tract. Propuls., 295 (1972).
342. Biegler, T., J. Electrochem. Soc., **116**, 1131 (1969).
343. Biegler, T., and R. Woods, J. Electroanal. Chem. Interfac. Electrochem., **20**, 73 (1969).
344. Biggs, A., D.F. Dailly, and B.E. Waye, Proc. Br. Ceram. Soc., **23**, 44 (1972).
345. Bilmes, S.A., N.R. De Tacconi, and A.J. Arvia, J. Electrochem. Soc., **127**, 2184 (1980).
346. Binder, H., A. Köhling, H. Krupp, K. Richter, and G. Sandstede, Electrochim. Acta, **8**, 781 (1963).
347. Binder, H., A. Kohling, and G. Sandstede, in "Hydrocarbon Fuel Cell Technology," B.S. Baker, (ed.), Academic Press, NY (1965), p. 91.
348. Binder, H., A. Köhling, and G. Sandstede, in "Fuel Cell Systems I," Advances in Chem. Series **47**, A.C.S., Washington, DC (1965), p. 283.
349. Binder, H., A. Kohling, and G. Sandstede, Adv. Energy Conv. Eng., **7**, 77 (1967).
350. Binder, H., A. Kohling, and G. Sandstede, Energy Conversion, **11**, 17 (1971).
351. Binder, H., A. Köhling, H. Krupp, K. Richter, and G. Sandstede, in "Fuel Cell Systems I," Advances in Chem. Series **47**, A.C.S., Washington, DC (1965), p. 269.
352. Binder, H., A. Köhling, W. Kuhn, W. Lindner, and G. Sandstede, Nature, **224**, 1299 (1969); also Energy Conversion, **10**, 25 (1970).
353. Binder, H., in "From Electrocatalysis to Fuel Cells," G. Sandstede, (ed.), Univ. of Washington Press, Seattle (1972), p. 15.
354. Binder, H., in "From Electrocatalysis to Fuel Cells," G. Sandstede, (ed.), Univ. of Washington Press, Seattle (1972), p. 43.

355. Binder, H., W.H. Kuhn, W. Lindner, and G. Sandstede, U.S. Patent No. 3,708,342 (1973).
356. Binder, H., D. Hartmann, A. Koehling, H.G. Schomann, and G. Walter, NTIS Rep. No. NP-20868 (1975).
357. Bindra, P., S. Clouser, and E.B. Yeager, J. Electrochem. Soc., **126**, 1631 (1979).
358. Bindra, P., S. Sarangpani, S. Clouser, and E. Yeager, National Fuel Cell Seminar Abstracts, 1979, p. 106.
359. Bishop, E., and P.H. Hitchcock, The Analyst, **98**, 475 (1973).
360. Bishop, W.S., Proc. 3rd Canadian Fuel Cell Seminar, 72 (1970).
361. Blackmer, R.H., U.S. Patent No. 3,014,976 (1961).
362. Blackmer, R.H., U.S. Patent No. 3,061,658 (1962).
363. Blake, AR., A.T. Kuhn, and J.G. Sunderland, J. Electrochem. Soc., **120**, 492 (1973).
364. Blanchart, A.P.O., G.J.L. Gilbert, C. DeBrandt, and G.J.F. Spaepen, Ger. Offen. 2,856,262 (1979).
365. Blander, M., and M.L. Saboungi, Proc. Int. Symp. Molten Salts, 93 (1976).
366. Bleikamp, R.H., Jr., U.S. Patent No. 4,194,960 (1980).
367. Bliton, J.L., H.L. Rechter, and Y. Harado, Ceramic Bulletin, **42**, 6 (1963).
368. Block, M., and J.A. Kitchener, J. Electrochem. Soc., **113**, 947 (1966).
369. Blomgren, E., and J.O'M. Bockris, Nature, **186**, 305 (1960).
370. Bloomfield, D.P., and R. Cohen, U.S. Patent No. 3,972,731 (1976).
371. Bloomfield, D.P., and M.B. Landau, U.S. Patent No. 3,973,993 (1976).
372. Bloomfield, D.P., U.S. Patent No. 3,976,507 (1976).
373. Bloomfield, D.P., Ger. Offen. 2,604,966 (1976).
374. Bloomfield, D.P., R. Cohen, M.B. Landau, and M.C. Menard, Ger. Offen. 2,604,981 (1976).
375. Bloomfield, D.P., E. Behrin and P.H. Ross, LBL-14500, Lawrence Berkeley Lab., Berkeley, CA, 1982.
376. Blurton, K.F., and J.M. Sedlak, J. Electrochem. Soc., **121**, 1315 (1974).
377. Blurton, K.F., L.G. Marianowski, and T.D. Claar, National Fuel Cell Seminar Abstracts, 1978, p. 123.
378. Blurton, K.F., and J.R. Peterson, Commercial Applications of Molten Carbonate Fuel Cell Systems, NTIS Rep. No. CONF-790213-4, (1979).
379. Bocciarelli, C.V., in "From Electrocatalysis to Fuel Cells," G. Sandstede, (ed.), Univ. of Washington Press, Seattle (1972), p. 33.
380. Bockris, J.O'M., and B.E. Conway (eds), Modern Aspects of Electrochemistry (a series of monographs), Plenum, NY.
381. Bockris, J.O'M., in "Electrochemical Constants," N.B.S. Circular 524 (1953), Chap. 24.
382. Bockris, J.O'M., and A.K.M.S. Huq., Proc. Royal Soc. (London), **A237**, 277 (1956).

383. Bockris, J.O'M., M.A.V. Devanthan, and A.K.N. Reddy, Proc. Roy. Soc. (London), **A279** (1378), 327 (1964).
384. Bockris, J.O'M., B.J. Piersma, E. Gileadi and B.D. Cahan, J. Electroanal. Chem., **7**, 487 (1964).
385. Bockris, J.O'M., and S. Srinivasan, J. Electroanal. Chem., **11**, 350 (1966).
386. Bockris, J.O'M., A.K.N. Reddy, and M.A. Genshaw, J. Chem. Phys., **48**, 671 (1968).
387. Bockris, J.O'M., and B.J. Cahan, J. Chem. Phys., **50**, 1307 (1969).
388. Bockris, J.O'M., and S. Srinivasan, Fuel Cells: Their Electrochemistry, McGraw-Hill, Inc., NY (1969).
389. Bockris, J.O'M., and A.K.N. Reddy, Modern Electrochemistry (2 vol), Plenum, NY (1970).
390. Bockris, J.O'M. (ed.), Electrochemistry of Cleaner Environments, Plenum, NY (1972).
391. Bockris, J.O'M., in "From Electrocatalysis to Fuel Cells," G. Sandstede, (ed.), Univ. of Washington Press, Seattle (1972), p. 385.
392. Bockris, J.O'M., and D. Drazic, Electrochemical Science, Taylor & Francis, London (1972).
393. Bockris, J.O'M., "Energy Options," Australia and New Zealand Book Co., Sydney, Australia (1980).
394. Bockris, J.O'M., and A.J. Appleby, Energy, **11**, 95 (1986).
395. Boehm, H., and L. Baudendistel, Ger. Offen. 2,106,599 (1972).
396. Boehm, H., and V. Hartmann, Ger. Offen. 2,305,627 (1974).
397. Boehm, H., and G. Louis, Ger. Offen. 2,243,211 (1975).
398. Boehm, H., and R. Fleischmann, Ger. Offen. 2,417,447 (1975).
399. Boehm, H., J. Power Sources, **1 (2)**, 177 (1976).
400. Boehm, H., R. Fleischmann, and L. Baudendistel, Ger. Offen. 2,512,363 (1976).
401. Boehm, H., and J. Heffler, Ger. Offen. 2,533,215 (1977).
402. Boehm, H., R. Fleischmann, and J. Heffler, Ger. Patent No. 2,713,308 (1978).
403. Boehm, H., R. Fleischmann, and J. Heffler, U.S. Patent No. 4,172,808 (1979).
404. Bogdanovskaya, V.A., M.R. Tarasevich, and R.Kh. Burshtein, Elektrokhimiya, **8**, 1206 (1972).
405. Böhm, H., and F.A. Pohl, Wiss. Ber. AEG-Telefunken, **41**, 46 (1968).
406. Böhm, H., and F.A. Pohl, 3rd Internat. Symp. on Fuel Cells, Presses Académiques Européennes, Brussels (1969), p. 180.
407. Böhme, H.J., H. Bysel, W. Fischer, H. Kelinschmager, F.J. Rohr, and R. Steiner, Proc. 5th Intersoc. Energy Conv. Eng. Conf., 5–59 (1970).
408. Boies, D.B., and Dravnieks, Electrochem. Tech., **2**, 351 (1964).
409. Boies, D.B., and A. Dravnieks, in "Fuel Cell Systems I," Advances in Chem. Series **47**, A.C.S., Washington, DC (1965), p. 262.

410. Boikova, G.V., G.V. Zhutaeva, M.R. Tarasevich, V.S. Bagotskii, and N.A. Shumilova, Elektrokhimiya, **17**, 847 (1980).
411. Boilot, J.P., G. Collin, R. Comes, J. Thery, R. Collongues, and A. Cuinier, Superionic Conduc., (Proc. Conf.), 243 (1976).
412. Boilot, J.P., and J. Thery, Mater. Res. Bull., **11**, 407 (1976).
413. Boilot, J.P., A. Kahn, J. Thery, R. Collongues, J. Antoine, D. Vivien, C. Chevrette, and D. Gourier, Electrochim. Acta, **22**, 741 (1977).
414. Bol'shikh, A.E., and A.G. Kineev, Tr. Mosk. Energ. Inst., **365**, 51 (1978).
415. Bolan, P., and J.W. Staniunas, National Fuel Cell Seminar Abstracts, 1978, p. 33.
416. Bold, W., and M. Breiter, Electrochim. Acta, **5**, 145 (1961).
417. Bold, W., and M. Breiter, Electrochim. Acta, **5**, 169 (1961).
418. Bollenbacher, G., National Fuel Cell Seminar Abstracts, 1982, p. 65.
419. Bond, G.C., Disc. Faraday Soc., **41**, 200 (1966).
420. Bond, J.A., National Fuel Cell Seminar Abstracts, 1982, p. 95.
421. Bonds, T.L., M.H. Dawes, A.W. Schnake, and L.W. Spradin, "Fuel Cell Integrated Systems Evaluation," EPRI Rep. EM-1670, Palo Alto, CA (1981).
422. Bonnemay, M., G. Bronoel, L. Angely, and G. Peslerbe, Ger. Offen. 2,600,213 (1976).
423. Bono, P., P. Groult, G. Breban, and P. Blondeau, Ger. Offen. 2,416,094 (1974).
424. Boreskov, G.K., Disc. Faraday Soc., **41**, 263 (1966).
425. Borisoff, B., Proc. Symp. Batt. Tract. Propuls., 255 (1972).
426. Borucka, A., in "Fuel Cell Systems II," Advances in Chem. Series **90**, A.C.S., Washington, DC (1969), p. 242.
427. Borys, S.S., and J.P. Ackerman, Proc. 14th Intersoc. Energy Conv. Eng. Conf., Vol. 1, 563 (1979).
428. Boudart, M., J. Am. Chem. Soc., **72**, 1040 (1949).
429. Boudart, M., and L.D. Ptak, J. Catalysis, **16**, 90 (1970).
430. Bovin, J.O., Fast Ion Transp. Solids: Electrodes Electrolytes, Proc. Int. Conf., 315 (1979).
431. Bowman, R.M., B.J. Jody, K.C. Lu, and K.F. Blurton, National Fuel Cell Seminar Abstracts, 1980, p. 66.
432. Boyd, D.W., O.E. Buckley, C.E. Clark, Jr., R.B. Fancher, and J.R. Spelman, "EPRI Roles in Fuel Cell Commercialization," EPRI Rep. RP1677-15, Palo Alto, CA (1987).
433. Brainina, Kh.Z., E.V. Bazarova, and V.L. Volkov, Elektrokhimiya, **16**, 1203 (1980).
434. Brake, J.M., Biochemical Fuel Cells, Final Rep., U.S. Govt. Contract No. DA36-039-SC-90866 (1965).
435. Brand, M.J., J.J. Early, A. Kaufman, A. Stawsky, and J. Werth, National Fuel Cell Seminar Abstracts, 1983, p. 115.

436. Braunshtein, D., and D.S. Tannhauser, J. Phys. E., **2**, 921 (1979).
437. Braunstein, J., and C.E. Vallet, Electrochemical Studies of Mass Transport in High-Temperature Electrolytes, NTIS Rep. CONF-760312-I (1975).
438. Braunstein, J., and C.E. Vallet, Migrational Polarization in High-Current Density Molten Salt Electrochemical Devices, NTIS Rep. CONF-770531-6 (1977).
439. Braunstein, J., and C.E. Vallet, Proc.–Electrochem. Soc., 77–6 (Proc. Symp. Electrode Mater. Processes Energy Convers. Storage), 559 (1977).
440. Braunstein, J., H.R. Bronstein, S. Cantor, D. Heatherly, and C.E. Vallet, Molten Carbonate Fuel Cell Research at ORNL, NTIS Rep. ORNL/TM-5886 (1977).
441. Braunstein, J., H.R. Bronstein, S. Cantor, D.E. Heatherly, L.D. Hullet, R.L. Sherman, C.E. Vallet, and G. Watts, Molten Carbonate Fuel Cell Research at ORNL II. Theoretical and Experimental Transport Studies, Thermochemistry and Electron Microscopy, Oak Ridge Nat. Lab. Rep. 1977, ORNL/TM-6168/V2 (1977).
442. Braunstein, J., H.R. Bronstein, S. Cantor, D.E. Heatherly, J.I. Padova, T.M. Thomas, and C.E. Vallet, Composition Gradients Induced by Current Flow in Lithium Carbonate-Potassium Carbonate Mixtures, NTIS Rep. CONF-7810130-2 (1978).
443. Breault, R.D., U.S. Patent No. 3,972,735 (1976).
444. Breault, R.D., Ger. Offen. 2,638,988 (1977).
445. Breault, R.D., R.P. Harding, and F.S. Kemp, U.S. Patent No. 4,043,933 (1977).
446. Breault, R.D., U.S. Patent No. 4,185,145 (1980).
447. Breault, R.D., "Improved FCG-1 Technology," EPRI Rep. EM-1566, Palo Alto, CA (1980).
448. Breault, R.D., M. Krasij, D.A. Landsman, R.R. Lesieur, W. Luoma, R.C. Nickols, and D.G. Young, "Electric Utility Phosphoric Acid Power Plant and Subsystems," Rep. No. 12, May–June 1986, Contract DEN3-364, FCR 8022, NASA-Lewis Research Center, Cleveland, OH, (1986).
449. Bredbenner, A.M., Ceramic/Metal Seals (for Aerospace Batteries), NASA Conf. Publ. NASA-CP-2041, 355 (1977).
450. Breelle, Y., J. Cheron, and A. Grehier, Proc. 7th Intersoc. Energy Conv. Eng. Conf., 1 (1972).
451. Breelle, Y., J. Cheron, P. Degobert, and A. Grehier, Rev. Gen. Electr., **86**, 24 (1977).
452. Breelle, Y., and P. Degobert, Inf. Chim., **186**, 125–130, 133 (1979).
453. Breelle, Y., J. Cheron, P. Degobert, E. Goldenberg and A. Grehier, Rev. Inst. Français du Pétrole, **36** (4), 485 (1981).
454. Breelle, Y., "Performance d'un Bloc de Pile Hydrogene-Air, Institut Français du Pétrole, Rep. No. 29,889 (1982).
455. Bregoli, L.J., H.R. Kunz, and F.J. Luczak, U.S. Patent No. 4,206,270 (1980).

456. Bregoli, L.J., Electrochim. Acta, **23**, 489 (1978).
457. Bregoli, L., et. al., "Development of Improved Molten Carbonate Fuel Cell Technology," Final Rep., RP1085-4, EPRI, Palo Alto, CA (1983).
458. Breiter, M.W., J. Phys. Chem., **69**, 901 (1965).
459. Breiter, M.W., Electrochemical Processes in Fuel Cells, Springer-Verlag, NY (1969).
460. Breiter, M.W., and G.C. Farrington, Ionic Transport and Electronic Exchange at Solid Electrolyte Interfaces, U.S. Nat. Bur. Stand., Spec. Publ., 455, 323 (1976).
461. Breiter, M.W., G.C. Farrington, W.L. Toth, and J.L. Fuffy, Mater. Res. Bull., **12**, 895 (1977).
462. Breiter, M.W., Z. Phys. Chem. (Wiesbaden), **112**, 183 (1978).
463. Breiter, M.W., J. Electroanal. Chem. Interfac. Electrochem., **101**, 329 (1979).
464. Breiter, M.W., J. Electroanal. Chem. Interfac. Electrochem., **109**, 243 (1980).
465. Breiter, M.W., J. Electroanal. Chem. Interfac. Electrochem., **109**, 253 (1980).
466. Brenet, J.P., N.C. Hoan, M. Beley, and P. Chartier, Fr. Demande 2,356,286 (1978).
467. Brezina, M., and J. Riha, Chem. Listy, **74**, 737 (1980).
468. Brinkmann, J., Ger. Patent No. 1,596,278 (1975).
469. Britz, D., Anal. Chem., **52**, 1166 (1980).
470. Brodd, R.J., A. Kozawa, V.E. Zilionis, and T. Kalnoki-Kis, Proc. 5th Intersoc. Energy Conv. Cond., 3–37 (1970).
471. Broers, G.H.J., High Temperature Galvanic Fuel Cells, Ph.D. dissertation, Univ. of Amsterdam, Amsterdam, The Netherlands, (1958).
472. Broers, G.H.J., and J.A.A. Ketellaar, in "Fuel Cells I," G.J. Young (ed.), Reinhold, NY (1960), p. 78.
473. Broers, G.H.J., and M. Schenke, in "Hydrocarbon Fuel Cell Technology," B.S. Baker, (ed.), Academic Press, NY (1965), p. 225.
474. Broers, G.H.J., "Survey of Fundamental Research on Molten Carbonate Fuel Cells," Int. Rep. No. 69-0667/1272-7211, Centraal Technisch Instituut TNO, Apeldoorn, Netherlands (1969); translation in ORNL-tr-4663, Oak Ridge Nat. Lab.
475. Broers, G.H.J., in "Proceedings of the DOE/EPRI Workshop on Molten Carbonate Fuel Cells," J. Ackerman and A.J. Appleby, (eds.), EPRI WS-78-135, EPRI, Palo Alto, CA (1978).
476. Broglio, E.P., Proc. Symp. Batt. Tract. Propuls. 286 (1972).
477. Bronoel, G., Ann. Chim., (Paris), **1** (1-4), 209 (1976).
478. Bronoel, G., and J. Reby, Electrochim. Acta, **25**, 973 (1980).
479. Bronoel, G., J.C. Grenier, and J. Reby, Electrochim. Acta, **25**, 1015 (1980).
480. Brook, R.J., Microstruct. Ceram., Ext. Abstr. Pap. Jt. Meet., 12, 3 pp. (1975).
481. Brooman, E.W., and T.P. Hoar, Platinum Metals Rev., **9**, 122 (1966).

482. Brooman, E.W., K.R. Shillito, and W.K. Boyd, A Survey of the Use of Ceramics in Battery and Fuel Cell Applications, NTIS AD Rep. AD-A044888 (1977).
483. Browall, K.W., O. Muller, and R.H. Doremus, Mater. Res. Bull., 11, 1475 (1976).
484. Browall, K.W., and R.H. Doremus, J. Am. Ceram. Soc., 60, 262 (1977).
485. Browall, K.W., D.A. Shores, and P. Singh, National Fuel Cell Seminar Abstracts, 1981, p. 137.
486. Browall, K.W., General Electric Co. Rep. RP-1085-6.
487. Brown, C.T., Propr. Hydrazine Ses. Appl. Source Energy, Colloq. Int., 245 (1975).
488. Brown, C.T., Electrochemistry of Hydrazine-Hydrazine Azide Mixtures, NTIS Rep. AD-A031196 (1976).
489. Brown, D.E., and M. Mahmood, PCT Int. Appl. 79 00,709 (1979).
490. Brown, D.H., and J.C. Corman, Energy Conversion Alternatives Study (ECAS), General Electric Phase 1. Volume 2: Advanced Energy Conversion Systems. Part 1: Open Cycle Gas Turbines, NTIS NASA-CR-134948, Vol. 2, Pt. 1 (1976).
491. Brown, H.C., and C.A. Brown, J. Am. Chem. Soc., 84, 1493 (1962).
492. Brown, J., A. Murphy, P. Pietrogrande, and J. Tien, "Site-Specific Assessment of a 150 MW Coal Gasification Fuel Cell Power Plant," EPRI Rep. EM-3162, Palo Alto, CA (1983).
493. Brown, J., B. Nazar, and V. Varma, "Site-Specific Assessment of a 50 MW Coal-Gasification- Air-Cooled Fuel Cell Power Plant," EPRI Rep. RP1041-11, Palo Alto, CA (1983).
494. Brown, J.T., (Westinghouse Research Laboratories), personal communication, Jan. 1985.
495. Brown, J.T., Energy, 11, 209 (1986).
496. Brown, J.T., personal communication, 1987.
497. Broyde, R., J. Catal., 10, 13 (1968).
498. Brummer, S.B., in "Fuel Cell Systems II," Advances in Chem. Series 90, A.C.S., Washington, DC (1969), p. 223.
499. Brummer, S.B., M.J. Turner, S.D. Kirkland, and H. Feng, Proc. Symp. Electrocatal., 128 (1974).
500. Brummer, S., Spec. Rep.–Electr. Power Res. Inst., EPRI SR-13 (1975); Conf. Proc. Fuel Cell Catal. Workshop, PB-245-115, 131 (1975).
501. Brummer, S.B., and H.H. Horowitz, Proc. Workshop Electrocatalysis Fuel Cell Reactions, Brookhaven Nat. Lab. 1978, p. 93.
502. Brummer, S.B., J. McHardy, V.R. Koch, M. Turner, D. Toland, and J. McVeigh, National Fuel Cell Seminar Abstracts, 1978, p. 180.
503. Brummer, S.B., J.S. Foos, J. McHardy, J.B. McVeigh, D.E. Toland, and M. Turner, National Fuel Cell Seminar Abstracts, 1979, p. 48.

504. Brummer, S.B., Proc. Renewable Fuels and Advanced Power Sources for Transportation Workshop (June 1982), (H.L. Chum and S. Srinivasan, eds.), U.S. Govt. Rep. SERI/CP-234-1707, DE83011988, 127 (1983).
505. Buchanan, W., A. Kaufman, S. Pudick, C.L. Wang, J. Werth, and J.A. Whelan, National Fuel Cell Seminar Abstracts, 1983, p. 15.
506. Budgen, W.G., and J.H. Duncan, Sci. Ceram., **9**, 348 (1977).
507. Buggy, J.J., D.G. Hoover, and L. Christner, National Fuel Cell Seminar Abstracts, 1981, p. 41.
508. Buggy, J.J., and M.K. Wright, National Fuel Cell Seminar Abstracts, 1981, p. 56.
509. Bukreev, E.N., A.B. Gulyaenko, A.I. Kaninchak, and Yu.L. Tonkonogii, Inzh.-Fiz. Zh., **34**, 642 (1978).
510. Bump, C.M., Ph.D. dissertation, Pennsylvania State Univ., University Park, PA (1979).
511. Bunyagidj, C., H. Pietrowska, and M.H. Aldridge, The Preparation of Some Novel Electrolytes: Synthesis of Partially Fluorinated Alkane-sulfonic Acids and Potential Fuel Cell Electrolytes, NTIS Rep. No. AD-A078473 (1979).
512. Bunyagidj, C., The Preparation of Some Novel Fuel Cell Electrolytes: The Synthesis of Partially Fluorinated Alkanes-sulfonic Acids as Fuel Cell Electrolytes, Available Univ. Microfilms Int., Order No. 8008558 (1980).
513. Burke, L.D., and J.K. Mulcahy, J. Electroanal. Chem. Interfac. Electrochem., **73**, 207 (1976).
514. Burke, L.D., J.K. Mulcahy, and S. Venkatesan, J. Electroanal. Chem. Interfac. Electrochem., **81**, 339 (1977).
515. Burke, L.D., and O.J. Murphy, J. Electroanal. Chem. Interfac. Electrochem., **101**, 351 (1979).
516. Burke, L.D., and D.P. Whelan, J. Electroanal. Chem. Interfac. Electrochem., **103**, 179 (1979).
517. Burke, L.D., and O.J. Murphy, J. Electroanal. Chem. Interfac. Electrochem., **109**, 199 (1980).
518. Burlingame, M.V., in "Fuel Cell Systems II," Advances in Chem. Series **90**, A.C.S., Washington, DC (1969), p. 377.
519. Burns, R.K., Y.K. Choo, and S.N. Simons, National Fuel Cell Seminar Abstracts, 1980, p. 61.
520. Burr, J.F., R. Cooper, R.V. Homsy, and B.J. McKinley, Paper No. INDE 68, ACS Meeting, Las Vegas, NV (1982).
521. Burshtein, R.C., A.G. Pshenichnikow, and F.Z. Sabirov, in "Fuel Cell Systems II," Advances in Chem. Series **90**, A.C.S., Washington, DC (1969), p. 70.
522. Burshtein, R.Kh., M.R. Tarasevich, and V.A. Bogdanovskaya, Elektrokhimiya, **8**, 1542 (1972).

523. Bushnell, C.L., and H.R. Kunz, U.S. Patent No. 4,064,322 (1977).
524. Busi, J.D., Proc. Symp. Batt. Tract. Propuls., 195 (1972).
525. Busi, J.K., and L.R. Turner, J. Electrochem. Soc., **121**, 1830 (1974).
526. Buswell, R.F., Proc. 19th Power Sources Conf., 24 (1965).
527. Butherus, A.D., R.L. Cohen, and D.W. Maurer, Ger. Offen. 2,527,173 (1976).
528. Buzzelli, E.S., Ger. Offen. 2,547,491 (1976).
529. Caby, F., and J. Dauna, Ger. Offen. 2,534,725 (1976).
530. Caby, F., Fr. Demande 2,403,655 (1979).
531. Cahan, B., Ph.D. dissertation, University of Pennsylvania (1968).
532. Cahmus, H., M. Green, and J. Weber, Nature, **195**, 1310 (1962).
533. Cairns, E.J., D.L. Douglas, and L.W. Niedrach, A.I.Ch.E. Journal, **7**, 551 (1961).
534. Cairns, E.J., and D.C. Bartosik, J. Electrochem. Soc., **111**, 1205 (1964).
535. Cairns, E.J. and D.I. MacDonald, Electrochem. Tech., **2**, 65 (1964).
536. Cairns, E.J., in "Hydrocarbon Fuel Cell Technology," B.S. Baker, (ed.), Academic Press, NY (1965), p. 465.
537. Cairns, E.J., Nature, **210**, 161 (1966).
538. Cairns, E.J., J. Electrochem. Soc., **113**, 1200 (1966).
539. Cairns, E.J., and G.J. Holm, Extended Abstracts, Battery Division, Electrochem. Soc., **11**, 19 (1966).
540. Cairns, E.J., and E.J. McInerney, Technical Summary Report No1 9, ARPA Order No. 247, U.S. Govt. Contract No. DA-44-009-AMC-479(T), (1966).
541. Cairns, E.J., and E.J. McInerney, Extended Abstracts, Battery Division, Electrochem. Soc., **11**, 19 (1966).
542. Cairns, E.J. and A.M. Breitenstein, J. Electrochem. Soc., 114, 349 (1967).
543. Cairns, E.J., and E.J. McInerney, J. Electrochem. Soc., **114**, 980 (1967).
544. Cairns, E.J., and H. Shimotake, Science, **164**, 1347 (1969).
545. Cairns, E.J., and H. Shimotake, in "Fuel Cell Systems II," Advances in Chem. Series **90**, A.C.S., Washington, DC (1969), p. 392.
546. Cairns, E.J., in "Advances in Electrochemistry and Electrochemical Engineering," C.W. Tobias (ed.), Vol. 8, Wiley-Interscience, NY (1971).
547. Cairns, E.J., R.K. Stuenenberg, J.P. Ackerman, B.A. Feay, D.M. Gruen, M.L. Kyle, T.W. Latimer, J.N. Mundy, R. Rubischko, H. Shimotake, D.E. Walker, A.J. Zielen, and A.D. Tevebaugh, Development of High Energy Batteries for Electric Vehicles, NTIS PB 205254 (1971).
548. Cairns, E.J., and R.A. Murie, Corrosion Problems in Energy Conversion and Generation, (Symp. Paper), 3 (1974).
549. Cairns, E.J., and R.R. Witherspoon, Kirk-Othmer Encycl. Chem. Tech., 3rd ed., **3**, 545 (1978).
550. Callahan, M.A., Proc. 25th Power Sources Conf., 169 (1972).

551. Callahan, M.A., in "From Electrocatalysis to Fuel Cells," G. Sandstede, (ed.), Univ. of Washington Press, Seattle (1972), p. 189.
552. Callahan, M.A., Proc. 26th Power Sources Symp., 181 (1974).
553. Callahan, M.A., and W.R. Haas, Proc. 27th Power Sources Symp., 180 (1976).
554. Camara, E.H., G.A. Jarvi, A.L. Lee, and L.G. Marianowski, National Fuel Cell Seminar Abstracts, 1978, p. 141.
555. Camara, E.H., L.G. Marianowski, and R.A. Donado, U.S. Patent No. 4,160,067 (1979).
556. Camara E.H., and T.D. Claar, National Fuel Cell Seminar Abstracts, 1980, p. 153.
557. Camara, E.H., Status of Phosphoric Acid and Molten Carbonate Fuel Cell Technologies, paper presented at Ont.-Que. Section of Electrochem. Soc. Symposium Toronto, Feb. 1982.
558. Camara, E.H., and E.T. Ong, National Fuel Cell Seminar Abstracts, 1983, p. 46.
559. Camara, E.H., Institute of Gas Technology, Ongoing work under EPRI RP1085-10, 1986.
560. Cameron, D.S., Platinum Met. Rev., **22**, 38 (1978).
561. Cameron, D., Chemtech, **9**, 633 (1979).
562. Camp, R.N., and B.S. Baker, Matrices for Phosphoric Acid Fuel Cells, Final Rep., U.S. Govt. Contract No. DAAK02-72-C-0247 (1972).
563. Camp, R.N., and B.S. Baker, Proc. 7th Intersoc. Energy Conv. Eng. Conf., 7 (1972).
564. Canadian Defence Research Board, Proc. 3rd Canadian Fuel Cell Seminar, Sheridan Park, Ontario, Feb. 13–14, 1969 (1970).
565. Cannon, M.E., Power Conditioning and Control Module for Hydrazine-Air Battery, Type I, PP- 6204/U, NTIS Rep. AD732 339 (1971).
566. Carcia, P.F., R.D. Shannon, P.E. Bierstedt, and R.B. Flippen, J. Electrochem. Soc., **127**, 1974 (1980).
567. Carlin, W.W., U.S. Patent No. 4,184,941 (1980).
568. Cathey, W.N., L.J. Nicks, and D.J. Bauer, Platinum-substitute Materials as Electrocatalysts for Oxygen Reduction, U.S. Bur. Mines, Rep. Invest., RI-8341 (1979).
569. Crandall, R.S., and B.W. Faughnan, Phys. Rev. B., **16**, 1750 (1977).
570. Cathro, K.J., J. Electrochem. Soc., **116**, 1608 (1969).
571. Cavagnaro, D.M., NTIS Rep. NTIS/PS-78/0651 (1978).
572. Cavagnaro, D.M., Fuel Cells, Vol. 2: 1974–1978 (Citations from the NTIS Data Base), NTIS, Springfield, Va. (1978).
573. Cavagnaro, D.M., Fuel Cells, Vol. 3: 1977–June 1978 (Citations from the NTIS Data Base), NTIS, Springfield, Va. (1978).

574. Cavagnaro, D.M., NTIS Rep. NTIS/PS-79/0718 (1979).
575. Cavagnaro, D.M., Fuel Cells, Vol. 3: 1977–June 1979 (Citations from the NTIS Data Base), NTIS Rep. NTIS/PS-79-0717 (1979).
576. Cavallotti, P., D. Colombo, U. Ducati, and G. Jemmi, Riv. Combust., 33 (3–4), 93 (1979).
577. Centre Français du Commerce Extérieur (Economic Services of French Govt.), French Techniques, Transportation (1972).
578. Ceron, J., Ger. Offen. 2,620,933 (1976).
579. Ceynowa, J., and R. Wodzki, J. Power Sources, 1, 323 (1977).
580. Ceynowa, J., Polymer, 19, 73 (1978).
581. Ceynowa, J., R. Wodzki, and A. Narebska, Pol. Patent No. 100,706 (1979).
582. Cha, C.-S., Z.-D. Wang, Y.-C. Chu, and C.-T. Lo, Power Sources 7, J. Thompson (ed.), Academic Press, NY (1979), p. 769.
583. Cha, C.-S., in "Fuel Cells: Trends in Research and Applications," A.J. Appleby, (ed.), Hemisphere Publ., NY (1987), p. 95.
584. Chamberland, B.L., Inorg. Chem., 8, 1183 (1969).
585. Chamberlin, R.D., Ph. D. dissertation, Pennsylvania State University (1965).
586. Chapman, L.E., Proc. 7th Intersoc. Energy Conv. Eng. Conf., 466 (1972).
587. Chekhovskoi, V.Ya., I.A. Zhukova, and V.D. Tarasov, Teplofiz. Vys. Temp., (4), 754 (1979).
588. Chen, W.K.W., R.B. Mesrobian, D.S. Ballentine, D.J. Metz, and A. Glines, J. Polymer Sci., 23, 903 (1957).
589. Chenevey, E.C., Ultrafine Poly(benzimidazole) (PBI) Fibers, NASA (Contract Rep.) NASA-CR- 159644 (1979).
590. Chernousova, N.I., G.I. Elfimova, and G.A. Bodganovskii, Zh. Prikl. Khim. (Leningrad), 52, 2507 (1979).
591. Chernousova, N.I., G.A. Bogdanovskii, and G.D. Vovchenko, Vestn. Mosk. Univ., Ser. 2: Khim., 21, 389 (1980).
592. Chernyavskaya, A.M., E.P. Razgulyaev, and S.A. Telezhkin, Primen. Splavov Reniya, Dokl. Vses. Soveschch. Probl. Reniya (USSR), 4, 182 (1973).
593. Cheron, J., Ger. Offen. 2,454,183 (1976).
594. Cheron, J., Ger. Offen. 2,529,470 (1976).
595. Cheron, J., Ger. Offen. 2,545,098 (1976).
596. Cheron, J., Ger. Offen. 2,645,359 (1977).
597. Cheron, J., Ger. Offen. 2,947,288 (1980).
598. Chevrier, J., and G. Siclet, Bull. Soc. Chim. Fr. (7-8, Pt. 1), 1037 (1976).
599. Chiang, C.K., A.L. Dragoo, and A.D. Franklin, National Fuel Cell Seminar Abstracts, 1979,
600. Chierchie, T., P. Taberner, and C. Mayer, Rev. Latinoam. Ing. Quim. Quim. Apl., 9, 47 (1979).

601. Child, W.C., D.H. Smith, H.R. Bronstein, and J. Braunstein, National Fuel Cell Seminar Abstracts, 1980, p. 179.
602. Children's Hospital Medical Center, The, Boston, MA and Thermo Electron Engineering Corp., Waltham, MA, Artificial Heart Program–Final Rep.: Summary and Conclusions, U.S. Govt. Contract No. PH43-66-0 (1966).
603. Chillier-Duchatel, N., and B. Verger, Fr. Demande 2,292,343 (1976).
604. Ching, A.C., A.P. Gillis, and F.M. Plauche, Proc. 7th Intersoc. Energy Conv. Eng. Conf., 368 (1972).
605. Chirkov, Yu.G., Elektrokhimiya, **10**, 1804 (1974).
606. Chirkov, Yu.G., Elektrokhimiya, **10**, 1536 (1974).
607. Chizmadzhev, Yu.A., in "Fuel Cells: Their Electrochemical Kinetics," V.S. Bagotsky and Yu.B. Vasil'ev, (eds.), Consultants Bureau, NY (1966), p. 111.
608. Chludzinski, P.J., I.F. Canzig, A.P. Fickett, and D.W. Craft, Regenerative Fuel Cell Development for Satellite Secondary Power, NTIS Rep. AD-762923 (1973).
609. Chodosh, S.M., N.I. Palmer, and H.G. Oswin, in "Hydrocarbon Fuel Cell Technology," B.S. Baker, (ed.), Academic Press, NY (1965), p. 495.
610. Chong, V.M., and W.H. Seitzer, U.S. Patent No. 4,131,514 (1978).
611. Chottiner, J., U.S. Patent No. 4,152,489 (1979).
612. Christie, J.H., G. Lauer, and R.A. Osteryoung, J. Electroanal. Chem., **7**, 60 (1964).
613. Christie, J.H., R.A. Osteryoung, and F.C. Anson, J. Electroanal. Chem., **13**, 236 (1967).
614. Christner, L., and P.N. Ross, Proc. Workshop Electrocatalysis Fuel Cell Reactions, Brookhaven Nat. Lab. 1978, p. 40.
615. Christner, L.G., D.C. Nagle, and P.R. Watson, U.S. Patent No. 4,115,528 (1978).
616. Christner, L.G., and D.C. Nagle, U.S. Patent No. 4,115,627 (1978).
617. Christner, L.G., H.C. Maru, C.V. Chi, S.R. Perkari, and M.A. Lambrech, Proc. 14th Intersoc. Energy Conv. Eng. Conf., 554 (1979).
618. Christner, L.G., in "Proc. of the Workshop on the Electrochemistry of Carbon," S. Sarangapani, J.R. Akridge, and B. Schumm, (eds.), PV 84-5, The Electrochemical Society, Pennington, NJ (1984), p. 291.
619. Christopher, H.A., Proc. 20th Power Sources Conf., 18 (1966).
620. Chudoba, S., and Z. Kubas, High Temp. Sci., **8**, 377 (1976).
621. Chudoba, S., and Z. Kubas, High Temp. Sci., **8**, 385 (1976).
622. Chum, H.L., and S. Srinivasan, (eds.), Proc. Workshop on Renewable Fuels and Advanced Power Sources for Transportation (June 1982), U.S. Govt. Rep. No. SERI/CP-234-1707, DE 83011988 (1983).
623. Ciprios, G., Proc. 1st Intersoc. Energy Conv. Eng. Conf., 9 (1966).
624. Ciprios, G., Proc. 5th Intersoc. Energy Conv. Eng. Conf., 5–65 (1970).

625. Ciprios, G., Exxon Inc., Final Rep. RP1200-4, EPRI, Palo Alto, CA (1979).
626. Ciric, J., and W.F. Graydon, J. Phys. Chem., **66**, 1549 (1962).
627. Claar, T.D., R.A. Donado, and L.G. Marianowski, Effects of Sulfur-Containing Gases on Molten Carbonate Fuel Cell Behavior, NTIS Rep. CONF-781030-3 (1978).
628. Claar, T.D., L.G. Marianowski, and A.F. Sammells, Development of Sulfur-Tolerant Components or Second-Generation Molten Carbonate Fuel Cells, NTIS Rep. EPRI0EM01114 (1979).
629. Claar, T.D., E. Ong, and L.G. Marianowski, National Fuel Cell Seminar Abstracts, 1979, p. 126.
630. Clark, M.B., W.G. Darland, and K.V. Kordesch, Electrochem. Technol., **3**, 166 (1965).
631. Clark, W.N., and V.R. Shorter, "Cool Water Coal Gasification: A Mid-term Performance Assessment," Mining Engineering, Feb. 1987, p. 97.
632. Clausi, J.V., M.B. Landau, and G. Vartanian, U.S. Patent No. 3,748,180 (1973).
633. Clavilier, J., R. Faure, G. Guinet, and R. Durand, J. Electroanal. Chem. Interfac. Electrochem., **107**, 205 (1980).
634. Clavilier, J., J. Electroanal. Chem. Interfac. Electrochem., **107**, 211 (1980).
635. Clow, C.G., and J.G. Bannochie, Proc. 4th Intersoc. Energy Conv. Eng. Conf., 1057 (1969).
636. Clukh, A.M., V.M. Golub, Z.R. Karichev, L.I. Leonichev, Yu.M. Morozenskov, Yu.G. Protsenko, G.V. Samoilov, V.Kh. Stan'kov, and V.I. Yakovlev, USSR Patent No. 551,734 (1977).
637. Cnobloch, H., H. Nischik, and F. von Sturm, Chemie-Technik, **41** (4), 146 (1969).
638. Cnobloch, H., M. Marchetto, H. Nischik, G. Righter, and F. von Sturm, Proc. 3rd Internat. Symp. on Fuel Cells, Presses Académiques Européenes, Brussels (1969), p. 203.
639. Cnobloch, H., H. Kohlmüller, D. Kühl, and F. von Sturm, Hydrazin-Brennstoffzellen, Vortrag anlässlich der 5. Energie-Direkt-Umwandlung-Tagung in Essen, Nov. 1970.
640. Cnobloch, H., H. Kohlmüller, and D. Kühl, Energy Conversion, **14**, 75 (1975).
641. Cnobloch, H., and H. Kohlmuller, Paper No. 282, Int. Soc. Electrochem. Meeting (1976).
642. Cnobloch, H., D. Gröppel, H. Kohlmüller, D. Kühl, and G. Siemsen, Power Sources 7, Academic Press, NY (1979), pp. 389–404.
643. Cnobloch, H., D. Gröppel, H. Kohmüller, D. Kühl, H. Popoa, and G. Siemsen, Deutsche Gesellschaft für Wehrtechnik e.V. Forum, "Mobile Stromversorgung im Felde," Hamburg, Nov. 1979 (1980), p. 33.
644. Cochran, N.P., in "Fuel Cell Systems II," Advances in Chem. Series **90**, A.C.S., Washington, DC (1969), p. 383.

645. Coffin, D.L., Proc. 6th Intersoc. Energy Conv. Eng. Conf., 329 (1971).
646. Cohen, R., National Fuel Cell Seminar Abstracts, 1981, p. 48.
647. Cohn, E.M., in "Fuel Cell Systems I," Advan. in Chem. Series 47, A.C.S., Washington, DC (1965), p. 1.
648. Cohn, J.G.E., A.P. Hauel, and C.D. Keith, U.S. Patent No. 3,223,556 (1965).
649. Cole, T., N. Weber, and T.K. Hunt, Fast Ion Transp. Solids: Electrodes Electrolytes, Proc. Int. Conf., 277 (1979).
650. Colling, P.M., R.V. Norton, and J.E. Martin, Selection and Evaluation of Carbon-Resin Composites for Bipolar Plates for Hydrogen Fuel Cells, NTIS Rep. AD-A070185 (1978).
651. Collins, M.F., Proc. 25th Power Sources Conf., 162 (1972).
652. Collins, M.F., R. Michalek, and W. Brink, Proc. 7th Intersoc. Energy Conv. Eng. Conf., 32 (1972).
653. Collman, J.P., M. Marrocco, P. Denisevich, C. Koval, and F.C. Anson, J. Electroanal. Chem. Interfac. Electrochem., **101**, 117 (1979).
654. Collongues, R., Mater. Res. Bull., **13**, 1281 (1978).
655. Collongues, R., J. Thery, and J.P. Boilot, Solid Electrolytes, 253 (1978).
656. Collongues, R., A. Kahm, and D. Michel, Annu. Rev. Mater. Sci., **9**, 123 (1979).
657. Colman, W.P., D.Gershberg, J. Di Palma, and R.G. Haldeman, Proc. 19th Annual Power Sources Conf., May 18–20, 1965.
658. Colomban, P., J.P. Boilot, A. Kahn, J. Thery, G. Lucazeau, R. Mercier, G. Collin, and R. Comes, Colloq. Int. C.N.R.S. 1977, 277 (Ordre Desordre Solides), 197 (1978).
659. Compagnie Française Thomson-Houston, Brandstofcel, Netherlands Patent No. 6,508,067 (1965).
660. Compagnie Generale d'Electricite, Comm. Eur. Communities, (Rep.) EUR, EUR 6085, Semin. Hydrogen Energy Vector: Prod., Use, Transp., 166 (1978).
661. Conger, M.T., T.A. Evans, and H.F. Villemain, U.S. Patent No. 3,935,050 (1976).
662. Conner, W.J., B.J. Greenough, and G.B. Adams, in "Fuel Cell Systems II," Advances in Chem. Series **90**, A.C.S., Washington, DC (1969), p. 426,
663. Conroy, E., and G. Podolsky, Inorg. Chem., **7**, 614 (1968).
664. Constable, D.C., J. Electrochem. Soc., **118**, 1391 (1971).
665. Constantinescu, D., and I. Atanasiu, Rev. Chim. (Bucharest), **26**, 567 (1975).
666. Contractor, A.Q., and H. Lal, J. Electroanal. Chem. Interfac. Electrochem., **103**, 103 (1979).
667. Conway, B.E., Theory and Principles of Electrode Processes, Ronald Press, NY (1965).
668. Conway, B.E., Proc. 3rd Can. Fuel Cell Seminar, Can. Defence Research Board, Sheridan Park, Ontario, Feb. 13–14, 1969 (1970), p. 129.

669. Conway, B.E., and S. Gottesfeld, J. Chem. Soc., Faraday Trans. I, **69**, 1090 (1973).
670. Conway, B.E., H. Angerstein-Kozlowska, and L.H. Laliberté, J. Electrochem. Soc., **121**, 1596 (1974).
671. Conway, B.E., Proc. Symp. Electrode Mater. Processes Energy Convers. Storage, 441 (1977).
672. Conway, B.E., and J.A. Joebstl, Proc. Workshop Electrocatalysis Fuel Cell Reactions, Brookhaven Nat. Lab. 1978, p. 212.
673. Conway, B.E., in "Modern Aspects of Electrochemistry," Vol. 16, B.E. Conway, J.O'M. Bockris and R.E. White, (eds.), Plenum, NY (1986), p. 103.
674. Cook, A.R., Proc. Symp. Batt. Tract. Propuls., 65 (1972).
675. Cooper, J.F., Proc. 15th Intersoc. Energy Conv. Eng. Conf., Seattle, WA (1980), p. 1487.
676. Cooper, J.F., U.S. Dept. Commerce, Rep. UCLR-53536 (1984).
677. Corkum, R., and J. Milne, Can. J. Chem., **56**, 1832 (1978).
678. Corrigan, T.J., U.S. Patent No. 3,909,299 (1975).
679. Coughlin, R.W., and M. Farooque, J. Appl. Electrochem., **10**, 729 (1980).
680. Coughlin, R.W., and M. Farooque, Ind. Eng. Chem. Process Des. Dev., **19**, 211 (1981).
681. Crandall, R.S., P.J. Wojtowicz, and B.W. Faughnan, Solid State Commun., **18**, 1409 (1976).
682. Crandall, W.B., and Y. Harada, SAE (Tech. Pap.), 750466 (1975).
683. Criddle, E.E., and E.J. Casey, National Fuel Cell Seminar Abstracts, 1981, p. 112.
684. Criner, D.E., P. Steitz, and C.R. Rogers, National Fuel Cell Seminar Abstracts, 1979, p. 26.
685. Criner, D.E., and P. Steitz, National Fuel Cell Seminar Abstracts, 1980, p. 68.
686. Croset, M., J.P. Schnell, G. Gelasco, and J. Siejka, Proc. Workshop High Temperature Solid Oxide Fuel Cells, Brookhaven Nat. Lab., 1977, p. 139.
687. Crouse, D.N., F. Walsh, and M. Walsh, National Fuel Cell Seminar Abstracts, 1978, p. 94.
688. Crouse, D.N., ECO Inc., work performed under EPRI RP-1676-1, 1980.
689. Crowe, B.J., Fuel Cells: A Survey, NASA Rep. (SP-5115), Washington, DC (1973).
690. Cummings, C.R., and W.C. Racine, National Fuel Cell Seminar Abstracts, 1980, p. 104.
691. Cusamano, J.A. and R.B. Levy, "Assessment of Fuel Processing Alternatives for Fuel Cell Power Generation," EM-570, EPRI, Palo Alto, CA (1977).
692. Curry, B.E., National Fuel Cell Seminar Abstracts, 1981, p. 13.
693. Cutler, L.H., U.S. Patent No. 3,607,420 (1971).

694. Cutler, L.H., K.B. Keating, V. Mehra, and W.R. Wolfe, Proc. 8th Intersoc. Energy Conv. Eng. Conf., 86 (1973).
695. Cutlip, M.B., Electrochim. Acta, **20**, 767 (1975).
696. Cuttica, J.J., "Key Commercialization Issues Associated with Fuel Cell Power Plants," Proc. IAE International Symposium on Fuel Cells and Advanced Batteries, Institute of Applied Energy, Tokyo (1986).
697. D'Agostino, V.F., J.Y. Lee, and E.H. Cook, Jr., Ger. Offen. 2,558,393 (1976).
698. D'Agostino, V.F., J.Y. Lee, and E.H. Cook, Jr., U.S. Patent No. 4,012,303 (1977).
699. Dabrowski, R., and Z. Witkiewicz, Wiad. Chem., **28**, 821 (1975).
700. Dabrowski, R., and L. Kreja, Biul. Wojsk. Akad. Tech., **28**, 99 (1979).
701. Dalla Betta, R.A., Proc. Workshop Electrocatalysis Fuel Cell Reactions, Brookhaven Nat. Lab. 1978, p. 203.
702. Damiano, P.J., U.S. Patent No. 4,129,685 (1978).
703. Damien, D., and J.C. Sohm, The Use of Fuel Cell Ion Exchange Membranes, Rep P/539/77/23 (1977), Avail. NTIS.
704. Damjanovic, A., and J.O'M. Bockris, Electrochim. Acta, **11**, 376 (1966).
705. Damjanovic, A., and V. Brusic, Electrochim. Acta, **12**, 615 (1967).
706. Damjanovic, A., M.A. Genshaw, and J.O'M. Bockris, J. Electrochem. Soc., **114**, 466 (1967).
707. Damjanovic, A., D. Sepa, and J.O'M. Bockris, J. Res. Inst. Catal. Hokkaido Univ., **16**, 1 (1968).
708. Damjanovic, A., A.T. Ward, and M. O'Jea, J. Electrochem. Soc., **121**, 1186 (1974).
709. Damjanovic, A., and B. Jovanovic, J. Electrochem. Soc., **123**, 374 (1976).
710. Damjanovic, A., D.B. Sepa, and M.V. Vojnovic, Electrochim. Acta, **24**, 887 (1979).
711. Damjanovic, A., L.S.R. Yeh, and J.F. Wolf, J. Electrochem. Soc., **127**, 874 (1980).
712. Damjanovic, A., L.S.R. Yeh, and J.F. Wolf, J. Electrochem. Soc., **127**, 1945 (1980).
713. Damjanovic, A., L.S.R. Yeh, and J.F. Wolf, J. Electrochem. Soc., **127**, 1951 (1980).
714. Danielyan, G.L., Arm. Khim. Zh., **33**, 3 (1980).
715. Danilov, V.G., I.A. Kedrinskii, and Yu.G. Chirkov, Elektrokhimiya, **15**, 757 (1979).
716. Danilov, V.G., I.A. Kedrinskii, and Yu.G. Chirkov, Elektrokhimiya, **15**, 1296 (1979).
717. Darlington, W.B., J.D. Driskill, and D.W. DuBois, U.S. Patent No. 4,165,248 (1979).
718. Darrow, K.G., Jr., National Fuel Cell Seminar Abstracts, 1980, p. 48.

719. Davidson, C.R., Ph.D. dissertation, Univ. Virginia, Charlottesville, VA (1978).
720. Davidson, C.R., Brookhaven Nat. Lab. Rep. BNL 50951, UC-93, Energy Conversion TID-4500, Dec. 1978, p. 6.
721. Davidson, C.R., and S. Srinivasan, J. Electrochem. Soc., **127**, 1060 (1980).
722. Davis, D.G., in "Porphyrins," D. Dolphin (ed.), Academic Press, NY (1978), 5 (Part C), p. 127.
723. Davis, H.J., Proc. 3rd Can. Fuel Cell Seminar, 33 (1969).
724. Davis, S.M., and M.N. Hull, J. Electrochem. Soc., **125**, 1918 (1978).
725. Davtyan, O.K., in "Fuel Cells: Their Electrochemical Kinetics," V.S. Bagotsky and Yu.B. Vasil'ev, (eds.), Consultants Bureau, NY (1966), p. 35.
726. Davy, H., Nicholson's Journal, 144, (1802).
727. Dawes, M.H., and K.W. Browall, National Fuel Cell Seminar Abstracts, 1979, p. 140.
728. Dawes, M.H., National Fuel Cell Seminar Abstracts, 1980, p. 169.
729. De Jonghe, L.C., and M.Y. Hsieh, Substructure and Properties of Sodium Beta Alumina Solid Electrolytes, Argonne Nat. Lab., ANL Rep. ANL-76-8, Proc. Symp. Adv. Battery Res. Ees., B/12-B/24 (1976).
730. De Jonghe, L.C., J. Mater. Sci., **14**, 33 (1979).
731. De Nora, O., G. Modica, and L. Giuffre, Eur. Patent Appl. 4,029 (1979).
732. Deborski, G.A., U.S. Patent No. 4,179,350 (1979).
733. Degan, L.J., G.D. Rodriguez, and P.D. Hindley, Molten Carbonate Fuel Cell Powerplants, GE-PG & E.-SCE joint publication, Mar. 1980.
734. Delahay, P., New Instrumental Methods in Electrochemistry, Interscience, NY (1954).
735. Delahay, P., Double Layer and Electrode Kinetics, Interscience, N.Y. (1966).
736. Delahay, P., and C.W. Tobias, (eds.), Advances in Electrochemistry and Electrochemical Engineering, Interscience, NY (a series of monographs).
737. DeLevie, R., in "Advances in Electrochem. and Electrochem. Eng." Vol. 6, p. 329, P. Delahay (ed.), Interscience, NY (1967).
738. Dell, R.M., and A. Hooper, in "Solid Electrolytes 1978," P. Hagenmuller and W. Van Gool, (eds.), Academic Press, NY (1978), p. 291.
739. Delmas, C., C. Fouassier, and P. Hagenmuller, Ger. Offen. 2,614,676 (1977).
740. Denno, K., Proc. Symp. Eng. Probl. Fusion Res., 1975, **6**, 728 (1976).
741. Denny, J.E., U.S. Patent Appl. 544,716 (1975).
742. Dept. of Energy, Mines, and Resources (Canada), An Energy Policy for Canada, Vol. 2, I (1973).
743. DeRosa, J.F., Ph.D. dissertation, Drexel Univ., Philadelphia, PA (1974).
744. Despic, A.R., in "Electrochemistry: Past Thirty Next Thirty Years" (Proc. Int. Symp. 1975), Plenum, NY (1977), p. 9.

745. Dews, G.H., D. Eichner, and F.S. Kemp, U.S. Patent No. 3,859,139 (1975).
746. Dews, G.H., and F.S. Kemp, Ger. Offen. 2,500,302 (1975).
747. Dharia, D.J., J. Yasi, D. Kinnibrugh, and B.S. Baker, National Fuel Cell Seminar Abstracts, 1978, p. 105.
748. Dharia, D.J., C. Herscovici, H.C. Maru, and P. Patel, National Fuel Cell Seminar Abstracts, 1979, p. 121.
749. Dharia, D.J., H.C. Maru, and P.S. Patel, National Fuel Cell Seminar Abstracts, 1980, p. 148.
750. Dhooge, P.M., D.E. Stilwell and S.M. Park, J. Electrochem. Soc., **129**, 1719 (1982).
751. Diamond Shamrock Corp., Fr. Demande 2,413,418 (1979).
752. Diebenbusch, K., and U. Krey, Z. Phys. B., **34** (1), 17 (1979).
753. Dighe, S.V., M.B. Blinn, J.J. Buggy, B.R. Krasicki, and B.L. Pierce, Proc. 16th Intersoc. Energy Conv. Eng. Conf., 1059 (1981).
754. DiMarco, D.M., Ph.D. dissertation, Iowa State Univ., Ames, Iowa (1979).
755. Discussion Section, J. Electrochem. Soc., **116**, 874 (1969).
756. Dittman, H.M., E.W. Justi, and A.W. Winsel, in "Fuel Cells II," G.J. Young, (ed.), Reinhold, NY (1963), p. 133.
757. Divisek, J., and J. Mergel, Ger. Offen. 2,914,094 (1980).
758. Dixon, A.G., A.C. Houston, and J.K. Johnson, Proc. 7th Intersoc. Energy Conv. Eng. Conf., 1084 (1972).
759. Doktor, E., J. Havas, and M. Patko, Hung. Teljes 15,203 (1978).
760. Doniat, D., in "From Electrocatalysis to Fuel Cells," G. Sandstede, (ed.), Univ. of Washington Press, Seattle (1972), p. 205.
761. Dorrestijn, A., Ger. Offen. 2,935,485 (1980).
762. Dotto, L., Toronto Globe and Mail, Jan. 7, 1973.
763. Dribinskii, A.V., M.R. Tarasevich, and R.Kh. Burshtein, Elektrokhimiya, **9**, 1046 (1973).
764. Droshko, A.N., Gazov. Prom-st., **1974 (12)**, 45–48 (1974).
765. Dubinin, A.G., L.A. Mirkind, V.E. Kazarinov, and M.Ya. Fioshin, Elektrokhimiya, **15**, 1337 (1979).
766. Dubois, P., Les Piles à Combustible aux Laboratoires de Marcoussis, CR.CGE, Centre de Perfectionnement Technique (Jan. 1970).
767. Dubok, V.A., and V.V. Lashneva, Izv. Akad. Nauk SSSR, Neorg. Mater., **11**, 1250 (1975).
768. Dudley, G.J., and B.C.H. Steele, J. Solid State Chem., **31**, 233 (1980).
769. Dudley, R.F., Proc. Workshop Electrocatalysis Fuel Cell Reactions, Brookhaven Nat. Lab. 1978, p. 1.
770. Dumke, R.V., in "Alternative Energy Sources," CDA 1978 Fall Meeting, 114–133 (1978).

References

771. Dunbar, B.J., and S. Sarian, Solid State Commun., **21**, 729 (1977).
772. Dunlop, J.D., G. van Ommering, and J.F. Stockel, Proc. 6th Intersoc. Energy Conv. Eng. Conf., 906 (1971).
773. Dunlop, J.D., G. van Ommering, and M.W. Earl, in "Power Sources 6," D.H. Collins, (ed.), Academic Press, London (1977), p. 231.
774. Dunlop, J.D., Proc. DOE Battery Contractors Meeting, Washington, DC, Nov. 1985.
775. Durante, B., and L. Handley, Dual Mode Fuel Cell Powerplant, Proc. 5th Intersoc. Energy Conv. Eng. Conf., Las Vegas, NV (1970).
776. Durante, B., J.K. Stedman, and C.L. Bushnell, Proc. 4th Intersoc. Energy Conv. Eng. Conf., 1065 (1969).
777. Dzhambova, A., E. Sokolova, and S. Raicheva, Dokl. Bolg. Akad. Nauk, **33**, 223 (1980).
778. Edahiro, K., S. Kiga, T. Sakata, and N. Yoshida, U.S. Patent No. 4,208,246 (1980).
779. Edwards, R.N., M.E. Cohen, and J.R. Hill, Proc. 4th Intersoc. Energy Conv. Eng. Conf., 692 (1969).
780. Efremov, B.N., M.R. Tarasevich, G.I. Zakharkin, and S.R. Zhukov, Elektrokhimiya, **14**, 1304 (1978).
781. Ehara, R., T. Miwa, and M. Kamaya, Japan. Kokai 75 26,770 (1975).
782. Ehara, R., T. Miwa, and M. Kamaya, Japan. Kokai 75 26,771 (1975).
783. Eindhoven University of Technology, A New Type High Temperature Fuel Cell, Rev. Energ. Primaire, 1/2 (1965), paper read at the Brussels Fuel Cell Conf. (1965)
784. Electrolyte Matrix Evaluation Program, 3rd Report, U.S. Govt. Contract No. DA-44-009-AMC-1517(T), Aug. 1968.
785. Ellis, G.B., Proc. 3rd Canadian Fuel Cell Seminar, 70 (1970).
786. See Ref. 788.
787. Elzinga, E.R., Jr., Ger. Offen. 2,549,881 (1976).
788. Elzinga, E.R., J.G. Bannochie, R.J. Bellows, J.H. Correa, H.H. Horowitz, and C.W. Snyder, Application of the Alsthom/Exxon Alkaline Fuel Cell System to Utility Power Generation, EPRI Rep. EM-384 (Project 584-1) (1977).
789. Ember, L.R., P.L. Layman, W. Labkowski, and P.S. Zuner, Chemical & Engineering News, Nov. 26, 1986, p. 14.
790. Emery, A.T., National Fuel Cell Seminar Abstracts, 1983, p. 98.
791. Energy Research Corp., Hydrogen Generator Assemblies, U.S. Govt. Contract No. DAAB07-70-C-0153 (PIC-2417).
792. Engdahl, R.E., and A.J. Cassano, Proc. 24th Power Sources Conf., 192, (1970).
793. Engdahl, R.E., and E.S. Tillman, Hydrogen Generation Assemblies, Final

Rep., Energy Research Corp., U.S. Govt. Contract No. DAAB07-70-C-0153, Sept. 1971.

794. Engelhard Industries, Ammonia-Air Fuel Cell System, Tech. Bull. Vol. 10, No. 3, Dec. 1969.

795. Engelhard Industries, Evaluation of Phosphoric Acid Matrix Fuel Cell, 6th Rept., U.S. Govt. Contract No. DAAK02-67-C-0219, Feb. 1970.

796. Engelhard Industries, Fuel Conditioning Device Development Study, Final Rep., U.S. Govt. Contract No. DAAK-02-69-C-0454, Oct. 1970.

797. Engelhard Industries Div., Engelhard Minerals and Chemical Corp., Open Cycle Fuel Cell Powerplant, 1.5 kW, Phase I, Final Rep., U.S. Govt. Contract No. DAAK02-70-C-0517 (PIC 2313) (1971).

798. Entina, V.S., O.A. Petrii, and V.T. Rysikova, Elektrokhimiya, **3**, 758 (1967).

799. Enyo, M., and T. Maoka, J. Electroanal. Chem. Interfac. Electrochem., **108**, 277 (1980).

800. Epperly, W.R., Proc. 19th Power Sources Conf., 43 (1965).

801. Erb, G.O., and S.J. Skacel, Ger. Offen. 2,615,532 (1977).

802. Erusalimchik, I.G., B.M. Esel'son, A.F. Zhigach, D.M. Levin, and I.B. Shvets, Elektrokhimiya, **15**, 1454 (1979).

803. Erusalimchik, I.G., B.M. Esel'son, D.M. Levin, and I.B. Shvets, Elektrokhimiya, **16**, 458 (1980).

804. Esso Research Corp., Hydrocarbon-Air Fuel Cell, Rep. No. 8, U.S. Govt. Contract No. DA36-039-AMC-03743(E), Dec. 1965.

805. Esso Research Corp., Hydrocarbon-Air Fuel Cells, 10th Rep., U.S. Govt. Contract No. DA-36-039-AMC-03743(E) (1967).

806. Engelhard Industries, Evaluation of Phosphoric Acid Matrix Fuel Cells, 1st Rep., U.S. Govt. Contract No. DAAK02-67-C-0219, August 1967.

807. Esso Research and Engineering Co., Fuel Cell Studies, Final Rep., Contract No. DA-44-009-AMC 1484(T), Dec. (1968).

808. Esso Research Corp., Study of Thin Film Electrode Fuel Cells, 1st Quarterly Rep., U.S. Govt. Contract NAS8-21311, Jan. 1969.

809. Estrup, P.J., Surface Science, **25**, 1 (1971).

810. Etsell, T.H., Proc. Workshop High Temperature Solid Oxide Fuel Cells, Brookhaven Nat. Lab., 1977, p. 60.

811. Etsell, T.H., Proc. Workshop High Temperature Solid Oxide Fuel Cells, Brookhaven Nat. Lab., 1977, p. 66.

812. Etsel, T.H., and S.N. Flengas, Chem. Rev., **70**, 339 (1970).

813. Euler, K.J., and B. Seim, J. Appl. Electrochem., **8**, 49 (1978).

814. European Atomic Energy Community, Development of New Electrocatalysts for Use in Advanced Electrolysis, Mise Point Nouv. Electrocatal. Electrolyse Av. (lere Phase) CEC, Luxembourg (1978).

815. Evans, G.E., and K.V. Kordesch, Science, **158**, 1148 (1967).

816. Evans, U.R., Electrochim. Acta, **14**, 197 (1969).
817. Ewe, H.H., E.W. Justi, and A. Schmitt, Electrochim. Acta, **19**, 799 (1974).
818. Ewe, H.H., E.W. Justi, and H.R. Schroeder, Ger. Offen. 2,315,570 (1974).
819. Ewe, H.H,, E.W. Justi, and M. Pesditschek, Energy Convers., **15**, 9 (1975).
820. Eyring, H., D. Henderson, and W. Jost (eds), Physical Chemistry: An Advanced Treatise, Vol. IX B, Electrochemistry, Academic Press, NY (1970).
821. Fabjan, C., M.R. Kazemi, and A. Neckel, Ber. Bunsenges. Phys. Chem., **84**, 1026 (1980).
822. Fabry, P., E. Schouler, and M. Kleitz, Electrochim. Acta, **23**, 539 (1978).
823. Fadley, C.S., and R.A. Wallace, J. Electrochem. Soc., **115**, 1264 (1968).
824. Fagan, T.J., National Fuel Cell Seminar Abstracts, 1978, p. 30.
825. Fan, F-R., and L.R. Faulkner, J. Am. Chem. Soc., **101**, 4779 (1979).
826. Fanciullo, S., U.S. Patent No. 4,046,956 (1977).
827. Fanciullo, S., U.S. Patent No. 4,098,959 (1978).
828. Fang, J.C., U.S. Patent No. 4,189,369 (1980).
829. Farooque, M., and R.W. Coughlin, Fuel, **58**, 705 (1979).
830. Farooque, M., and T.Z. Fahidy, Electrochim. Acta, **24**, 547 (1979).
831. Farr, J.P.G., G.W. Greene, N.A. Hampson, S.A.G.R. Karunathilaka, S. Kelly, J.B. Lakeman, and R. Leek, Surface Technol., **9**, 401 (1979).
832. Farr, R.D., and C.G. Vayenas, J. Electrochem. Soc., **127**, 1478 (1980).
833. Farrington, G.C., J. Electrochem. Soc., **123**, 833 (1976).
834. Farrington, G.C., J.L. Duffy, M.W. Breiter, and W.L. Roth, Proc. Symp. Electrode Mater. Processes Energy Convers. Storage, 692 (1977).
835. Farrington, G.C., and J.L. Briant, Mater. Res. Bull., **13**, 763 (1978).
836. Farrington, G.C., J. Briant, H.S. Story, and W.C. Bailey, Electrochim. Acta, **24**, 769 (1979).
837. Farrington, G.C., Membr. Transp. Processes, **3** (Ion Permeation Membr. Channels), 43 (1979).
838. Farrington, G.C., and J.L. Weininger, U.S. Patent No. 4,183,988 (1980).
839. Federmann, E.F., National Fuel Cell Seminar Abstracts, 1981, p. 99.
840. Fedorchenko, I.M., V.S. Pugin, P.A. Kornienko, D.A. Tkalenko, and L.G. Voloshina, Poroshk. Metall. (Kiev), **(3)**, 38 (1979).
841. Fedorov, M.I., V.A. Shorin, and L.M. Fedorov, Izv. Vyssh. Uchebn. Saved., Fiz., **22**, 125 (1979).
842. Fedorova, T.V., and Z.P. Arkhangel'skaya, Sb. Rab. Khim. Istochnikam Toka, **10**, 305 (1975).
843. Fee, D.C., and J.P. Ackerman, National Fuel Cell Seminar Abstracts, 1983, p. 11.

844. Fee, D.C., R.K. Steunenberg, T.D. Claar, R.B. Peoppel, and J.P. Ackerman, National Fuel Cell Seminar Abstracts, 1983, p. 74.
845. Fee, G., and E. Storto, Proc. 23rd Annual Power Sources Conf., PSC Publications, Red Bank, NJ (1969), p. 8.
846. Fee, G.G., Proc. 23rd Annual Power Sources Conf., 8–10 (1969).
847. Feldberg, S.W., in "Electroanalytical Chemistry," Vol. 3, A.J. Bard, (ed.), Dekker, NY (1969) pp. 199–296.
848. Feltham, A.M., and M. Spiro, Chem. Rev., **71**, 177 (1971).
849. Felton, R.H., in "Porphyrins," D. Dolphin, (ed.), Academic Press, NY (1978), **5**, (Part C), p. 127.
850. Fen, E.K., Ya.S. Malakhov, and V.Ya. Malakhov, Konfiguratsionnye Predstavleniya Elektron. Str. Fix. Materialoved., (Materr. Nauchn. Semin. Konfiguratsionnoi Modeli Kondens. Sostoyaniya Veshchestva), 2nd, 1976, 101–105 (1977).
851. Feret, J.M., and L.G. Christner, National Fuel Cell Seminar Abstracts, 1982, p. 40.
852. Feret, J.M., Westinghouse Electric, Presentations at EPRI/DOE Fuel Cell Workshops, Palo Alto, CA and Clearwater, FL, May 1986 and Feb. 1987.
853. Fetterman, D.L., U.S. Patent No. 3,753,780 (1973).
854. Feuillade, G., R. Coffre and G. Outhier, Annales de Radioelectricite, **XXI**, 105 (1966).
855. Feuillade, G., Bouet, and Chenaux, Etude theorique et experimentale de l'electrocatalyse et application a l'oxydation en milieu acide, 3iemes Journees Internationales d"Etude des Piles à Combustible, Bruxelles, du 16 au 20 (June 1969).
856. Fickett, A.P., Proc. Symp. Electrode Mater. Processes Energy Convers. Storage, 542 (1977).
857. Fickett, A.P., Proc. Symp. Electrode Mater. Processes Energy Convers. Storage, 546 (1977).
858. Fickett, A.P., Sci. Am., **239**, 70 (1978).
859. Fickett, A.P., National Fuel Cell Seminar Abstracts, 1978, p. 6.
860. Fickett, A.P., National Fuel Cell Seminar Abstracts, 1979, p. 5.
861. Finn, P.A. K. Kinoshita, G.H. Kucera, and J.W. Sim, National Fuel Cell Seminar Abstracts, 1978, p. 110.
862. Fiore, V.B., National Fuel Cell Seminar Abstracts, 1978, p. 9.
863. Fiore, V.B., National Fuel Cell Seminar Abstracts, 1979, p. 10.
864. Fiori, G., C. Mandelli, C.M. Mari, and P.V. Scolari, Adv. Hydrogen Energy, 1 (Hydrogen Energy Syst., Vol. 1), 193 (1979).
865. Fischer, P., and J. Heitbaum, J. Electroanal. Chem. Interfac. Electrochem., **112**, 231 (1980).
866. Fischer, P., J. Heitbaum, and W. Vielstich, Chem.-Ing.-Tech., **52**, 423 (1980).

867. Fischer, W., and W. Siedlarek, Werkst. Korros., **30**, 695 (1979).
868. Fishman, H., J.F. Henry, and S. Tessore, Electrochim. Acta, **14**, 1314 (1969).
869. Fishtik, I.F., V.A. Kir'yanov, and V.S. Krylov, Elektrokhimiya, **16**, 416 (1980).
870. Fleischmann, R., H. Böhm, and J. Heffler, Proc. 14th Intersoc. Energy Conv. Eng. Conf., 559 (1979).
871. Filippov, E.L., E.I. Tanicheva, and G.I. Popova, Tr. Mosk. Energ. Inst., **248**, 87 (1975).
872. Foley, R.L., D.H. Bomkamp, and C.D. Thompson, NASA AD-740 210 (1972).
873. Foley, R.L., Properties of a Fuel Cell Electrolyte, NTIS Rep. AD-A072864 (1979).
874. Foley, R.T., National Fuel Cell Seminar Abstracts, 1978, p. 181.
875. Foley, R.T., National Fuel Cell Seminar Abstracts, 1979, p. 44.
876. Force, E.L., J.R. Joiner, and C.D. Crawford, National Fuel Cell Seminar Abstracts, 1979,
877. Fouchard, D.T., and C.L. Gardner, Defence Res. Establ., Ottawa, Rep. No. DREO-793 (1978).
878. Fouilloux, P., and J.P. Candy, Fr. Demande 2,404,312 (1979).
879. Foulkes, F.R., Membrane Fuel Cells, Ph.D. dissertation, Dept. Chemical Engineering and Applied Chemistry, Univ. of Toronto (1969).
880. Foulkes, F.R., and W.F. Graydon, Can. J. Chem. Eng., **47**, 171 (1969).
881. Foulkes, F.R., and W.F. Graydon, Electrochim. Acta, **16**, 1577 (1971).
882. Foulkes, F.R., and P.T. Choi, Can. J. Chem. Eng., **56**, 236 (1978).
883. Franklin, A.D., A.L. Dragoo, and K.F. Young, National Fuel Cell Seminar Abstracts, 1978, p. 167.
884. Frazer, E.J., and R. Woods, J. Electroanal. Chem. Interfac. Electrochem., **102**, 77 (1979).
885. Fredlein, R.A., and J. McHardy, Proc. 24th Power Sources Conf., 175 (1970).
886. Fredlein, R.A., Aust. J. Chem., **32**, 2343 (1979).
887. Fried, I., The Chemistry of Electrode Processes, Academic Press, NY (1973).
888. Friedman, D., ASME Adv. Energy Conv. Eng., 887 (1967).
889. Froberg, R.W., Ger. Offen. 2,526,036 (1976).
890. Frysinger, G.R., Proc. 19th Power Sources Conf., 11 (1965).
891. Frysinger, G.R., Proc. 20th Power Sources Conf., 14 (1966).
892. Frysinger, G.R., ASME Advan. Energy Conv. Eng., 897 (1967).
893. Fuchs, R., J. Chem. Phys., **42**, 3781 (1965).
894. Fujihira, M., and T. Kuwana, Electrochim. Acta, **20**, 565 (1975).
895. Fujihira, M., and T. Osa, Prog. Batteries Solar Cells, **2**, 244 (1979).

896. Fujinaga, T., and I. Sakamoto, J. Electroanal. Chem. Interfac. Electrochem., **85**, 185 (1977).
897. Fujita, Y., and H. Nakamura, GS News Tech. Rep., **39**, 37 (1980).
898. Fujita, Y., and H. Nakamura, Japan, Kokai Tokkyo Koho **80**, 35,462 (1980).
899. Fujiwara, K., Japan, Kokai **75**, 42,347 (1975).
900. Fukuda, F., and N. Wakabayashi, National Fuel Cell Seminar Abstracts, 1982, p. 8.
901. Fukuda, M., National Technical Report (Japan), **16**, 145 (1970).
902. Fukuda, M., T. Iwaki, and Y. Kobayashi, Japan, Kokai Tokkyo Koho **75** 21,225 (1975).
903. Fukuda, M., T. Iwaki, and I. Matsumoto, Japan, Kokai **75**, 21,237 (1975).
904. Fukuda, R., Moonlight Project Project Promotion Office, MITI, personal communication, Dec. 17, 1982.
905. Fung, S.C., and S.J. Tauster, Ger. Offen. 2,513,078 (1975).
906. Fung, S.C., and S.J. Tauster, U.S. Patent No. 3,964,933 (1976).
907. Fung, S.C., and Y-C. Pan, U.S. Patent No. 4,131,721 (1978).
908. Furman, N.H., J. Am. Chem. Soc., **44**, 2685 (1922).
909. Gabor, D., Nature, **161**, 777 (1948).
910. Gagnon, R.N., Ger. Offen. 2,756,651 (1978).
911. Galen, R.F., and F. Wrublewski, "Power Sources 2," D.H. Collins, (ed.), Pergamon Press, London (1970), p. 567.
912. Gambino, S., and G. Silvestri, Tetrahedron Letters No. **32**, 3025 (1973).
913. Gamburtsev, S., I. Iliev, A. Kaisheva, G.V. Shteinberg, and L.N. Mokrousov, Elektrokhimiya, **16**, 1069 (1980).
914. Gamze, M.G., and J.A. Orlando, National Fuel Cell Seminar Abstracts, 1980, p. 109.
915. Ganguly, P., Indian J. Chem., Sec. A, **15A**, 280 (1977).
916. Gannamuthu, D.S., and J.V. Petrocelli, J. Electrochem. Soc., **114**, 1036 (1967).
917. Garner, P.J., and K.R. Williams, Brit. Patent No. 844,584 (1960).
918. Garvie, R.D., J. Mater. Sci., **11**, 1365 (1976).
919. Gary, W.M., National Fuel Cell Seminar Abstracts, 1980, p. 12.
920. Gavrikova, A.E., A.L. Rotinyan, and V.N. Varypaev, Elektrokhimiya, **15**, 1519 (1979).
921. Gay, E.C., R.K. Steunenberg, J.E. Battles, and E.J. Cairns, Proc. 8th Intersoc. Energy Conv. Eng. Conf., 96 (1973).
922. Gelb, G.H., and E.L. Leventhal, National Fuel Cell Seminar Abstracts, 1979, p. 34.
923. Gelb, G.H., N.A. Richardson, E.T. Seo, and H.P. Silverman, Proc. Symp. Batt. Tract. Propuls., 178 (1972).
924. Gelb, G.H., National Fuel Cell Seminar Abstracts, 1980, p. 76.

925. Gelfond, L., National Fuel Cell Seminar Abstracts, 1982, p. 164.
926. Gelfond, L., National Fuel Cell Seminar Abstracts, 1983, p. 51.
927. Gendron, A.S., and E.E. Criddle, Proc. 3rd Can. Fuel Cell Seminar, Can. Defence Research Board, Sheridan Park, Ontario, Feb. 13–14, 1969 (1970), p. 1.
928. General Electric Co., Brit. Patent No. 794,471 (1958).
929. General Electric Co., Ion-Exchange Membrane Fuel Cell, Final Rep., U.S. Govt. Contract No. DA-36-039-SC-89140 (1962).
930. General Electric Co., Fuel Cell Technology Program, U.S. Govt. Contract No. NAS-9-11033.
931. General Electric Co., Hydrocarbon-Air Fuel Cells, 5th Report, U.S. Govt. Contract No. DA-44-009-AMC-479(T), Dec. 1964.
932. General Electric Co., Hydrocarbon-Air Fuel Cells, 9th Rep., U.S. Govt. Contract DA-44-009-AMC-479(T), June 1966.
933. General Electric Co., Fuel Cell System (30 Watts), U.S. Govt. Contract No. DAAB07-67-C-0403, (1967).
934. General Electric Co., Hydrocarbon-Air Fuel Cells, 13th Rep., U.S. Govt. Contract No. DAAK02-67-C-0080, June 1968.
935. General Electric Co., Presentation in Proceedings of Cornell Hydrogen Symposium, S. Linke, (ed.), Cornell Univ. Press, Ithaca, NY (1974).
936. George, J.H.B., (A.D. Little, Inc.) Electrochemical Power Sources for Electric Highway Vehicles, NTIS PB-216 622 (1972).
937. George, M.A., and R.C. Nickols, Jr., Can. Patent No. 969,234 (1975).
938. George, M., and S. Januszkiewicz, New Materials for Fluorosulfonic Acid Electrolyte Fuel Cells, U.S. NTIS AD Rep. AD-A044414 (1977).
939. George, M.A., National Fuel Cell Seminar Abstracts, 1979, p. 51.
940. George, M.A., National Fuel Cell Seminar Abstracts, 1980, p. 123.
941. Gerischer, H., and W. Vielstich, Z. Physik. Chem., **3**, 16 (1955).
942. Gerischer, H., D.M. Kolb and M. Przasnyski, Surface Science, **43**, 662 (1974).
943. Gerischer, H., D.M. Kolb and M. Przasnyski, J. Electroanal. Chem., **54**, 662 (1974).
944. Gershberg, D., W.P. Colman, K.E. Olson, and E.W. Schmitz, High Performance Light-Weight Electrodes for Hydrogen-Oxygen Fuel Cells, NASA Report CR-1216 (1968).
945. Gidaspow, D., and S.S. Sareen, Proc. 4th Intersoc. Energy Conv. Eng. Conf., 920 (1969).
946. Gidaspow, D., S.S. Sareen, R.W. Lyczkowski, and B.S. Baker, Proc. 5th Intersoc. Energy Conv. Eng. Conf., 5–87 (1970).
947. Gidaspow, D., R.W. Lyczkowski, and B.S. Baker, U.S. Patent No. 3,823,038 (1974).
948. Gidaspow, D., Brookhaven Nat. Lab. Rep. BNL 51144, UC-93, Energy Conversion TID-4500, FC-6, Oct. 1979, p. 41.

949. Gidaspow, G., Brookhaven Nat. Lab. Rep. BNL-51158, UC-98, Energy Conversion TID-4500, FC-7, Jan. 1980, p. 40.
950. Gieruszczak, T., Proc. 3rd Can. Fuel Cell Seminar, Can. Defence Research Board, Sheridan Park, Ontario, Feb. 13–14, 1969 (1970), p. 29.
951. Gilderman, V.K., A.D. Neuimin, S.F. Pal'guev, and Yu.S. Toropov, Elektrokhimiya, **10**, 1585 (1976).
952. Gillibrand, M.I., and J. Gray, Proc. 5th Int. Symp. on Res. and Dev. in Non-Mech. Electrical Power Sources, 565 (1966) (published 1967).
953. Gillibrand, M.I., J. Gray, and F. Twentyman, in "Power Sources 3," D.H. Collins, (ed.), Oriel Press, Newcastle Upon Tyne (1971), p. 405.
954. Gillis, E.A., Proc. 20th Power Sources Conf., 11 (1965).
955. Gillis, E.A., Proc. 20th Annual Power Sources Conf., Atlantic City, 41 (1966).
956. Gillis, E.A., Proc. 5th Intersoc. Energy Conv. Eng. Conf., Las Vegas, NV, 1970, Paper 709203 (1970).
957. Gillis, E.A., Proc. 25th Power Sources Conf., 159 (1972).
958. Gillis, E.A., Proc. 7th Intersoc. Energy Conv. Eng. Conf., 1111 (1972).
959. Gillis, E.A., Testimony to California Public Utilities Commission, Mar. 14, 1985.
960. Gilman, S., J. Phys. Chem., **68**, 2112 (1964).
961. Gilman, S., Trans. Faraday Soc., **61**, 2561 (1965).
962. Gilman, S., in "Electroanalytical Chemistry: A Series of Advances," A.J. Bard, (ed.), Vol. II, Chap. 3, Dekker, NY (1967).
963. Gilman, S., in "Fuel Cell Systems II," Advances in Chem. Series **90**, A.C.S., Washington, DC (1969), p. 144.
964. Gilman, S., in "From Electrocatalysis to Fuel Cells," G. Sandstede, (ed.), Univ. of Washington Press, Seattle (1972), p. 169.
965. Giner, J., J.M. Parry, and S.M. Smith, in "Fuel Cell Systems I," Advances in Chem. Series **47**, A.C.S., Washington, DC (1965), p. 151.
966. Giner, J., J.M Parry, and L. Swette, in "Fuel Cell Systems II," Advances in Chem. Series **90**, A.C.S., Washington, DC (1969), p. 102.
967. Giner, J., J.M. Parry, and S.M. Smith, Proc. 24th Power Sources Conf., 172 (1970).
968. Giner, J., G. Holleck, and P.A. Malachesky, Research on Rechargeable Oxygen Electrodes, U.S. Govt. Contract No. NASA-CR-72999 (1971).
969. Giner, J., G.L. Holleck, M. Turchan, and R. Fragala, Proc. 6th Intersoc. Energy Conv. Eng. Conf., 256 (1971).
970. Giner, J., J.M. Parry, C. Leung, S.M. Smith, and M. Turchan, Molecular Engineering of Pt-based Electrocatalysts for Hydrocarbon Fuel Cell Electrodes, NTIS Report AD730943 (1971).
971. Giner, J., A.H. Taylor, and F. Goebel, Lead/Acid Battery Development for Heat Engine/Electric Hybrid Vehicles, NTIS PB 213 257 (1972).

972. Giner, J., in "From Electrocatalysis to Fuel Cells," G. Sandstede, (ed.), Univ. of Washington Press, Seattle (1972), p. 215.
973. Giner, J., in "From Electrocatalysis to Fuel Cells," G. Sandstede, (ed.), Univ. of Washington Press, Seattle (1972), p. 283.
974. Giner, J., and L. Swette, "Evaluation of the Feasibility of Low-Cost Carbon Dioxide Removal/Transfer Methods for Fuel Cell Application," U.S. NTIS PB Rep. PB-251263 (1975); see also EPRI Rep. EM-391, Palo Alto, CA (1975).
975. Glass, J.T., C.L. Cahen, G. Stoner and E.J. Taylor, J. Electrochem. Soc., **134**, 58 (1987).
976. Glasser, K.F., and J. LaStella, National Fuel Cell Seminar Abstracts, 1978, p. 55.
977. Glasser, K.F., National Fuel Cell Seminar Abstracts, 1979, p. 77.
978. Glasser, K.F., National Fuel Cell Seminar Abstracts, 1980, p. 6.
979. Glasser, K.F., National Fuel Cell Seminar Abstracts, 1981, p. 5.
980. Glugla, P.G., and V.J. DeCarlo, J. Electrochem. Soc., **129**, 1745 (1982).
981. Gnusin, N.P., and N.P. Borisov, Izv. Sev.-Kavk. Nauchn. Tsentra. Vyssh. Shk., Ser. Estestv. Nauk, **3**, 15 (1975).
982. Godin, P., Electricite de France, personal communication, 1986.
983. Goebel, F., U.S. Patent No. 4,020,248 (1977).
984. Gold, T., J. Petroleum Geology, **1,3**, 3 (1979).
985. Goldman, W., Proc. 6th Intersoc. Energy Conv. Eng. Conf., 112 (1971).
986. Goldman, W.E., Proc. Symp. Batt. Tract. Propuls., 242 (1972).
987. Goldner, F., National Fuel Cell Seminar Abstracts, 1982, p. 8.
988. Goldstein, J.R., and A.C.C. Tseung, Nature, **222**, 869 (1969).
989. Golin, Yu.L., G.E. Golina, O.N. Pakulina, G.V. Rylova, and Yu.S. Sherstobitov, Elektrokhimiya, **14**, 27 (1978).
990. Golin, Yu.L., N.M. Danchenko, A.I. Lazarev, and O.V. Chumakovskii, Elektrokhimiya, **14**, 632 (1978).
991. Golin, Yu.L., G.F. Karaseva, and S.Yu. Serykh, Elektrokhimiya, **14**, 1233 (1978).
992. Golin, Yu.L., G.E. Golina, O.N. Pakulina, G.V. Rylova, A.N. Suvorov, and Yu.S. Sherstrobitov, Elektrokhimiya, **15**, 813 (1979).
993. Golin, Yu.L., G.E. Golina, O.N. Pakulina, G.V. Rylova, A.N. Suvorov, and Yu.S. Sherstobitov, Elektrokhimiya, **15**, 995 (1979).
994. Golin, Yu.L., V.A. Zhil'tsov, V.S. Karyakin, and A.A. Kosyakov, USSR Patent No. 697,870 (1979).
995. Goller, G.J., V.J. Petraglia, and G. Dews, U.S. Patent No. 4,185,131 (1980).
996. Goller, G.J., and J.R. Solonia, Brit. UK Patent Appl. 2,024,045 (1980).
997. Goodenough, J.B., in "Solid Electrolytes: General Principles, Characterization, Materials, Applications," P. Hagenmuller and W. Van Gool (eds.), Academic Press, NY (1978) pp. 393–415.

998. Goodenough, J.B., Bull. Soc. Chim. Fr., **1965**, 1200 (1965).
999. Goodenough, J.B., Comm. Eur. Communities, (Rep.), EUR, EUR 6350, Proc. Meet. Prospects Battery Appl. Subsequent R&D Requir., 185 (1979).
1000. Gormley, D.R., and J.H. Harrison, Proc. 6th Intersoc. Energy Conv. Eng. Conf., 520 (1971).
1001. Gossner, K., and E. Mizera, Z. Phys. Chem., **111**, 61 (1978).
1002. Goualard, J., and J.P. Harivel, in "Power Sources 2," D.H. Collins, (ed.), Pergamon Press, London (1970), p. 441.
1003. Gouet, M., B. Chappey, and M. Guillou, C.R. Hebd. Seances Acad. Sci., Ser. C., **283** (10), 367 (1976).
1004. Grachev, D.K., and Yu.M. Novak, Elektrokhimiya, **13**, 157 (1977).
1005. Grahame, D.C., Chem. Rev., **41**, 441 (1947).
1006. Grant, R.J., and M.D. Ingram, J. Electroanal. Chem. Interfac. Electrochem., **83**, 199 (1977).
1007. Gras, J.M., and M. Pernot, Proc. Symp. Electrode Mater. Processes Energy Convers. Storage, 425 (1977).
1008. Gras, J.M., and P. Blanc, Div. Convers. Electrochem., Electr. France, Paris, Rep. No. P/539/77/30 (1977).
1009. Gras, J.M., Bibliographical Study of Electrocatalysis Involving Mixed Oxide Compounds in an Alkaline Medium, NTIS Rep. No. P/539/77/20 (1977).
1010. Gras, J.M., M. Pernot, and C. Duricocq, Oxygen Overvoltage on Nickel and the 50-50 Nickel-Iron Alloy in 40% Potassium Hydroxide at 80°C, NTIS Rep. No. P/539/77/13 (1978).
1011. Gras, J.M., and J.J. LeCoz, Adv. Hydrogen Energy, 1 (Hydrogen Energy Syst., Vol. 1), 255 (1979).
1012. Grasso, A.P., U.S. Patent Appl. B 574,128 (1976).
1013. Grave, B., and W. Shulz, Ger. Offen. 2,542,207 (1977).
1014. Grave, B., H. Gutbier, and M. Deinzer, Ger. Offen. 2,610,427 (1977).
1015. Graydon, W.F., and R.J. Stewart, J. Phys. Chem., **59**, 86 (1955).
1016. Gregory, D.P., and H. Heilbronner, in "Hydrocarbon Fuel Cell Technology," B.S. Baker, (ed.), Academic Press, N.Y. (1965), p. 509.
1017. Gregory, D.P., "A Hydrogen Energy System," American Gas Assoc., Cat. No. L21173 (1973).
1018. Gregory, D.P., D.Y.C. Ng and G.M. Long, in "The Electrochemistry of Cleaner Environments," Plenum, NY (1972), chap. 8.
1019. Grehier, A., Ger. Offen. 2,549,250 (1976).
1020. Grehier, A., and J. Cheron, Ger. Offen. 2,607,690 (1976).
1021. Grehier, A., Ger. Offen. 2,607,691 (1976).
1022. Grehier, A., Ger. Offen. 2,639,695 (1977).

1023. Grevstad, P.E., Development of Advanced Fuel Cell Systems, Final Rep. (Pratt and Whitney Aircraft), U.S. Govt. Contract No. NAS3-15339, NTIS NASA CR-121136.
1024. Grevstad, P.E., and R.L. Gelting, U.S. Patent No. 3,969,145 (1976).
1025. Grevstad, P.E., Ger. Offen. 2,736,884 (1978).
1026. Grevstad, P.E., National Fuel Cell Seminar Abstracts, 1981, p. 58.
1027. Grevstad, P.E., National Fuel Cell Seminar Abstracts, 1982, p. 59.
1028. Griffen, J.E., Proc. 24th Power Sources Conf., 66 (1970).
1029. Griffith, L.R., R.P. Buck, R.T. McDonald, and M.J. Schlatter, Proc. 15th Annual Power Sources Conf., 85 (1961).
1030. Grisso, J.R., National Fuel Cell Seminar Abstracts, 1983, p. 70.
1031. Griswold, J.W., and J. Wilborn, National Fuel Cell Seminar Abstracts, 1983, p. 61.
1032. Grot, W., Chem.-Ing.-Tech. **50**, 299 (1978).
1033. Grove, W.R., Phil. Mag., **14**, 127 (1839).
1034. Grove, W.R., Phil. Mag., **15**, 287 (1839).
1035. Grove, W.R., Phil. Mag., **21**, 417 (1842).
1036. Grubb, W.T., U.S. Patent No. 2,861,116 (1958).
1037. Grubb, W.T., U.S. Patent No. 2,913,511 (1959).
1038. Grubb, W.T., J. Phys. Chem., **63**, 55 (1959).
1039. Grubb, W.T., and L.W. Niedrach, J. Electrochem. Soc., **107**, 131 (1960).
1040. Grubb, W.T., Nature, **198**, 883 (1963).
1041. Grubb, W.T., and L.W. Niedrach, Proc. 17th Power Sources Conf., 69 (1963).
1042. Grubb, W.T., and C.J. Michalske, Proc. 18th Power Sources Conf., 17 (1964).
1043. Grubb, W.T., and L.H. King, Technical Summary Rep. No. 9, NTIS AD 640,521 (1966).
1044. Gruene, H., Ger. Offen. 2,510,078 (1976).
1045. Grüneberg, G., and M. Jung, U.S. Patent No. 3,198,666 (1965).
1046. Grüneberg, G., and F. Weddeling, paper presented at 29th meeting of the Propulsion and Energetics Panel, AGARD, Liège (1967).
1047. Gurevich, I.G., and V.S. Bagotskii, in "Fuel Cells: Their Electrochemical Kinetics," V.S. Bagotsky and Yu.B. Vasil'ev, (eds.), Consultants Bureau, NY (1966), p. 65.
1048. Gurevich, I.G., and A.M. Skundin, Vesti Akad. Navuk B. SSR, Ser. Fiz.-Energ. Navuk, **1976** (3), 46–54 (1976).
1049. Gurevich, I.G., J. Power Sources, **4**, 145 (1979).
1050. Gurevich, I.G., Vestsi Akad. Navuk BSSR, Ser. Fix-Energ. Navuk, (**2**), 110 (1979).
1051. Gurney, R.W., Ionic Processes in Solution, Dover, NY (1953).

1052. Gutbier, H., Ger. Patent No. 1,671,879 (1975).
1053. Gutjahr, M., and W. Vielstich, Chim.-Ing.-Techn., **40**, 180 (1968).
1054. Gutjahr, M.A., in "From Electrocatalysis to Fuel Cells," G. Sandstede, (ed.), Univ. of Washington Press, Seattle (1972), p. 143.
1055. Hadzi-Jordanov, S., H. Angerstein-Kozlowska, M. Vukovic, and B.E. Conway, Proc. Symp. Electrode Mater. Processes Energy Convers. Storage, 185 (1977).
1056. Hadzi-Jordanov, S., H. Angerstein-Kozlowska, M. Vukovic, and B.E. Conway, J. Phys. Chem., **81**, 2271 (1977).
1057. Haecker, W., H. Jahnke, and M. Schonborn, in Aufbau und Eigenschaften einer luftatmenden Brennstoffzelle mit Formaldehyd als Brennstoff und Schwefelsaure als Elektrolyt, Tagung Energie-Direkt-Umwandlung, Essen (Oct. 14, 1970).
1058. Hafer, P.R., Proc. Symp. Batt. Tract. Propuls., 271 (1972).
1059. Hagans, P., and E. Yeager, Catalysts for Electrochemical Generation of Oxygen, NASA Contract Rep. NASA-CR-162757 (1979).
1060. Hagenmuller, P., A. Levasseur, C. Lucat, J.M. Reau, and G. Villeneuve, Ion Transp. Solids: Electrodes Electrolytes, Proc. Int. Conf., 637 (1979).
1061. Hagey, G., National Fuel Cell Program Overview, National Fuel Cell Seminar, Newport Beach, Nov. 14–18, 1982.
1062. Hagiwara, Z., H. Ohki, and S. Inoue, Japan, Kokai **74** 110,707 (1974).
1063. Hagler, H., R. Shanker, L. Koppelman, D. Greenstein, W.R. Mison, J.E. Christian, W.L. Jackson, and G.D. Pine, National Fuel Cell Seminar Abstracts, 1978, p. 18.
1064. Hahn, P.B., Ph.D. dissertation, Iowa State University (1973).
1065. Halacy, D.S., Jr., Fuel Cells: Power for Tomorrow, World Publ. (1966).
1066. Halbch, C.R., R.J. Page, and O.J. McCaughey, Rockets **10**, 415 (1973).
1067. Halberstadt, H.J., Proc. 8th Intersoc. Energy Conv. Eng. Conf., 63 (1973).
1068. Haldeman, R.G., W.P. Colman, S.H. Langer, and W.A. Barber, in "Fuel Cell Systems I," Advances in Chem. Series **47**, A.C.S., Washington, DC (1965), p. 106.
1069. Hale, J.M., and M.L. Hitchman, J. Electroanal. Chem. Interfac. Electrochem., **107**, 281 (1980).
1070. Hall, D.C., National Fuel Cell Seminar Abstracts, 1978, p. 81.
1071. Hamann, C.H., and P. Schmöde, J. Power Sources, **1**, 141 (1976/77).
1072. Hamlen, R.P., J.W. Maar, and C.B. Murphy, Ind. Eng. Chem. Prod. Res. Devel., **4**, 251 (1965).
1073. Hammerli, M., personal communication, Nov. 16, 1982.
1074. Hammou, A., Actual. Chim., (8), 7 (1978).
1075. Hammou, A., JFE, J. Four Electr. Ind. Electrochim., **9**, 28-31 (1979).
1076. Hampson, N.A., M.J. Willars, and B.D. McNicol, J. Chem. Soc., Faraday Trans. 1, **75**, 2535 (1979).

1077. Hampson, N.A., M.J. Willars, and B.D. McNicol, J. Power Sources, 4, 191 (1979).
1078. Handley, L.M., and A.P. Meyer, Proc. 24th Power Sources Conf., 188 (1970).
1079. Handley, L.M., Study of Fuel Cell System for Powered Balloon, NTIS Rep. AD 766253 (1973).
1080. Handley, L.M., A.P. Meyer, and W.F. Bell, Development of Advanced Fuel Cell System. Phase 3, NASA Contract. Rep. NASA-CR-134818 (1975).
1081. Handley, L.M., P.F. Grevstat, and D.R. McVay, Improvements in Phosphoric Acid Fuel Cell Powerplant Technology, UTC publication, Aug. 1977.
1082. Handley, L.M., National Fuel Cell Seminar Abstracts, 1978, p. 52.
1083. Handley, L.M., National Fuel Cell Seminar Abstracts, 1979, p. 73.
1084. Handley, L.M., "Integrated Cell Scale-Up and Performance Verification," EPRI Rep. EM-1134 (1979).
1085. Handley, L.M., National Fuel Cell Seminar Abstracts, 1980, p. 17.
1086. Handley, L.M., and R. Cohen, National Fuel Cell Seminar Abstracts, 1982, p. 75.
1087. Handley, L.M., "Description of a Generic 11-MW Fuel Cell Power Plant for Utility Applications," EPRI Rep. EM-3161, Palo Alto, CA (1983).
1088. Handley, L.M., M. Kobayashi, and D.M. Rastler, "Operational Experience with Tokyo Electric Power Company's 4.5 MW Fuel Cell Demonstration Plant," Report, Electric Power Research Institute, Palo Alto, CA (1986).
1089. Hanneken, J.W., Ph. D. dissertation, Rice University, Houston, Texas (1979).
1090. Harada, Y., U.S. Patent Appl. 404,091 (1973).
1091. Harada, H., Japan, Kokai Tokkyo Koho 79 101,790 (1979).
1092. Harada, H., Japan, Kokai Tokkyo Koho 79 10,791 (1979).
1093. Harman, T.C., "Gas Turbine Engineering," Macmillan Press Ltd., London (1981).
1094. Harrison, J.W., (General Electric Co.), The Potential of Fuel Cells for Generation of Electricity from Hydrogen, G.E. Direct Energy Conversion Programs, Rep. (undated).
1095. Harrison, J.W., and K.M. Parker, National Fuel Cell Seminar Abstracts, 1981, p. 95.
1096. Harry, I.L., Proc. Workshop High Temperature Solid Oxide Fuel Cells, Brookhaven Nat. Lab., 1977, p. 88.
1097. Hart, A.B., J.H. Powerll, and C.H. DeWhalley, Proc. 3rd Internat. Fuel Cell Symp., Bournemouth, 1962, D.H. Collins, (ed.), Pergamon Press, NY, 1963, p. 277.
1098. Hart, A., U.S. Patent 3,228,798 (1966).
1099. Hart, A.B., and G.F. Womack, Fuel Cells: Theory and Application, Chapman & Hall, London (1967).

1100. Hart, A., Energy from Electrochemical Reactions, Spec. Publ.-Chem. Soc., 26 (Chem. Needs Soc., Symp.), 41 (1974).
1101. Hart, A.B., Chem. Br., 11 (2), 55 (1975).
1102. Hart, A.B., Materials in Fuel Cells, Mater. Power Plant, Spring Resid. Course, 177 (1975).
1103. Hartner, A.J., M.A. Vertes, V.E. Medina, and H.G. Oswin, in "Fuel Cell Systems I," Advances in Chem. Series 47, A.C.S., Washington, DC (1965), p. 141.
1104. Hartung, R., U. Kuerschner, M. Nicolaou, and H.H. Moebius, Z. Phys. Chem.
1105. Hauffe, W., J. Heitbaum, and W. Vielstich, Ext. Abstr., Meet–Int. Soc. Electrochem., 28th, 2, 210 (1977).
1106. Hauffe, W., J. Heitbaum, and W. Vielstich, Proc. Symp. Electrode Mater. Processes Energy Convers. Storage, 77 (1977).
1107. Hausmann, E., Comm. Eur. Communities, (Rep.) EUR, EUR 6085, Semin. Hydrogen Energy Vector: Prod., Use, Transp., 229 (1978).
1108. Hawley, G.G., The Condensed Chemical Dictionary, 8th Ed., Reinhold, NY (1971).
1109. Hayami, R. and T. Yabuki, Osaka Kogyo Gijutsu Shikensho Kiho, 28, 98 (1977).
1110. Hayes, M., A.T. Kuhn, and W. Patefield, J. Power Sources, 2, 121 (1977).
1111. Headridge, J.B., Electrochemical Techniques for Inorganic Chemists, Academic Press, NY (1969), pp. 42–49.
1112. Healy, H.C., J.M. King, A.H. Levy, L.L. Van Dine, and R.J. Wertheim, National Fuel Cell Seminar Abstracts, 1981, p. 62.
1113. Healy, H.C., A.P. Meyer, and R.F. Buswell, National Fuel Cell Seminar Abstracts, 1981, p. 91.
1114. Heath, C.E., E.H. Okrent, M. Beltzer, and G. Ciprios, Symp. on Power Sources for Electric Vehicles, Columbia Univ., 307 (1967).
1115. Hechtl, W., Wacker RTV-S69L: A Silicone Adhesive With Low Outgassing Rate, Eur. Space Agency, (Spec. Publ.) ESA-SP-145, Spacecr. Mater. Space Environ., 305 (1979).
1116. Heed, B., A. Lunden, and K. Schroeder, Proc. 10th Intersoc. Energy Conv. Eng. Conf., 613 (1975).
1117. Heed, B., A. Lunden, and K. Schroeder, Electrochim. Acta, 22, 705 (1977).
1118. Heffler, J., Ger. Offen. 2,556,731 (1977).
1119. Heffner, W.H., A.C. Veverka, and G.T. Skaperdas, in "Fuel Cell Systems I," Advances in Chem. Series 47, A.C.S., Washington, DC (1965), Chap. 23.
1120. Heflinger, L., R. Wuerber, and R. Broaks, J. Appl. Phys., 37, 642 (1964).
1121. Heikel, H.R., U.S. Patent No. 4,180,449 (1979).
1122. Heintz, E.A., R.W. Marek, and W.E. Parker, in "Fuel Cell Systems I," Advances in Chem. Series 47, A.C.S., Washington, DC (1965), p. 18.

1123. Heise, C.J., Proc. 7th Intersoc. Energy Conv. Eng. Conf., 1103 (1972).
1124. Heltemes, C.J., and C.A. Vansant, ASME Advan. Energy Conv. Eng., 905 (1967).
1125. Henkel, H.J., and H. Stamm, Ger. Offen. 2,443,016 (1976).
1126. Henry, J.F., and J.H. Fishman, Proc. 5th Intersoc. Energy Conv. Eng. Conf., 3–49 (1970).
1127. Herceg, J.E., and J.P. Ackerman, National Fuel Cell Seminar Abstracts, 1980, p. 143.
1128. Hibbert, D.B., J. Chem. Soc., Chem. Commun., (5), 202 (1980).
1129. Hietbrink, E.H., J. McBreen, S.M. Selis, S.B. Tricklebank, and R.R. Witherspoon, in "Electrochemistry of Cleaner Environments," J.O'M. Bockris, (ed.), Plenum, NY (1972).
1130. Higashi, K., H. Fukushima, K. Takamatsu, and H. Ohashi, Nippon Kogyo Kaishi, **95** (1098), 467 (1979).
1131. Hirschenhofer, J.H., J.F. Sheehan, J. Zemkoski, and L.G. Eklund, National Fuel Cell Seminar Abstracts, 1979, p. 29.
1132. Hirschenhofer, J.H., J.C. Cutting, J. Zemkoski, and T.M. Piascik, National Fuel Cell Seminar Abstracts, 1980, p. 72.
1133. Hitachi Chemical Co., Ltd., Japan, Kokai Tokkyo Koho **80** 56,372 (1980).
1134. Hitachi Chemical Co., Ltd., Japan, Kokai Tokkyo Koho **80**, 56,373 (1980).
1135. Hitachi Chemical Co., Ltd., Japan, Kokai Tokkyo Koho **80** 96,570 (1980).
1136. Hitchman, M.L., J. Electroanal. Chem. Interfac. Electrochem., **85**, 135 (1977).
1137. Hoar, T.P., Proc. Roy. Soc. (London), **A142**, 628 (1933).
1138. Hoar, T.P., and E.W. Brooman, Electrochim. Acta, **11**, 545 (1966).
1139. Hoare, J.P., J. Electrochem. Soc., **109**, 858 (1962).
1140. Hoare, J.P., J. Electrochem. Soc., **110**, 1019 (1963).
1141. Hoare, J.P., Electrochim. Acta, **9**, 599 (1964).
1142. Hoare, J.P., J. Electrochem. Soc., **112**, 602 (1965).
1143. Hoare, J.P., J. Electrochem. Soc., **112**, 849 (1965).
1144. Hoare, J.P., in "Advances in Electrochemistry and Electrochemical Engineering," Vol. 6, P. Delahay, (ed.), Interscience, NY (1967).
1145. Hoare, J.P., J. Electrochem. Soc., **116**, 612 (1969).
1146. Hoare, J.P., Electrochim. Acta, **14**, 797 (1969).
1147. Hoare, J.P., J. Electrochem. Soc., **121**, 872 (1974).
1148. Hoare, J.P., J. Electrochem. Soc., **125**, 1768 (1978).
1149. Hoare, J.P., J. Electrochem. Soc., **126**, 1502 (1979).
1150. Hobbs, B.S., and A.C.C. Tseung, J. Electrochem. Soc., **119**, 580 (1972).
1151. Hobbs, B.S., and A.C.C. Tseung, J. Electrochem. Soc., **120**, 766 (1973).
1152. Hodgdon, M., Ion Exchange Membrane Fuel Cell, Final Rep., U.S. Govt. Contract No. DA-36-039-SC-89140 (1962).

1153. Hoehne, K., Ger. Offen. 2,713,855 (1978).
1154. Hollax, E., J. Power Sources, **4**, 11 (1979).
1155. Holt, N.A., "Cool Water Project Update," EPRI Journal, **11**, 34-36 (1986).
1156. Holthusen, H., Electrochim. Acta, **21**, 107 (1976).
1157. Hooie, D.T., and E.H. Camara, (EPRI), National Fuel Cell Seminar Abstracts, 1985, p. 182.
1158. Hooper, A., J. Phys. D, **10**, 1487 (1977).
1159. Hoover, D.Q., and S.G. Abens, National Fuel Cell Seminar Abstracts, 1979, p. 95.
1160. Hoover, D.Q., and A.R. Jones, National Fuel Cell Seminar Abstracts, 1983, p. 103.
1161. Hopfenberg, H.B., Polym. Sci. Technol., 8 (Polym. Med. Surg.), 99 (1975).
1162. Hora, C.J., and A.D. Babinsky, Ger. Offen. 2,546,205 (1976).
1163. Horanyi, G., Electrochim. Acta, **25**, 43 (1980).
1164. Hori, Y., K. Onishi, O. Iwamoto, K. Machida, Y. Morita, and M. Miyashita, Japan, Kokai Tokkyo Koho 79 43,880 (1979).
1165. Horiuchi, N., N. Itoh, and S. Takeshita, National Fuel Cell Seminar Abstracts, 1983, p. 173.
1166. Horkans, J., J. Electroanal. Chem. Interfac. Electrochem., **106**, 245 (1980).
1167. Horowitz, H.H., personal communication, Dec. 5, 1980.
1168. Horowitz, H.S., J.M. Longo, and J.T. Lewandowski, Ger. Offen. 2,852,084 (1979).
1169. Horton, J.H., Proc. 25th Power Sources Symp., 92 (1972).
1170. Houghtby, W.E., F. Luczak, and R. Coykendall, National Fuel Cell Seminar Abstracts, 1980, p. 34.
1171. Houghtby, W.E., United Technologies Corp., personal communication, 1980.
1172. Houghtby, W.E., et. al., "Development of the Adiabatic Reformer to Process No. 2 Fuel Oil and Coal-Derived Liquid Fuels," EM-1701, EPRI, Palo Alto, CA (1981).
1173. Houghtby, W.E., IFC, in "Assessment of Research Needs for Advanced Fuel Cells," S.S. Penner, (ed.), Energy, **12** (1–2), Jan./Feb. (1986), p. 12.
1174. Houseman, J., National Fuel Cell Seminar Abstracts, 1978, p. 134.
1175. Houseman, J., and G. Voecks, National Fuel Cell Seminar Abstracts, 1979, p. 149.
1176. Houseman, J., and G. Voecks, National Fuel Cell Seminar Abstracts, 1979, p. 156.
1177. Houseman, J., G. Voecks and R. Shah, "Autothermal Reforming of No. 2 Fuel Oil" EM-1126, EPRI, Palo Alto, CA (1979).
1178. Howe, A.T., Energy World, **67**, 3 (1980).

1179. Hsieh, S.Y., and K.M. Chen, J. Electrochem. Soc., **124**, 1171 (1977).
1180. Hsu, M.S., and T.B. Reed, Proc. 11th IECEC Meeting, State Line, NC, September 1976, Paper 769074, p. 443-446, AIChE, NY (1976).
1181. Hsu, M.S., Ext. Abstr., Fuel Cell Seminar, Tucson, AZ, May 1985, pp. 115–118, EPRI, Palo Alto, CA (1985).
1182. Hubbard, A.T., and F.C. Anson, Anal. Chem., **38**, 58 (1966).
1183. Hubbard, A.T., and F.C. Anson, Anal. Chem., **38**, 1601 (1966).
1184. Huening, R., Energie, **28**, 296 (1978).
1185. Huff, J.R., and J.C. Orth, in "Fuel Cells II," Reinhold, NY (1963), p. 328.
1186. Huff, J.R., in "From Electrocatalysis to Fuel Cells," G. Sandstede, (ed.), Univ. of Washington Press, Seattle (1972), p. 3.
1187. Huff, J.R., Fuel Cells, NTIS (AD-755 106) (1972).
1188. Huff, J.R., J.B. O'Sullivan, E. Dowigiallo, I.R. Snellings, and R. Anderson, Proc. Symp. Batt. Tract. Propuls., 65 (1972).
1189. Huff, J.R., National Fuel Cell Seminar Abstracts, 1978, p. 10.
1190. Huff, J.R., and E.J. Dowgiallo, Jr., National Fuel Cell Seminar Abstracts, 1978, p. 45.
1191. Huff, J.R., National Fuel Cell Seminar Abstracts, 1979, p. 14.
1192. Huff, J.R., J.B. McCormick, D.K. Lynn, R.E. Bobbett, S. Srinivasan, and C. Derouin, National Fuel Cell Seminar Abstracts, 1981, p. 37.
1193. Huff, J.R., J. B. McCormick, D.K. Lynn, R.E. Bobbett, G.R. Dooley, C.R. Derouin, and H.S. Murray, National Fuel Cell Seminar Abstracts, 1982, p. 19.
1194. Huff, J.R., Presentation at DOE Battery Contractors Meeting (Technology Base Research Project), Cleveland, OH, April 14–15, 1986.
1195. Huggins, R.A., W.H. Flygare, E.J. Cairns, and C.W. Tobias, Prelim. Rep., Memo. Tech. Notes Mater. Res. Counc. Summer Conf., **1** (AD-777 743), 579 (1973).
1196. Huggins, R.A., and R.L. Coble, Prelim. Rep., Memo. Tech. Notes Mater. Res. Counc. Summer Conf., **2** (AD-777 737) 241 (1973).
1197. Huggins, R.A., Mater. Sci. Res., **9** (Mass Transp. Phenom. Ceram.), 155 (1975).
1198. Huggins, R.A., Solid Electrolyte Battery Materials, U.S. NTIS AD Rep., AD-A024658 (1975).
1199. Huggins, R.A., Adv. Electrochem. Electrochem. Eng., **10**, 323 (1977).
1200. Hughes, V.B., B.D. McNicol, M.R. Andrews, R.B. Jones, and R.T. Short, J. Appl. Electrochem., **7**, 161 (1977).
1201. Hughes, W.L., R. Ramakumar, and H.J. Allison, Proc. 5th Intersoc. Energy Conv. Eng. Conf., 5-105 (1970).
1202. Hull, M.N., and H.I. James, J. Electrochem. Soc., **124**, 332 (1977).

1203. Hunger, H.F., and J.E. Wynn, U.S. Patent No. 3,013,098 (1961).

1204. Hunt, T.K., N. Weber, and T. Cole, Fast Ion Transp. Solids: Electrodes Electrolytes, Proc. Int. Conf., 95 (1979).

1205. Huschka, H., W. Heit, F.J. Herrmann, and G. Spener, Ger. Offen. 2,348,282 (1975).

1206. Iacovangelo, C.D., C.E. Baumgartner, B.R. Karas, and W.D. Pasco, National Fuel Cell Seminar Abstracts, 1982, p. 83.

1207. Iacovangelo, C., E. Jerabek, R. Reinstrom, A. Schnacke, and J.C. Dart, "Advanced Anode Internal Reforming Program, Molten Carbonate Fuel Cell Development," Final Rep., Contract No. Z12-5-091-83, Pacific Gas and Electric Co., San Francisco, CA (1984).

1208. Iammartino, N.R., Chemical Engineering, 81, 62 (1974).

1209. Ibl, N., and R. Muller, Z. Electrochem., 59, 644 (1955).

1210. Ichenhower, D.E., and H.B. Urbach, Proc. 8th Intersoc. Energy Conv. Eng. Conf., 67 (1973).

1211. Ichenhower, D.E., and H.B. Urbach, High Power Density Hydrazine Fuel Cells, NTIS AD 764530 (1973).

1212. Iemmi, G., and D. Macerata, Brit. UK Patent Appl. 2,023,916 (1980).

1213. Ihrig, H.K., Paper No. S-253, 11th Annual Earthmoving Industry Conference, SAE, Illinois (1960).

1214. Iijima, T., K. Kishimoto, T. Komabashira, and T. Kano, Ger. Offen. 2,932,197 (1980).

1215. Ikeda, H., M. Kumeda, and Y. Ishikura, Japan. Patent No. 73, 84,792 (1973).

1216. Ikeda, H., T. Sakai, M. Kumeta, and Y. Ishikura, Denki Kagaku Oyobi Kogyo Butsuri Kagaku, 45, 37 (1977).

1217. Ikeda, H., M. Kumeda, and Y. Ishikura, Japan Tokkyo Koho 80, 5,815 (1980).

1218. Ikesawa, K., H. Takao, K. Matoba, S. Ishitani, and S. Kimura, Ger. Offen. 2,920,268 (1979).

1219. Ilatovskii, V.A., I.B. Dmietriev, and G.G. Domissarov, Zh. Fiz. Khim., 52, 2551 (1978).

1220. Iliev, I., S. Gamburtsev, A. Kaisheva, E. Bakanova, I. Mukhovski, and E. Budevski, Izv. Otd. Khim. Nauki, Bulg. Akad. Nauk., 7, 223 (1974).

1221. Iliev, I., S. Gamburtsev, A. Kaisheva, E. Bakanova, I. Mukhovski, and E. Budevski, Izv. Otd. Khim. Nauki, Bulg. Akad. Nauk., 8, 233 (1974).

1222. Iliev, I., S. Gamburtsev, A. Kaisheva, and E. Budevski, Izv. Khim., 8, 359 (1975).

1223. Imai, S., S. Fujii, and M. Asada, Japan. Kokai Tokkyo Koho 79 38,285 (1979).

1224. Inai, M., C. Iwakura, and H. Tamura, Denki Kagaku Oyobi Kogyo Butsuri Kagaku, 48, 173 (1980).

1225. Inai, M., C. Iwakura, and H. Tamura, Denki Kagaku Oyobi Kogyo Butsuri Kagaku, 48, 229 (1980).

1226. Inglis, D.R., Nuclear Energy: Its Physics and Its Social Challenge, Addison-Wesley, Reading, MA (1973).

1227. Ingold, J.H., and R.C. DeVries, Acta Met., **6**, 736 (1958).

1228. Inozemtsev, M.V., M.V. Perfil'ev, and A.S. Lipilin, Elektrokhimiya, **10**, 1471 (1974).

1229. Inozemtsev, M.V., M.V. Perfil'ev, and V.P. Gorelov, Elektrokhimiya, **12**, 1231 (1976).

1230. Inozemtsev, M.V., and M.V. Perfil'ev, Elektrokhimiya, **12**, 1236 (1976).

1231. Institute of Gas Technology, JP-4 Fueled Molten Carbonate Fuel Cells, 1st Rep., U.S. Govt. Contract DA-44-009-AMC-1456(T), July 1966.

1232. Institute of Gas Technology, JP-4 Fuelled Molten Carbonate Fuel Cells, Final Rept., U.S. Gov't contract DA-44-009-AMC-1456(T), Jan. 1968.

1233. Inzelt, G., and G. Horanyi, Acta Chim. Acad. Sci. Hung., **99**, 393 (1979).

1234. Inzelt, G., and G. Horanyi, Acta Chim. Acad. Sci. Hung., **101**, 229 (1979).

1235. Ioffe, A.I., M.V. Inozemtzev, A.S. Lipilin, M.V. Perfil'ev, and S.V. Karpachev, Phys. Status Solidi A., **30** (1), 87 (1975).

1236. Isaacs, H.S., and L.J. Olmer, Proc. Workshop High Temperature Solid Oxide Fuel Cells, Brookhaven Nat. Lab., 1977, p. 153.

1237. Isaacs, H.S., and L.J. Olmer, Brookhaven Nat. Lab. Rep. BNL 50951, UC-93, Energy Conversion TID-4500, Dec. 1978, p. 10.

1238. Isaacs, H.S. L.J. Olmer, and S. Srinivasan, National Fuel Cell Seminar Abstracts, 1978, p. 154.

1239. Isaacs, H.S., and L.J. Olmer, Brookhaven Nat. Lab. Rep. BNL 51038, UC-93, Energy Conversion TID-4500, Apr. 1979, p. 13.

1240. Isaacs, H.S., and L.J. Olmer, Brookhaven Nat. Lab. Rep. BNL 51144, UC-93, Energy Conversion TID-4500, FC-6, Oct. 1979, p. 11.

1241. Isaacs, H.S., L.J. Olmer, and S. Srinivasan, National Fuel Cell Seminar Abstracts, 1979, p. 60.

1242. Isaacs, H.S., and L.J. Olmer, Brookhaven Nat. Lab. Rep. BNL 51053, UC-93, Energy Conversion TID-4500, FC-4, June 1979, p. 9.

1243. Isaacs, H.S., and L.J. Olmer, Brookhaven Nat. Lab. Rep. BNL 51198, UC-93, Energy Conversion TID-4500, FC-8, Mar. 1980, p. 11.

1244. Isaacs, H.S., L.J. Olmer, E.J.L. Schouler, and C.Y. Yang, National Fuel Cell Seminar Abstracts, 1980, p. 131.

1245. Isaacs, H.S., G. Kissel, W.E. O'Grady, E.J.L. Schouler, and G.P. Wirtz, National Fuel Cell Seminar Abstracts, 1981, p. 72.

1246. Isenberg, A.O., 6th Biennial Fuel Cell Symposium, A.C.S., 158th National Meeting, NY Sept. 1969.

1247. Isenberg, A.O., Proc. Symp. Electrode Mater. Processes Energy Convers. Storage, 572 (1977).

1248. Isenberg, A.O., Proc. Symp. on Electrode Materials and Processes for Energy Conversion and Storage, 682 (1977).

1249. Isenberg, A.O., National Fuel Cell Seminar Abstracts, 1978, p. 163.
1250. Isenberg, A.O., and R.J. Ruka, National Fuel Cell Seminar Abstracts, 1979, p. 65.
1251. Isenberg, A.O., National Fuel Cell Seminar Abstracts, 1980, p. 135.
1252. Isenberg, A.O., National Fuel Cell Seminar Abstracts, 1983, p. 78.
1253. Ishibashi, T., Y. Oye, and T. Sugimoto, TRDI Tech. Bull., **8** (76), Dec. 1969.
1254. Ishibashi, T., T. Hara, and T. Sugimoto, TRDI Tech. Bull., **10** (88), Feb. 1971.
1255. Ishikawajima-Harima Heavy Industries, ongoing work, Tokyo, 1986.
1256. Ito, Y., F.R. Foulkes, and S. Yoshizawa, J. Electrochem. Soc., **129**, 1936 (1982).
1257. IUPAC Analytical Chemistry Div., Pure Appl. Chem., **52** (5), 1387 (1980).
1258. Ives, D.J.G., and G.J. Janz, "Reference Electrodes: Theory and Practice," Academic Press, NY (1961).
1259. Iwai, K., Japan. Patent No. **73** 26,514 (1973).
1260. Iwase, K., T. Takada, and H. Bando, Japan. Kokai Tokkyo Koho **78** 43,153 (1978).
1261. Iwase, K., T. Takada, and N. Bando, Japan. Kokai Tokkyo Koho **78** 47,792 (1978).
1262. Iwase, M., and T. Mori, Metall. Trans., B., **98** (4), 653 (1978).
1263. Izawa, K. and I. Aramaki, Japan Kokai **77**, 66,876 (1977).
1264. Izuchi, S., and K. Murata, Japan. Kokai Tokkyo Koho **79** 133,566 (1979).
1265. Jackson and Moreland, Div. of United Engineers and Constructors Inc., Review and Evaluation of Project Fuel Cell, O.C.R. Contract No. 14-01-0001-500, O.C.R. R&D, Rep. No. 27, Feb. 1967.
1266. Jackson, S.B., National Fuel Cell Seminar Abstracts, 1981, p. 52.
1267. Jacquelin, J., and J.P. Pompon, "Power Sources 3," D.H. Collins, (ed.), Oriel Press, Newcastle Upon Tyne, England (1971), p. 391.
1268. Jaeger, P., Ger. Offen. 2,519,098 (1976).
1269. Jahnke, H., and M. Schonborn, Electrocatalysis on Organic Semiconductors, 19th Meeting of CITCE, Detroit (1968).
1270. Jahnke, H., H. Rehin, and G. Schnepf, Construction and Functioning of a Methanol Fuel Cell of 100 W Power, 134th Meeting of the Electrochemical Society, Montreal (1968).
1271. Jahnke, H., and M. Schonborn, Troisiemes Journees Internationalles d'Etude des Piles à Combustible, Comptes Rendus S60 (1969).
1272. Jahnke, H., and M. Schonborn, Kathodische Reduktion von Sauerstoff an Phthalocyaninen, Ber. Bunsenges. Physik. Chem. (1970).
1273. Jahnke, H., and M. Schonborn, Untersuchungen zum Mechniamsus der Kathodichen Sauerstoffreduktion an Phthalocyaninen, Bunsentagung für Physikalische Chemie, Heidelberg (1970).
1274. Jahnke, H., M. Schoenborn, and G. Zimmermann, Proc. Symp. Electrocatal. **1974**, 303 (1974).

1275. Jahnke, H., M. Schoenborn, and G. Zimmermann, Top. Curr. Chem., **61**, 133 (1976).
1276. Jahnke, H., M. Schoenborn, and G. Zimmermann, Photosynth. Oxygen Evol. (Symp.), 1977, 439 (1978).
1277. Jahnke, H., Chimia, **34** (2), 58 (1980).
1278. Jakubowski, W., and D.H. Whitmore, J. Am. Ceram. Soc., **62**, 381 (1979).
1279. Jalan, V.M., D.A. Landsman, and J.M. Lee, Belg. Patent No. 877,413 (1979).
1280. Jalan, V.M., and C.L. Bushnell, U.S. Patent No. 4,136,056, Jan. 1979.
1281. Jalan, V.M., and C.L. Bushnell, U.S. Patent No. 4,137,373, Jan. 1979.
1282. Jalan, V., and D. Wu, National Fuel Cell Seminar Abstracts, 1980, p. 157.
1283. Jalan, V.M., and D.A. Landsman, U.S. Patent No. 4,186,110, Jan. 1980.
1284. Jalan, V.M., D.A. Landsman and J.M. Lee, U.S. Patent No. 4,192,907, Mar. 1980.
1285. Jalan, V.M., U.S. Patent No. 4,202,934, May 1980.
1286. Jalan, V., National Fuel Cell Seminar Abstracts, 1981, p. 87.
1287. Jalan, V., E.J. Taylor, D. Frost, and B. Morriseau, National Fuel Cell Seminar Abstracts, 1983, p. 127.
1288. Jalan, V., and E.J. Taylor, J. Electrochem. Soc., **130**, 2299 (1983).
1289. Jalan, V., and E.J. Taylor, in "Proceedings of the Symposium on the Chemistry and Physics of Electrocatalysis," J.D.E. McIntyre, (ed.), 84–12, Electrochemical Society, Pennington, NJ, 1984.
1290. Jalan, V., Quarterly Reports Under Contract DEN3-294, 1982–1986, NASA-Lewis Research Center, Cleveland, OH.
1291. James, G.S., B.I. Dewar, and W.R. Moergeli, Ger. Offen. 2,437,271 (1975).
1292. James, S.D., J. Electrochem. Soc., **114**, 1113 (1967).
1293. Jandera, J., CKD Praha, o.p. Semiconductor Div., personal communication, 1980.
1294. Janjua, M.B.I., and R.L. LeRoy, U.S. Patent No. 4,183,790 (1980).
1295. Jankowska, H., and B. Szczesniak, Zesz. Naik. Politech. Slask. Chem. (Poland), **631** (91), 273 (1979).
1296. Janssen, M.M.P., and J. Moolhuysen, Ger. Offen. 2,506,568 (1975).
1297. Janssen, M.M.P., and J. Moolhuysen, Electrochim. Acta, **21**, 869 (1976).
1298. Janssen, M.M.P., and J. Moolhuysen, J. Catalysis, **46**, 289 (1977).
1299. Janz, G.J., Proc. Bienn. Int. CODATA Conf., 5th 1976, 383 (1977).
1300. Janz, G.J., C.B. Allen, N.P. Bansal, R.M. Murphy, and R.P.T. Tomkins, Physical Properties Data Compilations Relevant to Energy Storage II. Molten Salts: Data on Single and Multicomponent Salt Systems, Rep., NSRDS-NBS-61-PT-2, (1979); NTIS Order No. PB-295406.
1301. See Ref. 1267.
1302. Jaques, W.W., Harper's Magazine, **94**, 114 (1896).

1303. Jarvi, G.A., A.L. Lee, and E.H. Camara, National Fuel Cell Seminar Abstracts, 1979, p. 159.
1304. Jasem, S., and A.C.C. Tseung, Proc. Symp. Electrode Mater. Processes Energy Convers. Storage, 414 (1977).
1305. Jasem, S.M., and A.C.C. Tseung, J. Electrochem. Soc., **126**, 1353 (1979).
1306. Jasinski, R.J., paper presented at 123rd Meeting of Electrochemical Society, Pittsburgh, PA, Apr. 1963.
1307. Jasinski, R.J., Abstracts of 145th A.C.S. meeting, NY (1963).
1308. Jasinski, R., Nature, **201**, 1212 (1964).
1309. Jasinski, R., Proc. 18th Power Sources Conf., 9 (1964).
1310. Jasinski, R.J., in "Fuel Cell Systems I," Advances in Chem. Series 47, A.C.S., Washington, DC (1965), p. 95.
1311. Jasinski, R.J., Electrochem. Tech., **3**, 129 (1965).
1312. Jasinski, R., J. Electrochem. Soc., **112**, 526 (1965).
1313. Jensen, J., and P. McGeehin, J. Mater. Sci., **13**, 909 (1978).
1314. Jensen, J., and P. McGeehin, Silic. Ind., **44**, 191 (1979).
1315. Jensen, J., in "Fuel Cells: Trends in Results and Applications," A.J. Appleby, (ed.), Hemisphere, NY (1987).
1316. Jindra, J., M. Svata, and J. Mrha, "Power Sources 2," D.H. Collins, (ed.), Pergamon Press, London (1970), p. 505.
1317. Joebstl, J.A., G.W. Walker, and W.A. Adams, NTIS Rep. No. 785638/8GA (1973).
1318. Joebstl, J.A., A.A. Adams, A.J. Coleman, and L.S. Joyce, National Fuel Cell Seminar Abstracts, 1981, p. 81.
1319. Johansson, L.Y., J. Mrha, M. Musilova, and R. Larsson, J. Power Sources, **2**, 183 (1977).
1320. Johnsen, C.I., U.S. Patent No. 3,847,670 (1974).
1321. Johnson, G.L., National Fuel Cell Seminar Abstracts, 1978, p. 16.
1322. Johnson, D.L., National Fuel Cell Seminar Abstracts, 1978, p. 113.
1323. Johnson, G.L., National Fuel Cell Seminar Abstracts, 1979, p. 20.
1324. Johnson, G.L., National Fuel Cell Seminar Abstracts, 1979, p. 38.
1325. Johnson, G.K., and A. Kaufman, National Fuel Cell Seminar Abstracts, 1981, p. 43.
1326. Johnson, I.B., U.S. Patent No. 3,963,593 (1976).
1327. Johnson, J.K., in "Power Sources 3," D.H. Collins, (ed.), Oriel Press, Newcastle Upon Tyne (1971), p. 351.
1328. Johnson, L., Northwestern University, Evanston, IL, ongoing work, EPRI RP 2278-4 (1986).
1329. Johnson, R.D., R.D. Coykendall, L.M. Handley, D.L. Maricle, and A.P.

Mientek, Advances in Lower Cost Phorphoric Acid Fuel Cells, UTC publication HP-97, Aug. 1978.

1330. Johnson, R.T., Jr., Fuel Cells. Direct Conversion of Electrochemical Energy Into Electricity, NTIS Rep. SAND-74-0125 (1974).
1331. Johnson, W.C., and L.A. Heldt, J. Electrochem. Soc., **121**, 34 (1974).
1332. Johnson, W.H., National Fuel Cell Seminar Abstracts, 1978, p. 62.
1333. Johnson, W.H., National Fuel Cell Seminar Abstracts, 1979, p. 82.
1334. Johnson, W.H., and T.G. Schiller, National Fuel Cell Seminar Abstracts, 1982, p. 45.
1335. Jones, A.R., National Fuel Cell Seminar Abstracts, 1980, p. 30.
1336. Jones, J.C., and J.E. Cox, Energy Conversion, **8**, 113 (1968).
1337. Jones, M.S., Jr., National Fuel Cell Seminar Abstracts, 1983, p. 89.
1338. Jones, R.E.G., and T.H. Tio, National Fuel Cell Seminar Abstracts, 1978, p. 137.
1339. Joo, L., Great Lakes Inst., ongoing work under EPRI RP1041-21, 1986.
1340. Juda, W., Proc. Electrochem. Soc. Meeting, Montreal, Oct. 1968.
1341. Juda, W., and R.J. Allen, ongoing work under EPRI, RP1041-17, 1986.
1342. Jung, M., Ger. Patent No. 1,933,778 (1972).
1343. Justi, E.W., and A.W. Winsel, Cold Combustion Fuel Cells, Franz Steiner, Wiesbaden (1962).
1344. Justi, E.W., and A.W. Kalberlah, in "Fuel Cell Systems II," Advances in Chem. Series **90**, A.C.S., Washington, DC (1969), p. 1.
1345. Kadija, I.V., and H.M. Patel, U.S. Patent No. 4,169,774 (1979).
1346. Kadija, I.V., Ger. Offen. 2,939,222 (1980).
1347. Kaisheva, A., Elektrokhimiya, **15**, 1539 (1979).
1348. Kaisheva, A., I. Iliev, and S. Gamburzev, J. Appl. Electrochem., **9**, 511 (1979).
1349. Kamalian, N., National Fuel Cell Seminar Abstracts, 1981, p. 125.
1350. Kan, K-T., Neng Yuan Chi Kan, **6** (3), 74 (1976).
1351. Kaneki, N., H. Hara, K. Shimada, and Y. Jomoto, Kagaku Kogaku Ronbunshu, **6**, 172 (1980).
1352. Karpova, R.A., and N.V. Luzanova, Tezisy Dokl.-Vses. Soveshch. Elektrokhim., 5th, **2**, 329 (1974).
1353. Karube, I., T. Matsunaga, S. Tsuru, and S. Suzuki, Biotechnol. and Bioeng., **19**, 1727 (1977).
1354. Koch, D.F.A., D.A.J. Rand, and R. Woods, J. Electroanal. Chem., **70**, 73 (1976).
1355. Kasper, J.S., and W.L. Roth, Structure and Mechanism Studies of High Conductivity Solid Ionic Conductors, U.S. NTIS AD Rep. No. 787885/3GA (1974).

1356. Katan, J., Factors Affecting the Performance of Porous Structures Used as Oxygen Fuel Cell Electrodes, NASA CR-1623 (1970).
1357. Katan, T., and H.F. Bauman, J. Electrochem. Soc., **122**, 77 (1975).
1358. Katayama, A., J. Phys. Chem., **84**, 376 (1980).
1359. Katz, M., U.S. Patent No. 3,909,206 (1975).
1360. Katz, M., S.W. Smith, and D. Reitsma, U.S. Patent No. 3,923,546 (1975).
1361. Katz, M., Can. Patent No. 990,783 (1976).
1362. Katz, M., Proc. Symp. Electrode Mater. Processes Energy Convers. Storage, 621 (1977).
1363. Katz, M., J. Electrochem. Soc., **125**, 515 (1978).
1364. Kaufman, A., National Fuel Cell Seminar Abstracts, 1979, p. 91.
1365. Kaufman, A., P.L. Terry, A.K.P. Chu, H. Feigenbaum, R.M. Yarrington, B.S. Beshty, and J.A. Whelan, National Fuel Cell Seminar Abstracts, 1980, p. 21.
1366. Kaufman, A., National Fuel Cell Seminar Abstracts, 1982, p. 42.
1367. Kaufman, A., and J. Werth, Rep. under NASA DEN3-241, Lewis Research Center, Cleveland, OH, 1985–1985.
1368. Kaun, T.D., U.S. Patent No. 4,011,374 (1977).
1369. Kaun, T.D., paper presented at Electrochem. Soc. 4th Int. Symp. on Molten Salts, May (1983).
1370. Kaun, T.D., T.M. Fannon, and B.A. Baumert, paper presented at Electrochem. Soc. Meet., New Orleans, LA, Oct. 7–12 (1984).
1371. Kawana, H., T. Horiba, T. Kahara, and K. Tamura, Proc. 15th Intersoc. Energy Conv. Eng. Conf., 892 (1980).
1372. Kawasaki, K., J. Coll. Sci., **16**, 405 (1961).
1373. Kawasaki, K., and T. Takeshita, Japan. Kokai Tokkyo Koho 79 71,084 (1979).
1374. Kazarinov, V.E., M.R. Tarasevich, K.A. Radyushkina, and V.N. Andreev, J. Electroanal. Chem. Interfac. Electrochem., **100**, 225 (1979).
1375. Kellen, F.J., and S.S. Tomter, U.S. Patent No. 3,879,218 (1975).
1376. Keller, E.K., A.I. Leonov, V.P. Popov, V.E. Shvaiko-Shvaikovskii, J. Majauskas, and A.B. Ivanov, Ogneupory, **(9)**, 45 (1978).
1377. Keller, H., J. Bregler, H. Rhein, and H. Jahnke, U.S. Patent No. 3,850,695 (1974).
1378. Keller, R., Electrochim. Acta, **25**, 303 (1980).
1379. Kellogg, M.W. Co., Applicability of SO_2-Control Processes to Power Plants, NTIS Rep. PB-213-421, Nov. 1972.
1380. Kellomaki, A., Kem.-Kemi, **3** (11), 536 (1976).
1381. Kemp, F.S., and M.A. George, U.S. Patent No. 3,867,737 (1974).
1382. Kennedy, J.H., Superionic Conduc., (Proc. Conf.), 335 (1976).
1383. Kennedy, J.H., Top. Appl. Phys., **21** (Solid Electrolytes), 105 (1977).
1384. Keren, E., and A. Soffer, J. Catalysis, **50**, 43 (1977).

1385. Kettler, J., Proc. Symp. Batt. Tract. Propuls., 213 (1972).
1386. Khandar, A., Ceramatec, Inc., ongoing work under EPRI RP2278-7, 1986.
1387. Khanova, L.A., E.V. Kasatkin, and V.I. Veselovskii, Elektrokhimiya, **8** (2), 451 (1972).
1388. Khanova, L.A., E.V. Kasatkin, and V.I. Veselovskii, Elektrokhimiya, **9**, 562 (1973).
1389. Khomskaya, E.A., N.F. Gorbacheva, and N.B. Tolochkov, Elektrokhimiya, **16**, 56 (1980).
1390. Khristoforova, N.K., Uch. Zap. Dal'nevost. Univ., **42**, 46 (1970).
1391. Khrushcheva, E.I., O.V. Morovskaya, N.A. Shumilova, and V.S. Bagotskii, Elektrokhimiya, **8**, 205 (1972).
1392. Khrushcheva, E.I., V.V. Karonik, O.V. Moravskaya, N.A. Shumilova, V.S. Bagotskii, and D.I. Lainer, Elektrokhimiya, **9**, 915 (1973).
1393. Khrushcheva, E.I., V.V. Karonik, O.V. Moravskaya, N.A. Shumilova, V.S. Bagotskii, and D.I. Lainer, Elektrokhimiya, **9**, 1028 (1973).
1394. Khrushcheva, E.I., N.A. Shumilova, and O.V. Moravskaya, Ext. Abstr., Meet.–Int. Soc. Electrochem., **28**, 226 (1977).
1395. Kicheev, A.G., and A.E. Vol'shikh, Elektrokhimiya, **15**, 706 (1979).
1396. Kihara, K., and T. Ishibashi, Japan. Kokai Tokkyo Koho **79** 95,990 (1979).
1397. Kikuchi, T., H. Sasaki, and S. Toshima, Chem. Lett., (1), 5 (1980).
1398. Kimikuza, H., K. Kaibara, E. Kumamoto, and M. Shirozu, J. Membr. Sci., **4**, 81 (1978).
1399. Kimura, S.G., and G.E. Walmet, Separation Sci. Technol., **15**, 1115 (1980).
1400. King, J.M., Jr., Proc. 8th Intersoc. Energy Eng. Conf., 111 (1973).
1401. King, J.M., and S.H. Folstad, Project Engineers, United Aircraft Corp., Pratt and Whitney Aircraft Division, personal communication (1974).
1402. King, J.M., Advanced Fuel Cell Technology for Utility Applications, UTC publication, Aug. 1975.
1403. King, J.M., W.E. Houghtby, and R.A. Sederquist, Advanced Fuel Cell Technology and Applications, UTC publication, June 1977.
1404. King, J.M., "Advanced Technology Fuel Cell Program," EPRI Rep. EM-335, Palo Alto, CA (1976).
1405. King, J.M., "Advanced Technology Fuel Cell Program," EPRI Rep. EM-577, Palo Alto, CA (1977).
1406. King, J.M., "Advanced Technology Fuel Cell Program," EPRI Rep. EM-1328, Palo Alto, CA (1980).
1407. King, J.M., National Fuel Cell Seminar Abstracts, 1983, p. 23.
1408. King, J.M., A.P. Meyer, C.A. Reiser and C.R. Schroll, "Molten Carbonate Fuel Cell Verification and Scale-Up," EM-4129, EPRI, Palo Alto, CA (1985).
1409. Kinoshita, K., and P.N. Ross, J. Electroanal. Chem. Interfac. Electrochem., **78**, 313 (1977).

1410. Kinoshita, K., Effects of Sintering on Porous Fuel Cell Electrodes, Rep., CONF-770531-3, 14 pp, available NTIS (1977).

1411. Kinoshita, K., Proc.–Electrochem. Soc., 77–6 (1977)

1412. Kinoshita, K., Proc. Workshop Electrocatalysis Fuel Cell Reactions, Brookhaven Nat. Lab. 1978, p. 123.

1413. Kinoshita, K., and J.P. Ackerman, U.S. Patent No. 4,115,632 (1978).

1414. Kinoshita, K., in "Proc. of the Workshop on the Electrochemistry of Carbon," S. Sarangapani, J.R. Akridge and B. Schumm, (eds.), PV 84-5, The Electrochemical Society, Pennington, NJ (1984), p. 273.

1415. Kinza, H., Z. Phys. Chem. (Leipzig), **255**, 517 (1974).

1416. Kir'yanov, V.A., Elektrokhimiya, **15**, 1405 (1979).

1417. Kirkland, T.G., and W.G. Smoke, Proc. 19th Power Sources Conf., 20 (1965).

1418. Kirkland, T.G., Proc. 20th Power Sources Conf., 35 (1966).

1419. Kirschbaum, H.S., B.F. Habron, and P.W. Kolody, National Fuel Cell Seminar Abstracts, 1980, p. 95.

1420. Kirtland, J.A., Vehicle Electric Drive Development in the Military, Preprints, National Electric Car Symp., San Jose, CA, (1967).

1421. Kishida, K., E. Nishiyama, I. Hirata, and Y. Hamasaki, National Fuel Cell Seminar Abstracts, 1982, p. 53.

1422. Kishida, K., E. Nishiyama, and Y. Hamasaki, National Fuel Cell Seminar Abstracts, 1983, p. 96.

1423. Kishida, K., and R. Anahara, National Fuel Cell Seminar Abstracts, 1983, p. 149.

1424. Kita, H., and K. Shimazu, Shokubai, **20**, 361 (1978).

1425. Kiyota, K., and A. Shimizu, Japan. Kokai Tokkyo Koho **79** 4,290 (1979).

1426. Kiyota, T., K. Takahashi, S. Asami, and A. Shimizu, Japan. Kokai Tokkyo Koho **79** 158,378 (1979).

1427. Klein, H.A., Fuel Cells: An Introduction to Electrochemistry, Lippincott, Philadelphia and New York.(1966).

1428. Klein, M., and R.L. Costa, "Electrolytic Regenerative Hydrogen-Oxygen Secondary Fuel Cells," Proc. Space Technology and Heat Transfer Conf., Los Angeles, CA (1970).

1429. Klein, M., and R.L. Costa, American Society of Mechanical Engineers, Rep. 70-Av/SpT-39, NY (1970).

1430. Klein, M., Proc. 7th Intersoc. Energy Conv. Eng. Conf., 79 (1972).

1431. Klein, Y., and J. Goldstein, Israeli Patent No. 50,217 (1980).

1432. Klinedinst, K., W. Vogel, and P. Stonehart, J. Mater. Sci. **11**, 794 (1976).

1433. Klinedinst, K.A., W.M. Vogel and P. Stonehart, J. Mat. Sci., **11**, 214, 794 (1979).

1434. Knox, C., R. Sayano, E. Seo, and H. Silverman, J. Phys. Chem., **71**, 3102 (1967).

1435. Kobayashi, M., National Fuel Cell Seminar Abstracts, 1981, p. 9.
1436. Kobayashi, M., National Fuel Cell Seminar Abstracts, 1983, p. 55.
1437. Kobayashi, N., M. Fujihira, K. Sunakawa, and T. Osa, J. Electroanal. Chem. Interfac. Electrochem., **101**, 269 (1979).
1438. Kobayashi, N., T. Matsue, M. Fujihira, and T. Osa, J. Electroanal. Chem. Interfac. Electrochem., **103**, 427 (1979).
1439. Kobayashi, T., Nippon Kikai Gakkaishi, **91** (714), 458 (1978).
1440. Kobayashi, Y., T. Eguchi, and T. Iwashiro, Japan Kokai Tokkyo Koho **79**, 153,247 (1979).
1441. Kober, F.P., Proc. 25th Power Sources Conf., 186 (1972).
1442. Kobosev, N., and W. Monblanowa, Acta Physicochim. URSS, **1**, 611 (1934).
1443. Kobosev, N., and W. Monblanowa, Acta Physicochim. URSS, **4**, 395 (1936).
1444. Kobosev, N., and W. Monblanowa, Zh. Fiz. Khim., **7**, 645 (1936).
1445. Kobussen, A.G.C., F.R. Van Buren, and G.H.J. Broers, J. Electroanal. Chem. Interfac. Electrochem., **91**, 211 (1978).
1446. Kobussen, A.G.C., and H.J.A. Van Wees, Mater. Sci. Monogr., **6** (Energy Ceram.), 1019 (1980).
1447. Koch, D.F.A., Proc. 1st Austral. Conf. Electrochem., Sydney-Hobart, Austral., 439 (1963).
1448. Koczorowski, R., L. Malanowski, and K. Spelt, Pol. Patent No. 85,743 (1976).
1449. Kodama, T., K. Tamura, K. Murata, and N. Horiuchi, National Fuel Cell Seminar Abstracts, 1983, p. 161.
1450. Kohlmayr, G.M., and P. Stonehart, U.S. Patent No. 4,163,811 (1979).
1451. Kohlmüller, H., Chemical Instrumentation, **2**, 321 (1970).
1452. Kohlmüller, H., Energy Conversion, **10**, 201 (1970).
1453. Kohlmüller, H., H. Cnobloch, and F. v. Sturm, in "Power Sources 3," D.H. Collins, (ed.), Oriel Press, Newcastle Upon Tyne (1971), pp. 373-390.
1454. Kohlmüller, H., Sonderdruck aus Messtechnik, Jahrgang 79, 1971, Heft 12, pp. 290-293.
1455. Kohlmüller, H., and W. Naschwitz, Metalloberflaeche-Angew. Elektrochem., **27**, 317 (1973).
1456. Kohlmüller, H., Ger. Offen. 2,226,665 (1973).
1457. Kohlmüller, H., Messtechnik, **3**, 84 (1973).
1458. Kohlmüller, H., and W. Naschwitz, Sonderdruck Aus "Metalloberflache. AngewandteElektrochemie," 27 Jahrgang 1973, Heft 9, p. 317.
1459. Kohlmüller, H., Beeinfussung der anodischen Oxydation von Hydrazin durch Elektrolytzusätze, Siemens Forsch.-u. Entiwickle.-Ber. Bd. 2 (1973), Nr. 3.
1460. Kohlmüller, H., and K. Strasser, Ger. Offen. 2,258,482 (1974).
1461. Kohlmüller, H., Ger. Offen. 2,325,287 (1974).

1462. Kohlmüller, H., Can. Patent No. 965,477 (1975).
1463. Kohlmüller, H., J. Power Sources, **1**, 249 (1976–1977).
1464. Koifman, O.I., N.G. Zemlyanaya, M.I. Al'yanov, and V.R. Larionov, Izv. Vysssh. Uchbn. Saved., Khim. Khim. Tekhnol., **18**, 1644 (1975).
1465. Kojima, K., Japan. Kokai Tokkyo Koho **79** 142,179 (1979).
1466. Kolb, D.M., and G. Lehmpfuhl, J. Electrochem. Soc., **172**, 243 (1980).
1467. Kolbenev, I.L., Energomashinostroenie, **18**, 40 (1972).
1468. Kolyadko, E.A., R. Vettzel, B.I. Podlovchenko, and L. Mueller, Elektrokhimiya, **16**, 1096 (1980).
1469. Komaki, A., I. Aramaki, H. Tsubota, Y. Sho, and H. Hatano, Japan. Kokai **74** 113,139 (1974).
1470. Komatsu, M., Japan. Kokai **76** 150,692 (1976).
1471. Komatsu, M., Ger. Offen. 2,610,699 (1977).
1472. Komoroski, R.A., and K.A. Mauritz, J. Am. Chem. Soc., **100**, 7487 (1978).
1473. Konstantinova, G.S., Tr. Mosk. Energ. Inst., **365**, 42 (1978).
1474. Korczynski, A., A. Malachowski, G. Nawrat, and E. Luszczynska, Zesz. Nauk. Politech. Slask., Chem., **631** (91), 45 (1979).
1475. Kordesch, K.V., in "Hydrocarbon Fuel Cell Technology," B.S. Baker, (ed.), Academic Press, NY (1965), p. 17.
1476. Kordesch, K.V., Proc. 21st Power Sources Conf., Red Bank, NJ, 14 (1967).
1477. Kordesch, K.V., in "Handbook of Fuel Cell Technology," C. Berger, (ed.), Prentice-Hall, Englewood Cliffs, NJ (1968), p. 361.
1478. Kordesch, K.V., and M.B. Clark, Proc. 24th Power Sources Conf., 207 (1970).
1479. Kordesch, K.V., Abstract No. 10, Electrochemical Society Meeting, Oct. 1970.
1480. Kordesch, K.V., Proc. 6th Intersoc. Energy Conv. Eng. Conf., 103 (1971).
1481. Kordesch, K.V., and R.F. Scarr, Proc. 7th Intersoc. Energy Conv. Eng. Conf., 12 (1972).
1482. Kordesch, K., Carbon Electrodes, Electric Power Research Inst. Spec. Rep., EPRI SR-13, Conf. Proc.: Fuel Cell Catal. Workshop, 101 (1975).
1483. Kordesch, K.V., J. Electrochem. Soc., **125** (3), 77C (1978).
1484. Kordesch, K.V., in "Prog. Batteries Solar Cells," **2**, 19 (1979).
1485. Kordesch, K.V., Survey About Carbon and its Role in Phosphoric Acid Fuel Cells, U.S. DOE Final Rep. BNL 464459-S (1979).
1486. Kordesch, K.V., and K. Tomantschger, paper presented at Electric Vehicle Exposition, Aug. 25–29 (1980), Adelaide, Australia.
1487. Kordesch, K., National Fuel Cell Seminar Abstracts, 1983, p. 1.
1488. Kordesch, K., and J. Gsellman, Hydrogen Energy Progress, p. 1657, V.T.N. Veziroglu and J.B. Taylor, (eds.), Proc. 5th World Hydrogen Energy Conf., Toronto, Ont., 1984, Pergamon, NY (1984).

1489. Kornienko, V.L., N.V. Kalinichenko, G.V. Kornienko, I.A. Kedrinskii, and Yu. G. Chirkov, Elektrokhimiya, **15** (5), 756 (1979).

1490. Korovin, N.V., and B.N. Yanchuk, Elektrokhimiya, **6**, 1117 (1970).

1491. Korovin, N.V., and E.I. Grigorev, Proc. 7th Intersoc. Energy Conv. Eng. Conf., 37 (1972).

1492. Korovin, N.V., G.A. Kalinovskaya, and V.K. Luzhin, Elektrokhimiya, **8** (3), 374 (1972).

1493. Korovin, N.V., L.A. Kayutenko, and N.I. Kudinova, Elektrokhimiya, **8**, 1058 (1972).

1494. Korovin, N.V., G.S. Konstantinova, N.G. Ul'ko, and N.G. Ryzhova, Elektrokhimiya, **8** (9), 1358 (1972).

1495. Korovin, N.V., N.V. Kuleshov, and E.L. Filippov, USSR Patent No. 568,989 (1977).

1496. Korovin, N.V., G.N. Maksimov, and A.V. Modestov, Izv. Vyssh. Uchebn. Zaved., Energ., **21**, 86 (1978).

1497. Korovin, N.V., G.N. Voloshchenko, and A.I. Strikitsa, Elektrokhimiya, **14**, 44 (1978).

1498. Korovin, N.V., N.I. Kozlova, and O.N. Savel'eva, Elektrokhimiya, **14**, 491 (1978).

1499. Korovin, N.V., N.I. Kozlova, and O.N. Savel'eva, Elektrokhimiya, **16**, 583 (1980).

1500. Korovin, N.V., O.N. Savel'eva, and N.I. Kozlova, Elektrokhimiya, **16**, 585 (1980).

1501. Korovin, N.V., N.I. Kozlova, O.N. Savel'eva, and T.V. Lapshina, USSR Patent No. 715,646 (1980).

1502. Koryta, J., L. Nemee, J. Pivonka, and L. Posposil, J. Electroanal. Chem., **20**, 327 (1969).

1503. Koseki, K., Japan Kokai **75**, 45,249 (1975).

1504. Koshel, N.D., Elektrokhimiya, **15**, 1324 (1979).

1505. Koshel, N.D., Elektrokhimiya, **15**, 1599 (1979).

1506. Kozlov, A.G., and N.A. Nerozin, USSR Patent No. 732,412 (1980).

1507. Kravestskii, G.A., A.P. Zaikina, Yu.I. Kyrlov, V.S. Dergunova, and L.V. Lyutyagina, Tsvetn. Met., (8), 71 (1979).

1508. Kreja, L., Chem. Stosow., **23** (2), 242 (1979).

1509. Kretzschmar, C., and K. Wiesiner, Z. Phys. Chem. (Leipzig), **257**, 39 (1976).

1510. Kretzschmar, D., and K. Wiesiner, Elektrokhimiya, 14, 1330 (1978).

1511. Kring, E.V., U.S. Patent No. 4,076,899 (1978).

1512. Kroger, F.A., J. Electrochem. Soc. **120**, 75 (1973).

1513. Kronenberg, M.L., J. Electrochem. Soc., **109**, 753 (1962).

1514. Kronenberg, M.L., and K.V. Kordesch, Electrochem. Tech., **4**, 460 (1966).

1515. Kruissink, C.A., Electrolysis of Water in an Alkaline Medium at Higher Temperatures and Pressures, Comm. Eur. Communities, (Rep.) EUR 6601 (1979).

1516. Krumpelt, M., J.P. Ackerman, J. Herceg, S. Zwick, C. Slack, and Y. Lwin, National Fuel Cell Seminar Abstracts, 1982, p. 127.

1517. Krumpelt, M., J.P. Ackerman, D.C. Fee, J.E. Herceg, C. Slack, and V. Lwin, paper presented at Am. Power Conf., Chicago, IL, Apr. 18–20 (1983).

1518. Krumpelt, M., D.C. Fee, R.D. Pierce, and J.P. Ackerman, paper presented at IECEC Conf., Orlando, FL, Aug. (1983).

1519. Krumpelt, M., Argonne Nat. Lab., personal communication, Nov. 12, 1984.

1520. Krumpelt, M., G.M. Cook, R.D. Pierce, and J.P. Ackerman, J. Am. Soc. Mech. Eng., Paper 84-AES-10 (1984).

1521. Krumpelt, M., V. Minkov, J.P. Ackerman, and R.D. Pierce, "Fuel Cell Power Plant Designs: A Review," DOE/CC/49941-1833, U.S. Dept. of Energy, Washington, DC (1985).

1522. Kszenhek, O.S., in "Fuel Cells: Their Electrochemical Kinetics," V.S. Bagotsky and Yu.B. Vasil'ev, (eds.), Consultants Bureau, NY (1966), p. 1.

1523. Ksenzhek, O.S., E.M. Shembel, and V.Z. Moskovskii, Elektrokhimiya, **14**, 510 (1978).

1524. Ku, W.S., and V.T. Sulzberger, Proc. Am. Power Conf., 35, 485 (1973).

1525. Kucera, G.H., J.W. Sim, and J.L. Smith, paper presented at Electrochem. Soc. Meet., May (1983).

1526. Kucherenko, L.D., A.V. Lizogub, and B.N. Afanas'ev, Elektrokhimiya, **16**, 291 (1980).

1527. Kudela, V., J. Vacik, and J. Koepcek, J. Membr. Sci., **6**, 123 (1980).

1528. Kudo, T., M. Yoshida, and O. Okamoto, U.S. Patent No. 3,804,674 (1974).

1529. Kudo, T., and Y. Maki, Japan. Patent No. 75 12,566 (1975).

1530. Kudo, T., and H. Obayashi, Ger. Offen. 2,924,678 (1979).

1531. Kuehl, D., Ger. Offen. 2,326,070 (1974).

1532. Kuhn, W.H., in "From Electrocatalysis to Fuel Cells," G. Sandstede, (ed.), Univ. of Washington Press, Seattle (1972), p. 131.

1533. Kukushkina, I.A., G.V. Shteinberg, and V.S. Bagotskii, Elektrokhimiya, **11**, 1135 (1975).

1534. Kuliev, S.A., N.V. Osetrova, V.S. Bagotskii, and Yu. B. Vasil'ev, Elektrokhimiya, **16**, 1091(1980).

1535. Kozlova, N.I., and T.V. Lapshina, Tr. Mosk. Energ. Inst., **365**, 103 (1978).

1536. Kumano, H., H. Manabe, and N. Yanagihara, Japan Kokai **74**, 51,189 (1974).

1537. Kumm, W.H., National Fuel Cell Seminar Abstracts, 1981, p. 124.

1538. Kummer, J.T., and N. Weber, U.S. Patent No. 3,475,223 (1969).

1539. Kummer, J.T., U.S. Patent No. 4,195,119 (1980).

1540. Kummer, J.T., J. Electrochem. Soc., **127**, 364 (1980).

1541. Kunieda, Y., and T. Oki, Denki Kagaku Oyobi Kogyo Butsuri Kagaku, 48, 180 (1980).
1542. Kunz, H.R., and C.A. Reiser, Ger. Offen. 2,617,686 (1976).
1543. Kunz, H.R., Proc. Symposium Elect. Mat. Proc. Energy Conv. and Storage, The Electrochem. Soc. Inc., Princeton, NJ, (1977), p. 607.
1544. Kunz, H.R., Molten Carbonate Fuel Cell Performance, UTC publication HP-106, July 1978.
1545. Kunz, H.R., Proc. Workshop Electrocatalysis Fuel Cell Reactions, Brookhaven Nat. Lab. 1978, p. 14.
1546. Kunz, H.R., National Fuel Cell Seminar Abstracts, 1978, p. 96.
1547. Kunz, H.R., National Fuel Cell Seminar Abstracts, 1982, p. 116.
1548. Kuo, H.C., R.L. Dotson, and K.E. Woodard, Jr., U.S. Patent No. 4,105,531 (1978).
1549. Kuramoto, N., S. Matsuura, Y. Takasaki, and K. Motani, Japan. Kokai 78 37,598 (1978).
1550. Kuribayashi, K., and T. Sato, Ceramurgia Int., 5, 23 (1979).
1551. Kurpit, S.S., National Fuel Cell Seminar Abstracts, 1982, p. 168.
1552. Kutsumi, S., and T. Fukuda, Japan. Kokai Tokkyo Koho 74 40,795 (1974).
1553. Kuwajima, K., K. Kobayashi, T. Masaki, and S. Kawakita, Japan. Kokai 74 99,305 (1974).
1554. Kuwana, T., Spectrophotometric Techniques for the Study of Electrode Processes, Final Rep., U.S. Govt. Contract DAAB07-68-C-0278, June 1971.
1555. Kuz'ko, V.S., I.V. Kudryashov, and A.V. Ismailov, Izv. Vyssh. Uchebn. Zaved., Khim. Khim. Tekhnol., 22, 329 (1979).
1556. Kwasnik, J., H. Purol, M. Cyrankowska, and L. Bullert, Polich Pat. No. 104,119 (1979).
1556a. Kyle, M.L., H. Shimotake, R.K. Steunenberg, F.J. Martino, R. Rubischko, and E.J. Cairns, Proc. 6th Intersoc. Energy Conv. Eng. Conf., 80 (1971).
1557. La Conti, A.B., (General Electric Co.), Introduction to SPE Coal Technology, G.E. Direct Energy Conversion Programs, Rep. (undated).
1558. La Conti, A.B., and A.R. Fragala, Ger. Offen. 2,665,070 (1977).
1559. La Conti, A.B., and A.R. Fragala, Belg. Patent No. 848,997 (1977).
1560. La-Thanh, H., Ph.D. dissertation, University of Florida (1979).
1561. Lamarine, J.H., R.C. Stewart, Jr., and R.W. Vine, U.S. Patent No. 4,038,463 (1977).
1562. Lamarine, J.H., R.C. Stewart, Jr., and R.W. Vine, Ger. Offen. 2,736,883 (1978).
1563. Lambert, C., P. Bono, B. Pichon, and J. Daunay, Ger. Offen. 2,417,300 (1974).
1564. Landau, M.B., U.S. Patent No. 3,976,506 (1976).
1565. Landau, M.B., U.S. Patent No. 4,037,024 (1977).

1566. Lander, J.J., Progress in Solid State Chemistry, Vol. 2, Pergamon Press, NY (1965), p. 26.
1567. Lander, J.J., E. Findle, K. Kordesch, and D.L. Douglas, Proc. Symp. Batt. Tract. Propuls., 307 (1972).
1568. Landgrebe, A.R., and K. Klunder, paper presented at Electrochem. Soc. Meeting, Dallas (1975).
1569. Landsberg, R., Z. Chem., **20**, 203 (1980).
1570. Landi, H.P., J.D. Voorhies, and W.A. Barber, in "Fuel Cell Systems II," Advances in Chem. Series **90**, A.C.S., Washington, DC (1969), p. 13.
1571. Landsman, D.A., and E.I. Thiery, U.S. Patent No. 3,956,014 (1976).
1572. Landsman, D.A., and F.J. Luczak, United Technologies Power Systems, South Windsor, CT, U.S. Govt. Contract No. DAAK70-80-C-0049, Final Rep., (1979).
1573. Landsman, D.A., and F.J. Luczak, U.S. Patent No. 4,316,944, Feb. 1982.
1574. Landsman, D.A., and F.J. Luczak, U.S. Patent No. 4,373,014, Feb. 1983.
1575. Larson, E.S., National Fuel Cell Seminar Abstracts, 1981, p. 20.
1576. Larson, E.S., Keynote Address, National Fuel Cell Seminar, Newport Beach, CA, Nov. 14–18, 1982.
1577. Lauer, G., R. Able, and F.C. Anson, Anal. Chem., **39**, 765 (1967).
1578. Lave, L.B., Proc. 6th Intersoc. Energy Conv. Eng. Conf., 337 (1971).
1579. Lawrance, R.L., and J.H. Russel, U.S. Patent No. 4,210,512 (1980).
1580. Lawrence, L.R., Jr., Proc. Symp. Electrode Mater. Proc. Energy Conv. and Storage, 593 (1977).
1581. Lay, L.A., Brit. Patent No. 1,455,428 (1976).
1582. Layden, G.K., R.A. Pike, and M.A. DeCrescente, U.S. Patent No. 4,080,413 (1978).
1583. Lazarz, C.A., E.H. Cook, Jr., and L.V. Scripa, U.S. Patent No. 4,170,540 (1979).
1584. Lecoanet, A., Rev. Gen. Therm., **16** (188-189), 587 (1977).
1585. Lecoz, J.J., Thermoplastic Materials: Water Electrolysis, NTIS Rep. P/539/77/06 (1977).
1586. LeDuc, J.A.M., U.S. Patent 3,235,473 (1966).
1587. Lee, J.A., W.C. Maskell, and F.L. Tye, Membr. Sep. Processes, 399 (1976).
1588. Lee, J.M., National Fuel Cell Seminar Abstracts, 1983, p. 56.
1589. Lee, W.D., S. Mathias, and R.P. Stickles, National Fuel Cell Seminar Abstracts, 1980, p. 115.
1590. Leites, S.Sh., V.I. Luk'yanycheva, V.S. Bagotskii, A.V. Yuzhannina, and V.F. Konanykina, Elektrokhimiya, **9**, 620 (1973).
1591. Leites, S.Sh., V.I. Luk'yanycheva, and V.S. Bagotskii, Elektrokhimiya, **13**, 1093 (1977).

1592. Leitz, F.B., W. Glass, and D.K. Fleming, in "Hydrocarbon Fuel Cell Technology," B.S. Baker, (ed.), Academic Press, NY (1965), p. 37.
1593. Lenfant, P., Proc. 3rd Can. Fuel Cell Seminar, Can. Defence Research Board, Sheridan Park, Ontario, Feb. 13-14, 1969 (1970), p. 25.
1594. Lenfant, P., Proc. Symp. Batt. Tract. Propuls., 164 (1972).
1595. Lennon, J.F., E. Luksha, and E.Y. Weissman, J. Electrochem. Soc., 116, 122 (1969).
1596. Leo, A., "Zinc Bromide Battery Development," EPRI Rep. EM-4425, Palo Alto, CA (1986).
1597. Lessing, P.A., and J.J. Rasmussen, National Fuel Cell Seminar Abstracts, 1980, p. 161.
1598. Leveasseur, A., J.M. Reau, G. Villeneuve, P. Echegut, J.C. Brethous, and M. Couzi, Comm. Eur. Communities, (Rep.), EUR, EUR 6350, Proc. Meet. Prospects Battery Appl. Subsequent R&D Requir., 225 (1979).
1599. Levy, R.B., and K.K. Ushiba, National Fuel Cell Seminar Abstracts, 1978, p. 131.
1600. Lewis, D.L., L.A. Schenke, M.R. Suchanski, and C.R. Franks, U.S. Patent No. 4,125,449 (1978).
1601. Lewis, F.A., Adv. Hydrogen Energy, 1 (Hydrogen Energy Syst., Vol. 1), 279 (1979).
1602. Lewis, F.A., Surf. Technol., 11, 1 (1980).
1603. Lewis, P., and J. Cronin, NTIS Rep. PSI-TR-191 (1979).
1604. Lewis, W.B., Nuclear Energy Based Synthetic Fuels: Basic Technology and Economics, Atomic Energy of Canada Ltd., Rep., Chalk River, Ontario (1973).
1605. Leyen, J.J.P., Ger. Offen. 2,844,695 (1979).
1606. Leyerle, R.W., L.B. Welsh, M.A. George, and B.S. Baker, Proc. 28th Power Sources Symp., 26 (1978).
1607. Leysen, R.F.R., P. Vermeiren, L.H.J.M. Baetsle, G.F.J. Spaepen, and J.B.H. Vanderborre, Belg. Patent No. 874,961 (1979).
1608. Li, C.T., T.D. Claar, and L.G. Marianowski, National Fuel Cell Seminar Abstracts, 1981,
1609. Liang, C.C., B. McDonald, and W.F. Vierow, Ger. Offen. 2,411,568 (1974).
1610. Liang, C.C., and A.V. Joshi, Ger. Offen. 2,624,941 (1976).
1611. Licentia Patent-Verwaltungs-G.m.b.H., Belg. Patent No. 874,643 (1979).
1612. Liebhafsky, H.A., E.J. Cairns, W.T. Grubb, and L.W. Niedrach, in "Fuel Cell Systems I," Advances in Chem. Series 47, A.C.S., Washington, DC (1965), p. 116.
1613. Liebhafsky, H.A., and E.J. Cairns, Fuel Cells and Fuel Batteries, Wiley, NY (1968).
1614. Liebhafsky, H.A., and W.T. Grubb, in "Fuel Cell Systems II," Advances in Chem. Series 90, A.C.S., Washington, DC (1969), p. 162.

1615. Lieth, E., and J. Upatnichs, J. Am. Opt. Soc., **52**, 1123 (1962).
1616. Lihach, N., Electric Power Research Inst. Journal, Sept. 1984.
1617. Limaye, D.R., National Fuel Cell Seminar Abstracts, 1981, p. 26.
1618. Linden, H., and E. Schultz, U.S. Patent No. 3,145,131 (1964).
1619. Lindholm, I., Proc. Internat. Meeting on the Study of Fuel Cells, SERAI, Brussels, June, 1965.
1620. Lindström, O., Elec. Rev., **179**, 243 (1966).
1621. Lindström, O., in "Fuel Cell Systems II," Advances in Chem. Series **90**, A.C.S., Washington, DC (1969), p. 24.
1622. Lindström, O.B., Swedish Patent Appl. No. 7602575 (1976).
1623. Lindström, O.B., U.S. Patent No. 4,049,878 (1977).
1624. Lindström, O.B., Swedish Patent Appl. No. 780245 (1978).
1625. Lindström, O., T. Nilsson, M. Bursell, C. Hoernell, C. Karlsson, C. Sylwan, and B. Ahgren, Proc. 13th Intersoc. Energy Conv. Eng. Conf., San Diego, CA, 1978, Society of Automotive Engineers, Warrendale, PA (1978), p. 1178.
1626. Lindström, O., T. Nilsson, M. Bursell, C. Hoernell, G. Karlsson, C. Sylwan, and B. Ahgren, Power Sources 1978, **7**, 419 (1979).
1627. Lindström, O., in "Fuel Cells: Trends in Research and Applications," A.J. Appleby, (ed.), Hemisphere Publ., NY (1987), p. 191.
1628. Littauer, E.L., and K.C. Tsai, J. Electrochem. Soc., **126**, 1924 (1979).
1629. Litz, L.M., and K.V. Kordesch, in "Fuel Cell Systems I," Advances in Chem. Series **47**, A.C.S., Washington, DC (1965), p. 166.
1630. Liu, G., National Fuel Cell Seminar Abstracts, 1982, p. 7.
1631. Liu, G., National Fuel Cell Seminar Abstracts, 1983, p. 106.
1632. Liventsev, V.P., V.P. Luzgin, A.G. Frolov, and V.I. Yavoiskii, Izv. Vyssh. Uchebn. Zaved., Chern. Metall., (11), 14 (1974).
1633. Lodi, G., G. Zucchini, A. De Battisti, E. Sivieri, and S. Trasatti, Mater. Chem., **3**, 179 (1978).
1634. Loftin, R.B., Ph. D. dissertation, Rice University, Houston, TX (1975).
1635. Lohrberg, K., Ger. Offen. 2,829,901 (1980).
1636. Lomax, G.R., and F. Twentyman, in "Power Sources 2," D.H. Collins, (ed.), Pergamon Press, London (1970), p. 493.
1637. Lopez, M., B. Kipling, and H.L. Yeager, Anal. Chem., **48**, 1120 (1976).
1638. Lopez, M., B. Kipling, and H.L. Yeager, Anal. Chem., **49**, 629 (1977).
1639. Lotvin, B.M., and Yu.B. Vasil'ev, Elektrokhimiya, **16**, 1050 (1980).
1640. Louis, G., Ger. Offen. 2,306,896 (1974).
1641. Loutfy, R.O., and P.P.K. Ho, U.S. Patent No. 4,107,025 (1978).
1642. Lowry, S.R., and K.A. Mauritz, J. Am. Chem. Soc., **102**, 4665 (1980).
1643. Lu, P.W.T., and S. Srinivasan, Proc. Symp. Electrode Mater. Processes Energy Convers. Storage, 396 (1977).

1644. Luciani, C., Proc. 3rd Canadian Fuel Cell Seminar, 17 (1970).
1645. Luczak, F.J., and D.A. Landsman, U.S. Patent No. 4,447,586, May 1984.
1646. Lueckel, W.J., and R.J. Farris, The FCG-1 Fuel Cell Powerplant for Electric Utility Use, paper presented at July meeting of IEEE Power Eng. Soc., (1974).
1647. Luk'yanycheva, V.I., A.V. Yuzhanina, N.A. Shumilova, and V.S. Bagotskii, Elektrokhimiya, **12**, 952 (1976).
1648. Luk'yanycheva, V.I., L.A. Fokina, and N.A. Shumilova, Elektrokhimiya, **15**, 1620 (1979).
1649. Lundquist, J.T., and W.M. Vogel, J. Electrochem. Soc., **116**, 1066 (1969).
1650. Lunkwitz, K., A. Ferse, D. Handte, B. Klatt, and G. Gruenzig, East German Patent No. 133,257 (1978).
1651. Luskha, E., and E. Weissman, in "Fuel Cell Systems II," Advances in Chem. Series **90**, A.C.S., Washington, DC (1969), p. 200.
1652. Lutz, H.D., H. Haeuseler, and W. Schmidt, Ger. Offen. 2,838,924 (1980).
1653. Lux, H., R. Kuhn and T. Niedermeier, Z. Anorg. Allg. Chem., **298**, 285 (1959).
1654. Luzhin, V.K., A.G. Kicheev, and N.V. Korovin, Tr. Mosk. Energ. Inst., **248**, 49 (1975).
1655. Lyons, E.H., Jr., Eur. Patent Appl. 4,847 (1979).
1656. Lyubomov, V.N., and V.A. Fedotov, Izv. Sib. Otd. Akad. Nauk SSSR, Ser. Khim. Nauk, **(2)**, 43 (1977).
1657. Macdonald, J.P., and P. Stonehart, Ext. Abstr. Program–Bienn. Conf. Carbon, **14**, 183 (1979).
1658. Mack, R., and S. Bruckenstein, Non-wetting Materials for Fabrication of Fuel Cell Electrodes, NTIS Rep. No. AD-A066602 (1977).
1659. Mackintosh, A.R., J. Chem. Phys., **38**, 1991 (1963).
1660. MacKnight, W.J., E. Baer, R.D. Nelson, (eds.), Polymer Materials Basic Research Needs for Energy Applications. Proceedings of a Workshop Recommending Future Directions in Energy-Related Polymer Research, NTIS, Springfield, VA (1978).
1661. MacMullin, R.B., J. Electrochem. Soc., **120**, 135C (1973).
1662. Madariaga, H.A., National Fuel Cell Seminar Abstracts, 1983, p. 19.
1663. Maekawa, M., TRDI Tech. Bull., **9 (80)**, 409 (1970).
1664. Maget, H.J.R., and E.A. Oster, in "Fuel Cell Systems I," Advances in Chem. Series **47**, A.C.S., Washington, DC (1965), p. 83.
1665. Maget, H.J.R., and R. Roethlein, J. Electrochem. Soc., **112**, 1034 (1965).
1666. Maget, H.J.R., in "Handbook of Fuel Cell Technology," C. Berger, (ed.), Prentice-Hall, Englewood Cliffs, NJ (1968), p. 425.
1667. Maggiore, C., ongoing work, Los Alamos Nat. Lab., Los Alamos, NM, (1986).

1668. Magner, G., M. Savey, and G. Scarbeck, J. Electrochem. Soc., **127**, 1076 (1980).

1669. Makles, M., H. Przywarska-Boniecka, and W. Wojciechowski, Rocz. Chem. **49**, 1647 (1975).

1670. Makrides, A.C., J. Electrocham. Soc., **113**, 1158 (1966).

1671. Maksimov, G.N., and S.I. Zaitsev, Tr. Mosk. Energ. Inst., **365**, 56 (1978).

1672. Malachesky, P.A., C. Leung, and H. Feng, "A Study of the Degradation of Platinum Black Fuel Cell Cathodes," U.S. Army Mobility Equipment R&D Center, Final Rep., U.S. Govt. Contract No. DAAK02-71-C-0380 (1972).

1673. Male, J.A., "The Development of a Technology for Producing Ionically Conductive Membranes of Complex Shape," Ph.D. dissertation, Rutgers, State Univ., New Brunswick, NJ (1977).

1674. Manage, H., 50-Watt Hydrazine-Air Fuel Cell Power System, National Technical Rep. (Japan), No. **16** (2), April 1970.

1675. Manet, R.G. and D. Warren, "Assessment of Fuel Processing Systems for Dispersed Fuel Cell Power Plants," EM-1487, EPRI, Palo Alto, CA (1980).

1676. Manet, R.G. and D. Warren, "Evaluation of Hybrid THR-ATR Fuel Processor," EM-2096.

1677. Manion, J.P., U.S. Patent No. 3,758,339 (1973).

1678. Mansour, M.N., National Fuel Cell Seminar Abstracts, 1980, p. 1.

1679. Mansour, M.N., ongoing work under EPRI RP2706-4 (1987).

1680. Maoka, T., and M. Enyo, Surf. Technol., **9**, 147 (1979).

1681. Marchetti, C., and N. Nakicenovic, "The Dynamics of Energy Systems and the Logistic Substitution Model," The International Institute for Systems Analysis, Rep. RR-79-13, Laxemburg, Austria (1979).

1682. Marianowski, L.G., J. Meek, E.B. Shultz, Jr., and B.S. Baker, Proc. 17th Power Sources Conf. (1963).

1683. Marianowski, L.G., E.H. Camara, and H.C. Maru, U.S. Patent No. 4,079,171 (1978).

1684. Marianowski, L.G., E.H. Camara, and H.C. Maru, Ger. Offen. 2,804,318 (1978).

1685. Marianowski, L.G., R.A. Donado, and H.C. Maru, Ger. Offen. 2,945,565 (1980).

1686. Marianowski, L.G., R. Donado, and H. Maru, U.S. Patent No. 4,247,604 (1981).

1687. Marianowski, L.G., and R. Griffin, National Fuel Cell Seminar Abstracts, 1982, p. 120.

1688. Maricle, D.L. and D.C. Nagle, U.S. Patent No. 4,125,676 (1978).

1689. Markin, V.S., A.A. Chernenko, Yu.A. Chizmadzhev, and Yu.G. Chirkov, in "Fuel Cells: Their Electrochemical Kinetics," V.S. Bagotsky and Yu.B. Vasil'ev, (eds.), Consultants Bureau, NY (1966), p. 21.

1690. Maroie, S., M. Savy, and J.J. Verbist, Inorg. Chem., **18**, 2560 (1979).
1691. Marschoff, C.M., and C.E. Pena, Acta Client. Venez., **30**, 279 (1979).
1692. Marshall, E., Science, **224**, 268 (1984).
1693. Martin, C.R., T.A. Rhodes and J.A. Ferguson, Anal. Chem., **54**, 1693 (1982).
1694. Martin, G.E., I.D. Cockshott, and K.T. McAloon, S. African Patent No. 75 06,118 (1976).
1695. Martin, R.E., Advanced Technology Lightweight Fuel Cell Program, NASA Contract Rep. NASA-CR-159653, FCR-1017 (1978).
1696. Martin, R.G., J. Powers, and J.C. Trocciola, U.S. Patent No. 4,157,327 (1979).
1697. Maru, H.C., C. Chi, D. Patel, and D. Burns, Proc. 13th Intersoc. Energy Conv. Eng. Conf., 723 (1978).
1698. Maru, H.C., C. Chik and B.S. Baker, National Fuel Cell Seminar Abstracts, 1978, p. 70.
1699. Maru, H.C., and L. Christner, National Fuel Cell Seminar Abstracts, 1978, p. 74.
1700. Maru, H.C., L.G. Christner, S.G. Abens, and B.S. Baker, National Fuel Cell Seminar Abstracts, 1979, p. 86.
1701. Maru, H.C., Brit. U.K. Patent Appl. 2,025,119 (1980).
1702. Maru, H.C., L.G. Christner, and B.S. Baker, National Fuel Cell Seminar Abstracts, 1980, p. 25.
1703. Maru, H.C., A. Pigeaud, and L. Paetsch, National Fuel Cell Seminar Abstracts, 1981, p. 133.
1704. Marui, T., Japan. Patent No. **73**, 18,299 (1973).
1705. Marks, C., E.A. Rishavy, and F.A. Wyczalek, SAE Trans., **76**, 992 (1967).
1706. Martin, R.E., "Advanced Technology Lightweight Fuel Cell Program," NASA Final Rep. CR-169653 FCR-1017, NASA, Washington, DC (1979).
1707. Masaki, T., A. Okada, H. Kuwajima, and K. Kobayashi, Koen Yoshishu–Kotai Ionikusu Toronkai, 7th, 27 (1979).
1708. Mashikian, M.S., T.A. Alessi, and E. Hines, ASME Advanced Energy Conv. Eng., 875 (1967).
1709. Mason, D., and C. Van Drunen, U.S. Patent No. 3,998,939 (1976).
1710. Mason, D.M., and C.J. Wen, Proc. Workshop High Temperature Solid Oxide Fuel Cells, Brookhaven Nat. Lab., 1977, p. 160.
1711. Mathur, P.B., and N. Muniyandi, Indian Patent No. 141,680 (1977).
1712. Matsuda, Y., and T. Ishino, Denki Kagaku, **32**, 48 (1964).
1713. Matsuda, Y., J. Shiokawa, H. Tamura, and T. Ishino, Electrochim. Acta, **12**, 1435 (1967).
1714. Matsumoto, H., Y. Saito, and Y. Yoneda, J. Catalysis, **19**, 101 (1970).
1715. Matsumoto, T., T. Sasaki, and K. Hijikata, Nippon Daigaku Bunrigakubu Shizenkagaku Kenkyusho Kenkyu Kiyo, **10**, 28 (1975).

1716. Matsumoto, Y., and E. Sato, Electrochim. Acta, **24**, 421 (1979).
1717. Matsumoto, Y., J. Kurimoto, and E. Sato, J. Electroanal. Chem. Interfac. Electrochem., **102**, 77 (1979).
1718. Matsumoto, Y., H. Yoneyama, and H. Tamura, Denchi Toronkai, (Koen Yoshishu), 18th, 45 (1979).
1719. Matsumoto, Y., H. Manabe, and E. Sato, J. Electrochem. Soc., **127**, 811 (1980).
1720. Matsumoto, Y., S. Yamada, T. Nishida, and E. Sato, J. Electrochem. Soc., **127**, 2360 (1980).
1721. Matsushita Electric Industrial Co., Ltd., Japan. Kokai Tokkyo Koho **80**, 07,666 (1980).
1722. Matsushita Electric Industrial Co., Ltd., Japan. Kokai Tokkyo Koho **80**, 12,704 (1980).
1723. Matsuura, S., and Y. Ozaki, Japan. Kokai Tokkyo Koho **80**, 24,970 (1980).
1724. Mavrodin-Tarabic, M., I. Onaca, M. Zaharescu, and I. Solacolu, Stud. Cercet. Chim., **22** (6), 719 (1974).
1725. May, J.W., R.J. Szostak, and L.H. Germer, Surface Science, **15**, 37 (1960).
1726. Mazulevskii, E.A., and V.Sh. Palanker, Inst. Org. Katal. Elektrokhim., Alma-Ata, USSR, Deposited Doc. VINITI 2159-79 (1979).
1727. McAdam, W., and R.M. Taylor, Ger. Offen. 2,927,346 (1980).
1728. McAlister, A.J., L.H. Bennett, and M.I. Cohen, National Fuel Cell Seminar Abstracts, 1978, p. 169.
1729. McAlister, A.J., M.I. Cohen, and L.H. Bennett, National Fuel Cell Seminar Abstracts, 1979, p. 59.
1730. McAlister, A.J., L.H. Bennett, and M.I. Cohen, National Fuel Cell Seminar Abstracts, 1980, p. 118.
1731. McAloon, K.T., Res. Discl., **145**, 63 (1976).
1732. McBreen, J., in "Electrochemistry: Past Thirty Next Thirty Years" (Proc. Int. Symp. 1975), Plenum, NY (1977), p. 35.
1733. McBreen, J., E.J. Taylor, K.V. Kordesch, G. Kissel, F. Kulesa, and S. Srinivasan, Development of Fuel Cell Technology for Vehicular Applications, Brookhaven Nat. Lab. Annual Rep., BNL51047, Energy Conversion TID-4500 FCT-1 (1979).
1734. McBreen, J., W.E. O'Grady, H. Olender, E.J. Taylor, and S. Srinivasan, National Fuel Cell Seminar Abstracts, 1979, p. 56.
1735. McBreen, J., G. Kissel, K.V. Kordesch, F. Kulesa, E.J. Taylor, E. Gannon, and S. Srinivasan, Proc. 15th Intersoc. Energy Conv. Eng. Conf., 886 (1980).
1736. McBryar, H., and H.H. Jamison, U.S. Patent No. 3,382,105 (1968).
1737. McBryar H., National Fuel Cell Seminar Abstracts, 1979, p. 16.
1738. McCain, G.H., Ger. Offen. 2,713,677 (1977).
1739. McDonnough, W.J., D.R. Flinn, K.H. Stern, and R.W. Rice, J. Mater. Sci., **13**, 2403 (1978).

1740. McDougall, A., Energy Alternative Series: Fuel Cells, MacMillan, London (1976).

1741. McElroy, J.F., (General Electric Co.), Status of Solid Polymer Electrolyte Fuel Cell Technology, G.E. Direct Energy Conversion Program, Rep. (undated).

1742. McElroy, J.F., Evaluation of Series "R" Fuel Cell Design, NTIS AD 854534 (1969).

1743. McElroy, J.F., National Fuel Cell Seminar Abstracts, 1978, p. 176.

1744. McElroy, J.F., National Fuel Cell Seminar Abstracts, 1983, p. 123.

1745. McEvoy, J.E., U.S. Patent 3,097,974 (1963).

1746. McFayden, J., Proc. 25th Power Sources Conf., 176 (1972).

1747. McGraw, L.D., Fuel Cells: Introduction, Preprint, SAE Summer Meeting, Chicago (1960).

1748. McHardy, J., and J.O'M. Bockris, J. Electrochem. Soc., **120**, 53 (1973).

1749. McHardy, J., and J.O'M. Bockris, J. Electrochem. Soc., **120**, 61 (1973).

1750. McHardy, J., Proc. Symp. Electrode Mater. Processes Energy Convers. Storage, 537 (1977).

1751. McIntyre, J.A., and R.F. Phillips, U.S. Patent No. 4,187,350 (1980).

1752. McNeill W., and L.E. Conroy, J. Chem. Phys., **36**, 87 (1962).

1753. McNicol, B.D., A.G. Chapman, and R.T. Short, J. Appl. Electrochem., **6**, 221 (1976).

1754. McNicol, B.D., R.T. Short, and A.G. Chapman, J. Chem. Soc. Faraday I, **72**, 2735 (1976).

1755. McNicol, B.D., R.T. Short and A.G. Chapman, J. Chem. Soc. Faraday Trans. 1 **72**, 2735 (1976).

1756. McNicol, B.D., and R.T. Short, Ger. Offen. 2,657,726 (1977).

1757. McNicol, B.D., C. Pinnington, and R.T. Short, Ger. Offen. 2,632,632 (1977).

1758. McNicol, B.D., Proc. Workshop Electrocatalysis Fuel Cell Reactions, Brookhaven Nat. Lab. 1978, p. 93.

1759. McNichol, B.D., in "Special Periodical Reports on Catalysis," Vol. 2, The Chemical Society, London (1979).

1760. McNicol, B.D., and D.A.J. Rand, (eds.), "Power Sources for Electric Vehicles," Elsevier, NY (1984).

1761. Meadowcroft, D.B., Nature, **226**, 847 (1970).

1762. Meek, J., and B.S. Baker, in "Fuel Cell Systems I," Adv. in Chem. Series **47**, A.C.S., Washington, DC (1965), Chap. 16.

1763. Mehta, R.M., and V.K. Venkatesan, Trans. Soc. Adv. Electrochem. Sci. Technol., **11**, 411 (1976).

1764. Meibuhr, S.G., Electrochim. Acta, **21**, 1059 (1967).

1765. Meibuhr, S.G., Electrochim. Acta, **13**, 1973 (1968).

1766. Meibuhr, S.G., J. Electrochem. Soc., **121**, 1264 (1974).
1767. Meibuhr, S.G. and E.J. Zeitner, Jr., U.S. Patent No. 3,852,116 (1974).
1768. Meibuhr, S.G., and R.F. Paluch, J. Electrochem. Soc., **122**, 164 (1975).
1769. Meibuhr, S.G., U.S. Patent No. 3,859,181 (1975).
1770. Meibuhr, S., G.M. Research Lab., personal communication, Jan. 19, 1981.
1771. Meier, H., The Use of Radiochemical Techniques in Fuel Cell Research, NTIS Rep. No. AED-Conf-75-404-021 (1975).
1772. Meier, H., U. Tschirwitz, E. Zimmerhackl, and W. Albrecht, Investigation of Acid Resistant Electrocatalysts for Fuel Cells, NTIS Rep. No. BMVg-FBWT-75-6 (1975).
1773. Meschter, P.J., Proc. Workshop High Temperature Solid Oxide Fuel Cells, Brookhaven Nat. Lab., 1977, p. 83.
1774. Messinger, S., and R.F. Drake, Proc. 4th Intersoc. Energy Conv. Eng. Conf., 361 (1969).
1775. Meyer, A.P., R.F. Buswell, and H.C. Healy, National Fuel Cell Seminar Abstracts, 1982.
1776. Meyer, A.P., C.A. Reiser, H.R. Kunz, and C.R. Schroll, National Fuel Cell Seminar Abstracts, 1983, p. 38.
1777. Meyer, T.J., R.W. Murray, S.-I. Choi, and T.L. Isenhour, Fuel Cells and Solid Electrolytes, U.S. NTIS AD Rep. AD-A033782 (1976).
1778. Michelin et Cie, Belg. Patent No. 861,785 (1978).
1779. Micka, K., in "Fuel Cell Systems I," Advances in Chem. Series **47**, A.C.S., Washington, DC (1965), p. 73.
1780. Micka, K., Electrochim. Acta., **21**, 293 (1976).
1781. Mikhailova, A.A., N.V. Osetrova, G.G. Kutyokov, A.B. Fasman, and Yu.B.Vasil'ev, Elektrokhimiya, **16**, 1283 (1980).
1782. Miles, R., J. Chem. Technol. Biotechnol., **30**, 35 (1980).
1783. Miller, B., J. Electrochem. Soc., **116**, 1117 (1969).
1784. Millet, J., and R. Buvet, Proc. Journées Internationales d'Etude des Piles à Combustible, 49, SERAI, Brussels, (1965).
1785. Minet, R.G., and R.E.G. Jones, National Fuel Cell Seminar Abstracts, 1979, p. 164.
1786. Minh, N.Q., and J.L. Smith, paper presented at Electrochem. Soc. Meet., New Orleans, Oct. 7–12 (1984).
1787. Minkov, V., E. Daniels, C. Dennis, and M. Krumpelt, National Fuel Cell Seminar Abstracts, 1986, p. 255.
1788. Mirkind, L.A., and G.L. Al'bertinskii, Elektrokhimiya, **16**, 121 (1980).
1789. Mirkind, L.A., A.G. Kornienko, and G.L. Al'bertinskii, Elektrokhimiya, **16**, 122 (1980).
1790. Mirkovich, V.V., High Temper.–High Pressures, **8**, 231 (1976).

1791. Mitchell, W., Jr., (ed), Fuel Cells, Academic Press, NY (1963).
1792. Mittelmeyer, H., K. Strasser and B. Stuve, Rep. BFMT-FB-T83-113, Siemens, A.G. Zentralbereich Technik, Postfach 3240, Erlangen, 8520, F.R.G. (June 1983).
1793. Miyoshi, N., Y. Tanabe, M. Tsuruoka, Y. Tachibana, and H. Hoshikawa, Ger. Offen. 2,415,253 (1974).
1794. Mohr, V.H. and G. Ranke, "Acid and Sour Gas Treatment Processes," Chem. Eng. Progr., Oct. 1984, p. 27.
1795. Molnar, C.J., E.H. Price, and P.R. Resnick, U.S. Patent No. 4,176,215 (1979).
1796. Momot, E., and G. Bronoel, C.R. Acad. Sci., Ser. C., 1972, **274** (17), 1485.
1797. Moncrief, T.I., National Fuel Cell Seminar Abstracts, 1978, p. 127.
1798. Mond, K., G. Richter, and M. Wenzel, German Offen. 2,108,396 (1972).
1799. Mond, L.L., and D. Langer, Proc. Roy. Soc. (London), Services, A, Vol.46, 296 (1889).
1800. Monsanto Corp., Prism™ Separators: Simple Separation Systems for Hydrogen Recovery, Tech. Bull., Separations Business Group (1980).
1801. Monsanto Corp., St. Louis, MO, Prism Separator Product Bulletin, (1980).
1802. Monsanto Research Corp., Research to Improve Electrochemical Catalysts, 1st Rep., U.S. Govt. Contract No. DA-44-009-AMC-202(T), January 1964.
1803. Monsanto Research Corp., Fuel Cell System (60 Watt), U.S. Govt. Contract No. DAAB07-67-C-1460
1804. Moore, R.B., and C.R. Martin, Anal. Chem., **58**, 2569 (1986).
1805. Moos, A.M., U.S. Patent No. 3,276,909 (1966).
1806. Moran, P.J., G.L. Cahen, Jr., and G.E. Stoner, Proc. DOE Chem./Hydrogen Energy Contract. Rev. Syst. 1977, (CONF-771131) 47 (1978).
1807. Morcos, I., Can. Patent No. 1,080,204 (1980).
1808. Morcos, I., Proc. Symp. Electrode Mater. Processes Energy Convers. Storage, 280 (1977).
1809. Morgan, P.E.E., Proc. Workshop High Temperature Solid Oxide Fuel Cells, Brookhaven Nat. Lab., 1977, p. 54.
1810. Moriarity, A.J., Proc. Int. Symp. Alcohol Fuel Technol.: Methanol/Ethanol, 1977, NTIS Rep. COLNF-771175, 8–1 (1978).
1811. Moritake, M., and M. Fujii, Japan. Kokai, **75**, 28,483 (1975).
1812. Morrison, S., Spec. Rep. EPRI-SR-13, Conf. Proc.: Fuel Cell Catal. Workshop, 179 (1975).
1813. Motani, K., and T. Oku, Japan. Kokai Tokkyo Koho **79** 112,382 (1979).
1814. Motoo, A., M. Watanabe, and Y. Mochizuki, Denchi Toronkai (Koen Yoshishu), 18th, 23 (1979).
1815. Motoo, S., and M. Watanabe, J. Electroanal. Chem., **69**, 429 (1976).
1816. Motoo, S., Denki Kagaku Oyobi Kogyo Butsuri Kagaku, **48**, 84 (1980).

1817. Motoo, S., M. Shibata, and M. Watanabe, J. Electroanal. Chem. Interfac. Electrochem., **110**, 103 (1980).

1818. Moyer, B.A., M.S. Thompson, and T.J. Meyer, J. Am. Chem. Soc., **102**, 2310 (1980).

1819. Muchnik, G.F., V.G. Shlekhanov, E.M. Chistov, V.M. Andrianov, L.I. Leonichev, Yu.A. Lomtev, and V.P. Postanogov, USSR Patent No. 493,838 (1975).

1820. Muchnik, G.F., O.G. Sarandinaki, and P.P. Sidorov, Vesti Akad. Navuk. B. SSR, Ser. Fiz.-Energ. Navuk, **1976** (3), 55-8 (1976).

1821. Mueller, R., Ger. Offen. 2,905,168 (1980).

1822. Muller, L., and L.N. Nekrasov, Electrochim. Acta, **9**, 1015 (1964).

1823. Muller, R., J. Electrochem. Soc., **113**, 943 (1966).

1824. Mullin, J.P., L.P. Randolph, W.R. Hudson, and J.H. Ambrus, Proc. 14th Intersoc. Energy Conv. Eng. Conf., Vol. 1, 544 (1979).

1825. Mund, K., G. Richter, and F. von Sturm, Comité International de Thermodynamiques et Cinétique Electrochimiques, Prague, Sept. 1970.

1826. Mund, K., G. Richter, and M. Wenzel, Ger. Offen. 2,108,457 (1972).

1827. Mund, K., and R. Schulte, Ger. Offen. 2,328,050 (1974).

1828. Mund, K., Siemens Forsch.–Entwicklungsber, **5**, 209 (1976).

1829. Mund, K., G. Richter, and F. von Sturm, Proc. Workshop Electrocatalysis Fuel Cell Reactions, Brookhaven Nat. Lab. 1978, p. 47.

1830. Murakami, T., Japan Kokai Tokkyo Koho **79**, 68,727 (1979).

1831. Murayama, N., T. Sakagami, and M. Fukuda, Ger. Offen. 2,614,058 (1976).

1832. Murayama, N., M. Fukuda, and T. Sakagami, Japan. Kokai **77** 8,997 (1977).

1833. Murayama, N., T. Sakagami, M. Fukuda, and S. Suzuki, Japan. Kokai **78** 46,489 (1978).

1834. Murayama, N., T. Sakagami, M. Fukuda, and S. Suzuki, Japan. Kokai Tokkyo Koho **79** 143,798 (1979).

1835. Murgia, L.S., National Fuel Cell Seminar Abstracts, 1982, p. 33.

1836. Murphy, A.J., Jr., National Fuel Cell Seminar Abstracts, 1982, p. 79.

1837. Murphy, O.J., J. O'M. Bockris and D.W. Later, Int. J. Hydrogen Energy, **10**, 453 (1985).

1838. Murray, J.N., and P.G. Graimed, "Fuel Cells," A.I.Ch.E. Progr. Tech. Manual, (1963), p. 57.

1839. Murray, J.N., Proc. ERDA Contract. Rev. Meet. Chem. Energy Storage Hydrogen Energy Syst., (CONF-761134) 29 (1976).

1840. Murray, J.N., and M.R. Yaffe, Proc. DOE Chem./Hydrogen Energy Contract. Rev. Syst. 1977 (CONF-771131) 42 (1978).

1841. Murray, R.W., and D.J. Gross, Anal. Chem., **38**, 392 (1966).

1842. Murray, R.W., Acc. Chem. Res., **13**, 135 (1980).

1843. Nadler, M., and R.P. Cahn, U.S. Patent No. 4,080,791 (1978).
1844. Nagai, M., S. Fujitsu, and T. Kanazawa, Koen Yoshishu - Kotai Ionikusu Toronkai, 7th, 67 (1979).
1845. Nagata, S., Y. Ohno, and H. Sato, in "Applications of Solid Electrolytes," JEC Press, Cleveland, OH (1980), p. 193.
1846. Nagata, S., Y. Ohno, Y. Kasuga, and H. Sato, National Fuel Cell Seminar Abstracts, 1983, p. 165.
1847. Nakamura, H., and Y. Fujita, Japan Kokai Tokkyo Koho 80, 35,461 (1980).
1848. Nakamura, H., and Y. Fujita, Japan Kokai Tokkyo Koho 80, 39,173 (1980).
1849. Nakamura, O., T. Kodama, I. Ogino, and Y. Miyake, Japan. Kokai 76, 89,135 (1976).
1850. Nakamura, O., T. Kodama, I. Ogino, and Y. Miyake, Japan. Kokai 76, 107,444 (1976).
1851. Nakamura, O., T. Kodama, I. Ogino, and Y. Miyake, Chem. Lett., (1), 17 (1979).
1852. Nanis, L., Brit. Patent No. 1,364,795 (1974).
1853. Narebska, A., and R. Wodzki, Angew. Makromol. Chem., 80, 105 (1979).
1854. Narimatsu, Y., N. Niguchi, M. Shimizu, and M. Nagashima, National Fuel Cell Seminar Abstracts, 1985, p. 6.
1855. Narsavage, S.T., R.W. Vine, and R.C. Emanuelson, U.S. Patent No. 3,859,138 (1975).
1856. Naumov, V.I., I.N. Starodubrovskaya, V.V. Izotova, and Yu.M. Tyurin, Elektrokhimiya, 16, 301 (1980).
1857. Neiggemann, M.F., National Fuel Cell Seminar Abstracts, 1982, p. 1.
1858. Nelson, A.D., and L.E. Espelien, Ger. Offen. 2,416,520 (1974).
1859. Nelson, A.D., and L.E. Espelien, U.S. Patent No. 4,007,058 (1977).
1860. Nelson, S.H., National Fuel Cell Seminar Abstracts, 1978, p. 39.
1861. Nernst, W., and W. Wald, Z. Elektrochem., 7, 373 (1900).
1862. Nesterov, B.P., N.V. Korovin, and V.N. Brodyanskii, Elektrokhimiya, 12, 750 (1976).
1863. Nesterov, B.P., V.M. Alashkin, E.S. Tverdokhlebov, M.S. Beletskii, and S.A. Zastavskaya, Tr. Mosk. Energ. Inst. 365, 16 (1978).
1864. Newby, D., National Fuel Cell Seminar Abstracts, 1983, p. 59.
1865. Newby, D., and J.M. Feret, "Phosphoric Acid Fuel Cell 7.5 MW Conceptual Design Report," Westinghouse Electric Co., Rep. WAESD TR-83-1002, Pittsburgh, PA (1983).
1866. Newman, J.S., Electrochemical Systems, Prentice-Hall, Englewood Cliffs, NJ (1973).
1867. Newman, J., Electrochim. Acta, 24, 223 (1979).
1868. Newman, S.A., (ed.), "Acid and Sour Gas Treating Processes," Gulf Publ., Houston, TX (1985).

1869. Ng, D.Y.C., and K.K. Fleming, in "Fuel Cell Systems II," Advances in Chem. Series **90**, A.C.S., Washington, DC (1969), p. 354.
1870. Ng, D.Y.C., H.C. Maru, H. Feldkirchner, and B.S. Baker, Proc. 5th Intersoc. Energy Conv. Eng. Conf., 5–46 (1970).
1871. Ng, D.Y.C., and A.J. Appleby, in "Power Sources 3," D.H. Collins, (ed.), Oriel Press, Newcastle Upon Tyne (1971), p. 265.
1872. Nguyen, D.H., Tap Chi Hao Hoc, **16**, 13 (1978).
1873. Nickols, R.C., Jr., and J.C. Trocciola, U.S. Patent No. 3,940,285 (1976).
1874. Nicolas, E., and L. Bourgeois, Eur. Patent Appl. 8,476 (1980).
1875. Nidola, A., and P.M. Spaziante, U.S. Patent No. 4,132,620 (1979).
1876. Niederreither, H., U.S. Patent No. 2,070,612 (1937).
1877. Niedrach, L.W., J. Electrochem. Soc., **111**, 1309 (1964).
1878. Niedrach, L.W., U.S. Patent No. 3,134,697 (1964).
1879. Niedrach, L.W., S. Gilman, and A. Weinstock, J. Electrochem. Soc., **112**, 1161 (1965).
1880. Niedrach, L.W., and H.R. Alford, J. Electrochem. Soc., **112**, 117 (1965).
1881. Niedrach, L.W., and I.B. Weinstock, Electrochem. Tech., **3**, 270 (1965).
1882. Niedrach, L.W., D.W. McKee, J. Paynter, and I.F. Danzig, Extended Abstracts, Battery Division, Electrochemical Soc., **11**, 32 (1966).
1883. Niedrach, L.W., D.W. McKee, J. Paynter, and I.F. Danzig, Electrochem. Tech., **5**, 318 (1967).
1884. Nikitin, V.G., G.I. Elfimova, and G.A. Bogdanovskii, Zh. Fiz. Khim., **53**, 3116 (1979).
1885. Nikolov, I., V. Nikolova, T. Vitanov, and M. Svata, J. Power Sources, **4**, 65 (1979).
1886. Nikolov, I., T. Vitanov, and V. Nikolova, J. Power Sources, **5**, 197 (1980).
1887. Nimmons, J.T., and K.D. Sheehy, National Fuel Cell Seminar Abstracts, 1981, p. 30.
1888. Noak, Yu.M., and D.K. Grachev, Elektrokhimiya, **16**, 65 (1980).
1889. Nohe, H., and H. Hannebaum, AIChE Symp. Ser., **75** (185), 109 (1979).
1890. Noninski, Kh., M. Vasileva-Dimova, and I. Ivanov, Ext. Abstr., Meet.–Int. Soc. Electrochem., 28 th, **2**, 238 (1977).
1891. Notoya. R., and A. Matsuda, J. Res. Inst. Catal., Hokkaido Univ., **27**, 95 (1980).
1892. Novak, M., and C. Visy, Acta Phys. Chem., **25**, 157 (1979).
1893. Novak, Yu.M., D.K. Grachev, L.A. L'vova, and L.I. Pen'kova, Issled. v Obl. Khim. Istochnikov Toka, (5), 49 (1977).
1894. Novikova, Z.N., and V.A. Druz, Zh. Fiz. Khim., **54**, 2293 (1980).
1895. Novosel'skii, I.M., Elektrokhimiya, **15**, 1285 (1979).
1896. Nowak, R., F.A. Schultz, M. Uamana, H. Abruna, and R.W. Murray,

Chemically Modified Carbon Electrodes. XIV. NTIS Rep. No. AD-A061427 (1978).

1897. Nowick, A.S., Proc. Workshop High Temperature Solid Oxide Fuel Cells, Brookhaven Nat. Lab., 1977, p. 24.

1898. Noyes, R., Fuel Cells for Public Utility and Industrial Power, Noyes Data Corp., Park Ridge, NJ (1977).

1899. Nuttall, L.J., paper presented at Electrochem. Soc. Ontario-Quebec Section Winter Symp., Toronto (1975).

1900. Nuttal, L.J., Proc. Workshop on Renewable Fuels and Advanced Power Sources for Transport. Workshop, p. 99 (1983).

1901. Nwalor, J.U., and A.R. Manner, Mass Transport in Molten Salt Electrochemical Systems, Oak Ridge Nat. Lab. Rep. ORNL/MIT-290 (1979).

1902. Obayashi, H., and T. Kudo, Adv. Chem. Ser., **163** (Solid State Chem. Energy Convers. Storage, Symp.), 316 (1977).

1903. Obayashi, H., and T. Kudo, Mater. Res. Bull., **13**, 1409 (1978).

1904. O'Brien, R., and C. Rosenfield, J. Phys. Chem., **67**, 643 (1963).

1905. Oda, Y., T. Morimoto, and K. Suzuki, Eur. Patent Appl. 14,896 (1980).

1906. Oda, Y., T. Morimoto, and K. Suzuki, Japan. Kokai Tokkyo Koho **80**, 28,216 (1980).

1907. O'Dom, G.W., and R.W. Murray, Anal. Chem., **39**, 51 (1967).

1908. Ogoshi, T., Japan. Kokai Tokkyo Koho **78**, 149,873 (1978).

1909. O'Grady, W.E., M.Y.C. Wood, P.L. Hagans, and E. Yeager, Proc. Symp. Electrode Mater. Processes Energy Convers. Storage, 172 (1977).

1910. O'Grady, W.E., S. Srinivasan, and R.F. Dudley, (eds.), Proc. of the Workshop on the Electrocatalysis of Fuel Cell Reactions, Brookhaven Nat. Lab., The Electrochemical Society, Princeton, NJ May 15–16 (1978).

1911. O'Grady, W.E., and S. Srinivasan, Proc. Workshop Electrocatalysis Fuel Cell Reactions, Brookhaven Nat. Lab., 1978, p. 5.

1912. O'Grady, W.E., H. Olender, H.S. Isaacs, and S. Srinivasan, National Fuel Cell Seminar Abstracts, 1978, p. 172.

1913. O'Grady, W.E., H. Olender, and H.S. Isaacs, Brookhaven National Lab. Rep. BNL 51038, UC-93, Energy Conversion TID-4500, Apr. 1979, p. 1.

1914. O'Grady, W.E., and K. Daube, Brookhaven National Lab. Rep. BNL 51053, UC-93, Energy Conversion TID-4500, FC-4, June 1979, p. 1.

1915. O'Grady, W.E., and E.J. Taylor, Brookhaven Nat. Lab. Rep. BNL 51144, UC-93, Energy Conversion TID-4500, FC-6, Oct. 1979, p. 5.

1916. O'Grady, W.E., and E.J. Taylor, Brookhaven Nat. Lab. Rep. BNL-51158, UC-98, Energy Conversion TID-4500, FC-7, Jan. 1980, p. 6.

1917. O'Grady, W.E., J. McBreen, S. Srinivasan, A. Goel, and H. Olender, National Fuel Cell Seminar Abstracts, 1981, p. 84.

1918. O'Grady, W.E., J. McBreen, and E. Findl, National Fuel Cell Seminar Abstracts, 1982, p. 142.

1919. Ogryzko-Zhukovskaya, S.G., N.A. Fedotov, and V.P. Belokopytov, Elektrokhimiya, **12**, 91 (1976).

1920. Ogura, H., and T. Mitani, Suzuka Kogyo Koto Semmon Gakko Kiyo, **12** (2), 263 (1979).

1921. Ogura, Y., and N. Itoh, National Fuel Cell Seminar Abstracts, 1983, p. 145.

1922. Ohbayashi, H., M. Yoshida, and T. Kudo, Japan. Kokai **77**, 120,286 (1977).

1923. Ohbayashi, H., M. Yoshida, and T. Kudo, Japan. Kokai **77**, 120,287 (1977).

1924. Okabe, S., K. Iwamoto, and H. Tamura, Japan. Kokai Tokkyo Koho, **79**, 29,027 (1979).

1925. Okabe, S., K. Iwamoto, and K. Tamura, Japan. Kokai Tokkyo Koho **80**, 14,602 (1980).

1926. Okamoto, Y., H. Tajima, and K. Koseki, Japan. Kokai **75**, 45,248 (1975).

1927. Okita, K., Japan. Tokkyo Koho **79**, 19,909 (1979).

1928. Okrent, E.H., and C.E. Heath, in "Fuel Cell Systems I," Advances in Chem. Series **47**, A.C.S., Washington, DC (1965), p. 328.

1929. Okuma, S., Denki Gijutsu, **221**, 5 (1977).

1930. Okuyama, M., G. Belanger, and D.L. Piron, Int. J. Hydrogen Energy, **3**, 297 (1978).

1931. Olah, G.A., G.K.S. Prakash and J. Sommer, Science, **206**, 13 (1979).

1932. Olender, H., H.S. Isaacs, A.C.C. Tseung, and W.E. O'Grady, Brookhaven Nat. Lab. Rep. BNL50951, UC-93, Energy Conversion TID-4500, Dec. 1978, p. 1.

1933. Olender, H., and W.E. O'Grady, Brookhaven Nat. Lab. Rep. BNL 51072, UC-93, Energy Conversion TID-4500, FC-5, July 1979, p. 1.

1934. Olender H., and W.E. O'Grady, Brookhaven Nat. Lab. Rep. BNL 51144, UC-93, Energy Conversion TID-4500, FC-6, Oct. 1979, p. 1.

1935. Olender, H., and J. McBreen, Brookhaven Nat. Lab. Rep. BNL-51158, UC-98, Energy Conversion TID-4500, FC-7, Jan. 1980, p. 1.

1936. Olesen, O.L., and W.H. Johnson, "Fuel Processor Development for 11-MW Fuel Cell Power Plants," EM-4123 EPRI, Palo Alto, CA (1985).

1937. Olmer, L.J., and H.S. Isaacs, Brookhaven Nat. Lab. Rep. BNL-51158, UC-98, Energy Conversion TID-4500, FC-7, Jan. 1980, p. 10.

1938. Ong, E.T., R.A. Donado, C.T. Li, and T.D. Claar, Alternate Electrolyte Compositions for Molten Carbonate Fuel Cells, NTIS Rep. CONF-7810130-4 (1979).

1939. Ong, E.T., R.A. Donado, C.T. Li, and D.D. Claar, in "Proceedings of the DOE/EPRI Workshop on Molten Carbonate Fuel Cells," WS-78-135, EPRI, Palo Alto, CA (1979), pp. 4–13.

1940. Ong, E.T., T.D. Claar, and L.G. Marianowski, National Fuel Cell Seminar Abstracts, 1981, p. 77.

1941. Ong, E.T., and T.D. Claar, in "Molten Carbonate Fuel Cell Technology," J.R.

Selman and T.D. Claar, (eds.), PV 84-13, p. 54, The Electrochemical Society, Inc., Pennington, NJ, 1984.

1942. Oniciu, L., S. Agachi, E. Schmidt, E. Suciu, and V.A. Topan, Rev. Roum. Chim., **24**, 145 (1970).

1943. Oniciu, L., I. Mitrache, and V.A. Topan, Stud. Univ. Babes-Bolyai, Ser. Chem., **20**, 77 (1975).

1944. Oniciu, L., Fuel Cells, Abacus Press, Kent, England (1976).

1945. Oniciu, L., in "Fuel Cells," Abacus Press, Kent, England (1976), p. 62.

1946. Oniciu, L., E.M. Suciu, and T. Dogaru, Stud. Univ. Babes-Bolyai, Ser. Chem., **21**, 39 (1976).

1947. Oniciu, L., E.M. Suciu, I. Mitrache, and D.M. Constantin, Stud. Univ. Babes-Bolyai, Ser. Chem., **21**, 44 (1976).

1948. Oniciu, L., E.M. Rus, and I. Bochis, Rev. Chim. (Bucharest), 30, 62 (1979).

1949. Onoue, Y., T. Sata, A. Nakahara, and J. Itoh, Ger. Offen. 2,818,128 (1978).

1950. Ontario Govt. Advisory Committee on Energy, Energy in Ontario: The Outlook and Policy Implications, 2 Vols., Ontario Govt. Publication, Toronto (1973).

1951. Ooue, M., and I. Izumi, Kenkyu Kiyo–Nara Kogyo Koto Senmon Gakko, **14**, 57 (1979).

1952. Orlando, J.A., D.R. Limaye, R.A. Wakefield, and S. Karamschetty, National Fuel Cell Seminar Abstracts, 1978, p. 49.

1953. Ormrod, S.E., and D.L. Kirk, J. Phys. D., 10, 1497 (1977).

1954. Ormrod, S.E., and D.L. Kirk, J. Phys. D., 10, 1769 (1977).

1955. Osteryoung, R.A., and J.H. Christie, J. Electroanal. Chem., **25**, 157 (1970).

1956. Ostrovidova, G.V., G.V. Vivitskaya, and V.B. Alezkovskii, Elektrokhimiya, **6**, 1119 (1970).

1957. Ostwald, W., Z. Elektrochem., **1**, 122 (1894).

1958. O'Sullivan, J.B., Historical Review of Fuel Cell Technology, Proc. 25th Power Sources Symp., 149 (1972).

1959. Oyama, N., and F.C. Anson, J. Electrochem. Soc., **127**, 247 (1980).

1960. Oyama, N., and F.C. Anson, J. Electrochem. Soc., **127**, 249 (1980).

1961. Ozawa, T., and I. Oshida, Kagaku no Jikken, **30** (2), 171 (1979).

1962. Ozeryanaya, I.N., T.I. Manukhina, and B.D. Antonov, Fiz. Khim. Elektrokhim. Rasplavl. Tverd. Elektrolitov, Tezisy Dokl. Vses. Konf. Fiz. Khim. Ionnykh Rasplavov Tverd. Elektrolitov, 7th, **2**, 98 (1979).

1963. Padova, J.I., H.R. Bronstein, and J. Braunstein, National Fuel Cell Seminar Abstracts, 1978, p. 115.

1964. Paetsch, L., A. Pigeaud, and H.C. Maru, National Fuel Cell Seminar Abstracts, 1983, p. 32.

1965. Paetsch, L., P.S. Patel, H.C. Maru and B.S. Baker, ongoing work under

RP1273-1, EPRI, Palo Alto, CA (1986); see also National Fuel Cell Seminar Abstracts, 1986, p. 143.

1966. Pal'guev, S.F., and A.D. Neuimin, Proc. 10th Intersoc. Energy Conv. Eng. Conf., 46 (1975).

1967. Palanker, V.Sh., D.V. Sokol'skii, E.A. Mazulevskii, and E.N. Baibatyrov, J. Power Sources, **1**, 169 (1976).

1968. Pan, Y-C., Carbon Oxidation Catalyst Mechanism Study for Fuel Cells, NTIS Rep. No. PB-256420 (1976).

1969. Pan, Y.C., and G. Ciprios, National Fuel Cell Seminar Abstracts, 1978, p. 78.

1970. Pandya, J.D., in "Fuel Cells: Trends in Research and Applications," Hemisphere Publ., NY (1987), p. 203.

1971. Pang, C.K., S.T. Lee, and J. Kekela, National Fuel Cell Seminar Abstracts, 1982, p. 3.

1972. Pang, C.K., S.T. Lee, K. Lee, and D.T. Imamura, "Applications of Fuel Cells in Utility Systems. Vol. 1, Study Results," EPRI Rep. EM-3205, Vol. 1, EPRI, Palo Alto, CA (1983).

1973. Pangborn, J., Inst. Gas. Tech., Fuel Cell Technology, Report CONF-741266-1 (1974).

1974. Panich, R.U., N.M. Voznesenskii, and V.I. Korolev, Tr. Mosk. Energ. Inst., **365**, 81 (1978).

1975. Paquot, C., Bull. Soc. Chim., **8**, 695 (1941).

1976. Paquot, C., and F. Goursac, Bull. Soc. Chim., **12**, 450 (1945).

1977. Parker, K.M., "Development of Molten Carbonate Fuel Cell Power Plant," Interim Rep., DOE/ET/17019-T3, U.S. Dept. of Energy, Washington, DC (1985).

1978. Parkinson, B., F. Decker, J.F. Juliao, M. Abramovich, and H.C. Chagas, Electrochim. Acta, **25**, 521 (1980).

1979. Patel, P.S., C.H. Lee, and H.C. Maru, National Fuel Cell Seminar Abstracts, 1981, p. 106.

1980. Patel, P.S., H.C. Maru, and B.S. Baker, National Fuel Cell Seminar Abstracts, 1982, p. 88.

1981. Patel, P.S., R. Bharvani, and M. Matsumura, National Fuel Cell Seminar Abstracts, 1983, p. 25.

1982. Patel, P.S., "Assessment of a 6500 BTU/kWh Heat Rate Dispersed Generator," EM-3307 (1983).

1983. Patel, P.S., and M.G. Monn, "Parametric Analysis of a 6500 BTU/kWh Heat Rate Dispersed Generator, Rep. EM-4179, EPRI, Palo Alto, CA (1985).

1984. Paterson, R., Pontif. Acad. Sci. Scr. Varia, **40** (Sem. Etude Theme: Membr. Biol. Artif. Desalin. Eng., 1975), 517 (1976).

1985. Patterson, J.W., Proc. Workshop High Temperature Solid Oxide Fuel Cells, Brookhaven Nat. Lab., 1977, p. 8.

1986. Pauling, L., Phys. Rev., **54**, 899 (1938); see also Proc. Royal Soc. (Lond.), **A196**, 343 (1949).
1987. Paynter, J., U.S. Patent No. 3,793,083 (1974).
1988. Peattie, C.G., IEEE Spectrum, June 1966, pp. 69–76.
1989. Penner, S.S., (ed.), "Assessment of Research Needs for Advanced Fuel Cells," Energy, **12 (1-2)** (1986); also (hardcover), Pergamon Press, NY (1986).
1990. Penner, S.S., et. al., "Coal Gasification," Final Rep., DE-AC01-85ER30076 (1987); Energy, **12**, 623–903 (1987).
1991. Pennsylvania Research Assoc. Inc., Advanced Fuel Cell Catalysts, U.S. Govt. Contract No. NAS-7-574, Interagency Advanced Power Group Project Brief No. 1710, Aug. 1970.
1992. Penny, M., and S. Bourgeois, Development Status and Environmental Hazards of Several Candidate Advanced Energy Systems, NTIS Rep. PB-272759 (1977).
1993. Peopperling, R., and W. Schwenk, Electrochim. Acta, **22**, 121 (1977).
1994. Perfil'ev, M.V., and M.V. Inozemtsev, Tezisy Dok.–Vses. Soveshch. Elektrokhim., 5th 1974, **1**, 78 (1975).
1995. Perfil'ev, M.V., Tr. Inst. Elektrokhim., Ural. Nauchn. Tsentr, Akad., Naukl. SSSR, **26**, 81 (1978).
1996. Perfil'ev, M.V., and M.V. Inozemtsev, Elektrokhimiya, **14**, 1724 (1978).
1997. Perkins, F.G., Proc. 24th Power Sources Conf., 202 (1970).
1998. Perretta, A.T., and A. Kaufman, National Fuel Cell Seminar Abstracts, 1978, p. 47.
1999. Perroud, P., and G. Terrier, Adv. Hydrogen Energy, **1** (Hydrogen Energy Syst., Vol. 1), 241(1979).
2000. Perry, J., Proc. 23rd Power Sources Conf., (1969).
2001. Perry, J., Jr., U.S. Patent No. 3,844,839 (1974).
2002. Perry, J., Jr., Investigation of Hydrazine System Reaction Product Formation, NTIS Rep. No. AD-A028587 (1976).
2003. Perry, T.L., Nature Canada, **3** (2), 3 (1974).
2004. Peterson, J.R., and R.M. Reinstrom, National Fuel Cell Seminar Abstracts, 1980, p. 86.
2005. Peterson, J.R., R.M. Reinstrom, and A.W. Schnacke, National Fuel Cell Seminar Abstracts, 1981, p. 67.
2006. Petrii, O.A., V.A. Safonov, and I.G. Shchigorev, Elektrokhimiya, **9**, 125 (1973).
2007. Petrii, O.A., A.V. Ushmaev, and E.Yu. Alekseeva, Dvoinoi Sloi Adsorbtsiya Tverd. Elektrodakh, **5**, 191 (1978).
2008. Petrii, O.A., A.V. Ushmaev, and E.Yu. Alekseeva, Elektrokhimiya, **15**, 1329 (1979).

2009. Petrii, O.A., S.Ya. Vasina, and M.I. Zhirnova, Elektrokhimiya, **16**, 348 (1980).
2010. Petrii, O.A., A.V. Ushmaev, and I.L. Gogichadze, Elektrokhimiya, **16**, 891 (1980).
2011. Petrova, S.A., and O.S. Ksenzhek, Vopr., Khim. Khim. Tekhnol., **40**, 3 (1975).
2012. Petrow, H.G., and R.J. Allen, Belg. Patent No. 831,297 (1975).
2013. Petrow, H.G., and R.J. Allen, U.S. Patent No. 3,992,331 (1976).
2014. Petrow, H.G., and R.J. Allen, U.S. Patent No. 3,992,512 (1976).
2015. Petrow, H.G., and R.J. Allen, U.S. Patent No. 4,044,193 (1977).
2016. Petrow, H.G., and R.J. Allen, U.S. Patent No. 4,059,541 (1977).
2017. Petrow, H.G., and R.J. Allen, U.S. Patent No. 4,082,695 (1978).
2018. Petrow, H.G., and R.J. Allen, U.S. Patent No. 4,166,143 (1979).
2019. Piasecki, R., National Fuel Cell Seminar Abstracts, 1981, p. 22.
2020. Pierce, R.D., P.A. Finn, K. Kinoshita, G.H. Kucera, and J.W. Sim, National Fuel Cell Seminar Abstracts, 1979, p. 133.
2021. Pierce, R.D., G.H. Kucera, J.W. Sim, J.L. Smith, and R.B. Poeppel, National Fuel Cell Seminar Abstracts, 1980, p. 176.
2022. Pierce, R.D., J.L. Smith, and R.B. Poeppell, Proc. Electrochem. Soc. Symp. on Molten Carbonate Fuel Cells, Montreal, May 9–14 (1982).
2023. Pierce, R.D., and J.P. Ackerman, National Fuel Cell Seminar Abstracts, 1983, p. 6.
2024. Piersma, B.J., and E. Gileadi, Modern Aspects of Electrochemistry, Vol. 4, Chap. 2, J.O'M. Bockris, (ed.), Plenum Press, NY (1966).
2025. Pigeaud, E., ongoing work, Energy Research Corporation, 1986.
2026. Pilla, A.A., and G.J. Di Masi, in "Fuel Cell Systems II," Advances in Chem. Series **90**, A.C.S., Washington, DC (1969), p. 171.
2027. Pilla, A.A., J.A. Christopulos, and G.J. Di Masi, in "Fuel Cell Systems II," Advances in Chem. Series **90**, A.C.S., Washington, DC (1969), p. 231.
2028. Pismen, L.M., Yu.M. Vol'fkovich, and V.S. Bagotskii, J. Appl. Electrochem., **6**, 485 (1976).
2029. Pivnik, E.D., A.A. Lanin, Z.L. Koposova, T.V. Efimovskaya, E.A. Ukshe, and N.G. Bukun, USSR Patent No. 490,212 (1975).
2030. Plass, G.N., in "Electrochemistry of Cleaner Environments," Bockris, J.O'M., (ed), Plenum Press, NY (1972), Chap. 2.
2031. Platner, J., D. Ghere, and P. Hess, Proc. 19th Power Sources Conf., 32 (1965).
2032. Pobedinskii, S.N., A.A. Trofimenko, and K.N. Belonogov, Izv. Vyssh. Uchebn. Zaved., Khim. Khim. Tekhnol., **21**, 696 (1978).
2033. Pober, R.L., U.S. Patent No. 4,166,159 (1979).
2034. Podlovchenko, V.I., and R. Petrukhova, Elektrokhimiya, **6**, 198 (1970).
2035. Podlovchenko, B.I., T.D. Gladysheva, V.F. Stenin, and V.I. Levina, Elektrokhimiya, **9**, 1680 (1973).

2036. Poirier, A.R., Engineering Development of a Direct Hydrocarbon-Air Fuel Cell System, Proc. 1st Intersoc. Energy Conv. Eng. Conf., Sept. 1966.
2037. Pollak, F.H., Optical Investigation of the Oxides of Ruthenium and Iridium in Relation to their Electrocatalytic Activity, Proc. DOE Chem./Hydrogen Energy Contract. Rev. Syst. 1977 (CONF-771131) (1978).
2038. Pollard, R., and J. Newman, Electrochim. Acta, **25**, 315 (1980).
2039. Polyanskii, A.B., Tr. Molodykh Uch., Saratov Univ., 231 (1971).
2040. Pons, S., J. Electroanal. Chem., **150**, 481 (1983).
2041. Post, R.E., Evaluation of Potassium Titanate as a Component of Alkaline Fuel Cell Matrices, NASA Tech. Note NASA-TN-D-8341 (1977).
2042. Potter, E.C., Electrochemistry: Principles and Applications, Cleaver-Hume, London (1961).
2043. Pouchot, W.D., National Fuel Cell Seminar Abstracts, 1982, p. 36.
2044. Pourbaix, M., "Atlas d'Equilbres Electrochimiques," Gautier-Villars, Paris (1963), p. 449.
2045. Powers, R.W., Superionic Conduct., (Proc. Conf.), 351 (1976).
2046. Powers, R.W., and S.P. Mitoff, in "Solid Electrolytes," P. Hagenmuller and W. Van Gool, (eds.), Academic Press, NY (1978), p. 123.
2047. Prater, K.B., and A.J. Bard, J. Electrochem. Soc., **117**, 207 (1970).
2048. Pratt and Whitney Aircraft, Evaluation of Fuel Cell Electrodes, 4th Rep., U.S. Govt. Contract DA-44-009-AMC-1651(T), Nov. 1968.
2049. Pratt and Whitney Aircraft, Cost Effectiveness Analysis of Military Hydrocarbon Fuel Cell Power Plants, Final Rept., U.S. Gov't Contract No. DAAK-2-68-C-0524, Mar. 1970.
2050. Pratt and Whitney Aircraft, Open-Cycle Fuel Cell Powerplant Development, U.S. Govt. Contract No. DAAK-02-70-C-0518, June 1971.
2051. Presbry, C.H., and S. Schuldiner, J. Electrochem. Soc., **108**, 985 (1961).
2052. Presnov, V.A., Uu.A. Fadorin, and A.P. Yutrov, Elektrokhimiya, **11**, 450 (1975).
2053. Prigent, M., A. Sugier, and O. Bloch, Comptes Rendus, Troisièmes Journées Internat. d'Etude des Piles à Combustibles, Presses Académiques Européenes, Bruxelles (1969), p. 263.
2054. Prigent, M., C. Dezael, and Y. Breele, Proc. 5th Intersoc. Energy Conv. Eng. Conf., 5–111 (1970).
2055. Prigent, M., Proc. 5th Intersoc. Energy Conv. Eng. Conf., Sept. 1970.
2056. Proengin, S.a.f.L., Fr. Demande 2,349,221 (1977).
2057. Prokopius, P.R., and N.H. Hagedorn, NASA Tech. Note, NASA-TN-D-7757 (1974).
2058. Prokopius, P.R., Internal Voltage Control of Hydrogen-Oxygen Fuel Cells. Feasibility Study. NASA Tech. Note, NASA-TN-D-7956 (1975).
2059. Prokopius, P.R., M. Warshay, S.N. Simons, and R.B. King, Commercial

Phosphoric Acid Fuel Cell System Technology Development, NASA Tech. Memo. NASA-TM-79169, DOE/NASA/11272-79/1, E-034 (1979).

2060. Prokopius, P.R., M. Warshay, S.N. Simons, and R.B. King, Proc. 14th Intersoc. Energy Conv. Eng. Conf., 538 (1979).

2061. Prototech, Inc., Newton Highlands, MA, commercial product.

2062. Przybyla, F., Proc. 3rd Canadian Fuel Cell Seminar, p. 107 (1969).

2063. Pshenichnikov, A.G., in "Fuel Cells: Their Electrochemical Kinetics," V.S. Bagotsky and Yu.B. Vasil'ev, (eds.), Consultants Bureau, NY (1966), p. 11.

2064. Pshenichnikov, A.G., in "Power Sources 2," D.H. Collins, (ed.), Pergamon Press, London (1970), p. 276.

2065. Quehl, H.E., U.S. Patent No. 4,165,404 (1979).

2066. Raabe, J., A. Szymanski, W. Wlosinski, and W. Tomassi, Electrochim. Acta, **24**, 31 (1979).

2067. Radovici, O., M. Mavro-Tarabic, I. Onaca, and I. Solacolu, Rev. Roum. Chim., **21**, 2 (1976).

2068. Radovici, O., A. Bandi, and C. Capat, Rev. Roum. Chim., **23**, 1513 (1978).

2069. Radyushkina, K.A., M.R. Tarasevich, and S.I. Andruseva, Dvoino Sloi Adsorbtsiya Tverd. Elektrodakh, **5**, 208 (1978).

2070. Radyushkina, K.A., M.R. Tarasevich, and E.A. Akhundov, Elektrokhimiya, **15**, 1884 (1979).

2071. Raicheva, S., and E. Sakolova, Khim. Ind. (Sofia), **50** (9), 402 (1978).

2072. Raistrick, I.D., and R.A. Huggins, Considerations in the Use of Electrical Measurement Techniques to Evaluate Potential New Electrolytes and Mixed Conductors, Argonne Nat. Lab. Rep., ANL, ANL-76-8, Proc. Symp. Workshop. Adv. Battery Res. Des. B/277-B/292 (1976).

2073. Ramanarayanan, T.A., and W.L. Worrell, J. Electrochem. Soc., **121**, 1530 (1974).

2074. Ramp, F.L., Ger. Offen. 2,624,203 (1976).

2075. Rand, D.A.J., R. Woods, and D. Michell, Proc. Symp. Electrode Mater. Processes Energy Convers. Storage, 217 (1977).

2076. Randin, J-P., A.K. Vijh, and A.B. Chughtai, J. Electrochem. Soc., **120**, 1174 (1973).

2077. Randin, J-P., J. Electrochem. Soc., **120**, 1325 (1973).

2078. Randin, J-P., Electrochim. Acta, **19**, 83 (1974).

2079. Randin, J-P., Electroanal. Chem. Interfac. Electrochem., **51**, 471 (1974).

2080. Randin, J-P., J. Electrochem. Soc., **121**, 1029 (1974).

2081. Ranney, M., Fuel Cells, Noyes Data Corp., NY (1969).

2082. Rao, K.V.N., Proc. 3rd Can. Fuel Cell Seminar, Can. Defence Research Board, Sheridan Park, Ontario, Feb. 13–14, 1969 (1970), p. 51.

2083. Rapp, K.I., National Fuel Cell Seminar Abstracts, 1983, p. 24.

2084. Rapp, R.A., Proc. Workshop High Temperature Solid Oxide Fuel Cells, Brookhaven Nat. Lab., 1977, p. 6.
2085. Rapp, R.A., personal communication, 1985.
2086. Ratcliff, M., F.L. Posey, D.K. Johnson and H.L. Chum, J. Electrochem. Soc., **132**, 577 (1985).
2087. Ravaine, D., J. Non-Crystall. Solids, **38-39** (1), 353 (1980).
2088. Ray, S.P., and V.S. Stubican, Proc. Workshop High Temperature Solid Oxide Fuel Cells, Brookhaven Nat. Lab., 1977, p. 18.
2089. Razina, N.F., Izv. Akad. Nauk Kaz. SSR, Ser. Khim. (1), 75 (1980).
2090. Reau, J.M., J. Portier, and C. Lucat, Prog. Batteries Solar Cells, **2**, 67 (1979).
2091. Record, R.G.H., Ger. Offen. 2,554,999 (1976).
2092. Refer to Japanese developers, 1986.
2093. Regnaut, B., Fr. Demande 2,300,423 (1976).
2094. Regnaut, B., Fr. Demande 2,300,425 (1976).
2095. Regnaut, B., Fr. Demande 2,300,426 (1976).
2096. Reid, W.T., in "From Electrocatalysis to Fuel Cells," G. Sandstede, (ed.), Univ. of Washington Press, Seattle (1972), p. 369.
2097. Reinstrom, R.M., and K.M. Parker, National Fuel Cell Seminar Abstracts, 1982, p. 134.
2098. Reiser, C.A., C.R. Schroll, S.J. Szymanski and R.F. Buswell, National Fuel Cell Seminar Abstracts, 1979, p. 130.
2099. Reiser, C.A., and C.R. Schroll, National Fuel Cell Seminar Abstracts, 1980, p. 173.
2100. Reiser, C.A., and M.B. Landau, Ger. Offen. 2,941,514 (1980).
2101. Reiser, C.A., "Molten Carbonate Fuel Cell Verification and Scale-Up," EPRI Rep. EM-4129, Palo Alto, CA (1985).
2102. Reiser, C.A., IFC, ongoing work, 1986.
2103. Reiss, H., J.L. Katz, and O.J. Kleppa, J. Chem. Phys., **36**, 144 (1962).
2104. Remick, R.J., "Effects of Hydrogen Sulfide on Molten Carbonate Fuel Cells," Rep. DOE/MC/20212-1739 (1985).
2105. Remick, R.J., and D.R. Vasil, National Fuel Cell Seminar Abstracts, 1985, p. 173.
2106. Rentz, R.L., A.U. Hira, A.J. Parker, and J.S. Moore, National Fuel Cell Seminar Abstracts, 1983, p. 84.
2107. Repka, V.V., O.S. Ksenzhek, and N.D. Koshel, Vopr. Khim. Khim. Tekhnol., **41**, 149 (1975).
2108. Repka, V.V., Inst. Geotek. Mekh., Dnepropetrovsk, USSR, Deposited Doc. VINITI 2449-78 (1978).
2109. Reshetnikova, A.K., V.A. Shaposhnik, and G.M. Mamedov, Sint. Issled. Svoistv Kompleksn. Soedin., **1**, 103 (1973).

2110. Reshetnikova, A.K., N.I. Isaev, V.A. Shaposhnik, and A.I. Pleshkov, Teor. Prakt. Sorbtsionnykh Protsessou, **9**, 136 (1974).
2111. Resier, D.A., and C.R. Schroll, National Fuel Cell Seminar Abstracts, 1981, p. 44.
2112. Resnick, P.R., and W.G. Grot, U.S. Patent No. 4,085,071 (1978).
2113. Reul, S., and U. Wieland, Brennst.-Waerme-Kraft, **31**, 244 (1979).
2114. Rhodes, W.H., and R.E. Carter, Ionic Self-Diffusion in Calcia-Stabilized Zirconia, Abstract in Bull. Amer. Chem. Soc. (April 1962).
2115. Rice, W.E., and D. Bell, Proc. 7th Intersoc. Energy Conv. Eng. Conf., 390 (1972).
2116. Richards, W.T., J. Phys. Chem., **32**, 990 (1928).
2117. Richardson, N.A., G.H. Gelb, T.C. Wang, and J.A. Licari, Intersoc. Energy Conv. Eng. Conf., 1968 Record, p. 789.
2118. Rigney, D.M., National Fuel Cell Seminar Abstracts, 1981, p. 1.
2119. Rishavy, E.A., W.D. Bond, and T.A. Zechin, SAE Trans., 78, 981 (1967).
2120. Ritschel, M., and W. Vielstich, Electrochim. Acta, **24**, 885 (1979).
2121. Roberts, C.A., National Fuel Cell Seminar Abstracts, 1982, p. 171.
2122. Roberts, R., in "The Primary Battery," G.W. Heise and N.C. Cahoon, (eds.), Wiley, NY (1971) Chap. 9.
2123. Robertson, P.M., J. Electroanal. Chem. Interfac. Electrochem., **111**, 97 (1980).
2124. Rocklin, R.D., and R.W. Murray, Chemically Modified Carbon Electrodes. XVII. NTIS Rep. No. AD-A066632 (1979).
2125. Roe, O.K., Anal. Chem., **44**, 85R (1972).
2126. Rogers, L.J., Proc. 23rd Power Sources Conf., 1–4 (1969).
2127. Rogers, L.J., Proc. 25th Power Sources Conf., 179 (1972).
2128. Rohr, F.J., Proc. Workshop High Temperature Solid Oxide Fuel Cells, Brookhaven Nat. Lab., 1977, p. 122.
2129. Rohr, F.J., in "Applications of Solid Electrolytes," JEC Press, Cleveland, OH (1980), p. 196.
2130. Roiter, V.A., G.I. Golodetz, and Yu. Pyatnitzky, 4th Int. Congr. on Catalysis, Vol. **2**, 62 (1968).
2131. Rolison, K.E., Proc. 3rd Canadian Fuel Cell Seminar, 79 (1970).
2132. Rolla, A., A. Sadkowski, J. Wild, and P. Zoltowski, J. Power Sources, **5**, 189 (1980).
2133. Romashin, O.P., M.M. Fioshin, R.G. Erenburg, E.F. Ryabov, V.L. Kubasov, and L.I. Krishtalik, Elektrokhimiya, **15**, 653 (1979).
2134. Ross, D.K., and A.A. Law, Power Sources 2, D.G. Collins, (ed.), Pergamon Press, London (1970).
2135. Ross, P.N., Proc. Symp. Electrode Mater. Processes Energy Convers. Storage, 290 (1977).

2136. Ross, P.N., Proc. Workshop Electrocatalysis Fuel Cell Reactions, Brookhaven Nat. Lab. 1978, p. 169.
2137. Ross, P.N., Jr., Oxygen Reduction on Supported Pt Alloys and Intermetallic Compounds in Phosphoric Acid, Electric Power Research Inst. Rep. EPRI-EM-1553, Project 1200-5, Final Rep., Sept. 1980.
2138. Ross, P.N., National Fuel Cell Seminar Abstracts, 1980, p. 42.
2139. Ross, P.N., Proc. 16th Intersoc. Energy Conv. Eng. Conf., 726 (1981).
2140. Ross, P.N., National Fuel Cell Seminar Abstracts, 1982, p. 147.
2141. Ross, P.N., National Fuel Cell Seminar Abstracts, 1983, p. 141.
2142. Ross, P.N., National Fuel Cell Seminar Abstracts, 1983, p. 183.
2143. Ross, P.N., J. Electrochem. Soc., **130**, 882 (1983).
2144. Ross, P.N., "Deactivation and Poisoning of Fuel Cell Electrocatalysts," LBL-19766, Lawrence Berkeley Laboratory, Berkeley, CA (1985).
2145. Ross, P.N., Jr., Ext. Abstr. Fall Meeting Electrochemical Society, Las Vegas, NV (1985).
2146. Ross, P.N., Lawrence Berkeley Laboratory, Work under EPRI RP-1676-2, 1986.
2147. Ross, P.N., personal communication, 1986.
2148. Roth, R.S., T. Negas, H.S. Parker, D.B. Minor, C.D. Olson, and C. Skarda, Rare Earths Mod. Sci. Technol., (Rare Earths Res. Conf.), 13th 1977, 163 (1978).
2149. Roth, W.L., F. Reidinger, and S. LaPlaca, Superionic Conduc., (Proc. Conf.), 223 (1976).
2150. Roth, W.L., and R.C. DeVries, J. Solid State Chem., **20**, 111 (1977).
2151. Rouget, R., in "Fuel Cells: Trends in Research and Applications," A.J. Appleby, (ed.), Hemisphere Publ., NY (1987), p. 219.
2152. Rudd, D.A., Ger. Offen. 2,554,997 (1976).
2153. Ruka, R.J., Proc. Workshop High Temperature Solid Oxide Fuel Cells, Brookhaven Nat. Lab., 1977, p. 56.
2154. Ruka, R.J., National Fuel Cell Seminar Abstracts, 1978, p. 159.
2155. Ruka, R.J., National Fuel Cell Seminar Abstracts, 1980, p. 139.
2156. Russel, D.G., and J.B. Senior, Can. J. Chem., **52**, 2975 (1974).
2157. Russel, P.G., H.S. Isaacs, and S. Srinivasan, Proc. Workshop High Temperature Solid Oxide Fuel Cells, Brookhaven Nat. Lab., 1977, p. 96.
2158. Russell, P.G., H.S. Isaacs, and A.C.C. Tseung, Brookhaven Nat. Lab. Rep. BNL 50951, UC-93, Energy Conversion TID-4500, Dec. 1978, p. 17.
2159. Sabbioni, F., "Impact of Fuel Cells on Electric Power Generation Systems in Industrial Countries," in "Fuel Cells: Trends in Research and Applications," A.J. Appleby, (ed.), Hemisphere Publ., NY (1987).
2160. Sachter, W.M.H., and N.H. DeBoer, Proc. 3rd Int. Congr. on Catalysis, Vol. **1**, p. 252 (1965).

2161. Sadaoka, Y., N. Yamazoe, and T. Seiyama, Denki Kagaku Oyobi Kogyo Butsuri Kagaku, **46**, 597 (1978).
2162. Safonov, V.A., V.D. Dishel, and O.A. Petrü, Elektrokhimiya, **9** (2), 264 (1973).
2163. Saharay, S.K., and A.S. Basu, Indian J. Technol., **17**, 423 (1979).
2164. Saito, S., A. Sawaoka, K. Kondo, and K. Itoh, Process. Prop. Electron. Magn. Ceram., Proc. U.S.–Japan Semin. Basic Sci. Ceram. 1975, 113 (1976). See also references 2320 and 2321.
2165. Sakai, T., M. Kumeda, and N. Nishizawa, Denchi Toronkai, (Koen Yoshishu), 18th, 31 (1979).
2166. Sakai, T., K. Tsukamoto, R. Saito, and M. Ide, National Fuel Cell Seminar Abstracts, 1983, p. 157.
2167. Salathe, R.E., J.O. Smith, J.P. Gallagher, P.L. Terry, J. Kozloff, L.V. Athearn, and P. Dantowitz, Extended Abstracts, A.C.S. Meeting, Chicago, p. 257 (1967).
2168. Salathe, R.E., and P. Terry, Monsanto Research Corp., U.S. Govt. Contract No. DA-44-009-AMC-983(T) (1967).
2169. Salathe, R.E., Fuel Cell Battery: 60-Watt Advanced Development Model, Final Report, U.S. Govt. Contract No. DA28-043-AMC-01460(E), (ECOM), Monsanto Research Corp., Sept. 1969.
2170. Salathe, R.E., Proc. 24th Power Sources Conf., 204 (1970).
2171. Salathe, R.E., Fuel Cell Electrodes (Hydrogen-Air), NTIS AD-742 732 (1972).
2172. Sale, B., Rev. Gen. Therm., **16** (181), 29 (1977).
2173. Salihi, J.T., Proc. 25th Power Sources Conf., 81 (1972).
2174. Salisbury, J.D., and E. Behrin, Proc. 15th Intersoc. Energy Conv. Eng. Conf., 2080 (1980).
2175. Salisbury, J.D., E. Behrin, M.K. Kong, and D.J. Whisler, U.S. Dept. of Commerce, Rep. No. UCRL-52933, Feb. 1980.
2176. Salisbury, J.K., "Steam Turbines and Their Cycles," Wiley, NY (1950).
2177. Samarev, A., Z. Physik. Chem., **A168**, 45 (1934).
2178. Sammells, A.F., and P.G.P. Ang, Anode Electrochemistry in Selected Molten Carbonate Melts, Inst. of Gas Technol. Rep. CONF-7810130 (1978).
2179. Sammells, A.F., S.B. Nicholson, and P.G.P. Ang, J. Electrochem. Soc., **127**, 350 (1980).
2180. Samsonov, G.V., E.K. Fen, Ya.S. Malakhov, and V.Ya. Malakhov, Izv. Adad. Nauk SSSR, Neorg. Mater., **12**, 1404 (1976).
2181. Sandelli, G.J., U.S. Patent No. 4,165,349 (1979).
2182. Sanderson, R.A., C.L. Bushnell, and T.F. McKiernan, in "Fuel Cell Systems II," Advances in Chem. Series **90**, A.C.S., Washington, DC (1969), p. 60.
2183. Sandler, Y.L., J. Electrochem. Soc., **109**, 1115 (1962).

2184. Sandstede, G., 5th Symposium, Energie-Direkt-Umwandlung, Essen, Oct. 1970.
2185. Sandstede, G., (ed), From Electrocatalysis to Fuel Cells, Univ. of Washington Press, Seattle (1972).
2186. Sandstede, G., in "From Electrocatalysis to Fuel Cells," G. Sandstede, (ed.), Univ. of Washington Press, Seattle (1972), p. 59.
2187. Sandstede, G., in "From Electrocatalysis to Fuel Cells," G. Sandstede, (ed.), Univ. of Washington Press, Seattle (1972), p. 87.
2188. Sandstede, G., in "From Electrocatalysis to Fuel Cells," G. Sandstede, (ed.), Univ. of Washington Press, Seattle (1972), p. 113.
2189. Santo, Y., and K. Kikuchi, National Fuel Cell Seminar Abstracts, 1982, p. 160.
2190. Sarada, T., R.D. Granata, and R.T. Foley, Materials of Construction and Electrode Materials for a Fuel Cell Using a Perfluoroalkane Sulfonic Acid Electrolyte, NTIS Rep. COO-2879-2 (1977).
2191. Sarada, T., R.D. Granata, and R.T. Foley, J. Electrochem. Soc., **125**, 1899 (1978).
2192. Sarangapani, S., P. Bindra and E.B. Yeager, "The Physical and Chemical Properties of Phosphoric Acid," Brookhaven Nat. Lab., Upton, NY (1979).
2193. Sata, T., A. Nakahara, S. Baba, and M. Shiromizu, Japan. Kokai Tokkyo Koho, **79**, 107,88 (1979).
2194. Sato, H., T. Shimamune, T. Goto, and H. Nitta, Belg. Patent No. 878,691 (1979).
2195. Satoh, T., K. Kudo, and T. Tsushima, Denshi Gijutsu Sogo Kenkyusho Iho, **42**, 652 (1978).
2196. Sauer, H., and W. Kloss, Ger. Offen. 2,812,040 (1979).
2197. Savage, R.L., L. Blank, T. Cady, K. Cox, R. Murray, and R.D. Williams, (eds.), A Hydrogen Energy Carrier (2 vols), Systems Design Institute, NASA Grant NGT 44-005-114, Sept. (1973).
2198. Savel'eva, O.N., Tr. Mosk. Energ. Inst., **365**, 100 (1978).
2199. Savitz, M.L., and R.L. Carreras, in "Fuel Cell Systems II," Advances in Chem. Series 90, American Chemical Society, Washington, DC (1969), p. 188.
2200. Sawyer, D.T., and J.L. Roberts, Jr., Experimental Electrochemistry for Chemists, Wiley-Interscience, NY (1974).
2201. Sawyer, J.W., (ed.), "Sawyer's Gas Turbine Engineering Handbook," 2nd edition, Vol. 2, Gas Turbine Publications, Inc., Stamford, CT (1972).
2202. Sayano, R.R., R.A. Mendelson, M.E. Kirkpatrick, E.T. Seo, and H.P. Silverman, Proc. 23rd Annual Power Sources Conf., 32–34 (1969).
2203. Sayano, R.R., R.A. Mendelson, M.E. Kirkpatrick, E.T. Seo, and H.P. Silverman, Proc. 4th Intersoc. Energy Conv. Eng. Conf., 898 (1969).

2204. Sayano, R.R., R.A. Mendelson, E.T. Seo, and H.P. Silverman, Proc. 24th Power Sources Conf., 178 (1970).
2205. Sbrockey, N.M., "Sintering and Surface Smoothing Studies on Nickel and Nickel with Dispersed Oxide Particles," Ph.D. dissertation, Northwestern University, June 1982.
2206. Scarr, R.F., Proc. 25th Power Sources Conf., 173 (1972).
2207. Schenke, M., and G.H.J. Broers, High Temperature Galvanic Fuel Cells, Final Rep., U.S. Govt. Contract No. DA-91-591-EUC-1398 (1961).
2208. Scherson, D., A. Tanaka, S. Gupta, B. Tryk, C. Fierro, E. Yeager, and R. Latimer, Electrochim. Acta, **31**, 1247 (1986).
2209. Schlatter, M.J., in "Fuel Cell Systems I," Advances in Chem. Series **47**, A.C.S., Washington, DC (1965), p. 292.
2210. Schlatter, M.J., SAE Trans. **78**, 1576 (1967).
2211. Schmickler, W., and J.W. Schultze, Z. Phys. Chem. (Wiesbaden), **110**, 277 (1978).
2212. Schmidberger, R., and W. Doenitz, Ger. Offen. 2,735,934 (1979).
2213. Schmidberger, R., and W. Doenitz, Ger. Offen. 2,824,408 (1979).
2214. Schmidberger, R., and W. Doenitz, Ger. Offen. 2,837,118 (1980).
2215. Schmidt, J.E., U.S. Patent No. 3,843,413 (1974).
2216. Schmitt, J.W., personal communication, Dec. 9, 1980.
2217. Schneider, F.A., Chem. Week **67** (45) 13 (1971).
2218. Schouler, E., A. Hammour, and M. Kleitz, Mater. Res. Bull., **11**, 1137 (1976).
2219. Schouler, E.J.L., and H.S. Isaacs, Brookhaven Nat. Lab. Rep. BNL-51158, UC-98, Energy Conversion TID-4500, FC-7, Jan. 1980, p. 16.
2220. Schouler, E.J.L., and H.S. Isaacs, Brookhaven Nat. Lab. Rep. BNL 51198, UC-93, Energy Conversion TID-4500, FC-8, Mar. 1980, p. 19.
2221. Schroeder, J.E., D. Pouli, and H.J. Seim, in "Fuel Cell Systems II," Advances in Chem. Series **90**, A.C.S., Washington, DC (1969), p. 93.
2222. Schroeder, K.H., ETZ, Elektrotech. Z., **100**, 1266 (1979).
2223. Schroll, C.R., U.S. Patent No. 3,855,002 (1974).
2224. Schroll, C.R., and C.A. Reiser, National Fuel Cell Seminar Abstracts, 1982, p. 85.
2225. Schubert, F.H., D.B. Heppner, T.M. Hallick, and R.R. Woods, Technology Advancement of the Electrochemical Carbon Dioxide-Concentrating Process, NASA CR-152250, LSI-ER-258-7 (1979).
2226. Schuldiner, S., T.B. Warner, and B.J. Piersma, J. Electrochem. Soc., **114**, 343 (1967).
2227. Schuldiner, S., M. Rosen, and D.R. Flinn, Electrochim. Acta, **18**, 19 (1973).
2228. Schuldiner, S., Spec. Rep. EPRI-SR-13, Conf. Proc.: Fuel Cell Catal. Workshop, PB-245-115, 171 (1975).
2229. Schulman, F., Proc. 3rd Canadian Fuel Cell Seminar, 75 (1970).

2230. Schultze, J.W., M. Lohrengel, D. Ross, and U. Stimmig, Chem.-Ing.-Tech., **52**, 447 (1980).
2231. Schulz, D.A., U.S. Patent No. 3,960,601 (1976).
2232. Schwartz, A., L.L. Holbrook, and H. Wise, J. Catalysis, **21**, 199 (1971).
2233. Schwarz, H.J., Proc. 6th Intersoc. Energy Conv. Eng. Conf., 690 (1971).
2234. Schwarz, J.A., and S.R. Kelemen, Surf. Sci., **87**, 510 (1979).
2235. Schwarz, J.A., Surf. Sci., **87**, 525 (1979).
2236. Sederquist, R.A., U.S. Patent No. 4,128,700 (1978).
2237. Sederquist, R.A., U.S. Patent No. 4,200,682 (1980).
2238. Segura, M.A., C.L. Aldridge, K.L. Riley, and L.A. Pine, U.S. Patent No. 4,054,644 (1977).
2239. Seida, T., K. Takahashi, S. Asaumi, and A. Shimizu, Japan, Kokai, **78**, 28,589 (1978).
2240. Seita, T., T. Sato, and A. Shimizu, Japan, Kokai Tokkyo Koho, **79**, 106,087 (1979).
2241. Seko, M., Y. Yamakoshi, H. Miyauchi, M. Kukumoto, K. Kimoto, T. Hane, and M. Hamada, Japan, Kokai Tokkyo Koho, **78**, 125,986 (1978).
2242. Seko, M., Y. Yamakoshi, H. Miyauchi, M. Fukumoto, K. Kimoto, T. Hane, and M. Hamada, Japan, Kokai Tokkyo Koho, **79**, 38,284 (1979).
2243. Selman, J.R., Energy, **11**, 153 (1986).
2244. Semkina, N.V., V.P. Obrosov, and E.A. Nechaev, USSR Patent No. 459,448 (1975).
2245. Sen, R.K., E. Yeager, and W.E. O'Grady, Annu. Rev. Phys. Chem., **26**, 287 (1975).
2246. Sepa, D.B., A. Damjanovic, and J.O'M. Bockris, Electrochim. Acta, **12**, 746 (1967).
2247. Sepa, D.B., A. Damjanovic and J.O'M Bockris, J. Res. Inst. Catalysis, Hokkaido Univ., **16**, 1 (1968).
2248. Sepa, D.B., M.V. Vojnovic, D.S. Ovcin, and N.D. Pavlovic, J. Electroanal. Chem. Interfac. Electrochem., **51**, 99 (1974).
2249. Seth, R.L., D.M. Bhatt, S.P. Ghosh, and P.N. Mukherjee, Trans. Soc. Adv. Electrochem. Sci. Technol., **11**, 425 (1976).
2250. Setzer, H.J., and W.R. Standley, U.S. Patent No. 3,962,411 (1973).
2251. Shafer, M.W., R.A. Figat, R. Johnson, and R.A. Pollak, Ext. Abstr., Meet.–Int. Soc. Electrochem., 30th, 313 (1979).
2252. Shair, R.C., Intersoc. Energy Conv. Eng. Conf., 1968 Record, 785.
2253. Sharaevskii, A.P., and A.P. Baranov, Issled. v Obl. Khim. Istochnikov Toka, (4), 126 (1976).
2254. Sheblovinskii, V.M., A.G. Kicheev, and B.N. Yanchuk, Elektrokhimiya, **15**, 1279 (1979).
2255. Sheibley, D.W., J.M. Bozek, and D.G. Soltis, U.S. Patent Appl. 665,034 (1976).

2256. Sheibley, D.W., and H. McBryar, National Fuel Cell Seminar Abstracts, 1982, p. 29.

2257. Sheibley, D.W., J.D. Denais, and L.S. Murgia, National Fuel Cell Seminar Abstracts, 1983,

2258. Shepelev, V.Ya., M.R. Tarasevich, and R.Kh. Burshtein, Elektrokhimiya, 9, 189 (1973).

2259. Sherson, D.A., S.L. Gupta, C. Fierro, E.B. Yeager, M.E. Kordesch, J. Eldridge, R.W. Hoffman and J. Blue, Electrochim. Acta, **28**, 1205 (1983).

2260. Sherson, D.A., C. Fierro, E. Yeager, M. Kordesch, J. Eldridge, R. Hoffman, and A. Burns, J. Electroanal. Chem., **169**, 287 (1984).

2261. Sherson, D.A., R.W. Grimes, R. Holze, A. Tanaka, C. Fierro, E.B. Yeager, and R. Lattimer, Ext. Abs. Electrochem. Soc. Fall Meeting, New Orleans, LA (1984), p. 810.

2262. Shimokobe, I., and M. Nakamura, Japan, Kokai Tokkyo Koho, **79**, 82,042 (1979).

2263. Shinn, B.H., Ger. Offen. 2,430,614 (1975).

2264. Shipps, P.R., D.V. Ragone, and A.C. Eulberg, ASME Advan. Energy Conv. Eng., 879 (1967).

2264a. Shreeve, J., see reference 2322.

2265. Shropshire, J.A., and B.L. Tarmy, in "Fuel Cell Systems I," Advances in Chem. Series **47**, A.C.S., Washington, DC (1965), p. 153.

2266. Shumilova, N.A., V.N. Evko, and V.I. Luk'yanycheva, Tezisy Dokl.–Vses. Soveshch. Polyarogr., 7th, 103 (1978).

2267. Shumov, Y.S., S.S. Chakhmakhchyan, V.I. Mityaev, and G.G. Komissarov, Zh. Fiz. Khim., **53**, 2834 (1979).

2268. Sicafus, E.N., and H.P. Bonzel, Recent Progress in Surface Science, K.G.A. Paukhurst and A.C. Riddiford, (eds.), Academic Press, NY (1970).

2269. Sidheswaran, P., J. Electrochem. Soc. India, **28**, 27 (1979).

2270. Siemens A.-G., Brit. Patent No. 1,363,999 (1974).

2271. Siemens A.-G., Fr. Demande 2,198,786 (1974).

2272. Siemens A.-G., Brit. Patent No. 1,397,487 (1975).

2273. Sienko, M.J, J. Am. Chem. Soc., **81**, 5556 (1959).

2274. Silverman, H.P., Development of High Charge and Discharge Rate Lead/Acid Battery Technology, NTIS PB 213 257 (1972).

2275. Sim, J.W., G.H. Kucera, J.L. Smith, A.M. Breindel, and R.D. Pierce, National Fuel Cell Seminar Abstracts, 1982, p. 108.

2276. Simbeck, D.R., R.L. Dickenson and E.D. Oliver, "Coal Gasification Systems: A Guide to Status, Applications, and Economics," AP-3109, EPRI, Palo Alto, CA (1983).

2277. Simon, W.E., and J.E. Cox, ASME Advanced Energy Convers. Eng., 389 (1967).

2278. Simons, H.M., Ger. Offen. 2,855,049 (1979).
2279. Simons, R., Nature (London), **280** (5725), 824 (1979).
2280. Singer, R.M., U.S. Patent No. 4,177,159 (1979).
2281. Singh, G., M.H. Miles, and S. Srinivasan, Mixed Oxides as Oxygen Electrodes, Brookhaven Nat. Lab. Rep. No. BNL-20984 (1975).
2282. Singh, P., L. Paetsch and H.C. Maru, "Corrosion Prevention in Molten Carbonate Fuel Cells," Paper No. 87, "Corrosion 86," National Association of Corrosion Engineers, Houston, TX (1986).
2283. Sitnikova, V.V., G.G. Chuvileva, and N.I. Nikolaev, Zh. Fiz. Khim., **54**, 1320 (1980).
2284. Skewes, H.R., Proc. 1st Austral. Conf. Electrochem., Sydney-Hobart, Australia, 634 (1963) (published 1965).
2285. Skowronski, J.M., Fuel, **56**, 385 (1977).
2286. Smale, F.J., Z. Phys. Chem., **14**, 577 (1894).
2287. Smirnov, A.V., M.V. Simonova, and E.G. Shibanova, Elektrokhimiya, **11**, 1836 (1975).
2288. Smith, J.L., R.M. Arons, J.T. Dusek, A.V. Fraioli, G.H. Kucera, J.W. Sim, and R.D. Pierce, National Fuel Cell Seminar Abstracts, 1981, p. 130.
2289. Smith, J.L., T.D. Kaun, N.Q. Minh, and R.D. Pierce, National Fuel Cell Seminar Abstracts, 1983, p. 42.
2290. Smith, J.L., T.D. Kaun, N.Q. Minh, E.H. VanDeventer, F.C. Mrazek, and R.D. Pierce, paper presented at Int. Soc. Electrochem. Meet., Berkeley, CA, Aug. 5-10 (1984).
2291. Smith, J., ongoing work, Argonne Nat. Lab., 1987.
2292. Smith, R., H. Urbach, J. Harrison, and N. Hatfield, J. Phys. Chem., **71**, 1250 (1967).
2293. Smith, S.W., H.R. Kunz, W.M. Vogel, and S.J. Szymanski, "Effects of Sulfur on Molten Carbonate Fuel Cells: A Survey," in "Molten Carbonate Fuel Cell Technology," J.R. Selman and T.D. Claar, (eds.), Proceedings Vol. 84-13, The Electrochemical Society, Inc., Pennington, NJ (1984), p. 246.
2294. Smrcek, K., Elektrotech. Obz., **65** (4), 213 (1976).
2295. Smrcek, K., J. Beran, and J. Jandera, Kniznice Odb. Ved. Spisu Vys. Uceni Tech. Brne B, B-67, 161 (1976).
2296. Smrcek, K., and Z. Ministr, Porozim. Jeji Pousiti (Czech.), **5**, 156 (1979).
2297. Snell, K.D., and A.G. Keenan, Chem. Soc. Rev. **8**, 259 (1979).
2298. Snow, G.S., and P.D. Wilcox, Novel Zirconium Dioxide-Platinum Seal for Molecular Oxygen Sensors, NTIS Rep. SAND-76-0329 (1976).
2299. Sobkowski, J., A. Wieckowski, P. Zelanay, and A. Czerwinski, J. Electroanal. Chem. Interfac. Electroçhem., **100**, 781 (1979).
2300. Soffer, A., and M. Folman, Electroanal. Chem. and Interfac. Electrochem., **38**, 25 (1972).

2301. Sokolova, E., S. Kalcheva, M. Khristov, and S. Raicheva, Dokl. Bolg. Akad. Nauk, **32**, 189 (1979).

2302. Solacolu, I., M. Mavrodin-Tarabic, I. Onaca, and Gh. Vasu, Electrochim. Acta, **21**, 115 (1976).

2303. Solomon, F., Ext. Abstr., Electrochemical Society Spring Meeting, Toronto, Canada, 1985.

2304. Solomon, M., Proc. Symp. Batteries for Traction and Propulsion, 26 (1972).

2305. Somers, E.V., E.J. Vidt, and M.B. Blinn, National Fuel Cell Seminar Abstracts, 1979, p. 99.

2306. Somers, E.V., R.E. Grimble, and E.J. Vidt, National Fuel Cell Seminar Abstracts, 1980, p. 82.

2307. Sooki-Toth, G., P. David, J. Lukacs, I. Molnar, P. Horvath, F. Krajcsovics, J. Keszler, S. Gal, E. Pungor, and K. Tomor, Hung. Teljes 9,936 (1975).

2308. Sorensen, J.C., and G.T. McLandrich, National Fuel Cell Seminar Abstracts, 1980, p. 78.

2309. Sosenkin, V.E., and Yu.M. Vok'fkovich, Elektrokhimiya, **14**, 326 (1978).

2310. Spacil, H.S., and C.S. Tedman, Jr., General Electric TIS Report No. 69-C-176, 11 (1969).

2311. Spacil, H.S., and C.S. Tedman, Jr., General Electric TIS Report No. 69-C-177, 9 (1969).

2312. Spacil, H.S., National Fuel Cell Seminar Abstracts, 1978, p. 102.

2313. Spahrbier, D., Ger. Patent No. 2,115,310 (1974).

2314. Specht, S.J., and J.O. Adams, U.S. Patent No. 4,076,571 (1978).

2315. Specht, S.J., U.S. Patent No. 4,191,627 (1980).

2316. Specht, S.J., and J.A. Wood, U.S. Patent No. 4,196,071 (1980).

2317. Spedding, P.L., Proc. 1st Austral. Congress Electrochem., 1963, Sydney-Hobart, Australia, p. 688 (1965).

2318. Spitsyn, V.I., T.A. Drobasheva, T.S. Skoropad, G.E. Shakh, A.A. Gavrish, and A.N. Khodan, Zh. Neorg. Khim., **24**, 1738 (1979).

2319. Spitsyn, V.I., T.I. Drobasheva, and L.P. Kazanskii, Khim. Soedin. Mo (VI) W (VI), 3 (1979).

2320. Saito, M., T. Kishi, and T. Nagai, Denki Kagaku Oyobi Kogyo Butsuri Kagaku, **47**, 298 (1979).

2321. Saito, M., T. Kishi, and T. Nagai, Asahi Garasu Kogyo Gijutsu Sorreikai Kenkyu Hokoku, **32**, 99 (1978).

2322. Shreeve, J., Gas Research Institute Contractor's Meeting, Apr. 1986.

2323. Sprengel, D., Chemie Ingenieur Tecchnik, **46**, 967 (1974).

2324. Srinivasan, S., and E. Gileadi, in "Handbook of Fuel Cell Technology," C. Berger, (ed.), Prentice-Hall, Englewood Cliffs, NJ (1968), p. 332.

2325. Srinivasan, S., and R.H. Wiswall, Proc. Symp. Energy Storage, 1975, 82 (1976).

2326. Srinivasan, S., P.W.T. Lu, G. Kissel, F. Kulesa, C.R. Davidson, H. Huang, S. Gottesfeld, and J. Orehotsky, Proc. DOE Chem./Hydrogen Energy Contract. Rev. Syst. 1977 (CONF-771131) 34 (1978).

2327. Srinivasan, S., and H.S. Isaacs, Proc. 14th Intersoc. Energy Conv. Eng. Conf., 568 (1979).

2328. Srinivasan, S., R.S. Yeo, G. Kissel, E. Gannon, F. Kulesa, J. Orehotsky, and W. Visscher, Selection and Evaluation of Materials for Advanced Water Electrolyzers, Brookhaven Nat. Lab. Rep. BNL-26807 (1979).

2329. Srinivasan, S., J. McBreen, W.E. O'Grady, H. Olender, E.J. Taylor, and R.R. Adzic, National Fuel Cell Seminar Abstracts, 1980, p. 118.

2330. Srinivasan, S, E.R. Gonzalez, K-L. Hsueh, D-T. Chin, C.J. Maggiore, R. Naranjo, S. Gottesfeld, P.J. Hyde, A. Carubelli, C.R. Derouin, K.V. Kordesch, N. Vanderborgh, and H. Chang, National Fuel Cell Seminar Abstracts, 1982, p. 150.

2331. Srinivasan, S., P.J. Hyde, A. Carubelli, M.T. Paffett, C.R. Derouin, R.E. Bobbett, C.T. Campbell, C.J. Maggiore, T.N. Taylor, E.R. Gonzalez, S. Gottesfeld, S-M. Park, K-L. Hseuh, H. Chang, and D-T. Chin, National Fuel Cell Seminar Abstracts, 1983, p. 131.

2332. Srinivasan, S., Los Alamos Nat. Lab., personal communication.

2333. Stanley, J.S., National Fuel Cell Seminar Abstracts, 1982, p. 157.

2334. Starkman, E.S., SAE Trans., **77**, 2176 (1968).

2335. Staschewski, D., in "Proc. 5th World Hydrogen Energy Conf.," Toronto, Ont., 1984, V.N. Veziroglu and J.B. Taylor, (eds.), Pergamon Press, NY (1984), p. 1677.

2336. Stead, H., Jr., Proc. 16th Intersoc. Energy Conv. Eng. Conf., 2096 (1981).

2337. Steck, A., and H.L. Yeager, Anal. Chem., **52**, 1215 (1980).

2338. Stedman, J.K., U.S. Patent No. 3,761,316 (1973).

2339. Steele, B.C.H., Mater. Sci. Res., **9** (Mass Transp. Phenom. Ceram.), 269 (1975).

2340. Steele, R.V., G.L. Johnson, and G. Ciprios, Alternative Energy Sources, (Proc. Miami Int. Conf.), 1977, **9**, 4375 (1978).

2341. Steele, R.V., National Fuel Cell Seminar Abstracts, 1978, p. 37.

2342. Steele, R.V., D.C. Bomberger, K.M. Clark, R.F. Goldstein, and R.L. Hays, Comparative Assessment of Residential Energy Supply Systems that use Fuel Cells, NTIS Rep. EPA/600/7-79/105B (1979).

2343. Steitz, P., and G. Mayo, National Fuel Cell Seminar Abstracts, 1978, p. 22.

2344. Stepanov, G.K., and A.M. Trunov, Electrochemistry of Molten and Solid Electrolytes, Vol. **3**, A.N. Baraboshkin, (ed.), Consultants Bureau, NY (1966), p. 73.

2345. Stepanov, G.K., E.N. Rodingina, and A.P. Sinel'nikova, USSR Patent No. 465,681 (1975).

2346. Stewart, R.C., Jr., Ger. Offen. 2,852,051 (1979).

2347. Stewart, R.C., Jr., U.S. Patent No. 4,173,662 (1979).
2348. Stickles, R.P., and J.K. O'Neill, National Fuel Cell Seminar Abstracts, 1978, p. 26.
2349. Stickles, R.P., and C.T. Breuer, National Fuel Cell Seminar Abstracts, 1982, p. 22.
2350. Stockel, J.F., Proc. 5th Intersoc. Energy Conv. Eng. Conf., 5–95 (1970).
2351. Stockel, J.F., Proc. 7th Intersoc. Energy Conv. Eng. Conf., 729019 (1972).
2352. Stoenciu, A., Mater. Plast. (Bucharest), **16**, 172 (1979).
2353. Stonehart, P., Spec. Rep. EPRI-SR-13, Conf. Proc.: Fuel Cell Catal. Workshop, 19 (1975).
2354. Stonehart, P., and G.M. Kohlmayr, Ger. Offen. 2,720,529 (1976).
2355. Stonehart, P., and A.C.C. Tseung, Proc. Workshop Electrocatalysis Fuel Cell Reactions, Brookhaven Nat. Lab. 1978, p. 165.
2356. Stonehart, P., and J. MacDonald, National Fuel Cell Seminar Abstracts, 1978, p. 84.
2357. Stonehart, P., J. MacDonald, V. Jalan, and J. Baris, National Fuel Cell Seminar Abstracts, 1979, p. 102.
2358. Stonehart, P., J. MacDonald, and J. Baris, National Fuel Cell Seminar Abstracts, 1980, p. 37.
2359. Stonehart, P., and J.P. MacDonald, National Fuel Cell Seminar Abstracts, 1983, p. 135.
2360. Stonehart, P., and J.P. MacDonald, in "Proc. of the Workshop on the Electrochemistry of Carbon," S. Sarangapani, J.R. Akridge and B. Schumm, (eds.), PV 84-5, The Electrochemical Society, Pennington, NJ (1984), p. 292.
2361. Stonehart, P., and J. MacDonald, "Stability of Acid Fuel Cell Cathode Materials," EPRI EM-1664, 1981; Final Rep., RP1200-2, EPRI, Palo Alto, CA (1984).
2362. Stonehart, P., ongoing work under EPRI RP1200-2, 1986.
2363. Stortnik, H.J., Spez. Ber. Kernforschungsanlage Jeulich, Juel-Spez-31 (1979).
2364. Stortnik, H., R. Stockmeyer, and M. Monkenbusch, J. Mol. Struct., **60**, 443 (1980).
2365. Stover, F.S., and R.P. Buck, J. Electroanal. Chem. Interfac. Electrochem., **107**, 165 (1980).
2366. Stoy, A., U.S. Patent No. 3,953,545 (1976).
2367. Strasser, K., Ger. Offen. 2,729,640 (1979).
2368. Strasser, K., J. Electrochem. Soc., **127**, 2172 (1980).
2369. Strasser, K., H. Gruner and H.B. Gutbier, Power Sources 6, 569 (1980).
2370. Strom, U., and P.C. Taylor, J. Appl. Phys., **50**, 5761 (1979).
2371. Strombotne, R.L., Proc. 24th Power Sources Conf., 77 (1970).
2372. Suchanski, M.R., Eur. Patent Appl. 14,596 (1980).

2373. Sugimoto, T., K. Sokura, and I. Aramaki, Denki Kagaku, **50**, 422 (1982).
2374. Sulkes, M.J., Characteristics of Conventional Design Nickel-Zinc Batteries for Hybrid Vehicle Operation, NTIS AD 756 431 (1972).
2375. Sulkes, M., Proc. 25th Power Sources Conf., 77 (1972).
2376. Summers, W.A., National Fuel Cell Seminar Abstracts, 1981, p. 119.
2377. Summers, W.S., and S. Fanciullo, U.S. Patent No. 3,915,747 (1975).
2378. Sunderland, J.G., Electr. Counc. Res. Cent., (memo), ECRC/M1065 (1977).
2379. Sundheim, B.R., M.H. Waxman, and H.P. Gregor, J. Phys. Chem., **57**, 974 (1953).
2380. Sundheim, B.R., Chem. Phys. Lett., **60**, 427 (1979).
2381. Sustersic, M.G., O.R. Cordova, W.E. Triaca, and A.J. Arvia, J. Electrochem. Soc., **127**, 1242 (1980).
2382. Sutyagina, A.A., N.I. Il'chenko, M.N. Semenenko, and G.D. Vovchenko, Elektrokhimiya, **15**, 1094 (1979).
2383. Sutyagina, A.A., N.I. Il'chenko, M.N. Semenenko, and G.D. Vovchenko, Zh. Fiz. Khim., **53**, 1865 (1979).
2384. Sverdrup, E.F., D.H. Archer, and A.D. Glasser, in "Fuel Cell Systems II," Advances in Chem. Series **90**, A.C.S., Washington, DC (1969), p. 301.
2385. Sverdrup, E.F., in "From Electrocatalysis to Fuel Cells," G.Sandstede, (ed.), Univ. of Washington Press, Seattle (1972), p. 255.
2386. Swaroope, R.B., J.W. Sim, and K. Kinoshita, J. Electrochem. Soc., **125**, 1799 (1978).
2387. Swe, B., Ger. Offen. 2,836,464 (1980).
2388. Swinkels, D.A.J., SAE Trans. 680452 (1968).
2389. Sylwan, C.L., Energy Convers., **15**, 137 (1976).
2390. Sylwan, C.L., Energy Convers., **17**, 67 (1977).
2391. Symons, P.C., and P. Carr, Proc. 8th Intersoc. Energy Conv. Eng. Conf., 72 (1973).
2392. Symons, P.C., "Advanced Technology Zinc/Chlorine Batteries for Utility Load Levelling," EPRI Final Rep. RP-1198-14, Palo Alto, CA (1985).
2393. Szabo, S., and F. Nagy, Isr. J. Chem., **18**, 162 (1979).
2394. Szczesniak, B., H. Jankowska, and J. Downarowicz, Ext. Abstr., Meeting of Internat. Soc. Electrochem., 30th, 386 (1979).
2395. Szpak, S., and T. Katan, Proc.–Electrochem. Soc., 77-6 (1977).
2396. Taberner, P., J. Heitbaum, and W. Vielstich, Electrochim. Acta, **21**, 39 (1976).
2397. Tafel, J., Z. Phys. Chem., **50**, 541 (1905).
2398. Tagawa, H., J. of Appl. Electrochem., **2**, 251 (1972).
2399. Tajima, H., M. Sakurai, K. Mizukami, and R. Endo, Denchi Toronkai (Koen Yoshishu), 18th, 27 (1979).
2400. Tajima, H., S. Maruyama, and S. Watanabe, Japan, Kokai Tokkyo Koho, **80**, 11,131 (1980).

2401. Tajima, H., M. Sakurai, and K. Mizukami, Japan, Kokai Tokkyo Koho, **80**, 19,713 (1980).
2402. Tajima, H., and H. Kaneko, National Fuel Cell Seminar Abstracts, 1983, p. 93.
2403. Takagi, J., Kaotsu Gasu, **13**, 447 (1976).
2404. Takahashi, T., and H. Ikeda, Japan, Kokai, **76**, 90,435 (1976).
2405. Takahashi, T., H. Iwahara and T. Esaka, J. Electrochem. Soc., **124**, 1563 (1977), and references quoted therein.
2406. Takahashi, K., K. Hiratsuka, H. Sasaki and S. Toshima, Chem. Lett. (**4**), 305 (1979).
2407. Takahashi, T., and Y. Suzuki, Ger. Offen. 2,837,593 (1979).
2408. Takahashi, T., S. Tanase, O. Yamamoto, S. Kamauchi, and H. Kabeya, Int. J. Hydrogen Energy, **4**, 327 (1979).
2409. Takamatsu, T., and A. Eisenberg, J. Appl. Polym. Sci., **24**, 2221 (1979).
2410. Takamatsu, T., M. Hashiyama, and A. Eisenberg, J. Appl. Polym. Sci., **24**, 2199 (1979).
2411. Takashi, K., T. Kiyota, S. Asaumi, and A. Shimizu, Japan, Kokai Tokkyo Koho, **80**, 50,033 (1980).
2412. Takashio, H., Japan. Patent No. 74 40,845 (1974).
2413. Takashio, H., Japan. Patent No. 74 40,846 (1974).
2414. Takashio, H., Japan. Patent No. 74 40,847 (1974).
2415. Takenaka, M., Y. Kakihara, and E. Tonegi, Japan, Kokai Tokkyo Koho, **80**, 22,311 (1980).
2416. Tamura, H., Japan, Kokai Tokkyo Koho, **74**, 45,069 (1974).
2417. Tamura, K., and K. Iwamoto, Ger. Offen. 2,725,743 (1977).
2418. Tamura, H., H. Yoneyama, and Y. Matsumoto, Stud. Phys. Theor. Chem., **11** (Electrodes Conduct. Met. Oxides, Part A), 261 (1980).
2419. Tamura, K., and T. Kahara, J. Electrochem. Soc., **123**, 776 (1976).
2420. Tamura, K., and K. Iwamoto, Japan, Kokai Tokkyo Koho, **80**, 33,794 (1980).
2421. Tamura, K., Hitachi, Ltd., personal Communication.
2422. Tan, S.R., and G.J. May, Fast Ion Transp. Solids: Electrodes Electrolytes, Proc. Int. Conf., 91 (1979).
2423. Tanaka, H., T. Ueno, and T. Yamamoto, Japan. Patent No. 74 31,636 (1974).
2424. Tanase, S., Kotai Butsuri, **14**, 406 (1979).
2425. Tang, T.E., T.D. Claar, and L.G. Marianowski, "Effects of Sulfur-Containing Gases on the Performance of Molten Carbonate Fuel Cells," EPRI Rep. EM-1699, Palo Alto, CA (1981).
2426. Tannenberger, H., and H. Siegert, in "Fuel Cell Systems II," Advances in Chem. Series **90**, A.C.S., Washington, DC (1969), p. 281.
2427. Tannenberger, H., in "From Electrocatalysis to Fuel Cells," G. Sandstede, (ed.), Univ. of Washington Press, Seattle (1972), p. 235.

2428. Tanso, S., and K. Murata, Japan. Kokai 76 6,883 (1976).
2429. Tant, V.E., Proc. 3rd Canadian Fuel Cell Seminar, 13 (1970).
2430. Tantram, A.D.S., A.C.C. Tseung, and B.S. Harris, in "Hydrocarbon Fuel Cell Technology," B.S. Baker, (ed.), Academic Press, NY (1965), p. 187.
2431. Tarasevich, M.R., R.Kh. Burshtein, and Yu.A. Chismadzhev, in "Fuel Cell Systems II," Advances in Chem. Series 90, A.C.S., Washington, DC (1969), p. 81.
2432. Tarasevich, M.R., and V.A. Bogdanovskaya, Elektrokhimiya, 7, 1072 (1971).
2433. Tarasevich, M.R., Elektrokhimiya, 9, 599 (1973).
2434. Tarasevich, M.R., G.I. Zakharkin, and R.M. Smirnova, Elektrokhimiya, 9, 645 (1973).
2435. Tarasevich, M.R., V.Ya. Shepelev, and R.Kh. Burshtein, Elektrokhimiya, 9, 1593 (1973).
2436. Tarasevich, M.R., V.S. Bagotskii, and B.N. Efremov, Geterog. Katal., 4th, Pt. 1, 367 (1979).
2437. Tarasevich, M.R., and B.N. Efremov, Stud. Phys. Theor. Chem., 11 (Electrodes Conduct. Met. Oxides, Part A), 221 (1980).
2438. Tarasevich, M.R., and K.A. Radyushkina, Usp. Khim., 49, 1498 (1980).
2439. Tarbet, G.W., Proc. 3rd Canadian Fuel Cell Seminar, p. 83 (1969).
2440. Tarmy, B.L., Proc. 19th Power Sources Conf., 41 (1965).
2441. Taschek, W.G, U.S. Patent No. 4,155,712 (1979).
2442. Taylor, A.H., J. Giner, and F. Goebel, Proc. Symp. Batt. Tract. Propuls., 102 (1972).
2443. Taylor, E.J., and W.E. O'Grady, Brookhaven Nat. Lab. Rep. BNL 51072, UC-93, Energy Conversion TID-4500, FC-5, July 1979, p. 7.
2444. Taylor, N.J., Vacuum Sci. and Tech., 6, 241 (1968).
2445. Taylor, O.C., Proc. Intersoc. Energy Conv. Eng. Conf., 1971, 323.
2446. Tedman, C.S., H.S. Spacil, and M. Mitoff, General Electric TIS Rep. No. 69-C-056 (1969).
2447. Tedmon, C.S., (ed.), Corrosion Problems in Energy Conversion and Generation, Part I: Corrosion Problems in Electrochemical Energy Conversion, The Electrochemical Society, Princeton, NJ (1974), pp. 1–62.
2448. Tempo, G.C.L., Proc. 4th Intersoc. Energy Conv. Eng. Conf., 933 (1969).
2449. Terry, P., J. Galagher, R. Salathe, and J.O. Smith, Proc. 20th Power Sources Conf., 39 (1966).
2450. Texas Instruments, 15 KW Hydrocarbon-Air Fuel Cell Electric Power Plant Design, Final Rep., U.S. Govt. Contract DA-44-009-AMC-1806T, Jan. 1968.
2451. Texas Instruments, Fuel Cell Power Plant, Final Rep., U.S. Govt. Contract No. DA-44-009-AMC-1806(T), Nov. 1968.
2452. Thacker, R., Nature, 206, 186 (1965).
2453. Thacker, R., and J.P. Hoare, J. Electrochem. Soc., 113, 862 (1966).

2454. Thacker, T., Nature, **212**, 182 (1966).
2455. Thaller, L.H., R.E. Post, and R.W. Easter, Proc. 5th Intersoc. Energy Conv. Eng. Conf., 5–72 (1970).
2456. Thaller, L.H., Proc. 25th Power Sources Symp., 152 (1972).
2457. Thompson, C.E., U.S. Patent No. 3,231,428 (1966).
2458. Thompson, J., (ed.), Power Sources 7: Research and Development in Non-mechanical Electrical Power Sources, Academic Press, London (1979).
2459. Tilak, B.V., B.E. Conway, and H. Angerstein-Kozlowska, J. Electroanal. Chem., **48**, 1 (1973).
2460. Tilak, B.V., and S.K. Rangrajan, Trans. SAEST, **13**, 261 (1978).
2461. Tillman, E.S., Proc. 25th Power Sources Conf., 165 (1972).
2462. Tio, T., R. Jones, and R. Minet, Assessment of Fuel Processing Systems for Dispersed Fuel Cell Power Plants, EPRI Rep. EPRI-EM-1010 (1979).
2463. Tirrell, C.E., U.S. Patent No. 3,117,034 (1964).
2464. Tirrell, C.E., U.S. Patent No. 3,152,015 (1964).
2465. Titterington, W.A., and A.P. Fickett, Proc. 8th Intersoc. Energy Conv. Eng. Conf., 574 (1973).
2466. Toedbeide, K., Proc. Int. Symp. Molten Salts, 20 (1976).
2467. Togawa, T., T. Nakamoto, and K. Kuwata, Japan, Kokai, **77**, 52,877 (1977).
2468. Tokovoi, O.K., D.Ya. Povolotskii, B.M. Zolotarevskii, L.S. Alekseenko, and S.N. Govalev, Tr., Chelyab. Politekh. Inst., **163**, 64-67 (1975).
2469. Tokuyama Soda Co., Ltd., Japan, Kokai Tokkyo Koho, **80**, 58,228 (1980).
2470. Tomashov, N.D., G.P. Chernova, and E.G. Manskii, USSR Patent 664,379 (1979).
2471. Tombalakian, A.S., H.J. Barton, and W.F. Graydon, J. Phys. Chem., **66**, 1006 (1962).
2472. Tombalakian, A.S., C.Y. Yeh, and W.F. Graydon, Can. J. Chem. Eng., **42**, 61 (1964).
2473. Tombalakian, A.S., M. Worsely, and W.F. Graydon, J. Am. Chem. Soc., **88**, 661 (1966).
2474. Tomter, S.S., and A.P. Antony, in "Fuel Cells: A Technical Manual," Am. J. Chem. Eng., NY (1963), p. 22.
2475. Tonami, M., M. Takeuchi, and F. Nakajima, National Fuel Cell Seminar Abstracts, 1982, p. 99.
2476. Toni, J.E.A., and R.L. Zwaagstra, U.S. Govt. Contract DA-44-009-AMC-1552(T), Final Rep. (1967).
2477. Tonomura, T., S. Imagane, Y. Imanaka, K. Tonomori, and S. Hayashi, Kinki Daigaku Rikugakubu Kenkyu Hokoku (**15**), 45 (1980).
2478. Torikai, E., Y. Kawami, M. Namba, and S. Shuhei, Ger. Offen. 2,717,512 (1977).

2479. Toshima, S., and H. Tanabe, Denchi Toronkai (Koen Yoshishu), 18th, 47 (1979).
2480. Townley, D., J. Winnick, and H.S. Huang, J. Electrochem. Soc., **127**, 1104 (1980).
2481. Toyo Soda Mfg. Co., Ltd., Japan, Kokai Tokkyo Koho, 80, 52,328 (1980).
2482. Tracey, V.A., Powder Metall., **22**, 45 (1979).
2483. Trachtenberg, I., J. Electrochem. Soc., **111**, 110 (1964).
2484. Trachtenberg, I., in "Fuel Cell Systems I," Advan. in Chem. Series 47, A.C.S., Washington, DC (1965), p. 232.
2485. Trachtenberg, I., in "Hydrocarbon Fuel Cell Technology," B.S. Baker, (ed.), Academic Press, NY (1965), p. 25.
2486. Trachtenberg, I., and D.F. Cole, in "Fuel Cell Systems II," Advan. in Chem. Series **90**, A.C.S., Washington, DC, (1969), p. 269.
2487. Trassati, S., and G. Lodi, Stud. Phys. Theor. Chem., **11** (Electrodes Conduct. Met. Oxides, Part A), 301 (1980).
2488. Tret'yakov, Yu.D., Tezisy Dokl.–Vses. Soveshch. Elektrokhim., 5th, 1974, **1**, 98 (1975).
2489. Triaca, W.E., A.M. Castro Luna, and A.J. Arvia, Simp. Bras. Electroquim. Eletroanal. (An.), 1st, 74 (1978).
2490. Triaca, W.E., A.M. Castro Luna, and A.J. Arvia, J. Electrochem. Soc., **127**, 827 (1980).
2491. Trocciola, J.C., Can. Patent No. 856,363 (1974).
2492. Trocciola, J.C., C.R. Schroll, and D.E. Elmore, U.S. Patent No. 3,867,206 (1975).
2493. Trocciola, J.C., and D.E. Elmore, U.S. Patent No. 4,000,006 (1976).
2494. Trocciola, J.C., D.E. Elmore, and R.J. Stosak, U.S. Patent No. 4,001,042 (1977).
2495. Truitt, J., Proc. Journées Internationales d'Etudes des Piles à Combustible, **88**, SERAI, Brussels (1965).
2496. Trunov, A.M., M.V. Uminskii, V.A. Prenov, and F.V. Makordei, Izv. Vyssh. Uchebn. Zaved., Khim. Khim. Tekhnol., **19**, 498 (1976).
2497. Trunov, A.M., A.M. Kotseruba, N.M. Yakovleva, and V.E. Polishchuk, Elektrokhimiya, **14**, 1614 (1978).
2498. Trunov, A.M., A.A. Domnikov, G.L. Reznikov, and F.R. Ypuuets, Elektrokhimiya, **15**, 783 (1979).
2499. TRW Systems Group, Research Investigation of Corrosion-Resistant Materials for Electrochemical Cells, 6th Rep., U.S. Govt. Contract No. DA-44-009-AMC-1452(T), May 1970.
2500. TRW Inc., Energy Systems Planning Div., Coal-Derived Fuel Gases for Molten Carbonate Fuel Cells, NTIS Rep. METC-8085-T2 (1979).

2501. Tseung, A.C.C., and H.L. Bevan, German Federal Republic Patent No. 2,119,702 (1971).
2502. Tseung, A.C.C., and H.L. Bevan, Electroanal. Chem. Interfac. Electrochem., 45, 429 (1973).
2503. Tseung, A.C.C., Spec. Rep.–Electr. Power Res. Inst., EPRI SR-13, Conf. Proc.: Fuel Cell Catal. Workshop; PB-245 115, pp. 97–99 (1975).
2504. Tseung, A.C.C., Proc. Symp. Electrode Mater. Processes Energy Convers. Storage, 205 (1977).
2505. Tseung, A.C.C., and K.L.K. Keung, J. Electrochem. Soc., 125, 1003 (1978).
2506. Tseung, A.C.C., Educ. Chem., 15 (1), 27 (1978).
2507. Tseung, A.C.C., S. Jasem, and M.N. Mahmood, Adv. Hydrogen Energy, 1 (Hydrogen Energy Syst., Vol. 1), 215 (1979).
2508. Tsien, H.C., J. A. Shropshire, and A.F. Venero, U.S. Patent No. 4,124,478 (1978).
2509. Tsuburaya, Y., Ger. Offen. 2,024,353 (1971).
2510. Tuller, H.L., Proc. Workshop High Temperature Solid Oxide Fuel Cells, Brookhaven Nat. Lab., 1977, p. 104.
2511. Tunstall, D.P., Phys. Rev. B., 11, 2821 (1975).
2512. Turkevich, J., Proc. Workshop Electrocatalysis Fuel Cell Reactions, Brookhaven Nat. Lab. 1978, p. 123.
2513. Turner, G.F., A.K. Johnson, and M.G. Gandel, Proc. 7th Intersoc. Energy Conv. Eng. Conf., 451 (1972).
2514. Tuschner, J.G., U.S. Patent No. 3,990,913 (1976).
2515. Tuschner, J.G., and R.C. Nickols, Jr., U.S. Patent No. 4,028,324 (1977).
2516. Twarog, T.J., Jr., National Fuel Cell Seminar Abstracts, 1982, p. 175.
2517. Tyco Laboratories Inc., Development of Cathodic Electrocatalysts for Use in Low Temperature Hydrogen/Oxygen Fuel Cells with an Alkaline Electrolyte, 11th Quarterly Rep., U.S. Govt. Contract No. NASW-1233, Mar. 1968.
2518. Tyco Laboratories, Inc., Research on Electrochemical Oxidation of Hexane and Its Isomers, 7th Interim Rep., U.S. Govt. Contract No. DA-44-009-AMC-1408(T), May 1969.
2519. Tyurin, V.S., A.G. Pshenichnikov, and R.Kh. Burshtein, Elektrokhimiya, 7, 708 (1971).
2520. Tyurin, Yu.M., G.F. Volodin, and Yu.V. Battalova, Elektrokhimiya, 16, 35 (1980).
2521. U.S. Dept. of the Interior, Office of Coal Research, Project Fuel Cell, Final Rep., R&D Rep. No. 57 (1970).
2522. U.S. Energy and Development Administration, Fuel Cells: A Bibliography, ERDA Rep. TID-3359 (1977).
2523. Uchida, S., S. Kutsumi, O. Ishida, and A. Shimizu, Japan, Kokai Tokkyo Koho, 74, 29,075 (1974).

2524. Uemura, T., M. Yoshida, M. Fujita, and T. Gejo, Japan, Kokai, **74**, 87,600 (1974).
2525. Ukihashi, H., T. Asawa, and T. Korishima, Japan, Kokai Tokkyo Koho, **79**, 157,777 (1979).
2526. Union Carbide Corp., Secondary Zinc-Oxygen Cell for Spacecraft Applications, 3rd Quarterly Rep., U.S. Govt. Contract No. NAS-5-10247, Mar. 1967.
2527. Union Carbide Corp., Hydrazine-Air Fuel Cell Simplification Studies, Final Rep., U.S. Govt. Contract DAAB07-68-C-0163 (1968).
2528. Union Carbide Corp., Austrian Patent No. 320,769 (1975).
2529. United Aircraft Corp., Pratt and Whitney Div., Pulse Power Fuel Cell, U.S. Govt. Contract No. AF33(615) (PIC 1496).
2530. United Aircraft Corp., Pratt and Whitney Div., Fuel Cell Technology Program, U.S. Govt. Contract No. NAS-9-11034.
2531. United Aircraft Corp., Pratt and Whitney Div., Fuel Cell Assemblies for Hydrocarbon-Air Fuel Cell System, U.S. Govt. Contract No. DAAB07-70-C-0215 (PIC 2259).
2532. United Aircraft Corp., Pratt and Whitney Div., Hydrocarbon-Air Fuel Cell System Component Improvement, U.S. Govt. Contract No. DAAB07-70-C-0125 (PIC-2260).
2533. United Aircraft Corp., Pratt and Whitney Div., Hydrocarbon-Air Fuel Cell System, U.S. Govt. Contract No. DAAB07-67-C-0376.
2534. United Aircraft Corp., Pratt and Whitney Div., High Power Density Fuel Cell, U.S. Govt. Contract No. F33-615-67-C-1935 (1967).
2535. United Aircraft Corp., Pratt and Whitney Div., High Power Density Fuel Cell, U.S. Govt. Contract No. F33-615-69-C-1267 (PIC 2053) (1969).
2536. United Aircraft Corp., Pratt and Whitney Div., 1.5 KW Fuel Cell Power Plant (PWA 4210), Phase I Technical Rep., U.S. Govt. Contract No. DAAK02-70-C-0518 (PIC-2314) (1970).
2537. United Aircraft Corp., Pratt and Whitney Div., Dual Mode High Power Density Fuel Cell, U.S. Govt. Contract No. F33-615-70-C-1134 (PIC 2179) (1970).
2538. United Nations, Small Scale Power Generation, (Sales No. 66.11.B.7) (1967).
2539. United Technologies Corp., Neth. Patent Appl. 78 10,871 (1979).
2540. United Technologies Corp., Brit. Patent No. 1,537,713 (1979).
2541. United Technologies Corp., Improvement in Fuel Cell Technology Base, Phase II, U.S. Govt. Contract No. DOE-AC-03-79-ET-11301 (1979).
2542. Urbach, H.B., D.E. Icenhower, and R.J. Bowen, Proc. 5th Intersoc. Energy Conv. Eng. Conf., 5–101 (1970).
2543. Urbach, H.B., Proc. 25th Power Sources Conf., 182 (1972).
2544. Urbach, H.B., R.J. Bowen, and D.E. Icenhower, Proc. 7th Intersoc. Energy Conv. Eng. Conf., 23 (1972).
2545. Urbach, H.B., L.G. Adams, and R.E. Smith, J. Electrochem. Soc., **121**, 233 (1974).

2546. Uri, N., Nature, **177**, 1177 (1956).
2547. Ushiba, K.U., I. Mahawili, and T.H. Tio, Proc. 14th Intersoc. Energy Conv. Eng. Conf., 550 (1979).
2548. Ushida, I., Y. Mugikura, T. Nishina and K. Itaya, J. Electroanal. Chem., **206**, 229; 241 (1986).
2549. Vahldieck, N.P., U.S. Patent No. 3,012,086 (1961).
2550. Vallet, C.E., and J. Braunstein, Steady-State Composition Profiles in Mixed Molten Salt Battery and Fuel Cell Analogs, NTIS Rep. No. CONF-771109-67 (1977).
2551. Vallet, C.E., and J. Braunstein, J. Phys. Chem. **81**, 2438 (1977).
2552. Vallet, C.E., and J. Braunstein, J. Electrochem. Soc., **125**, 1193 (1978).
2553. Vallet, C.E., and J. Braunstein, J. Electrochem. Soc., **126**, 527 (1979).
2554. Vallet, C.E., D.E. Heatherly, and J. Braunstein, Ext. Abstr., Meet.–Int. Soc. Electrochem., 30th, 62 (1979).
2555. Van den Brink, F., E. Barendrecht, and W. Visscher, J. Electrochem. Soc., **127**, 2003 (1980).
2556. Van den Brink, F., E. Barendrecht, and W. Visscher, Recl. Trav. Chim. Pays-Bas, **99**, 253 (1980).
2557. Van den Broeck, H., M. Alfenaar, G. Hovestreydt, A. Blanchard, G. van Bogeart, M. Bomke, and L. van Pouke, in "Hydrogen Energy Systems," T.N. Veziloglu and W. Seifritz, (eds.), Proc. 2nd World Hydrogen Energy Conf., Zurich, Switzerland, 1978, Pergamon Press, NY (1978).
2558. Van den Broeck, H., Progr. Batt. Solar Cells, JEC Press Inc., Cleveland, OH (1979), p. 204.
2559. Van den Broeck, H., personal communication, 1980.
2560. Van den Broeck, H., personal communication, 1981.
2561. Van den Broeck, H., National Fuel Cell Seminar Abstracts, 1983, p. 89.
2562. Van den Broeck, H., L. Adriensen, M. Alfenaar, A. Beekman, A. Blanchard, and G. Vanneste, in "Hydrogen Energy Progress," T.N. Veziloglu and J.B. Taylor, (eds.), Proc. 5th World Hydrogen Energy Conf., Toronto, Ontario, 1984, Pergamon Press, NY (1984), p. 1669.
2563. Van den Broeck, H., in "Fuel Cells: Trends in Research and Applications," A.J. Appleby, (ed.), Hemisphere Publ., NY (1987), p. 281.
2564. Van Duin, P.J., and C.A. Kruissink, Rev. Energ. Primaire, **2**, 19 (1966).
2565. Van Effen, R.M., Ph.D. dissertation, Univ. Wisconsin, Madison, WI (1979).
2566. Van Effen, R.M., and D.H. Evans, J. Electroanal. Chem. Interfac. Electrochem., **103**, 383 (1979).
2567. Van Effen, R.M., and D.H. Evans, J. Electroanal. Chem. Interfacial Electrochem., **107**, 405 (1980).
2568. Van Gool, W., (ed.), Fast Ion Transport in Solids: Solid State Batteries and Devices, American Elsevier, NY (1973).

2569. Van Gool, W., Chem. Weekbl. Mag., (10), 541 (1976).
2570. Van Gool, W., Solid Electrolytes, 9–26 (1978).
2571. Van Lang, H., Fr. Demande 2,408,630 (1979).
2572. Van Lier, J.A., U.S. Patent No. 4,175,167 (1979).
2573. Van Linden, J., Ger. Offen. 2,828,397 (1979).
2574. Van Tillborg, W.J.M., Chem. Weekbl. Mag. (Mar.), 97 (1977).
2575. Van Veen, J.A.R., and C. Visser, Electrochim. Acta, **24**, 921 (1979).
2576. Van Veen, J.A.R., and C.J. Korese, Ger. Offen. 2,855,117 (1979).
2577. Vannatta, D.W., NTIS AD Report 871299 (1970).
2578. Vanoiv, M., in "From Electrocatalysis to Fuel Cells," G. Sandstede, (ed.), Univ. of Washington Press, Seattle (1972), p. 101.
2579. Vaseen, V.A., U.S. Patent No. 4,195,118 (1980).
2580. Vaseen, V.A., U.S. Patent No. 4,218,518 (1980).
2581. Vasil'ev, Yu.B., and V.S. Bagotskii, in "Fuel Cells: Their Electrochemical Kinetics," V.S. Bagotsky and Yu.B. Vasil'ev, (eds.), Consultants Bureau, NY (1966), p. 77.
2582. Venkatasan, V.K., H.V. Gupa, K.A. Ghanasekharan, R.S. Pattabiraman, R. Tiruvidaimarudhur, C.J.A. Indira, and R. Chandrasekaran, Indian Patent No. 143,695 (1978).
2583. Venkatesan, V.K., H.V. Udupa, F. Pattabiraman, T.R. Jayaraman, and C.P.J.A. Indira, Indian Patent No. 143,696 (1978).
2584. Verger, M.B., D. Chillier, and N. Melle, Proper. Hydrazine Sess., Appl. Source Energy., Inter. Colloq., 233 (1975).
2585. Vernamonti, L., D.A.R.P.A., personal communication (AJA), Jan. 1987.
2586. Verstraete, J., D. Lefevre, R. Lefort, and J. Henry, in "Handbook of Fuel Cell Technology," C. Berger, (ed.), Prentice-Hall, Englewood Cliffs, NJ (1968), p. 496.
2587. Vert, Zh.L., T.V. Lipets, G.F. Muchnik, V.T. Serebryanskii, and I.P. Tverdovskii, Elektrokhimiya, **15**, 1595 (1979).
2588. Vertes, M.A., and A.J. Hartner, Proc. Journées Internationales d'Etudes des Piles à Combustible, 63, SERAI, Brussels (1965).
2589. Vetter, K.J., Electrochemical Kinetics, Academic Press, NY (1967).
2590. Vielstich, W., Chem.-Ing.-Tech., **35**, 362 (1963).
2591. Vielstich, V., Fuel Cells, (trans. D.J.G. Ives), Wiley-Interscience, London (1970).
2592. Vielstich, W., in "Fuel Cells," (trans., D.J.G. Ives), Wiley-Interscience, London, (1970), p. 83.
2593. Vielstich, W., in "Fuel Cells," (trans., D.J.G. Ives), Wiley-Interscience, London, (1970), p. 93.
2594. Vielstich, W., Proc. Workshop Electrocatalysis Fuel Cell Reactions, Brookhaven Nat. Lab. 1978, p. 67.

2595. Vielstich, W., Nachr. Chem. Tech. Lab., **27** (8), 491 (1979).
2596. Vielstich, W., Recent Advances in Electrocatalysis and their Implications for Fuel Cells, privately communicated report (1981).
2597. Vignaud, R., Fr. Demande 2,224,881 (1974).
2598. Vijh, A.K., Electrochemistry of Metals and Semiconductors, Dekker, NY (1973).
2599. Vilinskaya, V.S., N.G. Bulavina, V.Ya. Shepelev, and R.Kh. Burshtein, Elektrokhimiya, **15**, 932 (1979).
2600. Vine, R.W., R.G. Emanuelson, and C.A. Reiser, Can. Patent No. 954,934 (1975).
2601. Vinnikov, Yu.Ya., V.A. Shepelin, and V.I. Veselovskii, Elektrokhimiya, **9**, 552 (1973).
2602. Vinnikov, Yu.Ya., V.A. Shepelin, and V.I. Veselovskii, Elektrokhimiya, **9**, 649 (1973).
2603. Virnik, A.M., E.V. Mel'nikov, G.P. Telegin, and B.V. Roshchin, Zharostoikie Pokrytiya Zashch. Konstr. Mater., Tr. Vses. Soveshch., 7th 1975, 151 (1977).
2604. Visscher, W., and E. Barendrecht, (Neth.) Chem. Weekbl. Mag., (Jan.), 29–31 (1980).
2605. Visscher, W., and E. Barendrecht, J. Appl. Electrochem., **10**, 269 (1980).
2606. Vitvitskaya, G.V., Elektrokhimiya, **6**, 1234 (1970).
2607. Voecks, G.E., National Fuel Cell Seminar Abstracts, 1978, p. 132.
2608. Voelker, G.E., National Fuel Cell Seminar Abstracts, 1978, p. 1.
2609. Voelker, G.E., National Fuel Cell Seminar Abstracts, 1979, p. 1.
2610. Vogel, W.M., and J.T. Lundquist, J. E.ectrochem. Soc., **117**, 1512 (1970).
2611. Vogel, W.M., J. Lundquist, and A. Bradford, Electrochim. Acta, 17, 1735 (1972).
2612. Vogel, W.M., and K.A. Klinedinst, Electrochim. Acta, **22**, 1385 (1977).
2613. Vogel, W.M., and K.A. Klinedinst, J. Adhesion, 9 (2), 123 (1978).
2614. Vogel, W.M., L.J. Bregoli, and S.W. Smith, J. Electrochem. Soc., **127**, 833 (1980).
2615. Vogel, W.M., L.J. Bregoli, and S.W. Smith, National Fuel Cell Seminar Abstracts, 1980, p. 181.
2616. Vogel, W.M., and S.W. Smith, National Fuel Cell Seminar Abstracts, 1981, p. 127.
2617. Vogel, W.M., and S.W. Smith, J. Electrochem. Soc., **129**, 1441 (1982).
2618. Vol'fkovich, Yu.M., Elektrokhimiya, **14**, 262 (1978).
2619. Vol'fkovich, Yu.M., Elektrokhimiya, **14**, 361 (1978).
2620. Vol'fkovich, Yu.M., Elektrokhimiya, **14**, 546 (1978).
2621. Vol'fkovich, Yu.M., V.N. Ponomarev, and V.S. Duvasova, Elektrokhimiya, **15** (5), 639 (1979).

2622. Voloshin, A.G., Zh. Fiz. Khim., **53**, 787 (1979).
2623. Voloshchenko, G.N., N.V. Korovin, and A.I. Strikitsa, Elektrokhimiya, **13**, 1899 (1977).
2624. Voloshin, A.G., I.P. Kolesnikova, and S.D. Korolenko, Zh. Prikl. Khim. (Leningrad), **49**, 1801 (1976).
2625. Voloshin, A.G., and S.D. Korolenko, USSR Patent No. 642,805 (1979).
2626. Voloshin, A.G., S.D. Korolenko, and L.S. Sheremetikova, Elektrokhimiya, **16**, 111 (1980).
2627. Von Benda, K., and G. Sandstede, in "From Electrocatalysis to Fuel Cells," G. Sandstede, (ed.), Univ. of Washington Press, Seattle (1972), p. 87.
2628. Von Benda, K., Ger. Offen. 2,649,931 (1978).
2629. Von Sturm, F., and F. Kozdon, Ger. Patent No. 1,496,230 (1974).
2630. Von Sturm, F., and F. Kozdon, Ger. Patent No. 1,596,221 (1975).
2631. Von Sturm, F., Chem.-Ing.-Tech., **48**, 91 (1976).
2632. Von Sturm, F., Proc. Symp. Electrode Mater. Processes Energy Convers. Storage, 247 (1977).
2633. Voorhies, J.D., J.S. Mayell, and H.P. Landi, in "Hydrocarbon Fuel Cell Technology," B.S. Baker, (ed.), Academic Press, NY (1965), p. 455.
2634. Wakabayashi, S., S. Saito, M. Yokoyama, K. Ohe, H. Miyama, E. Omi, and T. Kawasaki, Japan, Kokai, **74**, 123,481 (1974).
2635. Wakefield, R.A., D.R. Limaye, S. Karamchetty, and N.R. Friedman, National Fuel Cell Seminar Abstracts, 1979, p. 21.
2636. Wakefield, R.A., D.R. Limaye, S.D.S.R. Karamchetty, and N.R. Friedman, "An Analysis of the Application of Fuel Cells in Dual Energy Use Systems," 2 Vols., EPRI Reps. EM-981-SY, EM-981, Palo Alto, CA (1979).
2637. Wakefield, R.A., S. Karamchetty, R. Rand, W.S. Ku, and V. Tekumalla, National Fuel Cell Seminar Abstracts, 1980, p. 90.
2638. Waldbott, G.L., Health Effects of Environmental Pollutants, C.V. Mosby Co., St. Louis, MO (1973).
2639. Waldman, E.I., and J.R. Aylward, U.S. Patent No. 3,877,989 (1975).
2640. Waldman, E.I., U.S. Patent No. 4,002,805 (1977).
2641. Walker, G.W., A.A. Adams, A.J. Coleman, and J.A. Joebstl, Proc. 28th Power Sources Symp., 41 (1978).
2642. Walsh, L.B., R.W. Leyerle, B.S. Baker, and M.A. George, National Fuel Cell Seminar Abstracts, 1978, p. 90.
2643. Walsh, M.A., D.N. Crouse, and F. Walsh, National Fuel Cell Seminar Abstracts, 1979, p. 115.
2644. Walsh, M.A., D.N. Crouse, J. Eynone, and R.S. Morris, (Eco, Inc.), National Fuel Cell Seminar Abstracts, 1980, p. 127.
2645. Walsh, M.A., J. Egnon, and R.S. Morris, National Fuel Cell Seminar Abstracts, 1981, p. 86.

2646. Wang, D-Y., and A.S. Nowick, Proc. Workshop High Temperature Solid Oxide Fuel Cells, Brookhaven Nat. Lab., 1977, p. 169.
2647. Warner, T.B., and S. Schuldiner, J. Electrochem. Soc., **112**, 853 (1965).
2648. Warner, T.B., S. Schuldiner, and B.J. Piersma, J. Electrochem. Soc., **116**, 938 (1969).
2649. Warrok, D.R., Proc. 24th Power Sources Conf., 155 (1972).
2650. Warshay, M., Proc. Workshop High Temperature Solid Oxide Fuel Cells, Brookhaven Nat. Lab., 1977, p. 89.
2651. Warshay, M., J. Energy, **2**, 46 (1978).
2652. Warshay, M., National Fuel Cell Seminar Abstracts, 1980, p. 2.
2653. Warszawski, B., B. Verger, and J.C. Dumas, J. Marine Technol. Soc., **5**, 28 (1971).
2654. Warszawski, B., and B. Verger, Ger. Patent No. 1,671,932 (1976).
2655. Waryawski, B., Entropie, **14**, 33 (1967); see also Proc. of the 2nd Internat. Symp. on Fuel Cells, Bruxelles, p. 108 (1967).
2656. Watanabe, M., T. Suzuki, and S. Motoo, Denki Kagaku, **40**, 205 (1972).
2657. Watanabe, M., and S. Motoo, J. Electroanal. Chem., **60**, 267 (1975).
2658. Watanabe, N. and M.A.V. Devanathan, J. Electrochem. Soc., **111**, 615 (1964).
2659. Watanabe, S., Japan. Kokai **76** 84,039 (1976).
2660. Weaver, M.J., Electroanal. Chem. Interfac. Electrochem., **51**, 231 (1974).
2661. Weaver, R., EPRI, personal communication, 1986.
2662. Weber, J., and T. Loucka, "Power Sources 3," D.H. Collins, (ed.), Oriel Press, Newcastle Upon Tyne (1971), p. 455.
2663. Weber, M.F., H.R. Shanks, and A.J. Bevolo, Proc. Symp. Electrode Mater. Processes Energy Convers. Storage, 265 (1977).
2664. Weber, M.F., H.R. Shanks, A.J. Bevelo, and G.C. Danielson, J. Electrochem. Soc., **127**, 329(1980).
2665. Webman, I., J. Jortner, and M.H. Cohen, Phys. Rev. B, **13**, 713 (1976).
2666. Weinberg, M.I, SAE meeting preprint, May (1968).
2667. Weissbart, J., J. Chem. Ed., **38**, 267 (1961).
2668. Weissbart, J., and R. Ruka, J. Electrochem. Soc., **109**, 723 (1962).
2669. Welsh, L.B., R.W. Layerle, B.S. Baker, and M.A. George, Power Sources 1978, **7**, 659 (1979).
2670. Welsh, L.B., R.W. Leyerle, and D.S. Scarlata, National Fuel Cell Seminar Abstracts, 1979, p. 115.
2671. Wendt, R.P., E. Klein, and S. Lynch, J. Membr. Sci., **1**, 165 (1976).
2672. Weppner, W., Proc. Workshop High Temperature Solid Oxide Fuel Cells, Brookhaven Nat. Lab., 1977, p. 114.
2673. Westinghouse Electric Corp., Project Fuel Cell, 1970 Final Rep., U.S. Govt. Contract 14-01-0001-303, Jan. 1971.

2674. Westinghouse Research and Development Center, Thin-Film Battery/Fuel Cell Power Generating System, NTIS Rep. CONS/1197-9 (1978).
2675. Weston, J.E., and B.C.H. Steele, J. Apl. Electrochem., **10**, 49 (1980).
2676. Wetzel, R., L. Mueller, and R.D. Schilling, Z. Chem., **15**, 358 (1975).
2677. Wetzel, R., H. Guenther, and L. Mueller, J. Electroanal. Chem. Interfac. Electrochem., **103**, 271 (1979).
2678. Wetzel, R., L. Mueller, and I. Babanskaja, Z. Phys. Chem. (Leipzig), **260**, 719 (1979).
2679. White, D.H., General Electric TIS Rep. No. 68-C-254 (1968).
2680. White, D.W., J. Intern. d'Etud. Piles Combust., Brussels, 1965, Ed. Serai, Burssels, Vol. III, 10 (1966).
2681. White, D.W., and H.S. Spacil, Systems for Economical Hydrogen Production, paper presented at Fall Meeting, Electrochemical Society, Montreal, 1968.
2682. White, E.R., and H.J.R. Maget, Proc. 19th Power Sources Conf., 46 (1965).
2683. White, S.H., J. McHardy, S.B. Brummer, M.M. Bower, G. Wilemski, G. Simons, and J. Mitteldorf, National Fuel Cell Seminar Abstracts, 1980, p. 165.
2684. Whitmore, D.H., J. Cryst. Growth, **39**, 160 (1977).
2685. Whittingham, M.S., and R.A. Huggines, in "Fast Ion Transport in Solids: Solid State Batteries and Devices," W. van Gool, (ed.), American Elsevier, NY (1973), p. 109.
2686. Whittingham, M.S., Annu. Rep., Conf. Electr. Insul. Dielectri. Phenom., 148 (1974).
2687. Wieckowski, A., V.N. Andreev, P. Zelenay, J. Sobkowski, and V.E. Kazarinov, Elektrokhimiya, **16**, 668 (1980).
2688. Wilberg, K.B., Computer Programming for Chemists, Benjamin Inc. (1965).
2689. Wilemski, G., P.F. Lewis, M.L. Finson, G.A. Simons, and K.L. Wray, National Fuel Cell Seminar Abstracts, 1978, p. 119.
2690. Wilemski, G., and T. Wolf, National Fuel Cell Seminar Abstracts, 1979, p. 136.
2691. Wilemski, G., J. Chem. Phys. **72**, 369 (1980).
2692. Wilemski, G., and J. Mitteldorf, National Fuel Cell Seminar Abstracts, 1981, p. 110.
2693. Wilemski, G., and A. Gelb, National Fuel Cell Seminar Abstracts, 1983, p. 85.
2694. Will, F., and C.A. Knorr, Z. Electrochem., **64**, 258 (1960).
2695. Will, F., and C.A. Knorr, Z. Electrochem., **64**, 270 (1960).
2696. Will, F., Electric Power Research Inst., Spec. Rep. EPRI-SR-13, Conf. Proc.: Fuel Cell Catal. Workshop; PB-245, 71 (1975).
2697. Will, F.G., J. Electrochem. Soc., **123**, 834 (1976).

2698. Will, F.G., in "Power Sources for Electric Vehicles," B.D. McNichol and D.A.J. Rand, (eds.), p. 573, Elsevier, NY (1984).
2699. Williams, K.R., and Gregory, D.P., J. Electrochem. Soc., **110**, 209 (1963).
2700. Williams, K.R., M.R. Andrew, and F. Jones, in "Hydrocarbon Fuel Cell Technology," B.S. Baker, (ed.), Academic Press, NY (1965), p. 143.
2701. Williams, K.R., (ed), An Introduction to Fuel Cells, Elsevier, Amsterdam (1966).
2702. Williams, K.R., and A.G. Dixon, in "Fuel Cell Systems II," Advances in Chem. Series **90**, A.C.S., Washington, DC (1969), p. 366.
2703. Williams, K.R., in "Power Sources 2," D.H. Collins, (ed.), Pergamon Press, London (1970), p. 521.
2704. Williams, L.F.G., and R.J. Taylor, J. Electroanal. Chem. Interfac. Electrochem., **108**, 293 (1980).
2705. Williams, R.M., T.E. Ranney, and J.W. Warren, Jr., Fr. Demande 2,214,976 (1974).
2706. Williams, R.M., T.E. Ranney, and J.W. Warren, Ger. Offen. 2,402,890 (1974).
2707. Williams, T.S., U.S. Patent No. 3,881,956 (1975).
2708. Wilsmore, N.T.M., Z. Phys. Chem., **35**, 291 (1900).
2709. Wilstater, R., and E.W. Leitz, Ber., 54, 115 (1921).
2710. Winsel, A., Electrochimica Acta, **14**, 961 (1969).
2711. Winsel, A., and R. Wendtland, Ger. Patent No. 1,771,570 (1974).
2712. Winsel, A., Ber. Bunsenges. Phys. Chem., **79**, 827 (1975).
2713. Winsel, A., and H.J. Schwartz, Ger. Patent No. 1,667,334 (1975).
2714. Winsel, A., Chem.-Ing.-Tech., **48**, 103 (1976).
2715. Winsel, A., in "Ullmanns Encykl. Tech. Chem., 4. Aufl., **12**, 113 (1976), Verlag, Weinheim, Germany.
2716. Wirtz, G.P., Proc. of the Workshop on High Temperature Solid Oxide Fuel Cells, Brookhaven National Lab., Upton, N.Y. (1977), p. 30.
2717. Wirtz, G.P., Proc. of the Workshop on High Temperature Solid Oxide Fuel Cells, Brookhaven Nat. Lab., Upton, NY (1977), p. 33.
2718. Wirtz, G.P., and H.S. Isaacs, Brookhaven Nat. Lab. Rep. BNL 51072, UC-93, Energy Conversion TID-4500, FC-5, July 1979, p. 17.
2719. Witherspoon, R.R., E.J. Zeitner, and H.A. Schulte, Proc. 6th Intersoc. Energy Conv. Eng. Conf., 96 (1971).
2720. Witherspoon, R.R., E.M. Domanski, and J.A. Davis, U.S. Patent No. 4,007,059 (1977).
2721. Wodzki, R., A. Narebska, and J. Ceynowa, Angew. Makromol. Chem., **78**, 145 (1979).
2722. Woerner, J.A., and J.H. Harrison, Proc. 24th Power Sources Conf., 195 (1970).

References

2723. Wolf, G., "Aktuelle Batteriforschung," Vart A-G., Frankfurt/Main, (1966), p. 38.
2724. Wolfe, W.R., Jr., U.S. Patent No. 3,492,164 (1970).
2725. Wolfe, W.R., Jr., K.B. Keating, V. Mehra, and L.H. Cutler, Proc. 8th Intersoc. Energy Conv. Eng. Conf., 91 (1973).
2726. Wolfe, W.R., Jr. and K.B. Keating, J. Electrochem. Soc., 121, 1125 (1974).
2727. Woods, R.R., R.D. Marshall, F.H. Schubert, and D.B. Heppner, Electrochemically Regenerable Carbon Dioxide Absorber, NASA (Contract Rep.) CR NASA-CR-152099, LSI-ER-290-3 (1970).
2728. Woods, R.W., Jr., Gas Research Inst. Digest, 7, 4 (1984).
2729. Worrell, W.L., Top. Appl. Phys., 21 (Solid Electrolytes), 143 (1977).
2730. Worsely, M., A.S. Tombalakian, and W.F. Graydon, J. Phys. Chem., 69, 883 (1965).
2731. Wragg, A.A., A. Einersson, and J.L. Dawson, Electrochim. Acta, 19, 503 (1974).
2732. Wright, M.K., and L.E. VanBibber, National Fuel Cell Seminar Abstracts, 1982, p. 69.
2733. Wyczalek, F.A., L.F. Daniel, and G.E. Smith, SAE Trans., 670181 (1967).
2734. Wyczalek, F.A., D.L. Frank, and G.E. Smith, SAE Trans., 670182 (1967).
2735. Wynn, J.E., and H. Knapp, Proc. 24th Power Sources Conf., 88 (1970).
2736. Wynn, J.E., Proc. 24th Power Sources Conf., (1970).
2737. Wynn, J.R., Proc. 24th Power Sources Conf. (1970).
2738. Wynn, J.W., Proc. 24th Power Sources Conf., 198 (1970).
2739. Wynveen, R.A., and T.G. Kirkland, Proc. 16th Power Sources Conf., 24 (1962).
2740. Wynveen, R.A., Proc. 1st Austral. Congr. Electrochem., 1963, Sydney-Hobart Australia, p. 642 (1965).
2741. Wynveen, R.A., and F.H. Schubert, Proc. 8th Intersoc. Energy Conv. Eng. Conf., 104 (1973).
2742. Yamamoto, H., and M. Igarashi, Japan, Kokai, 77, 122,276 (1977).
2743. Yamamoto, Y., K. Yoshino, and Y. Inuishi, Technol. Rep. Osaka Univ., 28 (1430-1458), 427(1978).
2744. Yamamoto, H., and Y. Kakihara, Japan, Kokai, 79, 29,892 (1979).
2745. Yamanaka, A., Japan, Kokai, 78, 72,021 (1978).
2746. Yamanaka, A., Japan, Kokai, 78, 72,198 (1978).
2747. Yamasaki, H., T. Yamane, Y. Omukai, and T. Hayashi, Japan, Kokai, 75, 43,428 (1975).
2748. Yanagihara, N., and H. Manabe, Japan, Kokai, 75, 161,649 (1975).
2749. Yanagihara, N., Japan, Kokai, 76, 96,037 (1976).

2750. Yang, C.H., S.F. Lin, H.L. Chen, and C.T. Chang, Inorg. Chem., **19**, 3541 (1980).
2751. Yang, C.Y., and H.S. Isaacs, Brookhaven Nat. Lab. Rep. BNL 51072, UC-93, Energy Conversion TID-4500, FC-5, July 1979, p. 16.
2752. Yang, C.Y., and H.S. Isaacs, Brookhaven Nat. Lab. Rep. BNL-51158, UC-98, Energy Conversion TID-4500, FC-7, Jan. 1980, p. 9.
2753. Yao, S.J., in "From Electrocatalysis to Fuel Cells," G. Sandstede, (ed.), Univ. of Washington Press, Seattle (1972), p. 291.
2754. Yargeau, B.A., U.S. Patent No. 4,120,787 (1978).
2755. Yarrington, R.M., H.S. Hwang, I.R. Feins, and C.P. Mayer, National Fuel Cell Seminar Abstracts, 1979, p. 163.
2756. Yarrington, R.M., I.R. Feins, and H.S. Hwang, National Fuel Cell Seminar Abstracts, 1980, p. 51.
2757. Yarrington, R.M., and J. Werth, National Fuel Cell Seminar Abstracts, 1982, p. 61.
2758. Yasuda, M., R.D. Weaver, and L. Nanis, National Fuel Cell Seminar Abstracts, 1978, p. 151.
2759. Yeager, E., and A.J. Salkind, (eds.), Techniques of Electrochemistry, (a series of monographs), Wiley, NY (1971-73).
2760. Yeager, E., Mechanisms of Electrochemical Reactions on Non-metallic Surfaces, U.S. Nat. Bur. Stand. Spec. Publ. **455**, 203 (1976).
2761. Yeager, E., Proc. Symp. Electrode Mater. Processes Energy Convers. Storage, 149 (1977).
2762. Yeager, E., and P. Bindra, National Fuel Cell Seminar Abstracts, 1978, p. 87.
2763. Yeager, E., J. Zagal, B.Z. Nikolic, and R.R. Adzic, Proc.–Electrochem. Soc. 1979, 80-3, 436 (1980).
2764. Yeager, E., D. Scherson and B. Simic-Glavaski, in "Proc. Symp. Chem. Phys. Electrocatal.," J.D.E. McIntyre, M.J. Weaver and E.B. Yeager, (eds.), The Electrochemical Society, Pennington, NJ (1984), p. 247.
2765. Yeager, E., Presentation at DOE Contractors Meeting (Technology Base Research Project), Cleveland, OH, April 14–15, 1986.
2766. Yeager, E., J. Molecular Catalysis, **38**, 5 (1986).
2766a. Yeager, E., Gas Research Inst. Contractors' Conf., Chicago, IL, Apr. 1986.
2767. Yeager, H.L., and A. Steck, Anal. Chem. **51**, 862 (1979).
2768. Yeager, H.L., and B. Kipling, J. Phys. Chem., **83**, 1836 (1979).
2769. Yeager, H.L., B. Kipling, and R.L. Doston, J. Electrochem. Soc., **127**, 303 (1980).
2770. Yeager, H.L., and A. Eisenberg, in "Perfluorinated Ionomer Membranes," ACS Symposium Series, American Chemical Society, Washington, DC (1982), p. 1.
2771. Yeo R.S., Polymer, **21**, 432 (1980).

2772. Yeo, S.C., and A. Eisenberg, J. Appl. Polym. Sci., **21**, 875 (1977).
2773. Yeung, K.L.K., and A.C.C. Tseung, J. Electrochem. Soc., **125**, 878 (1978).
2774. Yonehara, M., R. Anahara, and K. Suzuki, National Fuel Cell Seminar Abstracts, 1982, p. 49.
2775. Yonehara, M., R. Anahara, and K. Suzuki, National Fuel Cell Seminar Abstracts, 1983, p. 169.
2776. Yoshida, K., and M. Watanabe, Japan, Kokai Tokkyo Koho, **80**, 25,915 (1980).
2777. Yoshida, M., H. Obayashi, T. Kudo, and M. Kitada, Japan, Kokai, **77**, 120,347 (1977).
2778. Yoshihisa, H., and K. Fuke, Japan. Patent No. 75 09,066 (1975).
2779. Yoshizawa, S., Kagaku (Kyoto), **32**, 172 (1977).
2780. Young, A.R., Brit. Patent No. 1,449,233 (1976).
2781. Young, C.G., Battery Electrolytes: 1970–April 1980 (Citations from the Engineering Index data base), NTIS Rep. PB80-808637 (1980).
2782. Young, C.G., Battery Electrolytes: 1972-April 1980 (Citations from the NTIS data Base), NTIS Rep. PB80-808629 (1980).
2783. Young, G.E., National Fuel Cell Seminar Abstracts, 1980, 49.
2784. Young, G.J., (ed.), Fuel Cells I and II, Reinhold, NY, I (1960), II (1963).
2785. Young, J.A., R.C. Clark, and K.D. Lawrence, British N.R.L. Rep. 5995, Sept. 1963.
2786. Young, K.F., and A.D. Franklin, Proc. Workshop High Temperature Solid Oxide Fuel Cells, Brookhaven Nat. Lab., 1977, p. 172.
2787. Yuzhanina, A.V., V.I. Kuk'yanycheva, B.I. Lentsner, L.I. Knots, N.A. Shumilova, and V.S. Bagotskii, Elektrokhimiya, **8**, 877 (1972).
2788. Zagal, J., P. Bindra, and E. Yeager, A Mechanistic Study of Oxygen Reduction on Water Soluble Phthalocyanines Adsorbed on Graphite Electrodes, NTIS Rep. No. AD-A081235 (1980).
2789. Zagal, J., P. Bindra, and E. Yeager, J. Electrochem. Soc., **127**, 1506 (1980).
2790. Zagal, J.H., J. Electroanal. Chem. Interfac. Electrochem., **109**, 389 (1980).
2791. Zahn, M., P.G. Grimes and R.J. Bellows, U.S. Patent No. 4,197,169, April 1980.
2792. Zahradnik, R.L., L. Elikan, and D.H. Archer, in "Fuel Cell Systems I," Advances in Chem. Series **47**, A.C.S., Washington, DC (1965), p. 357.
2793. Zaromb, S., U.S. Patent No. 4,150,197 (1979).
2794. Zeh, C.M., F.D. Gmeindl, and R.A. Bajura, National Fuel Cell Seminar Abstracts, 1983, p. 82.
2795. Zeliger, H.I., J. Electrochem. Soc., **114**, 236 (1967).
2796. Zhirnova, M.I., S.Ya. Vanina, G.S. Mamankova, and O.A. Petrii, Elektrokhimiya, **15**, 1334 (1979).

2797. Zhuravleva, V.N., and A.G. Pshenichnikov, Elektrokhimiya, **12**, 851 (1976).
2798. Zielke, C.W., and E. Gorin, U.S. Patent No. 4,137,298 (1979).
2799. Zimmermann, G., M. Schoenborn, H. Magenau, H. Jahnke, and B. Becker, Ger. Offen. 2,326,667 (1974).
2800. Zimmermann, G., H. Jahnke, H. Magenau, and D. Baresel, Ger. Offen. 2,334,709 (1975).
2801. Zito, R., Jr., and L.J. Kunz, U.S. Patent No. 4,053,684 (1977).
2802. Zymboly, G.E., National Fuel Cell Seminar Abstracts, 1981, p. 74.

Index

Acetlyene black, 376
Activation polarization, 21, 22, 26, 251, 598
Adiabatic reactor, 213, 215
Adiabatic reforming, 213, 217
AEG Telefunken, 146, 419
AEM-Milan, 617
Aerospace Programs, 163–175
Agency of Industrial Science and Technology (AIST), 137, 311
Air pollution, 29–31
Alkaline fuel cell (AFC), 137–139, 144–147, 164, 165, 168, 171, 172, 174, 175, 185–189, 193, 194, 196–198, 200, 201, 240, 246, 250, 251, 253, 256, 258, 259, 266, 275, 277, 279, 392, 414, 425, 446, 449, 452, 453, 456, 457, 469, 492, 532, 613
Allis-Chalmers Manufacturing Company, 11, 186, 264, 446, 474
Alsthom, 127, 147, 148, 240, 266, 274, 275, 340, 349, 445, 453, 470, 471
American Cyanamid Corporation, 246, 256, 329
American electric utility policy, 613–616
American Gas Association (AGA), 13
American Norit Polycarbon C, 373
American on-site PAFC units, 620
American Petroleum Institute, 184
American Public Power Association (APPA), 59, 615
American University, 282, 339
Annual Reviews, Inc., 633
Ansaldo (Genoa), 617, 630
Anode creep, 568–570
Apollo, 4, 12, 34, 64, 149, 163–165, 244, 262, 264, 413, 465, 467, 477
Appleby, A. J., 624
Argonne National Laboratory (ANL), 83, 86, 89, 94, 96, 101, 102, 282, 301, 439, 570, 574, 585, 586, 607, 608, 631
Ariane satellite launcher, 457

Aromax T/12 foaming agent, 471
Arthur D. Little, Inc., 107, 129
Asahi Glass, 293
ASEA, Sweden, 146, 186
Atmospheric fluidized bed combustor (AFBC), 616
Auger electron spectroscopy (AES), 322
Automatic extraction noncondensing turbines, 495
Autothermal reformer (ATR), 215
Autothermal reforming, 208, 209, 213, 214
Auxiliary power unit (APU), 175
Axorb additive process, 269

Bacon, Francis T., 10, 12, 164, 244, 245, 262, 264, 330, 387, 413
Ballard, 295, 621, 622
Battelle Institute, Geneva, 404, 436, 439
Battelle-Frankfurt, 439
Baur, E., 8, 9, 579
Baymal, 247
Bechtel Corporation, 75, 613, 615
Becquerel, A. C., 8, 9
Bekaert, 146, 196
Bell Telephone System, 112
Belleli, 617
Belval, Ron, 615
Bergman, Michael, 615
Bipolar plate, 519, 525, 560
Blowers, 486, 487
Bockris, J. O'M., 10, 228
Boston Edison Co., 64, 614
Bottoming cycle systems, 234, 493–500
  choice of systems, 495, 496
  fuel cell system examples, 496–500
  gas turbines, 493, 494
  steam turbines, 494, 495
Boudouard reaction, 209, 541
Breda, 617
Breelle, 175
British Columbia Hydro, 620

British Gas Council, 82
British Gas, 145, 632
Broers, G. H. J., 196, 298, 430
Brookhaven National Laboratory (BNL), 198, 275, 347, 352, 362, 364, 369, 373, 377, 384, 435, 438, 439
Brown, Boveri, and Cie, AG, 102, 146, 306, 579, 587, 599, 610
Bubble-pressure barrier (BPB), 452, 557
Bureau of Mines, 395
Bureau of Labor Statistics, 184
Burns and McDonnell Engineering Company, 126
Burshtein, R. C., 246

Cabot carbon, 362, 363, 373
Cabot Corporation, 367, 374, 402, 522
Cahan, B., 245
Cairns, E. J., 4, 296
Canadian Department of Energy, Mines and Resources, 15
Canadian Western Natural Gas Co., Ltd., 62
Carbon Black Pearls 2000 furnace black, 374
Carbonization, 250
Carboxane TW-100 foaming agent, 471
Carnot, Sadi, vii, 15–17, 20, 21, 28, 32
Case Western Reserve University, 369, 626
Catacarb additive process, 269
Catalytica Associates, 214
Cathode dissolution, 570–574
Cell-stack assemblies (CSAs), 70, 71
Center for Industrial Research, Oslo, 632
Central Electric Power Council, 137, 138
Central Research Institute of the Electric Power Industry (CRIEPI), 135, 137, 628
Ceramaseal, Inc., 440
Ceramatec, 570, 626, 631
CH_4 emission, 623
Chemical Engineering, xii
Chemical vapor deposition (CVD), 593, 594, 603
Chemisorption, 315–318, 324, 327, 331, 332, 351, 360, 385, 400, 425
Chinese Apollo-type system, 175
Chita power plant, 617
Chubu Electric Power Co. (CHEPCO), 137, 143, 144, 617
CISE, Italy, 617, 630, 632
CNR IRTEC (Faenza), 630
CNR/ITAE, Messina, 630
Co-flow, 449, 548, 550, 597, 598, 609, 627
Coal gasification, 223–225
Coal-assisted electrolysis, 225–228
Coal-burning molten carbonate fuel cells, 81–86, 555

Coal-gasifier combined cycle (CGCC), 555, 556
Cogeneration Technology Alternatives Study (CTAS), 127
Cogeneration, 36, 40, 47, 48, 51, 52, 78–80, 83, 93, 101, 105, 109, 110, 120–129, 131, 132, 151, 488, 495, 527
Cold combustion, 6
Columbia Gas Systems, 63, 64
Combustion Engineering, 624, 631
Combustion turbine combined cycle (CTCC), 40, 43–46, 49–51, 57, 81, 93, 513, 538
Compagnie Generale d'Electricité (CGE), 147, 269
Composite fuels, 233
Concentration polarization, 22–25, 203, 236, 240, 256, 273, 296, 320, 349, 358, 555, 598, 599
Condensing turbine, 495
Configuration B, 535, 615, 620
Consolidated Edison (Con Ed), 46, 59, 64–66, 69, 72, 75, 489, 506, 512, 513, 614–616
Consumers Power Co., 64, 118
Contact resistance, 243, 257, 288, 307, 437, 508, 522, 535, 551, 553, 554, 561, 568, 569
Continental Oil, 456
Cool Water (power plant), 44, 87, 223, 556
Cooling plate, 118, 119, 173, 229, 237, 416, 417, 447, 472, 477, 478, 482, 507, 514, 515, 521, 522, 525, 526, 532, 535, 618
Cooling, 475–484
 active, 476
 comparison of active techniques, 480–484
 DIGAS®,, 477, 478
 liquid, 478–480
 passive, 475, 476
 process gas, 476, 477
Cost of electricity (COE), 577, 586, 587
Couglin, 225
Cross-flow, 99, 448–450, 465, 503, 505, 507, 548–551, 625, 628
CSZ 98 carbon catalyst, 373
Current efficiency, 27, 226, 339, 441
Cyclic reformer, 213

Davy, Sir Humphrey, 7
De Nora, 617, 622
Decision Focus Incorporated (DFI), 49, 52
Deep Submergence Search Vehicle System, 275
Defense Advanced Research Projects Agency (DARPA), 607

Department of Commerce, Bureau of Economic Analysis/Statistics, xi, xii
Department of Commerce, Bureau of Labor Statistics, 184
Department of Defense (DOD), 116, 151, 161
Department of Energy (DOE), 46, 60, 65, 66, 75, 77, 78, 81-83, 93-100, 103, 107, 109-111, 115, 116, 119, 126, 127, 161, 196-198, 213, 215, 216, 224, 284, 300, 379, 390, 429, 514-516, 518, 558, 563, 587, 604, 607, 624, 625
  Morgantown Energy Research Center, (METC), 95, 96, 99
  Program in Vehicular Fuel Cells, 196-198
Depew, C. A., 623
DEUS (Dual Energy Use Systems), 129
Deutsche Automobilgesellschaft, 310
Diamond Shamrock Technologies S.A., 408
Difluoromethanediphosphonic acid (DFMDPA), 365
DIGAS®,, 78, 158, 159, 225, 447, 477, 478, 480-483, 519, 521
Direct fuel cell (DFC), 96, 544
Direct internal reforming (DIR), 228, 229
Distant Early Warning (DEW), 153
Dornier System G.m.b.H., 102, 439, 632
Dow Chemical Company, 294, 295, 310, 401, 621-623
Du Pont de Nemours and Co., Inc., 188, 237, 246, 253, 286, 287, 335, 393, 394
Duke Power, 183
Dutch State Mines (DMS), 146, 196

ECO, Inc., 283, 365, 373, 401
Edison Electric Institute (EEI), 59, 65
EIC Corp., 94
Electrical demand, 44
Electric Power Research Institute (EPRI), x, 39, 45, 46, 49, 52, 59, 60, 65, 66, 73, 78, 81-83, 93, 95, 96, 98-100, 122, 125, 129, 137, 148, 169, 183, 214, 216, 224, 228, 229, 283, 300, 302, 365, 373, 379, 518, 523, 530, 547, 570, 613-615, 617, 623, 625, 631, 633
Electric vehicles (EVs), 10, 14, 177-202, 275, 341
  fuel cell-powered, 178-182, 185-202
    air quality, 180
    Department of Energy (DOE) Program in Vehicular Fuel Cells, 196-198
    ease of maintenance, 181
    efficiency, 178-180
    ELENCO Alkaline Fuel Cell (AFC) Program, 196-198
    fuel cell-battery hybrids, 191-196
    fuel supply, 181
    General Motors (GM) Corporation Electrovan, 190
    noise pollution, 180, 181
    outlook, 200-202
    raw material implications, 199, 200
    reliability, 181
    requirements, 181, 182
    types best suited for applications, 186-190
    Union Carbide Hybrid City Car, 193-196
    vs. batteries, 185, 186
  fuel costs, 183-185
  performance, 183-185
Electrocatalysis: definition, 313
Electrocatalysis of hydrazine, 335-337
  miscellaneous, 337
  nickel, 336
  nickel boride,336
  noble metals, 336, 337
Electrocatalysis of hydrocarbons, 337-354
  acetaldehyde, 352
  butane, 339
  ethane, 338
  ethanol, 350
  ethylene, 340
  ethylene glycol, 354
  formaldehyde, 351
  formate, 352, 353
  formic acid, 352, 353
  methane, 338
  methanol, 341-350
    acid solution, 341-348
    alkaline solution, 348, 349
    buffer solutions, 349
  octane, 340
  propane, 339
  propanol, 350, 351
  propylene, 340
Electrocatalysis of hydrogen, 322-335
  miscellaneous, 335
  mixed metal, 332, 333
  nickel, 330-332
  noble metal, 323-330
    poisoning, 326-330
  spinel, 332, 333
  tungsten-based, 333-335
Electrocatalysis of oxidants other than oxygen, 409, 410
Electrocatalysis of oxygen reduction, 357-409
  alloys of noble metals and platinum, 382
  carbon, 387-390
  metal chelates, 398-407
  miscellaneous, 407-409
  mixed metal oxides, 392-395
  nickel, 387

(Electrocatalysis: *cont'd*)
  other noble metals, 382–384
  perovskites, 390–392
  silver, 384–386
  spinels, 392–395
  tungsten bronzes, 395–398
  tungsten carbide, 395
  use of platinum, 359–382
    acid systems, 362–366
    alkaline systems, 381, 382
    corrosion of carbon catalyst support, 366–381
    oxide film, 360–362
    surface area loss, 366–381
Electrocatalysis, principles, 313–322
  adsorption, 321
  electrocatalytic activity, 318, 319
  electronic conductivity, 319
  electronic factor, 317, 318
  evaluation, 318–322
  function, 314, 315
  geometric factor, 317, 318
  precautions, 319, 320
  requirements for fuel cell, 315–318
  stability, 318
  surface study methods, 321, 322
Electrocatalysts, miscellaneous, 354, 355
Electrochemical Society, Inc., 633
Electrochemical vapor deposition (EVD), 593, 594, 596, 603
Electrochemische Energieconversie N.V., 453
Electrode cost, 258–260
Electrode evaluation, 257, 258
Electrodes, nonporous, 256, 257
Electrodes, porous, 242–256, 299, 385, 458, 474, 485, 603
  carbon, 247–255
  composite, 256
  metal, 244–246
  plastic microporous, 255, 256
  screen, 246, 247
Electrolyte feed and distribution, 470, 471
Electrolyte layer, 557, 558, 567, 594, 603, 605, 607
Electrolyte loss, 562–567, 625
Electrolyte matrix, 299, 301, 302, 429, 430, 452, 504, 505, 529, 553, 554, 557, 558, 561, 569, 572–574
Electrolyte reservoir plate (ERP), 165, 173, 414, 441, 472
Electrolytes, aqueous acid, 277–296
  solid polymer acid, 284–296
Electrolytes, aqueous alkaline, 261–277
  carbonation, 264–266

  carbon dioxide removal, 266–277
  electrochemical removal, 270–274
  expendable absorbers, 267–269
  membrane removal, 274–277
  regenerable absorbers, 269, 270
Electrolytes, molten carbonate, 297–303
Electrolytes, organic, 297
Electrolytes, other aqueous, 296
Electrolytes, solid, 303–311
  beta-alumina, 308–310
  other ionic conductors, 310–312
  oxide, 304–307
Electron spectroscopy for chemical analysis (ESCA), 322, 630
Electrotechnical Laboratory in Ibaraki Prefecture, 136, 137, 605
Electrotechnical Laboratory of the Agency of Industrial Science and Technology (AIST), 587
Electrotechnical Laboratory, Scuba, 631
Electrovair II, 182
ELENCO, 146, 147, 186, 196–198, 275
Ellipsometry, 321
Empire State Electric Energy Research Co. (ESEERCO), 78
ENEA, Italy, 629
ENEA-Casaccia, 630
ENEL, Italy, 617
Energy Center Netherlands (ECN), 629, 632
Energy Conversion Alternatives Study (ECAS), 81, 82
Energy conversion efficiency, 16–28, 35, 223, 358, 441, 533, 553
Energy Conversion, Ltd. (ECL), 63
Energy Information Administration Annual Reports to Congress, 184
Energy loss mechanisms, 19–26
Energy Management Associates (EMA), 45, 48, 52
Energy Research and Development Administration (ERDA), 60, 65, 93, 587
Energy Research Corporation (ERC), 60, 78–80, 93–97, 106, 118, 119, 143, 145, 154, 157–161, 163, 216, 217, 225, 229, 282, 302, 303, 363, 377, 419, 431, 447, 448, 450, 476–478, 480–483, 519–527, 534, 544, 545, 549, 551, 557, 560, 562, 565, 569, 570, 575, 576, 617, 619, 624, 625, 628
  DIGAS®,, 78, 158, 159, 225, 447, 476–478, 480–483, 519, 521
  Mark II 7.5 MW power plant, 519–527
Energy Systems Research Group Inc., 198
Engel-Brewer intermetallic platinum catalysts, 378, 379

Index

Engelhard Minerals and Chemicals Corporation, 119, 120, 370, 418, 422, 447, 448, 480, 519, 528–535, 619
Engelhard Industries Division, 122, 193, 215, 216, 419
Engineering News Record, xii
ENI, Italy 617
Environmental Policy Act, 30
Environmental Protection Agency (EPA), 30, 60, 82
EPRI Technical Assessment Guide (EPRI-TAG,), x, 40, 43
Ergenics, Inc., 621
Esso Research and Engineering Company, 245
ESTN/Hoogovens, 629
Ethylene propylene copolymer (EPC), 417
Ethylene-tetrafluoroethylene copolymer (ETFE-Tefzel), 422
European Economic Community (EEC), 311, 622, 629, 632
European Space Agency, 175, 449, 457
Evans, U. R., 390
Exchange current density, 22, 319, 323, 353, 357, 360, 363
Experience curve, 46
Extended X-ray absorption fine structure (EXAFS), 322
External reforming, 89, 229, 544, 627
Exxon Research and Engineering Co., 216, 255, 269, 273, 274, 340, 349, 372, 388, 393, 408, 471, 567

Faradaic efficiency, 27
Faraday, 7, 21, 278
Farooque, 225
Fermi level, 314, 315, 369, 398
Fermi, E., 314
Ferrugas Process, 146
Fiat, 617
Flemion, membrane, 293
Fluor, Inc., 96, 97
Fluor Solvent Process, 269
Forward Air Controller Radar Power, 161
Fourier-transform infra-red spectrometry (FTIR), 321
French Institute of Petroleum, 453, 468, 470, 471
French Petroleum Institute (IFP), 147
Frumkin, A. N., 186
Fuel Cells
  air pollution, 29–31
  definition, 3
  economic benefits, 38–40
  electrical operating characteristics, 28, 29
  energy loss mechanisms, 19–26
  energy conversion efficiency, 16–28, 35, 223, 358, 441, 533, 553
  future transportation needs, 622–624
  history, 7–14
  maintenance, 33, 34
  manpower requirements, 33, 34
  market, 40–54
  materials, 34
  modularity, 37–38
  multi-fuel ability, 34–36
  noise pollution, 32, 180, 181
  operation, 4–7
  Power Plant Applications Study, 59
  reliability, 34
  safety, 33
  siting flexibility, 36, 37
  thermal pollution, 32
  thermodynamics, 26–28
  visual aesthetics, 32
Fuel Cell Generator-1 (FCG-1), 46, 64, 67, 73–78, 84, 85, 500, 510–514, 519
Fuel cell programs, European, 145–148
  AEG Telefunken, 146
  Alsthom, 147, 148
  ASEA, Sweden, 146
  Bekaert, 146
  British Gas, 145
  Brown, Boveri, and Cie, AG, 146
  Compagnie Generale d'Electricité (CGE), 147
  Dutch State Mines (DMS), 146
  ELENCO, 146, 147
  Johnson Matthey, 147
  Royal Institute of Technology, Sweden, 146
  SCK/CEN (Belgian Atomic Energy Commission), 146
  Siemens AG, 145
  Sorapec, 148
Fuel cell programs, Japanese, 131–145
  Agency of Industrial Science and Technology (AIST), 137
  Central Electric Power Council, 137, 138
  Central Research Institute of the Electric Power Industry (CRIEPI), 135, 137
  Chubu Electric Power Co. (CHEPCO), 137, 143, 144
  Electrotechnical Laboratory in Ibaraki Prefecture, 136, 137
  Government Industrial Research Institute, Osaka (GIRIO), 135–137, 144
  Hitachi Ltd., 134–137, 139, 143–145
  Ishikawajima-Harima Heavy Industries Co. Ltd., (IHI), 136, 140
  Japan Gasoline Co., 138

Kansai Electric Power Co., Inc.
  (KEPCO), 138, 140, 142, 143
Kawasaki Heavy Industries, 136
Kureha KES-1 integrated substrate, 140
Matsushita Electric Industrial Co., Ltd.,
  136
Ministry of International Trade and Industry (MITI), 132, 140
Mitsui-Toatsu Chemical, Inc., 143
Moonlight Project, 132–145
New Energy Development Organization (NEDO), 134, 138–140, 142–144
Osaka Gas Co., 135
Sanyo Electric Co., Ltd., 136
Shin-Kobe Electric Machinery Co., Ltd., 144
Sunshine Project, 132
Takenaka Engineering, 140
Tohoku Electric Company, 140
Tokyo Electric Power Company (TEPCO), 131, 134, 135, 140, 145
Tokyo Gas Co., 135
Toshiba Corporation, 134–137, 139, 143, 144
Toyo Engineering Corporation (TEC), 145
Fuel Cell Programs, other, 148, 149
Fuel cell stack design, 440–454
  design concepts, 453, 454
  design parameters and correlations, 441, 442
  electrical configuration, 442–445
  manifolding arrangements, 445–449
  pressure control, 449–453
  pressure seals, 449–453
Fuel cell stack materials selection, 411–440
  aqueous alkaline systems, 412–415
    matrixes, 414, 415
  aqueous acid systems, 415–419
    electrode assemblies, 417, 418
    sealants, 417
    separators and matrixes, 418, 419
  molten carbonate systems, 426–433
    anodes, 429–431
    cathodes, 431, 432
    current collectors, 432, 433
    electrocatalysts, 432, 433
    electrolyte matrixes, 429
    separators (bipolar plates), 428, 429
  solid oxide systems, 433–440
    anodes, 435, 436
    cathodes, 434, 435
    electrode materials, 434
    electrolytes, 433
    interconnect material (ICM), 436–439
    miscellaneous cell components and sealants, 439, 440
  use of plastics for aqueous components, 419–426
    chemical stability, 420–422
    electrical conductivity, 425
    gas permeability, 425, 426
    heat build-up, 422
    mechanical stability, 420
    thermal expansion, 422
Fuel Cell Symposium, Tokyo '88, 613
Fuel Cell Users Group (FCUG), 39, 57, 59, 614, 620
Fuel cell-battery hybrids, 193, 198–202
Fuel Cells and Fuel Batteries, 4
Fuel processing, 25, 31, 50, 67, 69, 75, 101, 139, 142, 185, 455–457, 512
Fuels, 205–239
  ammonia, 238–239
  formaldehyde, 239
  formic acid, 239
  hydrazine, 233–236
  hydrogen, 205–233
    adiabatic reforming, 213, 217
    coal gasification, 223–225
    coal-assisted electrolysis, 225–228
    external hydrocarbon processing, 208–223
    from coal, 223–228
    internal hydrocarbon processing studies, 228–233
    miscellaneous generation processes, 233
    partial oxidation, 208, 214, 215, 218–220, 223
    pyrolysis, 208, 219–223
    steam reforming, 208–218
    steam reforming, catalyzed, 209–217
    steam reforming from methanol, 216, 217
    steam reforming, uncatalyzed, 217, 218
  methanol, 236–238
  oxidants, 239, 240
Fuji Electric Co., Ltd., 134, 136–140, 142, 143, 416, 448, 449, 473, 478, 489, 492, 534, 551, 562, 565, 617, 618, 622, 628
Fuji-Mitsubishi Kansai unit, 618

Galfac RE-610 foaming agent, 471
Galvani, 7
Garrett, 631
Gas clean-up systems, 458–464
  high-temperature vs. low-temperature sulfur removal for molten carbonate systems, 461, 462
  hydrodesulfurization, 459, 462, 463
  other contaminants, 464
  other methods of sulfur removal, 464

Index

sulfur contaminants, 458–460
Gas Research Institute (GRI), 86, 93, 110, 111, 115, 116, 135, 229, 626, 633
Gaseous voltaic battery, 8
Gate-turn-off (GTO) thyristors, 72, 492
Gautier-Villars, 633
GEC, United Kingdom, 632
Gedol, 617
Gellings, C. W., x
Gemini, 4, 12, 163–165, 188, 286, 292 465, 466, 474
General Dynamics, Electric Boat Division, 11
General Electric (GE), 12, 49, 50, 82, 84, 85, 93, 95, 97–100, 127, 146, 164, 166, 169, 188, 189, 274, 285, 288, 295, 301, 303, 310, 429, 459, 464, 471, 474, 496, 498, 548, 557, 558, 562, 569, 570, 577, 625
   Advanced Energy Systems Division, 99
   SP100 unit, 166
General Motors (GM) Corporation, 66, 186, 190
   Electrovair II, 182
   Electrovan, 190, 191, 194
   Handivan, 190
   Harrison Radiator Division, 66
General purpose power sources, 152, 153
Geosynchronous orbit (GEO), 165, 173
Gilbert/Commonwealth, 126
Gillis, Ed, 614
Giner, Inc., 379
Goodyear Tire and Rubber Co., 419
Government Industrial Research Institute, Osaka (GIRIO), 135—137, 144, 628
Grande Paroisse fluidized bed process, 214, 215
Graphitization, 250, 251, 374, 535
Great Lakes Research Corporation, 530
Greenhouse effect, 16, 30, 206, 410, 536, 622, 623
Gross National Product (GNP), 109, 177
Grotthus chain conduction mechanism, 283, 311
Grove cell, 8
Grove, Sir William, 8
Grubb, W. T., 313
Gulf acetylene black (GAB), 530

Haldor-Topsøe, 523, 617
Hall, E. W., 613
Handley, L. M., 613
Hansen, James, 622
Hanser, P., x
Hass, S. M., x
Heat rate, 26, 28, 36, 40, 43, 45, 50–52, 71, 75–77, 79, 80, 84, 85, 88, 89, 96, 147, 214, 215, 225, 258, 366, 499, 511, 513, 521, 523, 533, 535, 536, 576
Heat removal, 471, 472, 475–484
Heat transfer, 465–500
   bottoming cycle systems, 493–500
   electrical design, 489–491
   heat removal, 471, 472, 475–484
   mechanical components, 486–489
   power conditioning, 489–492
   start-up and shutdown, 484, 485
Heat treatment, 250, 251, 303, 367, 374, 418, 522, 531, 533
Heating, ventilating, and air-conditioning (HVAC) system, 117, 119
Hermes refurbishable mini-orbiter, 175
HIDRILA, Spain, 620
HIDROLA, Spain, 630
High flow turbocompressor, 67
High pressure-high temperature (HTHP), 139
High temperature steam reforming (HTSR) process, 215, 218
Higher-heating value (HHV), 17, 26, 28, 35, 36, 40, 41, 50, 71, 75, 84, 85, 88, 97, 101, 138, 143, 158, 179, 231–234, 277, 495, 499, 533, 576, 614, 616, 619, 620
Hitachi Ltd., 134–137, 139, 143–145, 237, 260, 311, 341, 348, 445, 453, 478, 534, 551, 562, 574, 617, 622, 626, 627
Hitco Corporation, 474
Holmes and Narver, Inc., 125
Homans, James E., 177
Hoogovens, 624, 629
Hoskins 815 alloy, 426
House Subcommittee on Science, Space and Technology, 624
House Subcommittees on Natural Resources, Agricultural Research and the Environment, and Science Research and Technology, 624
Hybrid reformer, 213
Hydro-Quebec, 63
Hydrodesulfurization, 67, 85, 89, 459, 462, 463, 512, 585
Hydrogasification, 213
Hydrogen beta-alumina, 309, 310
Hydrogen economy, 206, 455
Hydrogen enrichment, 455–457

Ihrig, Harry Karl, 11
Incremental break-even capital cost (DBECC), 123
Indian Institute of Technology, 225
Indirect internal reforming (IIR), 229, 237
Industrial cogeneration, 52, 83, 101, 105, 109, 120–129, 495, 527

(Industrial cogeneration: *cont'd*)

   economic dispatch, 123
   molten carbonate fuel cell, 127, 128
   phosphoric acid fuel cells, 122–127
   thermal dispatch, 123
Industrial Fuel Cell Users Group, 614
Infra-red (IR) spectrophotometry, 322
Infra-red reflectance Raman spectroscopy, 322
Institute for Applied Energy, Tokyo, 622
Institute of Chemical Technology, Germany, 235
Institute of Gas Technology (IGT), 13, 62, 94, 100, 128, 211, 213, 260, 298, 301, 459, 464, 469, 545, 547, 551, 559, 563, 570, 573, 624, 629–631
Integrated gasifier combined cycle (IGCC), 45, 86, 223
Internal combustion (IC) engine, 177, 178, 180–185, 191, 201, 311, 337, 340, 441, 610
Internal reflection spectroscopy, 322
Internal-reforming molten-carbonate fuel cell (IRMCFC), 40, 42, 51–53, 87–93, 96–98, 100, 101, 223, 224, 228–232, 234, 344, 497, 498, 538, 539, 544, 545, 548, 575–577, 583, 625, 627
Internally manifolded heat exchanger (IMHEX), 625
International Energy Agency Fuel Cell Workshop, 613
International Fuel Cell (IFC), 75, 77, 78, 279, 422, 450, 503, 510, 534, 566, 533, 537, 613, 617, 620, 622, 624, 625, 628, 633
International Society for Precious Metals, 633
IR-drop, 243, 257, 273, 283, 285, 290, 292, 295, 299, 300, 303, 432, 442, 443, 446, 467, 504, 517, 518, 528, 530, 536, 553, 554, 557, 561, 583, 589, 599, 603, 607, 609
Ishikawajima-Harima Heavy Industries Co. Ltd., (IHI), 136, 140, 449–451, 489, 551, 562, 624, 626, 628, 633
ITAE-CNR, 617
Italgas, 617
Italian 1 MW PAFC Demonstrator, 617
Italian Fuel Cell Working Group, 617
Italian Ministry of Defense, 617

Jablochkoff, 9
Jacques, W. W., 9
Japan Gasoline Co., 138
Japanese 1 MW PAFC Demonstrators, 617, 618
Japanese Fine Ceramics Center, 628
Japanese Fuel Cell Information Center, 622
Jet Propulsion Laboratory (JPL), 214, 216
Johnson Matthey, 147, 329

Kansai Electric Power Co., Inc. (KEPCO), 138, 140, 142, 143, 619
Kanthal A alloy, 426
Kanthal A1 alloy, 426
Kanthal Corporation, 300, 301, 426
Kawasaki Heavy Industries, 136
Keio University, 397
KEL-F®, $(CF_2\text{-}CFCl)$, 422
Kent, George, Ltd., 440
Ketelaar, 298, 430
KGN gasifier, 84
Kinetics Technology International Corporation (KTI), 84, 214, 498, 617
Kintex Corp., 551
Kipp-type generator, 233
Kobe Steel, 628
Kolbe process, 228
Kordesch, Karl, 193, 195, 196, 199, 206, 248, 255, 390
Kureha KES-1 integrated substrate, 140, 535, 530
Kynar®, $(CH_2\text{-}CF_2)_n$, 422
Kynol, 419
Kyushu University, 632

Lawrence Berkeley Laboratory, 189, 283, 365, 378
Le Chatelier's principle, 210
Learning curve, 46–49, 52, 54, 118
Licentia Patent-Verwaltungs-G.m.b.H., 233
Liebhafsky, H. A., 4
Life cycle management, 153
Life Systems, Inc., 170, 273
Limits to Growth, ix
Linear a.c. electrode polarization impedance techniques, 322
Liquid natural gas (LNG), 69, 70, 135, 632
Liquid petroleum gas (LPG), 140, 489
Los Alamos National Laboratory, 126, 159, 189, 198, 216, 217, 379, 621
Low earth orbit (LEO), 165, 168, 170–173
Low energy electron diffraction (LEED), 321, 322
Low pressure-low temperature (LPLT), 138–140, 142, 143
Low-energy ion scattering spectroscopy (ISS), 322
Low-temperature fuel cell systems, 620–622
Lower-heating value (LHV), 17, 41, 50, 158

Index 757

M-C Power, 624–626
Marangoni effect, 413
Marchetti's functions, 48
Mark II 7.5 MW power plant, 519–527
Martin, C. R., 295, 296, 621
Mass transfer, 465–500
  bottoming cycle systems, 493–500
  electrical design, 489–491
  electrolyte feed and distribution, 470, 471
  mechanical components, 486–489
  power conditioning, 489–492
  reactant supply and distribution, 465–470
  start-up and shutdown, 484, 485
  water removal, 471, 472, 475–484
Matchack, T. A., 623
Mathtech, Inc., 107, 122, 127
Matsushita Electric Industrial Co., Ltd., 136, 336, 434, 622
May, G. W., 613
McGraw Edison Co., 490
McGuire-2 plant, 183
Mechanical components, 486–489
  blowers, 486, 487
  motors, 488
  pumps, 487, 488
  turbocompressors, 488, 489
Meidensha, 619
Messner, Bill, 615
MHD Component Development and Integration Facility, 526
Milan Polytechnic, 617
Ministry of International Trade and Industry (MITI), 102, 132, 140
  Agency of Industrial Science and Technology (AIST), 137, 311
Mitsubishi Electric Corporation (MELCO), 134, 136, 138–140, 142, 143, 145, 450, 478, 504, 534, 619, 626, 628
Mitsubishi Metal Corporation, 628
Mitsui Engineering and Shipbuilding Company, 632
Mitsui-Toatsu Chemical, Inc., 143
Mogul L carbon catalyst, 373
Molten carbonate fuel cell (MCFC), 27, 52, 53, 60, 62, 81–100, 101, 123, 127, 128, 132, 136, 137, 143, 144, 192, 193, 196, 219, 223, 224, 228, 229, 237, 238, 244, 246, 257, 260, 297–300, 303, 426, 429, 430, 433, 440, 446–453, 458–460, 462, 464, 472, 496, 497, 498, 499, 501, 539–578, 590, 609, 613, 620, 624–632
  coal-burning, 81–86, 555
  cost, 574–578
  description, 545–551
  development program, 93–100
  internal reforming, 40, 42, 51–53, 87–93, 96–98, 100, 101, 223, 224, 228–231, 234, 344, 497, 498, 538, 539, 544, 545, 548, 575–577, 583, 625, 627
  operating principles, 539–545
    internal reforming, 544, 545
  performance, 551–556
    lifetime goals, 555, 556
    losses, 553–555
    output, 551–553
  problem areas, 556–574
    anode creep, 568–570
    cathode dissolution, 570–574
    contaminants, 567, 568
    design and fabrication, 556, 557
    electrolyte layer, 557, 558, 567
    electrolyte loss, 562–567
    other cathode problems, 574
    seals, 562
    separator plate corrosion, 559–561
  stack materials selection, 426–433
  use with coal, 83–86
  with fuels other than coal, 86–93
Molten Carbonate Fuel Cell Power Generation System Technology Research Association, 628
Monarch 1300 carbon catalyst, 373
Mond, Ludwig, 8–10, 336
Mond-gas process, 9
Monolithic fuel cell, 607
Monolithic solid oxide fuel cells, 584, 607–610
Monsanto Co., 274, 456, 480, 490
Montana Energy and MHD Research and Development Institute (MERDI), 94
Moonlight Project, 132–145, 587, 605, 626
Morgantown Energy Research Center, (METC), 95, 96, 99
Mossbauer characteristics, 405
Mueller Associates, Inc., 75
Multi-fuel ability, 34–36

N_2O emission, 623
NAFION®,, 188, 189, 237, 283, 286–288, 292, 294, 295, 345, 350, 621
Nalge Co., 422
NASA-Goddard, 622
National Aeronautics and Space Administration (NASA), 11, 34, 64, 78, 79, 81, 106, 127, 163–168, 171, 173, 187, 275, 379, 453, 483, 492, 514–516
  Johnson Space Center, 168–170
  Lewis Research Center, 115, 119, 127, 168, 169
National Aerospace Plane, 175

National Bureau of Standards, 394, 402, 437
National Chemical Laboratory for Industry, Japan, 632
National Energy Act and the Energy Tax Act of 1979, 120
National Fuel Cell Coordinating Committee, 633
National Fuel Cell Coordinating Group (NFCCG), 60
National Fuel Cell Program, 55, 59–61, 161
National Geographic Magazine, ix
National Research and Development Corporation (NRDC), 63, 440
National Research Council, Canada, 111
National Rural Electric Cooperative Association, 59
National Space Transport System (NSTC), 175
Neosepta ion exchange membranes, 293
Nernst equation, 23, 326, 391, 528, 551, 580, 598
Nernst glower, 304, 579
Nernst potential, 555
Nernst, 23, 364, 548, 554, 579, 585, 597, 598, 609
Nernstian Hiatus, 10
New Energy Development Organization (NEDO), 134, 138–144, 618, 626–629
New England Electric System, 64
New Mexico School of Mines, 631
New York (Con Ed) 4.5 MW A.C. Demonstrator, 29, 37, 46, 59, 64–70, 72, 75, 512, 514
New York City (NYC) Fire Department, 66, 67
New York University, 282
Niagara Mohawk Power Corp. of New York, 64, 77, 99, 518
Nippon Kokan K.K. (NKK), 628, 632
Nisshin Steel Corporation, 628
Noise pollution, 32, 180, 181
North American P-108, 373
Northeast Utilities of Connecticut, 64, 77, 112, 518, 614
Northwest Natural Gas Company, 112
NO_x emissions, 30, 31, 44, 50, 82, 125, 126, 135, 138, 555, 617, 618–620, 631
Nusselt number, 472

Oak Ridge National Laboratory (ORNL), 94, 300, 301, 303
Occidental Petroleum, 127, 148, 275
Office of Coal Research, 579
Ohio State University, 404
Ohmic polarization, 21, 22, 25, 26, 554
Oil-canning effect, 548
Oil and Gas Journal, xii

On-Site Fuel Cell Users' Group, 110, 117, 135
On-Site Industrial Users Group, 52
On-site integrated energy systems (OS/IES), 105–120, 145, 161, 528
 Engelhard Program, 119, 120
 gas industry interest, 109
 Operational Feasibility Program, 109–118
 Westinghouse/ERC Program, 118, 119
Onada Cement, 632
Ontario Provincial Government, 15
OPEC, 77, 110
Operating and maintenance (O&M) costs, 42, 555, 556
Orbital transfer vehicles (OTV), 173
Osaka Gas Co., 111, 135, 233, 619, 628, 631
Ostwald ripening, 370
Ostwald, W., 9, 370
OTB Partecipazioni, 617
Oxidant processing, 457, 458
Oxygen enrichment, 457, 458
Oxygen film, 360

Pacific Gas and Electric (PG&E), 94, 98, 184, 496, 498, 625
Partial oxidation, 208, 214, 215, 218–220, 223, 237, 342, 455
Particle migration, 370
Patent-Verwaltungs-G.m.b.H., 233
Pauling's theory of metals, 317
PC-11 (Power Cell-11), 62, 63, 135
PC-17C (Power Cell-17C), 164–166, 174, 622
PC-18 (Power Cell-18), 65, 111, 135, 620
PC-19 (Power Cell-19), 65
PC-23 (Power Cell-23), 73, 77, 510, 511, 613, 614, 616, 617
PC-25 (Power Cell-25), 620
Pennsylvania State University, 283
Perfluoralkoxy (PFA), 416, 417, 422, 466, 478, 530
Perfluorodisulfone imide acids, 366
Pergamon Press, 633
Perovskites, 390–392, 611
Philadelphia Electric Co., 64
Phosphoric acid fuel cell (PAFC), 13, 22–24, 26, 27, 30, 35, 37, 40, 42, 43, 45, 46, 48–54, 59–81, 83–89, 95, 96, 99, 105, 107, 109, 110, 118, 119, 121—129, 132, 134, 135, 137, 140, 142–148, 154, 165, 187–189, 193, 196, 198, 214–217, 219, 224, 225, 229, 231, 237, 238, 246, 250–260, 264, 275, 277, 279–281, 284, 285, 292, 298, 303, 311, 328, 329, 359, 363, 366–369, 372, 375, 381, 383, 389, 394, 405, 412, 414–419, 441,

446–450, 452, 453, 458, 459, 462, 465, 477, 478, 480, 482–484, 488, 492, 498, 499, 501, 503–539, 543, 544, 546, 551, 553, 554, 563, 574, 575, 616–621, 624, 626, 629
cost, 536–538
Engelhard Minerals and Chemicals Corporation, 528–533
 bipolar plates, 530–532
 cooling plates, 532
 description, 528
 electrodes and catalysts, 529
 electrolyte matrix, 529, 530
 fuel processor, 532
 operating conditions, 528
Energy Research Corporation, 78–80, 519–527
required improvements, 533–536
FCG-1 power plant, 73–78
New York (Con Ed) 4.5 MW A.C. demonstrator, 29, 37, 46, 59, 64–70, 72, 75, 512, 514
TARGET Program, 61–63
Tokyo Electric Power Company 4.5 MW A.C. (Goi) demonstrator, 29, 30, 37, 46, 59, 65, 66, 68–76, 134, 135, 479, 489, 508, 512–514, 615
United Technologies Corporation, 63–65, 73–78, 503–519
 performance and technology improvements, 513–519
 proposed 11 MW power plant, 510–513
 Utility Program, 63–65
 use with coal, 83–86
Westinghouse/ERC, 78–80, 519–527
Physical Sciences Inc., 94, 225
Polarization, 7, 9, 21–26, 83, 85, 86, 97, 174, 194, 203, 235, 236, 240, 242, 243, 256, 257, 261, 262, 265, 266, 273, 293, 296, 303, 307, 318–320, 322–325, 328, 330–332, 334, 335, 339, 346, 349, 359, 362, 365, 386, 388, 392, 402, 408, 539, 554, 555, 559, 594, 598, 599, 601, 603
Polydyne, Inc., 184
Polyelectrode theory, 361
Polyether sulfone (PES), 422
Polyetheretherketone (PEEK), 422, 530
Polyethersulfone (PES), 530
Polytechnic of Milan, 630
Potential Industrial Fuel Cell Users Group (PIFCUG), 129
Pourbaix diagram, 248, 249
Power conditioning, 67, 489–492
Power Plant and Industrial Fuel Use Act of 1978 (PIFUA), 57
Pratt & Whitney Aircraft Division, United Aircraft Corp., 12, 61–63, 272, 273

Pregraphitization, 533
Preis, 579
Pressure swing absorption (PSA), 269
Prism™ separators, 456
Process and control (PAC) test, 66–68, 70, 142, 143, 617, 618,
Progetto Volta, 617, 632
Prototech, 255, 378, 407, 619, 621
Public Service Electric and Gas Co. of New Jersey, 64, 126, 614
Public Utility Regulatory Policies Act of 1978, 110
Puget Sound Power and Light, 614
Pumps, 487, 488
Pyrolysis, 208, 219–223, 250, 251, 405, 455

R.M. Parsons Company, 125
Ragone plot, 192
Raman microprobe, 322
Ranathan, V., 623
Raney nickel, 245, 255, 262, 330, 331, 336, 348, 349, 354, 387, 464
Raney palladium, 348, 353
Raney platinum, 338–340, 345, 348, 350, 351, 377, 381
Raney platinum-palladium alloy, 349
Raney silver, 384, 385
Rawlinson's New System Laundry, 112
Reactant supply and distribution, 465–470
 examples of feed distribution designs, 470
 feed modes, 466, 467
 reactant conditioning and control, 467–470
Rectosol process, 270
Reflected high-energy electron diffraction (RHEED), 321
R≈flexions Sur La Puissance Motrice du Feu, vii
Refrasil, 474
Regal 660R carbon catalyst, 362, 373
Regenerative energy storage (RES), 167
Regenerative fuel cell (RFC), 168–172
Relative energy usage (REU), 228
Resistance polarization, 25, 26
Reynolds Metals Company, 125
Ribbed substrate stack, 68, 533, 535
Risø University, 632
Roa, A. D., 623
Robert Bosch G.m.b.H., 409
Royal Institute of Technology, Sweden, 146, 233, 273
Rutherford backscattering spectroscopy, 379

Sander, M. T., 623
Sandia Laboratories, 440

Sanyo Electric Co., Ltd., 136, 476, 482, 534, 619, 622, 628
Saran, (polyvinylidene chloride), 422
Science Applications Inc. (SAI), 116
SCK/CEN (the Belgian Atomic Energy Commission), 146, 196, 197
Scotchcast resin, 490
Screening curve, 42, 43
Seals, 562, 565, 567, 574, 603, 609, 630, 632
Secondary ion mass spectroscopy (SIMS), 322
Selexol gas purification process, 97, 270, 462, 496, 631
Self-propelled vehicles, 177
Sensible heat reforming system, 90–92
Separator plate corrosion, 559–561
Shawinigan acetylene, 367, 373, 374, 529, 530
Shell Thornton, Ltd., (Research Center) 186, 237, 255, 328, 343, 382, 389, 404
Shimazu turbcompressor, 489
Shimazu, 140
Shin-Kobe Electric Machinery Co., Ltd., 144
Shin-Tokyo power plant, 619
Shoreham, 183
Siemens AG, 145, 175, 264, 265, 268, 269, 333, 447, 453, 454, 470, 471, 473, 474
Silent, lightweight electrical energy plants (SLEEP), 154, 159
Siting flexibility, 36, 37
Snyder, W. G., 623
Sol-gel powder processing, 632
Solid oxide electrolyte fuel cell (SOFC), 27, 60, 87, 96, 100–103, 137, 146, 175, 196, 201, 205, 223, 224, 238, 260, 304, 411, 433, 434, 436, 439, 440, 447, 449, 450, 452, 453, 501, 539, 553, 579–611, 613, 624, 626, 629–632
  components, 588–596
  anode, 588
  cathode, 588
  electrochemical vapor deposition, 594–596
  electrolyte, 588
  interconnection material, 588
  thin film technology, 589, 590, 594, 596
  thin wall concept, 590–594
  features, 583–586
  tolerance to contaminants, 584, 585
  typical system configuration, 585, 586
  future research directions, 610, 611
  performance, 596–600
  principles of operation, 579–583
  programs, 586, 587
  problem areas, 610, 611

stack materials selection, 433–440
state-of-the-art, 600–610
  Japanese, 605, 606
  monolithic, 607–610
  Westinghouse, 600–605
Solid polymer electrolyte (SPE) cell, 145, 146, 159–161, 164, 168–170, 175, 188, 189, 219, 228, 237, 260, 280, 284–286, 288, 292, 293, 295, 310, 329, 446, 457, 466, 469, 471, 474, 492, 621
Sorapec, 148
Southern California Edison (SCE) Co., 64, 94, 523
Southern California Gas, 117
Southern New England Telephone, 112
SO_x emissions, 30, 125, 126, 460, 555, 618, 620
Space Shuttle Orbiter, 164–166, 170, 171, 173, 174, 246, 247, 251, 263, 384, 412, 414, 492, 621, 622
Srinivasan, S., 622
Stackpole, 530
Stamicarbon B.V., 453
Standard Industrial Classification (SIC) groups, 123
Stanford University, 311, 366
Start-up and shutdown, 484, 485
Steam reforming, 208, 211, 219, 225, 237, 430, 455, 463, 464, 473, 532
Steam turbines, 494, 495
Stirling engine, 441
Stonehart Associates, 373, 374, 379
Straight noncondensing turbine, 494
Strategic Defense Initiative (SDI), 173
Stretford plant, 499
Sumitomo Precision Product Company, 628
Sun Oil Company, 458
Sunshine Project, 132
Superacid, 365, 366
Support tube, 590, 591, 594, 599, 601, 603–605, 610, 632
Surface acoustic spectroscopy, 322
Surface alloy, 360
Survey of Current Business, xii
Sydkraft, Sweden, 620
Synthetic natural gas (SNG), 57, 83

Tactical Aircraft Ground Support Power, 161
Tactical power sources, 152
Tafel, J., 22–24, 275, 278, 279, 314, 316, 319, 320, 364, 371, 374–376, 398, 408, 514, 517, 522
Takenaka Engineering, 140
Tapley drag data, 183

Index

Team to Advance Research on Gas Energy Transformation (TARGET), 13, 61–65, 100, 109, 110, 112, 135, 279, 536
Teflon®, (Polytetrafluorethylene [PTFE]), 68, 70, 71, 118, 148, 165, 171, 173, 175, 197, 243, 245–247, 251–253, 255, 256, 259, 263, 269, 282–284, 286, 288, 293, 294, 296, 323, 331, 362, 364, 365, 367, 369, 377, 378, 384, 386, 389, 390, 400, 401, 408, 412–419, 422, 453, 466, 478, 480, 503, 507, 521, 530–532
Tennessee Valley Authority (TVA), 35, 36, 68, 77, 85, 630
Tetrafluorinated ethylene (TFE), 417
Tetrafluoroethane-1,2-disulfonic acid (TFEDSA), 365, 407
Texaco coal gasifier, 97, 631
Texaco entrained-bed oxygen-blown gasifier, 82, 223, 224, 496, 577
Texas A&M University, 621, 622, 631
Thermal (flash) desorption (TDS), 322
Thermal Management System (TMS), 67, 68, 511, 512
Thermal pollution, 32
Thin film technology, 589, 590, 594, 596–598, 600, 610, 611
Thin wall concept, 590–594
TNO, The Netherlands, 632
Toa Nenyo-Kogo, 632
Tobler, J. Z., 8
Toho Gas, 628
Tohoku Electric Company, 140, 489
Tohoku University, 404
Tokuyama Soda Company, 293
Tokyo Electric Power Company (TEPCO), 29, 30, 37, 59, 65, 66, 68–76, 118, 131, 134, 135, 140, 145, 479, 483, 489, 508, 512–514, 518, 615–617, 631, 632
Tokyo Gas Co., 111, 135, 628, 631
Toray Industries, 530
Toshiba Corporation, 134–137, 139, 143, 144, 440, 478, 503, 534, 535, 551, 562, 613, 617, 628
Total hydrocarbon reforming (THR) process, 215
Total suspended particulates (TSP), 57, 82, 337
Toyo Engineering Corporation (TEC), 145, 215, 619
Transfer coefficient, 278, 526, 542
Tree pattern, 522, 525, 526, 535
Trifluoroacetic acid (TFAc), 365
Trifluoromethylsulfonic acid (TFMSA), 160, 226, 281, 282, 343–345, 364, 365

Trimble, Karen, 620
Triple contact, 8–10
TRW Energy Systems Group, 117, 123, 125, 225

U.S. Air Force Programs, 160–163, 173, 175, 275
   Aero Propulsion Laboratory, 161
   McClellan Air Force Base, 117
U.S. Army Mobility Equipment Research and Development Command (USAMERADCOM), 154, 158
U.S. Army Programs, 62, 153–161, 198, 214, 216, 220, 223, 275, 282–284, 336, 363
   Electronics Command, 64
   Mobility Research and Development Command, 198
U.S. Army Regulation 70-64, 153
U.S. Navy, 275, 473
   Deep Submergence Search Vehicle System, 275
Ultraviolet (UV) spectrophotometry, 322
Ultraviolet photoemission spectroscopy (UPS), 322
Underpotential deposition (UPD), 347, 352
Union Carbide Corporation, 186, 190, 193–196, 198, 253, 254, 269, 333
   Hybrid City Car, 193–196
   pressure swing absorption (PSA), 269
United Oil Products Co. (UOP), 363
United Technologies Corporation (UTC), 12, 13, 28–30, 35–37, 40, 46, 51, 59–61, 63–70, 73–78, 80, 81, 84, 85, 88, 93, 95, 96, 98–100, 110, 111, 113–115, 117, 119, 122, 125–127, 134, 135, 139, 140, 143, 146, 154, 157, 160, 161, 164–166, 169, 170, 173, 178, 212, 213, 216, 244, 254, 264, 269, 277, 279, 301, 346, 366, 372, 374, 377–379, 381–384, 407, 416, 417, 419, 422, 447, 448, 453, 454, 465–469, 471–473, 478–480, 483, 491, 497, 498, 504, 505, 507, 508, 510, 514–516, 521–523, 529, 530, 534, 535, 548, 549, 551, 558, 562, 563, 565–568, 570, 571, 573, 613, 617, 631
   Hamilton Standard Division, 99, 169
   FCG-1 Utility Fuel Cell 73–78
   Power Systems Division, 503
   proposed 11 MW power plant, 510–513
   Utility PAFC Program, 63–65
   Utility MCFC Development Program, 93–100
University of Connecticut, 225
University of Delft, 632
University of Illinois, 439

University of Messina, 617
University of Milan, 617
University of Missouri-Rolla, 439
University of Oslo, 632
University of Ottawa, 282
University of Pennsylvania, 245
University of Trondheim, 632

Varta Battery, 470, 474
Vibration spectroscopy via low energy electron loss (EES), 322
Visual aesthetics, 32
Viton, 417, 532
Volta, 7
Voltage efficiency, 27, 172
Voltaic pile, 7
Vulcan, 377, 379
Vulcan-6, 402
Vulcan XC-72, 371, 373, 374, 376, 401, 407, 522
Vulcan XC-72R, 367, 378
Vulcan XC-T2R, 346

Wacker-Chemie GmbH, 449
Wall Street Journal, 184
Water removal, 471, 472, 475–484

Wellman-Galusha coal gasifier, 84
Westinghouse Electric Corporation, 60, 78–80, 84, 85, 102, 106–108, 118–120, 125, 153, 161, 224, 225, 260, 359, 366, 394, 395, 436, 439, 440, 447, 449, 452, 453, 476, 482, 488, 519–527, 530, 534, 535, 579, 584, 587, 590–593, 596, 598–605, 609, 610, 617, 630–632
  Advanced Energy Systems Division, 161
  Mark II 7.5 MW power plant, 519–527
  Synthetic Fuels Division, 224
Westinghouse Electric Corporation-Sanyo Electric Co., Ltd. cooling method, 476
Westinghouse gasifier, 225
Westvaco Nuchar, 373
Window-frame gasket, 565

X-ray photoemission spectroscopy (XPS), 322

Yogogawa Electric, 619

Zee pattern, 522, 524, 525